T0178158

Introduction to Logic Circuits & Logic Design with Verilog

Introduction to Logic Circuits & Logic Design with Verilog

2nd Edition

Brock J. LaMeres

Springer

Brock J. LaMeres
Department of Electrical & Computer Engineering
Montana State University
Bozeman, MT, USA

ISBN 978-3-030-13607-9 ISBN 978-3-030-13605-5 (eBook)
https://doi.org/10.1007/978-3-030-13605-5

Library of Congress Control Number: 2019934718

© Springer Nature Switzerland AG 2019
This work is subject to copyright. All rights are reserved by the Publisher, whether the whole or part of the material is concerned, specifically the rights of translation, reprinting, reuse of illustrations, recitation, broadcasting, reproduction on microfilms or in any other physical way, and transmission or information storage and retrieval, electronic adaptation, computer software, or by similar or dissimilar methodology now known or hereafter developed.
The use of general descriptive names, registered names, trademarks, service marks, etc. in this publication does not imply, even in the absence of a specific statement, that such names are exempt from the relevant protective laws and regulations and therefore free for general use.
The publisher, the authors, and the editors are safe to assume that the advice and information in this book are believed to be true and accurate at the date of publication. Neither the publisher nor the authors or the editors give a warranty, express or implied, with respect to the material contained herein or for any errors or omissions that may have been made. The publisher remains neutral with regard to jurisdictional claims in published maps and institutional affiliations.

This Springer imprint is published by the registered company Springer Nature Switzerland AG
The registered company address is: Gewerbestrasse 11, 6330 Cham, Switzerland

Preface

The overall goal of this book is to fill a void that has appeared in the instruction of digital circuits over the past decade due to the rapid abstraction of system design. Up until the mid-1980s, digital circuits were designed using *classical* techniques. Classical techniques relied heavily on manual design practices for the synthesis, minimization, and interfacing of digital systems. Corresponding to this design style, academic textbooks were developed that taught classical digital design techniques. Around 1990, large-scale digital systems began being designed using hardware description languages (HDL) and automated synthesis tools. Broad-scale adoption of this *modern design* approach spread through the industry during this decade. Around 2000, hardware description languages and the modern digital design approach began to be taught in universities, mainly at the senior and graduate level. There were a variety of reasons that the modern digital design approach did not penetrate the lower levels of academia during this time. First, the design and simulation tools were difficult to use and overwhelmed freshman and sophomore students. Second, the ability to implement the designs in a laboratory setting was infeasible. The modern design tools at the time were targeted at custom integrated circuits, which are cost- and time-prohibitive to implement in a university setting. Between 2000 and 2005, rapid advances in programmable logic and design tools allowed the modern digital design approach to be implemented in a university setting, even in lower-level courses. This allowed students to learn the modern design approach based on HDLs and prototype their designs in real hardware, mainly field-programmable gate arrays (FPGAs). This spurred an abundance of textbooks to be authored, teaching hardware description languages and higher levels of design abstraction. This trend has continued until today. While abstraction is a critical tool for engineering design, the rapid movement toward teaching only the modern digital design techniques has left a void for freshman- and sophomore-level courses in digital circuitry. Legacy textbooks that teach the classical design approach are outdated and do not contain sufficient coverage of HDLs to prepare the students for follow-on classes. Newer textbooks that teach the modern digital design approach move immediately into high-level behavioral modeling with minimal or no coverage of the underlying hardware used to implement the systems. As a result, students are not being provided the resources to understand the fundamental hardware theory that lies beneath the modern abstraction such as interfacing, gate-level implementation, and technology optimization. Students moving too rapidly into high levels of abstraction have little understanding of what is going on when they click the "compile and synthesize" button of their design tool. This leads to graduates who can model a breadth of different systems in an HDL but have no depth into how the system is implemented in hardware. This becomes problematic when an issue arises in a real design and there is no foundational knowledge for the students to fall back on in order to debug the problem.

This new book addresses the lower-level foundational void by providing a comprehensive, bottoms-up, coverage of digital systems. The book begins with a description of lower-level hardware including binary representations, gate-level implementation, interfacing, and simple combinational logic design. Only after a foundation has been laid in the underlying hardware theory is the Verilog language introduced. The Verilog introduction gives only the basic concepts of the language in order to model, simulate, and synthesize combinational logic. This allows the students to gain familiarity with the language and the modern design approach without getting overwhelmed by the full capability of the language. The book then covers sequential logic and finite-state machines at the structural level. Once this secondary foundation has been laid, the remaining capabilities of Verilog are presented that allow sophisticated, synchronous systems to be modeled. An entire chapter is then dedicated to examples of sequential system modeling, which allows the students to learn by example. The second part of the textbook introduces the details of programmable logic, semiconductor memory, and arithmetic circuits. The book culminates with a discussion of computer system design, which incorporates all of the knowledge gained in the previous chapters. Each component of a computer system is described with an accompanying Verilog implementation, all while continually reinforcing the underlying hardware beneath the HDL abstraction.

Written the Way It Is Taught

The organization of this book is designed to follow the way in which the material is actually learned. Topics are presented only once sufficient background has been provided by earlier chapters to fully understand the material. An example of this *learning-oriented* organization is how the Verilog language is broken into two chapters. Chapter 5 presents an introduction to Verilog and the basic constructs to model combinational logic. This is an ideal location to introduce the language because the reader has just learned about combinational logic theory in Chap. 4. This allows the student to begin gaining experience using the Verilog simulation tools on basic combinational logic circuits. The more advanced constructs of Verilog, such as sequential modeling and test benches, are presented in Chap. 8 only after a thorough background in sequential logic is presented in Chap. 7. Another example of this learning-oriented approach is how arithmetic circuits are not introduced until Chap. 12. While technically the arithmetic circuits in Chap. 12 are combinational logic circuits and could be presented in Chap. 4, the student does not have the necessary background in Chap. 4 to fully understand the operation of the arithmetic circuitry, so its introduction is postponed.

This incremental, *just-in-time* presentation of material allows the book to follow the way the material is actually taught in the classroom. This design also avoids the need for the instructor to assign sections that move back and forth through the text. This not only reduces course design effort for the instructor but allows the student to know where they are in the sequence of learning. At any point, the student should know the material in prior chapters and be moving toward understanding the material in subsequent ones.

An additional advantage of this book's organization is that it supports giving the student hands-on experience with digital circuitry for courses with an accompanying laboratory component. The flow is designed to support lab exercises that begin using discrete logic gates on a breadboard and then move into HDL-based designs implemented on off-the-shelf FPGA boards. Using this approach to a laboratory experience gives the student experience with the basic electrical operation of digital circuits, interfacing, and HDL-based designs.

Learning Outcomes

Each chapter begins with an explanation of its learning objective followed by a brief preview of the chapter topics. The specific learning outcomes are then presented for the chapter in the form of concise statements about the measurable knowledge and/or skills the student will be able to demonstrate by the end of the chapter. Each section addresses a single, specific learning outcome. This eases the process of assessment and gives specific details on student performance. There are over 1000 assessment tools in the form of exercise problems and concept check questions that are tied directly to specific learning outcomes for both formative and summative assessment.

Teaching by Example

With nearly 250 worked examples, concept checks for each section, 200+ supporting figures, and 1000+ assessment problems, students are provided with multiple ways to learn. Each topic is described in a clear, concise written form with accompanying figures as necessary. This is then followed by annotated worked examples that match the form of the exercise problems at the end of each chapter. Additionally, concept check questions are placed at the end of each section in the book to measure the student's general understanding of the material using a concept inventory assessment style. These features provide the student multiple ways to learn the material and build an understanding of digital circuitry.

Course Design

The book can be used in multiple ways. The first is to use the book to cover two, semester-based college courses in digital logic. The first course in this sequence is an *introduction to logic circuits* and covers Chaps. 1, 2, 3, 4, 5, 6, and 7. This introductory course, which is found in nearly all accredited electrical and computer engineering programs, gives students a basic foundation in digital hardware and interfacing. Chapters 1, 2, 3, 4, 5, 6, and 7 only cover relevant topics in digital circuits to make room for a thorough introduction to Verilog. At the end of this course, students have a solid foundation in digital circuits and are able to design and simulate Verilog models of concurrent and hierarchical systems. The second course in this sequence covers *logic design* using Chaps. 8, 9, 10, 11, 12, and 13. In this second course, students learn the advanced features of Verilog such as procedural assignments, sequential behavioral modeling, system tasks, and test benches. This provides the basis for building larger digital systems such as registers, finite-state machines, and arithmetic circuits. Chapter 13 brings all of the concepts together through the design of a simple 8-bit computer system that can be simulated and implemented using many off-the-shelf FPGA boards.

This book can also be used in a more accelerated digital logic course that reaches a higher level of abstraction in a single semester. This is accomplished by skipping some chapters and moving quickly through others. In this use model, it is likely that Chap. 2 on number systems and Chap. 3 on digital circuits would be quickly referenced but not covered in detail. Chapters 4 and 7 could also be covered quickly in order to move rapidly into Verilog modeling without spending significant time looking at the underlying hardware implementation. This approach allows a higher level of abstraction to be taught but provides the student with the reference material so that they can delve into the details of the hardware implementation if interested.

All exercise and concept problems that do not involve a Verilog model are designed so that they can be implemented as a multiple-choice or numeric entry question in a standard course management system. This allows the questions to be automatically graded. For the Verilog design questions, it is expected that the students will upload their Verilog source files and screenshots of their simulation waveforms to the course management system for manual grading by the instructor or teaching assistant.

Instructor Resources

Instructors adopting this book can access a growing collection of supplementary learning resources including *YouTube* videos created by the author, a solutions manual, a laboratory manual, and Verilog test benches for all problems. Additional resources are made available as demand grows. The growing library of YouTube videos can provide supplementary learning materials for students or facilitate fully online or flipped delivery of this material. The videos are found at https://www.youtube.com/c/DigitalLogicProgramming_LaMeres. The solutions manual contains a graphic-rich description of select exercise problems. A complementary lab manual has also been developed to provide additional learning activities based on both the 74HC discrete logic family and an off-the-shelf FPGA board. This manual is provided separately from the book in order to support the ever-changing technology options available for laboratory exercises.

What's New in the Second Edition

The most common request from adopters of the first edition of this book was more assessment problems and accompanying videos. As a result, the second edition now contains over 1000 assessment questions and a growing library of YouTube videos. Additionally, more worked examples have been added so that every section has abundant examples of how to apply the content to designing and analyzing digital circuits.

Bozeman, MT, USA Brock J. LaMeres

Acknowledgment

For JoAnn, Alexis, and Kylie. Thank you for your endless support of this project. You are my inspiration.

Contents

Chapter 1: Introduction: Analog Versus Digital

We often hear that we live in a digital age. This refers to the massive adoption of computer systems within every aspect of our lives from smartphones to automobiles to household appliances. This statement also refers to the transformation that has occurred to our telecommunications infrastructure that now transmits voice, video, and data using 1's and 0's. There are a variety of reasons that digital systems have become so prevalent in our lives. In order to understand these reasons, it is good to start with an understanding of what a digital system is and how it compares to its counterpart, the analog system. The goal of this chapter is to provide an understanding of the basic principles of analog and digital systems.

Learning Outcomes—After completing this chapter, you will be able to:

1.1 Describe the fundamental differences between analog and digital systems.
1.2 Describe the advantages of digital systems compared to analog systems.

1.1 Differences Between Analog and Digital Systems

Let's begin by looking at signaling. In electrical systems, signals represent information that is transmitted between devices using an electrical quantity (voltage or current). An analog signal is defined as a continuous, time-varying quantity that corresponds directly to the information it represents. An example of this would be a barometric pressure sensor that outputs an electrical voltage corresponding to the pressure being measured. As the pressure goes up, so does the voltage. While the range of the input (pressure) and output (voltage) will have different spans, there is a direct mapping between the pressure and voltage. Another example would be sound striking a traditional analog microphone. Sound is a pressure wave that travels through a medium such as air. As the pressure wave strikes the diaphragm in the microphone, the diaphragm moves back and forth. Through the process of inductive coupling, this movement is converted to an electric current. The characteristics of the current signal produced (e.g., frequency and magnitude) correspond directly to the characteristics of the incoming sound wave. The current can travel down a wire and go through another system that works in the opposite manner by inductively coupling the current onto another diaphragm, which in turn moves back and forth forming a pressure wave and thus sound (i.e., a speaker or earbud). In both of these examples, the electrical signal represents the *actual* information that is being transmitted and is considered *analog*. Analog signals can be represented mathematically as a function with respect to time.

In digital signaling the electrical signal itself is not directly the information it represents; instead, the information is encoded. The most common type of encoding is binary (1's and 0's). The 1's and 0's are represented by the electrical signal. The simplest form of digital signaling is to define a threshold voltage directly in the middle of the range of the electrical signal. If the signal is above this threshold, the signal is representing a 1. If the signal is below this threshold, the signal is representing a 0. This type of signaling is not considered continuous as in analog signaling; instead, it is considered to be *discrete* because the information is transmitted as a series of distinct values. The signal transitions between a 1 to 0 and 0 to 1 are assumed to occur instantaneously. While this is obviously impossible, for the purposes of information transmission, the values can be interpreted as a series of discrete values. This is a *digital* signal and is not the actual information, but rather the binary encoded representation of the original information. Digital signals are not represented using traditional mathematical functions; instead, the digital values are typically held in tables of 1's and 0's.

© Springer Nature Switzerland AG 2019
B. J. LaMeres, *Introduction to Logic Circuits & Logic Design with Verilog*,
https://doi.org/10.1007/978-3-030-13605-5_1

Examples of analog signals and systems The human interface to a computer, the electrical signal representing sound that travels down the wire of a set of headphones, the actual sound coming out of headphones, electricity coming out of a wall outlet, and the voltage from a battery or solar cell.

Examples of digital signals and systems Information stored on a computer; the computer hardware that processes information; devices such as computers, tablets, and smartphones are also considered digital systems because their hardware and information being processed is digital; and the information being transmitted over the Internet and wireless networks is digital.

Figure 1.1 shows an example of analog signal (left) and an example of digital signal (right). While the digital signal is in reality continuous, it represents a series of discrete 1 and 0 values.

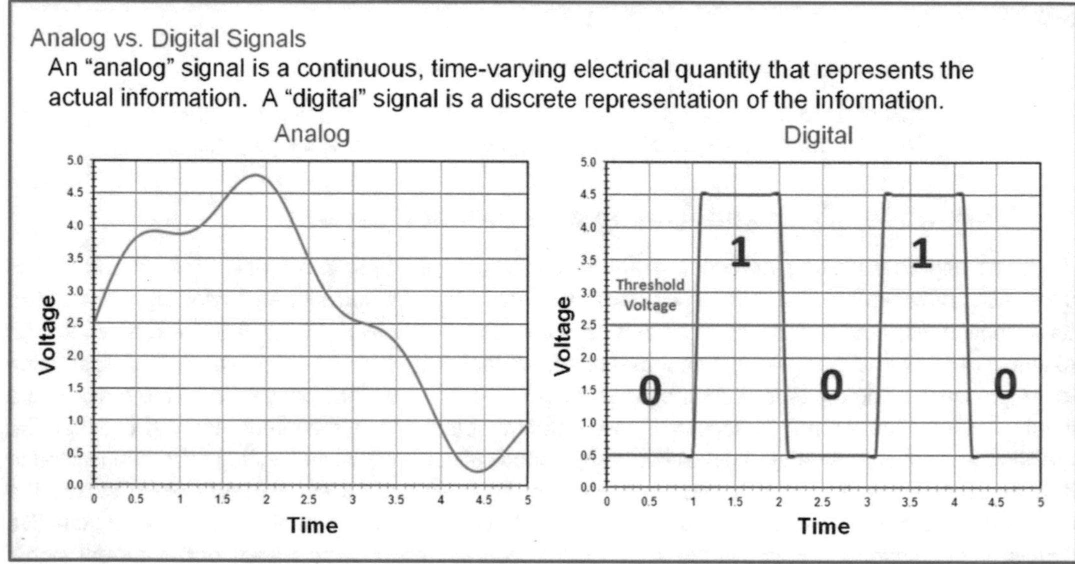

Fig. 1.1
Analog (left) vs. digital (right) signals

Concept Check

CC1.1 If a digital signal is only a discrete representation of real information, how is it possible to produce high-quality music without hearing "gaps" in the output due to the digitization process?

A) The gaps are present, but they occur so quickly that the human ear can't detect them.

B) When the digital music is converted back to analog sound, the gaps are smoothed out since an analog signal is by definition *continuous*.

C) Digital information is a continuous, time-varying signal so there aren't gaps.

D) The gaps can be heard if the music is played slowly, but at normal speed, they can't be.

1.2 Advantages of Digital Systems over Analog Systems

There are a variety of reasons that digital systems are preferred over analog systems. First is their ability to operate within the presence of noise. Since an analog signal is a direct representation of the physical quantity it is transmitting, any noise that is coupled onto the electrical signal is interpreted as noise on the original physical quantity. An example of this is when you are listening to an AM/FM radio and you hear distortion of the sound coming out of the speaker. The distortion you hear is not due to actual distortion of the music as it was played at the radio station, but rather electrical noise that was coupled onto the analog signal transmitted to your radio prior to being converted back into sound by the speakers. Since the signal in this case is analog, the speaker simply converts it in its entirety (noise + music) into sound. In the case of digital signaling, a significant amount of noise can be added to the signal while still preserving the original 1's and 0's that are being transmitted. For example, if the signal is representing a 0, the receiver will still interpret the signal as a 0 as long as the noise doesn't cause the level to exceed the threshold. Once the receiver interprets the signal as a 0, it stores the encoded value as a 0, thus ignoring any noise present during the original transmission. Figure 1.2 shows the exact same noise added to the analog and digital signals from Fig. 1.1. The analog signal is distorted; however, the digital signal is still able to transmit the 0's and 1's that represent the information.

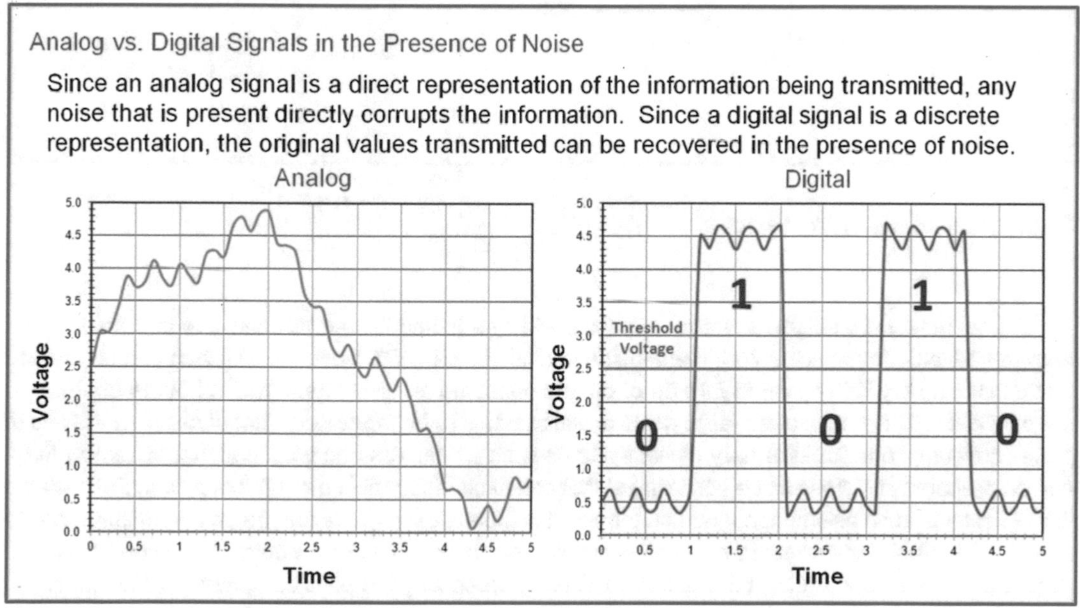

Fig. 1.2
Noise on analog (left) and digital (right) signals

Another reason that digital systems are preferred over analog ones is the simplicity of the circuitry. In order to produce a 1 and 0, you simply need an electrical switch. If the switch connects the output to a voltage below the threshold, then it produces a 0. If the switch connects the output to a voltage above the threshold, then it produces a 1. It is relatively simple to create such a switching circuit using modern transistors. Analog circuitry, however, needs to perform the conversion of the physical quantity it is representing (e.g., pressure, sound) into an electrical signal all the while maintaining a direct correspondence between the input and output. Since analog circuits produce a direct, continuous representation of information, they require more complicated designs to achieve linearity in the presence of environmental variations (e.g., power supply, temperature, fabrication differences). Since digital circuits only produce a discrete representation of the information, they can be implemented with simple switches that are only altered when information is produced or retrieved. Figure 1.3 shows an example comparison between an analog-inverting amplifier and a digital inverter. The analog amplifier uses dozens of transistors (inside the triangle) and two resistors to perform the inversion of the input. The digital inverter uses two transistors that act as switches to perform the inversion.

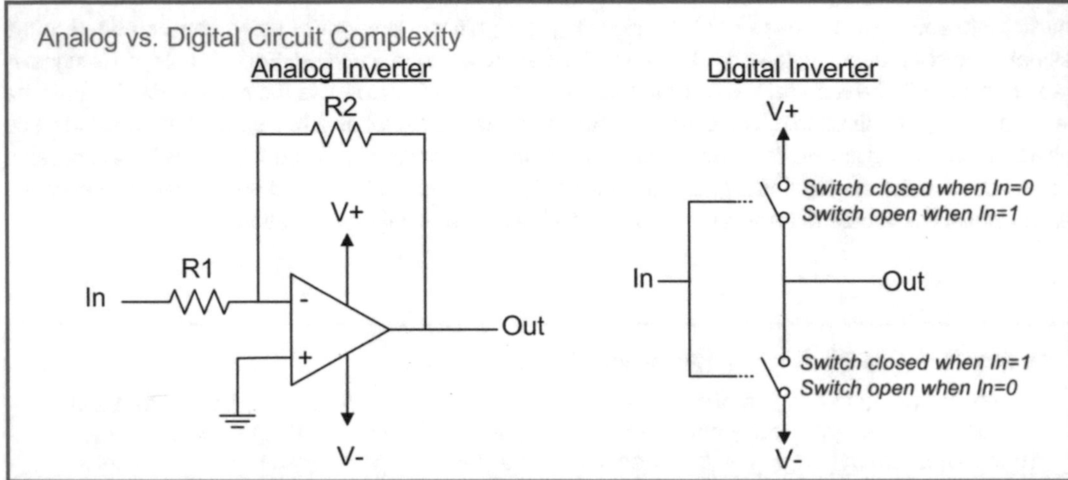

Fig. 1.3
Analog (left) vs. digital (right) circuits

A final reason that digital systems are being widely adopted is their reduced power consumption. With the advent of complementary metal-oxide transistors (CMOS), electrical switches can be created that consume very little power to *turn* on or off and consume relatively negligible amounts of power to *keep* on or off. This has allowed large-scale digital systems to be fabricated without excessive levels of power consumption. For stationary digital systems such as servers and workstations, extremely large and complicated systems can be constructed that consume reasonable amounts of power. For portable digital systems such as smartphones and tablets, this means useful tools can be designed that are able to run on portable power sources. Analog circuits, on the other hand, require continuous power to accurately convert and transmit the electrical signal representing the physical quantity. Also, the circuit techniques that are required to compensate for variances in power supply and fabrication processes in analog systems require additional power consumption. For these reasons, analog systems are being replaced with digital systems wherever possible to exploit their noise immunity, simplicity, and low-power consumption. While analog systems will always be needed at the transition between the physical (e.g., microphones, camera lenses, sensors, video displays) and the electrical world, it is anticipated that the push toward digitization of everything in between (e.g., processing, transmission, storage) will continue.

Concept Check

CC1.2 When does the magnitude of electrical noise on a digital signal prevent the original information from being determined?

A) When it causes the system to draw too much power.

B) When the shape of the noise makes the digital signal look smooth and continuous like a sine wave.

C) When the magnitude of the noise is large enough that it causes the signal to inadvertently cross the threshold voltage.

D) It doesn't. A digital signal can withstand any magnitude of noise.

Summary

- An analog system uses a direct mapping between an electrical quantity and the information being processed. A digital system, on the other hand, uses a discrete representation of the information.
- Using a discrete representation allows the digital signals to be more immune to noise in addition to requiring simple circuits that require less power to perform the computations.

Exercise Problems

Section 1.1: Differences Between Analog and Digital Systems

1.1.1 If an electrical signal is a direct function of a physical quantity, is it considered analog or digital?

1.1.2 If an electrical signal is a discrete representation of information, is it considered analog or digital?

1.1.3 What part of any system will always require an analog component?

1.1.4 Is the sound coming out of earbuds analog or digital?

1.1.5 Is the MP3 file stored on an iPod analog or digital?

1.1.6 Is the circuitry that reads the MP3 file from memory in an iPod analog or digital?

1.1.7 Is the electrical signal that travels down earphone wires analog or digital?

1.1.8 Is the voltage coming out of the battery in an iPod analog or digital?

1.1.9 Is the physical interface on the touch display of an iPod analog or digital?

1.1.10 Take a look around right now, and identify two digital technologies in use.

1.1.11 Take a look around right now, and identify two analog technologies in use.

1.1.12 Using the terminology from this chapter, describe the difference between an analog clock with a smooth moving second hand and a digital clock.

1.1.13 Using the terminology from this chapter, describe the difference between an analog and digital thermometer.

Section 1.2: Advantages of Digital Systems over Analog Systems

1.2.1 Give three advantages of using digital systems over analog.

1.2.2 Name a technology or device that has evolved from analog to digital in your lifetime.

1.2.3 Name an analog technology or device that has become obsolete in your lifetime.

1.2.4 Name an analog technology or device that has been replaced by digital technology but is still in use due to nostalgia.

1.2.5 Name a technology or device *invented* in your lifetime that could not have been possible without digital technology.

1.2.6 In this chapter it was discussed that in digital circuits, a transistor is used to create a simple switch. Doing a quick Internet search, about how many transistors are in a modern microprocessor?

1.2.7 In this chapter it was discussed that in analog circuits, the circuit design is more complex, and as a result, fewer transistors are used. The analog circuit shown in Fig. 1.3 is based on an analog device called an *operational amplifier*. Doing a quick Internet search, about how many transistors are in an operational amplifier device?

Chapter 2: Number Systems

Logic circuits are used to generate and transmit 1's and 0's to compute and convey information. This two-valued number system is called *binary*. As presented earlier, there are many advantages of using a binary system; however, the human brain has been taught to count, label, and measure using the *decimal* number system. The decimal number system contains ten unique symbols (0 → 9) commonly referred to as the *Arabic numerals*. Each of these symbols is assigned a relative magnitude to the other symbols. For example, 0 is less than 1, 1 is less than 2, etc. It is often conjectured that the 10-symbol number system that we humans use is due to the availability of our ten fingers (or *digits*) to visualize counting up to 10. Regardless, our brains are trained to think of the real world in terms of a decimal system. In order to bridge the gap between the way our brains think (decimal) and how we build our computers (binary), we need to understand the basics of number systems. This includes the formal definition of a positional number system and how it can be extended to accommodate any arbitrarily large (or small) value. This also includes how to convert between different number systems that contain different numbers of symbols. In this chapter, we cover 4 different number systems: decimal (10 symbols), binary (2 symbols), octal (8 symbols), and hexadecimal (16 symbols). The study of decimal and binary is obvious as they represent how our brains interpret the physical world (decimal) and how our computers work (binary). Hexadecimal is studied because it is a useful means to represent large sets of binary values using a manageable number of symbols. Octal is rarely used but is studied as an example of how the formalization of the number systems can be applied to all systems regardless of the number of symbols they contain. This chapter will also discuss how to perform basic arithmetic in the binary number system and represent negative numbers. The goal of this chapter is to provide an understanding of the basic principles of binary number systems.

Learning outcomes—After completing this chapter, you will be able to:

2.1 Describe the formation and use of positional number systems.
2.2 Convert numbers between different bases.
2.3 Perform binary addition and subtraction by hand.
2.4 Use two's complement numbers to represent negative numbers.

2.1 Positional Number Systems

A positional number system allows the expansion of the original set of symbols so that they can be used to represent any arbitrarily large (or small) value. For example, if we use the 10 symbols in our decimal system, we can count from 0 to 9. Using just the individual symbols, we do not have enough symbols to count beyond 9. To overcome this, we use the same set of symbols but assign a different value to the symbol based on its position within the number. The *position* of the symbol with respect to other symbols in the number allows an individual symbol to represent greater (or lesser) values. We can use this approach to represent numbers larger than the original set of symbols. For example, let's say we want to count from 0 upward by 1. We begin counting 0, 1, 2, 3, 4, 5, 6, 7, 8 to 9. When we are out of symbols and wish to go higher, we bring on a symbol in a different position with that position being valued higher and then start counting over with our original symbols (e.g., …, 9, 10, 11,... 19, 20, 21,...). This is repeated each time a position runs out of symbols (e.g., …, 99, 100, 101… 999, 1000, 1001,…).

First, let's look at the formation of a number system. The first thing that is needed is a set of symbols. The formal term for one of the symbols in a number system is a *numeral*. One or more numerals are used to form a *number*. We define the number of numerals in the system using the terms *radix* or *base*. For

© Springer Nature Switzerland AG 2019

B. J. LaMeres, *Introduction to Logic Circuits & Logic Design with Verilog*,
https://doi.org/10.1007/978-3-030-13605-5_2

example, our decimal number system is said to be *base 10* or have a *radix of 10* because it consists of ten unique numerals or symbols.

Radix = Base ≡ the number of numerals in the number system

The next thing that is needed is the relative value of each numeral with respect to the other numerals in the set. We can say $0 < 1 < 2 < 3$, etc. to define the relative magnitudes of the numerals in this set. The numerals are defined to be greater or less than their neighbors by a magnitude of 1. For example, in the decimal number system, each of the subsequent numerals is greater than its predecessor by exactly 1. When we define this relative magnitude, we are defining that the numeral 1 is greater than the numeral 0 by a magnitude of 1, the numeral 2 is greater than the numeral 1 by a magnitude of 1, etc. At this point we have the ability to count from 0 to 9 by 1's. We also have the basic structure for mathematical operations that have results that fall within the numeral set from 0 to 9 (e.g., $1 + 2 = 3$). In order to expand the values that these numerals can represent, we need to define the rules of a positional number system.

2.1.1 Generic Structure

In order to represent larger or smaller numbers than the lone numerals in a number system can represent, we adopt a positional system. In a positional number system, the relative position of the numeral within the overall number dictates its value. When we begin talking about the position of a numeral, we need to define a location to which all of the numerals are positioned with respect to. We define the *radix point* as the point within a number to which numerals to the left represent whole numbers and numerals to the right represent fractional numbers. The radix point is denoted with a period (i.e., "."). A particular number system often renames this radix point to reflect its base. For example, in the base 10 number system (i.e., decimal), the radix point is commonly called the *decimal point*; however, the term *radix point* can be used across all number systems as a generic term. If the radix point is not present in a number, it is assumed to be to the right of number. Figure 2.1 shows an example number highlighting the radix point and the relative positions of the whole and fractional numerals.

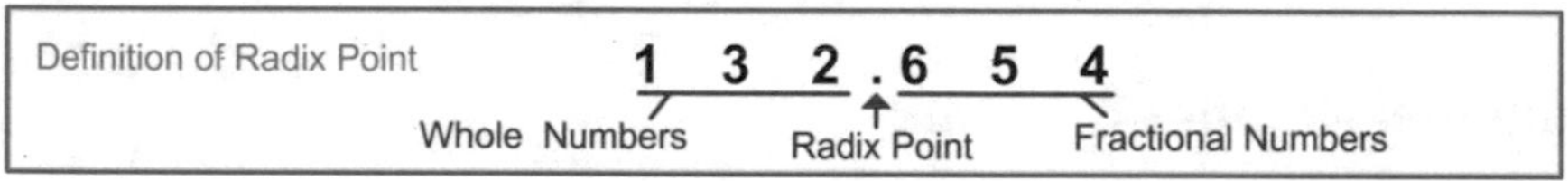

Fig. 2.1
Definition of radix point

Next, we need to define the position of each numeral with respect to the radix point. The position of the numeral is assigned a whole number with the number to the left of the radix point having a position value of 0. The position number increases by 1 as numerals are added to the left (2, 3, 4...) and decreased by 1 as numerals are added to the right (−1, −2, −3). We will use the variable *p* to represent position. The position number will be used to calculate the value of each numeral in the number based on its relative position to the radix point. Figure 2.2 shows the example number with the position value of each numeral highlighted.

Definition of Position

	1	3	2 .	6	5	4
Position (p)	2	1	0	-1	-2	-3

Fig. 2.2
Definition of position number (p) within the number

In order to create a generalized format of a number, we assign the term *digit* (*d*) to each of the numerals in the number. The term digit signifies that the numeral has a position. The position of the digit within the number is denoted as a subscript. The term *digit* can be used as a generic term to describe a numeral across all systems, although some number systems will use a unique term instead of digit which indicates its base. For example, the binary system uses the term *bit* instead of digit; however, using the term digit to describe a generic numeral in any system is still acceptable. Figure 2.3 shows the generic subscript notation used to describe the position of each digit in the number.

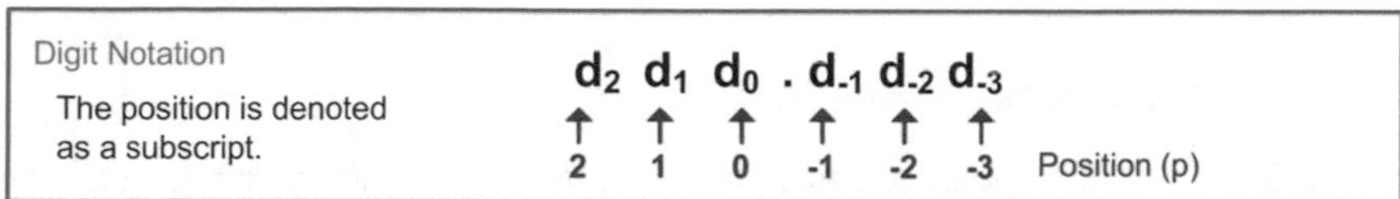

Fig. 2.3
Digit notation

We write a number from left to right starting with the highest position digit that is greater than 0 and ending with the lowest position digit that is greater than 0. This reduces the number of numerals that are written; however, a number can be represented with an arbitrary number of 0's to the left of the highest position digit greater than 0 and an arbitrary number of 0's to the right of the lowest position digit greater than 0 without affecting the value of the number. For example, the number 132.654 could be written as 0132.6540 without affecting the value of the number. The 0's to the left of the number are called *leading 0's* and the 0's to the right of the number are called *trailing 0's*. The reason this is being stated is because when a number is implemented in circuitry, the number of numerals is fixed, and each numeral must have a value. The variable *n* is used to represent the number of numerals in a number. If a number is defined with $n = 4$, that means 4 numerals are always used. The number 0 would be represented as 0000 with both representations having an equal value.

2.1.2 Decimal Number System (Base 10)

As mentioned earlier, the decimal number system contains ten unique numerals (0, 1, 2, 3, 4, 5, 6, 7, 8 and 9). This system is thus a base 10 or a radix 10 system. The relative magnitudes of the symbols are $0 < 1 < 2 < 3 < 4 < 5 < 6 < 7 < 8 < 9$.

2.1.3 Binary Number System (Base 2)

The binary number system contains two unique numerals (0 and 1). This system is thus a base 2 or a radix 2 system. The relative magnitudes of the symbols are $0 < 1$. At first glance, this system looks very limited in its ability to represent large numbers due to the small number of numerals. When counting up, as soon as you count from 0 to 1, you are out of symbols and must increment the $p + 1$ position in order to represent the next number (e.g., 0, 1, 10, 11, 100, 101, etc.); however, magnitudes of each position scale quickly so that circuits with a reasonable amount of digits can represent very large numbers. The term *bit* is used instead of *digit* in this system to describe the individual numerals and at the same time indicate the base of the number.

Due to the need for multiple bits to represent meaningful information, there are terms dedicated to describing the number of bits in a group. When 4 bits are grouped together, they are called a **nibble**. When 8 bits are grouped together, they are called a **byte**. Larger groupings of bits are called **words**. The size of the word can be stated as either an *n-bit word* or omitted if the size of the word is inherently implied. For example, if you were using a 32-bit microprocessor, using the term *word* would be interpreted as a *32-bit word*. For example, if there was a 32-bit grouping, it would be referred to as a 32-bit word. The leftmost bit in a binary number is called the ***Most Significant Bit*** (**MSB**). The rightmost bit in a binary number is called the ***Least Significant Bit*** (**LSB**).

2.1.4 Octal Number System (Base 8)

The octal number system contains eight unique numerals (0, 1, 2, 3, 4, 5, 6, 7). This system is thus a base 8 or a radix 8 system. The relative magnitudes of the symbols are 0 < 1 < 2 < 3 < 4 < 5 < 6 < 7. We use the generic term *digit* to describe the numerals within an octal number.

2.1.5 Hexadecimal Number System (Base 16)

The hexadecimal number system contains 16 unique numerals. This system is most often referred to in spoken word as "hex" for short. Since we only have ten Arabic numerals in our familiar decimal system, we need to use other symbols to represent the remaining six numerals. We use the alphabetic characters A–F in order to expand the system to 16 numerals. The 16 numerals in the hexadecimal system are 0, 1, 2, 3, 4, 5, 6, 7, 8, 9, A, B, C, D, E, and F. The relative magnitudes of the symbols are 0 < 1 < 2 < 3 < 4 < 5 < 6 < 7 < 8 < 9 < A < B < C < D < E < F. We use the generic term digit to describe the numerals within a hexadecimal number.

At this point, it becomes necessary to indicate the base of a written number. The number 10 has an entirely different value if it is a decimal number or binary number. In order to handle this, a subscript is typically included at the end of the number to denote its base. For example, 10_{10} indicates that this number is decimal "ten." If the number was written as 10_2, this number would represent binary "one zero." Table 2.1 lists the equivalent values in each of the four number systems just described for counts from 0_{10} to 15_{10}. The left side of the table does not include leading 0's. The right side of the table contains the same information but includes the leading zeros. The equivalencies of decimal, binary, and hexadecimal in this table are typically committed to memory.

Equivalency Between Different Number Systems

Decimal	Binary	Octal	Hex	Decimal	Binary	Octal	Hex
0	0	0	0	00	0000	00	0
1	1	1	1	01	0001	01	1
2	10	2	2	02	0010	02	2
3	11	3	3	03	0011	03	3
4	100	4	4	04	0100	04	4
5	101	5	5	05	0101	05	5
6	110	6	6	06	0110	06	6
7	111	7	7	07	0111	07	7
8	1000	10	8	08	1000	10	8
9	1001	11	9	09	1001	11	9
10	1010	12	A	10	1010	12	A
11	1011	13	B	11	1011	13	B
12	1100	14	C	12	1100	14	C
13	1101	15	D	13	1101	15	D
14	1110	16	E	14	1110	16	E
15	1111	17	F	15	1111	17	F
(Without Leading 0's)				(With Leading 0's)			

Table 2.1
Number system equivalency

Concept Check

CC2.1 The base of a number system is arbitrary and is commonly selected to match a particular aspect of the physical system in which it is used (e.g., base 10 corresponds to our ten fingers; base 2 corresponds to the two states of a switch). If a physical system contained three unique modes and a base of 3 was chosen for the number system, what is the base 3 equivalent of the decimal number 3?

A) $3_{10} = 11_3$ B) $3_{10} = 3_3$ C) $3_{10} = 10_3$ D) $3_{10} = 21_3$

2.2 Base Conversion

Now we look at converting between bases. There are distinct techniques for converting to and from decimal. There are also techniques for converting between bases that are powers of 2 (e.g., base 2, 4, 8, 16, etc.).

2.2.1 Converting to Decimal

The value of each digit within a number is based on the individual digit value and the digit's position. Each position in the number contains a different *weight* based on its relative location to the radix point. The weight of each position is based on the radix of the number system that is being used. The weight of each position in decimal is defined as:

$$\textbf{Weight} = (\textbf{Radix})^{p}$$

This expression gives the number system the ability to represent fractional numbers since an expression with a negative exponent (e.g., x^{-y}) is evaluated as one over the expression with the exponent change to positive (e.g., $1/x^y$). Figure 2.4 shows the generic structure of a number with its positional weight highlighted.

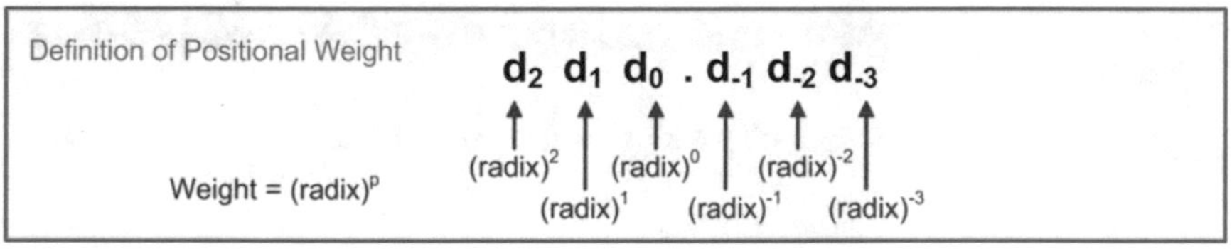

Fig. 2.4
Weight definition

In order to find the decimal value of each of the numerals in the number, its individual numeral value is multiplied by its positional weight. In order to find the value of the entire number, each value of the individual numeral-weight products is summed. The generalized format of this conversion is written as:

$$\textbf{\textit{Total Decimal Value}} = \sum_{i=p_{min}}^{p_{max}} d_i \cdot (radix)^i$$

In this expression, p_{max} represents the highest position number that contains a numeral greater than 0. The variable p_{min} represents the lowest position number that contains a numeral greater than 0. These limits are used to simplify the hand calculations; however, these terms theoretically could be $+\infty$ to $-\infty$ with no effect on the result since the summation of every leading 0 and every trailing 0 contributes nothing to the result.

As an example, let's evaluate this expression for a decimal number. The result will yield the original number but will illustrate how positional weight is used. Let's take the number 132.654_{10}. To find the decimal value of this number, each numeral is multiplied by its positional weight, and then all of the products are summed. The positional weight for the digit 1 is $(\text{radix})^p$ or $(10)^2$. In decimal this is called the hundred's position. The positional weight for the digit 3 is $(10)^1$, referred to as the ten's position. The positional weight for digit 2 is $(10)^0$, referred to as the one's position. The positional weight for digit 6 is $(10)^{-1}$, referred to as the tenth's position. The positional weight for digit 5 is $(10)^{-2}$, referred to as the hundredth's position. The positional weight for digit 4 is $(10)^{-3}$, referred to as the thousandth's position.

When these weights are multiplied by their respective digits and summed, the result is the original decimal number 132.654_{10}. Example 2.1 shows this process step-by-step.

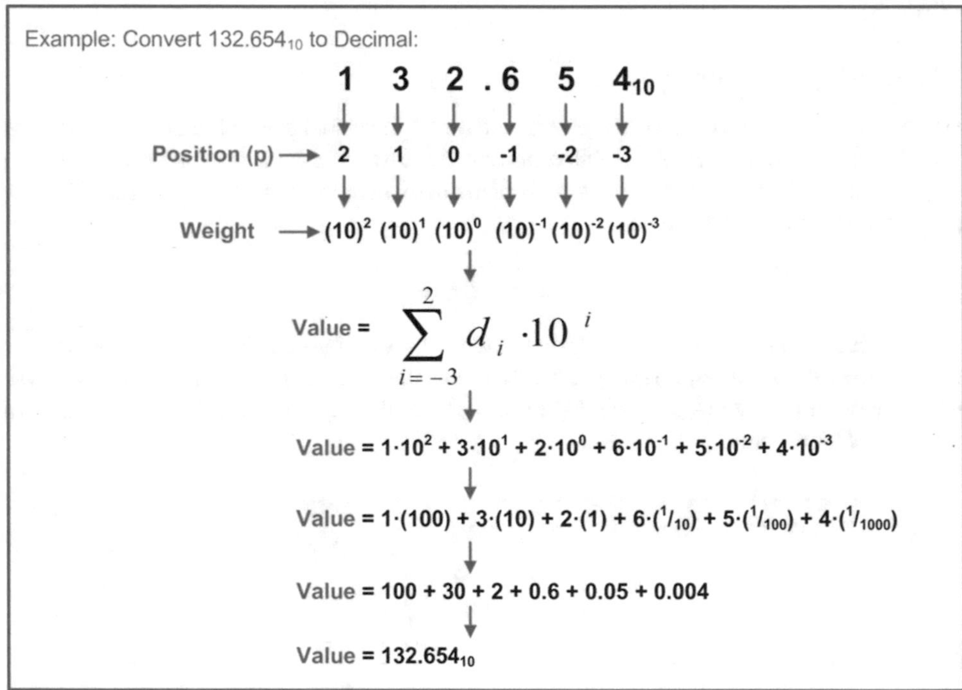

Example 2.1
Converting decimal to decimal

This process is used to convert between any other base to decimal.

2.2.1.1 Binary to Decimal

Let's convert 101.11_2 to decimal. The same process is followed with the exception that the base in the summation is changed to 2. Converting from binary to decimal can be accomplished quickly in your head due to the fact that the bit values in the products are either 1 or 0. That means any bit that is a 0 has no impact on the outcome and any bit that is a 1 simply yields the weight of its position. Example 2.2 shows the step-by-step process converting a binary number to decimal.

Example: Convert 101.11_2 to Decimal:

$$1 \quad 0 \quad 1 \,.\, 1 \quad 1_2$$

Position (p) → 2 1 0 -1 -2

Weight → $(2)^2$ $(2)^1$ $(2)^0$ $(2)^{-1}$ $(2)^{-2}$

$$\text{Value} = \sum_{i=-2}^{2} d_i \cdot 2^i$$

$$\text{Value} = 1 \cdot 2^2 + 0 \cdot 2^1 + 1 \cdot 2^0 + 1 \cdot 2^{-1} + 1 \cdot 2^{-2}$$

$$\text{Value} = 1 \cdot (4) + 0 \cdot (2) + 1 \cdot (1) + 1 \cdot ({}^1/_2) + 1 \cdot ({}^1/_4)$$

$$\text{Value} = 4 + 0 + 1 + 0.5 + 0.25$$

$$\text{Value} = 5.75_{10}$$

Example 2.2
Converting binary to decimal

2.2.1.2 Octal to Decimal

When converting from octal to decimal, the same process is followed with the exception that the base in the weight is changed to 8. Example 2.3 shows an example of converting an octal number to decimal.

Example: Convert 17.17_8 to Decimal:

$$1 \quad 7 \,.\, 1 \quad 7_8$$

Position (p) → 1 0 -1 -2

Weight → $(8)^1$ $(8)^0$ $(8)^{-1}$ $(8)^{-2}$

$$\text{Value} = \sum_{i=-2}^{1} d_i \cdot 8^i$$

$$\text{Value} = 1 \cdot 8^1 + 7 \cdot 8^0 + 1 \cdot 8^{-1} + 7 \cdot 8^{-2}$$

$$\text{Value} = 1 \cdot (8) + 7 \cdot (1) + 1 \cdot ({}^1/_8) + 7 \cdot ({}^1/_{64})$$

$$\text{Value} = 8 + 7 + 0.125 + 0.109375$$

$$\text{Value} = 15.234375_{10}$$

Example 2.3
Converting octal to decimal

2.2.1.3 Hexadecimal to Decimal

Let's convert $1AB.EF_{16}$ to decimal. The same process is followed with the exception that the base is changed to 16. When performing the conversion, the decimal equivalent of the numerals A–F needs to be used. Example 2.4 shows the step-by-step process converting a hexadecimal number to decimal.

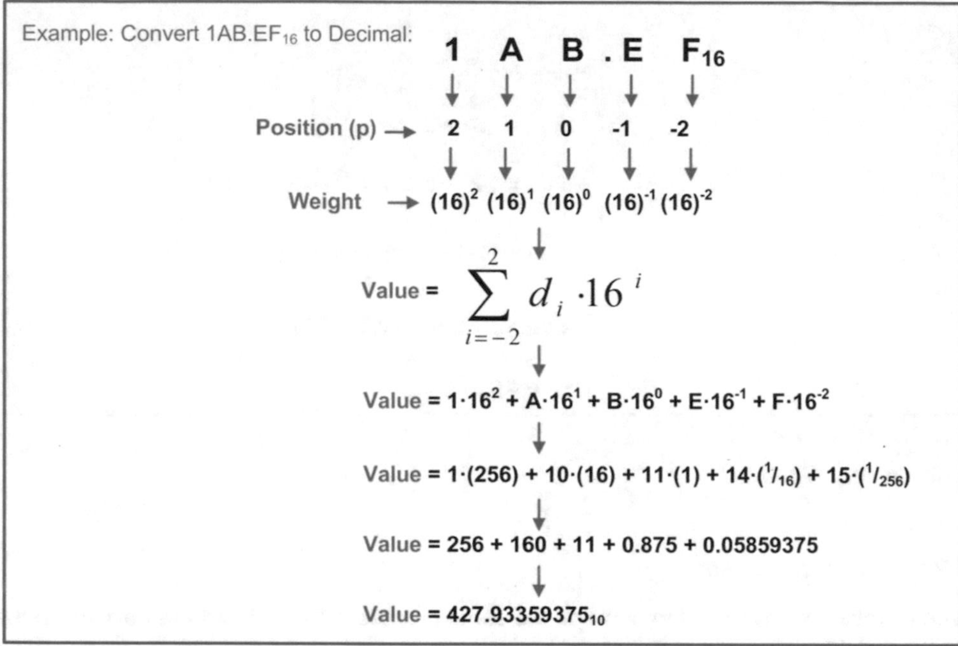

Example 2.4
Converting hexadecimal to decimal

In some cases, it is desired to specify a *level of accuracy* for the conversion in order to bound the number of fractional digits in the final result. For example, if the conversion in Example 2.4 was stated as "convert $1AB.EF_{16}$ to decimal with a fractional accuracy of 2 digits," the final result would be 427.93_{10}. How rounding is handled can also be specified with the two options being *with* or *without rounding*. In the case where the conversion is performed with rounding, additional fractional digits may need to be computed to determine if the least significant digit of the new decimal fraction needs to be altered. For example, let's say the conversion in Example 2.4 is stated as "convert $1AB.EF_{16}$ to decimal with a fractional accuracy of 4 digits with rounding." In this case, the final result would be 427.9336_{10}. Notice how rounding was applied to the digit in position $p = -3$ changing it from a 5 to a 6 based on the value in position $p = -4$. Now let's say the conversion in Example 2.4 is stated as "convert $1AB.EF_{16}$ to decimal with a fractional accuracy of 4 digits without rounding." In this case, the final result would be 427.9335_{10}. Notice how without rounding simply drops all of the digits beyond the specified level of accuracy.

2.2.2 Converting from Decimal

The process of converting from decimal to another base consists of two separate algorithms. There is one algorithm for converting the whole number portion of the number and another algorithm for converting the fractional portion of the number. The process for converting the whole number portion is to divide the decimal number by the base of the system you wish to convert to. The division will result in a quotient and a whole number remainder. The remainder is recorded as the *least significant numeral* in the converted number. The resulting quotient is then divided again by the base, which results in a new quotient and new remainder. The remainder is recorded as the next higher-order numeral in the new number. This process is repeated until a quotient of 0 is achieved. At that point the conversion is complete. The remainders will always be within the numeral set of the base being converted to.

The process for converting the fractional portion is to multiply just the fractional component of the number by the base. This will result in a product that contains a whole number and a fraction. The whole number is recorded as the *most significant digit* of the new converted number. The new fractional portion is then multiplied again by the base with the whole number portion being recorded as the next lower-order numeral. This process is repeated until the product yields a fractional component equal to zero or the desired level of accuracy has been achieved. The level of accuracy is specified by the number of numerals in the new converted number. For example, the conversion would be stated as "convert this decimal number to binary with a fractional accuracy of 4 bits." This means the final result would only have 4 bits in the fraction. In cases where the conversion does not yield exactly 4 fractional bits, there are two approaches that can be used. The first is to have *no rounding*, which means the conversion simply stops at the desired accuracy. The second is to apply *rounding*, which means additional bits beyond the desired accuracy are computed in order to determine whether the least significant bit reported.

2.2.2.1 Decimal to Binary

Let's convert 11.375_{10} to binary. Example 2.5 shows the step-by-step process converting a decimal number to binary.

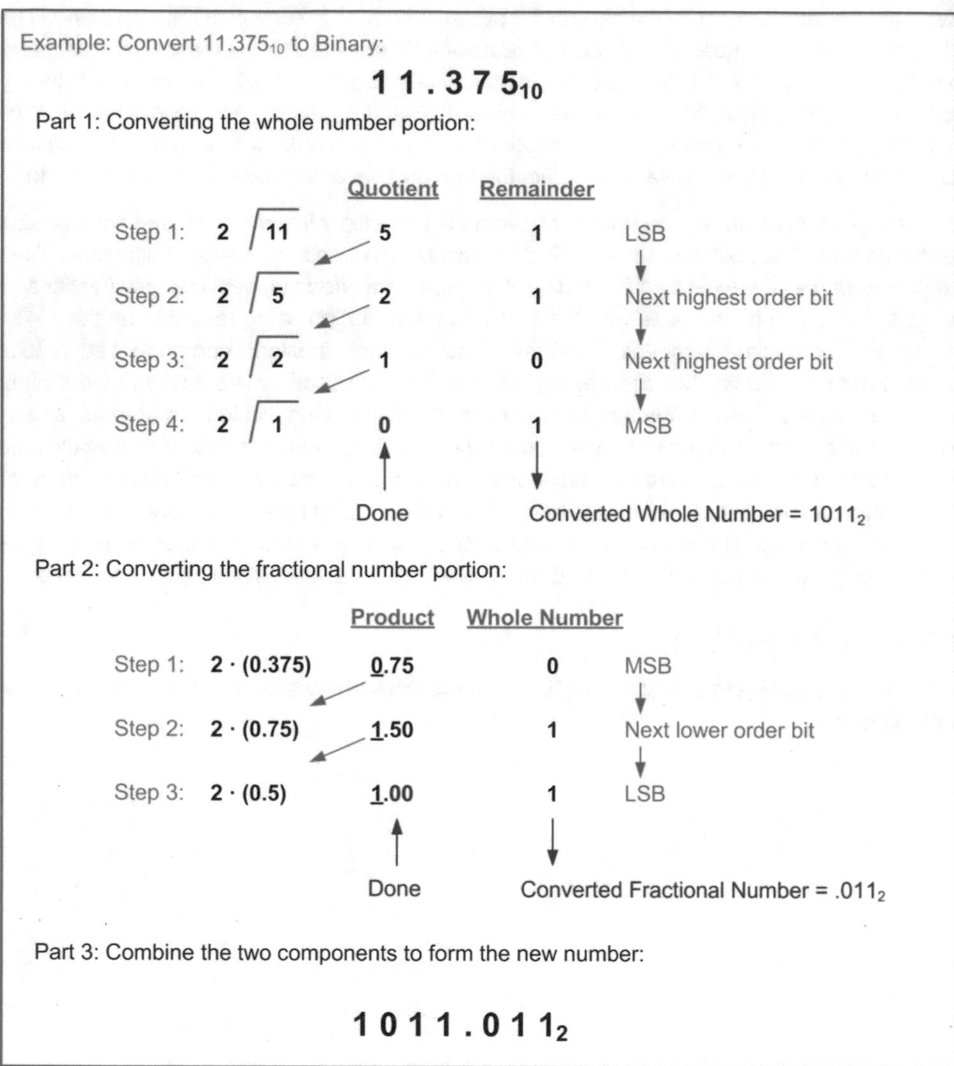

Example 2.5
Converting decimal to binary

Many times when converting to binary, the number of fractional bits that result from the conversion is more than which is needed. In this case, rounding is applied to limit the fractional accuracy. The simplest rounding approach for binary numbers is to continue the conversion for one more bit beyond the desired fractional accuracy. If the next bit is a 0, then you leave the fractional component of the number as is. If the next bit is a 1, you round the least significant bit of your number up. Often this rounding will result in a cascade of roundings from the LSB to the MSB. As an example, let's say that the conversion in Example 2.5 was specified to have a fractional accuracy of 2 bits. If the bit in position $p = -3$ was a 0 (which it is not, but let's just say it is for the sake of this example), then the number would be left as is, and the final converted number would be 1011.01_2. However, if the bit in position $p = -3$ was a 1 (as it actually is in Example 2.5), then we would need to apply rounding. We would start with the bit in position $p = -2$. Since it is a 1, we would round that up to a 0, but we would need to apply the overflow of this rounding to the next higher-order bit in position $p = -1$. That would then cause the value of $p = -1$ to go from a 0 to a 1. The final result of the conversion with rounding would be 1011.10_2.

2.2.2.2 Decimal to Octal

Let's convert 10.4_{10} to octal with an accuracy of 4 fractional digits. When converting the fractional component of the number, the algorithm is continued until 4 digits worth of fractional numerals have been achieved. Once the accuracy has been achieved, the conversion is finished even though a product with a zero fractional value has not been obtained. Example 2.6 shows the step-by-step process converting a decimal number to octal with a fractional accuracy of 4 digits.

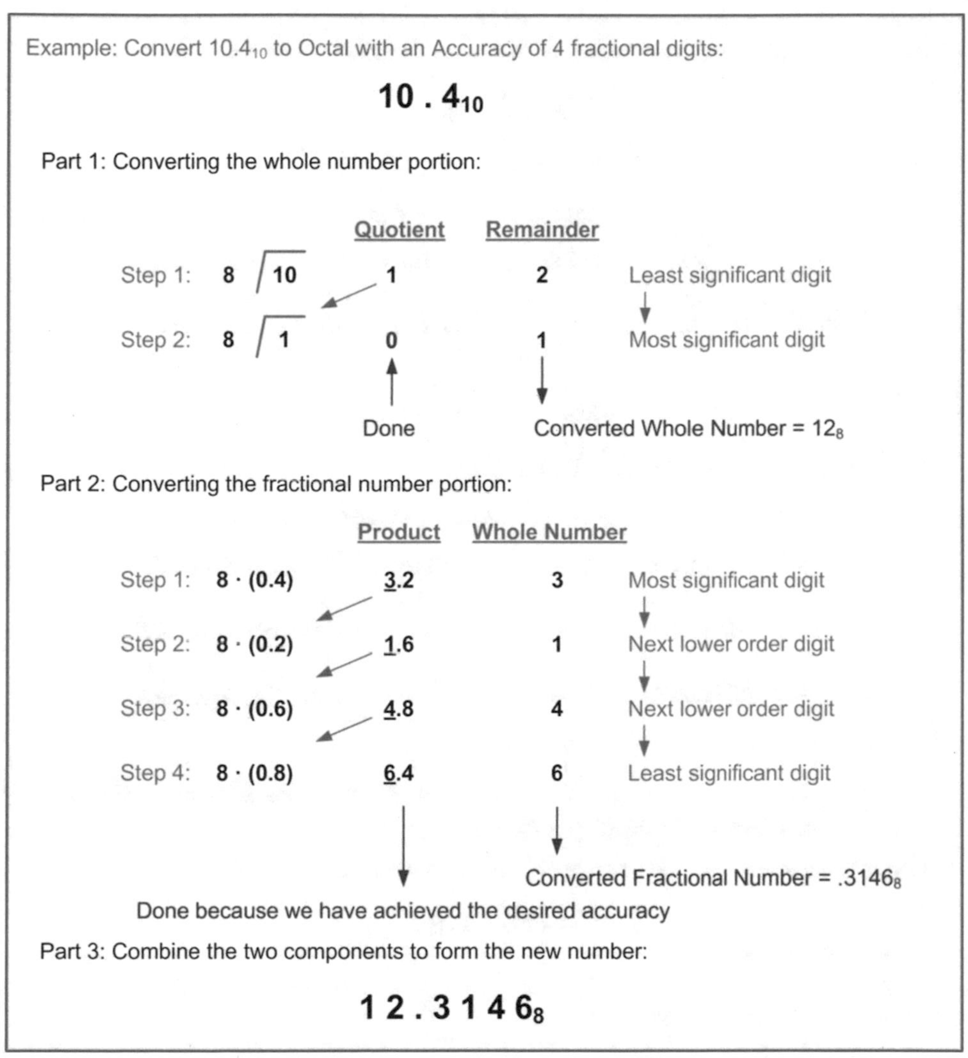

Example 2.6
Converting decimal to octal

Rounding of octal digits uses a similar approach as when rounding decimal numbers, with the exception that the middle of the range of the numbers lies between digits 3_8 and 4_8. This means that any number to be rounded that is 4_8 or greater will be rounded up. Numbers that are 3_8 or less will be rounded down, which means the fractional component of the converted number is left as in.

2.2.2.3 Decimal to Hexadecimal

Let's convert 254.655_{10} to hexadecimal with an accuracy of 3 fractional digits. When doing this conversion, all of the divisions and multiplications are done using decimal. If the results end up between 10_{10} and 15_{10}, then the decimal numbers are substituted with their hex symbol equivalent (i.e., A to F). Example 2.7 shows the step-by-step process of converting a decimal number to hex with a fractional accuracy of 3 digits.

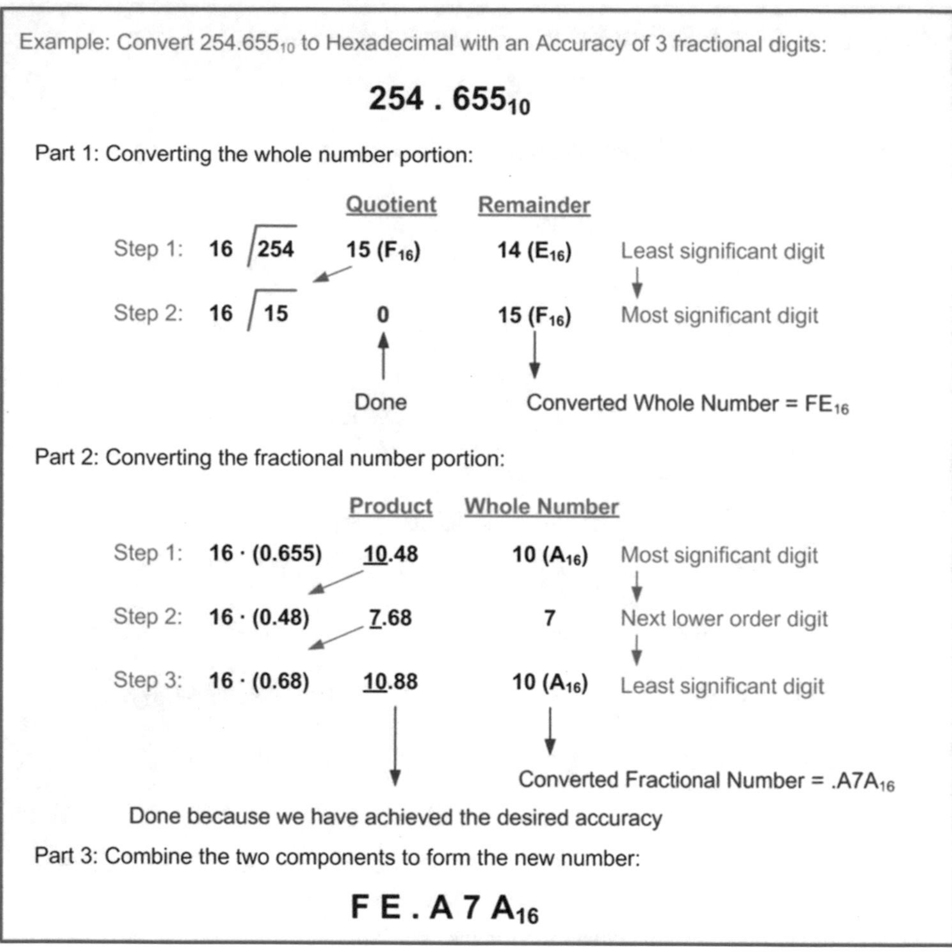

Example 2.7
Converting decimal to hexadecimal

Rounding of hexadecimal digits uses a similar approach as when rounding decimal numbers, with the exception that the middle of the range of the numbers lies between digits 7_{16} and 8_{16}. This means that any number to be rounded that is 8_{16} or greater will be rounded up. Numbers that are 7_{16} or less will be rounded down, which means the fractional component of the converted number is left as in.

2.2.3 Converting Between 2^n Bases

Converting between 2^n bases (e.g., 2, 4, 8, 16, etc.) takes advantage of the direct mapping that each of these bases has back to binary. Base 8 numbers take exactly 3 binary bits to represent all 8 symbols (i.e., $0_8 = 000_2$, $7_8 = 111_2$). Base 16 numbers take exactly 4 binary bits to represent all 16 symbols (i.e., $0_{16} = 0000_2$, $F_{16} = 1111_2$).

When converting *from* binary to any other 2^n base, the whole number bits are grouped into the appropriate-sized sets starting from the radix point and working left. If the final leftmost grouping does not have enough symbols, it is simply padded on left with leading 0's. Each of these groups is then directly substituted with their 2^n base symbol. The fractional number bits are also grouped into the appropriate-sized sets starting from the radix point, but this time working right. Again, if the final rightmost grouping does not have enough symbols, it is simply padded on the right with trailing 0's. Each of these groups is then directly substituted with their 2^n base symbol.

2.2.3.1 Binary to Octal

Example 2.8 shows the step-by-step process of converting a binary number to octal.

Example: Convert 10111.01_2 to Octal:

$$10111\,.\,01_2$$

Part 1: Form groups of 3 bits representing octal symbols.

$$(0\;1\;0)\;(1\;1\;1)\,.\,(0\;1\;0)_2$$

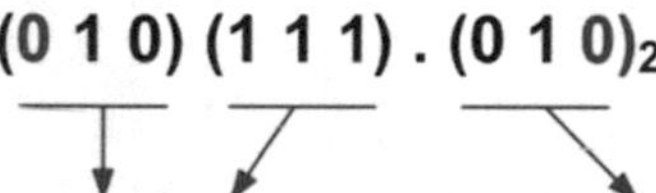

Whole number groupings start at the radix point and work left. Leading 0's are added as necessary.

Fractional number groupings start at the radix point and work right. Trailing 0's are added as necessary.

Part 2: Perform a direct substitution of the bit groupings with the equivalent octal symbol.

$$(0\;1\;0)\;(1\;1\;1)\,.\,(0\;1\;0)_2$$

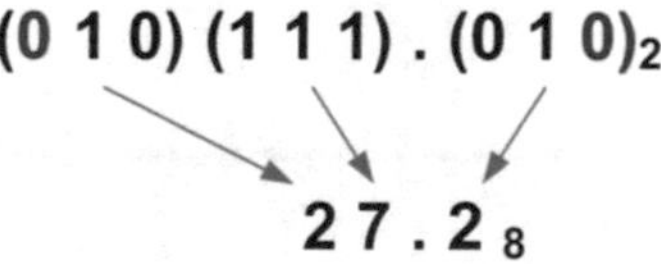

$$2\,7\,.\,2_8$$

Example 2.8
Converting binary to octal

2.2.3.2 Binary to Hexadecimal

Example 2.9 shows the step-by-step process of converting a binary number to hexadecimal.

Example: Convert 111011.11111_2 to Hexadecimal:

$$111011\,.\,11111_2$$

Part 1: Form groups of 4 bits representing hex symbols.

$$(0\,0\,1\,1)\,(1\,0\,1\,1)\,.\,(1\,1\,1\,1)\,(1\,0\,0\,0)_2$$

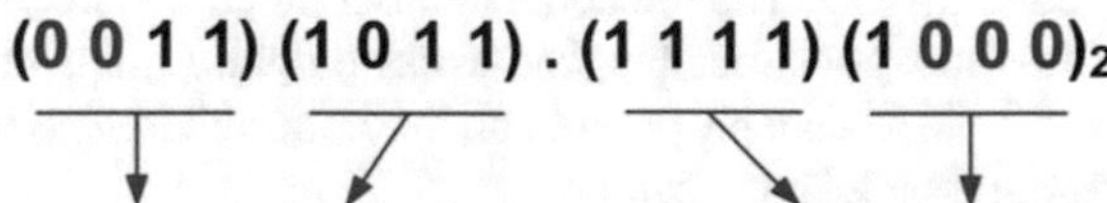

Whole number groupings start at the radix point and work left. Leading 0's are added as necessary.

Fractional number groupings start at the radix point and work right. Trailing 0's are added as necessary.

Part 2: Perform a direct substitution of the bit groupings with the equivalent hex symbol.

$$(0\,0\,1\,1)\,(1\,0\,1\,1)\,.\,(1\,1\,1\,1)\,(1\,0\,0\,0)_2$$

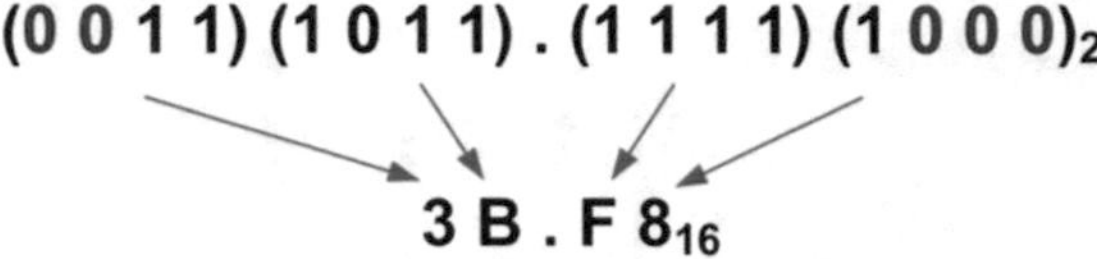

Example 2.9
Converting binary to hexadecimal

2.2.3.3 Octal to Binary

When converting *to* binary from any 2^n base, each of the symbols in the originating number are replaced with the appropriate-sized number of bits. An octal symbol will be replaced with 3 binary bits, while a hexadecimal symbol will be replaced with 4 binary bits. Any leading or trailing 0's can be removed from the converted number once complete. Example 2.10 shows the step-by-step process of converting an octal number to binary.

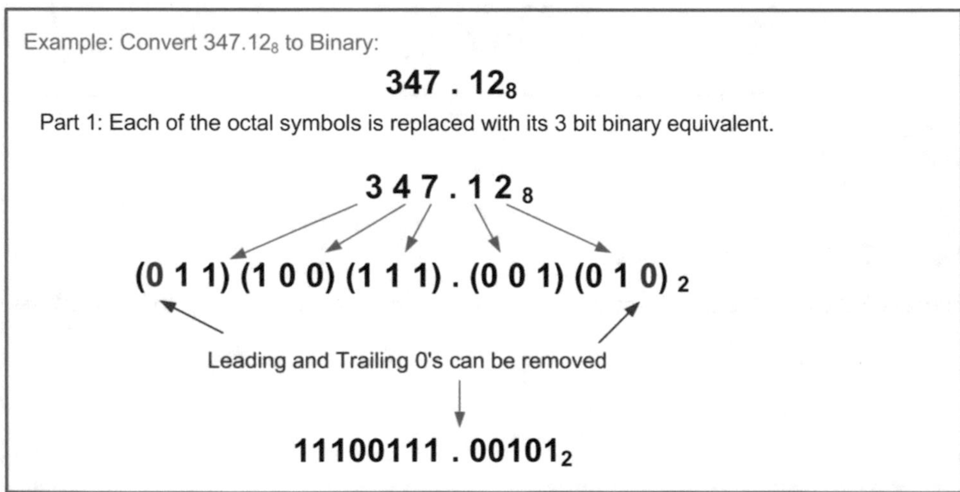

Example 2.10
Converting octal to binary

2.2.3.4 Hexadecimal to Binary

Example 2.11 shows the step-by-step process of converting a hexadecimal number to binary.

Example: Convert $1B.A_{16}$ to Binary:

Part 1: Each of the hex symbols is replaced with its 4 bit binary equivalent.

$$1\ B\ .\ A_{16}$$

$$(0\ 0\ 0\ 1)\ (1\ 0\ 1\ 1)\ .\ (1\ 0\ 1\ 0)_2$$

Part 2: Leading and trailing zeros can be removed.

$$11011\ .\ 101_2$$

Example 2.11
Converting hexadecimal to binary

2.2.3.5 Octal to Hexadecimal

When converting between 2^n bases (excluding binary), the number is first converted into binary and then converted from binary into the final 2^n base using the algorithms described before. Example 2.12 shows the step-by-step process of converting an octal number to hexadecimal.

Example: Convert 71.5_8 to Hexadecimal:

Part 1: Convert the octal number into binary. Each octal symbol is represented with 3 bits.

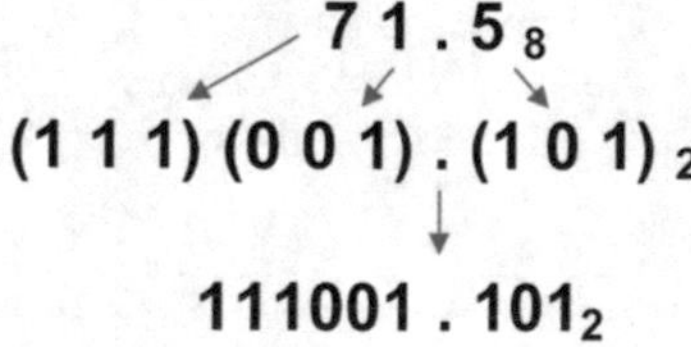

Part 2: Convert the binary number into hexadecimal. Form groups of 4 bits representing hex symbols.

Step 1: $(0\ 0\ 1\ 1)\ (1\ 0\ 0\ 1)\ .\ (1\ 0\ 1\ 0)_2$

Whole number groupings start at the radix point and work left. Leading 0's are added as necessary.

Fractional number groupings start at the radix point and work right. Trailing 0's are added as necessary.

Step 2: $(0\ 0\ 1\ 1)\ (1\ 0\ 0\ 1)\ .\ (1\ 0\ 1\ 0)_2$

$$3\ 9\ .\ A_{16}$$

Example 2.12
Converting octal to hexadecimal

2.2.3.6 Hexadecimal to Octal

Example 2.13 shows the step-by-step process of converting a hexadecimal number to octal.

Example: Convert $AB.C_{16}$ to Octal:

$$AB \,.\, C_{16}$$

Part 1: Convert the hex number into binary. Each hex symbol is represented with 4 bits.

$$AB \,.\, C_{16}$$

$$(1\,0\,1\,0)\,(1\,0\,1\,1)\,.\,(1\,1\,0\,0)_2$$

$$10101011\,.\,11_2$$

Part 2: Convert the binary number into octal. Form groups of 3 bits representing octal symbols.

Step 1: $(0\,1\,0)\,(1\,0\,1)\,(0\,1\,1)\,.\,(1\,1\,0)_2$

Step 2: $2\,5\,3\,.\,6_8$

Example 2.13
Converting hexadecimal to octal

Concept Check

CC2.2 A "googol" is the term for the decimal number 1e100. When written out manually, this number is a 1 with 100 zeros after it (e.g., 10,000,000,000,000,000,000,000, 000,000,000,000,000,000,000,000,000,000,000,000,000,000,000,000,000,000,000,0- 00,000,000,000,000,000,000). This term is more commonly associated with the search engine company Google, which uses a different spelling but is pronounced the same. How many bits does it take to represent a googol in binary?

A) 100 bits B) 256 bits C) 332 bits D) 333 bits

2.3 Binary Arithmetic

2.3.1 Addition (Carries)

Binary addition is a straightforward process that mirrors the approach we have learned for longhand decimal addition. The two numbers (or terms) to be added are aligned at the radix point, and addition begins at the least significant bit. If the sum of the least significant position yields a value with 2 bits (e.g., 10_2), then the least significant bit is recorded, and the most significant bit is *carried* to the next higher position. The sum of the next higher position is then performed including the potential *carry bit* from the prior addition. This process continues from the least significant position to the most significant position. Example 2.14 shows how addition is performed on two individual bits.

Example: Single Bit Binary Addition

There are four possible results when adding two bits.

$$\begin{array}{r} 0 \\ +\ 0 \\ \hline 0 \end{array} \qquad \begin{array}{r} 0 \\ +\ 1 \\ \hline 1 \end{array} \qquad \begin{array}{r} 1 \\ +\ 0 \\ \hline 1 \end{array} \qquad \begin{array}{r} 1 \\ +\ 1 \\ \hline \text{Carry} \rightarrow \underline{1}\,0 \end{array}$$

Example 2.14
Single-bit binary addition

When performing binary addition, the width of the inputs and output is fixed (i.e., *n*-bits). Carries that exist within the *n*-bits are treated in the normal fashion of including them in the next higher position sum; however, if the highest position summation produces a carry, this is a uniquely named event. This event is called a *carry out* or the sum is said to *generate a carry.* The reason this type of event is given special terminology is because in real circuitry, the number of bits of the inputs and output is fixed in hardware and the carry out is typically handled by a separate circuit. Example 2.15 shows this process when adding two 4-bit numbers.

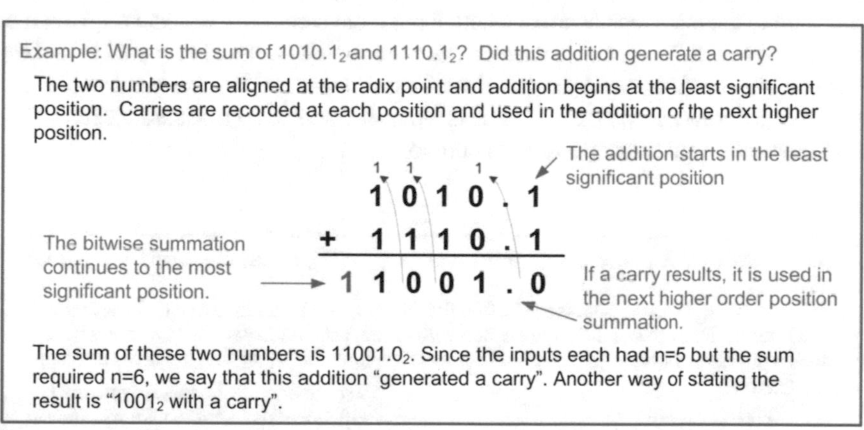

Example 2.15
Multiple-bit binary addition

The largest decimal sum that can result from the addition of two binary numbers is given by $2{\cdot}(2^n - 1)$. For example, two 8-bit numbers to be added could both represent their highest decimal value of $(2^n - 1)$ or 255_{10} (i.e., $1111\ 1111_2$). The sum of this number would result in 510_{10} or $(1\ 1111\ 1110_2)$. Notice that the largest sum achievable would only require one additional bit. This means that a single carry bit is sufficient to handle all possible magnitudes for binary addition.

2.3.2 Subtraction (Borrows)

Binary subtraction also mirrors longhand decimal subtraction. In subtraction, the formal terms for the two numbers being operated on are *minuend* and *subtrahend*. The subtrahend is subtracted from the minuend to find the *difference*. In longhand subtraction, the minuend is the top number and the subtrahend is the bottom number. For a given position, if the minuend is less than the subtrahend, it needs to *borrow* from the next higher-order position to produce a difference that is positive. If the next higher position does not have a value that can be borrowed from (i.e., 0), then it in turn needs to borrow from the next higher position, and so forth. Example 2.16 shows how subtraction is performed on two individual bits.

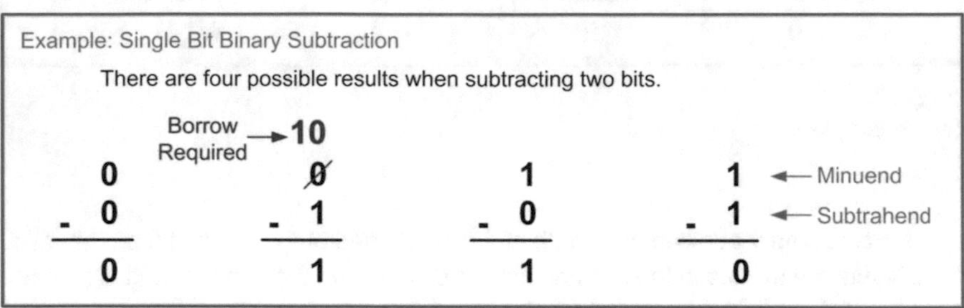

Example 2.16
Single-bit binary subtraction

As with binary addition, binary subtraction is accomplished on fixed widths of inputs and outputs (i.e., *n*-bits). The minuend and subtrahend are aligned at the radix point, and subtraction begins at the least significant bit position. Borrows are used as necessary as the subtractions move from the least significant position to the most significant position. If the most significant position requires a borrow, this is a uniquely named event. This event is called a *borrow in* or the subtraction is said to *require a borrow*. Again, the reason this event is uniquely named is because in real circuitry, the number of bits of the input and output is fixed in hardware and the borrow in is typically handled by a separate circuit. Example 2.17 shows this process when subtracting two 4-bit numbers.

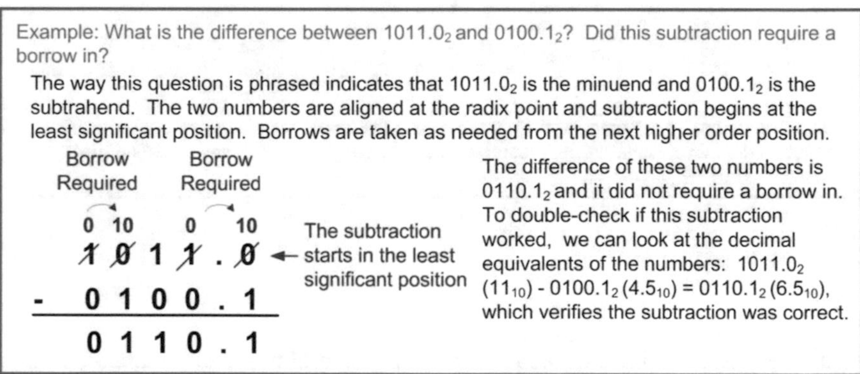

Example 2.17
Multiple-bit binary subtraction

Notice that if the minuend is less than the subtrahend, then the difference will be negative. At this point, we need a way to handle negative numbers.

CONCEPT CHECK

CC2.3 If an 8-bit computer system can only perform unsigned addition on 8-bit inputs and produce an 8-bit sum, how is it possible for this computer to perform addition on numbers that are larger than what can be represented with 8 bits (e.g., $1000_{10} + 1000_{10} = 2000_{10}$)?

A) There are multiple 8-bit adders in a computer to handle large numbers.

B) The result is simply rounded to the nearest 8-bit number.

C) The computer returns an error and requires smaller numbers to be entered.

D) The computer keeps track of the carry out and uses it in a subsequent 8-bit addition, which enables larger numbers to be handled.

2.4 Unsigned and Signed Numbers

All of the number systems presented in the prior sections were positive. We need to also have a mechanism to indicate negative numbers. When looking at negative numbers, we only focus on the mapping between decimal and binary since octal and hexadecimal are used as just another representation of a binary number. In decimal, we are able to use the negative *sign* in front of a number to indicate it is negative (e.g., -34_{10}). In binary, this notation works fine for writing numbers on paper (e.g., -1010_2), but we need a mechanism that can be implemented using real circuitry. In a real digital circuit, the circuits can only deal with 0's and 1's. There is no "−" in a digital circuit. Since we only have 0's and 1's in the hardware, we use a bit to represent whether a number is positive or negative. This is referred to as the *sign bit*. If a binary number is not going to have any negative values, then it is called an **unsigned** number, and it can only represent positive numbers. If a binary number is going to allow negative numbers, it is called a **signed** number. It is important to always keep track of the type of number we are using as the same bit values can represent very different numbers depending on the coding mechanism that is being used.

2.4.1 Unsigned Numbers

An unsigned number is one that does not allow negative numbers. When talking about this type of code, the number of bits is fixed and stated up front. We use the variable *n* to represent the number of bits in the number. For example, if we had an 8-bit number, we would say, "This is an 8-bit, unsigned number."

The number of unique codes in an unsigned number is given by 2^n. For example, if we had an 8-bit number, we would have 2^8 or 256 unique codes (e.g., $0000\ 0000_2$ to $1111\ 1111_2$).

The *range* of an unsigned number refers to the decimal values that the binary code can represent. If we use the notation $N_{unsigned}$ to represent any possible value that an *n*-bit, unsigned number can take on, the range would be defined as $0 < N_{unsigned} < (2^n - 1)$.

$$\textbf{Range of an UNSIGNED number} \Rightarrow \mathbf{0 \leq N_{unsigned} \leq (2^n - 1)}$$

For example, if we had an unsigned number with $n = 4$, it could take on a range of values from $+0_{10}$ (0000_2) to $+15_{10}$ (1111_2). Notice that while this number has 16 unique possible codes, the highest decimal value it can represent is 15_{10}. This is because one of the unique codes represents 0_{10}. This is the reason that the highest decimal value that can be represented is given by (2^n-1). Example 2.18 shows this process for a 16-bit number.

Example: What is the range of decimal numbers that an 16-bit, unsigned word can represent?

The term "16-bit word" means that the binary number has n=16. We can plug this into the equation for the range of an unsigned numbers directly.

$$0 \leq N_{unsigned} \leq (2^n - 1)$$

$$\downarrow$$

$$0 \leq N_{unsigned} \leq (2^{16} - 1)$$

$$\downarrow$$

$$0 \leq N_{unsigned} \leq (65{,}536 - 1)$$

$$\downarrow$$

$$0 \leq N_{unsigned} \leq 65{,}535$$

An unsigned 16-bit word can represent decimal numbers from 0 to 65,535.

Example 2.18
Finding the range of an unsigned number

2.4.2 Signed Numbers

Signed numbers are able to represent both positive and negative numbers. The most significant bit of these numbers is always the *sign bit*, which represents whether the number is positive or negative. The sign bit is defined to be a **0 if the number is positive** and **1 if the number is negative**. When using signed numbers, the number of bits is fixed so that the sign bit is always in the same position. There are a variety of ways to encode negative numbers using a sign bit. The encoding method used exclusively in modern computers is called *two's complement*. There are two other encoding techniques called *signed magnitude* and *one's complement* that are rarely used but are studied to motivate the power of two's complement. When talking about a signed number, the number of bits and the type of encoding are always stated. For example, we would say, "This is an 8-bit, two's complement number."

2.4.2.1 Signed Magnitude

Signed magnitude is the simplest way to encode a negative number. In this approach, the most significant bit (i.e., leftmost bit) of the binary number is considered the sign bit (0 = positive, 1 = negative). The rest of the bits to the right of the sign bit represent the magnitude or absolute value of the number. As an example of this approach, let's look at the decimal values that a 4-bit, signed magnitude number can take on. These are shown in Example 2.19.

Example: What decimal values can a 4-bit "Signed Magnitude" code represent?

Decimal	4-bit Signed Magnitude
-7	1111
-6	1110
-5	1101
-4	1100
-3	1011
-2	1010
-1	1001
-0	1000
0	0000
1	0001
2	0010
3	0011
4	0100
5	0101
6	0110
7	0111

↑ Sign bit

Example 2.19
Decimal values that a 4-bit, signed magnitude code can represent

There are drawbacks of signed magnitude encoding that are apparent from this example. First, the value of 0_{10} has two signed magnitude codes (0000_2 and 1000_2). This is an inefficient use of the available codes and leads to complexity when building arithmetic circuitry since it must account for two codes representing the same number.

The second drawback is that addition using the negative numbers does not directly map to how decimal addition works. For example, in decimal if we added (−5) + (1), the result would be −4. In signed magnitude, adding these numbers using a traditional adder would produce (−5) + (1) = (−6). This is because the traditional addition would take place on the magnitude portion of the number. A 5_{10} is represented with 101_2. Adding 1 to this number would result in the next higher binary code 110_2 or 6_{10}. Since the sign portion is separate, the addition is performed on |5|, thus yielding 6. Once the sign bit is included, the resulting number is −6. It is certainly possible to build an addition circuit that works on signed magnitude numbers, but it is more complex than a traditional adder because it must perform a different addition operation for the negative numbers versus the positive numbers. It is advantageous to have a single adder that works across the entire set of numbers.

Due to the duplicate codes for 0, the range of decimal numbers that signed magnitude can represent is reduced by 1 compared to unsigned encoding. For an *n*-bit number, there are 2^n unique binary codes available, but only $2^n - 1$ can be used to represent unique decimal numbers. If we use the notation N_{SM} to represent any possible value that an *n*-bit, signed magnitude number can take on, the range would be defined as:

$$\textbf{Range of a SIGNED MAGNITUDE number} \Rightarrow -\left(2^{n-1} - 1\right) \le N_{SM} \le +\left(2^{n-1} - 1\right)$$

Example 2.20 shows how to use this expression to find the range of decimal values that an 8-bit, signed magnitude code can represent.

Example: What is the range of decimal numbers that an 8-bit, signed magnitude number can represent?

The term "8-bit" means that n=8. We can plug this into the equation for the range of a signed magnitude number directly.

$$-(2^{n-1}-1) \leq N_{SM} \leq +(2^{n-1} - 1)$$

$$\downarrow$$

$$-(2^{8-1}-1) \leq N_{SM} \leq +(2^{8-1} - 1)$$

$$\downarrow$$

$$-127 \leq N_{SM} \leq +127$$

An 8-bit, signed magnitude number can represent decimal numbers from -127 to +127.

Example 2.20
Finding the range of a signed magnitude number

The process to determine the decimal value from a signed magnitude binary code involves treating the sign bit separately from the rest of the code. The sign bit provides the polarity of the decimal number (0 = positive, 1 = negative). The remaining bits in the code are treated as unsigned numbers and converted to decimal using the standard conversion procedure described in the prior sections. This conversion yields the magnitude of the decimal number. The final decimal value is found by applying the sign. Example 2.21 shows an example of this process.

Example: What is the decimal value of the 5-bit, signed magnitude code 11010_2?

The most significant bit of this 5-bit number is a 1, which indicates that the number is negative.

Sign Bit → 1 1010 ← Magnitude

The remaining 4-bits are the magnitude of the decimal number and are converted directly to decimal.

$$1\ 0\ 1\ 0_2$$

$$\downarrow$$

$$|\text{Value}| = \sum_{i=0}^{3} d_i \cdot 2^i$$

$$\downarrow$$

$$|\text{Value}| = 1 \cdot 2^3 + 0 \cdot 2^2 + 1 \cdot 2^1 + 0 \cdot 2^0$$

$$\downarrow$$

$$|\text{Value}| = 1 \cdot (8) + 0 \cdot (4) + 1 \cdot (2) + 0 \cdot (1)$$

$$\downarrow$$

$$|\text{Value}| = 8 + 0 + 2 + 0$$

$$\downarrow$$

$$|\text{Value}| = 10_{10}$$

The negative sign is then added back to the converted number giving a decimal value of -10_{10}

Example 2.21
Finding the decimal value of a signed magnitude number

2.4.2.2 One's Complement

One's complement is another simple way to encode negative numbers. In this approach, the negative number is obtained by taking its positive equivalent and flipping all of the 1's to 0's and 0's to 1's. This procedure of *flipping the bits* is called a **complement** (notice the two *e*'s). In this way, the most significant bit of the number is still the sign bit (0 = positive, 1 = negative). The rest of the bits represent the value of the number, but in this encoding scheme, the negative number values are less intuitive. As an example of this approach, let's look at the decimal values that a 4-bit, one's complement number can take on. These are shown in Example 2.22.

Example: What decimal values can a 4-bit "One's Complement" code represent?

Decimal	4-bit One's Complement
-7	1000
-6	1001
-5	1010
-4	1011
-3	1100
-2	1101
-1	1110
-0	1111
0	0000
1	0001
2	0010
3	0011
4	0100
5	0101
6	0110
7	0111

↳Sign bit

Example 2.22
Decimal values that a 4-bit, one's complement code can represent

Again, we notice that there are two different codes for 0_{10} (0000_2 and 1111_2). This is a drawback of one's complement because it reduces the possible range of numbers that can be represented from 2^n to (2^n-1) and requires arithmetic operations that take into account the gap in the number system. There are advantages of one's complement, however. First, the numbers are ordered such that traditional addition works on both positive and negative numbers (excluding the double 0 gap). Taking the example of $(-5) + (1)$ again, in one's complement the result yields -4, just as in a traditional decimal system. Notice in one's complement, -5_{10} is represented with 1010_2. Adding 1 to this entire binary code would result in the next higher binary code 1011_2 or -4_{10} from the above table. This makes addition circuitry less complicated, but still not as simple as if the double 0 gap was eliminated. Another advantage of one's complement is that as the numbers are incremented beyond the largest value in the set, they *roll over* and start counting at the lowest number. For example, if you increment the number 0111_2 (7_{10}), it goes to the next higher binary code 1000_2, which is -7_{10}. The ability to have the numbers roll over is a useful feature for computer systems.

If we use the notation N_{1comp} to represent any possible value that an n-bit, one's complement number can take on, the range is defined as:

$$\textbf{Range of a ONE'S COMPLEMENT number} \Rightarrow -\left(2^{n-1}-1\right) \leq N_{1's\ comp} \leq +\left(2^{n-1}-1\right)$$

Example 2.23 shows how to use this expression to find the range of decimal values that a 24-bit, one's complement code can represent.

Example: What is the range of decimal numbers that a 24-bit, one's complement number can represent?

The term "24-bit" means that n=24. We can plug this into the equation for the range of a one's complement number directly.

$$-(2^{n-1}-1) \le N_{1comp} \le +(2^{n-1}-1)$$

$$\downarrow$$

$$-(2^{24-1}-1) \le N_{1comp} \le +(2^{24-1}-1)$$

$$\downarrow$$

$$-8{,}388{,}607 \le N_{1comp} \le +8{,}388{,}607$$

A 24-bit, one's complement number can represent decimal numbers from -8,388,607 to +8,388,607.

Example 2.23
Finding the range of a 1's complement number

The process of finding the decimal value of a one's complement number involves first identifying whether the number is positive or negative by looking at the sign bit. If the number is positive (i.e., the sign bit is 0), then the number is treated as an unsigned code and is converted to decimal using the standard conversion procedure described in prior sections. If the number is negative (i.e., the sign bit is 1), then the number sign is recorded separately, and the code is complemented in order to convert it to its positive magnitude equivalent. This new positive number is then converted to decimal using the standard conversion procedure. As the final step, the sign is applied. Example 2.24 shows an example of this process.

Example: What is the decimal value of the 5-bit, one's complement code 11010_2?

The most significant bit of this 5-bit number is a 1, which indicates that the number is negative.

Sign Bit → **1**1010

To find the magnitude of the number, we first perform a complement on the entire number to find its positive equivalent.

$$1\ 1\ 0\ 1\ 0_2$$

$$\downarrow$$

$$0\ 0\ 1\ 0\ 1_2$$

A complement operation turns all 1's to 0's and all 0's to 1's

The number can now be converted into decimal to find its magnitude.

$$|Value| = \sum_{i=0}^{4} d_i \cdot 2^i$$

$$\downarrow$$

$$|Value| = 0\cdot2^4 + 0\cdot2^3 + 1\cdot2^2 + 0\cdot2^1 + 1\cdot2^0$$

$$\downarrow$$

$$|Value| = 0\cdot(16) + 0\cdot(8) + 1\cdot(4) + 0\cdot(2) + 1\cdot(1)$$

$$\downarrow$$

$$|Value| = 0 + 0 + 4 + 0 + 1 = 5_{10}$$

The negative sign is then added back to the converted number giving a decimal value of -5_{10}

Example 2.24
Finding the decimal value of a 1's complement number

2.4.2.3 Two's Complement

Two's complement is an encoding scheme that addresses the double 0 issue in signed magnitude and 1's complement representations. In this approach, the negative number is obtained by subtracting its positive equivalent from 2^n. This is identical to performing a complement on the positive equivalent and then adding one. If a carry is generated, it is discarded. This procedure is called *taking the two's complement of a number.* The procedure of complementing each bit and adding one is the most common technique to perform a two's complement. In this way, the most significant bit of the number is still the sign bit (0 = positive, 1 = negative), but all of the negative numbers are in essence *shifted up* so that the double 0 gap is eliminated. Taking the two's complement of a positive number will give its negative counterpart and vice versa. Let's look at the decimal values that a 4-bit, two's complement number can take on. These are shown in Example 2.25.

Example: What decimal values can a 4-bit "Two's Complement" code represent?

Decimal	4-bit Two's Complement
-8	1000
-7	1001
-6	1010
-5	1011
-4	1100
-3	1101
-2	1110
-1	1111
0	0000
1	0001
2	0010
3	0011
4	0100
5	0101
6	0110
7	0111

↑ Sign bit

Example 2.25
Decimal values that a 4-bit, two's complement code can represent

There are many advantages of two's complement encoding. First, there is no double 0 gap, which means that all possible 2^n unique codes that can exist in an *n*-bit number are used. This gives the largest possible range of numbers that can be represented. Another advantage of two's complement is that addition with negative numbers works exactly the same as decimal. In our example of (−5) + (1), the result is (−4). Arithmetic circuitry can be built to mimic the way our decimal arithmetic works without the need to consider the double 0 gap. Finally, the roll over characteristic is preserved from one's complement. Incrementing +7 by +1 will result in −8.

If we use the notation N_{2comp} to represent any possible value that an *n*-bit, two's complement number can take on, the range is defined as:

$$\textbf{Range of a TWO'S COMPLEMENT number} \Rightarrow -\left(2^{n-1}\right) \leq N_{2's\ comp} \leq +\left(2^{n-1}-1\right)$$

Example 2.26 shows how to use this expression to find the range of decimal values that a 32-bit, two's complement code can represent.

Example: What is the range of decimal numbers that a 32-bit, two's complement number can represent?

The term "32-bit" means that n=32. We can plug this into the equation for the range of a two's complement number directly.

$$-(2^{n-1}) \leq N_{2comp} \leq +(2^{n-1} - 1)$$

$$-(2^{32-1}) \leq N_{2comp} \leq +(2^{32-1} - 1)$$

$$-2,147,483,648 \leq N_{2comp} \leq +2,147,483,647$$

A 32-bit, two's complement number can represent decimal numbers from -2,147,483,648 to +2,147,483,647.

Example 2.26
Finding the range of a two's complement number

The process of finding the decimal value of a two's complement number involves first identifying whether the number is positive or negative by looking at the sign bit. If the number is positive (i.e., the sign bit is 0), then the number is treated as an unsigned code and is converted to decimal using the standard conversion procedure described in prior sections. If the number is negative (i.e., the sign bit is 1), then the number sign is recorded separately, and a two's complement is performed on the code in order to convert it to its positive magnitude equivalent. This new positive number is then converted to decimal using the standard conversion procedure. The final step is to apply the sign. Example 2.27 shows an example of this process.

Example: What is the decimal value of the 5-bit, 2's complement code 11010_2?

The most significant bit of this 5-bit number is a 1, which indicates that the number is negative.

Sign Bit → **1**1010

To find the magnitude of the number, we take the 2's complement of the entire number to find its positive equivalent.

Step 1 – Complement the number

$$1\,1\,0\,1\,0_2 \rightarrow 0\,0\,1\,0\,1_2$$

Step 2 – Add 1, ignore carry out if any

$$\begin{array}{r} 0\,0\,1\,0\,1 \\ +\quad\quad\;\; 1 \\ \hline 0\,0\,1\,1\,0_2 \end{array}$$

The number can now be converted into decimal to find its magnitude (i.e., $00110_2 = 6_{10}$). The negative sign is then added giving a final decimal value of -6_{10}.

Example 2.27
Finding the decimal value of a two's complement number

To convert a decimal number into its two's complement code, the range is first checked to determine whether the number can be represented with the allocated number of bits. The next step is to convert the decimal number into unsigned binary. The final step is to apply the sign bit. If the original decimal number was positive, then the conversion is complete. If the original decimal number was negative, then the two's complement is taken on the unsigned binary code to find its negative equivalent. Example 2.28 shows this procedure when converting -99_{10} to its 8-bit, two's complement code.

Example: What is the 8-bit, 2's complement code for -99_{10}?

Step 1 – Determine if -99_{10} can be represented within the 2's complement number range

An 8-bit, 2's complement number has a range of:

$$-(2^{n-1}) \leq N_{2comp} \leq +(2^{n-1} - 1)$$

$$\downarrow$$

$$-(2^{8-1}) \leq N_{2comp} \leq +(2^{8-1} - 1)$$

$$\downarrow$$

$$-128 \leq N_{2comp} \leq +127$$

Yes, the number -99_{10} falls within the range that an 8-bit, 2's complement number.

Step 2 – Find the positive binary code for -99_{10}

	Quotient	Remainder	
2 ⟌ 99	49	1	LSB
2 ⟌ 49	24	1	
2 ⟌ 24	12	0	
2 ⟌ 12	6	0	
2 ⟌ 6	3	0	
2 ⟌ 3	1	1	
2 ⟌ 1	0	1	MSB

Done

The converted 8-bit number is $0110\ 0011_2$.

Step 3 – Perform 2's Complement on the positive equivalent of 99_{10}

First, complement the number

$$0110\ 0011_2$$

$$\downarrow$$

$$1001\ 1100_2$$

Second, add 1, ignore carry out if any

$$\begin{array}{r} 1001\ 1100 \\ +\quad\quad\quad\quad 1 \\ \hline 1001\ 1101_2 \end{array}$$

The 8-bit, 2's complement code for -99_{10} is $1001\ 1101_2$

Example 2.28
Finding the two's complement code of a decimal number

2.4.2.4 Arithmetic with Two's Complement

Two's complement has a variety of arithmetic advantages. First, the operations of addition, subtraction, and multiplication are handled exactly the same as when using unsigned numbers. This means that duplicate circuitry is not needed in a system that uses both number types. Second, the ability to convert a number from positive to its negative representation by performing a *two's complement* means that an adder circuit can be used for subtraction. For example, if we wanted to perform the subtraction $13_{10} - 4_{10} = 9_{10}$, this is the same as performing $13_{10} + (-4_{10}) = 9_{10}$. This allows us to use a single adder circuit to perform both addition and subtraction as long as we have the ability to take the two's complement of a number. Creating a circuit to perform two's complement can be simpler and faster than building a separate subtraction circuit, so this approach can sometimes be advantageous.

There are specific rules for performing two's complement arithmetic that must be followed to ensure proper results. First, any carry or borrow that is generated is **ignored**. The second rule that must be followed is to always check if **two's complement overflow** occurred. Two's complement overflow refers to when the result of the operation falls outside of the range of values that can be represented by the number of bits being used. For example, if you are performing 8-bit, two's complement addition, the range of decimal values that can be represented is -128_{10} to $+127_{10}$. Having two input terms of 127_{10} ($0111\ 1111_2$) is perfectly legal because they can be represented by the 8 bits of the two's complement number; however, the summation of $127_{10} + 127_{10} = 254_{10}$ ($1111\ 1110_2$). This number does *not* fit within the range of values that can be represented and is actually the two's complement code for -2_{10}, which is obviously incorrect. Two's complement overflow occurs if any of the following occurs:

- The sum of like signs results in an answer with opposite sign (i.e., positive + positive = negative or negative + negative = positive).
- The subtraction of a positive number from a negative number results in a positive number (i.e., negative – positive = positive).
- The subtraction of a negative number from a positive number results in a negative number (i.e., positive – negative = negative).

Computer systems that use two's complement have a dedicated logic circuit that monitors for any of these situations and lets the operator know that overflow has occurred. These circuits are straightforward since they simply monitor the sign bits of the input and output codes. Example 2.29 shows how to use two's complement in order to perform subtraction using an addition operation.

Example: Use 4-bit, two's complement addition to find the differences between 6_{10} and 3_{10}.

The answer in decimal to this problem is $6_{10} - 3_{10} = 3_{10}$. Instead of using subtraction, we will use the two's complement representation of -3_{10} and add the two numbers.

$$\begin{array}{r} 6_{10} \\ -\ 3_{10} \\ \hline 3_{10} \end{array} \quad = \quad \begin{array}{r} 6_{10} \\ +\ (-3_{10}) \\ \hline 3_{10} \end{array}$$

Step 1 – Find the 4-bit, two's complement codes for $+6_{10}$ and -3_{10}.

Since 6 is positive, its code is simply its 4-bit binary equivalent ($+6_{10} = 0110_2$)

Since 3 is negative, we'll need to take the two's complement of its 4-bit positive binary equivalent ($+3_{10} = 0011_2$)

1) Complement the number

$$0011_2 \rightarrow 1100_2$$

2) Add 1, ignore carry out if any

$$\begin{array}{r} 1100 \\ +\quad 1 \\ \hline 1101_2 \end{array}$$

Step 2 – Add the two codes, ignore carry out if any

$$\begin{array}{r} 6_{10} \\ +\ (-3_{10}) \\ \hline 3_{10} \end{array} \quad = \quad \begin{array}{r} \scriptstyle 1 \\ 0\ 1\ 1\ 0_2 \\ +\ \ 1\ 1\ 0\ 1_2 \\ \hline 1\ 0\ 0\ 1\ 1_2 \end{array}$$

The sum resulted in a carry out, but in two's complement addition, this bit is ignored.

The result of the addition was 0011_2 or $+3_{10,}$ verifying that this approach was correct. Also, two's complement overflow did not occur because the result of this operation was within the range of possible values that a 4-bit, two's complement number can represent (e.g., -8_{10} to $+7_{10}$).

Example 2.29
Two's complement addition

Concept Check

CC2.4 A 4-bit, two's complement number has 16 unique codes and can represent decimal numbers between -8_{10} and $+7_{10}$. If the number of unique codes is even, why is it that the range of integers it can represent is not symmetrical about zero?

A) One of the positive codes is used to represent zero. This prevents the highest positive number from reaching $+8_{10}$ and being symmetrical.

B) It is asymmetrical because the system allows the numbers to roll over.

C) It isn't asymmetrical if zero is considered a positive integer. That way there are eight positive numbers and eight negatives numbers.

D) It is asymmetrical because there are duplicate codes for 0.

Summary

- The base, or radix, of a number system refers to the number of unique symbols within its set. The definition of a number system includes both the symbols used and the relative values of each symbol within the set.
- The most common number systems are base 10 (decimal), base 2 (binary), and base 16 (hexadecimal). Base 10 is used because it is how the human brain has been trained to treat numbers. Base 2 is used because the two values are easily represented using electrical switches. Base 16 is a convenient way to describe large groups of bits.
- A positional number system allows larger (or smaller) numbers to be represented beyond the values within the original symbol set. This is accomplished by having each position within a number have a different *weight*.
- There are specific algorithms that are used to convert any base to or from decimal. There are also algorithms to convert between number systems that contain a power-of-two symbols (e.g., binary to hexadecimal and hexadecimal to binary).
- Binary arithmetic is performed on a fixed width of bits (n). When an n-bit addition results in a sum that cannot fit within n-bits, it generates a *carry out* bit. In an n-bit subtraction, if the minuend is smaller than the subtrahend, a *borrow in* can be used to complete the operation.
- Binary codes can represent both unsigned and signed numbers. For an arbitrary n-bit binary code, it is important to know the encoding technique and the range of values that can be represented.
- Signed numbers use the most significant position to represent whether the number is negative (0 = positive, 1 = negative). The width of a signed number is always fixed.
- Two's complement is the most common encoding technique for signed numbers. It has an advantage that there are no duplicate codes for zero and that the encoding approach provides a monotonic progression of codes from the most negative number that can be represented to the most positive. This allows addition and subtraction to work the same on two's complement numbers as it does on unsigned numbers.
- When performing arithmetic using two's complement codes, the carry bit is ignored.
- When performing arithmetic using two's complement codes, if the result lies outside of the range that can be represented, it is called *two's complement overflow*. Two's complement overflow can be determined by looking at the sign bits of the input arguments and the sign bit of the result.

Exercise Problems

Section 2.1: Positional Number Systems

2.1.1 What is the radix of the binary number system?

2.1.2 What is the radix of the decimal number system?

2.1.3 What is the radix of the hexadecimal number system?

2.1.4 What is the radix of the octal number system?

2.1.5 What is the radix of a number system with *base 3*?

2.1.6 For the number 261.367, what position (p) is the number 2 in?

2.1.7 For the number 261.367, what position (p) is the number leftmost 6 in?

2.1.8 For the number 261.367, what position (p) is the number 1 in?

2.1.9 For the number 261.367, what position (p) is the number 3 in?

2.1.10 For the number 261.367, what position (p) is the number rightmost 6 in?

2.1.11 For the number 261.367, what position (p) is the number 7 in?

2.1.12 What is the name of the number system containing 10_2?

2.1.13 What is the name of the number system containing 10_{10}?

2.1.14 What is the name of the number system containing 10_{16}?

2.1.15 What is the name of the number system containing 10_8?

2.1.16 Which of the four number systems covered in this chapter (i.e., binary, decimal, hexadecimal, and octal) could the number 22 be part of? Give all that are possible.

2.1.17 Which of the four number systems covered in this chapter (i.e., binary, decimal, hexadecimal, and octal) could the number 99 be part of? Give all that are possible.

2.1.18 Which of the four number systems covered in this chapter (i.e., binary, decimal, hexadecimal, and octal) could the number 1F be part of? Give all that are possible.

2.1.19 Which of the four number systems covered in this chapter (i.e., binary, decimal, hexadecimal, and octal) could the number 88 be part of? Give all that are possible.

2.1.20 Which symbols could be used in all of the four number systems covered in this chapter (i.e., binary, decimal, hexadecimal, and octal)?

2.1.21 What is the only symbol that could be used in every number system from base 1 to base ∞?

Section 2.2: Base Conversions

2.2.1 If the number 101.111 has a radix of 2, what is the weight of the position containing the left-most 1?

2.2.2 If the number 101.111 has a radix of 2, what is the weight of the position containing the bit 0?

2.2.3 If the number 101.111 has a radix of 2, what is the weight of the position containing the 1 immediately to the left of the radix point?

2.2.4 If the number 101.111 has a radix of 2, what is the weight of the position containing the 1 immediately to the right of the radix point?

2.2.5 If the number 101.111 has a radix of 2, what is the weight of the position containing the 1 that second to the right of the radix point?

2.2.6 If the number 101.111 has a radix of 2, what is the weight of the position containing the right-most 1?

2.2.7 If the number 261.367 has a radix of 10, what is the weight of the position containing the numeral 2?

2.2.8 If the number 261.367 has a radix of 10, what is the weight of the position containing the left-most 6?

2.2.9 If the number 261.367 has a radix of 10, what is the weight of the position containing the numeral 1?

2.2.10 If the number 261.367 has a radix of 10, what is the weight of the position containing the numeral 3?

2.2.11 If the number 261.367 has a radix of 10, what is the weight of the position containing the right-most 6?

2.2.12 If the number 261.367 has a radix of 10, what is the weight of the position containing the numeral 7?

2.2.13 If the number 261.367 has a radix of 16, what is the weight of the position containing the numeral 2?

2.2.14 If the number 261.367 has a radix of 16, what is the weight of the position containing the left-most 6?

2.2.15 If the number 261.367 has a radix of 16, what is the weight of the position containing the numeral 1?

2.2.16 If the number 261.367 has a radix of 16, what is the weight of the position containing the numeral 3?

2.2.17 If the number 261.367 has a radix of 16, what is the weight of the position containing the right-most 6?

2.2.18 If the number 261.367 has a radix of 16, what is the weight of the position containing the numeral 7?

2.2.19 If the number 261.367 has a radix of 8, what is the weight of the position containing the numeral 2?

2.2.20 If the number 261.367 has a radix of 8, what is the weight of the position containing the left-most 6?

2.2.21 If the number 261.367 has a radix of 8, what is the weight of the position containing the numeral 1?

2.2.22 If the number 261.367 has a radix of 8, what is the weight of the position containing the numeral 3?

2.2.23 If the number 261.367 has a radix of 8, what is the weight of the position containing the right-most 6?

2.2.24 If the number 261.367 has a radix of 8, what is the weight of the position containing the numeral 7?

2.2.25 Convert $11\ 1111_2$ to decimal. Treat all numbers as unsigned.

2.2.26 Convert $10\ 1001_2$ to decimal. Treat all numbers as unsigned.

2.2.27 Convert $1100\ 1100_2$ to decimal. Treat all numbers as unsigned.

2.2.28 Convert $1001\ 1001_2$ to decimal. Treat all numbers as unsigned.

2.2.29 Convert 0.1111_2 to decimal. Provide the full answer without limiting its accuracy or rounding. Treat all numbers as unsigned.

2.2.30 Convert 0.1111_2 to decimal with a fractional accuracy of 2 digits without rounding. Treat all numbers as unsigned.

2.2.31 Convert 0.1111_2 to decimal with a fractional accuracy of 2 digits with rounding. Treat all numbers as unsigned.

2.2.32 Convert 11.01_2 to decimal. Provide the full answer without limiting its accuracy or rounding. Treat all numbers as unsigned.

2.2.33 Convert 11.01_2 to decimal with a fractional accuracy of 1 digit without rounding. Treat all numbers as unsigned.

2.2.34 Convert 11.01_2 to decimal with a fractional accuracy of 1 digit with rounding. Treat all numbers as unsigned.

2.2.35 Convert 1001.1001_2 to decimal. Provide the full answer without limiting its accuracy or rounding. Treat all numbers as unsigned.

2.2.36 Convert 1001.1001_2 to decimal with a fractional accuracy of 3 digits without rounding. Treat all numbers as unsigned.

2.2.37 Convert 1001.1001_2 to decimal with a fractional accuracy of 3 digits with rounding. Treat all numbers as unsigned.

2.2.38 Convert 1100.1101_2 to decimal. Provide the full answer without limiting its accuracy or rounding. Treat all numbers as unsigned.

2.2.39 Convert 1100.1101_2 to decimal with a fractional accuracy of 3 digits without rounding. Treat all numbers as unsigned.

2.2.40 Convert 1100.1101_2 to decimal with a fractional accuracy of 3 digits with rounding. Treat all numbers as unsigned.

2.2.41 Convert 72_8 to decimal. Treat all numbers as unsigned.

2.2.42 Convert 777_8 to decimal. Treat all numbers as unsigned.

2.2.43 Convert 123_8 to decimal. Treat all numbers as unsigned.

2.2.44 Convert 7654_8 to decimal. Treat all numbers as unsigned

2.2.45 Convert 0.777_8 to decimal. Provide the full answer without limiting its accuracy or rounding. Treat all numbers as unsigned.

2.2.46 Convert 0.777_8 to decimal with a fractional accuracy of 2 digits without rounding. Treat all numbers as unsigned.

2.2.47 Convert 0.777_8 to decimal with a fractional accuracy of 2 digits with rounding. Treat all numbers as unsigned.

2.2.48 Convert 12.57_8 to decimal. Provide the full answer without limiting its accuracy or rounding. Treat all numbers as unsigned.

2.2.49 Convert 12.57_8 to decimal with a fractional accuracy of 4 digits without rounding. Treat all numbers as unsigned.

2.2.50 Convert 12.57_8 to decimal with a fractional accuracy of 4 digits with rounding. Treat all numbers as unsigned.

2.2.51 Convert 123.123_8 to decimal. Provide the full answer without limiting its accuracy or rounding. Treat all numbers as unsigned.

2.2.52 Convert 123.123_8 to decimal with a fractional accuracy of 1 digit without rounding. Treat all numbers as unsigned.

2.2.53 Convert 123.123_8 to decimal with a fractional accuracy of 1 digit with rounding. Treat all numbers as unsigned.

2.2.54 Convert 7654.7654_8 to decimal. Provide the full answer without limiting its accuracy or rounding. Treat all numbers as unsigned.

2.2.55 Convert 7654.7654_8 to decimal with a fractional accuracy of 4 digits without rounding. Treat all numbers as unsigned.

2.2.56 Convert 7654.7654_8 to decimal with a fractional accuracy of 4 digits with rounding. Treat all numbers as unsigned.

2.2.57 Convert $F3_{16}$ to decimal. Treat all numbers as unsigned.

2.2.58 Convert FFF_{16} to decimal. Treat all numbers as unsigned.

2.2.59 Convert $FACE_{16}$ to decimal. Treat all numbers as unsigned.

2.2.60 Convert BEEF $FEED_{16}$ to decimal. Treat all numbers as unsigned.

2.2.61 Convert $0.FF_{16}$ to decimal. Provide the full answer without limiting its accuracy or rounding. Treat all numbers as unsigned.

2.2.62 Convert $0.FF_{16}$ to decimal with a fractional accuracy of 4 digits without rounding. Treat all numbers as unsigned.

2.2.63 Convert $0.FF_{16}$ to decimal with a fractional accuracy of 4 digits with rounding. Treat all numbers as unsigned.

2.2.64 Convert $EE.0F_{16}$ to decimal. Provide the full answer without limiting its accuracy or rounding. Treat all numbers as unsigned.

2.2.65 Convert $EE.0F_{16}$ to decimal with a fractional accuracy of 4 digits without rounding. Treat all numbers as unsigned.

2.2.66 Convert $EE.0F_{16}$ to decimal with a fractional accuracy of 4 digits with rounding. Treat all numbers as unsigned.

2.2.67 Convert $15B.CEF_{16}$ to decimal. Provide the full answer without limiting its accuracy or rounding. Treat all numbers as unsigned.

2.2.68 Convert $15B.CEF_{16}$ to decimal with a fractional accuracy of 2 digits without rounding. Treat all numbers as unsigned.

2.2.69 Convert $15B.CEF_{16}$ to decimal with a fractional accuracy of 2 digits with rounding. Treat all numbers as unsigned.

2.2.70 Convert $1ACE.E1F_{16}$ to decimal. Provide the full answer without limiting its accuracy or rounding. Treat all numbers as unsigned.

2.2.71 Convert $1ACE.E1F_{16}$ to decimal with a fractional accuracy of 4 digits without rounding. Treat all numbers as unsigned.

2.2.72 Convert $1ACE.E1F_{16}$ to decimal with a fractional accuracy of 4 digits with rounding. Treat all numbers as unsigned.

2.2.73 Convert 67_{10} to binary. Treat all numbers as unsigned.

2.2.74 Convert 100_{10} to binary. Treat all numbers as unsigned.

2.2.75 Convert 999_{10} to binary. Treat all numbers as unsigned.

2.2.76 Convert 1111_{10} to binary. Treat all numbers as unsigned.

2.2.77 Convert 0.875_{10} to binary. Provide the full answer without limiting its accuracy or rounding. Treat all numbers as unsigned.

2.2.78 Convert 0.875_{10} to binary with a fractional accuracy of 2 bits without rounding. Treat all numbers as unsigned.

2.2.79 Convert 0.875_{10} to binary with a fractional accuracy of 2 bits with rounding. Treat all numbers as unsigned.

2.2.80 Convert 1.4375_{10} to binary. Provide the full answer without limiting its accuracy or rounding. Treat all numbers as unsigned.

2.2.81 Convert 1.4375_{10} to binary with a fractional accuracy of 3 bits without rounding. Treat all numbers as unsigned.

2.2.82 Convert 1.4375_{10} to binary with a fractional accuracy of 3 bits with rounding. Treat all numbers as unsigned.

2.2.83 Convert 31.65625_{10} to binary. Provide the full answer without limiting its accuracy or rounding. Treat all numbers as unsigned.

2.2.84 Convert 31.65625_{10} to binary with a fractional accuracy of 3 bits without rounding. Treat all numbers as unsigned.

2.2.85 Convert 31.65625_{10} to binary with a fractional accuracy of 3 bits with rounding. Treat all numbers as unsigned.

2.2.86 Convert 252.987_{10} to binary. Provide the full answer without limiting its accuracy or rounding. Treat all numbers as unsigned.

2.2.87 Convert 252.987_{10} to binary with a fractional accuracy of 4 bits without rounding. Treat all numbers as unsigned.

2.2.88 Convert 252.987_{10} to binary with a fractional accuracy of 4 bits with rounding. Treat all numbers as unsigned.

2.2.89 Convert 67_{10} to octal. Treat all numbers as unsigned.

2.2.90 Convert 101_{10} to octal. Treat all numbers as unsigned.

2.2.91 Convert 777_{10} to octal. Treat all numbers as unsigned.

2.2.92 Convert 7654_{10} to octal. Treat all numbers as unsigned.

2.2.93 Convert $0.1875\ _{10}$ to octal. Provide the full answer without limiting its accuracy or rounding. Treat all numbers as unsigned.

2.2.94 Convert 0.1875_{10} to octal with a fractional accuracy of 1 digit without rounding. Treat all numbers as unsigned.

2.2.95 Convert 0.1875_{10} to octal with a fractional accuracy of 1 digit with rounding. Treat all numbers as unsigned.

2.2.96 Convert 4.5625_{10} to octal. Provide the full answer without limiting its accuracy or rounding. Treat all numbers as unsigned.

2.2.97 Convert 4.5625_{10} to octal with a fractional accuracy of 1 digit without rounding. Treat all numbers as unsigned.

2.2.98 Convert 4.5625_{10} to octal with a fractional accuracy of 1 digit with rounding. Treat all numbers as unsigned.

2.2.99 Convert 77.15625_{10} to octal. Provide the full answer without limiting its accuracy or rounding. Treat all numbers as unsigned.

2.2.100 Convert 77.15625_{10} to octal with a fractional accuracy of 1 digit without rounding. Treat all numbers as unsigned.

2.2.101 Convert 77.15625_{10} to octal with a fractional accuracy of 1 digit with rounding. Treat all numbers as unsigned.

2.2.102 Convert 22.2890625_{10} to octal. Provide the full answer without limiting its accuracy or rounding. Treat all numbers as unsigned.

2.2.103 Convert 22.2890625_{10} to octal with a fractional accuracy of 2 digits without rounding. Treat all numbers as unsigned.

2.2.104 Convert 22.2890625_{10} to octal with a fractional accuracy of 2 digits with rounding. Treat all numbers as unsigned.

2.2.105 Convert 67_{10} to hexadecimal. Treat all numbers as unsigned.

2.2.106 Convert 100_{10} to hexadecimal. Treat all numbers as unsigned.

2.2.107 Convert 999_{10} to hexadecimal. Treat all numbers as unsigned.

2.2.108 Convert 6789_{10} to hexadecimal. Treat all numbers as unsigned.

2.2.109 Convert 0.109375_{10} to hexadecimal. Provide the full answer without limiting its accuracy or rounding. Treat all numbers as unsigned.

2.2.110 Convert 0.109375_{10} to hexadecimal with a fractional accuracy of 1 digit without rounding. Treat all numbers as unsigned.

2.2.111 Convert 0.109375_{10} to hexadecimal with a fractional accuracy of 1 digit with rounding. Treat all numbers as unsigned.

2.2.112 Convert 10.6640625_{10} to hexadecimal. Provide the full answer without limiting its accuracy or rounding. Treat all numbers as unsigned.

2.2.113 Convert 10.6640625_{10} to hexadecimal with a fractional accuracy of 1 digit without rounding. Treat all numbers as unsigned.

2.2.114 Convert 10.6640625_{10} to hexadecimal with a fractional accuracy of 1 digit with rounding. Treat all numbers as unsigned.

2.2.115 Convert 186.66796875_{10} to hexadecimal. Provide the full answer without limiting its accuracy or rounding. Treat all numbers as unsigned.

2.2.116 Convert 186.66796875_{10} to hexadecimal with a fractional accuracy of 2 digits without rounding. Treat all numbers as unsigned.

2.2.117 Convert 186.66796875_{10} to hexadecimal with a fractional accuracy of 2 digits with rounding. Treat all numbers as unsigned.

2.2.118 Convert 57005.74560546875_{10} to hexadecimal. Provide the full answer without limiting its accuracy or rounding. Treat all numbers as unsigned.

2.2.119 Convert 57005.74560546875_{10} to hexadecimal with a fractional accuracy of 2 digits without rounding. Treat all numbers as unsigned.

2.2.120 Convert 57005.74560546875_{10} to hexadecimal with a fractional accuracy of 2 digits with rounding. Treat all numbers as unsigned.

2.2.121 Convert $11\ 1110_2$ to octal. Treat all numbers as unsigned.

2.2.122 Convert $10\ 1010.01_2$ to octal. Treat all numbers as unsigned.

2.2.123 Convert $1010\ 1010.0101_2$ to octal. Treat all numbers as unsigned.

2.2.124 Convert $1\ 0000\ 1111_2$ to octal. Treat all numbers as unsigned.

2.2.125 Convert $11\ 1110_2$ to hexadecimal. Treat all numbers as unsigned.

2.2.126 Convert $10\ 1010.01_2$ to hexadecimal. Treat all numbers as unsigned.

2.2.127 Convert $1010\ 1010.0101_2$ to hexadecimal. Treat all numbers as unsigned.

2.2.128 Convert $1\ 0000\ 1111.011_2$ to hexadecimal. Treat all numbers as unsigned.

2.2.129 Convert 77_8 to binary. Treat all numbers as unsigned.

2.2.130 Convert 77.7_8 to binary. Treat all numbers as unsigned.

2.2.131 Convert 123.4_8 to binary. Treat all numbers as unsigned.

2.2.132 Convert 261.367_8 to binary. Treat all numbers as unsigned.

2.2.133 Convert AB_{16} to binary. Treat all numbers as unsigned.

2.2.134 Convert $F.A_{16}$ to binary. Treat all numbers as unsigned.

2.2.135 Convert $AB.CD_{16}$ to binary. Treat all numbers as unsigned.

2.2.136 Convert 261.367_{16} to binary. Treat all numbers as unsigned.

2.2.137 Convert 66_8 to hexadecimal. Treat all numbers as unsigned.

2.2.138 Convert 66.7_8 to hexadecimal. Treat all numbers as unsigned.

2.2.139 Convert 261.367_8 to hexadecimal. Treat all numbers as unsigned.

2.2.140 Convert 1234.5678_8 to hexadecimal. Treat all numbers as unsigned.

2.2.141 Convert AB_{16} to octal. Treat all numbers as unsigned.

2.2.142 Convert $AB.D_{16}$ to octal. Treat all numbers as unsigned.

2.2.143 Convert $ABC.DE_{16}$ to octal. Treat all numbers as unsigned.

2.2.144 Convert $BABE.FACE_{16}$ to octal. Treat all numbers as unsigned.

Section 2.3: Binary Arithmetic

2.3.1 Compute $11_2 + 01_2$ by hand. Treat all numbers as unsigned. Provide the 2-bit sum, and indicate whether a *carry out* occurred.

2.3.2 Compute $1010_2 + 1011_2$ by hand. Treat all numbers as unsigned. Provide the 4-bit sum, and indicate whether a *carry out* occurred.

2.3.3 Compute $1111\ 1111_2 + 0000\ 0001_2$ by hand. Treat all numbers as unsigned. Provide the 8-bit sum, and indicate whether a *carry out* occurred.

2.3.4 Compute $1010.1010_2 + 1011.1011_2$ by hand. Treat all numbers as unsigned. Provide the 8-bit sum, and indicate whether a *carry out* occurred.

2.3.5 Compute $1111\ 1111.1011_2 + 0000\ 0001.1100_2$ by hand. Treat all numbers as unsigned. Provide the 12-bit sum, and indicate whether a *carry out* occurred.

2.3.6 Compute $10_2 - 01_2$ by hand. Treat all numbers as unsigned. Provide the 2-bit difference, and indicate whether a *borrow in* occurred.

2.3.7 Compute $1010_2 - 1011_2$ by hand. Treat all numbers as unsigned. Provide the 4-bit difference, and indicate whether a *borrow in* occurred.

2.3.8 Compute $1111\ 1111_2 - 0000\ 0001_2$ by hand. Treat all numbers as unsigned. Provide the 8-bit difference, and indicate whether a *borrow in* occurred.

2.3.9 Compute $1010.1010_2 - 1011.1011_2$ by hand. Treat all numbers as unsigned. Provide the 8-bit difference, and indicate whether a *borrow in* occurred.

2.3.10 Compute $1111\ 1111.1011_2 - 0000\ 0001.1100_2$ by hand. Treat all numbers as unsigned. Provide the 12-bit difference, and indicate whether a *borrow in* occurred.

Section 2.4: Unsigned and Signed Numbers

2.4.1 What range of decimal numbers can be represented by 8-bit, two's complement numbers?

2.4.2 What range of decimal numbers can be represented by 16-bit, two's complement numbers?

2.4.3 What range of decimal numbers can be represented by 32-bit, two's complement numbers?

2.4.4 What range of decimal numbers can be represented by 64-bit, two's complement numbers?

2.4.5 What is the 8-bit, two's complement code for $+88_{10}$?

2.4.6 What is the 8-bit, two's complement code for -88_{10}?

2.4.7 What is the 8-bit, two's complement code for -128_{10}?

2.4.8 What is the 8-bit, two's complement code for -1_{10}?

2.4.9 What is the decimal value of the 4-bit, two's complement code 0010_2?

2.4.10 What is the decimal value of the 4-bit, two's complement code 1010_2?

2.4.11 What is the decimal value of the 8-bit, two's complement code $0111\ 1110_2$?

2.4.12 What is the decimal value of the 8-bit, two's complement code $1111\ 1110_2$?

2.4.13 Compute $1110_2 + 1011_2$ by hand. Treat all numbers as 4-bit, two's complement codes. Provide the 4-bit sum and indicate whether *two's complement overflow* occurred.

2.4.14 Compute $1101\ 1111_2 + 0000\ 0001_2$ by hand. Treat all numbers as 8-bit, two's complement codes. Provide the 8-bit sum, and indicate whether *two's complement overflow* occurred.

2.4.15 Compute $1010.1010_2 + 1000.1011_2$ by hand. Treat all numbers as 8-bit, two's complement codes. Provide the 8-bit sum, and indicate whether *two's complement overflow* occurred.

2.4.16 Compute $1110\ 1011.1001_2 + 0010\ 0001.1101_2$ by hand. Treat all numbers as 12-bit, two's complement codes. Provide the 12-bit sum, and indicate whether *two's complement overflow* occurred.

2.4.17 Compute $4_{10} - 5_{10}$ using 4-bit two's complement addition. You will need to first convert each number into its 4-bit two's complement code and then perform binary addition (i.e., $4_{10} + (-5_{10})$). Provide the 4-bit result, and indicate whether two's complement overflow occurred. Check your work by converting the 4-bit result back to decimal.

2.4.18 Compute $7_{10} - 7_{10}$ using 4-bit two's complement addition. You will need to first convert each decimal number into its 4-bit two's complement code and then perform binary addition (i.e., $7_{10} + (-7_{10})$). Provide the 4-bit result, and indicate whether two's complement overflow occurred. Check your work by converting the 4-bit result back to decimal.

2.4.19 Compute $7_{10} + 1_{10}$ using 4-bit two's complement addition. You will need to first convert each decimal number into its 4-bit two's complement code and then perform binary addition. Provide the 4-bit result, and indicate whether two's complement overflow occurred. Check your work by converting the 4-bit result back to decimal.

2.4.20 Compute $64_{10}-100_{10}$ using 8-bit two's complement addition. You will need to first convert each number into its 8-bit two's complement code and then perform binary addition (i.e, $64_{10} + (-100_{10})$). Provide the 8-bit result and indicate whether two's complement overflow occurred. Check your work by converting the 8-bit result back to decimal.

2.4.21 Compute $(-99)_{10} - 11_{10}$ using 8-bit two's complement addition. You will need to first convert each decimal number into its 8-bit two's complement code and then perform binary addition (i.e., $(-99_{10}) + (-11_{10})$). Provide the 8-bit result, and indicate whether two's complement overflow occurred. Check your work by converting the 8-bit result back to decimal.

2.4.22 Compute $50_{10} + 100_{10}$ using 8-bit two's complement addition. You will need to first convert each decimal number into its 8-bit two's complement code and then perform binary addition. Provide the 8-bit result, and indicate whether two's complement overflow occurred. Check your work by converting the 8-bit result back to decimal.

Chapter 3: Digital Circuitry and Interfacing

Now we turn our attention to the physical circuitry and electrical quantities that are used to represent and operate on the binary codes 1 and 0. In this chapter we begin by looking at how logic circuits are described and introduce the basic set of gates used for all digital logic operations. We then look at the underlying circuitry that implements the basic gates including digital signaling and how voltages are used to represent 1's and 0's. We then look at interfacing between two digital circuits and how to ensure that when one circuit sends a binary code, the receiving circuit is able to determine which code was sent. Logic families are then introduced, and the details of how basic gates are implemented at the switch level are presented. Finally, interfacing considerations are covered for the most common types of digital loads (i.e., other gates, resistors, and LEDs). The goal of this chapter is to provide an understanding of the basic electrical operation of digital circuits.

Learning Outcomes—After completing this chapter, you will be able to:

3.1 Describe the functional operation of a basic logic gate using truth tables, logic expressions, and logic waveforms.
3.2 Analyze the DC and AC behavior of a digital circuit to verify it is operating within specification.
3.3 Describe the meaning of a logic family and the operation of the most common technologies used today.
3.4 Determine the operating conditions of a logic circuit when driving various types of loads.

3.1 Basic Gates

The term *gate* is used to describe a digital circuit that implements the most basic functions possible within the binary system. When discussing the operation of a logic gate, we ignore the details of how the 1's and 0's are represented with voltages and manipulated using transistors. We instead treat the inputs and output as simply ideal 1's and 0's. This allows us to design more complex logic circuits without going into the details of the underlying physical hardware.

3.1.1 Describing the Operation of a Logic Circuit

3.1.1.1 The Logic Symbol

A logic symbol is a graphical representation of the circuit that can be used in a schematic to show how circuits in a system interface to one another. For the set of basic logic gates, there are uniquely shaped symbols that graphically indicate their functionality. For more complex logic circuits that are implemented with multiple basic gates, a simple rectangular symbol is used. Inputs of the logic circuit are typically shown on the left of the symbol and outputs are on the right. Figure 3.1 shows two example logic symbols.

© Springer Nature Switzerland AG 2019
B. J. LaMeres, *Introduction to Logic Circuits & Logic Design with Verilog*,
https://doi.org/10.1007/978-3-030-13605-5_3

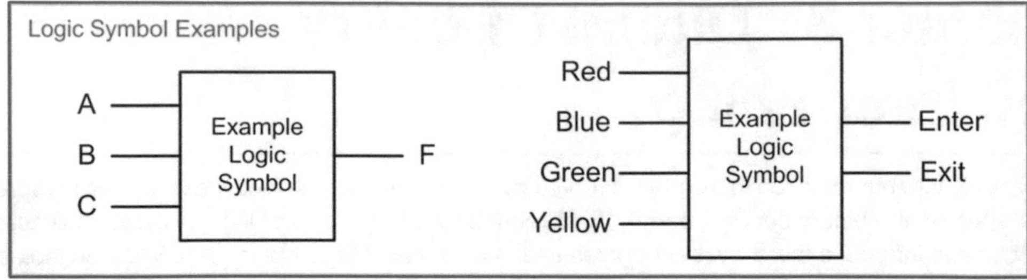

Fig. 3.1
Example logic symbols

3.1.1.2 The Truth Table

We formally define the functionality of a logic circuit using a *truth table*. In a truth table, each and every possible input combination is listed, and the corresponding output of the logic circuit is given. If a logic circuit has n inputs, then it will have 2^n possible input codes. The binary codes are listed in ascending order within the truth table mimicking a binary count starting at 0. By always listing the input codes in this way, we can assign a *row number* to each input that is the decimal equivalent of the binary input code. Row numbers can be used to simplify the notation for describing the functionality of larger circuits. Figure 3.2 shows the formation of an example 3-input truth table.

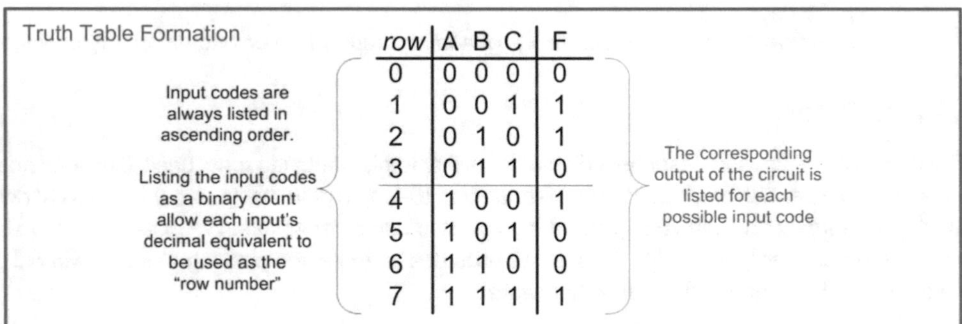

row	A	B	C	F
0	0	0	0	0
1	0	0	1	1
2	0	1	0	1
3	0	1	1	0
4	1	0	0	1
5	1	0	1	0
6	1	1	0	0
7	1	1	1	1

Fig. 3.2
Truth table formation

3.1.1.3 The Logic Function

A logic *expression*, (also called a *logic function*), is an equation that provides the functionality of each output in the circuit as a function of the inputs. The logic operations for the basic gates are given a symbolic set of operators (e.g., +, ·, ⊕), the details of which will be given in the next sections. The logic function describes the operations that are necessary to produce the outputs listed in the truth table. A logic function is used to describe a single output that can take on only the values 1 and 0. If a circuit contains multiple outputs, then a logic function is needed for each output. The input variables can be included in the expression description just as in an analog function. For example, "F(A,B,C) = . . ." would state that "F is a function of the inputs A, B, and C." This can also be written as "$F_{A,B,C}$ = ..." The input variables can also be excluded for brevity as in "F = ..." Figure 3.3 shows the formation of an example 3-input logic expression.

Logic Expression Formation

A	B	C	F
0	0	0	0
0	0	1	1
0	1	0	1
0	1	1	0
1	0	0	1
1	0	1	0
1	1	0	0
1	1	1	1

=

$$F(A,B,C) = A \oplus B \oplus C$$

or

$$F_{A,B,C} = A \oplus B \oplus C$$

or

$$F = A \oplus B \oplus C$$

Fig. 3.3
Logic function formation

3.1.1.4 The Logic Waveform

A logic *waveform* is a graphical depiction of the relationship of the output to the inputs with respect to time. This is often a useful description of behavior since it mimics the format that is typically observed when measuring a real digital circuit using test equipment such as an oscilloscope. In the waveform, each signal can only take on a value of 1 or 0. It is useful to write the logic values of the signal at each transition in the waveform for readability. Figure 3.4 shows an example logic waveform.

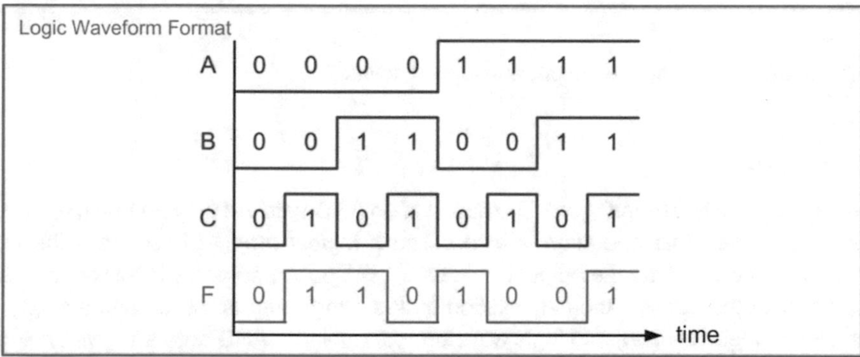

Fig. 3.4
Example logic waveform

3.1.2 The Buffer

The first basic gate is the *buffer*. The output of a buffer is simply the input. The logic symbol, truth table, logic function, and logic waveform for the buffer are given in Fig. 3.5.

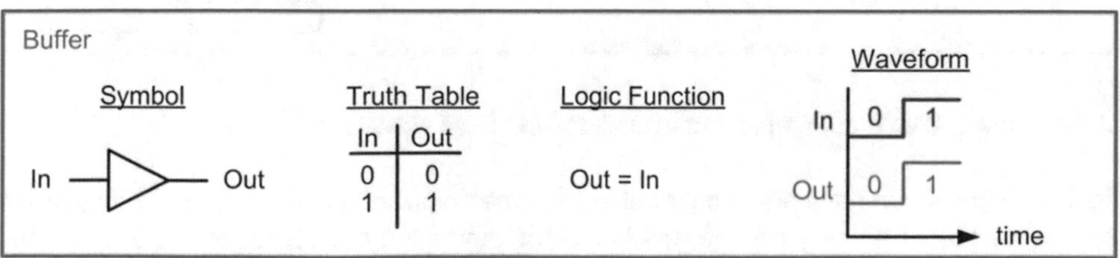

Fig. 3.5
Buffer symbol, truth table, logic function, and logic waveform

3.1.3 The Inverter

The next basic gate is the *inverter.* The output of an inverter is the complement of the input. Inversion is also often called the *not* operation. In spoken word, we might say "A is equal to *not* B;" thus this gate is also often called a *not* gate. The symbol for the inverter is the same as the buffer with the exception that an *inversion bubble* (i.e., a circle) is placed on the output. The inversion bubble is a common way to show inversions in schematics and will be used by many of the basic gates. In the logic function, there are two common ways to show this operation. The first way is by placing a prime (') after the input variable (e.g., Out = In′). This notation has the advantage that it is supported in all text editors but has the drawback that it can sometimes be difficult to see. The second way to indicate inversion in a logic function is by placing an *inversion bar* over the input variable (e.g., Out $= \overline{In}$). The advantage of this notation is that it is easy to see but has the drawback that it is not supported by many text editors. In this text, both conventions will be used to provide exposure to each. The logic symbol, truth table, logic function, and logic waveform for the inverter are given in Fig. 3.6.

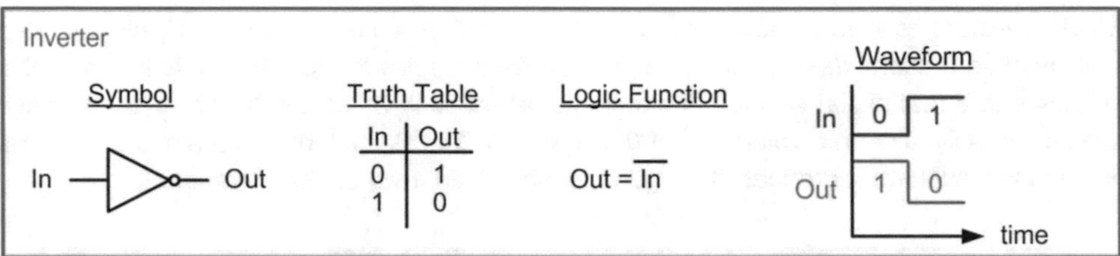

Fig. 3.6
Inverter symbol, truth table, logic function, and logic waveform

3.1.4 The AND Gate

The next basic gate is the *AND gate*. The output of an AND gate will only be true (i.e., a logic 1) if **all of the inputs are true**. This operation is also called a *logical product* because if the inputs were multiplied together, the only time the output would be a 1 is if each and every input was a 1. As a result, the logic operator is the dot (·). Another notation that is often seen is the ampersand (&). The logic symbol, truth table, logic function, and logic waveform for a 2-input AND gate are given in Fig. 3.7.

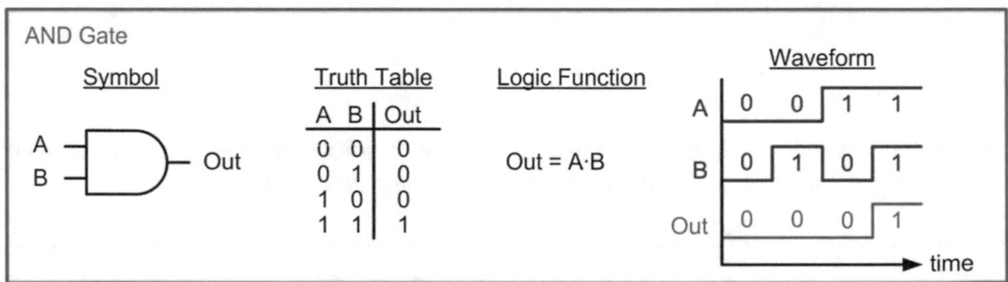

Fig. 3.7
2-input AND gate symbol, truth table, logic function, and logic waveform

Ideal AND gates can have any number of inputs. The operation of an n-bit, AND gates still follows the rule that the output will only be true when all of the inputs are true. Later sections will discuss the limitations on expanding the number of inputs of these basic gates indefinitely.

3.1.5 The NAND Gate

The NAND gate is identical to the AND gate with the exception that the output is inverted. The "N" in NAND stands for "NOT," which represents the inversion. The symbol for a NAND gate is an AND gate with an inversion bubble on the output. The logic expression for a NAND gate is the same as an AND gate but with an inversion bar over the entire operation. The logic symbol, truth table, logic function, and logic waveform for a 2-input NAND gate are given in Fig. 3.8. Ideal NAND gates can have any number of inputs with the operation of an n-bit, NAND gate following the rule that the output is always the inversion of an n-bit, AND operation.

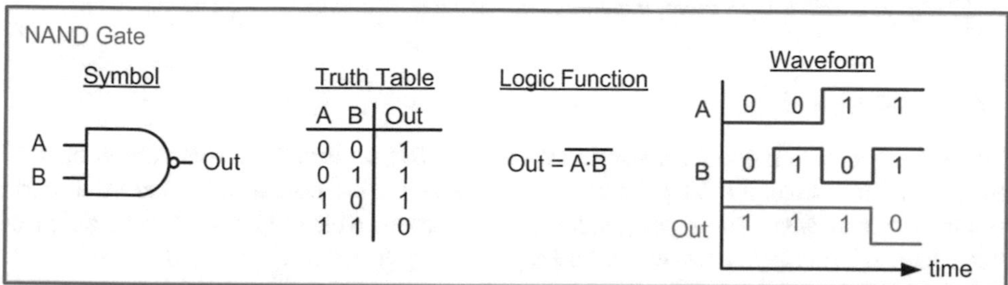

Fig. 3.8
2-input NAND gate symbol, truth table, logic function, and logic waveform

3.1.6 The OR Gate

The next basic gate is the *OR gate*. The output of an OR gate will be true when **any of the inputs are true**. This operation is also called a *logical sum* because of its similarity to logical disjunction in which the output is true if *at least one* of the inputs is true. As a result, the logic operator is the plus sign (+). The logic symbol, truth table, logic function, and logic waveform for a 2-input OR gate are given in Fig. 3.9. Ideal OR gates can have any number of inputs. The operation of an n-bit, OR gates still follows the rule that the output will be true if any of the inputs are true.

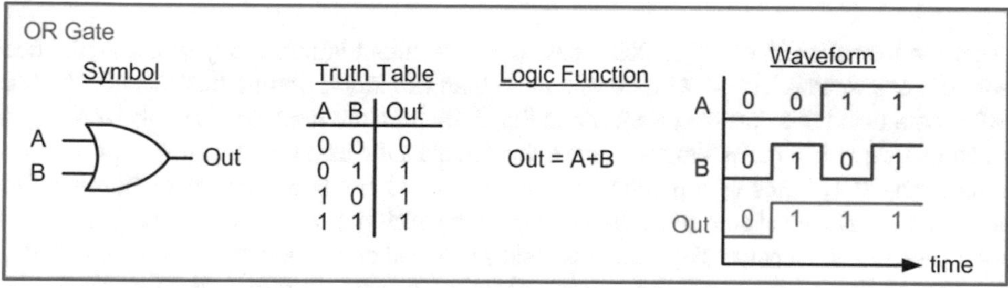

Fig. 3.9
2-input OR gate symbol, truth table, logic function, and logic waveform

3.1.7 The NOR Gate

The NOR gate is identical to the OR gate with the exception that the output is inverted. The symbol for a NOR gate is an OR gate with an inversion bubble on the output. The logic expression for a NOR gate is the same as an OR gate but with an inversion bar over the entire operation. The logic symbol, truth table, logic function, and logic waveform for a 2-input NOR gate are given in Fig. 3.10. Ideal NOR gates can have any number of inputs with the operation of an n-bit, NOR gate following the rule that the output is always the inversion of an n-bit, OR operation.

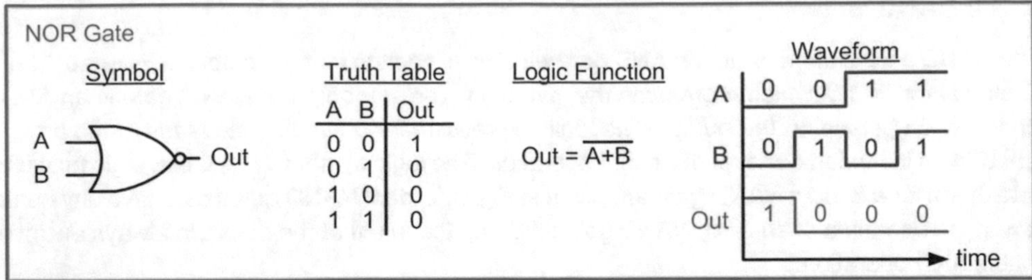

Fig. 3.10
2-input NOR gate symbol, truth table, logic function, and logic waveform

3.1.8 The XOR Gate

The next basic gate is the *exclusive-OR gate* or XOR gate for short. This gate is also called a *difference* gate because for the 2-input configuration, its output will be true when the **input codes are different from one another**. The logic operator is a circle around a plus sign ($\oplus$). The logic symbol, truth table, logic function, and logic waveform for a 2-input XOR gate are given in Fig. 3.11.

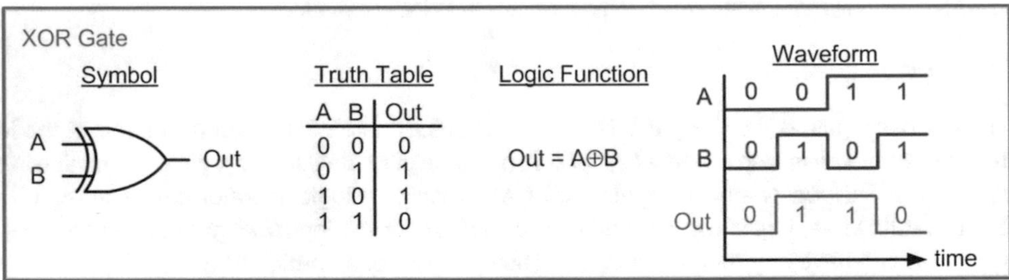

Fig. 3.11
2-input XOR gate symbol, truth table, logic function, and logic waveform

Using the formal definition of an XOR gate (i.e., the output is true if any of the input codes are different from one another), an XOR gate with more than two inputs can be built. The truth table for a 3-bit, XOR gate using this definition is shown in Fig. 3.12. In modern electronics, this type of gate has found little use since it is much simpler to build this functionality using a combination of AND and OR gates. As such, XOR gates with greater than two inputs do not implement the *difference* function. Instead, a more useful functionality has been adopted in which the output of the n-bit, XOR gate is the result of a cascade of 2-input XOR gates. This results in an ultimate output that is true when there is **an ODD number of 1's on the inputs**. This functionality is much more useful in modern electronics for error correction codes and arithmetic. As such, this is the functionality that is seen in modern n-bit, XOR gates. This functionality is also shown in Fig. 3.12.

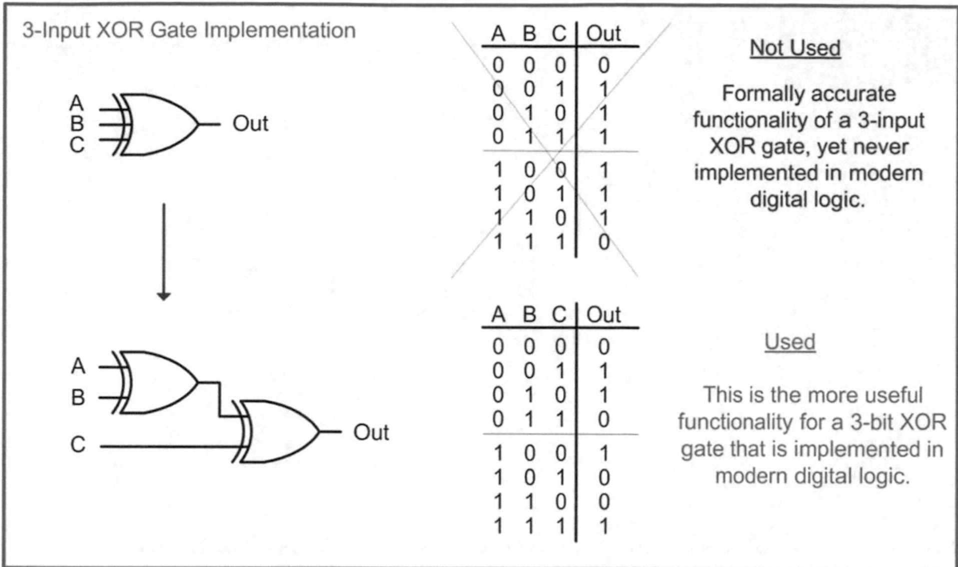

Fig. 3.12
3-input XOR gate implementation

3.1.9 The XNOR Gate

The exclusive-NOR gate is identical to the XOR gate with the exception that the output is inverted. This gate is also called an *equivalence* gate because for the 2-input configuration, its output will be true when the **input codes are equivalent to one another**. The symbol for an XNOR gate is an XOR gate with an inversion bubble on the output. The logic expression for an XNOR gate is the same as an XOR gate but with an inversion bar over the entire operation. The logic symbol, truth table, logic function, and logic waveform for a 2-input XNOR gate are given in Fig. 3.13. XNOR gates can have any number of inputs with the operation of an n-bit, XNOR gate following the rule that the output is always the inversion of an n-bit, XOR operation (i.e., the output is true if there is an ODD number of 1's on the inputs).

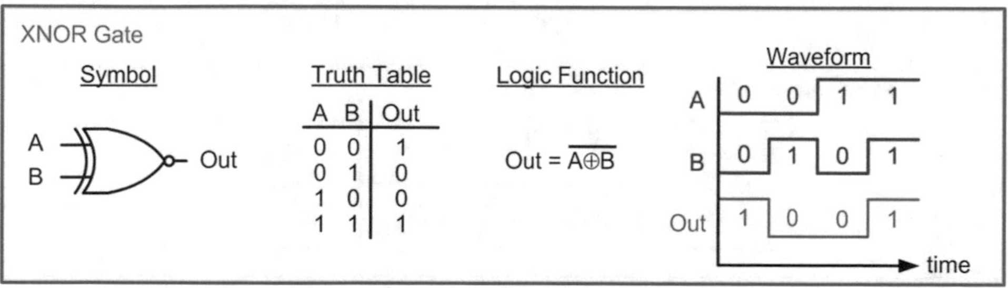

Fig. 3.13
2-input XNOR gate symbol, truth table, logic function, and logic waveform

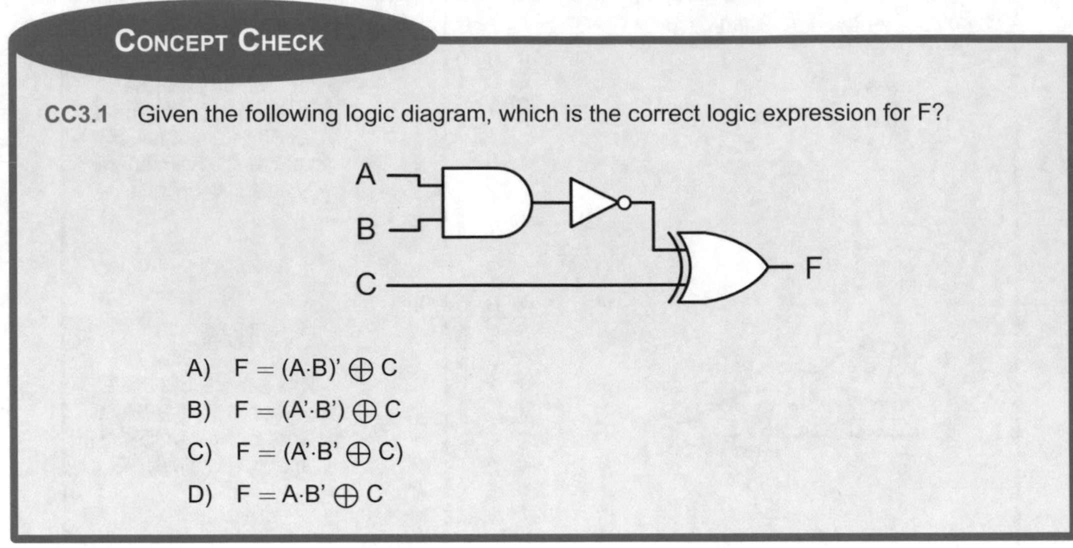

Concept Check

CC3.1 Given the following logic diagram, which is the correct logic expression for F?

A) F = (A·B)' ⊕ C
B) F = (A'·B') ⊕ C
C) F = (A'·B' ⊕ C)
D) F = A·B' ⊕ C

3.2 Digital Circuit Operation

Now we turn our attention to the physical hardware that is used to build the basic gates just described and how electrical quantities are used to represent and communicate the binary values 1 and 0. We begin by looking at digital signaling. Digital signaling refers to how binary codes are generated and transmitted successfully between two digital circuits using electrical quantities (e.g., voltage and current). Consider the digital system shown in Fig. 3.14. In this system, the sending circuit generates a binary code. The sending circuit is called either the *transmitter* (Tx) or *driver*. The transmitter represents the binary code using an electrical quantity such as voltage. The receiving circuit (Rx) observes this voltage and is able to determine the value of the binary code. In this way, 1's and 0's can be communicated between the two digital circuits. The transmitter and receiver are both designed to use the same digital signaling scheme so that they are able to communicate with each other. It should be noted that all digital circuits contain both inputs (Rx) and outputs (Tx) but are not shown in this figure for simplicity.

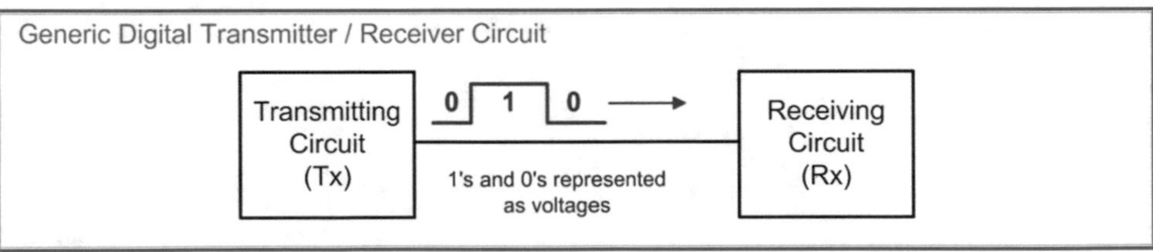

Fig. 3.14
Generic digital transmitter/receiver circuit

3.2.1 Logic Levels

A *logic level* is the term to describe all possible states that a signal can have. We will focus explicitly on circuits that represent binary values, so these will only have two finite states (1 and 0). To begin, we define a simplistic model of how to represent the binary codes using an electrical quantity. This model uses a voltage threshold (V_{th}) to represent the switching point between the binary codes. If the voltage of the signal (V_{sig}) is above this threshold, it is considered a *logic HIGH*. If the voltage is below this threshold, it is considered a *logic LOW*. A graphical depiction of this is shown in Fig. 3.15. The terms HIGH and LOW are used to describe which logic level corresponds to the *higher* or *lower* voltage.

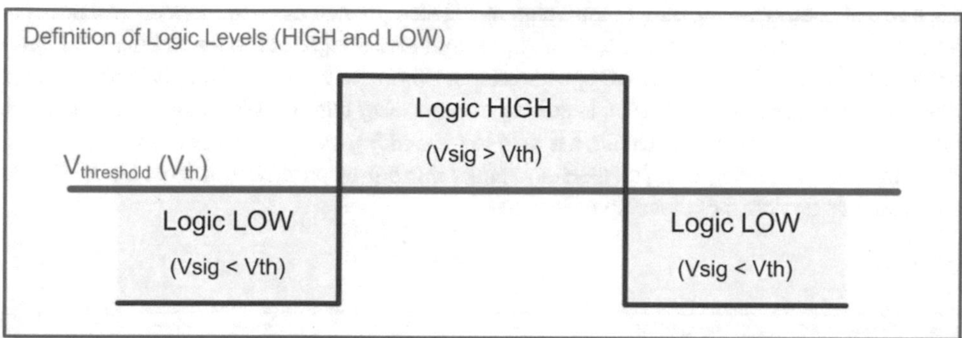

Fig. 3.15
Definition of logic HIGH and LOW

It is straightforward to have the HIGH level correspond to the binary code 1 and the LOW level correspond to the binary code 0; however, it is equally valid to have the HIGH level correspond to the binary code 0 and the LOW level correspond to the binary code 1. As such, we need to define how the logic levels HIGH and LOW map to the binary codes 1 and 0. We define two types of digital assignments: positive logic and negative logic. In **positive logic**, the logic HIGH level represents a binary 1, and the logic LOW level represents a binary 0. In n**egative logic**, the logic HIGH level represents a binary 0, and the logic LOW level represents a binary 1. Table 3.1 shows the definition of positive and negative logic. There are certain types of digital circuits that benefit from using negative logic; however, we will focus specifically on systems that use positive logic since it is more intuitive when learning digital design for the first time. The transformation between positive and negative logic is straightforward and will be covered in Chap. 4.

Definition of Positive and Negative Logic

Logic Level	Logic Value	
	Positive Logic	Negative Logic
LOW	0	1
HIGH	1	0

Table 3.1
Definition of positive and negative logic

3.2.2 Output DC Specifications

Transmitting circuits provide specifications on the range of output voltages (V_O) that they are guaranteed to provide when outputting a logic 1 or 0. These are called the DC output specifications. There are four DC voltage specifications that specify this range: $V_{OH\text{-}max}$, $V_{OH\text{-}min}$, $V_{OL\text{-}max}$, and $V_{OL\text{-}min}$.

The $V_{OH\text{-}max}$ and $V_{OH\text{-}min}$ specifications provide the range of voltages the transmitter is guaranteed to provide when outputting a logic HIGH (or logic 1 when using positive logic). The $V_{OL\text{-}max}$ and $V_{OL\text{-}min}$ specifications provide the range of voltages the transmitter is guaranteed to provide when outputting a logic LOW (or logic 0 when using positive logic). In the subscripts for these specifications, the "O" signifies "output" and the "L" or "H" signifies "LOW" or "HIGH," respectively.

The maximum amount of current that can flow through the transmitter's output (I_O) is also specified. The specification $I_{OH\text{-}max}$ is the maximum amount of current that can flow through the transmitter's output when sending a logic HIGH. The specification $I_{OL\text{-}max}$ is the maximum amount of current that can flow through the transmitter's output when sending a logic LOW. When the maximum output currents are violated, it usually damages the part. Manufacturers will also provide a *recommended* amount of current for I_O that will guarantee the specified operating parameters throughout the life of the part. Figure 3.16 shows a graphical depiction of these DC specifications. When the transmitter output is providing current to the receiving circuit (aka, the *load*), it is said to be ***sourcing*** current. When the transmitter output is drawing current from the receiving circuit, it is said to be ***sinking*** current. In most cases, the transmitter sources current when driving a logic HIGH and sinks current when driving a logic LOW. Figure 3.16 shows a graphical depiction of these specifications.

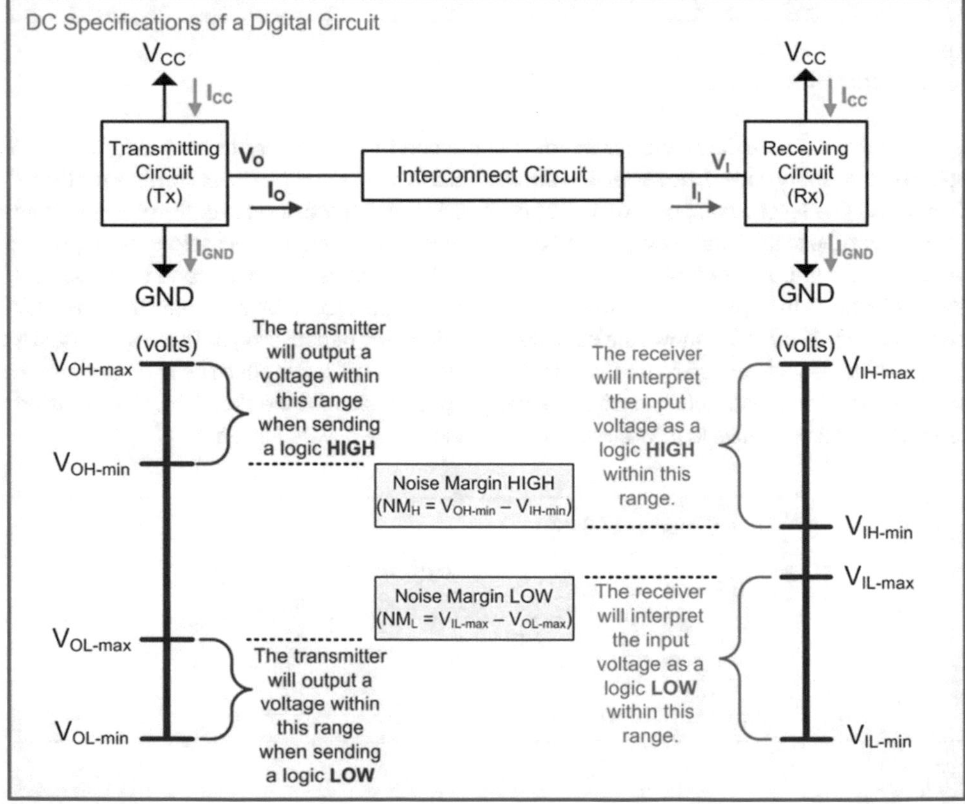

Fig. 3.16
DC specifications of a digital circuit

3.2.3 Input DC Specifications

Receiving circuits provide specifications on the range of input voltages (V_I) that they will interpret as either a logic HIGH or LOW. These are called the DC input specifications. There are four DC voltage specifications that specify this range: $V_{IH\text{-}max}$, $V_{IH\text{-}min}$, $V_{IL\text{-}max}$, and $V_{IL\text{-}min}$. The $V_{IH\text{-}max}$ and $V_{IH\text{-}min}$ specifications provide the range of voltages that the receiver will interpret as a logic HIGH (or logic 1 when using positive logic). The $V_{IL\text{-}max}$ and $V_{IL\text{-}min}$ specifications provide the range of voltages that the receiver will interpret as a logic LOW (or logic 0 when using positive logic). In the subscripts for these specifications, the "I" signifies "input."

The maximum amount of current that the receiver will draw, or take in, when connected is also specified I_I). The specification $I_{IH\text{-}max}$ is the maximum amount of current that the receiver will draw when it is being driven with a logic HIGH. The specification $I_{IL\text{-}max}$ is the maximum amount of current that the receiver will draw when it is being driven with a logic LOW. Again, Fig. 3.16 shows a graphical depiction of these DC specifications.

3.2.4 Noise Margins

For digital circuits that are designed to operate with each other, the $V_{OH\text{-}max}$ and $V_{IH\text{-}max}$ specifications have equal voltages. Similarly, the $V_{OL\text{-}min}$ and $V_{IL\text{-}min}$ specifications have equal voltages. The $V_{OH\text{-}max}$ and $V_{OL\text{-}min}$ output specifications represent the *best-case* scenario for digital signaling as the transmitter is sending the largest (or smallest) signal possible. If there is no loss in the interconnect between the transmitter and receiver, the full voltage levels will arrive at the receiver and be interpreted as the correct logic states (HIGH or LOW).

The *worst-case* scenario for digital signaling is when the transmitter outputs its levels at $V_{OH\text{-}min}$ and $V_{OL\text{-}max}$. These levels represent the furthest away from an *ideal* voltage level that the transmitter can send to the receiver and are susceptible to loss and noise that may occur in the interconnect system. In order to compensate for potential loss or noise, digital circuits have a predefined amount of margin built into their worst-case specifications. Let's take the worst-case example of a transmitter sending a logic HIGH at the level $V_{OH\text{-}min}$. If the receiver was designed to have $V_{IH\text{-}min}$ (i.e., the lowest voltage that would still be interpreted as a logic 1) equal to $V_{OH\text{-}min}$ and then if even the smallest amount of the output signal was attenuated as it traveled through the interconnect, it would arrive at the receiver below $V_{IH\text{-}min}$ and would not be interpreted as a logic 1. Since there will always be some amount of loss in any interconnect system, the specifications for $V_{IH\text{-}min}$ are always less than $V_{OH\text{-}min}$. The difference between these two quantities is called the **noise margin**. More specifically, it is called the noise margin HIGH (or NM_H) to signify how much margin is built into the Tx/Rx circuit when communicating a logic 1. Similarly, the $V_{IL\text{-}max}$ specification is always higher than the $V_{OL\text{-}max}$ specification to account for any voltage added to the signal in the interconnect. The difference between these two quantities is called the noise margin LOW (or NM_L) to signify how much margin is built into the Tx/Rx circuit when communicating a logic 0. Noise margins are always specified as positive quantities; thus the order of the subtrahend and minuend in these differences.

$$\mathbf{NM_H = V_{OH\text{-}min} - V_{IH\text{-}min}}$$
$$\mathbf{NM_L = V_{IL\text{-}max} - V_{OL\text{-}max}}$$

Figure 3.16 includes the graphical depiction of the noise margins. Notice in this figure that there is a region of voltages that the receiver will not interpret as either a HIGH or LOW. This region lies between the $V_{IH\text{-}min}$ and $V_{IL\text{-}max}$ specifications. This is the **uncertainty region** and should be avoided. Signals in this region will cause the receiver's output to go to an unknown voltage. Digital transmitters are designed to transition between the LOW and HIGH states quickly enough so that the receiver does not have time to react to the input being in the uncertainty region.

3.2.5 Power Supplies

All digital circuits require a power supply voltage and a ground. There are some types of digital circuits that may require multiple power supplies. For simplicity, we will focus on digital circuits that only require a single power supply voltage and ground. The power supply voltage is commonly given the abbreviations of either V_{CC} or V_{DD}. The "CC" and "DD" have to do with how the terminals of the transistors inside of the digital circuit are connected (i.e., "collector to collector" or "drain to drain"). Digital circuits will specify the required power supply voltage. Ground is considered an ideal 0v. Digital circuits will also specify the maximum amount of DC current that can flow through the V_{CC} (I_{CC}) and GND (I_{GND}) pins before damaging the part.

There are two components of power supply current. The first is the current that is required for the functional operation of the device. This is called the *quiescent current* (I_q). The second component of the power supply current is the output currents (I_O). Any current that flows out of a digital circuit must also flow into it. When a transmitting circuit sources current to a load on its output pin, it must bring in that same amount of current on another pin. This is accomplished using the power supply pin (V_{CC}). Conversely, when a transmitting circuit sinks current from a load on its output pin, an equal amount of current must exit the circuit on a different pin. This is accomplished using the GND pin. This means that the amount of current flowing through the V_{CC} and GND pins will vary depending on the logic states that are being driven on the outputs. Since a digital circuit may contain numerous output pins, the maximum amount of current flowing through the V_{CC} and GND pins can scale quickly and care must be taken not to damage the device.

The quiescent current is often specified using the term I_{CC}. This should not be confused with the specification for the maximum amount of current that can flow through the V_{CC} pin, which is often called $I_{CC\text{-}max}$. It is easy to tell the difference because I_{CC} (or I_q) is much smaller than $I_{CC\text{-}max}$ for CMOS parts. I_{CC} (or I_q) is specified in the uA to nA range, while the maximum current that can flow through the V_{CC} pin is specified in the mA range. Example 3.1 shows the process of calculating the I_{CC} and I_{GND} currents when sourcing multiple loads.

Example: Calculating I_{CC} and I_{GND} when Sourcing Multiple Loads

Given: The driver is specified to have a quiescent current of 1mA and is driving a logic HIGH on two of its output pins. Each of the two loads on the output pins is being sourced with 4mA of current from the driver.

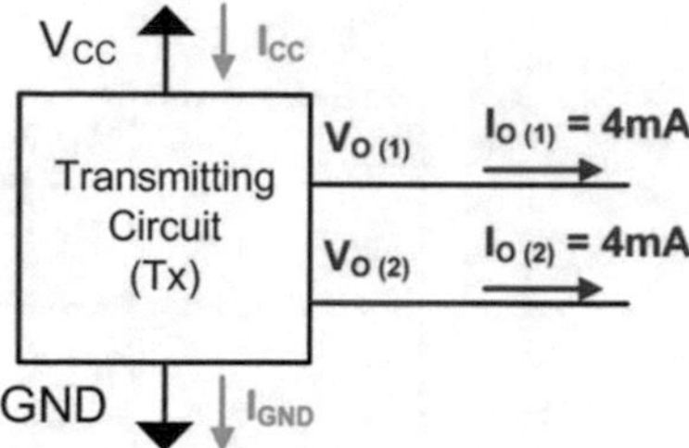

Find: I_{CC} and I_{GND}

Solution: The current into the device must equal the current out of the device. The quiescent current of 1mA is used for the functional operation of the transistors within the transmitter and will flow into the device through the V_{CC} pin and out of the device on the GND pin. The output currents that are being sourced by the driver exit the circuit on the two output pins $V_{O(1)}$ and $V_{O(2)}$. An equal amount of current must also flow into the device ($I_{O(1)} + I_{O(2)} = 8mA$), which enters the device on the V_{CC} pin. This means the total amount of current flowing into the circuit on the V_{CC} pin is:

$$I_{CC} = I_q + I_{O(1)} + I_{O(2)} = 1mA + 4mA + 4mA = 9mA$$

The total amount of current flowing out of the circuit on the GND pin is simply the quiescent current I_q.

$$I_{GND} = I_q = 1mA$$

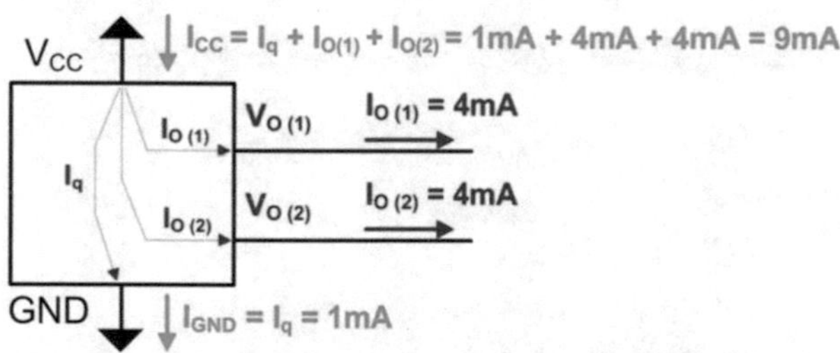

Check: Does the total amount of current entering the circuit equal the total amount of current exiting the circuit?

Yes, there is 9mA entering the circuit through the V_{CC} pin. There is also 9mA exiting the circuit using the $V_{O(1)}$, $V_{O(2)}$ and GND pins.

Example 3.1
Calculating I_{CC} and I_{GND} when sourcing multiple loads

Example 3.2 shows the process of calculating the I_{CC} and I_{GND} currents when both sourcing and sinking loads.

Example: Calculating I_{CC} and I_{GND} When Both Sourcing and Sinking Loads

Given: The driver is specified to have a quiescent current of 0.5mA and is driving a logic HIGH on one of its output pins and a logic LOW on two of its output pins. The driver is sourcing 1mA when driving a HIGH and sinking 2mA when driving a LOW.

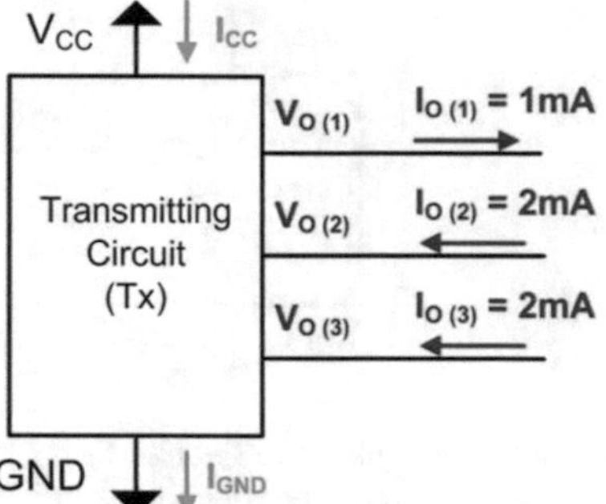

Find: I_{CC} and I_{GND}

Solution: The current into the device must equal the current out of the device. The quiescent current of 0.5mA enters the circuit on the V_{CC} pin and exits on the GND pin. The output current for $V_{O(1)}$ enters the circuit on the V_{CC} pin and exits the circuit on the $V_{O(1)}$ pin. The output current for $V_{O(2)}$ and $V_{O(3)}$ enters the circuit on the $V_{O(2)}$ and $V_{O(3)}$ pins and exits the circuit on the GND pin. This means the total amount of current flowing into the circuit on the V_{CC} pin is:

$$I_{CC} = I_q + I_{O(1)} = 0.5mA + 1mA = 1.5mA$$

The total amount of current flowing out of the circuit on the GND pin is the quiescent current I_q plus the current being sunk from the pins $V_{O(2)}$ and $V_{O(3)}$:

$$I_{GND} = I_q + I_{O(2)} + I_{O(3)} = 0.5mA + 2mA + 2mA = 4.5mA$$

Example 3.2
Calculating I_{CC} and I_{GND} when both sourcing and sinking loads

3.2.6 Switching Characteristics

Switching characteristics refer to the transient behavior of the logic circuits. The first group of switching specifications characterize the *propagation delay* of the gate. The propagation delay is the time it takes for the output to respond to a change on the input. The propagation delay is formally defined as the time it takes from the point at which the input has transitioned to 50% of its final value to the point at which the output has transitioned to 50% of its final value. The initial and final voltages for the input are defined to be GND and V_{CC}, while the output initial and final voltages are defined to be V_{OL} and V_{OH}. Specifications are given for the propagation delay when transitioning from a LOW to HIGH (t_{PLH}) and from a HIGH to LOW (t_{PHL}). When these specifications are equal, the values are often given as a single specification of t_{pd}. These specifications are shown graphically in Fig. 3.17.

The second group of switching specifications characterize how quickly the output switches between states. The *transition time* is defined as the time it takes for the output to transition from 10% to 90% of the output voltage range. The *rise time* (t_r) is the time it takes for this transition when going from a LOW to HIGH, and the *fall time* (t_f) is the time it takes for this transition when going from a HIGH to LOW. When these specifications are equal, the values are often given as a single specification of t_t. These specifications are shown graphically in Fig. 3.17.

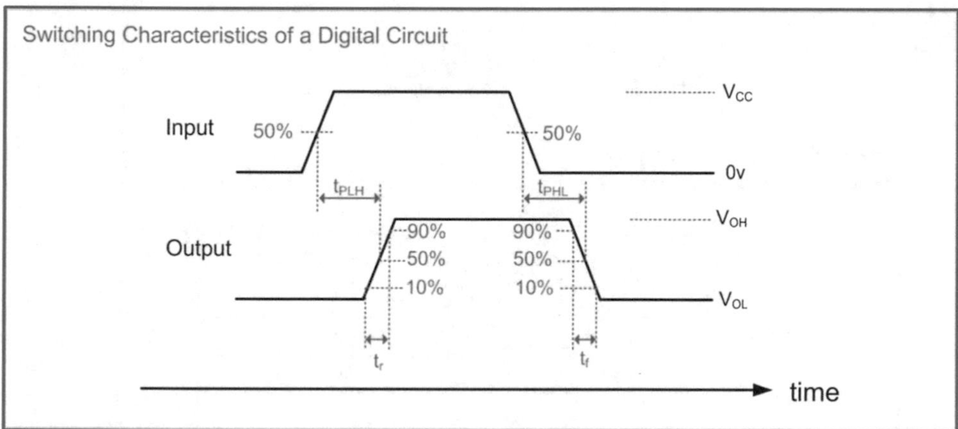

Fig. 3.17
Switching characteristics of a digital circuit

3.2.7 Data Sheets

The specifications for a particular part are given in its **data sheet**. The data sheet contains all of the operating characteristics for a part, in addition to functional information such as package geometries and pin assignments. The data sheet is usually the first place a designer will look when selecting a part. Figures 3.18, 3.19, and 3.20 show excerpts from an example data sheet highlighting some of the specifications just covered.

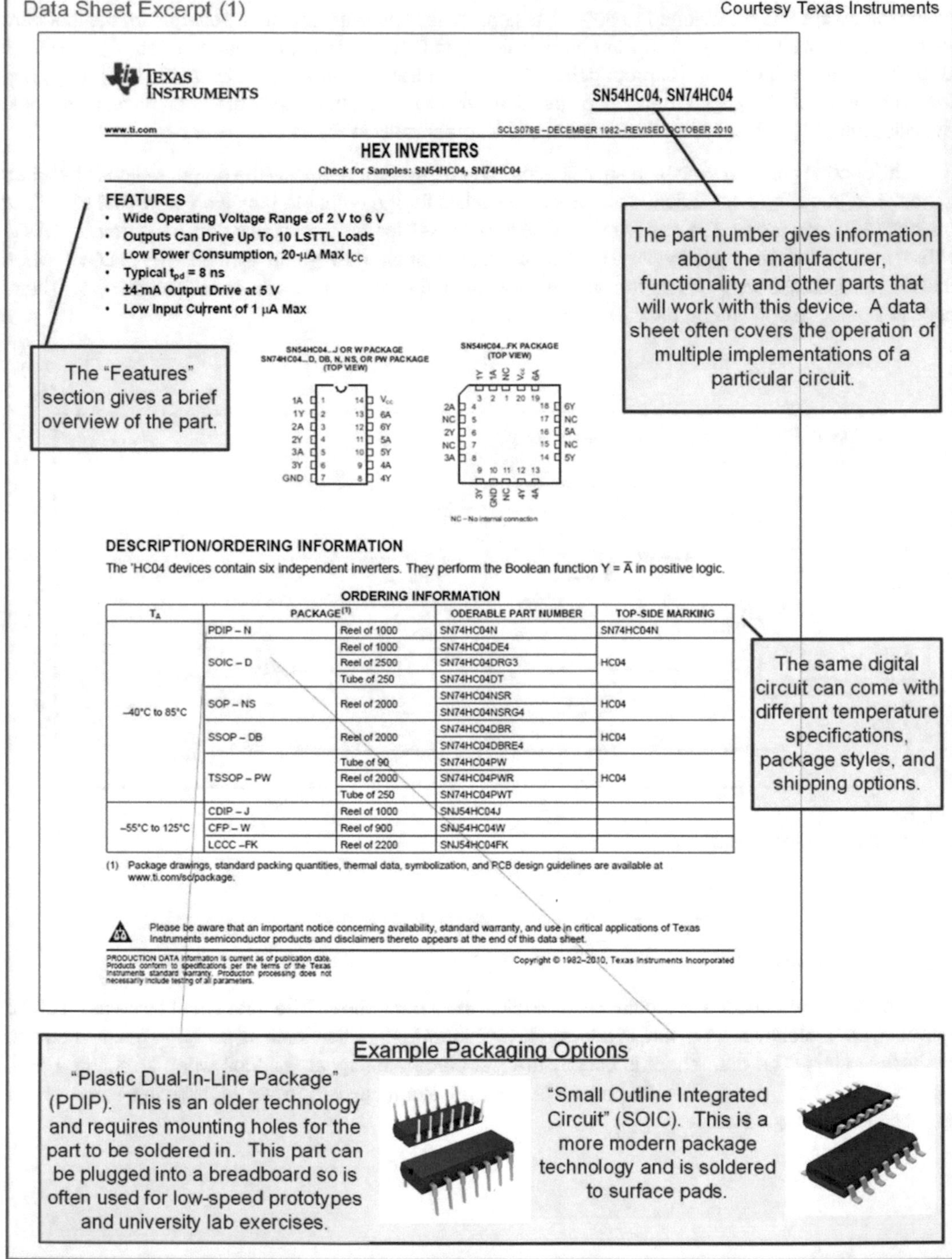
Data Sheet Excerpt (1)

Courtesy Texas Instruments

TEXAS INSTRUMENTS

SN54HC04, SN74HC04

www.ti.com

SCLS078E – DECEMBER 1982 – REVISED OCTOBER 2010

HEX INVERTERS

Check for Samples: SN54HC04, SN74HC04

FEATURES

- Wide Operating Voltage Range of 2 V to 6 V
- Outputs Can Drive Up To 10 LSTTL Loads
- Low Power Consumption, 20-μA Max I_{CC}
- Typical t_{pd} = 8 ns
- ±4-mA Output Drive at 5 V
- Low Input Current of 1 μA Max

DESCRIPTION/ORDERING INFORMATION

The 'HC04 devices contain six independent inverters. They perform the Boolean function $Y = \overline{A}$ in positive logic.

ORDERING INFORMATION

T_A	PACKAGE(1)		ODERABLE PART NUMBER	TOP-SIDE MARKING
–40°C to 85°C	PDIP – N	Reel of 1000	SN74HC04N	SN74HC04N
	SOIC – D	Reel of 1000	SN74HC04DE4	HC04
		Reel of 2500	SN74HC04DRG3	
		Tube of 250	SN74HC04DT	
	SOP – NS	Reel of 2000	SN74HC04NSR	HC04
			SN74HC04NSRG4	
	SSOP – DB	Reel of 2000	SN74HC04DBR	HC04
			SN74HC04DBRE4	
	TSSOP – PW	Tube of 90	SN74HC04PW	HC04
		Reel of 2000	SN74HC04PWR	
		Tube of 250	SN74HC04PWT	
–55°C to 125°C	CDIP – J	Reel of 1000	SNJ54HC04J	
	CFP – W	Reel of 900	SNJ54HC04W	
	LCCC – FK	Reel of 2200	SNJ54HC04FK	

(1) Package drawings, standard packing quantities, thermal data, symbolization, and PCB design guidelines are available at www.ti.com/sc/package.

Please be aware that an important notice concerning availability, standard warranty, and use in critical applications of Texas Instruments semiconductor products and disclaimers thereto appears at the end of this data sheet.

PRODUCTION DATA information is current as of publication date. Products conform to specifications per the terms of the Texas Instruments standard warranty. Production processing does not necessarily include testing of all parameters.

Copyright © 1982–2010, Texas Instruments Incorporated

Fig. 3.18
Example data sheet excerpt (1)

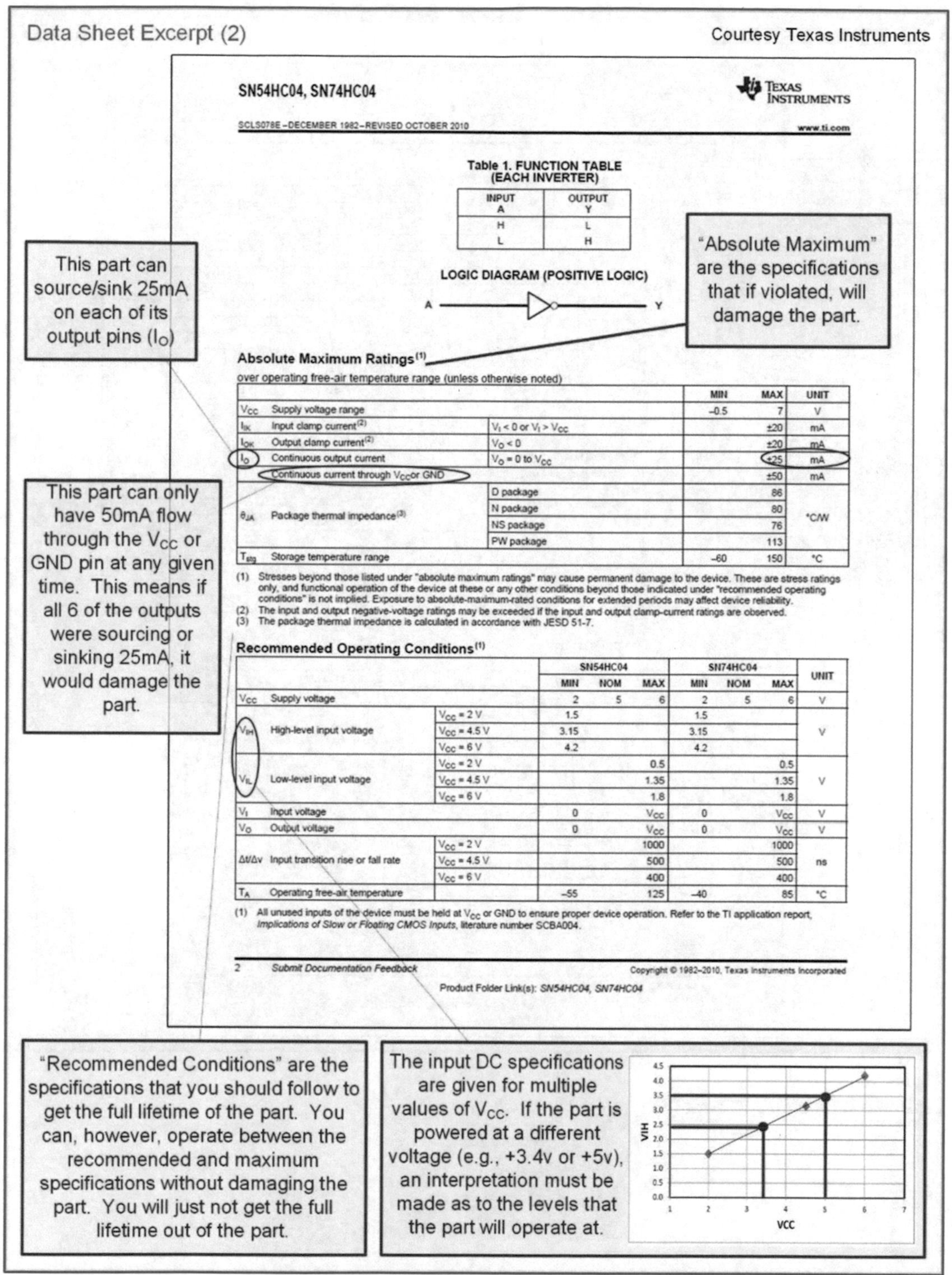
Data Sheet Excerpt (2)

Courtesy Texas Instruments

SN54HC04, SN74HC04

TEXAS INSTRUMENTS

SCLS078E – DECEMBER 1982 – REVISED OCTOBER 2010

www.ti.com

Table 1. FUNCTION TABLE (EACH INVERTER)

INPUT A	OUTPUT Y
H	L
L	H

LOGIC DIAGRAM (POSITIVE LOGIC)

A — Y

Absolute Maximum Ratings(1)

over operating free-air temperature range (unless otherwise noted)

			MIN	MAX	UNIT
V_{CC}	Supply voltage range		–0.5	7	V
I_{IK}	Input clamp current(2)	$V_I < 0$ or $V_I > V_{CC}$		±20	mA
I_{OK}	Output clamp current(2)	$V_O < 0$		±20	mA
I_O	Continuous output current	$V_O = 0$ to V_{CC}		±25	mA
	Continuous current through V_{CC} or GND			±50	mA
θ_{JA}	Package thermal impedance(3)	D package		86	°C/W
		N package		80	
		NS package		76	
		PW package		113	
T_{stg}	Storage temperature range		–60	150	°C

(1) Stresses beyond those listed under "absolute maximum ratings" may cause permanent damage to the device. These are stress ratings only, and functional operation of the device at these or any other conditions beyond those indicated under "recommended operating conditions" is not implied. Exposure to absolute-maximum-rated conditions for extended periods may affect device reliability.
(2) The input and output negative-voltage ratings may be exceeded if the input and output clamp-current ratings are observed.
(3) The package thermal impedance is calculated in accordance with JESD 51-7.

Recommended Operating Conditions(1)

			SN54HC04			SN74HC04			UNIT
			MIN	NOM	MAX	MIN	NOM	MAX	
V_{CC}	Supply voltage		2	5	6	2	5	6	V
V_{IH}	High-level input voltage	$V_{CC} = 2$ V	1.5			1.5			V
		$V_{CC} = 4.5$ V	3.15			3.15			
		$V_{CC} = 6$ V	4.2			4.2			
V_{IL}	Low-level input voltage	$V_{CC} = 2$ V			0.5			0.5	V
		$V_{CC} = 4.5$ V			1.35			1.35	
		$V_{CC} = 6$ V			1.8			1.8	
V_I	Input voltage		0		V_{CC}	0		V_{CC}	V
V_O	Output voltage		0		V_{CC}	0		V_{CC}	V
$\Delta t/\Delta v$	Input transition rise or fall rate	$V_{CC} = 2$ V			1000			1000	ns
		$V_{CC} = 4.5$ V			500			500	
		$V_{CC} = 6$ V			400			400	
T_A	Operating free-air temperature		–55		125	–40		85	°C

(1) All unused inputs of the device must be held at V_{CC} or GND to ensure proper device operation. Refer to the TI application report, *Implications of Slow or Floating CMOS Inputs*, literature number SCBA004.

2 *Submit Documentation Feedback* Copyright © 1982–2010, Texas Instruments Incorporated

Product Folder Link(s): *SN54HC04, SN74HC04*

Fig. 3.19
Example data sheet excerpt (2)

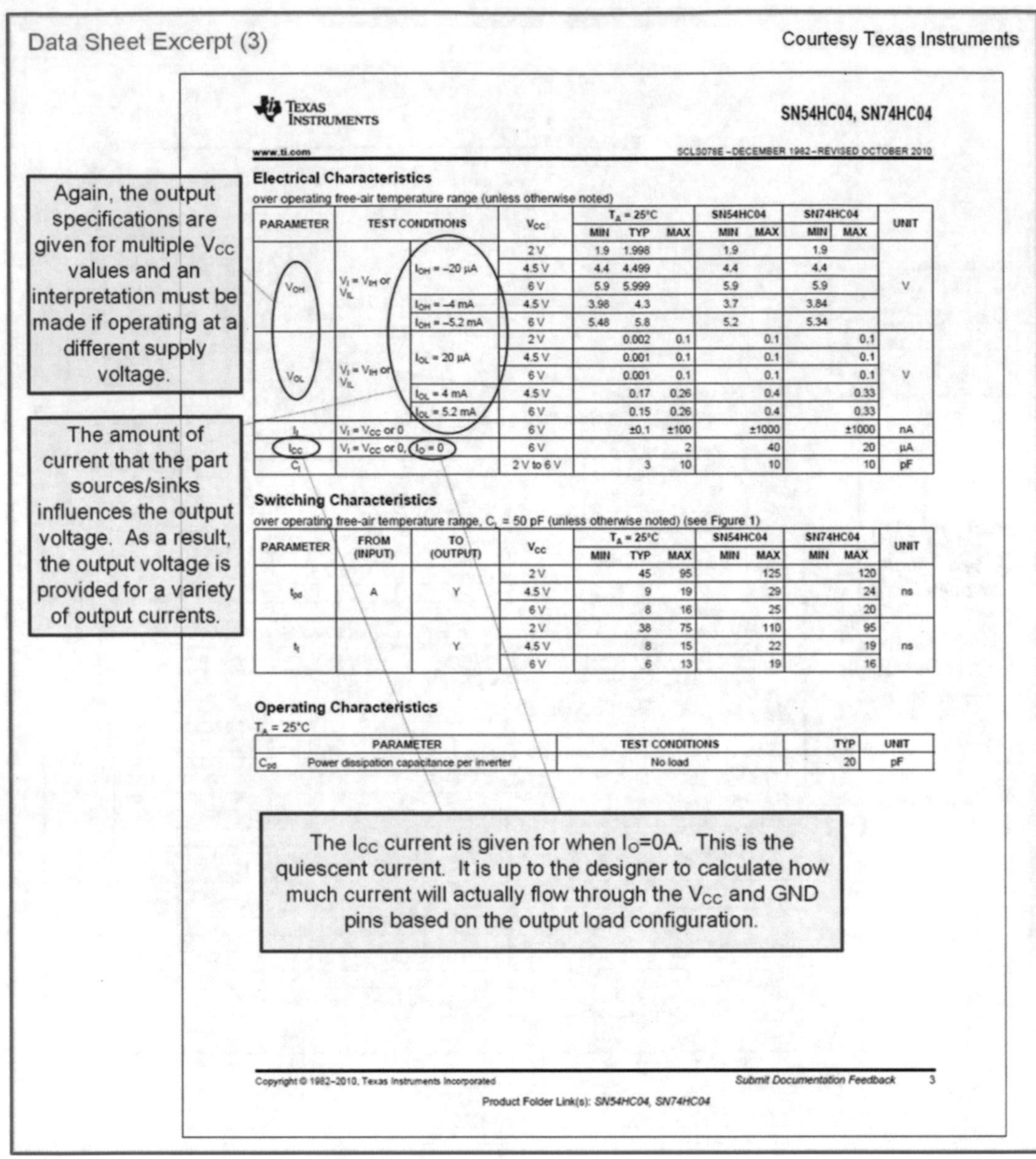

Data Sheet Excerpt (3)

Courtesy Texas Instruments

TEXAS INSTRUMENTS

SN54HC04, SN74HC04

www.ti.com

SCLS078E – DECEMBER 1982 – REVISED OCTOBER 2010

Electrical Characteristics

over operating free-air temperature range (unless otherwise noted)

PARAMETER	TEST CONDITIONS		V_{CC}	T_A = 25°C			SN54HC04		SN74HC04		UNIT
				MIN	TYP	MAX	MIN	MAX	MIN	MAX	
V_{OH}	V_I = V_{IH} or V_{IL}	I_{OH} = –20 µA	2 V	1.9	1.998		1.9		1.9		V
			4.5 V	4.4	4.499		4.4		4.4		
			6 V	5.9	5.999		5.9		5.9		
		I_{OH} = –4 mA	4.5 V	3.98	4.3		3.7		3.84		
		I_{OH} = –5.2 mA	6 V	5.48	5.8		5.2		5.34		
V_{OL}	V_I = V_{IH} or V_{IL}	I_{OL} = 20 µA	2 V		0.002	0.1		0.1		0.1	V
			4.5 V		0.001	0.1		0.1		0.1	
			6 V		0.001	0.1		0.1		0.1	
		I_{OL} = 4 mA	4.5 V		0.17	0.26		0.4		0.33	
		I_{OL} = 5.2 mA	6 V		0.15	0.26		0.4		0.33	
I_I	V_I = V_{CC} or 0		6 V		±0.1	±100		±1000		±1000	nA
I_{CC}	V_I = V_{CC} or 0,	I_O = 0	6 V			2		40		20	µA
C_i			2 V to 6 V		3	10		10		10	pF

Switching Characteristics

over operating free-air temperature range, C_L = 50 pF (unless otherwise noted) (see Figure 1)

PARAMETER	FROM (INPUT)	TO (OUTPUT)	V_{CC}	T_A = 25°C			SN54HC04		SN74HC04		UNIT
				MIN	TYP	MAX	MIN	MAX	MIN	MAX	
t_{pd}	A	Y	2 V		45	95		125		120	ns
			4.5 V		9	19		29		24	
			6 V		8	16		25		20	
t_t		Y	2 V		38	75		110		95	ns
			4.5 V		8	15		22		19	
			6 V		6	13		19		16	

Operating Characteristics

T_A = 25°C

PARAMETER		TEST CONDITIONS	TYP	UNIT
C_{pd}	Power dissipation capacitance per inverter	No load	20	pF

Copyright © 1982–2010, Texas Instruments Incorporated

Submit Documentation Feedback 3

Product Folder Link(s): *SN54HC04, SN74HC04*

Fig. 3.20
Example data sheet excerpt (3)

Concept Check

CC3.2(a) Given the following DC specifications for a driver/receiver pair, in what situation *may* a logic signal transmitted not be successfully received?

$V_{OH\text{-}max} = +3.4v$ $V_{IH\text{-}max} = +3.4v$
$V_{OH\text{-}min} = +2.5v$ $V_{IH\text{-}min} = +2.5v$
$V_{OL\text{-}max} = +1.5v$ $V_{IL\text{-}max} = +2.0v$
$V_{OL\text{-}min} = 0v$ $V_{IL\text{-}min} = 0v$

A) Driving a HIGH with $V_o = +3.4v$

B) Driving a HIGH with $V_o = +2.5v$

C) Driving a LOW with $V_o = +1.5v$

D) Driving a LOW with $V_o = 0v$

CC3.2(b) For the following driver configuration, which of the following is a valid constraint that could be put in place to prevent a violation of the maximum power supply currents ($I_{CC\text{-}max}$ and $I_{GND\text{-}max}$)?

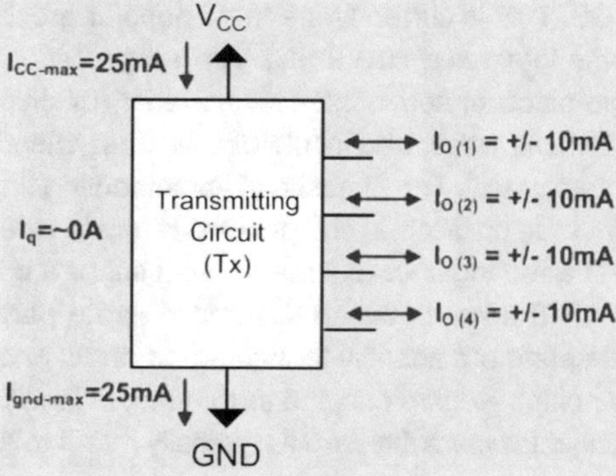

A) Modify the driver transistors so that they can't provide more than 5 mA on any output.

B) Apply a cooling system (e.g., a heat sink or fan) to the driver chip.

C) Design the logic so that no more than half of the outputs are HIGH at any given time.

D) Drive multiple receivers with the same output pin.

CC3.2(c) Why is it desirable to have the output of a digital circuit transition quickly between the logic LOW and logic HIGH levels?

A) So that the outputs are not able to respond as the input transitions through the uncertainty region. This avoids unwanted transitions.

B) So that all signals look like square waves.

C) To reduce power by minimizing the time spent switching.

D) Because the system can only have two states, a LOW and a HIGH.

3.3 Logic Families

It is apparent from the prior discussion of operating conditions that digital circuits need to have comparable input and output specifications in order to successfully communicate with each other. If a transmitter outputs a logic HIGH as +3.4v and the receiver needs a logic HIGH to be above +4v to be successfully interpreted as a logic HIGH, then these two circuits will not be able to communicate. In order to address this interoperability issue, digital circuits are grouped into *logic families*. A logic family is a group of parts that all adhere to a common set of specifications so that they work together. The logic family is given a specific name, and once the specifications are agreed upon, different manufacturers produce parts that work within the particular family. Within a logic family, parts will all have the same power supply requirements and DC input/output specifications such that if connected *directly*, they will be able to successfully communicate with each other. The phrase "connected directly" is emphasized because it is very possible to insert an interconnect circuit between two circuits within the same logic family and alter the output voltage enough so that the receiver will not be able to interpret the correct logic level. Analyzing the effect of the interconnect circuit is part of the digital design process. There are many logic families that exist (up to 100 different types!), and more emerge each year as improvements are made to circuit fabrication processes that create smaller, faster, and lower power circuits.

3.3.1 Complementary Metal-Oxide Semiconductors (CMOS)

The first group of logic families we will discuss is called complementary metal-oxide semiconductors, or CMOS. This is currently the most popular group of logic families for digital circuits implemented on the same integrated circuit (IC). An **integrated circuit** is where the entire circuit is implemented on a single piece of semiconductor material (or chip). The IC can contain transistors, resistors, capacitors, inductors, wires, and insulators. Modern integrated circuits can contain billions of devices and meters of interconnect. The opposite of implementing the circuit on an integrated circuit is to use **discrete components**. Using discrete components refers to where every device (transistor, resistor, etc.) is its own part and is wired together externally using either a printed circuit board (PCB) or jumper wires as on a breadboard. The line between ICs and discrete parts has blurred in the past decades because modern discrete parts are actually fabricated as an IC and regularly contain multiple devices (e.g., four logic gates per chip). Regardless, the term *discrete* is still used to describe components that only contain a *few* components where the term IC typically refers to a much larger system that is custom designed.

The term CMOS comes from the use of particular types of transistors to implement the digital circuits. The transistors are created using a metal-oxide semiconductor (MOS) structure. These transistors are turned on or off based on an electric field, so they are given the name metal-oxide semiconductor *field-effect* transistors, or MOSFETs. There are two transistors that can be built using this approach that operate *complementary* to each other, thus the term *complementary metal-oxide semiconductors*. To understand the basic operation of CMOS logic, we begin by treating the MOSFETs as ideal switches. This allows us to understand the basic functionality without diving into the detailed electronic analysis of the transistors.

3.3.1.1 CMOS Operation

In CMOS, there is a single power supply (V_{CC} or V_{DD}) and a single ground (GND). The ground signal is sometimes called V_{SS}. The maximum input and output DC specifications are equal to the power supply ($V_{CC} = V_{OH\text{-}max} = V_{IH\text{-}max}$). The minimum input and output DC specification are equal to ground (GND $= 0v = V_{OL\text{-}min} = V_{IL\text{-}min}$). In this way, using CMOS simplifies many of the specifications. If you state that you are using "CMOS with a +3.4v power supply," you are inherently stating that $V_{CC} = V_{OH\text{-}max} = V_{IH\text{-}max} =$ +3.4v and that $V_{OL\text{-}min} = V_{IL\text{-}min} =$ 0v. Many times, the name of the logic family will be associated with the power supply voltage. For example, a logic family may go by the name "+3.3v CMOS" or "+2.5v CMOS." These names give a first-level description of the logic family operation, but more details about the operation must be looked up in the data sheet.

There are two types of transistors used in CMOS. The transistors will be closed or open based on an input logic level. The first transistor is called an N-type MOSFET, or **NMOS**. This transistor will turn on, or close, when the voltage between the gate and source (V_{GS}) is greater than its *threshold voltage*. The threshold voltage (V_T) is the amount of voltage needed to create a conduction path between the drain and the source terminals. The threshold voltage of an NMOS transistor is typically between 0.2v and 1v and much less than the V_{CC} voltage in the system. The second transistor is called a P-type MOSFET, or **PMOS**. This transistor turns on, or closes, when the voltage between the gate and the source (V_{GS}) is less than V_T, where the V_T for a PMOS is a negative value. This means that to turn on a PMOS transistor, the gate terminal needs to be at a lower voltage than the source. The type of transistor (i.e., P-type or N-type) has to do with the type of semiconductor material used to conduct current through the transistor. An NMOS transistor uses negative charge to conduct current (i.e., *n*egative-type), while a PMOS uses positive charge (i.e., *p*ositive-type). Figure 3.21 shows the symbols for the PMOS and NMOS, the fabrication cross sections, and their switch-level equivalents.

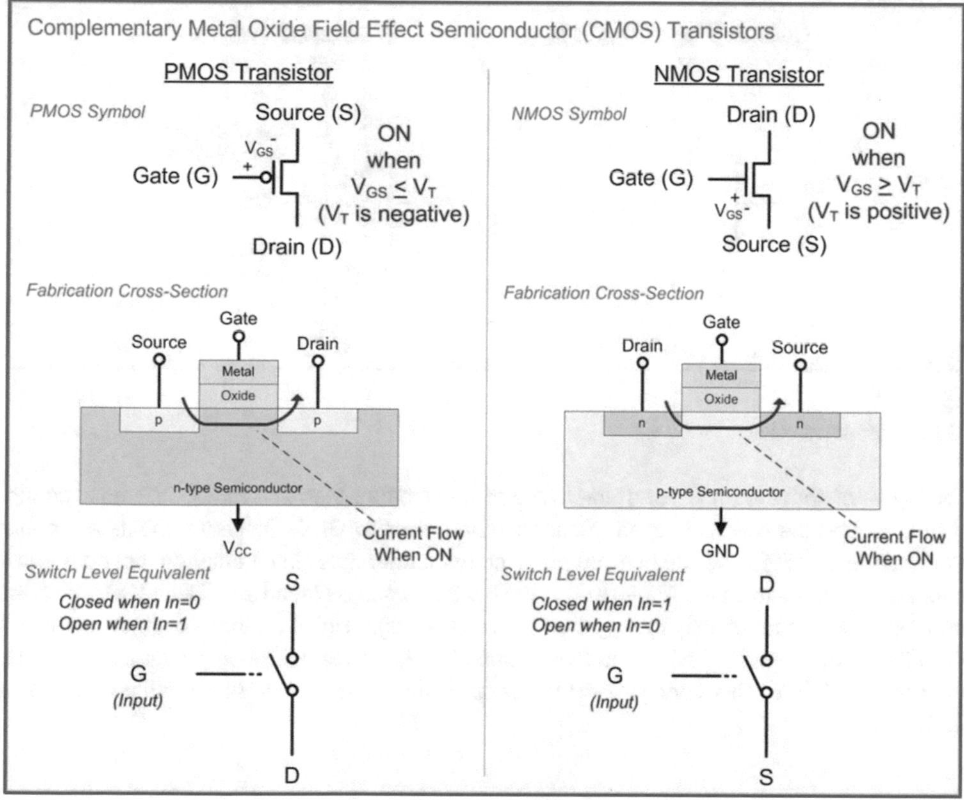

Fig. 3.21
CMOS transistors

The basic operation of CMOS is that when driving a logic HIGH, the switches are used to connect the output to the power supply (V_{CC}), and when driving a logic LOW, the switches are used to connect the output to GND. In CMOS, V_{CC} is considered an ideal logic HIGH and GND is considered an ideal logic

LOW. V_{CC} is typically much larger than V_T so using these levels can easily turn on and off the transistors. The design of the circuit must never connect the output to V_{CC} and GND at the same time or else the device itself will be damaged due to the current flowing directly from V_{CC} to GND through the transistors. Due to the device physics of the MOSFETS, PMOS transistors are used to form the network that will connect the output to V_{CC} (aka, the pull-up network), and NMOS transistors are used to form the network that will connect the output to GND (aka, the pull-down network). Since PMOS transistors are closed when the input is a 0 (thus providing a logic HIGH on the output) and NMOS transistors are closed when the input is a 1 (thus providing a logic LOW on the output), CMOS implements negative logic gates. This means CMOS can implement inverters, NAND and NOR gates, but not buffers, AND and OR gates, directly. In order to create a CMOS AND gate, the circuit would implement a NAND gate followed by an inverter and similarly for an OR gate and buffer.

3.3.1.2 CMOS Inverter

Let's now look at how we can use these transistors to create a CMOS inverter. Consider the transistor arrangement shown in Fig. 3.22.

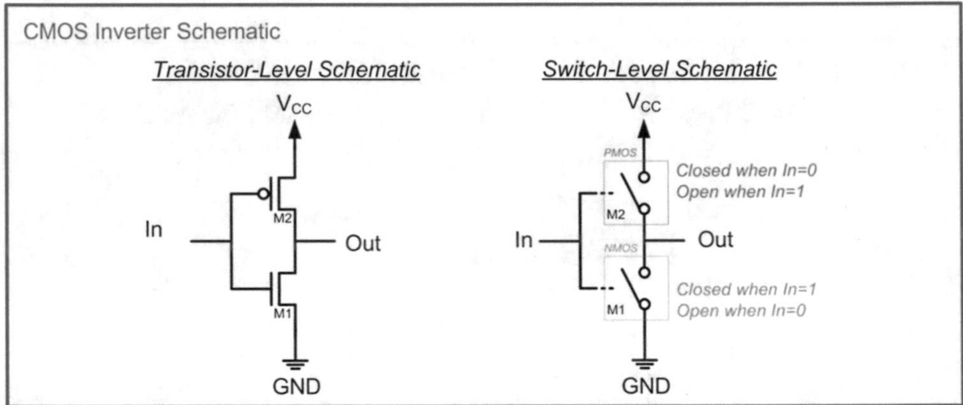

Fig. 3.22
CMOS inverter schematic

The inputs of both the PMOS and NMOS are connected together. The PMOS is used to connect the output to V_{CC}, and the NMOS is used to connect the output to GND. Since the inputs are connected together and the switches operate in a complementary manner, this circuit ensures that both transistors will never be on at the same time. When In = 0, the PMOS switch is closed and the NMOS switch is open. This connects the output directly to V_{CC}, thus providing a logic HIGH on the output. When In = 1, the PMOS switch is open and the NMOS switch is closed. This connects the output directly to GND, thus providing a logic LOW. This configuration yields an inverter. This operation is shown graphically in Fig. 3.23.

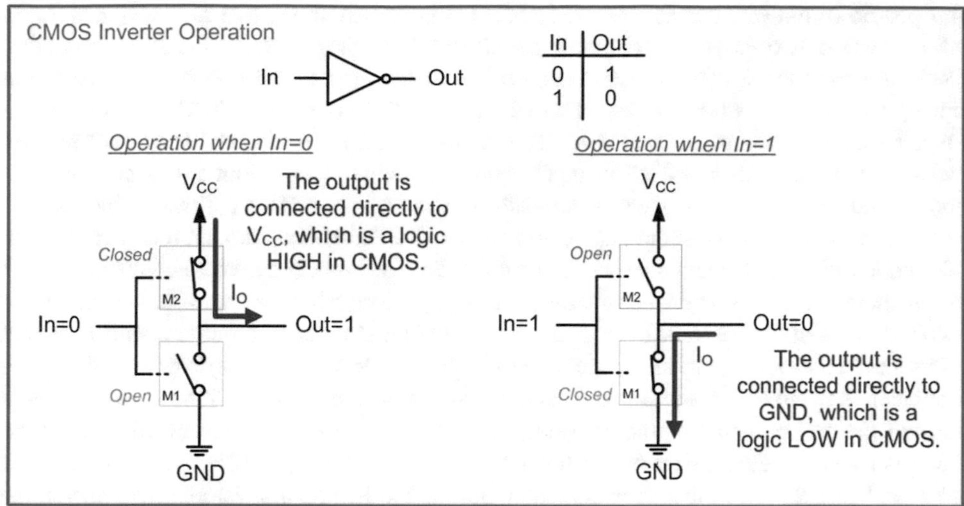

Fig. 3.23
CMOS inverter operation

3.3.1.3 CMOS NAND Gate

Let's now look at how we use a similar arrangement of transistors to implement a 2-input NAND gate. Consider the transistor configuration shown in Fig. 3.24.

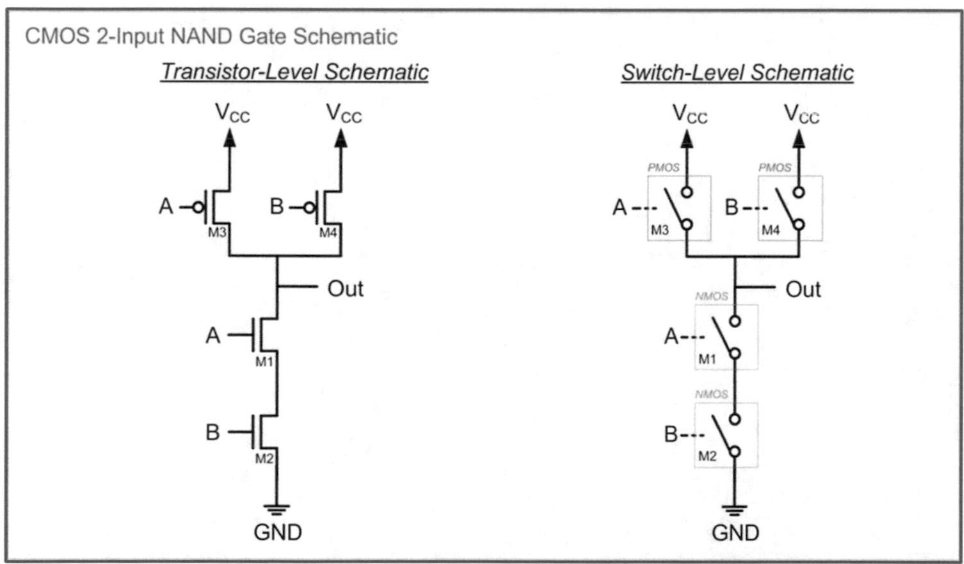

Fig. 3.24
CMOS 2-input NAND gate schematic

The pull-down network consists of two NMOS transistors in series (M1 and M2), and the pull-up network consists of two PMOS transistors in parallel (M3 and M4). Let's go through each of the input conditions and examine which transistors are on and which are off and how they impact the output. The first input condition is when A = 0 and B = 0. This condition turns *on* both M3 and M4 creating two parallel paths between the output and V_{CC}. At the same time, it turns *off* both M1 and M2 preventing a path between the output and GND. This input condition results in an output that is connected to V_{CC} resulting in a logic HIGH. The second input condition is when A = 0 and B = 1. This condition turns *on* M3 in the pull-up network and M2 in the pull-down network. This condition also turns *off* M4 in the pull-up network and M1 in the pull-down network. Since the pull-up network is a parallel combination of PMOS transistors, there is still a path between the output and V_{CC} through M3. Since the pull-down network is a series combination of NMOS transistors, both M1 and M2 must be on in order to connect the output to GND. This input condition results in an output that is connected to V_{CC} resulting in a logic HIGH. The third input condition is when A = 1 and B = 0. This condition again provides a path between the output and V_{CC} through M4 and prevents a path between the output and ground by having M2 open. This input condition results in an output that is connected to V_{CC} resulting in a logic HIGH. The final input condition is when A = 1 and B = 1. In this input condition, both of the PMOS transistors in the pull-up network (M3 and M4) are off preventing the output from being connected to V_{CC}. At the same time, this input turns on both M1 and M2 in the pull-down network connecting the output to GND. This input condition results in an output that is connected to GND resulting in a logic LOW. Based on the resulting output values corresponding to the four input codes, this circuit yields the logic operation of a 2-input NAND gate. This operation is shown graphically in Fig. 3.25.

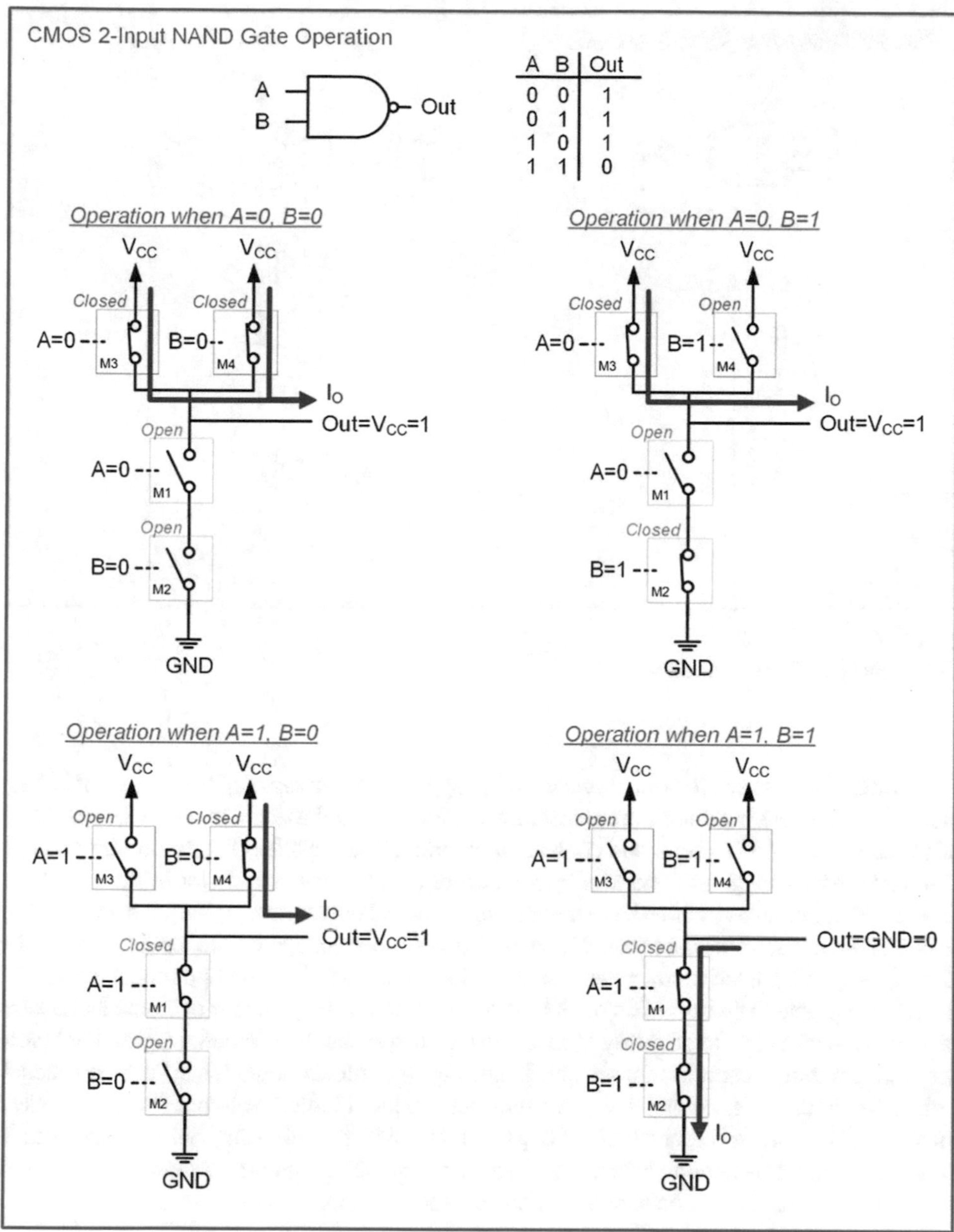

Fig. 3.25
CMOS 2-input NAND gate operation

Creating a CMOS NAND gate with more than two inputs is accomplished by adding additional PMOS transistors to the pull-up network in parallel and additional NMOS transistors to the pull-down network in series. Figure 3.26 shows the schematic for a 3-input NAND gate. This procedure is followed for creating NAND gates with larger numbers of inputs.

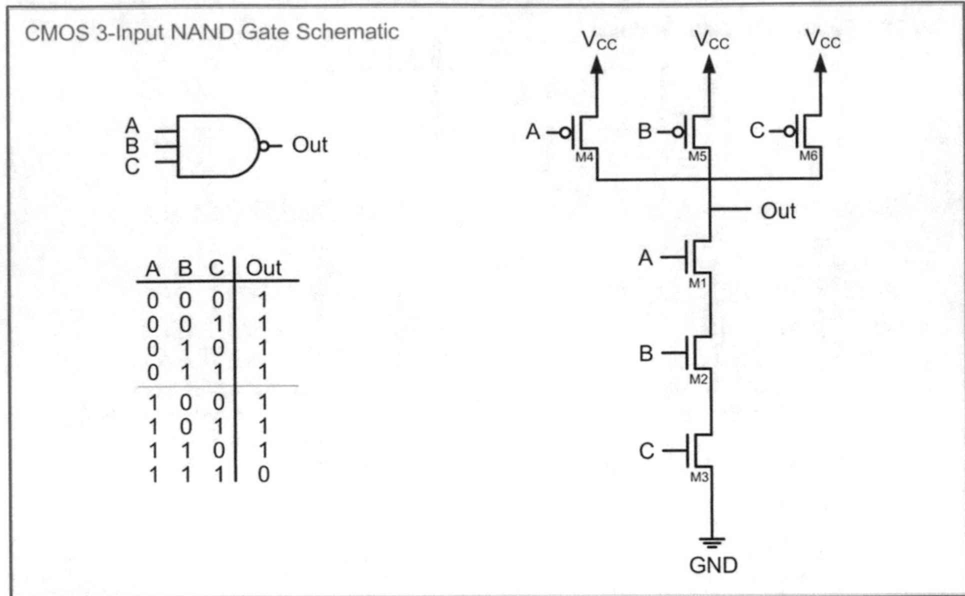

A	B	C	Out
0	0	0	1
0	0	1	1
0	1	0	1
0	1	1	1
1	0	0	1
1	0	1	1
1	1	0	1
1	1	1	0

Fig. 3.26
CMOS 3-input NAND gate schematic

If the CMOS transistors were ideal switches, the approach of increasing the number of inputs could be continued indefinitely. In reality, the transistors are not ideal switches, and there is a limit on how many transistors can be added in series and continue to operate. The limitation has to do with ensuring that each transistor has enough voltage to properly turn on or off. This is a factor in the series network because the drain terminals of the NMOS transistors are not all connected to GND. If a voltage develops across one of the lower transistors (e.g., M3), then it takes more voltage on the input to turn on the next transistor up (e.g., M2). If too many transistors are added in series, then the uppermost transistor in the series may not be able to be turned on or off by the input signals. The number of inputs that a logic gate can have within a particular logic family is called its **fan-in** specification. When a logic circuit requires a number of inputs that exceed the fan-in specification for a particular logic family, then additional logic gates must be used. For example, if a circuit requires a 5-input NAND gate but the logic family has a fan-in specification of 4, this means that the largest NAND gate available only has 4-inputs. The 5-input NAND operation must be accomplished using additional circuit design techniques that use gates with 4 or less inputs. These design techniques will be covered in Chap. 4.

3.3.1.4 CMOS NOR Gate

A CMOS NOR gate is created using a similar topology as a NAND gate with the exception that the pull-up network consists of PMOS transistors in series and the pull-down network that consists of NMOS transistors in parallel. Consider the transistor configuration shown in Fig. 3.27.

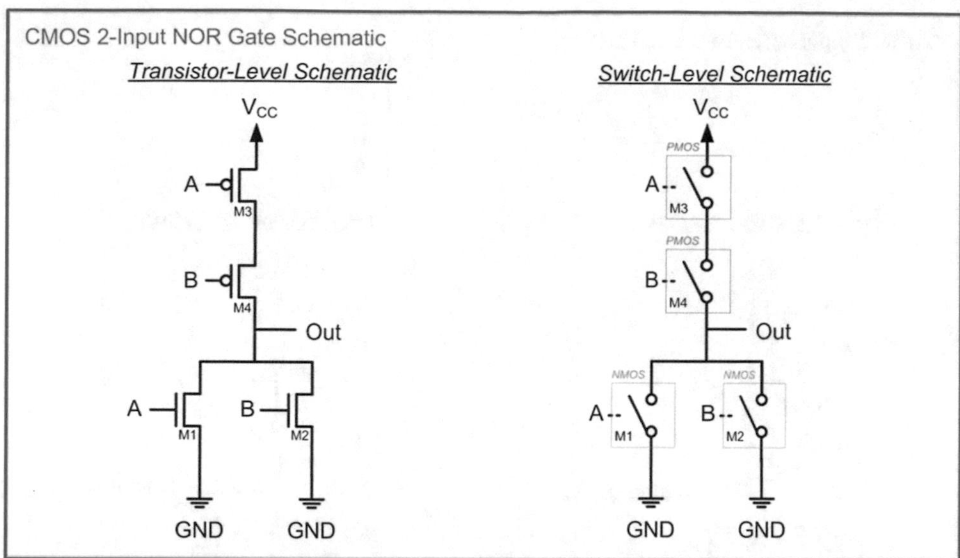

Fig. 3.27
CMOS 2-input NOR gate schematic

The series configuration of the pull-up network will only connect the output to V_{CC} when both inputs are 0. Conversely, the pull-down network prevents connecting the output to GND when both inputs are 0. When either or both of the inputs are true, the pull-up network is off, and the pull-down network is on. This yields the logic function for a NOR gate. This operation is shown graphically in Fig. 3.28. As with the NAND gate, the number of inputs can be increased by adding more PMOS transistors in series in the pull-up network and more NMOS transistors in parallel in the pull-down network.

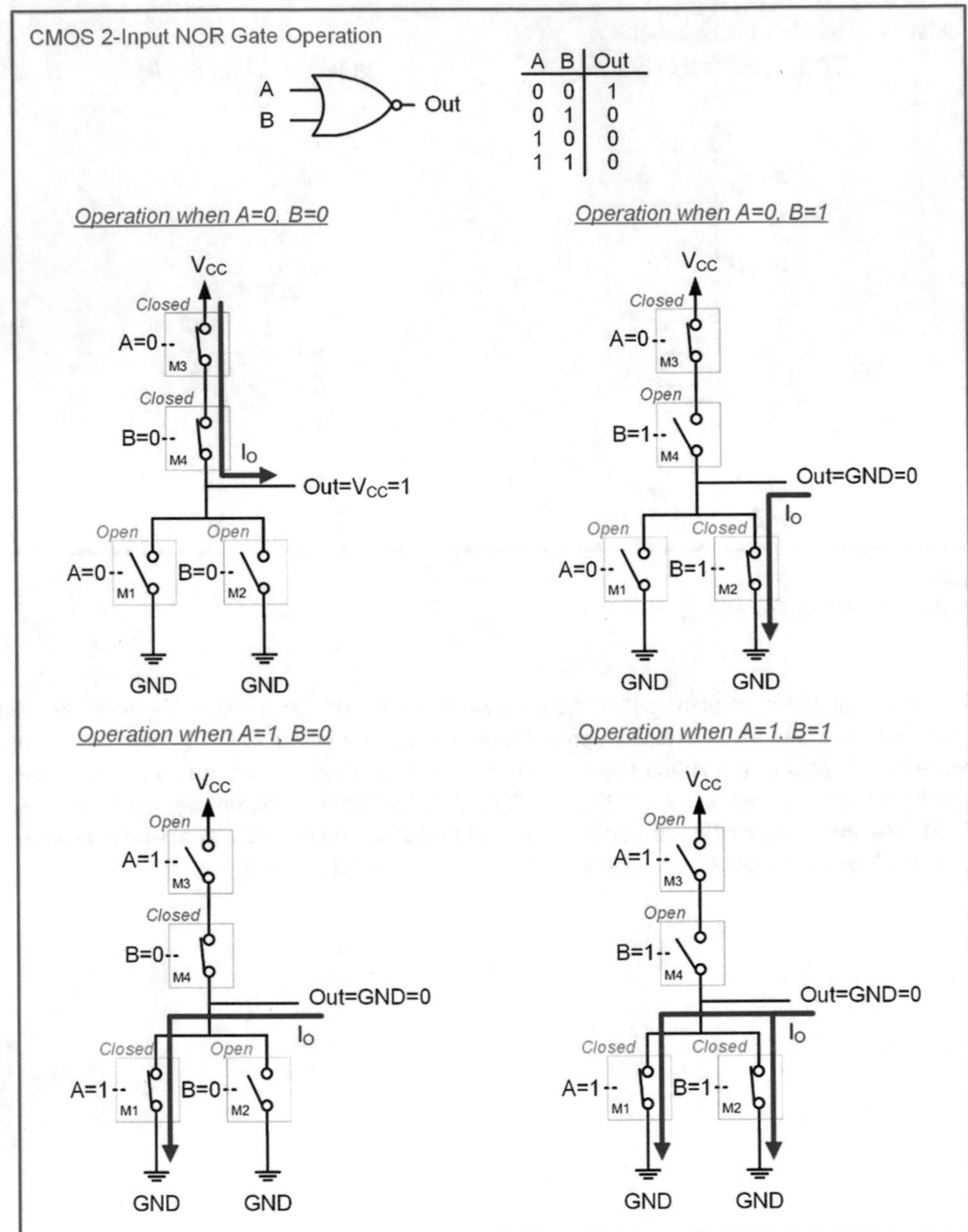

Fig. 3.28
CMOS 2-input NOR gate operation

The schematic for a 3-input NOR gate is given in Fig. 3.29. This approach can be used to increase the number of inputs up until the fan-in specification of the logic family is reached.

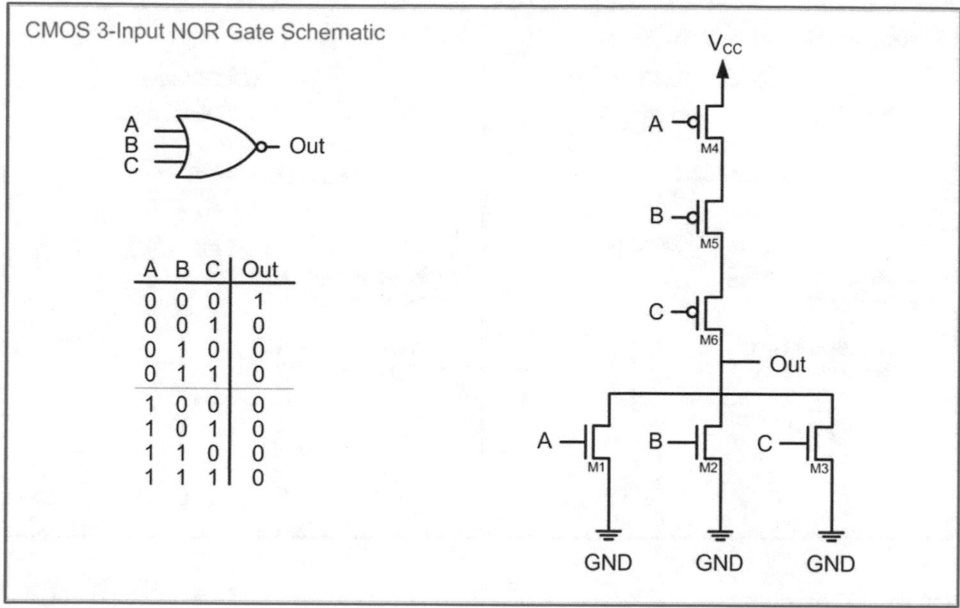

Fig. 3.29
CMOS 3-input NOR gate schematic

3.3.2 Transistor-Transistor Logic (TTL)

One of the first logic families that emerged after the invention of the integrated circuit was transistor-transistor logic (TTL). This logic family uses bipolar junction transistor (BJT) as its fundamental switching item. This logic family defined a set of discrete parts that contained all of the basic gates in addition to more complex building blocks. TTL was used to build the first computer systems in the 1960s. TTL is not widely used today other than for specific applications because it consumes more power than CMOS and cannot achieve the density required for today's computer systems. TTL is discussed because it was the original logic family based on integrated circuits, so it provides a historical perspective of digital logic. Furthermore, the discrete logic pin-outs and part-numbering schemes are still used today for discrete CMOS parts.

3.3.2.1 TTL Operation

TTL logic uses BJT transistors and resistors to accomplish the logic operations. The operation of a BJT transistor is more complicated than a MOSFET; however, it performs essentially the same switch operation when used in a digital logic circuit. An input is used to turn the transistor on, which in turn allows current to flow between two other terminals. Figure 3.30 shows the symbol for the two types of BJT transistors. The PNP transistor is analogous to a PMOS and the NPN is analogous to an NMOS. Current will flow between the emitter and collector terminals when there is a sufficient voltage on the base terminal. The amount of current that flows between the emitter and collector is related to the current flowing into the base. The primary difference in operation between BJTs and MOSFETs is that BJTs require proper voltage biasing in order to turn on and also draw current through the base in order to stay on. The detailed operation of BJTs is beyond the scope of this text, so an overly simplified model of TTL logic gates is given.

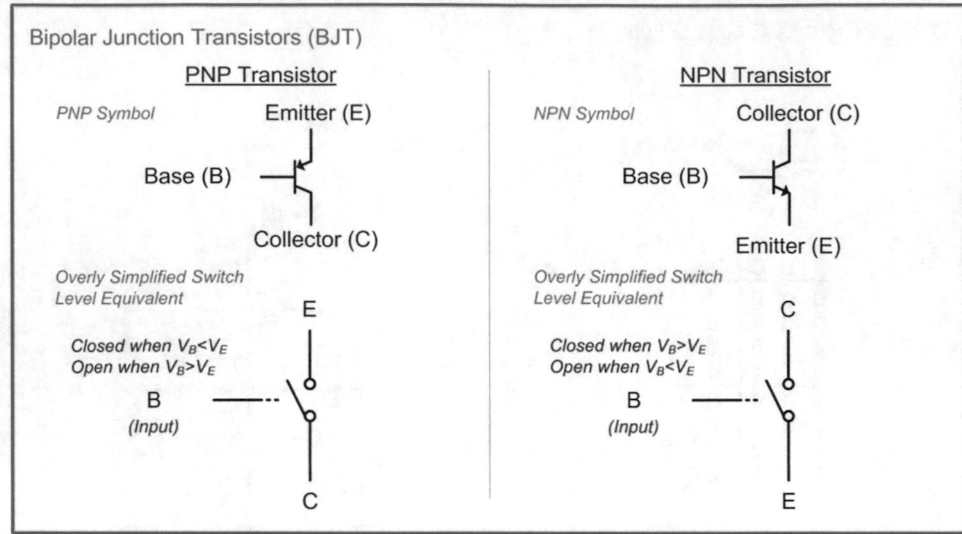

Fig. 3.30
PNP and NPN transistors

Figure 3.31 shows a simplified model of how TTL logic operates using BJTs and resistors. This simplified model does not show all of the transistors that are used in modern TTL circuits but instead is intended to provide a high-level overview of the operation. This gate is an inverter that is created with an NPN transistor and a resistor. When the input is a logic HIGH, the NPN transistor turns on and conducts current between its collector and emitter terminals. This in effect closes the switch and connects the output to GND providing a logic LOW. During this state, current will also flow through the resistor to GND through Q1, thus consuming more power than the equivalent gate in CMOS. When the input is a logic LOW, the NPN transistor turns off and no current flows between its collector and emitter. This, in effect, is an open circuit leaving only the resistor connected to the output. The resistor pulls the output up to V_{CC} providing a logic HIGH on the output. One drawback of this state is that there will be a voltage drop across the resistor, so the output is not pulled fully to V_{CC}.

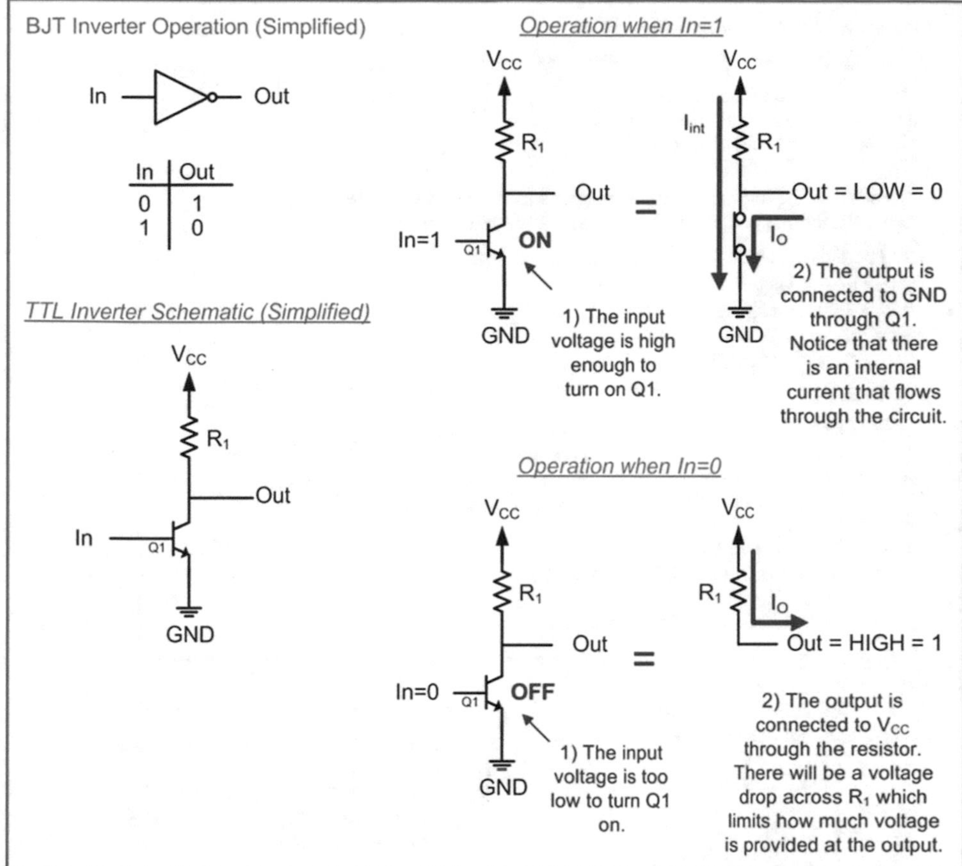

Fig. 3.31
TTL inverter

3.3.3 The 7400 Series Logic Families

The 7400 series of TTL circuits became popular in the 1960s and 1970s. This family was based on TTL and contained hundreds of different digital circuits. The original circuits came in either plastic or ceramic dual-in-line packages (DIP). The 7400 TTL logic family was powered off of a +5v supply. As mentioned before, this logic family set the pin-outs and part-numbering schemes for modern logic families. There were many derivatives of the original TTL logic family that made modifications to improve speed and reliability, decrease power, and reduce power supplies. Today's CMOS logic families within the 7400 series still use the same pin-outs and numbering schemes as the original TTL family. It is useful to understand the history of this series because these parts are often used in introductory laboratory exercises to learn how to interface digital logic circuits.

3.3.3.1 Part-Numbering Scheme

The part-numbering scheme for the 7400 series and its derivatives contains five different fields: (1) manufacturer, (2) temperature range, (3) logic family, (4) logic function, and (5) package type. The breakdown of these fields is shown in Fig. 3.32.

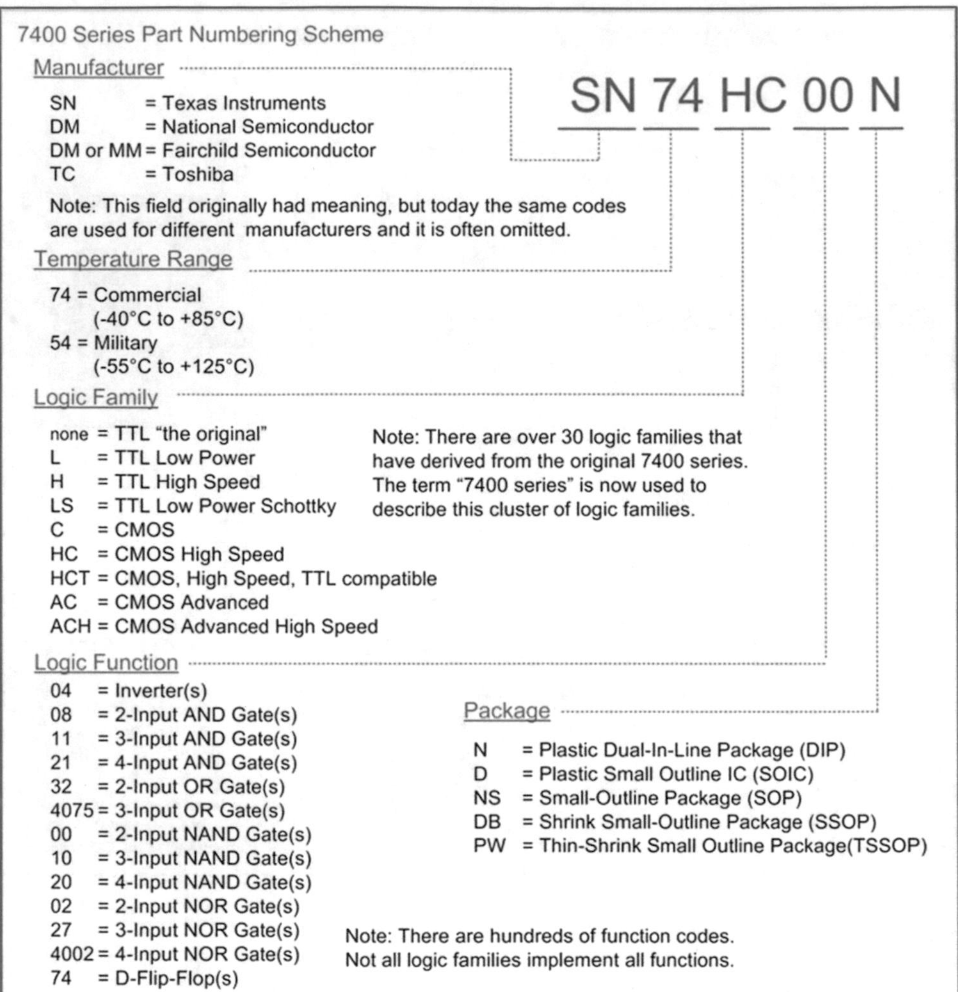

Fig. 3.32
7400 series part-numbering scheme

3.3.3.2 DC Operating Conditions

Table 3.2 gives the DC operating conditions for a few of the logic families within the 7400 series. Notice that the CMOS families consume much less power than the TTL families. Also notice that the TTL output currents are asymmetrical. The differences between the I_{OH} and I_{OL} within the TTL families have to do with the nature of the bipolar transistors and the resistors used to create the pull-up networks within the devices. CMOS has symmetrical drive currents due to using complementary transistors for the pull-up (PMOS) and pull-down networks (NMOS).

DC Operating Conditions for a Sample of 7400 Series Logic Families

Logic Family	Year	DC Operating Condition											Speed (MHz)
		V_{CC}	V_{OHmax}	V_{OHmin}	V_{OLmax}	V_{OLmin}	V_{IHmax}	V_{IHmin}	V_{ILmax}	V_{ILmin}	I_{CC}	$I_{Omax\ (H/L)}$	
Orig. (TTL)	1964	+5	+5	+2.4	+0.4	GND	+5	+2	+0.8	GND	40m	-4/+16m	25
LS (TTL)	1976	+5	+5	+2.4	+0.4	GND	+5	+2	+0.8	GND	8.8m	-4/+8m	40
HC (CMOS)	1982	+2-6	V_{CC}	$0.8 \cdot V_{CC}$	0.33	GND	V_{CC}	$0.7 \cdot V_{CC}$	$0.3 \cdot V_{CC}$	GND	40u	+/-25m	50
AC (CMOS)	1985	+2-6	V_{CC}	$0.8 \cdot V_{CC}$	0.33	GND	V_{CC}	$0.7 \cdot V_{CC}$	$0.3 \cdot V_{CC}$	GND	80u	+/-50m	125

Note 1: All voltage specifications have units of volts. All current specifications have units of amps.

Note 2: The V_O and V_I specifications for the AC and HC logic families are worst case and vary depending on the V_{CC} selection and the output current.

Note 3: All specifications are given for the commercial temperature range (74 series).

Table 3.2
DC operating conditions for a sample of 7400 series logic families

3.3.3.3 Pin-Out Information for the DIP Packages

Figure 3.33 shows the pin-out assignments for a subset of the basic gates from the 74HC logic family in the dual-in-line package form factor. Most of the basic gates within the 7400 series follow these assignments. Notice that each of these basic gates comes in a 14-pin DIP package, each with a single V_{CC} and single GND pin. It is up to the designer to ensure that the maximum current flowing through the V_{CC} and GND pins does not exceed the maximum specification. This is particularly important for parts that contain numerous gates. For example, the 74HC00 part contains four, 2-input NAND gates. If each of the NAND gates was driving a logic HIGH at its maximum allowable output current (i.e., 25 mA from Fig. 3.19), then a total of $4 \cdot 25$ mA + $I_q = \sim 100$ mA would be flowing through its V_{CC} pin. Since the V_{CC} pin can only tolerate a maximum of 50 mA of current (from Fig. 3.19), the part would be damaged since the output current of ~100 mA would also flow through the V_{CC} pin. The pin-outs in Fig. 3.33 are useful when first learning to design logic circuits because the DIP packages plug directly into a standard breadboard.

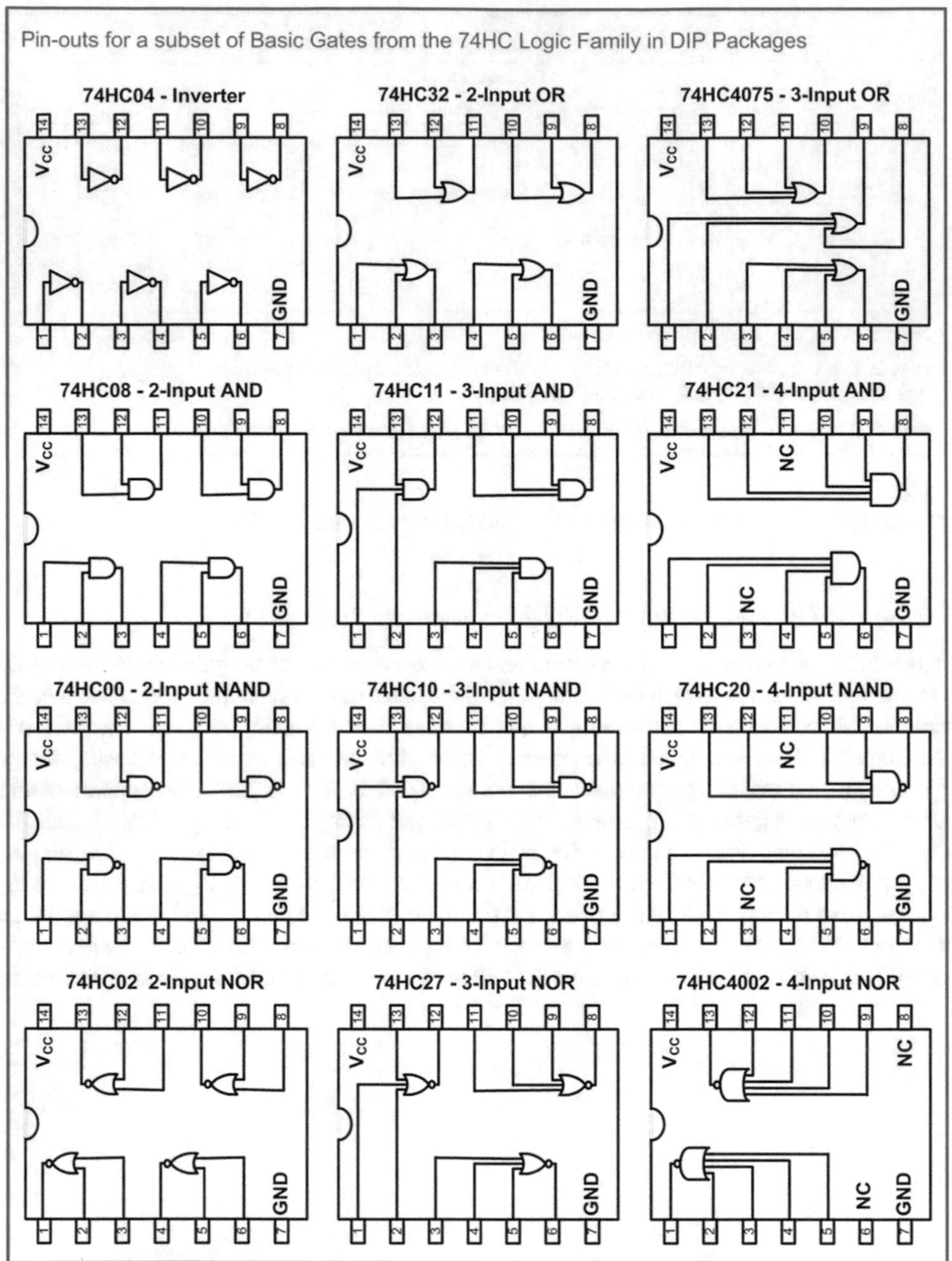

Fig. 3.33
Pin-outs for a subset of basic gates from the 74HC logic family in DIP packages

CONCEPT CHECK

CC3.3 Why doesn't the following CMOS transistor configuration yield a buffer?

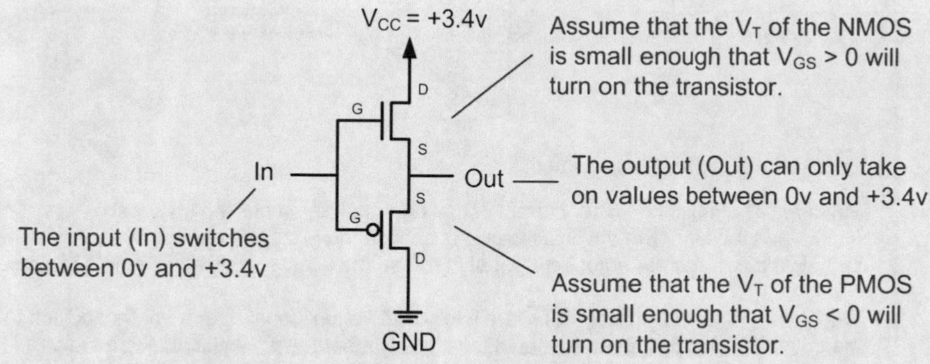

A) In order to turn on the NMOS transistor, V_{GS} needs to be greater than zero. In the given configuration, the gate terminal of the NMOS (G) needs to be driven above the source terminal (S). If the source terminal was at +3.4v, then the input (In) would never be able to provide a positive enough voltage to ensure the NMOS is on because "In" doesn't go above +3.4v.

B) There is no way to turn on both transistors in this configuration.

C) The power consumption will damage the device because both transistors will potentially be on.

D) The sources of the two devices can't be connected together without causing a short in the device.

3.4 Driving Loads

At this point we've discussed in depth how proper care must be taken to ensure that not only do the output voltages of the driving gate meet the input specifications of the receiver in order to successfully transmit 1's and 0's but that the output current of the driver does not exceed the maximum specifications so that the part is not damaged. The output voltage and current for a digital circuit depend greatly on the load that is being driven. The following sections discuss the impact of driving some of the most common digital loads.

3.4.1 Driving Other Gates

Within a logic family, all digital circuits are designed to operate with one another. If there is minimal loss or noise in the interconnect system, then 1's and 0's will be successfully transmitted, and no current specifications will be exceeded. Consider the example in Example 3.3 for an inverter driving another inverter from the same logic family.

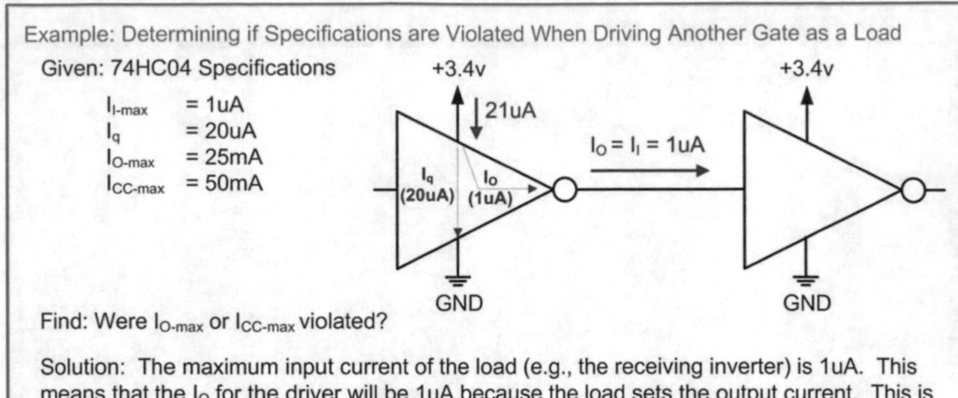

Solution: The maximum input current of the load (e.g., the receiving inverter) is 1uA. This means that the I_O for the driver will be 1uA because the load sets the output current. This is far below the maximum output current of 25mA so the I_{O-max} specification is <u>not</u> violated.

The driver will draw I_q through its V_{CC} pin to power its functional operation. In addition to I_q, the driver will also pull a current equal to I_O through the V_{CC} pin while driving a logic HIGH. This means the maximum current pulled through the V_{CC} pin is $I_q + I_O$ = 20uA + 1uA = 21uA. Again, this is well below the specification for the maximum amount of current that can flow through the V_{CC} pin (50mA) so the I_{CC}-max specification is also <u>not</u> violated.

Example 3.3
Determining if specifications are violated when driving another gate as a load

From this example, it is clear that there are no issues when a gate is driving another gate from the same family. This is as expected because that is the point of a logic family. In fact, gates are designed to drive multiple gates from within their own family. Based solely on the DC specifications for input and output current, it could be assumed that the number of other gates that can be driven is simply I_{O-max}/I_{I-max}. For the example in Example 3.3, this would result in a 74HC gate being able to drive 25,000 other gates (i.e., 25 mA/1uA = 25,000). In reality, the maximum number of gates that can be driven is dictated by the switching characteristics. This limit is called the **fan-out** specification. The fan-out specification states the maximum number of other gates from within the same family that can be driven. As discussed earlier, the output signal needs to transition quickly through the uncertainty region so that the receiver does not have time to react and go to an unknown state. As more and more gates are driven, this transition time is slowed down. The fan-out specification provides a limit to the maximum number of gates from the same family that can be driven while still ensuring that the output signal transitions between states fast enough to avoid the receivers from going to an unknown state. Example 3.4 shows the process of determining the maximum output current that a driver will need to provide when driving the maximum number of gates allowed by the fan-out specification.

Example: Determining the Output Current When Driving Multiple Gates as the Load

Given: 74HC04 Specifications

Fan-out = 3
$I_{I\text{-}max}$=1uA
Driving the maximum gates allowed by fan-out.

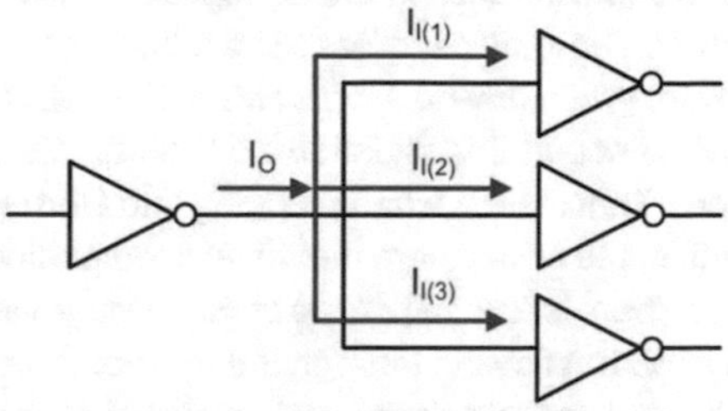

Find: I_O

Solution: The fan-out specification is 3, which means that the transmitting inverter can drive up to 3 other gates from its own logic family. Each of the receivers will draw their input current of I_I=1uA, which will be provided by the driver. The total amount of output current from the driver is 3·1uA = 3uA.

Example 3.4
Determining the output current when driving multiple gates as the load

3.4.2 Driving Resistive Loads

There are many situations where a resistor is the load in a digital circuit. A resistive load can be an actual resistor that is present for some other purpose such as a pull-up, pull-down, or for impedance matching. More complex loads such as buzzers, relays, or other electronics can also be modeled as a resistor. When a resistor is the load in a digital circuit, care must be taken to avoid violating the output current specifications of the driver. The electrical circuit analysis technique that is used to evaluate how a resistive load impacts a digital circuit is **Ohm's law**. Ohm's law is a very simple relationship between the current and voltage in a resistor. Figure 3.34 gives a primer on Ohm's law. For use in digital circuits, there are only a select few cases that this technique will be applied to, so no prior experience with Ohm's law is required at this point.

A Primer on Ohm's Law

Ohm's Law describes the relationship between current and voltage in a resistor. This simple equation is used in nearly all electrical circuit analysis. The equation is as follows:

$$V = I \cdot R$$

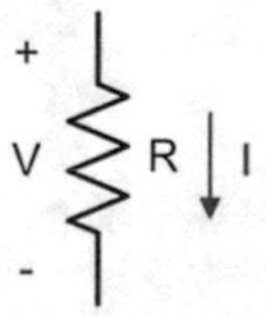

A resistor is characterized by its *resistance*, which describes how much current will flow when a voltage is present across its two terminals. The units for resistance are Ohms (Ω = Volts / Amp). The current in Ohm's Law is defined to flow from the + to – of the voltage.

Example: Use Ohm's Law to find the current flowing through the following resistor.

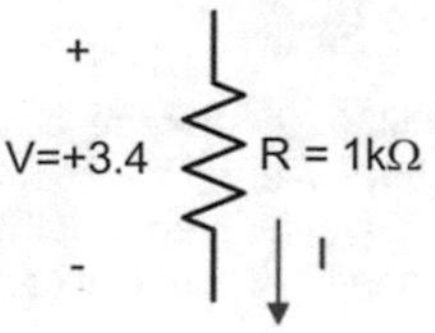

Solution: Plugging the parameters directly into Ohm's Law we find:

$$V = I \cdot R$$
$$3.4 = I \cdot (1k)$$
$$I = 0.0034\ A = 3.4\ mA$$

Fig. 3.34
A primer on Ohm's law

Let's see how we can use Ohm's law to analyze the impact of a resistive load in a digital circuit. Consider the circuit configuration in Example 3.5 and how we can use Ohm's law to determine the output current of the driver. The load in this case is a resistor connected between the output of the driver and the power supply (+5v). When driving a logic HIGH, the output level will be approximately equal to the power supply (i.e., +5v). Since in this situation both terminals of the resistor are at +5v, there is no voltage difference present. That means when plugging into Ohm's law, the voltage component is 0v, which gives 0 amps of current. In the case where the driver is outputting a logic LOW, the output will be approximately GND. In this case, there is a voltage drop of +5v across the resistor (5v–0v). Plugging this into Ohm's law yields a current of 50 mA flowing through the resistor. This can become problematic because the current flows through the resistor and then into the output of the driver. For the 74HC logic family, this would exceed the I_O max specification of 25 mA and damage the part. Additionally, as more current is drawn through the output, the output voltage becomes less and less ideal. In this example, the first-order analysis uses $V_O = GND$. In reality, as the output current increases, the output voltage will move further away from its ideal value and may eventually reach a value within the uncertainty region.

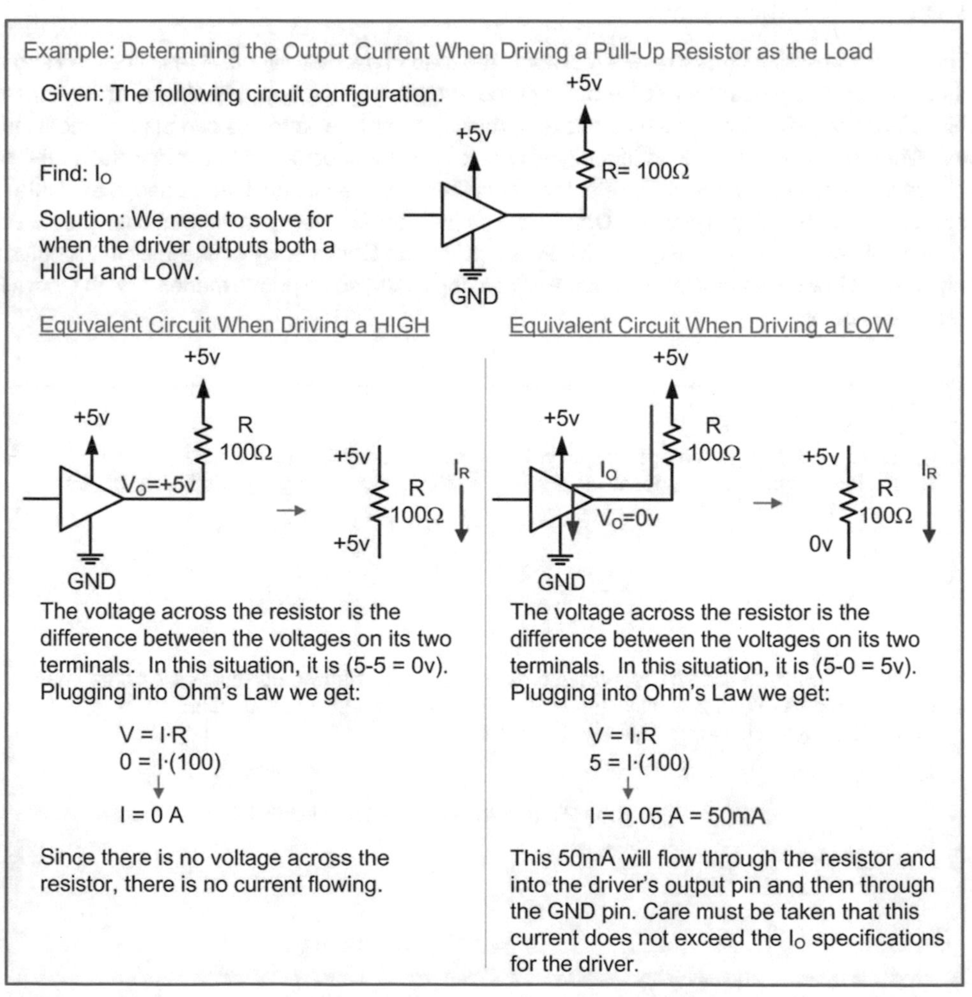

Example 3.5
Determining the output current when driving a pull-up resistor as the load

A similar process can be used to determine the output current when driving a resistive load between the output and GND. This process is shown in Example 3.6.

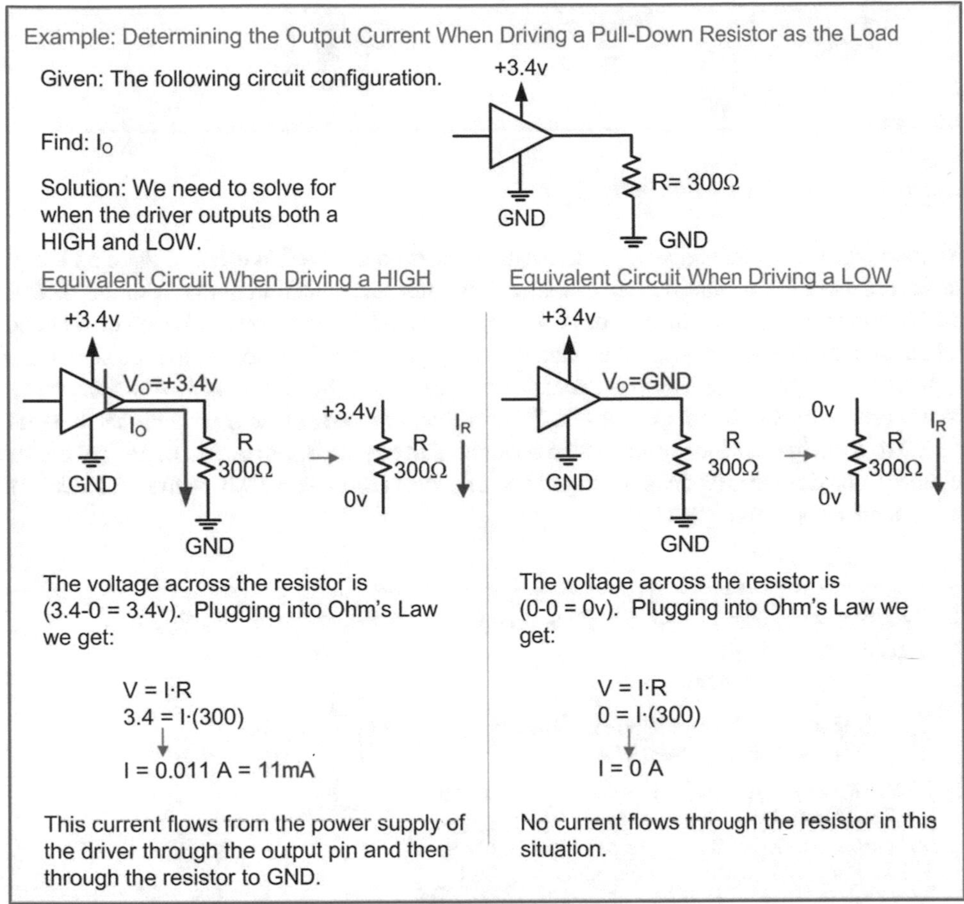

Example 3.6
Determining the output current when driving a pull-down resistor as the load

3.4.3 Driving LEDs

A light-emitting diode (LED) is a very common type of load that is driven using a digital circuit. The behavior of diodes is typically covered in an analog electronics class. Since it is assumed that the reader has not been exposed to the operation of diodes, the behavior of the LED will be described using a highly simplified model. A diode has two terminals, the anode and cathode. Current that flows from the anode to the cathode is called the *forward current*. A voltage that is developed across a diode from its anode to cathode is called the *forward voltage*. A diode has a unique characteristic that when a forward voltage is supplied across its terminal, it will only increase up to a certain point. The amount is specified as the LED's forward voltage (v_f) and is typically between 1.5v and 2v in modern LEDs. When a power supply circuit is connected to the LED, no current will flow until this forward voltage has been reached. Once it has been reached, current will begin to flow, and the LED will prevent any further voltage from developing across it. Once current flows, the LED will begin emitting light. The more current that flows, the more light that will be emitted up until the point that the maximum allowable current through the LED is reached and then the device will be damaged. When using an LED, there are two specifications of interest: the forward voltage and the recommended forward current. The symbols for a diode and an LED are given in Fig. 3.35.

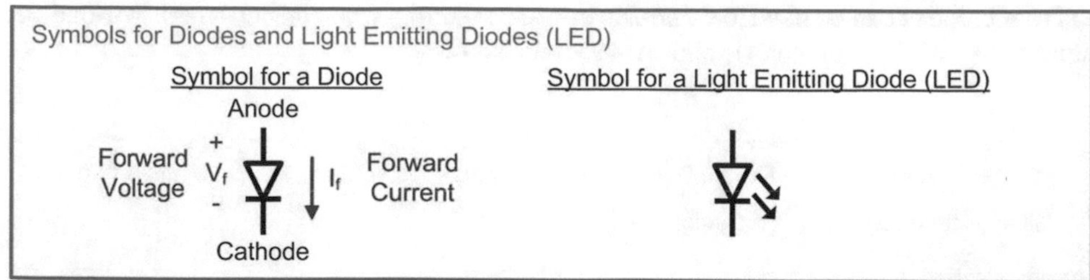

Fig. 3.35
Symbols for a diode and a light-emitting diode

When designing an LED driver circuit, a voltage must be supplied in order to develop the forward voltage across the LED so that current will flow. A resistor is included in series with the LED for two reasons. The first reason is to provide a place for any additional voltage provided by the driver to develop in the situation that $V_o > V_f$, which is most often the case. The second reason for the resistor is to set the output current. Since the voltage across the resistor will be a fixed amount (i.e., $V_o - V_f$), then the value of the resistor can be chosen to set the current. This current is typically set to an optimum value that turns on the LED to a desired luminosity while also ensuring that the maximum output current of the driver is not violated. Consider the LED driver configuration shown in Example 3.7 where the LED will be turned on when the driver outputs a HIGH.

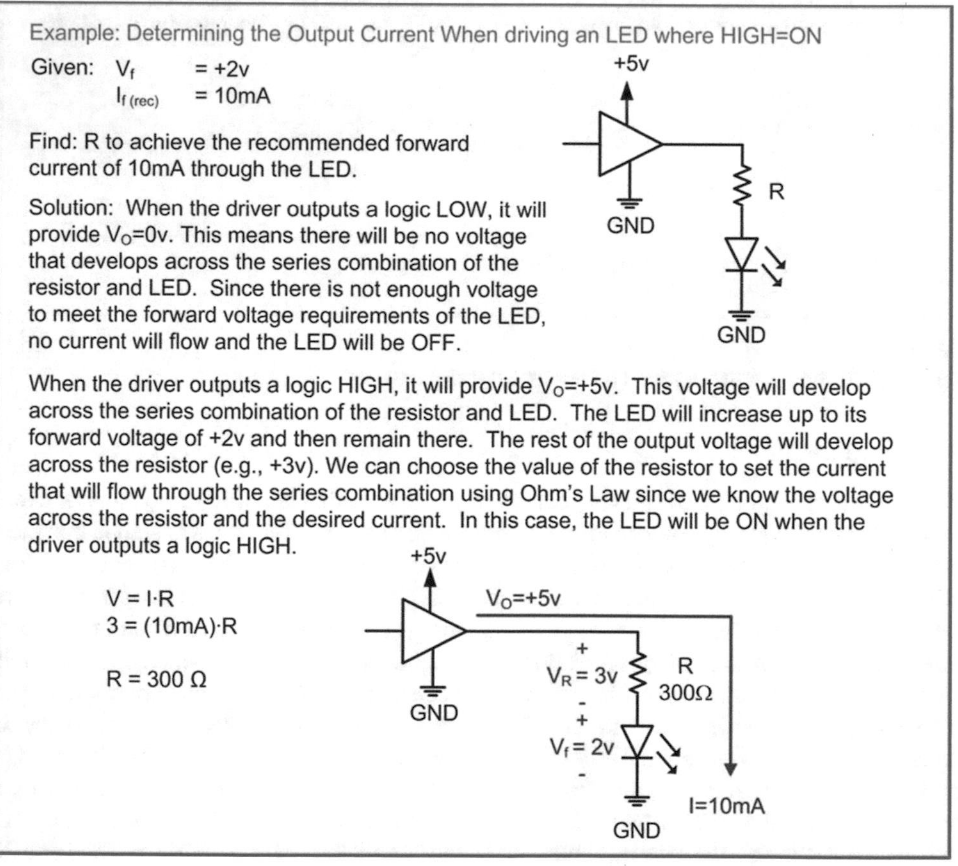

Example 3.7
Determining the output current when driving an LED where HIGH = ON

Example 3.8 shows another example of driving an LED, but this time using a different configuration where the LED will be on when the driver outputs a logic LOW.

Example: Determining the Output Current When Driving an LED where LOW=ON

Given: V_f = +1.8v
$I_{f\,(rec)}$ = 4mA

Find: R to achieve the recommended forward current of 4mA through the LED.

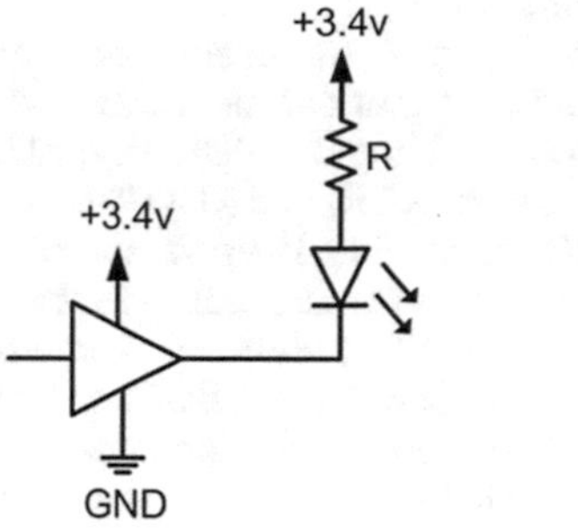

Solution: When the driver outputs a logic HIGH, it will provide V_O=+3.4v. This means there will be no voltage that develops across the series combination of the resistor and LED since the other end of the combination is also at +3.4. This means when driving a logic HIGH, the LED will be OFF.

When the driver outputs a logic LOW, it will provide V_O=0v. Since the resistor is tied to +3.4v, this voltage will develop across the series combination of the resistor and LED. The LED will increase up to its forward voltage of +1.8v and then remain there. The rest of the output voltage will develop across the resistor (e.g., +1.6v). We can choose the value of the resistor to set the current that will flow through the series combination using Ohm's Law since we know the voltage across the resistor and the desired current. In this case, the LED will be ON when the driver outputs a logic LOW.

$$V = I \cdot R$$
$$1.6 = (4mA) \cdot R$$
$$R = 400\ \Omega$$

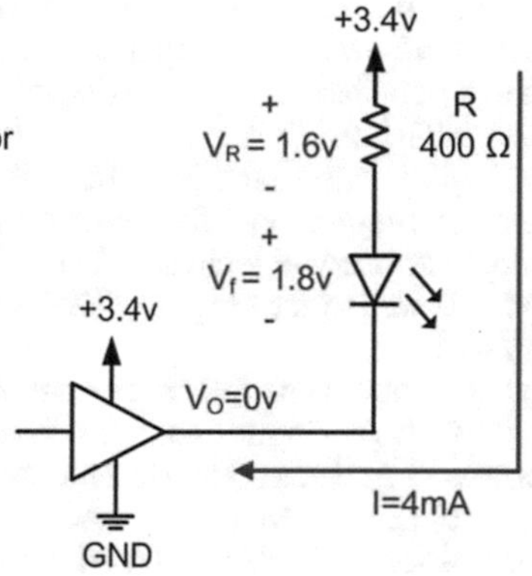

Example 3.8
Determining the Output Current When Driving an LED where HIGH=OFF

Concept Check

CC3.4 A fan-out specification is typically around 6–12. If a logic family has a maximum output current specification of $I_{O\text{-}max} = 25$ mA and a maximum input current specification of only $I_{I\text{-}max} = 1$uA, a driver could conceivably source up to 25,000 gates ($I_{O\text{-}max}/I_{I\text{-}max} = 25$ mA/1uA = 25,000) without violating its maximum output current specification. Why isn't the fan-out specification then closer to 25,000?

A) The fan-out specification has significant margin built into it in order to protect the driver.

B) Connecting 25,000 loads to the driver would cause significant wiring congestion and would be impractical.

C) The fan-out specification is in place to reduce power, so keeping it small is desirable.

D) The fan-out specification is in place for AC behavior. It ensures that the AC loading on the driver doesn't slow down its output rise and fall times. If too many loads are connected, the output transition will be too slow, and it will reside in the uncertainty region for too long leading to unwanted switching on the receivers.

Summary

- The operation of a logic circuit can be described using either a logic symbol, a truth table, a logic expression, or a logic waveform.
- Logic *gates* represent the most basic operations that can be performed on binary numbers. They are BUF, INV, AND, NAND, OR, NOR, XOR, and XNOR.
- XOR gates that have a number of inputs greater than two are created using a cascade of 2-input XOR gates. This implementation has more practical applications such as arithmetic and error detection codes.
- The logic *level* describes whether the electrical signal representing one of two states is above or below a switching threshold region. The two possible values that a logic level can be are HIGH or LOW.
- The logic *value* describes how the logic levels are mapped into the two binary codes 0 and 1. In positive logic a HIGH = 1 and a LOW = 0. In negative logic a HIGH = 0 and a LOW = 1.
- Logic circuits have DC specifications that describe how input voltage levels are interpreted as either HIGHs or LOWs ($V_{IH\text{-}max}$, $V_{IH\text{-}min}$, $V_{IL\text{-}max}$, and $V_{IL\text{-}min}$). Specifications are also given on what output voltages will be produced when driving a HIGH or LOW ($V_{OH\text{-}max}$, $V_{OH\text{-}min}$, $V_{OL\text{-}max}$, and $V_{OL\text{-}min}$).
- In order to successfully transmit digital information, the output voltages of the driver that represent a HIGH and LOW must arrive at the receiver within the voltage ranges that are *interpreted* as a HIGH and LOW. If the voltage arrives at the receiver outside of these specified input ranges, the receiver will not know whether a HIGH or LOW is being transmitted.
- Logic circuits also specify maximum current levels on the power supplies (I_{VCC}, I_{gnd}), inputs ($I_{I\text{-}max}$), and outputs ($I_{O\text{-}max}$) that may not be exceeded. If these levels are exceeded, the circuit may not operate properly or be damaged.
- The current exiting a logic circuit is equal to the current entering.
- When a logic circuit *sources* current to a load, an equivalent current is drawn *into* the circuit through its power supply pin.
- When a logic circuit *sinks* current from a load, an equivalent current flows *out of* the circuit through its ground pin.
- The type of load that is connected to the output of a logic circuit dictates how much current will be drawn from the driver.
- The *quiescent current* (I_q or I_{cc}) is the current that the circuit always draws independent of the input/output currents.
- Logic circuits have AC specifications that describe the delay from the input to the output (t_{PLH}, t_{PHL}) and also how fast the outputs transition between the HIGH and LOW levels (t_r, t_f).
- A *logic family* is a set of logic circuits that are designed to operate with each other.
- The *fan-in* of a logic family describes the maximum number of inputs that a gate may have.
- The *fan-out* of a logic family describes the maximum number of other gates from within the same family that can be driven simultaneously by one gate.
- Complementary metal-oxide semiconductor (CMOS) logic is the most popular family series in use today. CMOS logic uses two transistors (NMOS and PMOS) that act as complementary switches. CMOS transistors draw very low quiescent current and can be fabricated with extremely small feature sizes.
- In CMOS, only inverters, NAND gates, and NOR gates can be created directly. If it is desired to create a buffer, AND gate, or OR gate, an inverter is placed on the output of the original inverter, NAND, or NOR gate.

Exercise Problems

Section 3.1: Basic Gates

3.1.1 Give the truth table for a 3-input AND gate with the input variables A, B, C and output F.

3.1.2 Give the truth table for a 3-input OR gate with the input variables A, B, C and output F.

3.1.3 Give the truth table for a 3-input XNOR gate with the input variables A, B, C and output F.

3.1.4 Give the truth table for a 4-input XNOR gate with the input variables A, B, C and output F.

3.1.5 Give the logic expression for a 3-input AND gate with the input variables A, B, C and output F.

3.1.6 Give the logic expression for a 3-input OR gate with the input variables A, B, C and output F.

3.1.7 Give the logic expression for a 3-input XNOR gate with the input variables A, B, C and output F.

3.1.8 Give the logic expression for a 4-input XNOR gate with the input variables A, B, C and output F.

3.1.9 Give the logic waveform for a 3-input AND gate with the input variables A, B, C and output F.

3.1.10 Give the logic waveform for a 3-input OR gate with the input variables A, B, C and output F.

3.1.11 Give the logic waveform for a 3-input XNOR gate with the input variables A, B, C and output F.

3.1.12 Give the logic waveform for a 4-input XNOR gate with the input variables A, B, C and output F.

Section 3.2: Digital Circuit Operation

3.2.1 Using the DC operating conditions from Table 3.2, give the noise margin HIGH (NM_H) for the 74LS logic family.

3.2.2 Using the DC operating conditions from Table 3.2, give the noise margin LOW (NM_L) for the 74LS logic family.

3.2.3 Using the DC operating conditions from Table 3.2, give the noise margin HIGH (NM_H) for the 74HC logic family with $V_{CC} = +5v$.

3.2.4 Using the DC operating conditions from Table 3.2, give the noise margin LOW (NM_L) for the 74HC logic family with $V_{CC} = +5v$.

3.2.5 Using the DC operating conditions from Table 3.2, give the noise margin HIGH (NM_H) for the 74HC logic family with $V_{CC} = +3.4v$.

3.2.6 Using the DC operating conditions from Table 3.2, give the noise margin LOW (NM_L) for the 74HC logic family with $V_{CC} = +3.4v$.

3.2.7 Using the DC operating conditions from Table 3.2, give the noise margin HIGH (NM_H) for the 74HC logic family with $V_{CC} = +2.5v$.

3.2.8 Using the DC operating conditions from Table 3.2, give the noise margin LOW (NM_L) for the 74HC logic family with $V_{CC} = +2.5v$.

3.2.9 For the driver configuration in Fig. 3.36, give the current flowing through the V_{CC} pin.

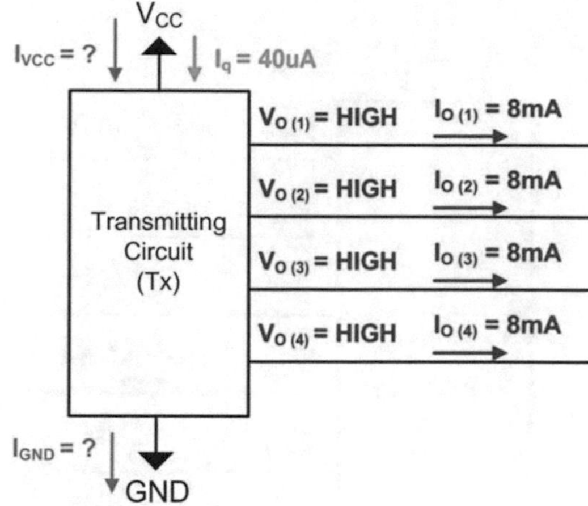

Fig. 3.36
Driver configuration 1

3.2.10 For the driver configuration in Fig. 3.36, give the current flowing through the GND pin.

3.2.11 For the driver configuration in Fig. 3.37, give the current flowing through the V_{CC} pin.

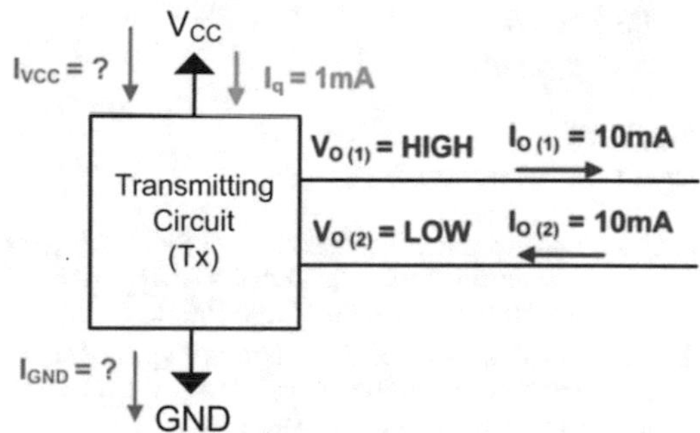

Fig. 3.37
Driver configuration 2

3.2.12 For the driver configuration in Fig. 3.37, give the current flowing through the GND pin.

3.2.13 For the driver configuration in Fig. 3.38, give the current flowing through the V_{CC} pin if all of the output pins are driving a HIGH.

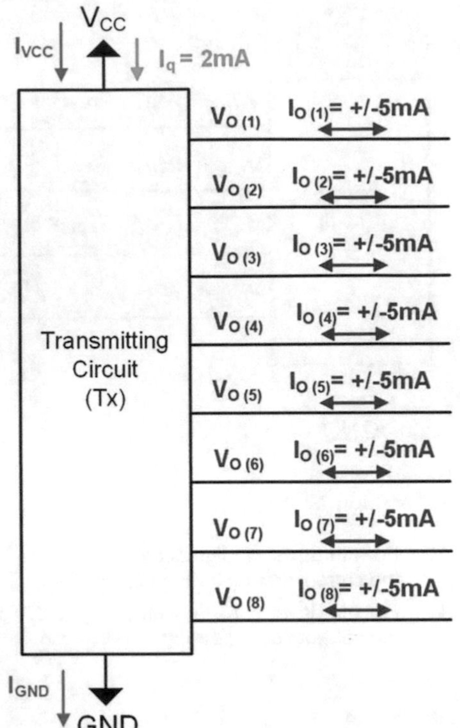

Fig. 3.38
Driver configuration 3

3.2.14 For the driver configuration in Fig. 3.38, give the current flowing through the GND pin if all of the output pins are driving a HIGH.

3.2.15 For the driver configuration in Fig. 3.38, give the current flowing through the V_{CC} pin if all of the output pins are driving a LOW.

3.2.16 For the driver configuration in Fig. 3.38, give the current flowing through the GND pin if all of the output pins are driving a LOW.

3.2.17 For the driver configuration in Fig. 3.38, give the current flowing through the V_{CC} pin if half of the output pins are driving a HIGH and the other half are driving a LOW.

3.2.18 For the driver configuration in Fig. 3.38, give the current flowing through the GND pin if half of the output pins are driving a HIGH and the other half are driving a LOW.

3.2.19 For the driver configuration in Fig. 3.38, give the current flowing through the V_{CC} pin if six of the output pins are driving a HIGH and two are driving a LOW.

3.2.20 For the driver configuration in Fig. 3.38, give the current flowing through the GND pin if six of the output pins are driving a HIGH and two are driving a LOW.

3.2.21 For the driver configuration in Fig. 3.38, give the current flowing through the V_{CC} pin if two of the output pins are driving a HIGH and six are driving a LOW.

3.2.22 For the driver configuration in Fig. 3.38, give the current flowing through the GND pin if two of the output pins are driving a HIGH and six are driving a LOW.

3.2.23 For the driver configuration in Fig. 3.38, if the maximum current allowed through the V_{CC} pin is 25 mA, is this specification violated if all of the outputs are driving a HIGH?

3.2.24 For the driver configuration in Fig. 3.38, if the maximum current allowed through the V_{CC} pin is 25 mA, is this specification violated if six of the outputs are driving a HIGH and two are driving a LOW?

3.2.25 For the driver configuration in Fig. 3.38, if the maximum current allowed through the V_{CC} pin is 25 mA, is this specification violated if half of the outputs are driving a HIGH and half are driving a LOW?

3.2.26 For the driver configuration in Fig. 3.38, if the maximum current allowed through the V_{CC} pin is 25 mA, is this specification violated if two of the outputs are driving a HIGH and six are driving a LOW?

3.2.27 For the driver configuration in Fig. 3.38, if the maximum current allowed through the V_{CC} pin is 25 mA, is this specification violated if all of the outputs are driving a LOW?

3.2.28 For the driver configuration in Fig. 3.38, if the maximum current allowed through the GND pin is 25 mA, is this specification violated if all of the outputs are driving a HIGH?

3.2.29 For the driver configuration in Fig. 3.38, if the maximum current allowed through the GND pin is 25 mA, is this specification violated if six of the outputs are driving a HIGH and two are driving a LOW?

3.2.30 For the driver configuration in Fig. 3.38, if the maximum current allowed through the GND pin is 25 mA, is this specification violated if half of the outputs are driving a HIGH and half are driving a LOW?

3.2.31 For the driver configuration in Fig. 3.38, if the maximum current allowed through the GND pin is 25 mA, is this specification violated if two of the outputs are driving a HIGH and six are driving a LOW?

3.2.32 For the driver configuration in Fig. 3.38, if the maximum current allowed through the GND pin is 25 mA, is this specification violated if all of the outputs are driving a LOW?

3.2.33 Using the data sheet excerpt from Fig. 3.20, give the maximum propagation delay (t_{pd}) for the 74HC04 inverter when powered with V_{CC} = +2v.

3.2.34 Using the data sheet excerpt from Fig. 3.20, give the maximum propagation delay from low to high (t_{PLH}) for the 74HC04 inverter when powered with V_{CC} = +2v.

3.2.35 Using the data sheet excerpt from Fig. 3.20, give the maximum propagation delay from high to low (t_{PHL}) for the 74HC04 inverter when powered with V_{CC} = +2v.

3.2.36 Using the data sheet excerpt from Fig. 3.20, give the maximum transition time (t_t) for the 74HC04 inverter when powered with V_{CC} = +2v.

3.2.37 Using the data sheet excerpt from Fig. 3.20, give the maximum rise time (t_r) for the 74HC04 inverter when powered with V_{CC} = +2v.

3.2.38 Using the data sheet excerpt from Fig. 3.20, give the maximum fall time (t_f) for the 74HC04 inverter when powered with V_{CC} = +2v.

3.2.39 Using the data sheet excerpt from Fig. 3.20, give the maximum propagation delay (t_{pd}) for the 74HC04 inverter when powered with V_{CC} = +4.5v.

3.2.40 Using the data sheet excerpt from Fig. 3.20, give the maximum propagation delay from low to high (t_{PLH}) for the 74HC04 inverter when powered with V_{CC} = +4.5v.

3.2.41 Using the data sheet excerpt from Fig. 3.20, give the maximum propagation delay from high to low (t_{PHL}) for the 74HC04 inverter when powered with V_{CC} = +4.5v.

3.2.42 Using the data sheet excerpt from Fig. 3.20, give the maximum transition time (t_t) for the 74HC04 inverter when powered with V_{CC} = +4.5v.

3.2.43 Using the data sheet excerpt from Fig. 3.20, give the maximum rise time (t_r) for the 74HC04 inverter when powered with V_{CC} = +4.5v.

3.2.44 Using the data sheet excerpt from Fig. 3.20, give the maximum fall time (t_f) for the 74HC04 inverter when powered with V_{CC} = +4.5v.

3.2.45 Using the data sheet excerpt from Fig. 3.20, give the maximum propagation delay (t_{pd}) for the 74HC04 inverter when powered with V_{CC} = +6v.

3.2.46 Using the data sheet excerpt from Fig. 3.20, give the maximum propagation delay from low to high (t_{PLH}) for the 74HC04 inverter when powered with V_{CC} = +6v.

3.2.47 Using the data sheet excerpt from Fig. 3.20, give the maximum propagation delay from high to low (t_{PHL}) for the 74HC04 inverter when powered with V_{CC} = +6v.

3.2.48 Using the data sheet excerpt from Fig. 3.20, give the maximum transition time (t_t) for the 74HC04 inverter when powered with V_{CC} = +6v.

3.2.49 Using the data sheet excerpt from Fig. 3.20, give the maximum rise time (t_r) for the 74HC04 inverter when powered with V_{CC} = +6v.

3.2.50 Using the data sheet excerpt from Fig. 3.20, give the maximum fall time (t_f) for the 74HC04 inverter when powered with V_{CC} = +6v.

3.2.51 Using the data sheet excerpt from Fig. 3.20, give the maximum propagation delay (t_{pd}) for the 54HC04 inverter when powered with V_{CC} = +2v.

3.2.52 Using the data sheet excerpt from Fig. 3.20, give the maximum propagation delay from low to high (t_{PLH}) for the 54HC04 inverter when powered with V_{CC} = +2v.

3.2.53 Using the data sheet excerpt from Fig. 3.20, give the maximum propagation delay from high to low (t_{PHL}) for the 54HC04 inverter when powered with V_{CC} = +2v.

3.2.54 Using the data sheet excerpt from Fig. 3.20, give the maximum transition time (t_t) for the 54HC04 inverter when powered with V_{CC} = +2v.

3.2.55 Using the data sheet excerpt from Fig. 3.20, give the maximum rise time (t_r) for the 54HC04 inverter when powered with V_{CC} = +2v.

3.2.56 Using the data sheet excerpt from Fig. 3.20, give the maximum fall time (t_f) for the 54HC04 inverter when powered with V_{CC} = +2v.

3.2.57 Using the data sheet excerpt from Fig. 3.20, give the maximum propagation delay (t_{pd}) for the 54HC04 inverter when powered with V_{CC} = +4.5v.

3.2.58 Using the data sheet excerpt from Fig. 3.20, give the maximum propagation delay from low to high (t_{PLH}) for the 54HC04 inverter when powered with V_{CC} = +4.5v.

3.2.59 Using the data sheet excerpt from Fig. 3.20, give the maximum propagation delay from high to low (t_{PHL}) for the 54HC04 inverter when powered with V_{CC} = +4.5v.

3.2.60 Using the data sheet excerpt from Fig. 3.20, give the maximum transition time (t_t) for the 54HC04 inverter when powered with V_{CC} = +4.5v.

3.2.61 Using the data sheet excerpt from Fig. 3.20, give the maximum rise time (t_r) for the 54HC04 inverter when powered with V_{CC} = +4.5v.

3.2.62 Using the data sheet excerpt from Fig. 3.20, give the maximum fall time (t_f) for the 54HC04 inverter when powered with V_{CC} = +4.5v.

3.2.63 Using the data sheet excerpt from Fig. 3.20, give the maximum propagation delay (t_{pd}) for the 54HC04 inverter when powered with V_{CC} = +6v.

3.2.64 Using the data sheet excerpt from Fig. 3.20, give the maximum propagation delay from low to high (t_{PLH}) for the 54HC04 inverter when powered with V_{CC} = +6v.

3.2.65 Using the data sheet excerpt from Fig. 3.20, give the maximum propagation delay from high to low (t_{PHL}) for the 54HC04 inverter when powered with V_{CC} = +6v.

3.2.66 Using the data sheet excerpt from Fig. 3.20, give the maximum transition time (t_t) for the 54HC04 inverter when powered with V_{CC} = +6v.

3.2.67 Using the data sheet excerpt from Fig. 3.20, give the maximum rise time (t_r) for the 54HC04 inverter when powered with V_{CC} = +6v.

3.2.68 Using the data sheet excerpt from Fig. 3.20, give the maximum fall time (t_f) for the 54HC04 inverter when powered with V_{CC} = +6v.

Section 3.3: Logic Families

3.3.1 Provide the transistor-level schematic for a 4-input NAND gate.

3.3.2 Provide the transistor-level schematic for a 4-input NOR gate.

3.3.3 Provide the transistor-level schematic for a 5-input NAND gate.

3.3.4 Provide the transistor-level schematic for a 5-input NOR gate.

3.3.5 Provide the transistor-level schematic for a NAND gate with a fan-in specification of 6.

3.3.6 Provide the transistor-level schematic for a NOR gate with a fan-in specification of 6.

3.3.7 Provide the transistor-level schematic for a 2-input AND gate.

3.3.8 Provide the transistor-level schematic for a 2-input OR gate.

3.3.9 Provide the transistor-level schematic for a buffer.

Section 3.4: Driving Loads

3.4.1 In the driver configuration shown in Fig. 3.39, the buffer is driving its maximum fan-out specification of 6. The maximum input current for this logic family is I_I = 1 nA. What is the maximum output current (I_O) that the driver will need to source?

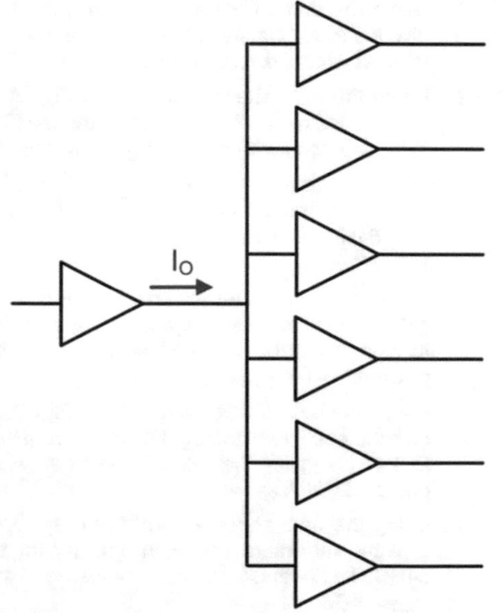

Fig. 3.39
Driver configuration 4

3.4.2 In the driver configuration shown in Fig. 3.39, the buffer is driving its maximum fan-out specification of 6. The maximum input current for this logic family is I_I = 4 nA. What is the maximum output current (I_O) that the driver will need to source?

3.4.3 For the pull-down driver configuration shown in Fig. 3.40, what is the output current (I_O) if V_{CC} = +5v and R = 50 Ω?

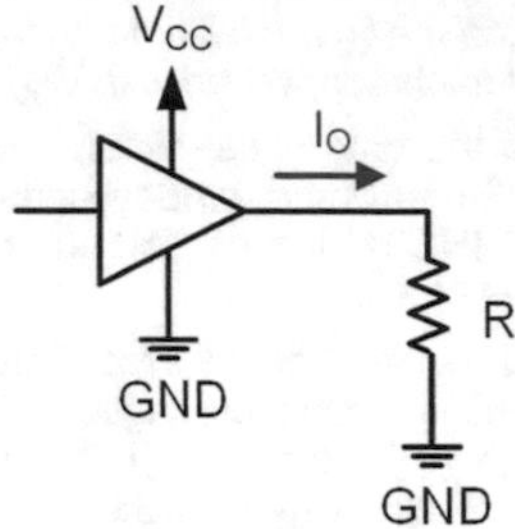

Fig. 3.40
Driver configuration 5

3.4.4 For the pull-down driver configuration shown in Fig. 3.40, what is the output current (I_O) if V_{CC} = +5v and R = 100 Ω?

3.4.5 For the pull-down driver configuration shown in Fig. 3.40, what is the output current (I_O) if V_{CC} = +5v and R = 1 kΩ?

3.4.6 For the pull-down driver configuration shown in Fig. 3.40, what is the output current (I_O) if V_{CC} = +3.4v and R = 50 Ω?

3.4.7 For the pull-down driver configuration shown in Fig. 3.40, what is the output current (I_O) if V_{CC} = +3.4v and R = 100 Ω?

3.4.8 For the pull-down driver configuration shown in Fig. 3.40, what is the output current (I_O) if V_{CC} = +3.4v and R = 1 kΩ?

3.4.9 For the pull-down driver configuration shown in Fig. 3.40, if V_{CC} = +5v, R = 75 Ω, and $I_{O\text{-}MAX}$ = 50 mA, is the maximum output current specification violated?

3.4.10 For the pull-down driver configuration shown in Fig. 3.40, if V_{CC} = +5v, R = 150 Ω, and $I_{O\text{-}MAX}$ = 50 mA, is the maximum output current specification violated?

3.4.11 For the pull-down driver configuration shown in Fig. 3.40, if V_{CC} = +5v, R = 500 Ω, and $I_{O\text{-}MAX}$ = 50 mA, is the maximum output current specification violated?

3.4.12 For the pull-down driver configuration shown in Fig. 3.40, if V_{CC} = +3.4v, R = 75 Ω, and $I_{O\text{-}MAX}$ = 25 mA, is the maximum output current specification violated?

3.4.13 For the pull-down driver configuration shown in Fig. 3.40, if V_{CC} = +3.4v, R = 150 Ω, and $I_{O\text{-}MAX}$ = 25 mA, is the maximum output current specification violated?

3.4.14 For the pull-down driver configuration shown in Fig. 3.40, if V_{CC} = +3.4v, R = 500 Ω, and $I_{O\text{-}}$

$_{MAX}$ = 25 mA, is the maximum output current specification violated?

3.4.15 For the pull-down driver configuration shown in Fig. 3.40 with V_{CC} = +5v, calculate the value of the pull-down resistor (R) in order to ensure that the output current does not exceed 20 mA.

3.4.16 For the pull-down driver configuration shown in Fig. 3.40 with V_{CC} = +5v, calculate the value of the pull-down resistor (R) in order to ensure that the output current does not exceed 25 mA.

3.4.17 For the pull-down driver configuration shown in Fig. 3.40 with V_{CC} = +5v, calculate the value of the pull-down resistor (R) in order to ensure that the output current does not exceed 50 mA.

3.4.18 For the pull-down driver configuration shown in Fig. 3.40 with V_{CC} = +3.4v, calculate the value of the pull-down resistor (R) in order to ensure that the output current does not exceed 20 mA.

3.4.19 For the pull-down driver configuration shown in Fig. 3.40 with V_{CC} = +3.4v, calculate the value of the pull-down resistor (R) in order to ensure that the output current does not exceed 25 mA.

3.4.20 For the pull-down driver configuration shown in Fig. 3.40 with V_{CC} = +3.4v, calculate the value of the pull-down resistor (R) in order to ensure that the output current does not exceed 50 mA.

3.4.21 For the pull-up driver configuration shown in Fig. 3.41, what is the output current (I_O) if V_{CC} = +5v and R = 50 Ω?

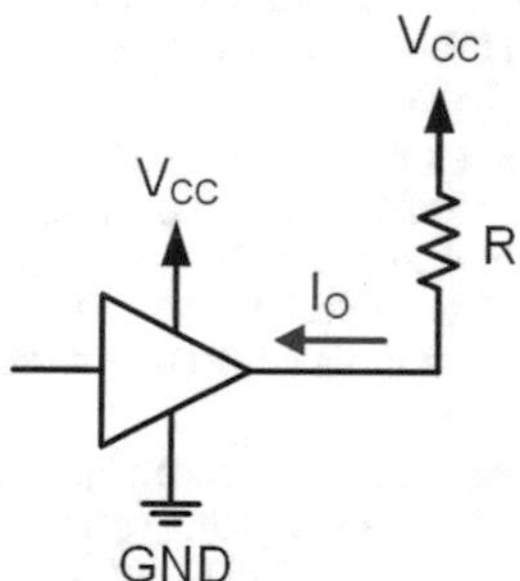

Fig. 3.41
Driver configuration 6

3.4.22 For the pull-up driver configuration shown in Fig. 3.41, what is the output current (I_O) if V_{CC} = +5v and R = 100 Ω?

3.4.23 For the pull-up driver configuration shown in Fig. 3.41, what is the output current (I_O) if V_{CC} = +5v and R = 1 kΩ?

3.4.24 For the pull-up driver configuration shown in Fig. 3.41, what is the output current (I_O) if V_{CC} = +3.4v and R = 50 Ω?

3.4.25 For the pull-up driver configuration shown in Fig. 3.41, what is the output current (I_O) if V_{CC} = +3.4v and R = 100 Ω?

3.4.26 For the pull-up driver configuration shown in Fig. 3.41, what is the output current (I_O) if V_{CC} = +3.4v and R = 1 kΩ?

3.4.27 For the pull-up driver configuration shown in Fig. 3.41, if V_{CC} = +5v, R = 75 Ω, and $I_{O\text{-}MAX}$ = 50 mA, is the maximum output current specification violated?

3.4.28 For the pull-up driver configuration shown in Fig. 3.41, if V_{CC} = +5v, R = 150 Ω, and $I_{O\text{-}MAX}$ = 50 mA, is the maximum output current specification violated?

3.4.29 For the pull-up driver configuration shown in Fig. 3.41, if V_{CC} = +5v, R = 500 Ω, and $I_{O\text{-}MAX}$ = 50 mA, is the maximum output current specification violated?

3.4.30 For the pull-up driver configuration shown in Fig. 3.41, if V_{CC} = +3.4v, R = 75 Ω, and $I_{O\text{-}MAX}$ = 25 mA, is the maximum output current specification violated?

3.4.31 For the pull-up driver configuration shown in Fig. 3.41, if V_{CC} = +3.4v, R = 150 Ω, and $I_{O\text{-}MAX}$ = 25 mA, is the maximum output current specification violated?

3.4.32 For the pull-up driver configuration shown in Fig. 3.41, if V_{CC} = +3.4v, R = 500 Ω, and $I_{O\text{-}MAX}$ = 25 mA, is the maximum output current specification violated?

3.4.33 For the pull-up driver configuration shown in Fig. 3.41 with V_{CC} = +5v, calculate the value of the pull-up resistor (R) in order to ensure that the output current does not exceed 20 mA.

3.4.34 For the pull-up driver configuration shown in Fig. 3.41 with V_{CC} = +5v, calculate the value of the pull-up resistor (R) in order to ensure that the output current does not exceed 25 mA.

3.4.35 For the pull-up driver configuration shown in Fig. 3.41 with V_{CC} = +5v, calculate the value of the pull-up resistor (R) in order to ensure that the output current does not exceed 50 mA.

3.4.36 For the pull-up driver configuration shown in Fig. 3.41 with V_{CC} = +3.4v, calculate the value of the pull-up resistor (R) in order to ensure that the output current does not exceed 20 mA.

3.4.37 For the pull-up driver configuration shown in Fig. 3.41 with V_{CC} = +3.4v, calculate the value of the pull-up resistor (R) in order to ensure that the output current does not exceed 25 mA.

3.4.38 For the pull-up driver configuration shown in Fig. 3.41 with V_{CC} = +3.4v, calculate the value of the pull-up resistor (R) in order to ensure that the output current does not exceed 50 mA.

3.4.39 For the LED driver configuration shown in Fig. 3.42 where an output of HIGH on the driver will turn on the LED, what is the output current (I_O) if V_{CC} = +5v, R = 50 Ω, and the LED has a forward voltage of Vf = 1.9v?

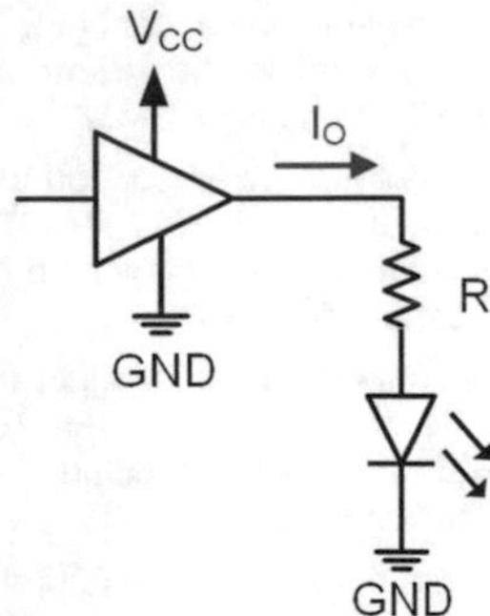

Fig. 3.42
Driver configuration 7

3.4.40 For the LED driver configuration shown in Fig. 3.42 where an output of HIGH on the driver will turn on the LED, what is the output current (I_O) if V_{CC} = +5v, R = 100 Ω, and the LED has a forward voltage of Vf = 1.9v?

3.4.41 For the LED driver configuration shown in Fig. 3.42 where an output of HIGH on the driver will turn on the LED, what is the output current (I_O) if V_{CC} = +5v, R = 1 kΩ, and the LED has a forward voltage of Vf = 1.9v?

3.4.42 For the LED driver configuration shown in Fig. 3.42 where an output of HIGH on the driver will turn on the LED, what is the output current (I_O) if V_{CC} = +3.4v, R = 50 Ω, and the LED has a forward voltage of Vf = 1.9v?

3.4.43 For the LED driver configuration shown in Fig. 3.42 where an output of HIGH on the driver will turn on the LED, what is the output current (I_O) if V_{CC} = +3.4v, R = 100 Ω, and the LED has a forward voltage of Vf = 1.9v?

3.4.44 For the LED driver configuration shown in Fig. 3.42 where an output of HIGH on the driver will turn on the LED, what is the output current (I_O) if V_{CC} = +3.4v, R = 1 kΩ, and the LED has a forward voltage of Vf = 1.9v?

3.4.45 For the LED driver configuration shown in Fig. 3.42 where an output of HIGH on the driver will turn on the LED, if V_{CC} = +5v, R = 75 Ω, Vf = 1.9v, and $I_{O\text{-}MAX}$ = 25 mA, is the maximum output current specification violated?

3.4.46 For the LED driver configuration shown in Fig. 3.42 where an output of HIGH on the driver will turn on the LED, if V_{CC} = +5v, R = 150 Ω, Vf = 1.9v, and $I_{O\text{-}MAX}$ = 25 mA, is the maximum output current specification violated?

3.4.47 For the LED driver configuration shown in Fig. 3.42 where an output of HIGH on the driver will turn on the LED, if V_{CC} = +5v, R = 500 Ω, Vf = 1.9v, and $I_{O\text{-}MAX}$ = 25 mA, is the maximum output current specification violated?

3.4.48 For the LED driver configuration shown in Fig. 3.42 where an output of HIGH on the driver will turn on the LED, if V_{CC} = +3.4v, R = 75 Ω, Vf = 1.9v, and $I_{O\text{-}MAX}$ = 15 mA, is the maximum output current specification violated?

3.4.49 For the LED driver configuration shown in Fig. 3.42 where an output of HIGH on the driver will turn on the LED, if V_{CC} = +3.4v, R = 150 Ω, Vf = 1.9v, and $I_{O\text{-}MAX}$ = 15 mA, is the maximum output current specification violated?

3.4.50 For the LED driver configuration shown in Fig. 3.42 where an output of HIGH on the driver will turn on the LED, if V_{CC} = +3.4v, R = 500 Ω, Vf = 1.9v, and $I_{O\text{-}MAX}$ = 15 mA, is the maximum output current specification violated?

3.4.51 For the LED driver configuration shown in Fig. 3.42 where an output of HIGH on the driver will turn on the LED with V_{CC} = +5v and Vf = 1.9v, calculate the value of the resistor (R) in order to set the current through the LED to be 5 mA.

3.4.52 For the LED driver configuration shown in Fig. 3.42 where an output of HIGH on the driver will turn on the LED with V_{CC} = +5v and Vf = 1.9v, calculate the value of the resistor (R) in order to set the current through the LED to be 12 mA.

3.4.53 For the LED driver configuration shown in Fig. 3.42 where an output of HIGH on the driver will turn on the LED with V_{CC} = +5v and Vf = 1.9v, calculate the value of the resistor (R) in order to set the current through the LED to be 15 mA.

3.4.54 For the LED driver configuration shown in Fig. 3.42 where an output of HIGH on the driver will turn on the LED with V_{CC} = +3.4v and Vf = 1.9v, calculate the value of the resistor (R) in order to set the current through the LED to be 5 mA.

3.4.55 For the LED driver configuration shown in Fig. 3.42 where an output of HIGH on the driver will turn on the LED with V_{CC} = +3.4v and Vf = 1.9v, calculate the value of the resistor (R) in order to set the current through the LED to be 12 mA.

3.4.56 For the LED driver configuration shown in Fig. 3.42 where an output of HIGH on the driver will turn on the LED with V_{CC} = +3.4v and Vf = 1.9v, calculate the value of the resistor (R) in order to set the current through the LED to be 15 mA.

3.4.57 For the LED driver configuration shown in Fig. 3.43 where an output of LOW on the driver will turn on the LED, what is the output current (I_O) if V_{CC} = +5v, R = 50 Ω, and the LED has a forward voltage of Vf = 1.9v?

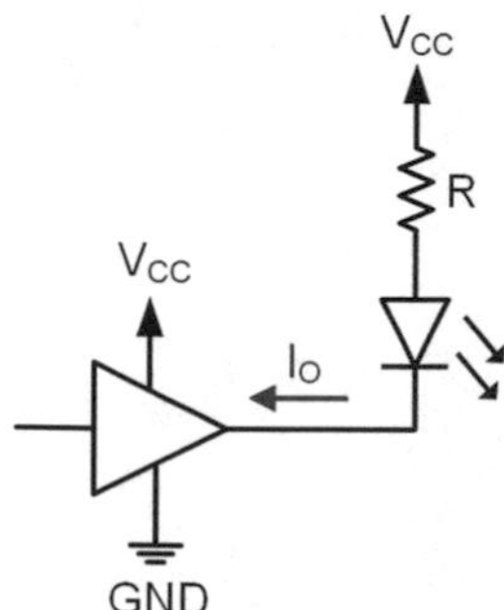

Fig. 3.43
Driver configuration 8

3.4.58 For the LED driver configuration shown in Fig. 3.43 where an output of LOW on the driver will turn on the LED, what is the output current (I_O) if V_{CC} = +5v, R = 100 Ω, and the LED has a forward voltage of Vf = 1.9v?

3.4.59 For the LED driver configuration shown in Fig. 3.43 where an output of LOW on the driver will turn on the LED, what is the output current (I_O) if V_{CC} = +5v, R = 1 kΩ, and the LED has a forward voltage of Vf = 1.9v?

3.4.60 For the LED driver configuration shown in Fig. 3.43 where an output of LOW on the driver will turn on the LED, what is the output current (I_O) if V_{CC} = +3.4v, R = 50 Ω, and the LED has a forward voltage of Vf = 1.9v?

3.4.61 For the LED driver configuration shown in Fig. 3.43 where an output of LOW on the driver will turn on the LED, what is the output current (I_O) if V_{CC} = +3.4v, R = 100 Ω, and the LED has a forward voltage of Vf = 1.9v?

3.4.62 For the LED driver configuration shown in Fig. 3.43 where an output of LOW on the driver will turn on the LED, what is the output current (I_O) if V_{CC} = +3.4v, R = 1 kΩ, and the LED has a forward voltage of Vf = 1.9v?

3.4.63 For the LED driver configuration shown in Fig. 3.43 where an output of LOW on the driver will turn on the LED, if V_{CC} = +5v, R = 75 Ω, Vf = 1.9v, and $I_{O\text{-}MAX}$ = 25 mA, is the maximum output current specification violated?

3.4.64 For the LED driver configuration shown in Fig. 3.43 where an output of LOW on the driver will turn on the LED, if V_{CC} = +5v, R = 150 Ω, Vf = 1.9v, and $I_{O\text{-}MAX}$ = 25 mA, is the maximum output current specification violated?

3.4.65 For the LED driver configuration shown in Fig. 3.43 where an output of LOW on the driver will turn on the LED, if V_{CC} = +5v, R = 500 Ω, Vf = 1.9v, and $I_{O\text{-}MAX}$ = 25 mA, is the maximum output current specification violated?

3.4.66 For the LED driver configuration shown in Fig. 3.43 where an output of LOW on the driver will turn on the LED, if V_{CC} = +3.4v, R = 75 Ω, Vf = 1.9v, and $I_{O\text{-}MAX}$ = 15 mA, is the maximum output current specification violated?

3.4.67 For the LED driver configuration shown in Fig. 3.43 where an output of LOW on the driver will turn on the LED, if V_{CC} = +3.4v, R = 150 Ω, Vf = 1.9v, and $I_{O\text{-}MAX}$ = 15 mA, is the maximum output current specification violated?

3.4.68 For the LED driver configuration shown in Fig. 3.43 where an output of LOW on the driver will turn on the LED, if V_{CC} = +3.4v, R = 500 Ω, Vf = 1.9v, and $I_{O\text{-}MAX}$ = 15 mA, is the maximum output current specification violated?

3.4.69 For the LED driver configuration shown in Fig. 3.43 where an output of LOW on the driver will turn on the LED with V_{CC} = +5v and Vf = 1.9v, calculate the value of the resistor (R) in order to set the current through the LED to be 5 mA.

3.4.70 For the LED driver configuration shown in Fig. 3.43 where an output of LOW on the driver will turn on the LED with V_{CC} = +5v and Vf = 1.9v, calculate the value of the resistor (R) in order to set the current through the LED to be 12 mA.

3.4.71 For the LED driver configuration shown in Fig. 3.43 where an output of LOW on the driver will turn on the LED with V_{CC} = +5v and Vf = 1.9v, calculate the value of the resistor (R) in order to set the current through the LED to be 15 mA.

3.4.72 For the LED driver configuration shown in Fig. 3.43 where an output of LOW on the driver will turn on the LED with V_{CC} = +3.4v and Vf = 1.9v, calculate the value of the resistor (R) in order to set the current through the LED to be 5 mA.

3.4.73 For the LED driver configuration shown in Fig. 3.43 where an output of LOW on the driver will turn on the LED with V_{CC} = +3.4v and Vf = 1.9v, calculate the value of the resistor (R) in order to set the current through the LED to be 12 mA.

3.4.74 For the LED driver configuration shown in Fig. 3.43 where an output of LOW on the driver will turn on the LED with VCC = +3.4v and Vf = 1.9v, calculate the value of the resistor (R) in order to set the current through the LED to be 15 mA.

Chapter 4: Combinational Logic Design

In this chapter we cover the techniques to synthesize, analyze, and manipulate logic functions. The purpose of these techniques is to ultimately create a logic circuit using the basic gates described in Chap. 3 from a truth table or word description. This process is called *combinational logic design*. Combinational logic refers to circuits where the output depends on the present value of the inputs. This simple definition implies that there is no storage capability in the circuitry and a change on the input immediately impacts the output. To begin, we first define the rules of Boolean algebra, which provide the framework for the legal operations and manipulations that can be taken on a two-valued number system (i.e., a binary system). We then explore a variety of logic design and manipulation techniques. These techniques allow us to directly create a logic circuit from a truth table and then to manipulate it to either reduce the number of gates necessary in the circuit or to convert the logic circuit into equivalent forms using alternate gates. The goal of this chapter is to provide an understanding of the basic principles of combinational logic design.

Learning Outcomes—After completing this chapter, you will be able to:

4.1 Describe the fundamental principles and theorems of Boolean algebra and how to use them to manipulate logic expressions.
4.2 Analyze a combinational logic circuit to determine its logic expression, truth table, and timing information.
4.3 Synthesis a logic circuit in canonical form (sum of products or product of sums) from a functional description including a truth table, minterm list, or maxterm list.
4.4 Synthesize a logic circuit in minimized form (sum of products or product of sums) through algebraic manipulation or with a Karnaugh map.
4.5 Describe the causes of timing hazards in digital logic circuits and the approaches to mitigate them.

4.1 Boolean Algebra

The term *algebra* refers to the rules of a number system. In Chap. 2 we discussed the number of symbols and relative values of some of the common number systems. Algebra defines the operations that are legal to perform on that system. Once we have defined the rules for a system, we can then use the system for more powerful mathematics such as solving for unknowns and manipulating into equivalent forms. The ability to manipulate into equivalent forms allows us to minimize the number of logic operations necessary and also put into a form that can be directly synthesized using modern logic circuits.

In 1854, English mathematician George Boole presented an abstract algebraic framework for a system that contained only two states, true and false. This framework essentially launched the field of computer science even before the existence of the modern integrated circuits that are used to implement digital logic today. In 1930, American mathematician Claude Shannon applied Boole's algebraic framework to his work on switching circuits at Bell Labs, thus launching the field of digital circuit design and information theory. Boole's original framework is still used extensively in modern digital circuit design and thus bears the name *Boolean algebra*. Today, the term Boolean algebra is often used to describe not only George Boole's original work but all of those that contributed to the field after him.

© Springer Nature Switzerland AG 2019
B. J. LaMeres, *Introduction to Logic Circuits & Logic Design with Verilog*,
https://doi.org/10.1007/978-3-030-13605-5_4

4.1.1 Operations

In Boolean algebra there are two valid states (true and false) and three core operations. The operations are conjunction ($\wedge$, equivalent to the AND operation), disjunction ($\vee$, equivalent to the OR operation), and negation ($\neg$, equivalent to the NOT operation). From these three operations, more sophisticated operations can be created including other logic functions (i.e., BUF, NAND, NOR, XOR, XNOR, etc.) and arithmetic. Engineers primarily use the terms AND, OR, and NOT instead of conjunction, disjunction, and negation. Similarly, engineers primarily use the symbols for these operators described in Chap. 3 (e.g., $\cdot$, +, and ') instead of $\wedge$, $\vee$, and $\neg$.

4.1.2 Axioms

An *axiom* is a statement of truth about a system that is accepted by the user. Axioms are very simple statements about a system but need to be established before more complicated theorems can be proposed. Axioms are so basic that they do not need to be proved in order to be accepted. Axioms can be thought of as the basic *laws* of the algebraic framework. The terms *axiom* and *postulate* are synonymous and used interchangeably. In Boolean algebra there are five main axioms. These axioms will appear redundant with the description of basic gates from Chap. 3 but must be defined in this algebraic context so that more powerful theorems can be proposed.

4.1.2.1 Axiom #1: Logical Values

This axiom states that in Boolean algebra, a variable A can only take on one of two values, 0 or 1. If the variable A is not 0, then it must be a 1, and conversely, if it is not a 1, then it must be a 0.

Axiom #1 – Boolean values: $A = 0$ if $A \neq 1$, conversely $A = 1$ if $A \neq 0$.

4.1.2.2 Axiom #2: Definition of Logical Negation

This axiom defines logical negation. Negation is also called the NOT operation or taking the *complement*. The negation operation is denoted using either a prime ('), an inversion bar or the negation symbol ($\neg$). If the complement is taken on a 0, it becomes a 1. If the complement is taken on a 1, it becomes a 0.

Axiom #2 – Definition of logical negation: if $A = 0$ then $A' = 1$, conversely, if $A = 1$ then $A' = 0$.

4.1.2.3 Axiom #3: Definition of a Logical Product

This axiom defines a logical product or multiplication. Logical multiplication is denoted using either a dot ($\cdot$), an ampersand (&), or the conjunction symbol ($\wedge$). The result of logical multiplication is true when *both* inputs are true and false otherwise.

Axiom #3 – Definition of a logical product: *$A \cdot B = 1$ if $A = B = 1$ and $A \cdot B = 0$ otherwise.*

4.1.2.4 Axiom #4: Definition of a Logical Sum

This axiom defines a logical sum or addition. Logical addition is denoted using either a plus sign (+) or the disjunction symbol ($\vee$). The result of logical addition is true when *any* of the inputs are true and false otherwise.

Axiom #4 – Definition of a logical sum: *$A + B = 1$ if $A = 1$ or $B = 1$ and $A + B = 0$ otherwise.*

4.1.2.5 Axiom #5: Logical Precedence

This axiom defines the order of precedence for the three operators. Unless the precedence is explicitly stated using parentheses, negation takes precedence over a logical product, and a logical product takes precedence over a logical sum.

Axiom #5 – Definition of logical precedence: *NOT precedes AND,* and *AND precedes OR.*

To illustrate Axiom #5, consider the logic function $F = A' \cdot B + C$. In this function, the first operation that would take place is the NOT operation on A. This would be followed by the AND operation of A' with B. Finally, the result would be OR'd with C. The precedence of any function can also be explicitly stated using parentheses such as $F = (((A') \cdot B) + C)$.

4.1.3 Theorems

A theorem is a more sophisticated truth about a system that is not intuitively obvious. Theorems are proposed and then must be proved. Once proved, they can be accepted as a truth about the system going forward. Proving a theorem in Boolean algebra is much simpler than in our traditional decimal system due to the fact that variables can only take on one of two values, true or false. Since the number of input possibilities is bounded, Boolean algebra theorems can be proved by simply testing the theorem using every possible input code. This is called **proof by exhaustion**. The following theorems are used widely in the manipulation of logic expressions and reduction of terms within an expression.

4.1.3.1 De Morgan's Theorem of Duality

Augustus De Morgan was a British mathematician and logician who lived during the time of George Boole. De Morgan is best known for his contribution to the field of logic through the creation of what have been later called the *De Morgan's theorems* (often called De Morgan's laws). There are two major theorems that De Morgan proposed that expanded Boolean algebra. The first theorem is named *duality.* Duality states that an algebraic equality will remain true if all 0's and 1's are interchanged and all AND and OR operations are interchanged. The new expression is called the *dual* of the original expression. Example 4.1 shows the process of proving duality using proof by exhaustion.

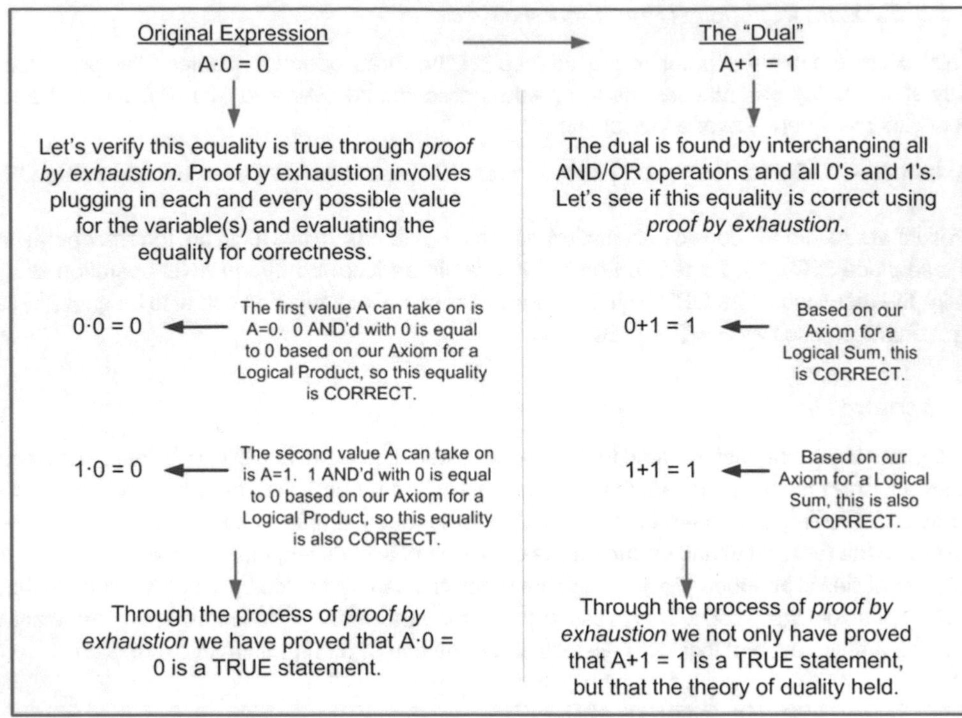

Example 4.1
Proving De Morgan's theorem of duality using proof by exhaustion

Duality is important for two reasons. First, it doubles the impact of a theorem. If a theorem is proved to be true, then the dual of that theorem is also proved to be true. This, in essence, gives twice the theorem with the same amount of proving. Boolean algebra theorems are almost always given in pairs, the original and the dual. That is why duality is covered as the first theorem.

The second reason that duality is important is because it can be used to convert between positive and negative logic. Until now, we have used positive logic for all of our examples (i.e., a logic HIGH = true = 1 and a logic LOW = false = 0). As mentioned earlier, this convention is arbitrary, and we could have easily chosen a HIGH to be false and a LOW to be true (i.e., negative logic). Duality allows us to take a logic expression that has been created using positive logic (F) and then convert it into an equivalent expression that is valid for negative logic (F_D). Example 4.2 shows the process for how this works.

Example: Converting Between Positive and Negative Logic Using Duality

Let's start with a logic expression that originates in positive logic convention.

$F = A \cdot B$

Positive Logic means that a HIGH=1 and a LOW=0. If we want to implement the equivalent function using negative logic, we instead assign a HIGH=0 and a LOW=1.

The Logic We Want

A	B	F
L	L	L
L	H	L
H	L	L
H	H	H

Mapping with Positive Logic

A	B	F
0	0	0
0	1	0
1	0	0
1	1	1

Mapping with Negative Logic

A	B	F
1	1	1
1	0	1
0	1	1
0	0	0

Let's use Duality to come up with the equivalent logic expression using negative logic.

$F_D = A+B$ ← The dual is found by interchanging all AND/OR operations and all 0's and 1's.

Does this give us what we want for a negative logic convention? Let's take the truth table of the "Mapping with Negative Logic" and rearrange the input codes into a more traditional format:

Mapping with Negative Logic

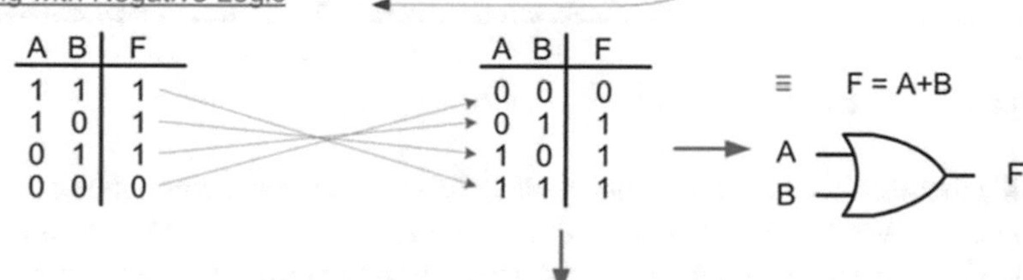

A	B	F
1	1	1
1	0	1
0	1	1
0	0	0

A	B	F
0	0	0
0	1	1
1	0	1
1	1	1

Yes, this truth table is the definition of a Logical Sum per our axioms (e.g., F=A+B). This means that the logic expression created using duality (F_D) created an equivalent function using negative logic.

Example 4.2
Converting between positive and negative logic using duality

One consideration when using duality is that the order of precedence follows the original function. This means that in the original function, the axiom for precedence states the order as NOT-AND-OR; however, this is not necessarily the correct precedence order in the dual. For example, if the original function was $F = A \cdot B + C$, the AND operation of A and B would take place first, and then the result would be OR'd with C. The dual of this expression is $F_D = A + B \cdot C$. If the expression for F_D was evaluated using traditional Boolean precedence, it would show that F_D does NOT give the correct result per the definition of a dual function (i.e., converting a function from positive to negative logic). The order of precedence for F_D must correlate to the precedence in the original function. Since in the original function A and B were operated on first, they must also be operated on first in the dual. In order to easily manage this issue, parentheses can be used to track the order of operations from the original function to the dual. If we put parentheses in the original function to explicitly state the precedence of the operations, it would take the form $F = (A \cdot B) + C$. These parentheses can be mapped directly to the dual yielding $F_D = (A + B) \cdot C$. This order of precedence in the dual is now correct.

Now that we have covered the duality operation, its usefulness and its pitfalls, we can formally define this theorem as:

De Morgan's Duality: An algebraic equality will remain true if all 0's and 1's are interchanged and all AND and OR operations are interchanged. Furthermore, taking the dual of a positive logic function will produce the equivalent function using negative logic if the original order of precedence is maintained.

4.1.3.2 Identity

An identity operation is one that when performed on a variable will yield itself regardless of the variable's value. The following is the formal definition of identity theorem. Figure 4.1 shows the gate-level depiction of this theorem.

Identity: OR'ing any variable with a logic 0 will yield the original variable. The dual: AND'ing any variable with a logic 1 will yield the original variable.

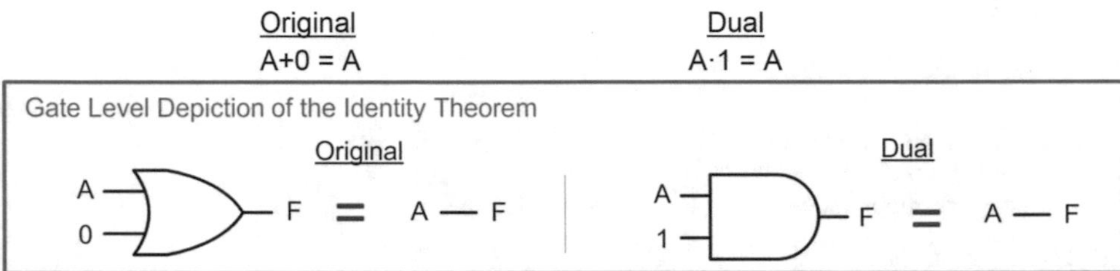

Fig. 4.1
Gate-level depiction of the identity theorem

The identity theorem is useful for reducing circuitry when it is discovered that a particular input will never change values. When this is the case, the static input variable can simply be removed from the logic expression making the entire circuit a simple wire from the remaining input variable to the output.

4.1.3.3 Null Element

A null element operation is one that, when performed on a constant value, will yield that same constant value regardless of the values of any variables within the same operation. The following is the formal definition of null element. Figure 4.2 shows the gate-level depiction of this theorem.

Null Element: OR'ing any variable with a logic 1 will yield a logic 1 regardless of the value of the input variable. The dual: AND'ing any variable with a logic 0 will yield a logic 0 regardless of the value of the input variable.

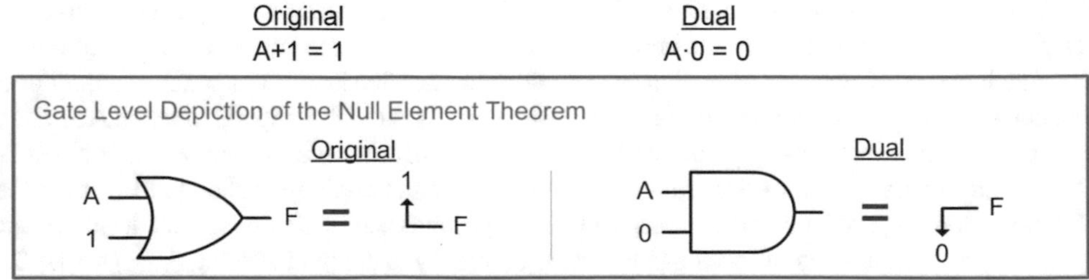

Fig. 4.2
Gate-level depiction of the null element theorem

The null element theorem is also useful for reducing circuitry when it is discovered that a particular input will never change values. It is also widely used in computer systems in order to set (i.e., force to a logic 1) or clear (i.e., force to a logic 0) the value of a storage element.

4.1.3.4 Idempotent

An idempotent operation is one that has no effect on the input, regardless of the number of times the operation is applied. The following is the formal definition of idempotence. Figure 4.3 shows the gate-level depiction of this theorem.

Idempotent: OR'ing a variable with itself results in itself. The dual: AND'ing a variable with itself results in itself.

Original	Dual
A+A = A	A·A = A

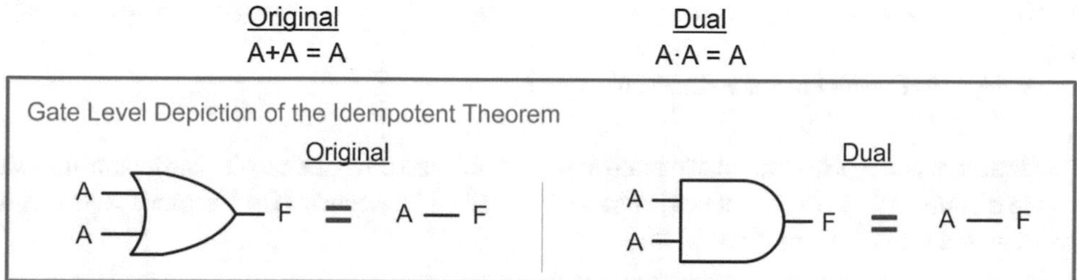

Fig. 4.3
Gate-level depiction of the idempotent theorem

This theorem also holds true for any number of operations such as A + A + A + ... + A = A and A·A·A·. . . ..·A = A.

4.1.3.5 Complements

This theorem describes an operation of a variable with the variable's own complement. The following is the formal definition of complements. Figure 4.4 shows the gate-level depiction of this theorem.

Complements: OR'ing a variable with its complement will produce a logic 1. The dual: AND'ing a variable with its complement will produce a logic 0.

Original	Dual
A+A' = 1	A·A' = 0

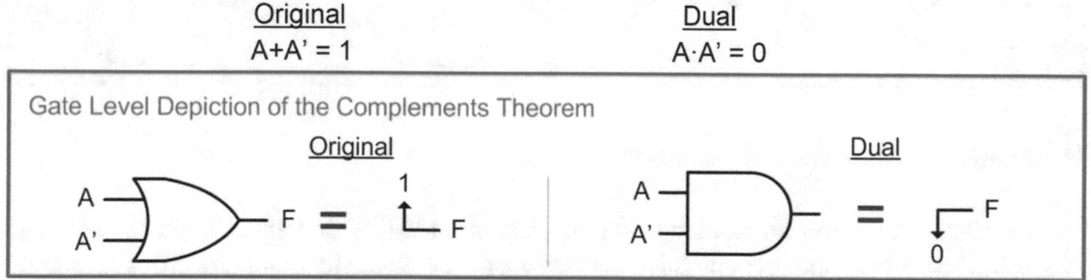

Fig. 4.4
Gate-level depiction of the complements theorem

The complement theorem is again useful for reducing circuitry when these types of logic expressions are discovered.

4.1.3.6 Involution

An involution operation describes the result of double negation. The following is the formal definition of involution. Figure 4.5 shows the gate-level depiction of this theorem.

Involution: Taking the double complement of a variable will result in the original variable.

Original
A'' = A

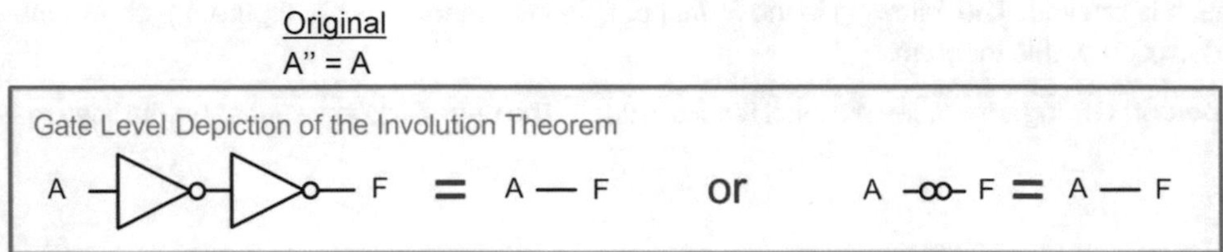

Fig. 4.5
Gate-level depiction of the involution theorem

This theorem is not only used to eliminate inverters but also provides us a powerful tool for *inserting* inverters in a circuit. We will see that this is used widely with the second of De Morgan's laws that will be introduced at the end of this section.

4.1.3.7 Commutative Property

The term commutative is used to describe an operation in which the order of the quantities or variables in the operation has no impact on the result. The following is the formal definition of the commutative property. Figure 4.6 shows the gate-level depiction of this theorem.

Commutative Property: Changing the order of variables in an OR operation does not change the end result. The dual: Changing the order of variables in an AND operation does not change the end result.

Original
A+B = B+A

Dual
A·B = B·A

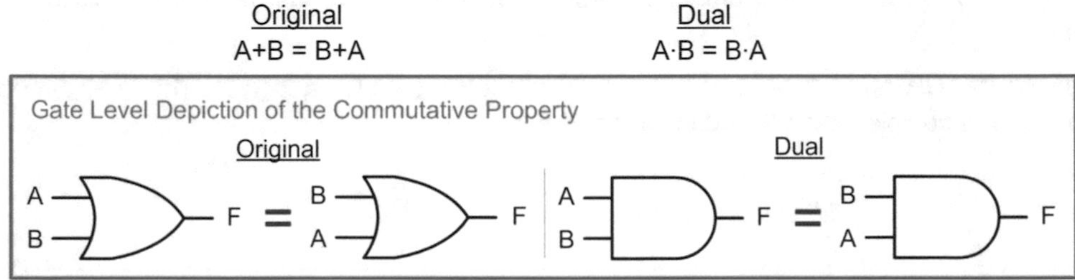

Fig. 4.6
Gate-level depiction of commutative property

One practical use of the commutative property is when wiring or routing logic circuitry together. Example 4.3 shows how the commutative property can be used to untangle crossed wires when implementing a digital system.

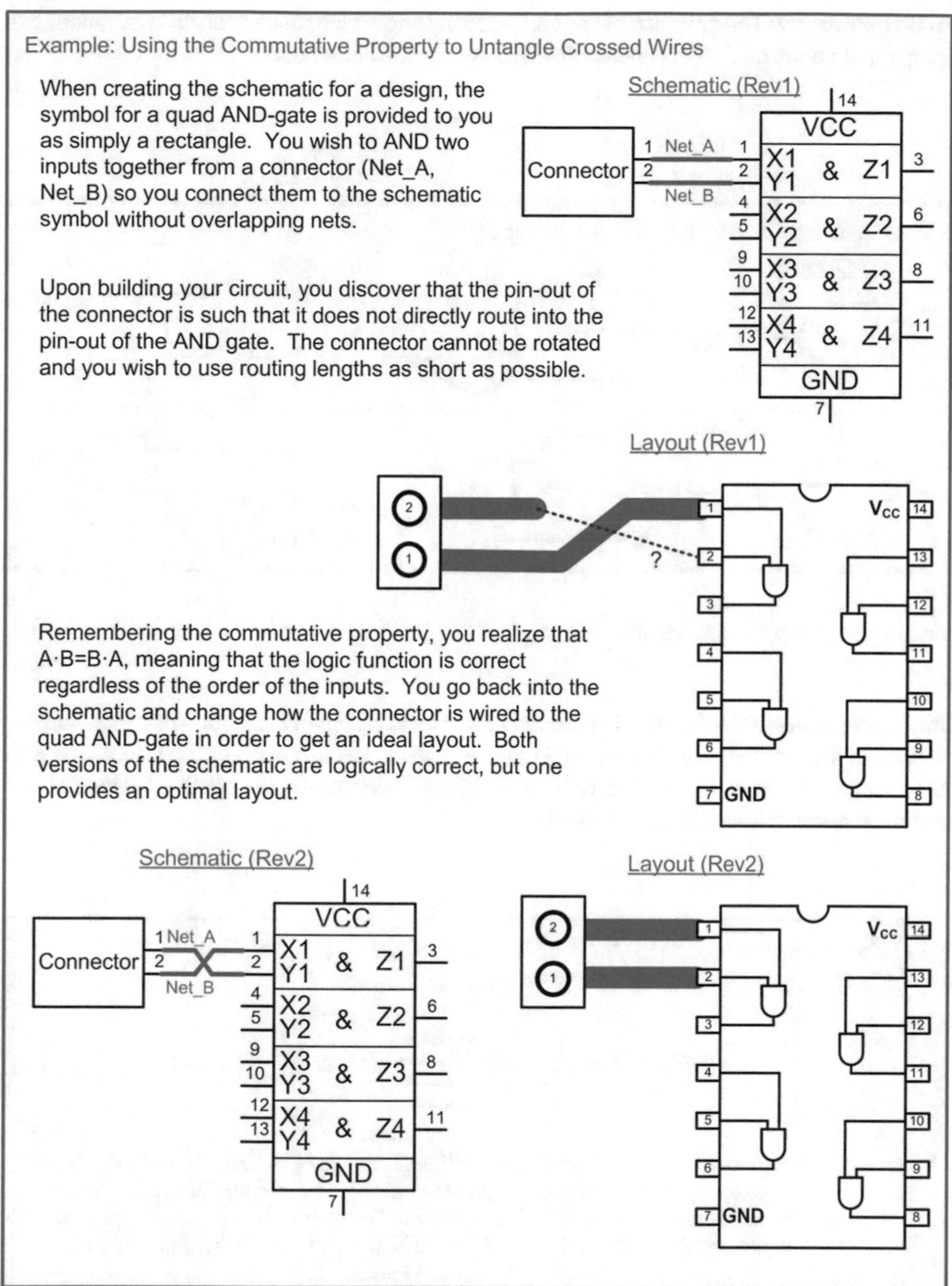

Example 4.3
Using the commutative property to untangle crossed wires

4.1.3.8 Associative Property

The term associative is used to describe an operation in which the grouping of the quantities or variables in the operation has no impact on the result. The following is the formal definition of the associative property. Figure 4.7 shows the gate-level depiction of this theorem.

Associative Property: The grouping of variables doesn't impact the result of an OR operation. The dual: The grouping of variables doesn't impact the result of an AND operation.

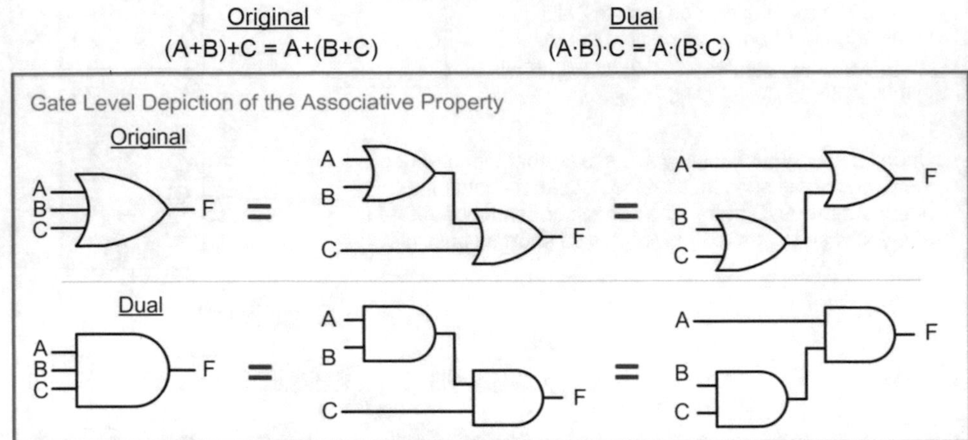

Fig. 4.7
Gate-level depiction of the associative property

One practical use of the associative property is addressing fan-in limitations of a logic family. Since the grouping of the input variables does not impact the result, we can accomplish operations with large numbers of inputs using multiple gates with fewer inputs. Example 4.4 shows the process of using the associative property to address a fan-in limitation.

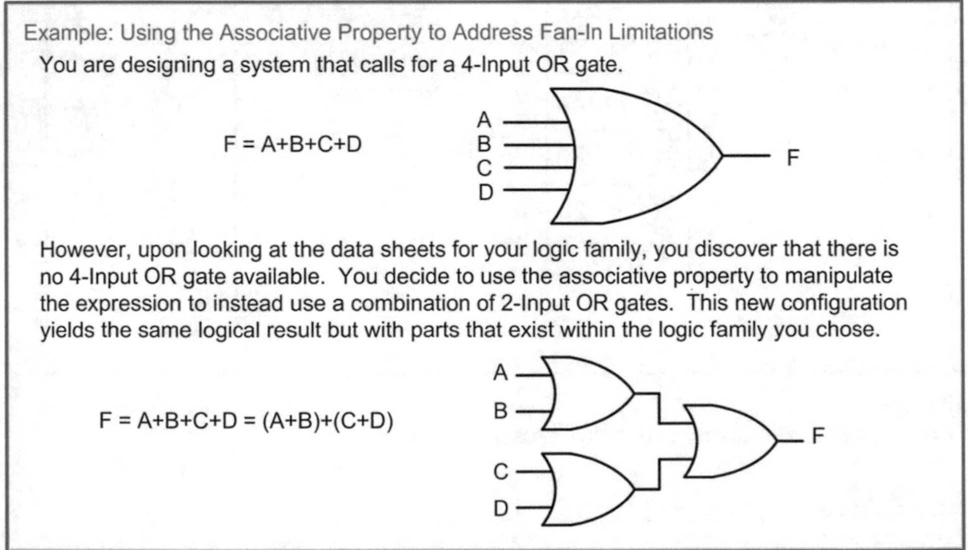

Example 4.4
Using the associative property to address fan-in limitations

4.1.3.9 Distributive Property

The term distributive describes how an operation on a parenthesized group of operations (or higher precedence operations) can be distributed through each term. The following is the formal definition of the distributive property. Figure 4.8 shows the gate-level depiction of this theorem.

Distributive Property: An operation on a parenthesized operation(s), or higher precedence operator, will distribute through each term.

Original	Dual
A·(B+C) = A·B + A·C	A+(B·C) = (A+B)·(A+C)

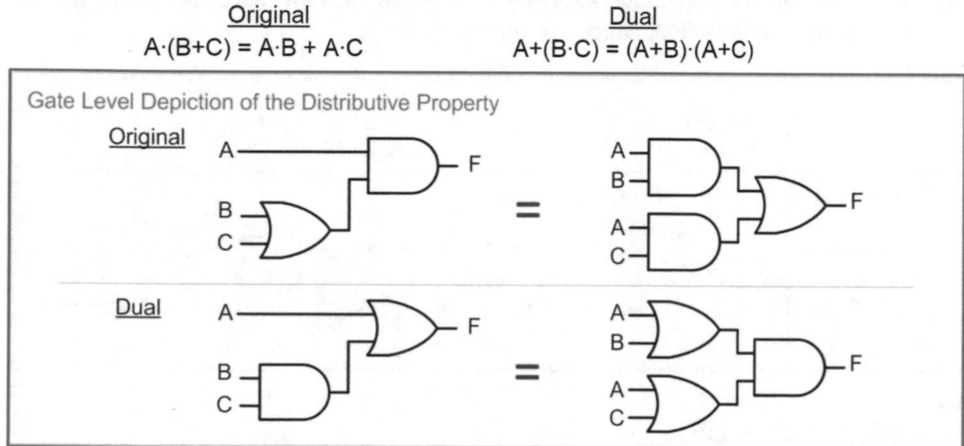

Fig. 4.8
Gate-level depiction of the distributive property

The distributive property is used as a logic manipulation technique. It can be used to put a logic expression into a form more suitable for direct circuit synthesis or to reduce the number of logic gates necessary. Example 4.5 shows how to use the distributive property to reduce the number of gates in a logic circuit.

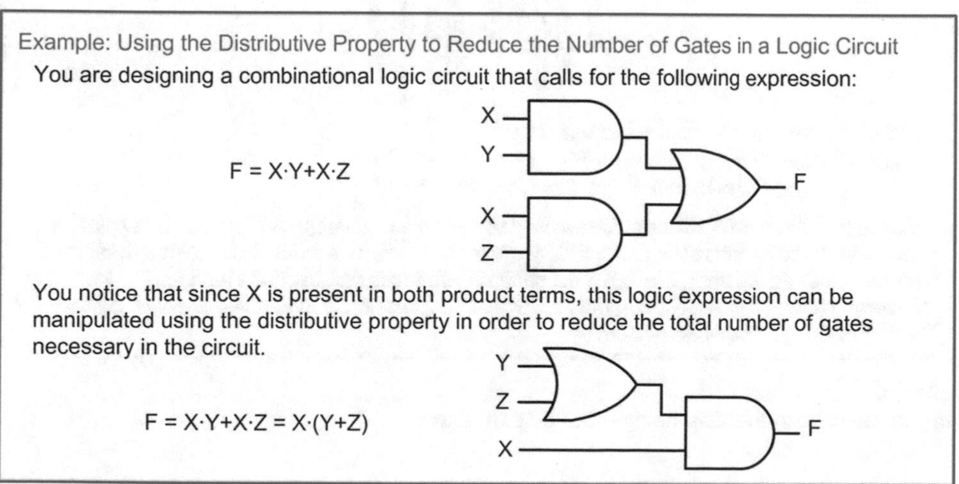

Example 4.5
Using the distributive property to reduce the number of logic gates in a circuit

4.1.3.10 Absorption

The term absorption refers to when multiple logic terms within an expression produce the same results. This allows one of the terms to be eliminated from the expression, thus reducing the number of logic operations. The remaining terms essentially *absorb* the functionality of the eliminated term. This theorem is also called *covering* because the remaining term essentially covers the functionality of both itself and the eliminated term. The following is the formal definition of the absorption theorem. Figure 4.9 shows the gate-level depiction of this theorem.

Absorption: When a term within a logic expression produces the same output(s) as another term, the second term can be removed without affecting the result.

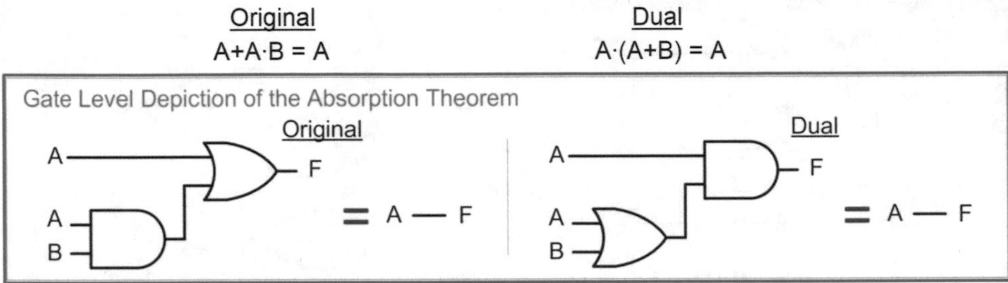

Fig. 4.9
Gate-level depiction of absorption

This theorem is better understood by looking at the evaluation of each term with respect to the original expression. Example 4.6 shows how the absorption theorem can be proven through proof by exhaustion by evaluating each term in a logic expression.

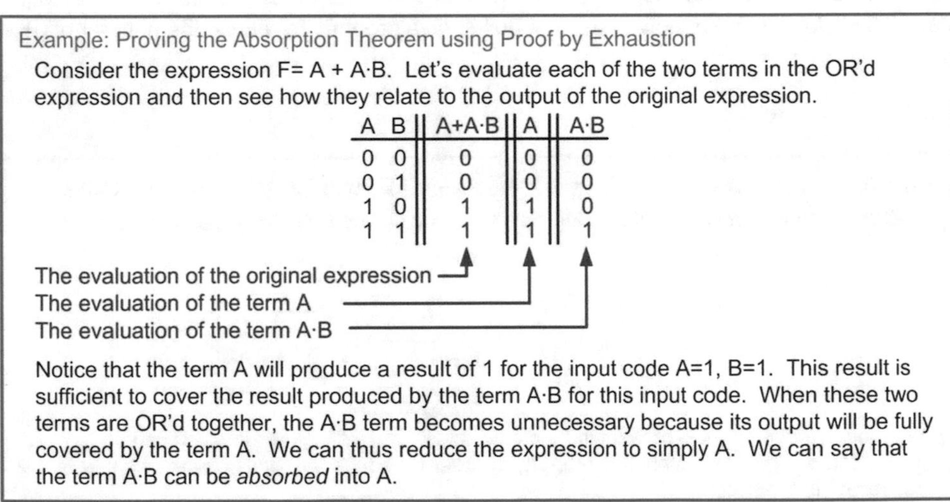

Example: Proving the Absorption Theorem using Proof by Exhaustion

Consider the expression F= A + A·B. Let's evaluate each of the two terms in the OR'd expression and then see how they relate to the output of the original expression.

A	B	A+A·B	A	A·B
0	0	0	0	0
0	1	0	0	0
1	0	1	1	0
1	1	1	1	1

The evaluation of the original expression
The evaluation of the term A
The evaluation of the term A·B

Notice that the term A will produce a result of 1 for the input code A=1, B=1. This result is sufficient to cover the result produced by the term A·B for this input code. When these two terms are OR'd together, the A·B term becomes unnecessary because its output will be fully covered by the term A. We can thus reduce the expression to simply A. We can say that the term A·B can be *absorbed* into A.

Example 4.6
Proving the absorption theorem using proof by exhaustion

4.1.3.11 Uniting

The uniting theorem, also called *combining* or *minimization*, provides a way to remove variables from an expression when they have no impact on the outcome. This theorem is one of the most widely used techniques for the reduction of the number of gates needed in a combinational logic circuit. The following is the formal definition of the uniting theorem. Figure 4.10 shows the gate-level depiction of this theorem.

Uniting: When a variable (B) and its complement (B') appear in multiple product terms with a common variable (A) within a logical OR operation, the variable B does not have any effect on the result and can be removed.

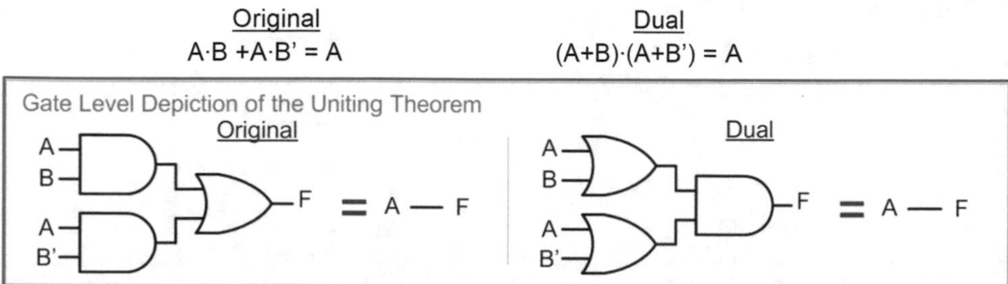

Fig. 4.10
Gate-level depiction of uniting

This theorem can be proved using prior theorems. Example 4.7 shows how the uniting theorem can be proved using a combination of the distributive property, the complements theorem, and the identity theorem.

Example: Proving the Uniting Theorem

Uniting theorem states that A·B + A·B' = A. Let's use the other Boolean algebra theorems to manipulate the original expression in order to prove this theorem.

The original expression: ⟶ F = A·B + A·B'

Using the distributive property, we can rewrite the expression as: ⟶ F = A·(B + B')

The "complements theorem" states that B+B'=1, so we can now rewrite the expression as: ⟶ F = A·1

The *identity theorem* states that A·1=A, so the expression can be written in its final form. ⟶ F = A

This proves that the uniting theorem holds true. Uniting theorem is also called *minimization* or *combining*.

Example 4.7
Proving of the uniting theorem

4.1.3.12 De Morgan's Theorem

Now we look at the second of De Morgan's laws. This second theorem is simply known as *De Morgan's theorem*. This theorem provides a technique to manipulate a logic expression that uses AND gates into one that uses OR gates and vice versa. It can also be used to manipulate traditional Boolean logic expressions that use AND-OR-NOT operators, into equivalent forms that uses NAND and NOR gates. The following is the formal definition of De Morgan's theorem. Figure 4.11 shows the gate-level depiction of this theorem.

De Morgan's Theorem: An OR operation with both inputs inverted is equivalent to an AND operation with the output inverted. The dual: An AND operation with both inputs inverted is equivalent to an OR operation with the output inverted.

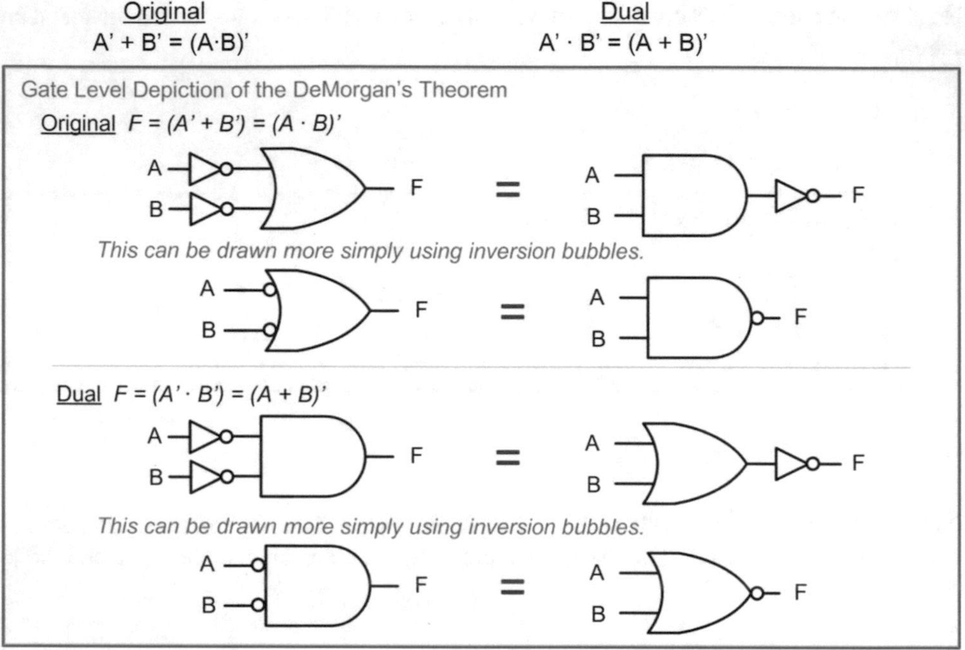

Fig. 4.11
Gate-level depiction of De Morgan's theorem

This theorem is used widely in modern logic design because it bridges the gap between the design of logic circuitry using Boolean algebra and the physical implementation of the circuitry using CMOS. Recall that Boolean algebra is defined for only three operations, the AND, the OR, and the inversion. CMOS, on the other hand, can only directly implement negative-type gates such as NAND, NOR, and NOT. De Morgan's theorem allows us to design logic circuitry using Boolean algebra and synthesize logic diagrams with AND, OR, and NOT gates and then directly convert the logic diagrams into an equivalent form using NAND, NOR, and NOT gates. As we'll see in the next section, Boolean algebra produces logic expressions in two common forms. These are the **sum of products (SOP)** and the **product of sums (POS)** forms. Using a combination of involution and De Morgan's theorem, SOP and POS forms can be converted into equivalent logic circuits that use only NAND and NOR gates. Example 4.8 shows a process to convert a sum of products form into one that uses only NAND gates.

Example: Converting a Sum of Products Form into One That Uses Only NAND Gates

You are designing a combinational logic circuit that will be implemented in CMOS. You use Boolean algebra to create a circuit in the form of a **sum of products (SOP)**.

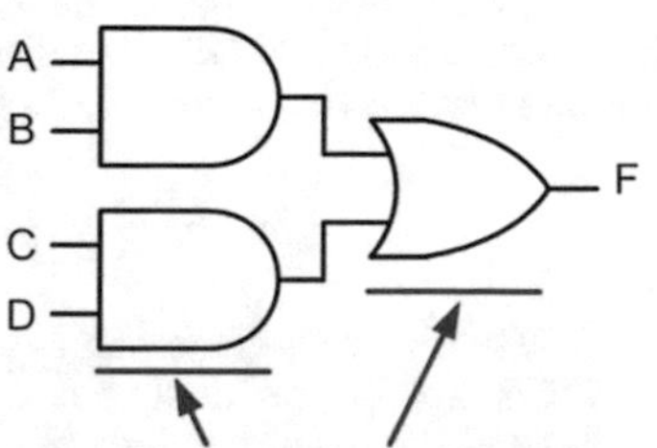

$$F = \underline{A \cdot B} + \underline{C \cdot D}$$

These two logical products (e.g., AND operations) are summed together (e.g., OR operation) to form a Sum of Product form.

A Sum of Products at the gate level always has a stage of AND gates feeding into a single OR gate.

Since this logic needs to be implemented in CMOS, you need to convert it into a form that uses only NAND, NOR or NOT gates. You know that DeMorgan's Theorem allows an OR gate with its inputs inverted to be converted to an AND gate with its output inverted (e.g., a NAND gate). To prepare for this manipulation, you take advantage of the theory of involution, which allows you to put double inversions on any net without affecting the result.

Double inverters are placed on these nodes in order to create an OR gate with its inputs inverted.

These inverters can also be denoted using inversion bubbles (e.g., double bubbles).

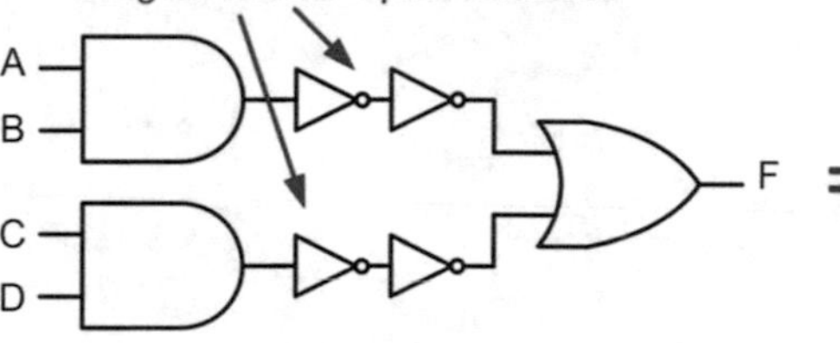

=

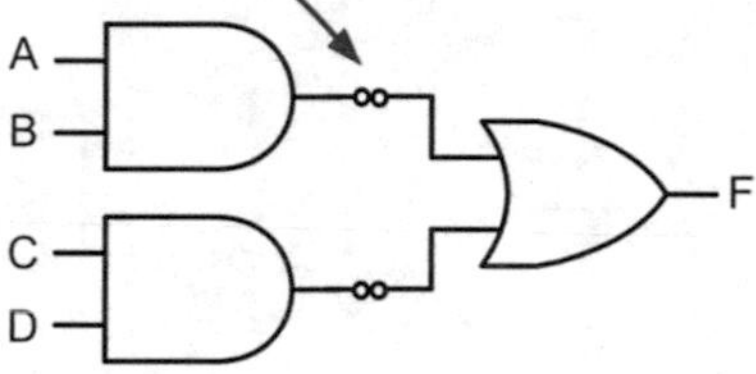

Moving the inversion bubbles to these locations on the wires highlights that the first stage of AND gates can be directly replaced with NAND gates and the OR gate is ready for DeMorgan's.

The final step is to convert the OR gate with its inputs inverted to an AND gate with its output inverted, which is a NAND gate.

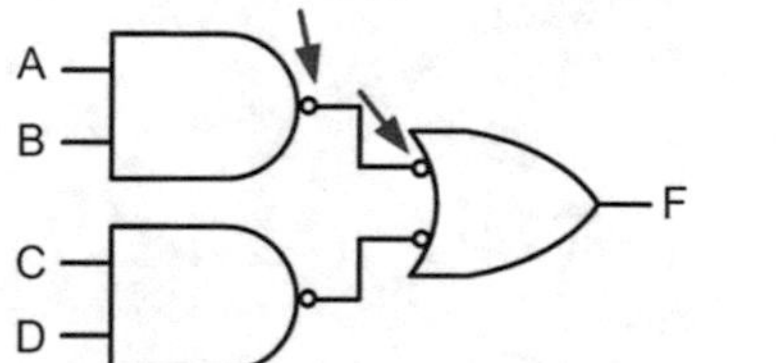

=

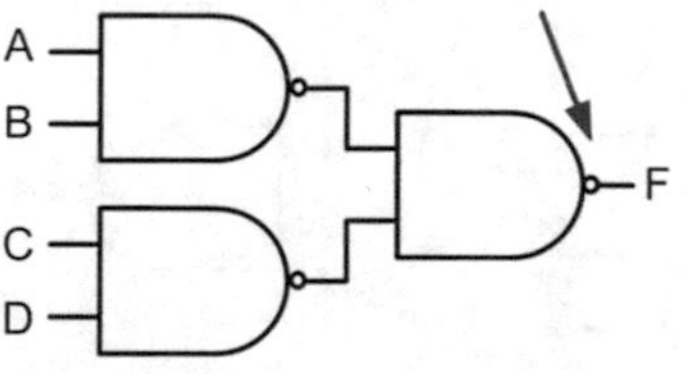

The original Sum of Products that was implemented with only AND/OR operations was replaced with an equivalent circuit that used only NAND gates. This replacement can be made directly anytime a Sum of Products form is present.

Example 4.8
Converting a sum of products form into one that uses only NAND gates

Example 4.9 shows a process to convert a product of sums form into one that uses only NOR gates.

Example: Converting a Product of Sums Form into One That Uses Only NOR Gates

You are designing a combinational logic circuit that will be implemented in CMOS. You use Boolean algebra to create a circuit in the form of a **product of sums (POS)**.

$F = (A+B) \cdot (C+D)$

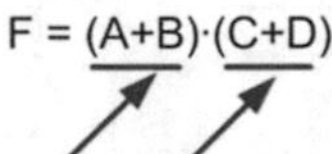

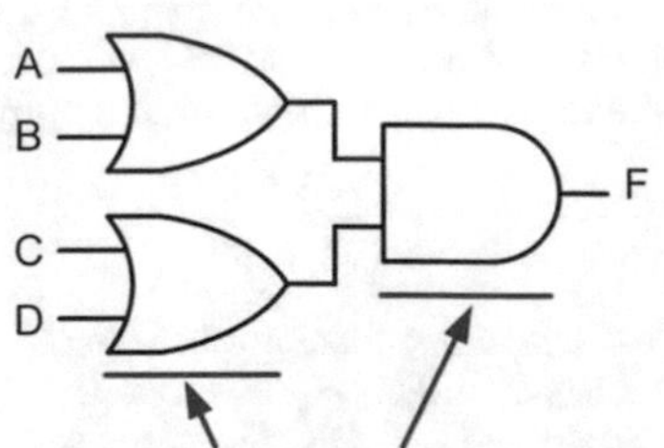

These two logical sums (e.g., OR operations) are multiplied together (e.g., AND operation) to form a Product of Sums form.

A Product of Sums at the gate level always has a stage of OR gates feeding into a single AND gate.

Since this logic needs to be implemented in CMOS, you need to convert it into a form that uses only NAND, NOR or NOT gates. You know that DeMorgan's Theorem allows an AND gate with its inputs inverted to be converted to an OR gate with its output inverted (e.g., a NOR gate). To prepare for this manipulation, you take advantage of the theory of involution, which allows you to put double inversions on any net without affecting the result.

Double inverters are placed on these nodes in order to create an AND gate with its inputs inverted.

These inverters can also be denoted using inversion bubbles (e.g., double bubbles).

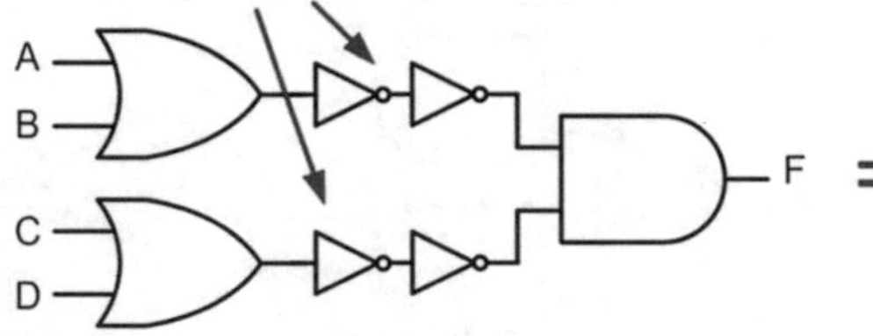

=

Moving the inversion bubbles to these locations on the wires highlights that the first stage of OR gates can be directly replaced with NOR gates and the AND gate is ready for DeMorgan's.

The final step is to convert the AND gate with its inputs inverted to an OR gate with its output inverted, which is a NOR gate.

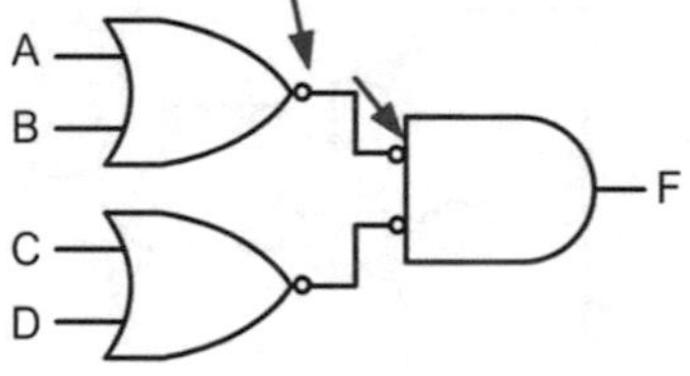

=

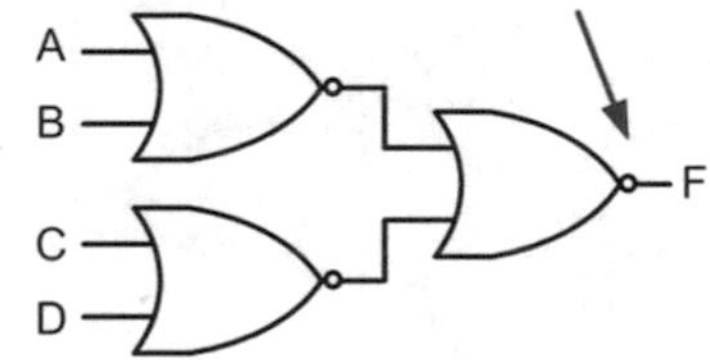

The original Product of Sums that was implemented with only OR/AND operations was replaced with an equivalent circuit that used only NOR gates. This replacement can be made directly anytime a Product of Sums form is present.

Example 4.9
Converting a product of sums form into one that uses only NOR gates

De Morgan's theorem can also be accomplished algebraically using a process known as *breaking the bar and flipping the operator*. This process again takes advantage of the involution theorem, which allows double negation without impacting the result. When using this technique in algebraic form, involution takes the form of a double inversion bar. If an inversion bar is *broken*, the expression will remain true as long as the operator directly below the break is flipped (AND to OR, OR to AND). Example 4.10 shows how to use this technique when converting an OR gate with its inputs inverted into an AND gate with its output inverted.

Example: Using DeMorgan's Theorem Algebraically, Breaking the Bar and Flipping the Sign (1)

DeMorgan's Theorem can be accomplished algebraically using a process called "breaking the bar and flipping the operator". Let's see if this approach works on an OR gate with its inputs inverted.

$F = \overline{A} + \overline{B}$ ← The original algebraic expression for an OR gate with both inputs inverted.

$F = \overline{\overline{\overline{A} + \overline{B}}}$ ← Involution allows double negation without impacting the result. This is accomplished with two inversion bars.

$F = \overline{\overline{\overline{A} + \overline{B}}}$ (+ to ·) ← An inversion bar can be "broken", but in order for the expression to remain true, the OR operator beneath the break must be flipped to an AND.

$F = \overline{\overline{\overline{A}} \cdot \overline{\overline{B}}}$ ← Involution can be used again to remove the double negations above A and B.

$F = \overline{A \cdot B}$ ← The resulting expression is an AND gate with its output inverted.

This technique upheld DeMorgan's Theorem that an OR gate with its inputs inverted is equivalent to an AND gate with its output inverted.

$$F = \overline{A} + \overline{B} = \overline{A \cdot B}$$

Example 4.10
Using De Morgan's theorem in algebraic form (1)

Example 4.11 shows how to use this technique when converting an AND gate with its inputs inverted into an OR gate with its output inverted.

Example: Using DeMorgan's Theorem Algebraically, Breaking the Bar and Flipping the Sign (2)

Let's see if the "breaking the bar and flipping the operator" approach works on an AND gate with its inputs inverted.

$F = \overline{A} \cdot \overline{B}$ ← The original algebraic expression for an AND gate with both inputs inverted.

$F = \overline{\overline{\overline{A} \cdot \overline{B}}}$ ← Involution allows double negation without impacting the result. This is accomplished with two inversion bars.

$F = \overline{\overline{\overline{A} \cdot \overline{B}}}$ (· to +) ← An inversion bar can be "broken", but in order for the expression to remain true, the AND operator beneath the break must be flipped to an OR.

$F = \overline{\overline{\overline{A}} + \overline{\overline{B}}}$ ← Involution can be used again to remove the double negations above A and B.

$F = \overline{A + B}$ ← The resulting expression is an OR gate with its output inverted.

This technique upheld DeMorgan's Theorem that an AND gate with its inputs inverted is equivalent to an OR gate with its output inverted.

$$F = \overline{A} \cdot \overline{B} = \overline{A+B}$$

Example 4.11
Using De Morgan's theorem in algebraic form (2)

Table 4.1 gives a summary of all the Boolean algebra theorems just covered. The theorems are grouped in this table with respect to the number of variables that they contain. This grouping is the most common way these theorems are presented.

Summary of Boolean Algebra Theorems

Single Variable Theorems	Original	Dual
Identity	$A+0 = A$	$A \cdot 1 = A$
Null Element	$A+1 = 1$	$A \cdot 0 = 0$
Idempotency	$A+A = A$	$A \cdot A = A$
Complements	$A+A' = 1$	$A \cdot A' = 0$
Involution	$A'' = A$	
Multiple Variable Theorems		
Commutative	$A+B = B+A$	$A \cdot B = B \cdot A$
Associative	$(A+B)+C = A+(B+C)$	$(A \cdot B) \cdot C = A \cdot (B \cdot C)$
Distributive	$A \cdot (B+C) = A \cdot B+A \cdot C$	$A+(B \cdot C) = (A+B) \cdot (A+C)$
Absorption (or Covering)	$A+A \cdot B = A$	$A \cdot (A+B) = A$
Uniting (or Combining)	$A \cdot B + A \cdot B' = A$	$(A+B) \cdot (A+B') = A$
DeMorgan's	$A' \cdot B' = (A + B)'$	$A'+B' = (A \cdot B)'$

Table 4.1
Summary of Boolean algebra theorems

4.1.4 Functionally Complete Operation Sets

A set of Boolean operators is said to be *functionally complete* when the set can implement all possible logic functions. The set of operators {AND, OR, NOT} is functionally complete because every other operation can be implemented using these three operators (i.e., NAND, NOR, BUF, XOR, XNOR). The De Morgan's theorem showed us that all AND and OR operations can be replaced with NAND and NOR operators. This means that NAND and NOR operations could be by themselves functionally complete if they could perform a NOT operation. Figure 4.12 shows how a NAND gate can be configured to perform a NOT operation. This configuration allows a NAND gate to be considered functionally complete because all other operations can be implemented.

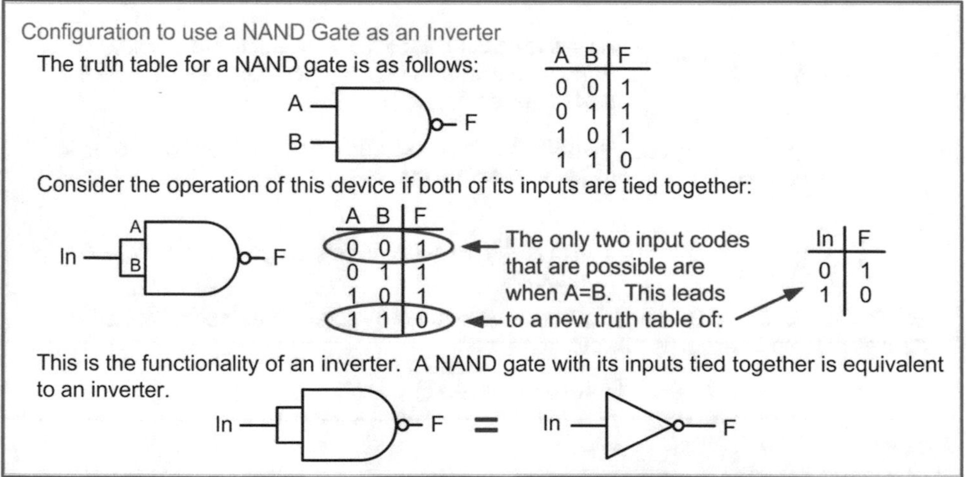

Fig. 4.12
Configuration to use a NAND gate as an inverter

This approach can also be used on a NOR gate to implement an inverter. Figure 4.13 shows how a NOR gate can be configured to perform a NOT operation, thus also making it functionally complete.

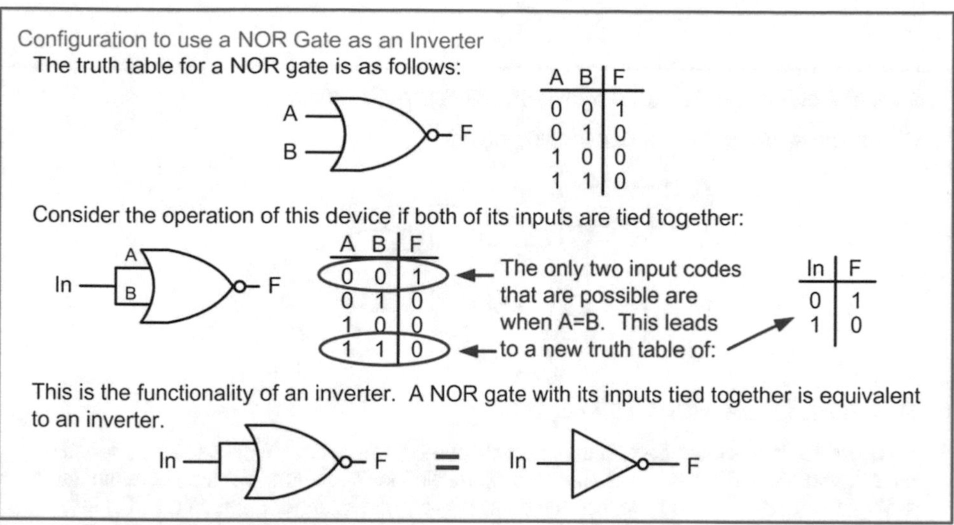

Fig. 4.13
Configuration to use a NOR gate as an inverter

CONCEPT CHECK

CC4.1 If the logic expression F = A·B·C·D·E·F·G·H is implemented with only 2-input AND gates, how many levels of logic will the final implementation have? Hint: Consider using the associative property to manipulate the logic expression to use only 2-input AND operations.

A) 2 B) 3 C) 4 D) 5

4.2 Combinational Logic Analysis

Combinational logic analysis refers to the act of deciphering the operation of a circuit from its final logic diagram. This is a useful skill that can aid designers when debugging their circuits. This can also be used to understand the timing performance of a circuit and to reverse-engineer an unknown design.

4.2.1 Finding the Logic Expression from a Logic Diagram

Combinational logic diagrams are typically written with their inputs on the left and their output on the right. As the inputs change, the intermediate *nodes*, or connections, within the diagram hold the interim computations that contribute to the ultimate circuit output. These computations propagate from left to right until ultimately the final output of the system reaches its final steady-state value. When analyzing the behavior of a combinational logic circuit, a similar *left-to-right* approach is used. The first step is to label each intermediate node in the system. The second step is to write in the logic expression for each node based on the preceding logic operation(s). The logic expressions are written working left-to-right until the output of the system is reached and the final logic expression of the circuit has been found. Consider the example of this analysis in Example 4.12.

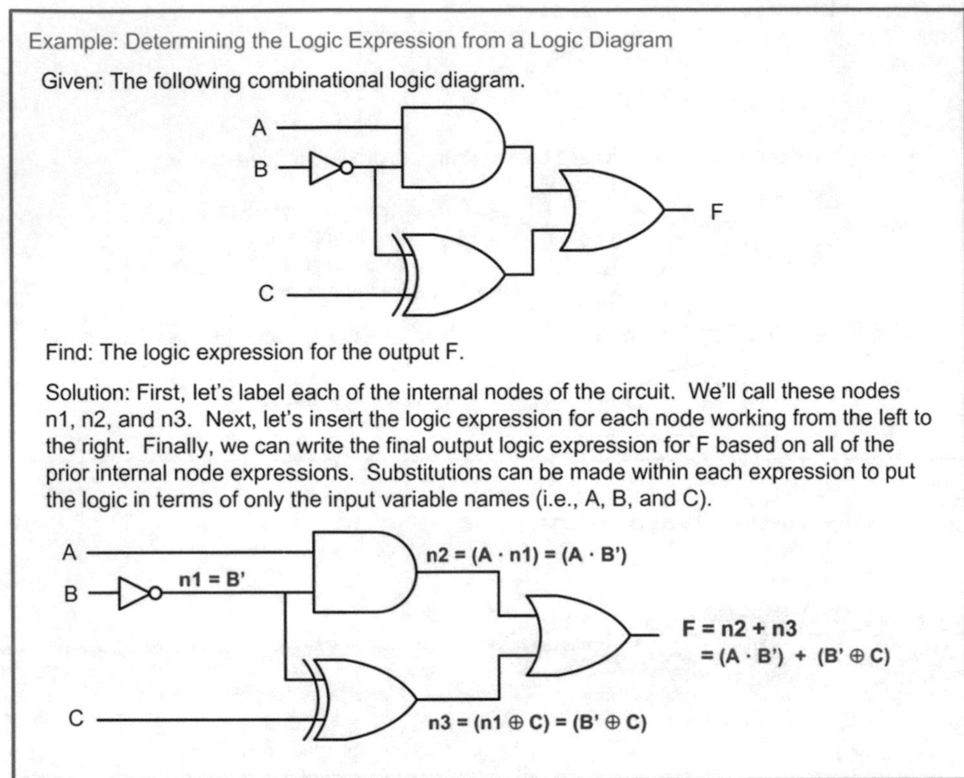

Example 4.12
Determining the logic expression from a logic diagram

4.2.2 Finding the Truth Table from a Logic Diagram

The final truth table of a circuit can also be found in a similar manner as the logic expression. Each internal node within the logic diagram can be evaluated working from the left to the right for each possible input code. Each subsequent node can then be evaluated using the values of the preceding nodes. Consider the example of this analysis is Example 4.13.

Example: Determining the Truth Table from a Logic Diagram

Given: The following combinational logic diagram.

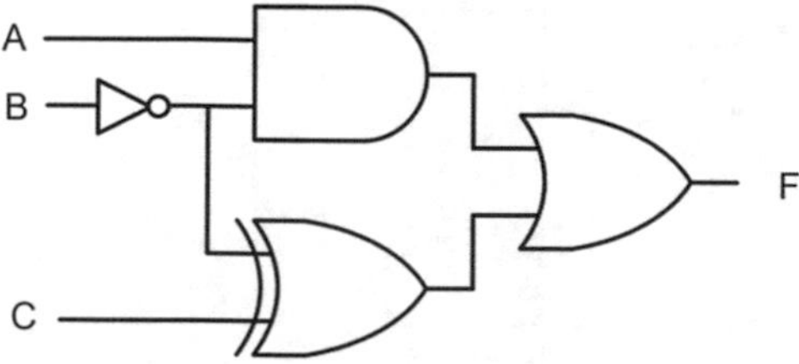

Find: The truth table for the output F.

Solution: First, we label each internal node and record the intermediate logic expressions.

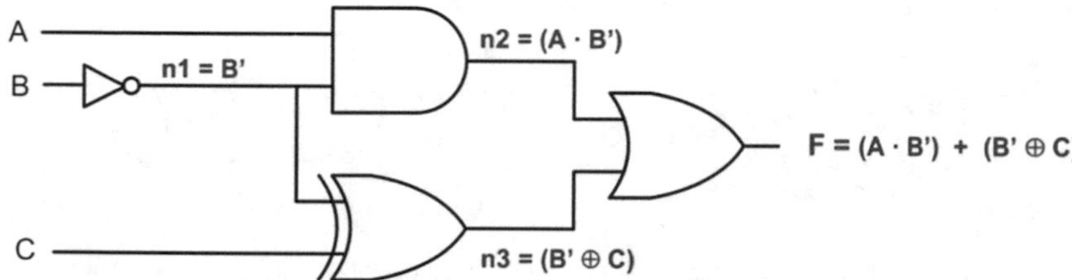

Next, we evaluate each node for all possible input codes working from the left to the right. This allows us to keep a record of the values of each intermediate node that can be used in the subsequent evaluations. We continue this process until we reach the final output F.

A	B	C	n1 = B'	n2 = A·B'	n3 =B'⊕C	F = (A·B') + (B'⊕C)
0	0	0	1	0	1	1
0	0	1	1	0	0	0
0	1	0	0	0	0	0
0	1	1	0	0	1	1
1	0	0	1	1	1	1
1	0	1	1	1	0	1
1	1	0	0	0	0	0
1	1	1	0	0	1	1

Notice that the intermediate computations can be used in the subsequent evaluations.

Example 4.13
Determining the truth table from a logic diagram

4.2.3 Timing Analysis of a Combinational Logic Circuit

Real logic gates have a propagation delay (t_{pd}, t_{PHL}, or t_{PLH}) as presented in Chap. 3. Performing a timing analysis on a combinational logic circuit refers to observing how long it takes for a change in the inputs to propagate to the output. Different paths through the combinational logic circuit will take different times to compute since they may use gates with different delays. When determining the delay of the entire combinational logic circuit, we always consider the longest delay path. This is because this delay represents the worst-case scenario. As long as we wait for the longest path to propagate through the circuit, then we are ensured that the output will always be valid after this time. To determine which signal path has the longest delay, we map out each and every path the inputs can take to the output of the circuit. We then sum up the gate delay along each path. The path with the longest delay dictates the delay of the entire combinational logic circuit. Consider this analysis shown in Example 4.14.

Example: Determining the Delay of a Combinational Logic Circuit

Given: The following combinational logic diagram with the associated gate delays.

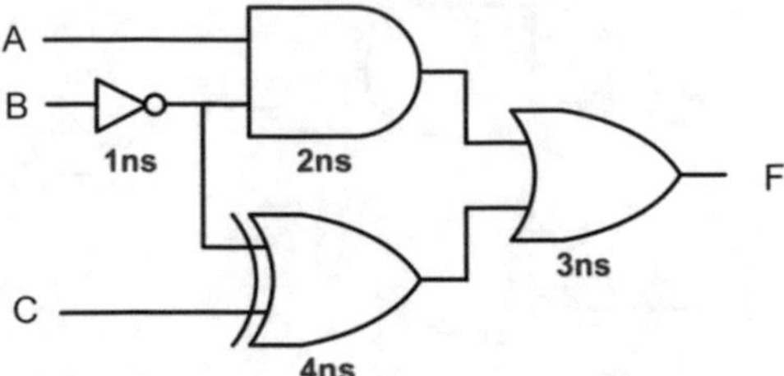

Find: The delay of the combinational logic circuit.

Solution: We begin by mapping the route of each and every path from the inputs to the output. For each path, we sum the delay through each gate that is used.

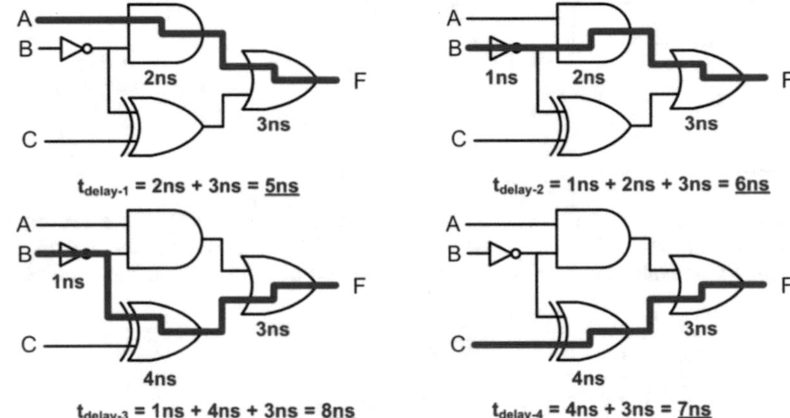

$t_{delay\text{-}1}$ = 2ns + 3ns = <u>5ns</u>

$t_{delay\text{-}2}$ = 1ns + 2ns + 3ns = <u>6ns</u>

$t_{delay\text{-}3}$ = 1ns + 4ns + 3ns = <u>8ns</u>

$t_{delay\text{-}4}$ = 4ns + 3ns = <u>7ns</u>

The longest delay path through this circuit is from B to F in which the signal traverses the inverter, XOR gate, and OR gate ($t_{delay\text{-}3}$). This path takes 8ns to compute. Since we must always consider the longest delay path when calculating how fast this circuit can operate, we can say that the delay of this combinational logic circuit is <u>8ns</u>.

Example 4.14

Determining the delay of a combinational logic circuit

Concept Check

CC4.2 Does the delay specification of a combinational logic circuit change based on the input values that the circuit is evaluating?

A) Yes. There are times when the inputs switch between inputs codes that use paths through the circuit with different delays.

B) No. The delay is always specified as the longest delay path.

C) Yes. The delay can vary between the longest delay path and zero. A delay of zero occurs when the inputs switch between two inputs codes that produce the same output.

D) No. The output is always produced at a time equal to the longest delay path.

4.3 Combinational Logic Synthesis

4.3.1 Canonical Sum of Products

One technique to directly synthesize a logic circuit from a truth table is to use a canonical sum of products topology based on **minterms**. The term *canonical* refers to this topology yielding potentially unminimized logic. A minterm is a product term (i.e., an AND operation) that will be true for one and only one input code. The minterm must contain every input variable in its expression. Complements are applied to the input variables as necessary in order to produce a true output for the individual input code. We define the word *literal* to describe an input variable which may or may not be complemented. This is a more useful word because if we say that a minterm "must include all variables," it implies that all variables are included in the term uncomplemented. A more useful statement is that a minterm "must include all literals." This now implies that each variable must be included, but it can be in the form of itself or its complement (e.g., A or A'). Figure 4.14 shows the definition and gate-level depiction of a minterm expression. Each minterm can be denoted using the lower case "m" with the row number as a subscript.

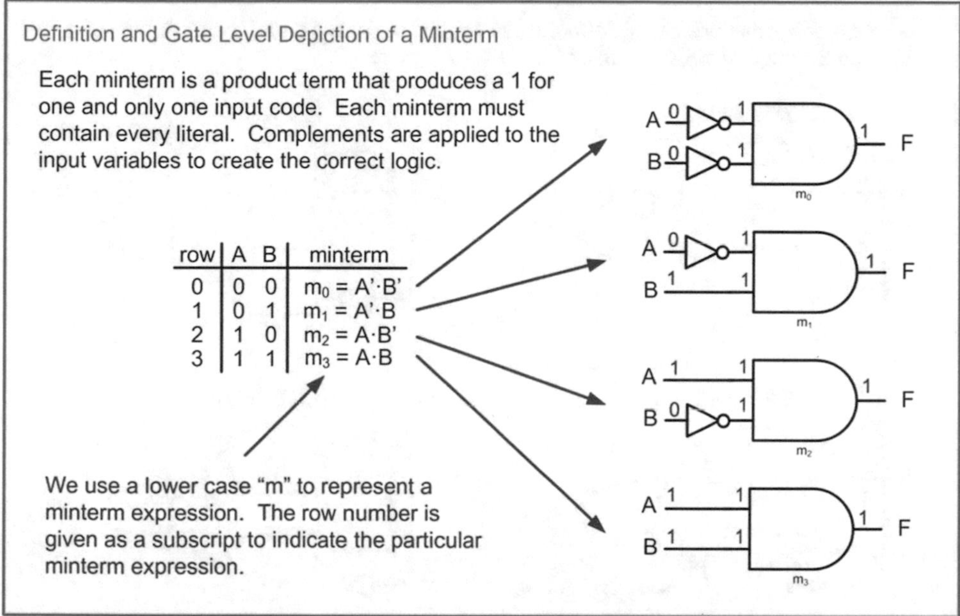

Fig. 4.14
Definition and gate-level depiction of a minterm

For an arbitrary truth table, a minterm can be used for each row corresponding to a true output. If each of these minterms' outputs are fed into a single OR gate, then a sum of products logic circuit is formed that will produce the logic listed in the truth table. In this topology, any input code that corresponds to an output of 1 will cause its corresponding minterm to output a 1. Since a 1 on any input of an OR gate will cause the output to go to a 1, the output of the minterm is passed to the final result. Example 4.15 shows this process. One important consideration of this approach is that no effort has been taken to minimize the logic expression. This unminimized logic expression is also called the **canonical sum**. The canonical sum is logically correct but uses the most amount of circuitry possible for a given truth table. This canonical sum can be the starting point for minimization using Boolean algebra.

Example: Creating a Canonical Sum of Products Logic Circuit using Minterms

Given: The following truth table.

A	B	F
0	0	0
0	1	1
1	0	1
1	1	0

Find: The Canonical SOP

Solution: Let's first start by writing the minterms for the rows that correspond to a 1 on the output. These can then be implemented using inverters and AND gates. The final step is to feed the outputs of each minterm circuit into a single OR gate.

row	A	B	minterm
0	0	0	-
1	0	1	$m_1 = A' \cdot B$
2	1	0	$m_2 = A \cdot B'$
3	1	1	-

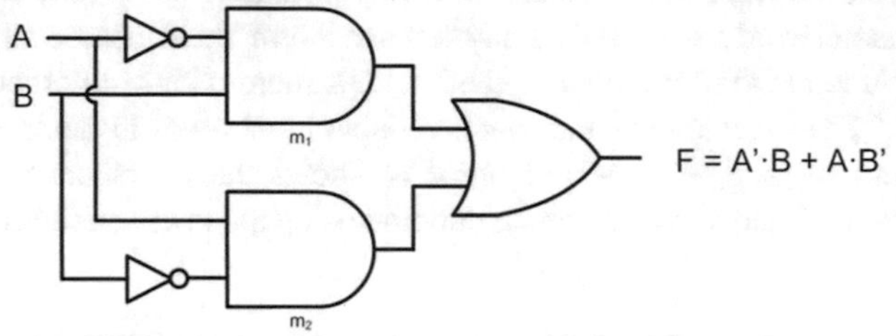

Let's now check that this circuit performs as intended by testing it under each input code for A and B and observing the output F.

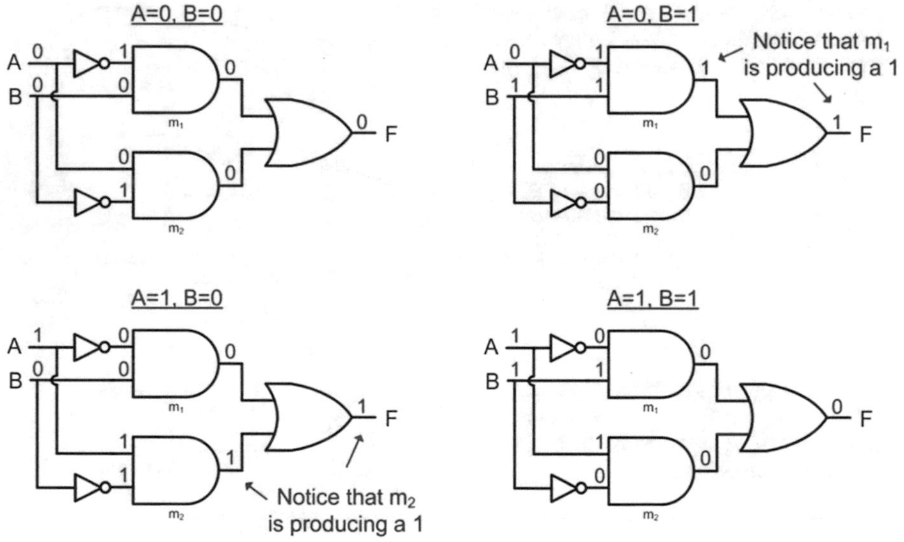

This circuit operates as intended.

Example 4.15
Creating a canonical sum of products logic circuit using minterms

4.3.2 The Minterm List (Σ)

A *minterm list* is a compact way to describe the functionality of a logic circuit by simply listing the row numbers that correspond to an output of 1 in the truth table. The ∑ symbol is used to denote a minterm list. All input variables must be listed in the order they appear in the truth table. This is necessary because since a minterm list uses only the row numbers to indicate which input codes result in an output of 1, the minterm list must indicate how many variables comprise the row number, which variable is in the most significant position, and which is in the least significant position. After the $\sum$ symbol, the row numbers corresponding to a true output are listed in a comma-delimited format within parentheses. Example 4.16 shows the process for creating a minterm list from a truth table.

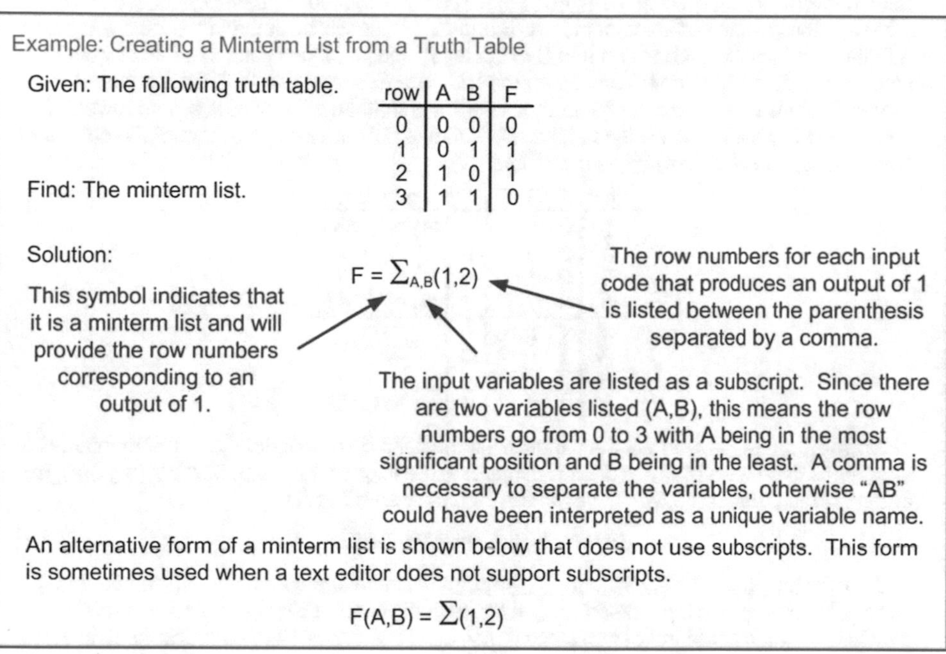

row	A	B	F
0	0	0	0
1	0	1	1
2	1	0	1
3	1	1	0

Example 4.16
Creating a minterm list from a truth table

A minterm list contains the same information as the truth table, the canonical sum and the canonical sum of products logic diagram. Since the minterms themselves are formally defined for an input code, it is trivial to go back and forth between the minterm list and these other forms. Example 4.17 shows how a minterm list can be used to generate an equivalent truth table, canonical sum, and canonical sum of products logic diagram.

Example: Creating Equivalent Functional Representations from a Minterm List

Given: The following minterm list. $F = \sum_{A,B,C}(0,3,7)$

Find: The truth table, canonical sum logic expression and the canonical sum of products logic diagram.

Solution: First, let's generate the **truth table**. From the minterm list subscripts, we know that there are three input variables named A, B and C. These will be listed in the truth table with A in the most significant position and C in the least significant position. We can fill in the input codes as a binary count and insert the row numbers. We can then list the output values that are true. From the minterm list we know that the true outputs are on rows 0, 3 and 7. Since we know we will need the minterm expressions for these rows in the canonical sum, we can also list them in the truth table.

row	A	B	C	F	minterm
0	0	0	0	1	$m_0 = A'{\cdot}B'{\cdot}C'$
1	0	0	1	0	-
2	0	1	0	0	-
3	0	1	1	1	$m_3 = A'{\cdot}B{\cdot}C$
4	1	0	0	0	-
5	1	0	1	0	-
6	1	1	0	0	-
7	1	1	1	1	$m_7 = A{\cdot}B{\cdot}C$

The **canonical sum** is simply the minterm expressions corresponding to a true output OR'd together. Since we already wrote the minterm expressions for rows 0, 3 and 7 (e.g., m_0, m_3 and m_7) in the truth table, we can write the canonical sum directly.

$$F = A'{\cdot}B'{\cdot}C' + A'{\cdot}B{\cdot}C + A{\cdot}B{\cdot}C$$

The **canonical sum of products logic diagram** is simply the gate level depiction of the canonical sum. When logic diagrams get larger, it is acceptable to indicate a variable's complement as a prime instead of placing individual inverters and drawing connection wires that cross each other. It is implied that multiple listings of a variable's complement (e.g., A' in m_0 and m_3) will come from the same inverter.

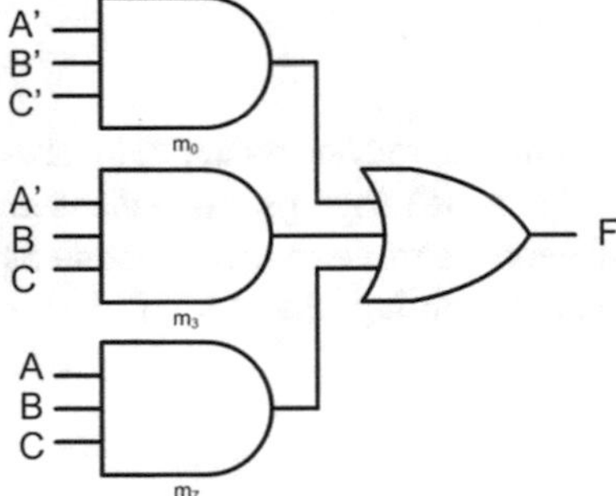

Example 4.17
Creating equivalent functional representations from a minterm list

4.3.3 Canonical Product of Sums (POS)

Another technique to directly synthesize a logic circuit from a truth table is to use a canonical product of sums topology based on **maxterms**. A maxterm is a sum term (i.e., an OR operation) that will be false for one and only one input code. The maxterm must contain every literal in its expression. Complements are applied to the input variables as necessary in order to produce a false output for the individual input code. Figure 4.15 shows the definition and gate-level depiction of a maxterm expression. Each maxterm can be denoted using the upper case "M" with the row number as a subscript.

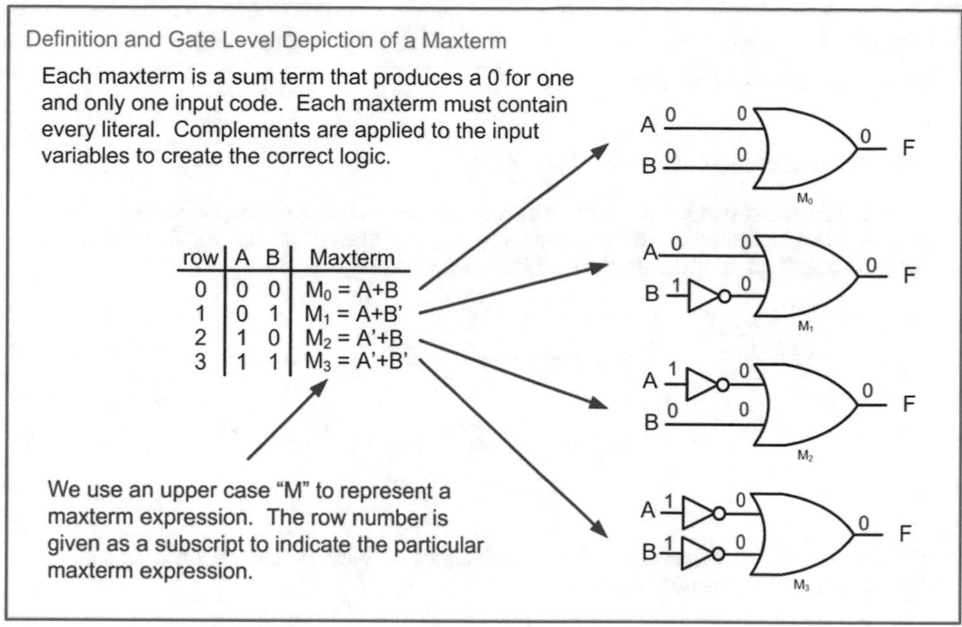

Fig. 4.15
Definition and gate-level depiction of a maxterm

For an arbitrary truth table, a maxterm can be used for each row corresponding to a false output. If each of these maxterms outputs are fed into a single AND gate, then a product of sums logic circuit is formed that will produce the logic listed in the truth table. In this topology, any input code that corresponds to an output of 0 will cause its corresponding maxterm to output a 0. Since a 0 on any input of an AND gate will cause the output to go to a 0, the output of the maxterm is passed to the final result. Example 4.18 shows this process. This approach is complementary to the sum of products approach. In the sum of products approach based on minterms, the circuit operates by producing 1's that are passed to the output for the rows that require a true output. For all other rows, the output is false. A product of sums approach based on maxterms operates by producing 0's that are passed to the output for the rows that require a false output. For all other rows, the output is true. These two approaches produce the equivalent logic functionality. Again, at this point no effort has been taken to minimize the logic expression. This unminimized form is called a **canonical product**. The canonical product is logically correct but uses the most amount of circuitry possible for a given truth table. This canonical product can be the starting point for minimization using the Boolean algebra theorems.

Example: Creating a Canonical Product of Sums Logic Circuit using Maxterms

Given: The following truth table.

A	B	F
0	0	0
0	1	1
1	0	1
1	1	0

Find: The Canonical POS.

Solution: Let's first start by writing the maxterms for the rows that correspond to a 0 on the output. These can then be implemented using inverters and OR gates. The final step is to feed the outputs of each maxterm circuit into a single AND gate.

row	A	B	Maxterm
0	0	0	M_0=A+B
1	0	1	-
2	1	0	-
3	1	1	M_3=A'+B'

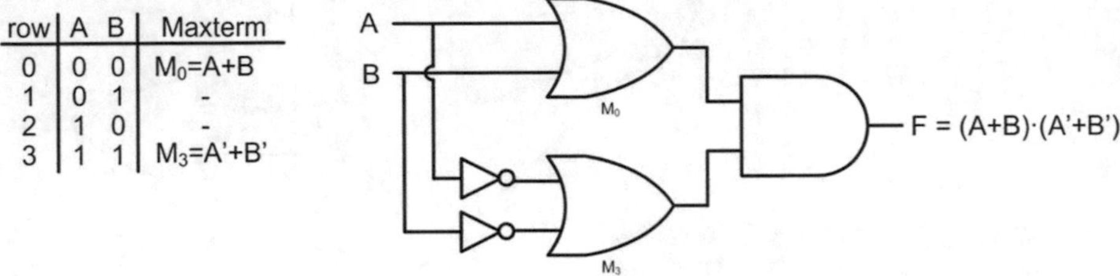

Let's now check that this circuit performs as intended by testing it under each input code for A and B and observing the output F.

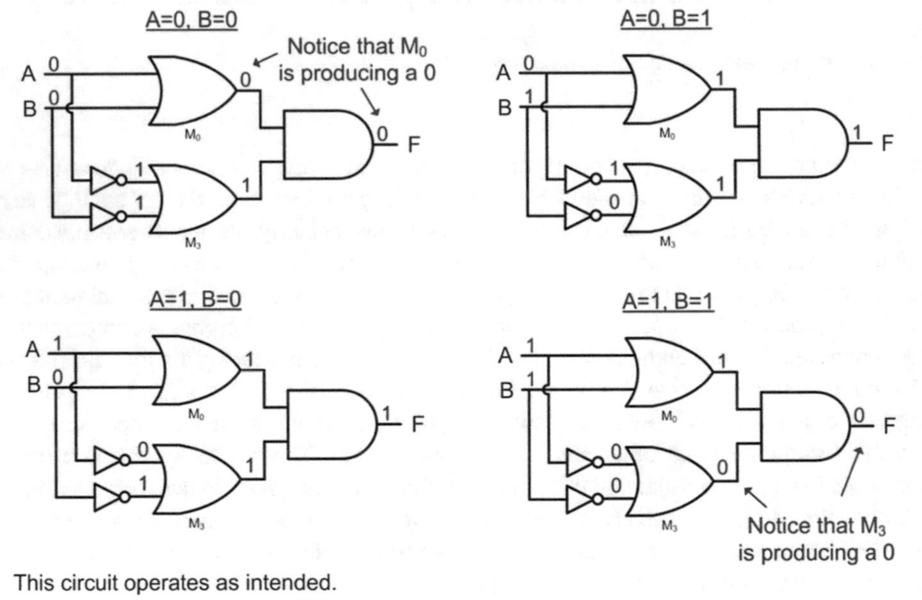

This circuit operates as intended.

Example 4.18
Creating a product of sums logic circuit using maxterms

4.3.4 The Maxterm List (Π)

A maxterm list is a compact way to describe the functionality of a logic circuit by simply listing the row numbers that correspond to an output of 0 in the truth table. The Π symbol is used to denote a maxterm list. All literals used in the logic expression must be listed in the order they appear in the truth table. After the Π symbol, the row numbers corresponding to a false output are listed in a comma-delimited format within parentheses. Example 4.19 shows the process for creating a maxterm list from a truth table.

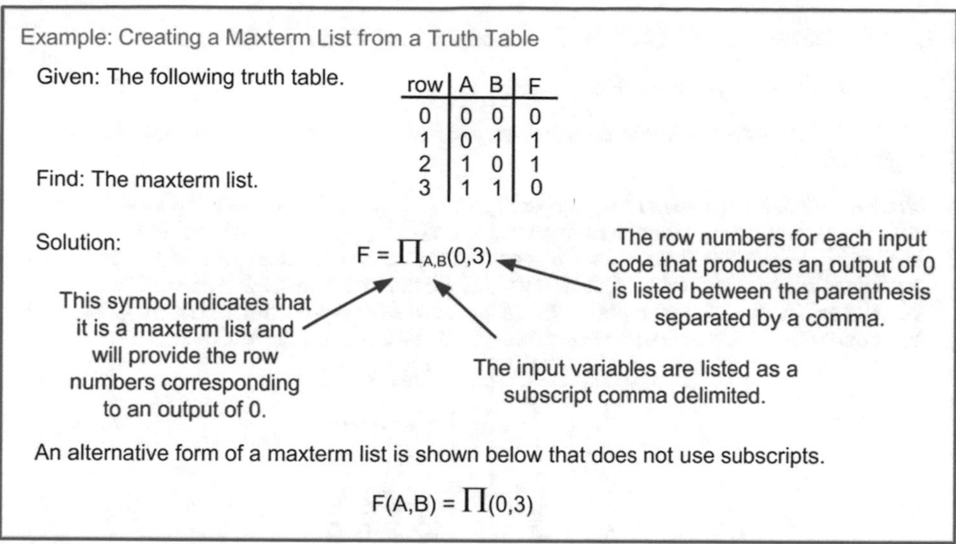

Example 4.19
Creating a maxterm list from a truth table

A maxterm list contains the same information as the truth table, the canonical product, and the canonical product of sums logic diagram. Example 4.20 shows how a maxterm list can be used to generate these equivalent forms.

Example: Creating Equivalent Functional Representations from a Maxterm List

Given: The following maxterm list. $F = \prod_{A,B,C}(1,2,4,5,6)$

Find: The truth table, canonical product logic expression and the canonical product of sums logic diagram.

Solution: First, let's generate the **truth table**. From the maxterm list subscripts, we know that there are three input variables named A, B and C that will be used in the truth table in that order. We can fill in the input codes as a binary count and insert the row numbers. We then can list the output values that are false. From the maxterm list we know that the false outputs are on rows 1, 2, 4, 5 and 6. Since we know we will need the maxterm expressions for these rows in the canonical product, we can also list them in the truth table.

row	A	B	C	F	Maxterm
0	0	0	0	1	-
1	0	0	1	0	M_1=A+B+C'
2	0	1	0	0	M_2=A+B'+C
3	0	1	1	1	-
4	1	0	0	0	M_4=A'+B+C
5	1	0	1	0	M_5=A'+B+C'
6	1	1	0	0	M_6=A'+B'+C
7	1	1	1	1	-

The **canonical product** is simply the maxterm expressions corresponding to a false output AND'd together. Since we already wrote these maxterm expressions in the truth table (M_1, M_2, M_4, M_5 and M_6) we can write the canonical product directly.

$$F = (A+B+C')\cdot(A+B'+C)\cdot(A'+B+C)\cdot(A'+B+C')\cdot(A'+B'+C)$$

The **canonical product of sums logic diagram** is simply the gate level depiction of the canonical product.

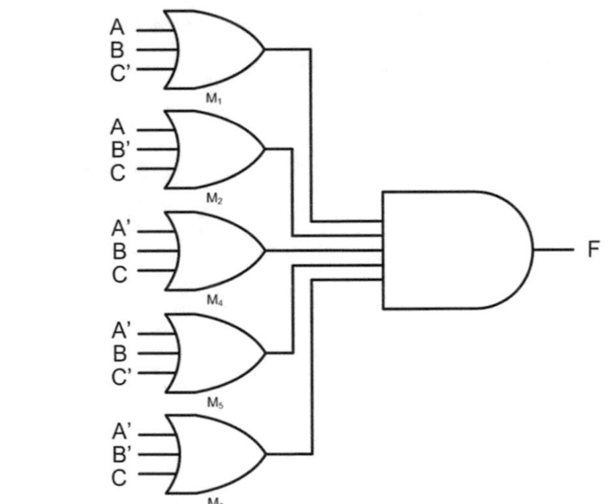

Example 4.20
Creating equivalent functional representations from a maxterm list

4.3.5 Minterm and Maxterm List Equivalence

The examples in Examples 4.17 and 4.20 illustrate how minterm and maxterm lists produce the exact same logic functionality but in a complementary fashion. It is trivial to switch back and forth between minterm lists and maxterm lists. This is accomplished by simply changing the list type (i.e., min to max, max to min) and then switching the row numbers between those listed and those not listed. Example 4.21 shows multiple techniques for representing equivalent logic functionality as a truth table.

Example: Creating Equivalent Forms to Represent Logic Functionality

Given: The following minterm list.

row	A	B	F
0	0	0	1
1	0	1	0
2	1	0	0
3	1	1	1

Find: All equivalent forms to describe the same functionality as the truth table.

Solution: Let's start by writing the **minterm list** and the **maxterm list**. These two lists are equivalent to each other. Remember that the minterm list provides the row numbers corresponding to an output of true while the maxterm list provides the row numbers corresponding to an output of false.

$$F = \sum_{A,B}(0,3) = \prod_{A,B}(1,2)$$

Let's write the minterm and maxterm expressions in the truth table. These will be used when creating the canonical sum and product expressions.

row	A	B	F	minterm	maxterm
0	0	0	1	$m_0 = A' \cdot B'$	-
1	0	1	0	-	$M_1 = A+B'$
2	1	0	0	-	$M_2 = A'+B$
3	1	1	1	$m_3 = A \cdot B$	-

Now let's write the **canonical sum** and **canonical product** logic expressions using these minterms and maxterms. Remember that a canonical sum is simply all of the minterms corresponding to an output of true OR'd together, and a canonical product is simply all of the maxterms corresponding to an output of false AND'd together.

$$F = A' \cdot B' + A \cdot B = (A+B') \cdot (A'+B)$$

Finally, let's draw the **canonical sum of products logic diagram** and the **canonical product of sums logic diagram**.

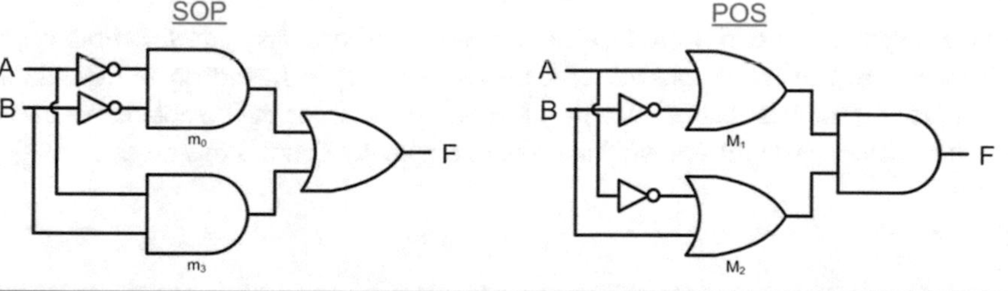

Example 4.21

Creating equivalent forms to represent logic functionality

Concept Check

CC4.3 All logic functions can be implemented equivalently using either a canonical sum of products (SOP) or canonical product of sums (POS) topology. Which of these statements is true with respect to selecting a topology that requires the least amount of gates?

A) Since a minterm list and a maxterm list can both be written to describe the same logic functionality, the number of gates in an SOP and POS will always be the same.

B) If a minterm list has over half of its row numbers listed, an SOP topology will require fewer gates than a POS.

C) A POS topology always requires more gates because it needs additional logic to convert the inputs from positive to negative logic.

D) If a minterm list has over half of its row numbers listed, a POS topology will require fewer gates than SOP.

4.4 Logic Minimization

We now look at how to reduce the canonical expressions into equivalent forms that use less logic. This minimization is key to reducing the complexity of the logic prior to implementing in real circuitry. This reduces the number of gates needed, placement area, wiring, and power consumption of the logic circuit.

4.4.1 Algebraic Minimization

Canonical expressions can be reduced algebraically by applying the theorems covered in prior sections. This process typically consists of a series of factoring based on the distributive property followed by replacing variables with constants (i.e., 0's and 1's) using the complements theorem. Finally, constants are removed using the identity theorem. Example 4.22 shows this process.

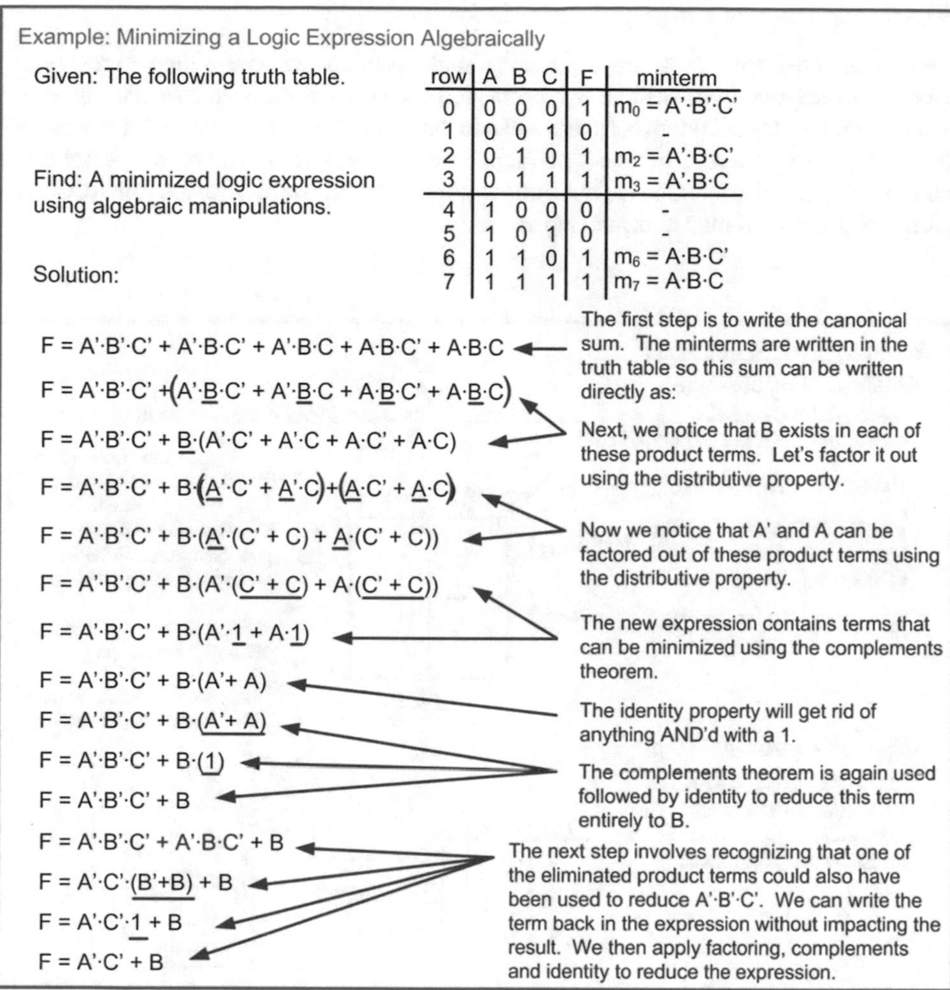

Example 4.22
Minimizing a logic expression algebraically

The primary drawback of this approach is that it requires recognition of where the theorems can be applied. This can often lead to missed minimizations. Computer automation is often the best mechanism to perform this minimization for large logic expressions.

4.4.2 Minimization Using Karnaugh Maps

A Karnaugh map is a graphical way to minimize logic expressions. This technique is named after Maurice Karnaugh, American physicist, who introduced the map in its latest form in 1953 while working at Bell Labs. The Karnaugh map (or K-map) is a way to put a truth table into a form that allows logic minimization through a graphical process. This technique provides a graphical process that accomplishes the same result as factoring variables via the distributive property and removing variables via the complements and identity theorems. K-maps present a truth table in a form that allows variables to be removed from the final logic expression in a graphical manner.

4.4.2.1 Formation of a K-Map

A K-map is constructed as a two-dimensional grid. Each *cell* within the map corresponds to the output for a specific input code. The cells are positioned such that neighboring cells only differ by one bit in their input codes. Neighboring cells are defined as cells immediately adjacent horizontally and immediately adjacent vertically. Two cells positioned diagonally next to each other are not considered neighbors. The input codes for each variable are listed along the top and side of the K-map. Consider the construction of a 2-input K-map shown in Fig. 4.16.

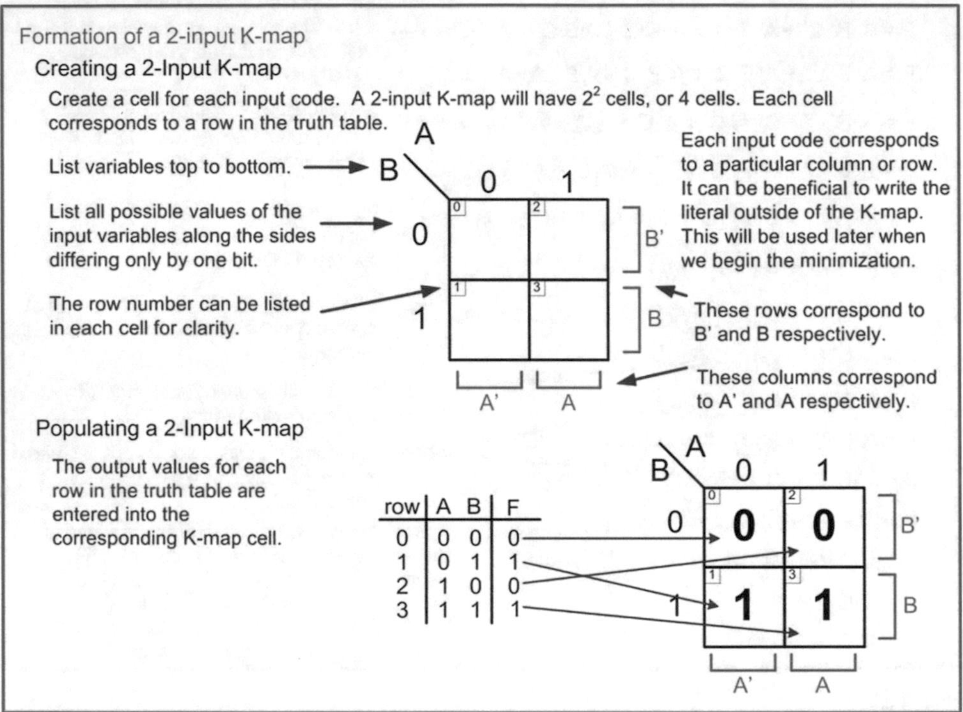

Fig. 4.16
Formation of a 2-input K-map

When constructing a 3-input K-map, it is important to remember that each input code can only differ from its neighbor by one bit. For example, the two codes 01 and 10 differ by two bits (i.e., the MSB is different and the LSB is different); thus they could not be neighbors; however, the codes 01-11 and 11-10 can be neighbors. As such, the input codes along the top of the 3-input K-map must be ordered accordingly (i.e., 00-01-11-10). Consider the construction of a 3-input K-map shown in Fig. 4.17. The rows and columns that correspond to the input literals can now span multiple rows and columns. Notice how in this 3-input K-map, the literals A, A', B, and B' all correspond to two columns. Also, notice that B' spans two columns, but the columns are on different edges of the K-map. The side edges of the 3-input K-map are still considered neighbors because the input codes for these columns only differ by one bit. This is an important attribute once we get to the minimization of variables because it allows us to examine an input literal's impact not only within the obvious adjacent cells but also when the variables *wrap* around the edges of the K-map.

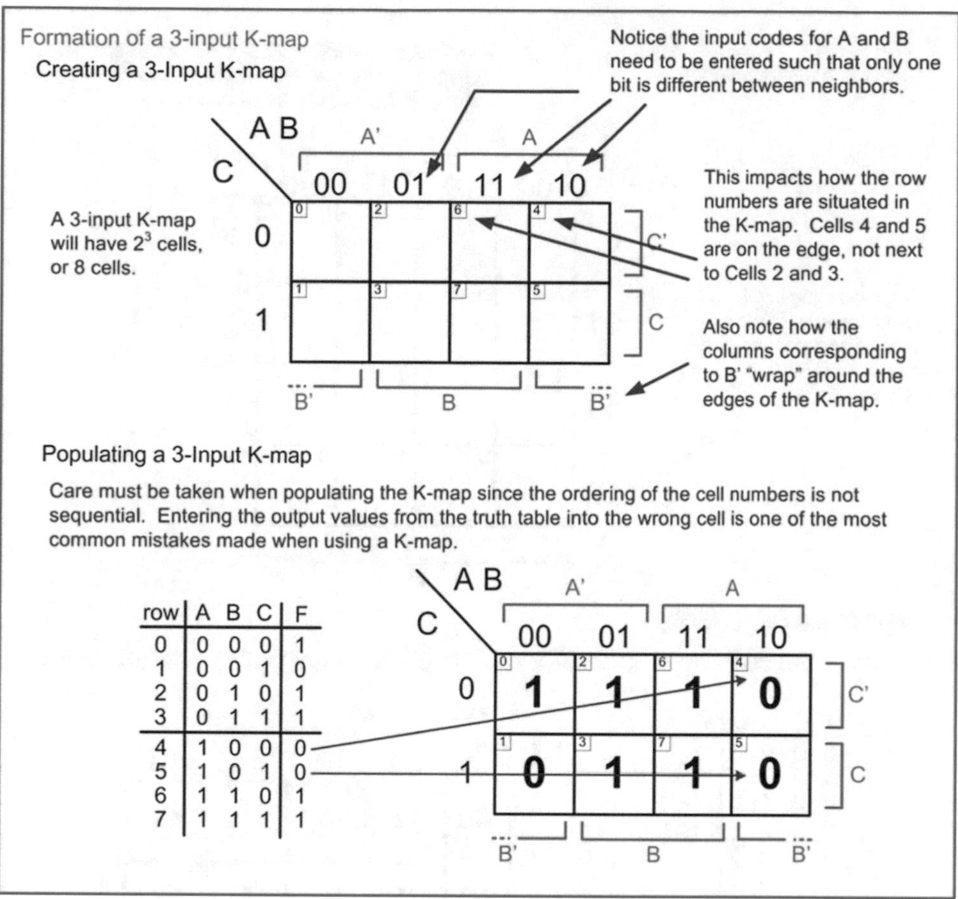

Fig. 4.17
Formation of a 3-input K-map

When constructing a 4-input K-map, the same rules apply that the input codes can only differ from their neighbors by one bit. Consider the construction of a 4-input K-map in Fig. 4.18. In a 4-input K-map, neighboring cells can wrap around both the top-to-bottom edges in addition to the side-to-side edges. Notice that all 16 cells are positioned within the map so that their neighbors on the top, bottom, and sides only differ by one bit in their input codes.

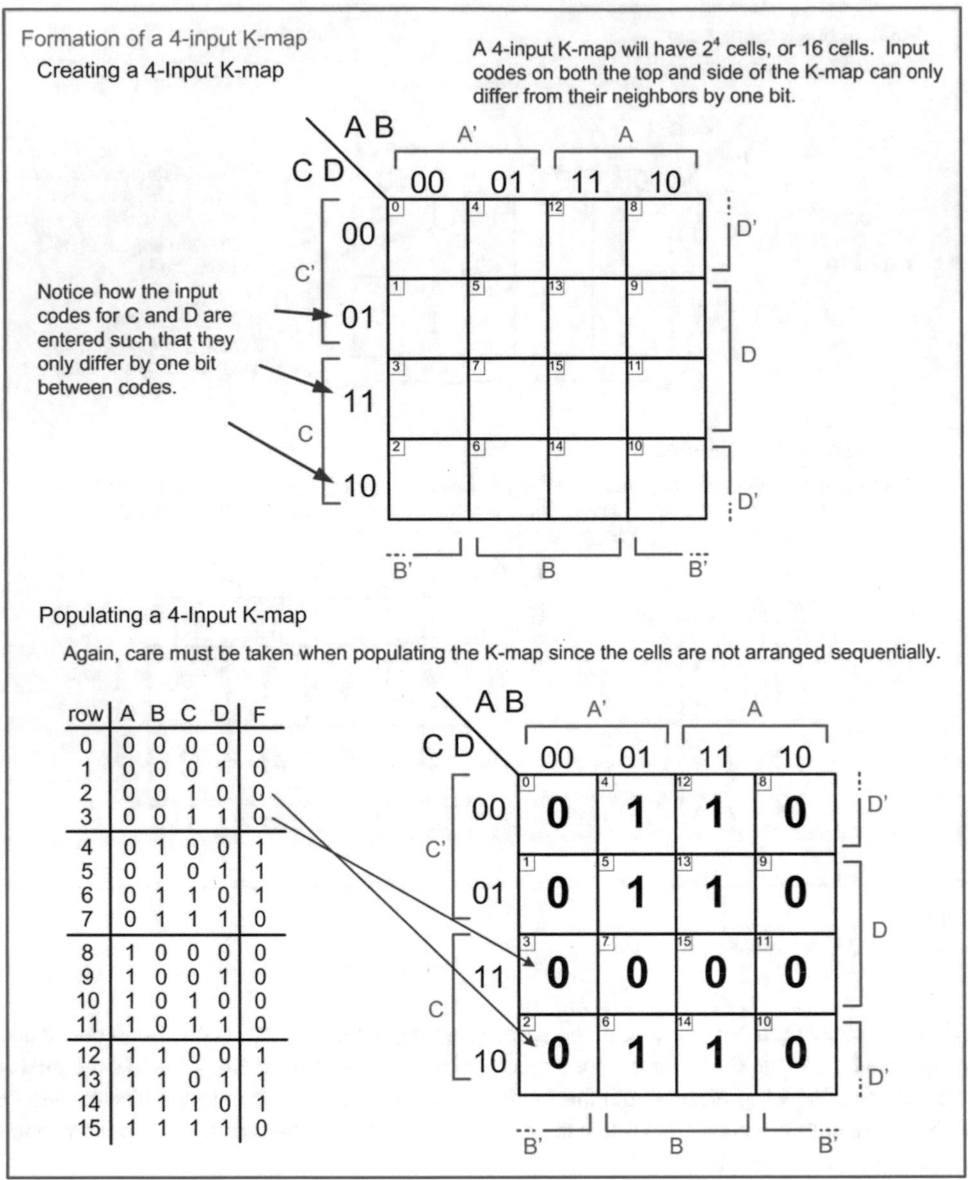

row	A	B	C	D	F
0	0	0	0	0	0
1	0	0	0	1	0
2	0	0	1	0	0
3	0	0	1	1	0
4	0	1	0	0	1
5	0	1	0	1	1
6	0	1	1	0	1
7	0	1	1	1	0
8	1	0	0	0	0
9	1	0	0	1	0
10	1	0	1	0	0
11	1	0	1	1	0
12	1	1	0	0	1
13	1	1	0	1	1
14	1	1	1	0	1
15	1	1	1	1	0

Fig. 4.18
Formation of a 4-input K-map

4.4.2.2 Logic Minimization Using K-Maps (Sum of Products)

Now we look at using a K-map to create a minimized logic expression in a SOP form. Remember that each cell with an output of 1 has a minterm associated with it, just as in the truth table. When two neighboring cells have outputs of 1, it graphically indicates that the two minterms can be reduced into a minimized product term that will cover both outputs. Consider the example given in Fig. 4.19.

Observing how K-Maps Visually Highlight Logic Minimizations

Let's look at how a K-map highlights minimizations. First, we put the truth table into K-map form.

row	A	B	F
0	0	0	0
1	0	1	1
2	1	0	0
3	1	1	1

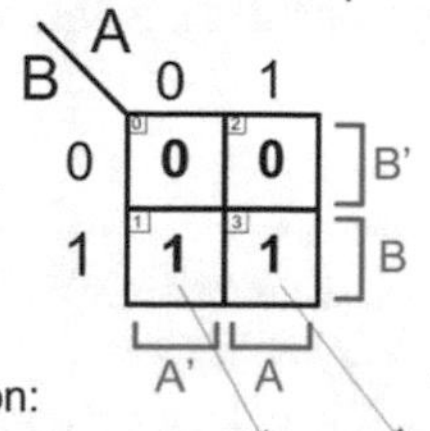

B \ A	0	1	
0	0	0	B'
1	1	1	B
	A'	A	

Let's first write the canonical SOP expression:

The canonical sum of products for this truth table is: F = A'·B + A·B ← Each of the outputs that are true have an associated minterm.

Next, let's minimize the canonical SOP algebraically to find the correct answer.

F = A'·B + A·B

F = B·(A' + A) ← Factor out the variable B using the distributive property.

F = B·(1) ← Replace (A' + A) = 1 using the complements theorem.

F = B ← Reduce to just B using the identity theorem.

Let's now look at the K-map. Notice that if we examine the grouping of cells 1 and 3, we can observe the dependence of the group on the input variables.

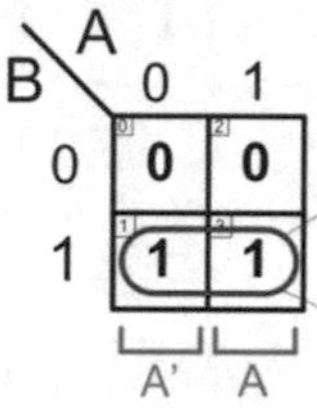

B \ A	0	1	
0	0	0	B'
1	(1	1)	B
	A'	A	

This group spans both A and A'. This means that if a single product term was created to produce these outputs, the variable A would not impact the result. This is a graphical way to notice a variable that can be factored through the distributive property, reduced to 1 through the complements theorem and removed from the product term using the identity theorem.

This group spans only the literal B. This means B must be included in the product term.

These two observations yield a product term that is associated with the grouping that is simply:

F=B

Fig. 4.19
Observing how K-maps visually highlight logic minimizations

These observations can be put into a formal process to produce a minimized SOP logic expression using a K-map. The steps are as follows:

1. Circle groups of 1's in the K-map following the rules:
   - Each circle should contain the largest number of 1's possible.
   - The circles encompass only neighboring cells (i.e., side-to-side sides and/or top and bottom).
   - The circles must contain a number of 1's that is a power of 2 (i.e., 1, 2, 4, 8, or 16).
   - Enter as many circles as possible without having any circles fully cover another circle.
   - Each circle is called a *prime implicant*.
2. Create a product term for each prime implicant following the rules:
   - Each variable in the K-map is evaluated one by one.
   - If the circle covers a region where the input variable is a 1, then include it in the product term uncomplemented.
   - If the circle covers a region where the input variable is a 0, then include it in the product term complemented.
   - If the circle covers a region where the input variable is both a 0 and 1, then the variable is excluded from the product term.
3. Sum all of the product terms for each prime implicant.

Let's apply this approach to our 2-input K-map example. Example 4.23 shows the process of finding a minimized sum of products logic expression for a 2-input logic circuit using a K-map. This process yielded the same SOP expression as the algebraic minimization and observations shown in Fig. 4.19, but with a formalized process.

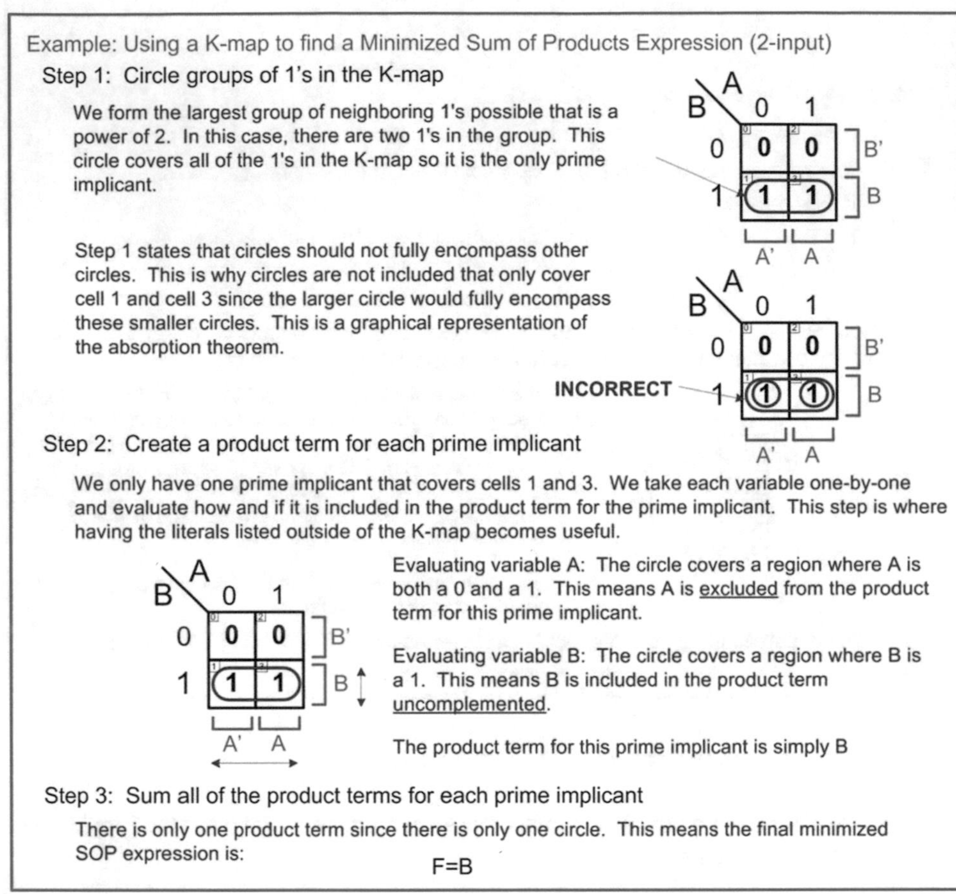

Example 4.23
Using a K-map to find a minimized sum of products expression (2-Input)

Let's now apply this process to our 3-input K-map example. Example 4.24 shows the process of finding a minimized sum of products logic expression for a 3-input logic circuit using a K-map. This example shows circles that overlap. This is legal as long as one circle does not fully encompass another. Overlapping circles are common since the K-map process dictates that circles should be drawn that group the largest number of ones possible as long as they are in powers of 2. Forming groups of ones using ones that have already been circled is perfectly legal to accomplish larger groupings. The larger the grouping of ones, the more chance there is for a variable to be excluded from the product term. This results in better minimization of the logic.

Example: Using a K-map to find a Minimized Sum of Products Expression (3-input)

Step 1: Circle groups of 1's in the K-map

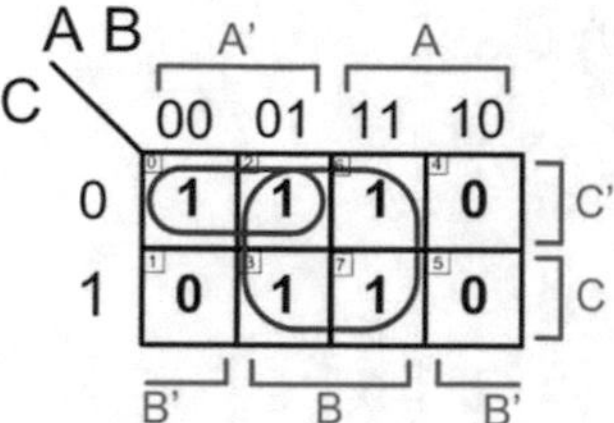

The two prime implicants overlap in cell 2, but this is legal because the larger circle does not fully encompass the smaller circle.

Step 2: Create a product term for each prime implicant

Variable A: The circle covers a region where A is a 0 so it is included in the product term <u>complemented</u>.

Variable B: The circle covers a region where B is both a 0 and 1, so it is <u>excluded</u> from the product term.

Variable C: The circle covers a region where C is a 0, so it is included in the product term <u>complemented</u>.

The product term for this prime implicant is: A'·C'

Variable A: The circle covers a region where A is both a 0 and 1, so it is <u>excluded</u> from the product term.

Variable B: The circle covers a region where B is a 1, so it is included in the product term <u>uncomplemented</u>.

Variable C: The circle covers a region where C is both a 0 and 1, so it is <u>excluded</u> from the product term.

The product term for this prime implicant is: B

Step 3: Sum all of the product terms for each prime implicant

There are two product terms, one for each circle. The final minimized SOP expression is:

F = A'·C' + B

Example 4.24
Using a K-map to find a minimized sum of products expression (3-input)

Let's now apply this process to our 4-input K-map example. Example 4.25 shows the process of finding a minimized sum of products logic expression for a 4-input logic circuit using a K-map.

Example: Using a K-map to find a Minimized Sum of Products Expression (4-input)

Step 1: Circle groups of 1's in the K-map

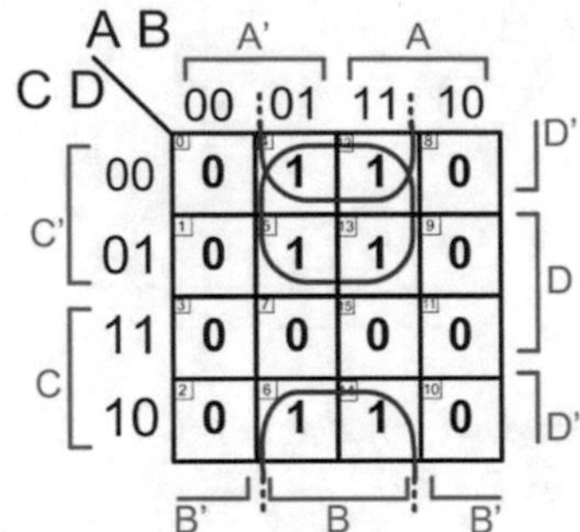

Circles can be drawn that "wrap" around the edges. Notice that the input codes for cells 4 and 12 only differ by 1 bit from cells 6 and 14. This makes them neighbors and grouping these 4 cells together is legal.

Again, circles that overlap are legal as long as one circle does not fully encompass another.

Step 2: Create a product term for each prime implicant

Variable A: The circle covers a region where A is both a 0 and 1, so it is <u>excluded</u> from the product term.

Variable B: The circle covers a region where B is a 1, so it is included in the product term <u>uncomplemented</u>.

Variable C: The circle covers a region where C is a 0, so it is included in the product term <u>complemented</u>.

Variable D: The circle covers a region where D is both a 0 and 1, so it is <u>excluded</u> from the product term.

The product term for this prime implicant is: B·C'

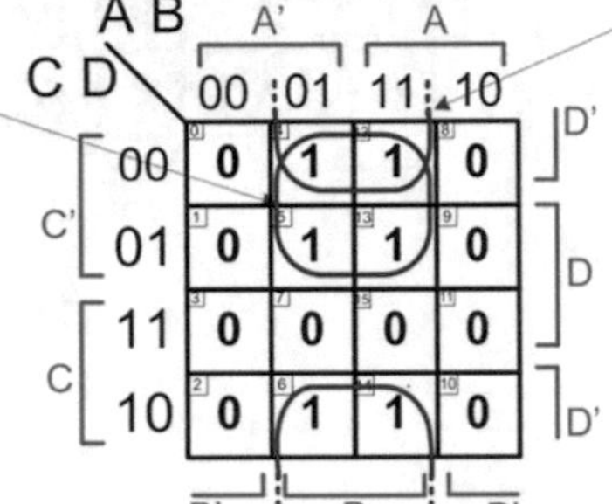

Variable A: The circle covers a region where A is both a 0 and 1, so it is <u>excluded</u> from the product term.

Variable B: The circle covers a region where B is a 1, so it is included in the product term <u>uncomplemented</u>.

Variable C: The circle covers a region where C is both a 0 and 1, so it is <u>excluded</u> from the product term.

Variable D: The circle covers a region where D is a 0, so it is included in the product term <u>complemented</u>.

The product term for this prime implicant is: B·D'

Step 3: Sum all of the product terms for each prime implicant

There are two product terms, one for each circle. The final minimized SOP expression is:

F = B·C' + B·D'

This expression could be further factored using the distributive property to F = B·(C' + D') to eliminate one more logic operation; however, since the problem asked for an SOP form, this last step was not necessary. Also, leaving a logic expression in an SOP form allows it to be directly converted into a NAND gate only implementation using DeMorgan's Theorem if the target logic family is CMOS.

Example 4.25
Using a K-map to find a minimized sum of products expression (4-input)

4.4.2.3 Logic Minimization Using K-Maps (Product of Sums)

K-maps can also be used to create minimized product of sums logic expressions. This is the same concept as how a minterm list and maxterm list each produce the same logic function, but in complementary fashions. When creating a product of sums expression from a K-map, groups of 0's are circled. For each circle, a sum term is derived with a negation of variables similar to when forming a maxterm (i.e., if the input variable is a 0, then it is included uncomplemented in the sum term and vice versa). The final step in forming the minimized POS expression is to AND all of the sum terms together. The formal process is as follows:

1. Circle groups of 0's in the K-map following the rules:
   - Each circle should contain the largest number of 0's possible.
   - The circles encompass only neighboring cells (i.e., side-to-side sides and/or top and bottom).
   - The circles must contain a number of 0's that is a power of 2 (i.e., 1, 2, 4, 8, or 16).
   - Enter as many circles as possible without having any circles fully cover another circle.
   - Each circle is called a *prime implicant*.
2. Create a sum term for each prime implicant following the rules:
   - Each variable in the K-map is evaluated one by one.
   - If the circle covers a region where the input variable is a 1, then include it in the sum term complemented.
   - If the circle covers a region where the input variable is a 0, then include it in the sum term uncomplemented.
   - If the circles cover a region where the input variable is both a 0 and 1, then the variable is excluded from the sum term.
3. Multiply all of the sum terms for each prime implicant.

Let's apply this approach to our 2-input K-map example. Example 4.26 shows the process of finding a minimized product of sums logic expression for a 2-input logic circuit using a K-map. Notice that this process yielded the same logic expression as the SOP approach shown in Example 4.23. This illustrates that both the POS and SOP expressions produce the correct logic for the circuit.

Example: Using a K-map to find a Minimized Product of Sums Expression (2-input)

Step 1: Circle groups of 0's in the K-map

We form the largest group of neighboring 0's possible that is a power of 2.

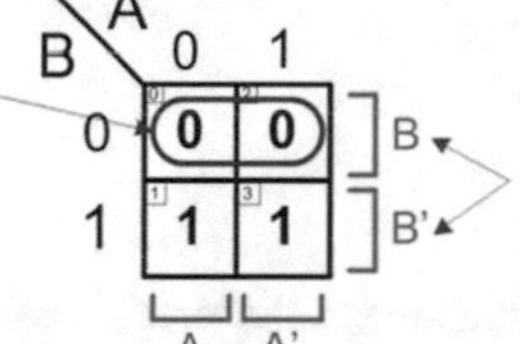

It is useful to change the variable polarities listed along the sides of the K-map to reflect how the variables are entered into the sum terms.

Step 2: Create a product term for each prime implicant

We take each variable one-by-one and evaluate how and if it is included in the sum term for the prime implicant.

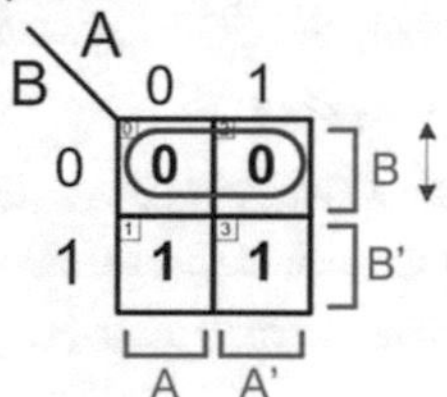

Evaluating variable A: The circle covers a region where A is both a 0 and a 1. This means A is excluded from the sum term for this prime implicant.

Evaluating variable B: The circle covers a region where B is a 0. This means B is included in the sum term uncomplemented.

The sum term for this prime implicant is simply B.

Step 3: Multiply all of the sums terms for each prime implicant

There is only one product term since there is only one circle. This means the final minimized POS expression is:

F=B ← This gives the exact same logic as the SOP form obtained by circling 1's.

Example 4.26
Using a K-map to find a minimized product of sums expression (2-input)

Let's now apply this process to our 3-input K-map example. Example 4.27 shows the process of finding a minimized product of sums logic expression for a 3-input logic circuit using a K-map. Notice that the logic expression in POS form is not identical to the SOP expression found in Example 4.24; however, using a few steps of algebraic manipulation shows that the POS expression can be put into a form that *is* identical to the prior SOP expression. This illustrates that both the POS and SOP produce equivalent functionality for the circuit.

Example: Using a K-map to find a Minimized Product of Sums Expression (3-input)

Step 1: Circle groups of 0's in the K-map

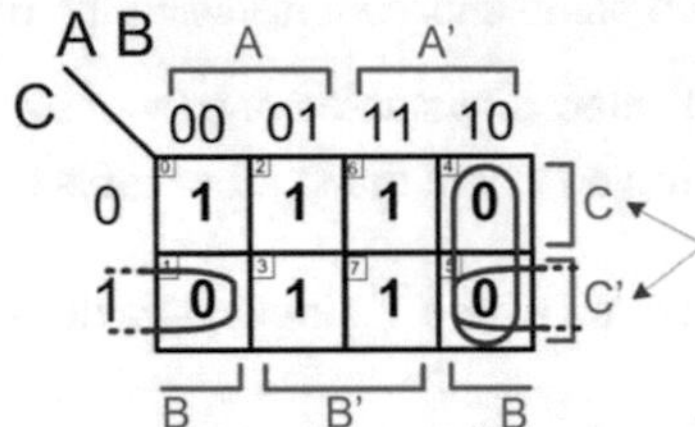

Again, the polarities of the variables along K-map are changed to reflect how the variables are entered into the sum terms.

Step 2: Create a sum term for each prime implicant

Variable A: The circle covers a region where A is both a 0 and 1, so it is excluded from the sum term.

Variable B: The circle covers a region where B is a 0, so it is included in the sum term uncomplemented.

Variable C: The circle covers a region where C is a 1, so it is included in the sum term complemented.

The sum term for this prime implicant is: B+C'

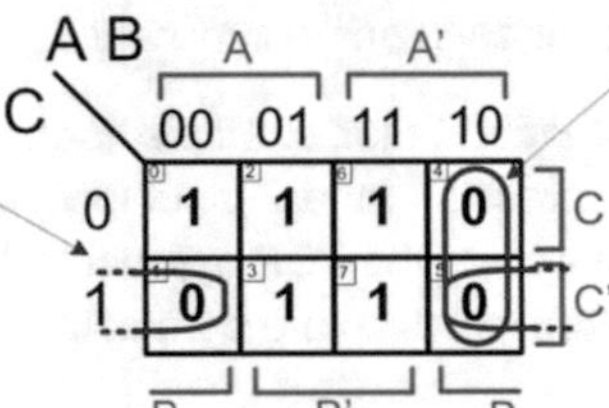

Variable A: The circle covers a region where A is a 1, so it is included in the sum term complemented.

Variable B: The circle covers a region where B is a 0, so it is included in the sum term uncomplemented.

Variable C: The circle covers a region where C is both a 0 and 1, so it is excluded from the sum term.

The sum term for this prime implicant is: A'+B

Step 3: Multiply all of the sum terms for each prime implicant

There are two sum terms, one for each circle. The final minimized POS expression is:

$$F = (B+C')\cdot(A'+B)$$

Check: Is this equivalent to the logic expression obtained using the SOP approach?

From the prior example, the minimized SOP expression was: $F = A'\cdot C' + B$

$F = (B+C')\cdot(A'+B)$ ← Let's use the Boolean algebra theorems to see if this is equal to $A'\cdot C' + B$

$F = B+(C'\cdot A')$ ← Using the distributive property on the POS expression, we can factor out B.

$F = A'\cdot C' + B$ ← The commutative property allows us to rearrange terms to match the SOP expression exactly.

Yes, its POS expression is equivalent to the SOP expression.

Example 4.27
Using a K-map to find a minimized product of sums expression (3-input)

Let's now apply this process to our 4-input K-map example. Example 4.28 shows the process of finding a minimized product of sums logic expression for a 4-input logic circuit using a K-map.

Example: Using a K-map to find a Minimized Product of Sums Expression (4-input)

Step 1: Circle groups of 0's in the K-map

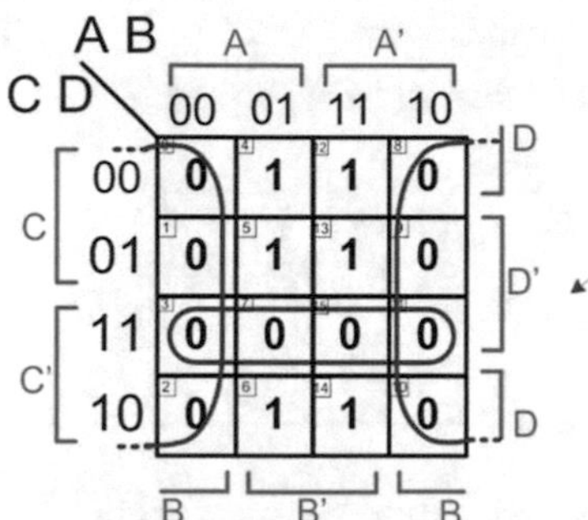

Again, the polarities of the variables along K-map are changed to reflect how the variables are entered into the sum terms.

Step 2: Create a sum term for each prime implicant

Variable A: The circle covers a region where A is both a 0 and 1, so it is <u>excluded</u> from the sum term.

Variable B: The circle covers a region where B is a 0, so it is included in the sum term <u>uncomplemented</u>.

Variable C: The circle covers a region where C is both a 0 and 1, so it is <u>excluded</u> from the sum term.

Variable D: The circle covers a region where D is both a 0 and 1, so it is <u>excluded</u> from the sum term.

The sum term for this prime implicant is: B

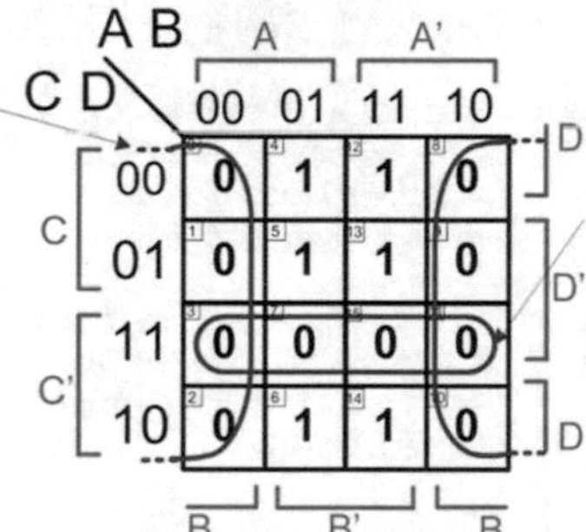

Variable A: The circle covers a region where A is both a 0 and 1, so it is <u>excluded</u> from the sum term.

Variable B: The circle covers a region where B is both a 0 and 1, so it is <u>excluded</u> from the sum term.

Variable C: The circle covers a region where C is a 1, so it is included in the sum term <u>complemented</u>.

Variable D: The circle covers a region where D is a 1, so it is included in the sum term <u>complemented</u>.

The sum term for this prime implicant is: C'+D'

Step 3: Multiply all of the sum terms for each prime implicant

There are two sum terms, one for each circle. The final minimized POS expression is:

F = (B)·(C'+D')

Check: Is this equivalent to the logic expression obtained using the SOP approach?

From the prior example, the minimized SOP expression was: F = B·C' + B·D'

F = (B)·(C'+D') ← Let's use the Boolean algebra theorems to see if this is equal to B·C' + B·D'

F = B·C' + B·D' ← Using the distributive property on the POS expression shows that this is equal to the minimized SOP expression.

Example 4.28
Using a K-map to find a minimized product of sums expression (4-input)

4.4.2.4 Minimal Sum

One situation that arises when minimizing logic using a K-map is that some of the prime implicants may be redundant. Consider the example in Fig. 4.20.

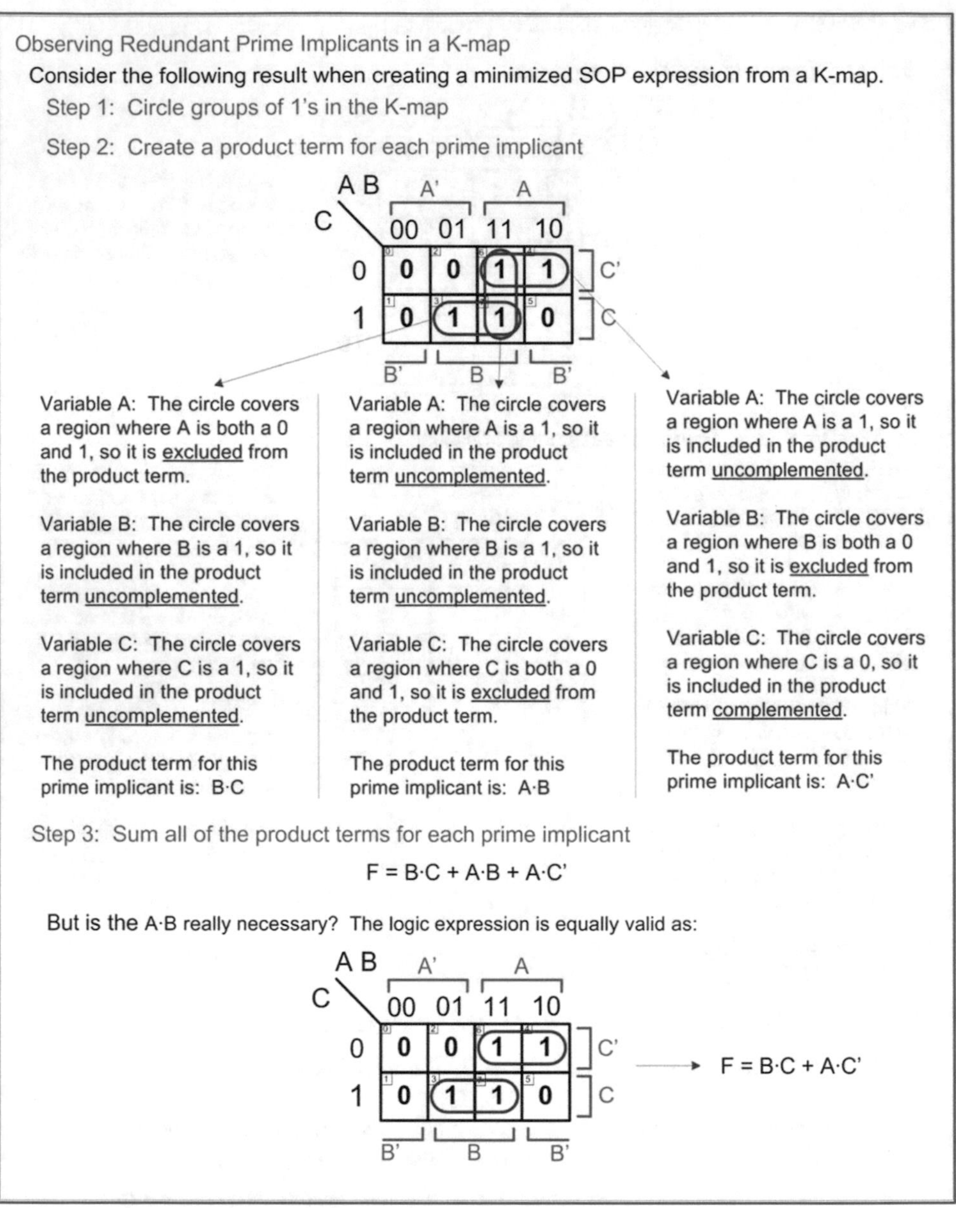

Fig. 4.20
Observing redundant prime implicants in a K-map

We need to define a formal process for identifying redundant prime implicants that can be removed without impacting the result of the logic expression. Let's start with examining the sum of products form. First, we define the term **essential prime implicant** as a prime implicant that *cannot* be removed from the logic expression without impacting its result. We then define the term **minimal sum** as a logic expression that represents the most minimal set of logic operations to accomplish a sum of products form. There may be multiple minimal sums for a given truth table, but each would have the same number of logic operations. In order to determine if a prime implicant is essential, we first put in each and every possible prime implicant into the K-map. This gives a logic expression known as the **complete sum**. From this point we identify any cells that have only one prime implicant covering them. These cells are called **distinguished one cells**. Any prime implicant that covers a distinguished one cell is defined as an essential prime implicant. All prime implicants that are not essential are removed from the K-map. A minimal sum is then simply the sum of all remaining product terms associated with the essential prime implicants. Example 4.29 shows how to use this process.

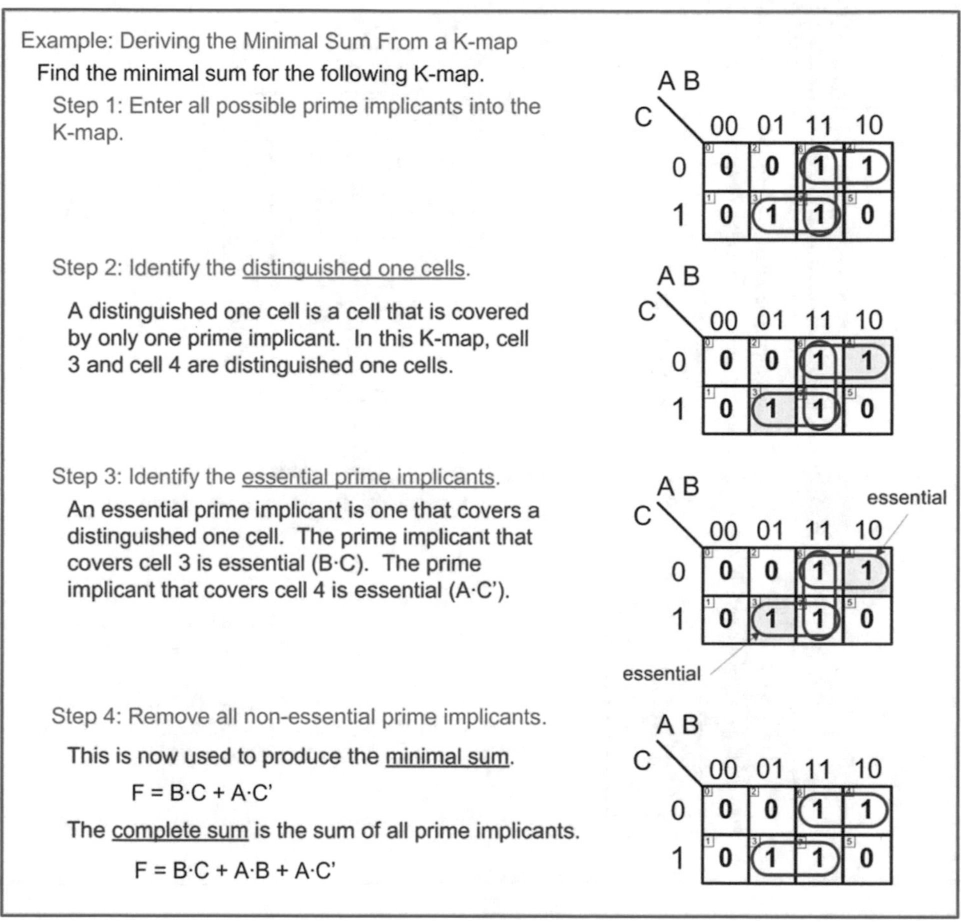

Example 4.29
Deriving the minimal sum from a K-map

This process is identical for the product of sums form to produce the **minimal product**.

4.4.3 Don't Cares

There are often times when framing a design problem that there are specific input codes that require exact output values, but there are other codes where the output value doesn't matter. This can occur for a variety of reasons, such as knowing that certain input codes will never occur due to the nature of the problem or that the output of the circuit will only be used under certain input codes. We can take advantage of this situation to produce a more minimal logic circuit. We define an output as a **don't care** when it doesn't matter whether it is a 1 or 0 for the particular input code. The symbol for a don't care is "X." We take advantage of don't cares when performing logic minimization by treating them as whatever output value will produce a minimal logic expression. Example 4.30 shows how to use this process.

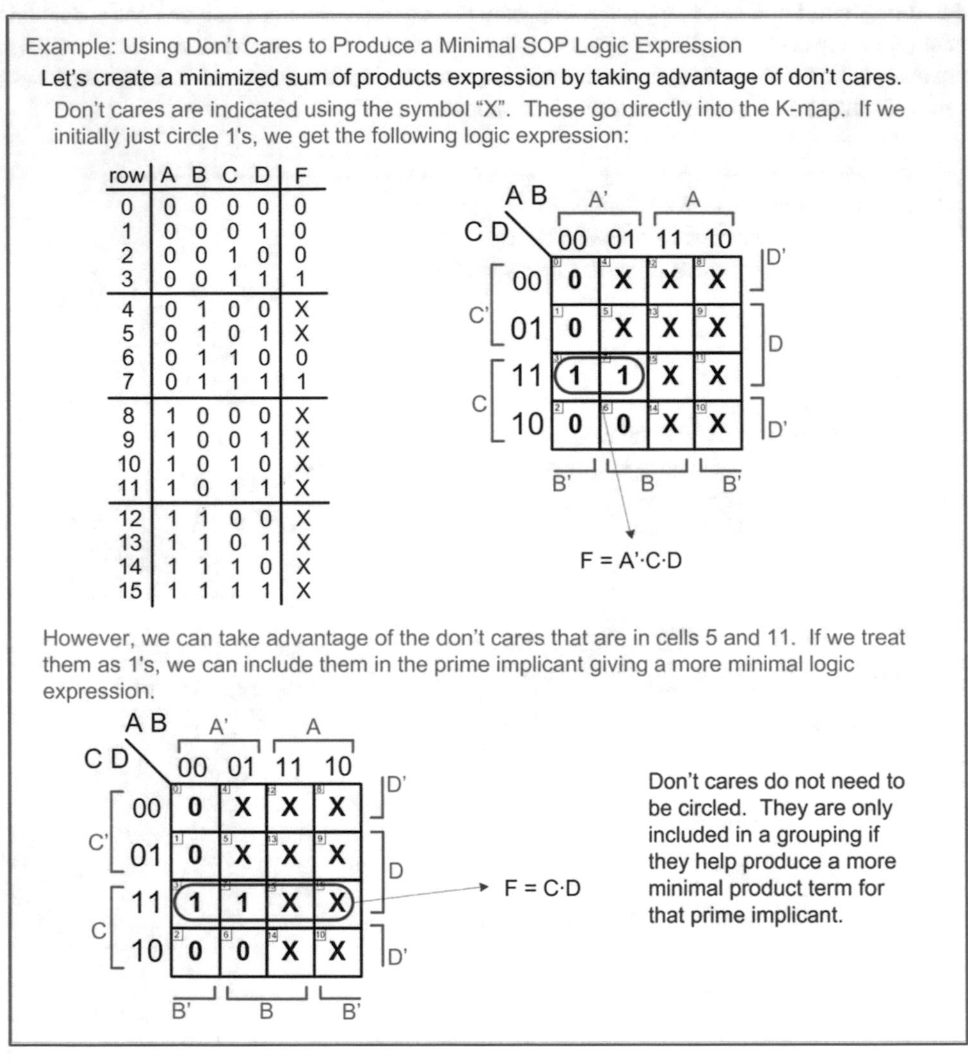

row	A	B	C	D	F
0	0	0	0	0	0
1	0	0	0	1	0
2	0	0	1	0	0
3	0	0	1	1	1
4	0	1	0	0	X
5	0	1	0	1	X
6	0	1	1	0	0
7	0	1	1	1	1
8	1	0	0	0	X
9	1	0	0	1	X
10	1	0	1	0	X
11	1	0	1	1	X
12	1	1	0	0	X
13	1	1	0	1	X
14	1	1	1	0	X
15	1	1	1	1	X

Example 4.30
Using don't cares to produce a minimal SOP logic expression

4.4.4 Using XOR Gates

While Boolean algebra does not include the exclusive-OR and exclusive-NOR operations, XOR and XNOR gates do indeed exist in modern electronics. They can be a useful tool to provide logic circuitry with less operations, sometimes even compared to a minimal sum or product synthesized using the techniques just described. An XOR/XNOR operation can be identified by putting the values from a truth table into a K-map. The XOR/XNOR operations will result in a characteristic checkerboard pattern in the K-map. Consider the following patterns for XOR and XNOR gates in Figs. 4.21, 4.22, 4.23, and 4.24. Anytime these patterns are observed, it indicates an XOR/XNOR gate.

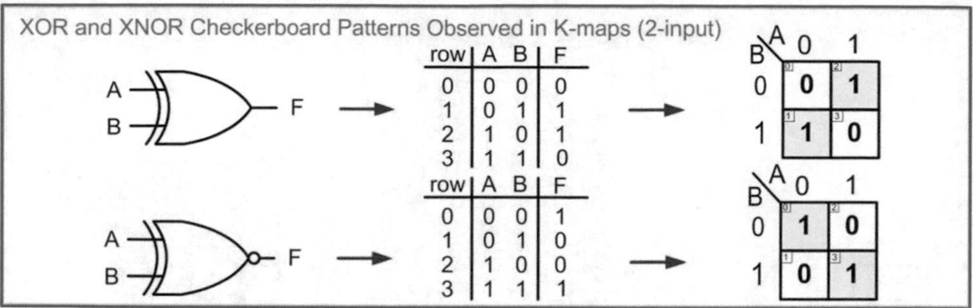

Fig. 4.21
XOR and XNOR checkerboard patterns observed in K-maps (2-input)

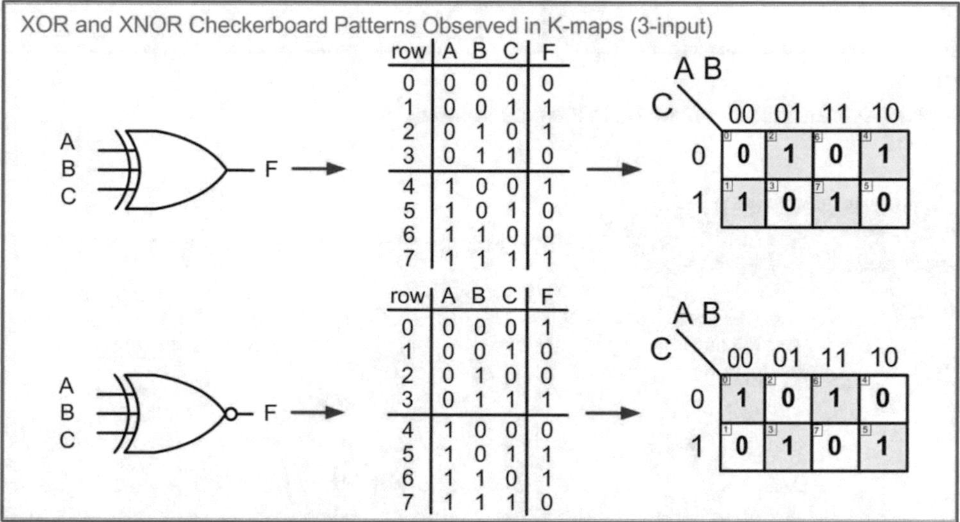

Fig. 4.22
XOR and XNOR checkerboard patterns observed in K-maps (3-input)

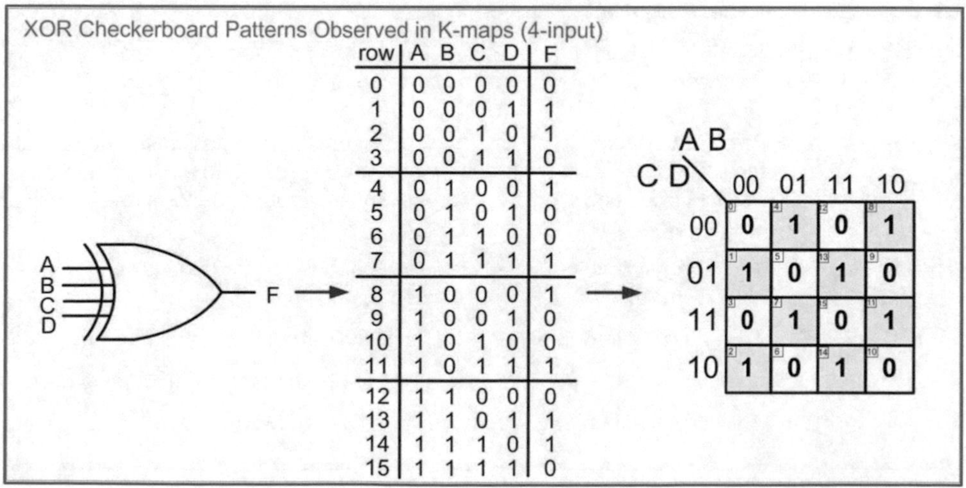

Fig. 4.23
XOR checkerboard pattern observed in K-maps (4-input)

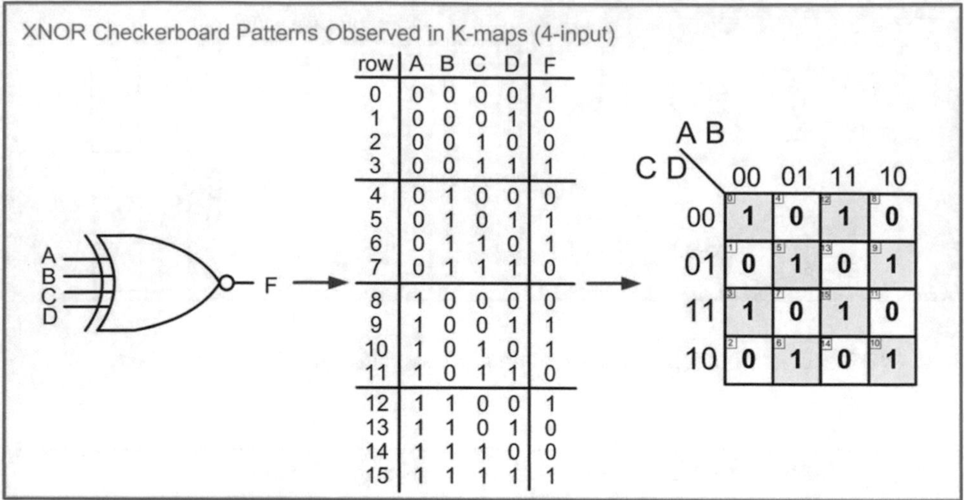

row	A	B	C	D	F
0	0	0	0	0	1
1	0	0	0	1	0
2	0	0	1	0	0
3	0	0	1	1	1
4	0	1	0	0	0
5	0	1	0	1	1
6	0	1	1	0	1
7	0	1	1	1	0
8	1	0	0	0	0
9	1	0	0	1	1
10	1	0	1	0	1
11	1	0	1	1	0
12	1	1	0	0	1
13	1	1	0	1	0
14	1	1	1	0	0
15	1	1	1	1	1

Fig. 4.24
XNOR checkerboard pattern observed in K-maps (4-input)

Concept Check

CC4.4(a) Logic minimization is accomplished by removing variables from the original canonical logic expression that don't impact the result. How does a Karnaugh map graphically show what variables can be removed?

A) K-maps contain the same information as a truth table but the data is formatted as a grid. This allows variables to be removed by inspection.

B) K-maps rearrange a truth table so that adjacent cells have one and only one input variable changing at a time. If adjacent cells have the same output value when an input variable is both a 0 and a 1, that variable has no impact on the interim result and can be eliminated.

C) K-maps list both the rows with outputs of 1's and 0's simultaneously. This allows minimization to occur for a SOP and POS topology that each has the same, but minimal, number of gates.

D) K-maps display the truth table information in a grid format, which is a more compact way of presenting the behavior of a circuit.

CC4.4(b) A "Don't Care" can be used to minimize a logic expression by assigning the output of a row to either a 1 or a 0 in order to form larger groupings within a K-map. How does the output of the circuit behave when it processes the input code for a row containing a don't care?

A) The output will be whatever value was needed to form the largest grouping in the K-map.

B) The output will go to either a 0 or a 1, but the final value is random.

C) The output can toggle between a 0 and a 1 when this input code is present.

D) The output will be driven to exactly halfway between a 0 and a 1.

4.5 Timing Hazards and Glitches

Timing hazards, or glitches, refer to unwanted transitions on the output of a combinational logic circuit. These are most commonly due to different delay paths through the gates in the circuit. In real circuitry there is always a finite propagation delay through each gate. Consider the circuit shown in Fig. 4.25 where gate delays are included and how they can produce unwanted transitions.

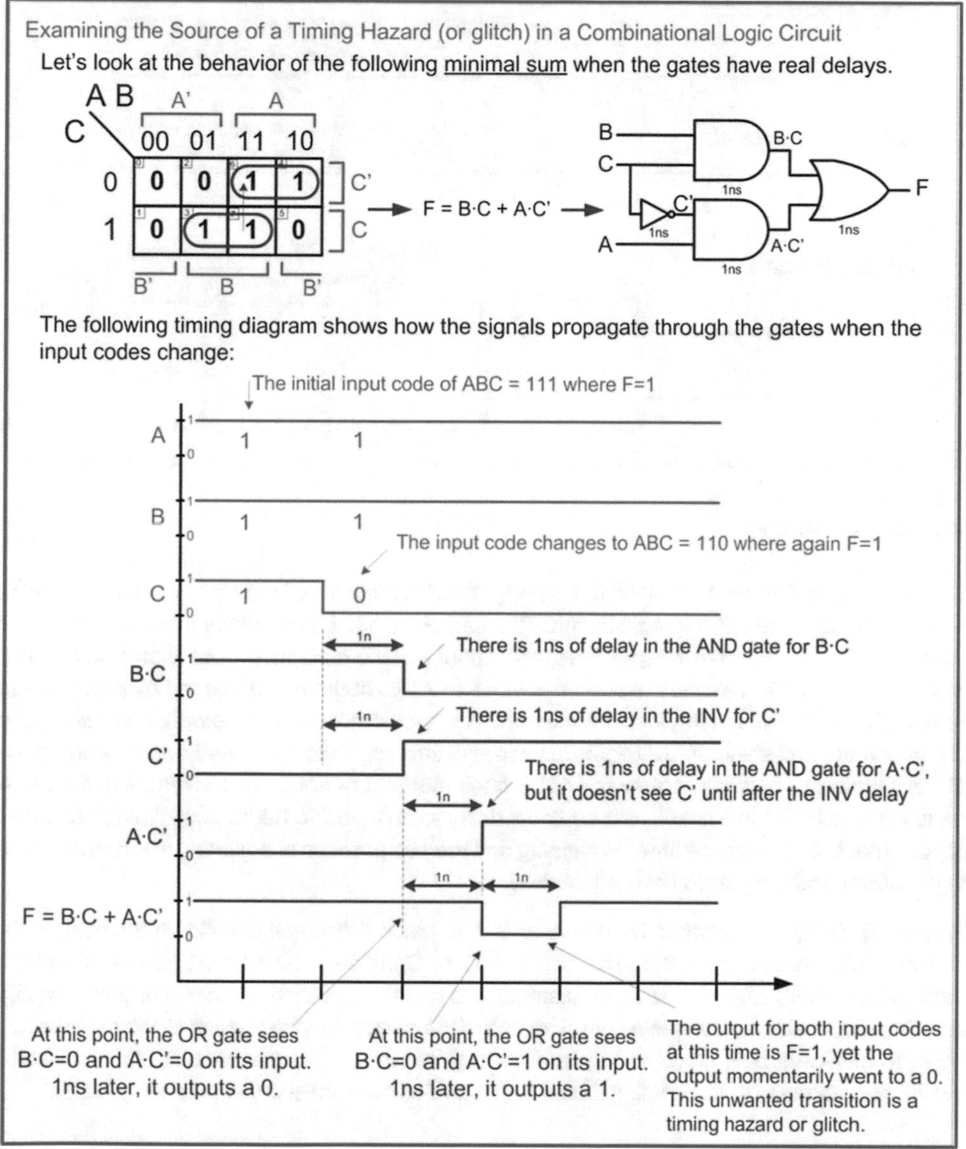

Fig. 4.25
Examining the source of a timing hazard (or glitch) in a combinational logic circuit

These timing hazards are given unique names based on the type of transition that occurs. A **static 0** timing hazard is when the input switches between two input codes that both yield an output of 0 but the output momentarily switches to a 1. A **static 1** timing hazard is when the input switches between two input codes that both yield an output of 1 but the output momentarily switches to a 0. A **dynamic hazard** is when the input switches between two input codes that result in a real transition on the output (i.e., 0 to 1 or 1 to 0), but the output has a momentary glitch before reaching its final value. These definitions are shown in Fig. 4.26.

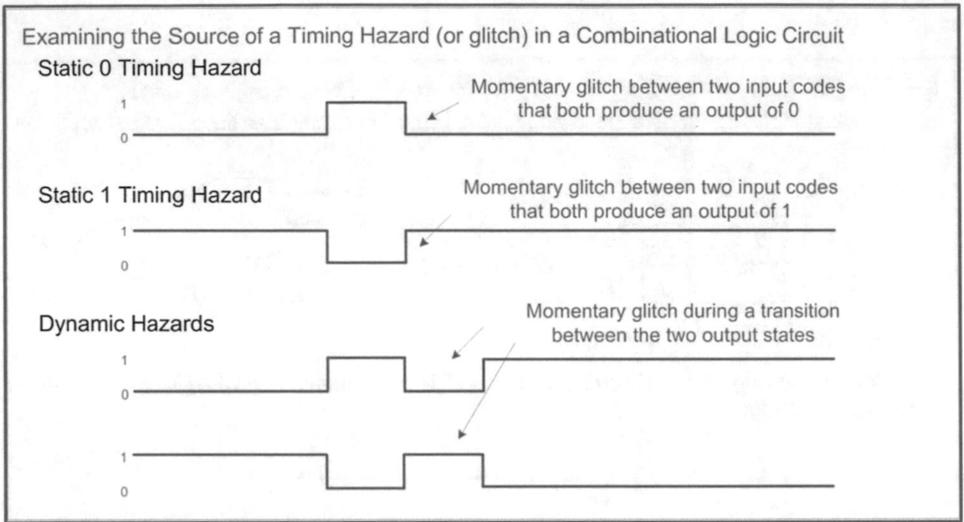

Fig. 4.26
Timing hazard definitions

Timing hazards can be addressed in a variety of ways. One way is to try to match the propagation delays through each path of the logic circuit. This can be difficult, particularly in modern logic families such as CMOS. In the example in Fig. 4.25, the root cause of the different propagation delays was due to an inverter on one of the variables. It seems obvious that this could be addressed by putting buffers on the other inputs with equal delays as the inverter. This would create a situation where all input codes would arrive at the first stage of AND gates at the same time regardless of whether they were inverted or not and eliminate the hazards; however, CMOS implements a buffer as two inverters in series, so it is difficult to insert a buffer in a circuit with an equal delay to an inverter. Addressing timing hazards in this way is possible, but it involves a time-consuming and tedious process of adjusting the transistors used to create the buffer and inverter to have equal delays.

Another technique to address timing hazards is to place additional circuitry in the system that will ensure the correct output while the input codes switch. Consider how including a nonessential prime implicant can eliminate a timing hazard in Example 4.31. In this approach, the minimal sum from Fig. 4.25 is instead replaced with the complete sum. The use of the complete sum instead of the minimal sum can be shown to eliminate both static and dynamic timing hazards. The drawback of this approach is the addition of extra circuitry in the combinational logic circuit (i.e., nonessential prime implicants).

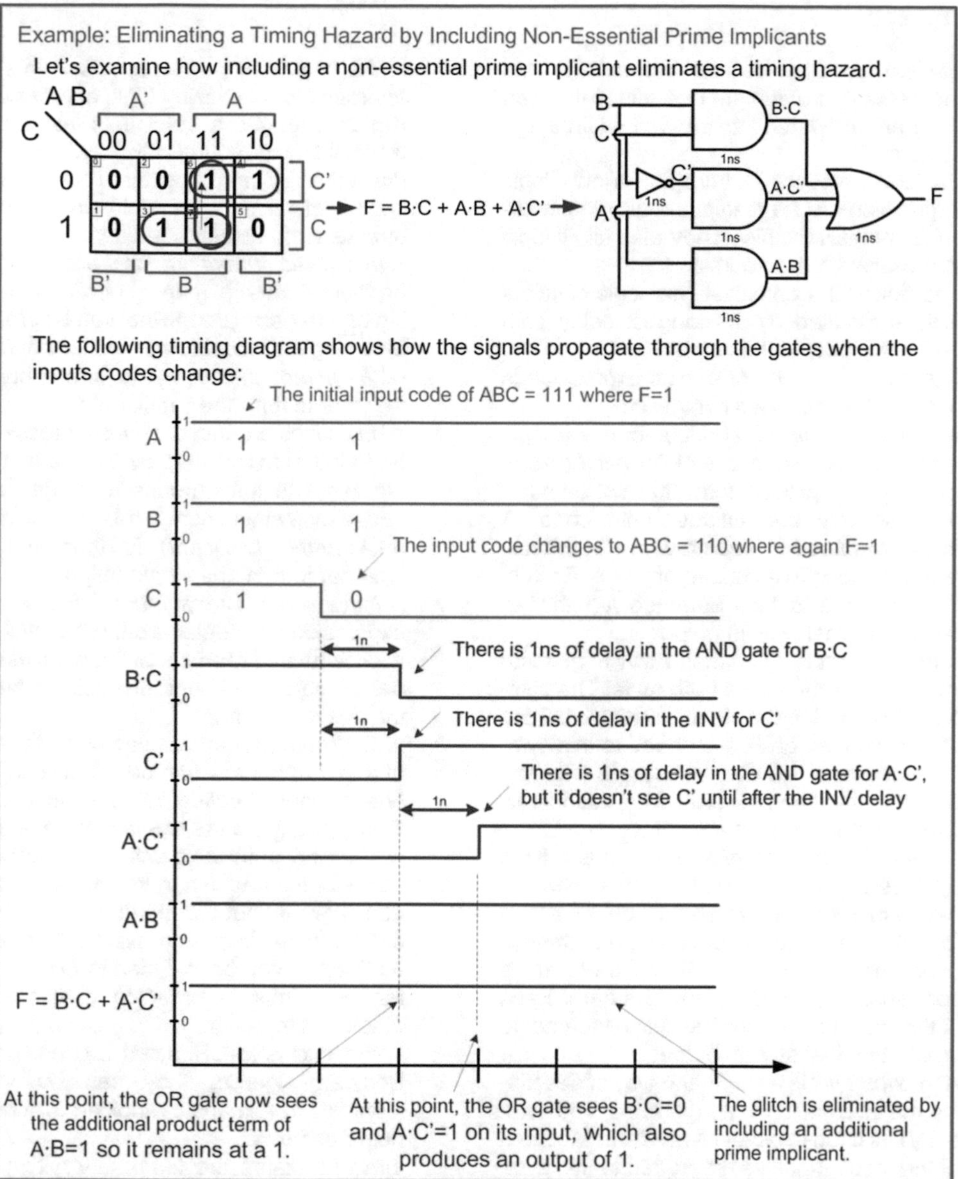

Example 4.31
Eliminating a timing hazard by including nonessential product terms

Concept Check

CC4.5 How long do you need to wait for all hazards to settle out?

A) The time equal to the delay through the nonessential prime implicants.

B) The time equal to the delay through the essential prime implicants.

C) The time equal to the shortest delay path in the circuit.

D) The time equal to the longest delay path in the circuit.

Summary

- Boolean algebra defines the axioms and theorems that guide the operations that can be performed on a two-valued number system.
- Boolean algebra theorems allow logic expressions to be manipulated to make circuit synthesis simpler. They also allow logic expressions to be minimized.
- The delay of a combinational logic circuit is always dictated by the longest delay path from the inputs to the output.
- The *canonical form* of a logic expression is one that has not been minimized.
- A *canonical sum of products* form is a logic synthesis technique based on *minterms*. A minterm is a product term that will output a one for only one unique input code. A minterm is used for each row of a truth table corresponding to an output of a one. Each of the minterms is then summed together to create the final system output.
- A *minterm list* is a shorthand way of describing the information in a truth table. The symbol "Σ" is used to denote a minterm list. Each of the input variables are added to this symbol as comma-delimited subscripts. The row number is then listed for each row corresponding to an output of a one.
- A *canonical product of sums* form is a logic synthesis technique based on *maxterms*. A maxterm is a sum term that will output a zero for only one unique input code. A maxterm is used for each row of a truth table corresponding to an output of a zero. Each of the maxterms is then multiplied together to create the final system output.
- A *maxterm list* is a shorthand way of describing the information in a truth table. The symbol "Π" is used to denote a maxterm list. Each of the input variables are added to this symbol as comma-delimited subscripts. The row number is then listed for each row corresponding to an output of a zero.
- Canonical logic expressions can be minimized through a repetitive process of factoring common variables using the *distributive* property and then eliminating remaining variables using a combination of the *complements* and *identity* theorems.
- A *Karnaugh map* (K-map) is a graphical approach to minimizing logic expressions. A K-map arranges a truth table into a grid in which the neighboring cells have input codes that differ by only one bit. This allows the impact of an input variable on a group of outputs to be quickly identified.
- A *minimized sum of products* expression can be found from a K-map by circling neighboring ones to form groups that can be produced by a single product term. Each product term (aka *prime implicant*) is then summed together to form the circuit output.
- A *minimized product of sums* expression can be found from a K-map by circling neighboring zeros to form groups that can be produced by a single sum term. Each sum term (aka *prime implicant*) is then multiplied together to form the circuit output.
- A *minimal sum* or *minimal product* is a logic expression that contains only essential prime implicants and represents the smallest number of logic operations possible to produce the desired output.
- A *don't care* (X) can be used when the output of a truth table row can be either a zero or a one without affecting the system behavior. This typically occurs when some of the input codes of a truth table will never occur. The value for the row of a truth table containing a don't care output can be chosen to give the most minimal logic expression. In a K-map, don't cares can be included to form the largest groupings in order to give the least amount of logic.
- While exclusive-OR gates are not used in Boolean algebra, they can be visually identified in K-maps by looking for checkerboard patterns.
- Timing hazards are temporary glitches that occur on the output of a combinational logic circuit due to timing mismatches through different paths in the circuit. Hazards can be minimized by including additional circuitry in the system or by matching the delay of all signal paths.

Exercise Problems

Section 4.1: Boolean Algebra

4.1.1 Which Boolean algebra theorem describes the situation where *any variable OR'd with itself will yield itself*?

4.1.2 Which Boolean algebra theorem describes the situation where *any variable that is double complemented will yield itself*?

4.1.3 Which Boolean algebra theorem describes the situation where *any variable OR'd with a 1 will yield a 1*?

4.1.4 Which Boolean algebra theorem describes the situation where a variable that exists in multiple product terms can be factored out?

4.1.5 Which Boolean algebra theorem describes the situation where when output(s) corresponding to a term within an expression are handled by another term the original term can be removed?

4.1.6 Which Boolean algebra theorem describes the situation where *any variable AND'd with its complement will yield a 0*?

4.1.7 Which Boolean algebra theorem describes the situation where *any variable AND'd with a 0 will yield a 0*?

4.1.8 Which Boolean algebra theorem describes the situation where an AND gate with its inputs inverted is equivalent to an OR gate with its outputs inverted?

4.1.9 Which Boolean algebra theorem describes the situation where *a variable that exists in multiple sum terms can be factored out*?

4.1.10 Which Boolean algebra theorem describes the situation where an OR gate with its inputs inverted is equivalent to an AND gate with its outputs inverted?

4.1.11 Which Boolean algebra theorem describes the situation where the grouping of variables in an OR operation does not affect the result?

4.1.12 Which Boolean algebra theorem describes the situation where *any variable AND'd with itself will yield itself*?

4.1.13 Which Boolean algebra theorem describes the situation where the order of variables in an OR operation does not affect the result?

4.1.14 Which Boolean algebra theorem describes the situation where *any variable AND'd with a 1 will yield itself*?

4.1.15 Which Boolean algebra theorem describes the situation where the grouping of variables in an AND operation does not affect the result?

4.1.16 Which Boolean algebra theorem describes the situation where *any variable OR'd with its complement will yield a 1*?

4.1.17 Which Boolean algebra theorem describes the situation where the order of variables in an AND operation does not affect the result?

4.1.18 Which Boolean algebra theorem describes the situation where *a variable OR'd with a 0 will yield itself*?

4.1.19 Use proof by exhaustion to prove that an OR gate with its inputs inverted is equivalent to an AND gate with its outputs inverted.

4.1.20 Use proof by exhaustion to prove that an AND gate with its inputs inverted is equivalent to an OR gate with its outputs inverted.

Section 4.2: Combinational Logic Analysis

4.2.1 For the ogic diagram given in Fig. 4.27, give the logic expression for the output F.

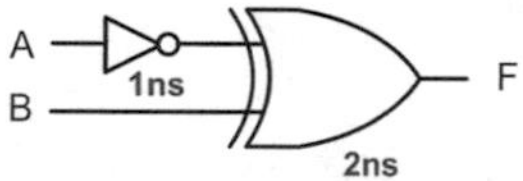

Fig. 4.27
Combinational logic analysis 1

4.2.2 For the logic diagram given in Fig. 4.27, give the truth table for the output F.

4.2.3 For the logic diagram given in Fig. 4.27, give the delay.

4.2.4 For the logic diagram given in Fig. 4.28, give the logic expression for the output F.

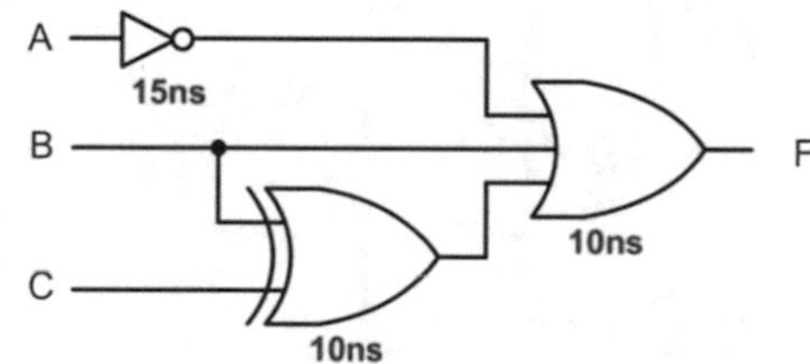

Fig. 4.28
Combinational logic analysis 2

4.2.5 For the logic diagram given in Fig. 4.28, give the truth table for the output F.

4.2.6 For the logic diagram given in Fig. 4.28, give the delay.

4.2.7 For the logic diagram given in Fig. 4.29, give the logic expression for the output F.

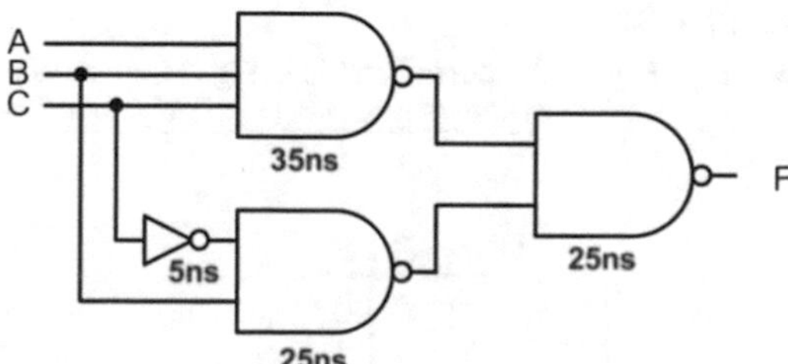

Fig. 4.29
Combinational logic analysis 3

4.2.8 For the logic diagram given in Fig. 4.29, give the truth table for the output F.

4.2.9 For the logic diagram given in Fig. 4.29, give the delay.

4.2.10 For the logic diagram given in Fig. 4.30, give the logic expression for the output F.

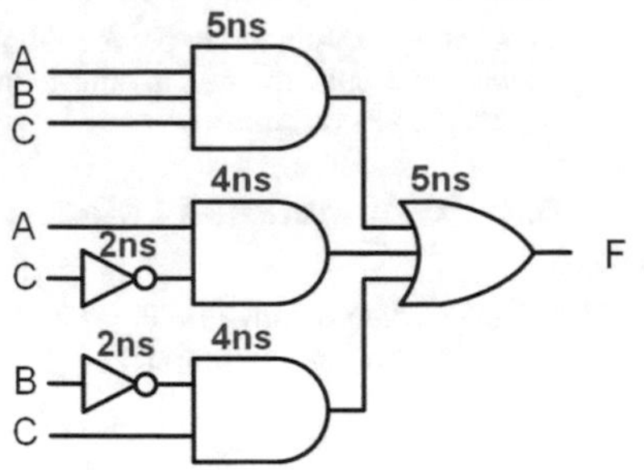

Fig. 4.30
Combinational logic analysis 4

4.2.11 For the logic diagram given in Fig. 4.30, give the truth table for the output F.

4.2.12 For the logic diagram given in Fig. 4.30, give the delay.

4.2.13 For the logic diagram given in Fig. 4.31, give the logic expression for the output F.

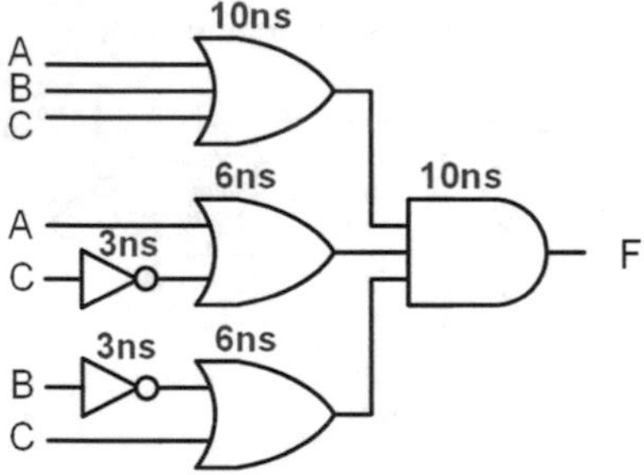

Fig. 4.31
Combinational logic analysis 5

4.2.14 For the logic diagram given in Fig. 4.31, give the truth table for the output F.

4.2.15 For the logic diagram given in Fig. 4.31, give the delay.

Section 4.3: Combinational Logic Synthesis

4.3.1 For the 2-input truth table in Fig. 4.32, give the canonical sum of products (SOP) logic expression.

A	B	F
0	0	0
0	1	1
1	0	0
1	1	1

Fig. 4.32
Combinational logic synthesis 1

4.3.2 For the 2-input truth table in Fig. 4.32, give the canonical sum of products (SOP) logic diagram.

4.3.3 For the 2-input truth table in Fig. 4.32, give the minterm list.

4.3.4 For the 2-input truth table in Fig. 4.32, give the canonical product of sums (POS) logic expression.

4.3.5 For the 2-input truth table in Fig. 4.32, give the canonical product of sums (POS) logic diagram.

4.3.6 For the 2-input truth table in Fig. 4.32, give the maxterm list.

4.3.7 For the 2-input minterm list in Fig. 4.33, give the canonical sum of products (SOP) logic expression.

$$F = \sum_{A,B}(1,2,3)$$

Fig. 4.33
Combinational logic synthesis 2

4.3.8 For the 2-input minterm list in Fig. 4.33, give the canonical sum of products (SOP) logic diagram.

4.3.9 For the 2-input minterm list in Fig. 4.33, give the truth table.

4.3.10 For the 2-input minterm list in Fig. 4.33, give the canonical product of sums (POS) logic expression.

4.3.11 For the 2-input minterm list in Fig. 4.33, give the canonical product of sums (POS) logic diagram.

4.3.12 For the 2-input minterm list in Fig. 4.33, give the maxterm list.

4.3.13 For the 2-input maxterm list in Fig. 4.34, give the canonical sum of products (SOP) logic expression.

$$F = \prod_{A,B}(1,2,3)$$

Fig. 4.34
Combinational logic synthesis 3

4.3.14 For the 2-input maxterm list in Fig. 4.34, give the canonical sum of products (SOP) logic diagram.

4.3.15 For the 2-input maxterm list in Fig. 4.34, give the minterm list.

4.3.16 For the 2-input maxterm list in Fig. 4.34, give the canonical product of sums (POS) logic expression.

4.3.17 For the 2-input maxterm list in Fig. 4.34, give the canonical product of sums (POS) logic diagram.

4.3.18 For the 2-input maxterm list in Fig. 4.34, give the truth table.

4.3.19 For the 2-input truth table in Fig. 4.35, give the canonical sum of products (SOP) logic expression.

A	B	F
0	0	1
0	1	1
1	0	0
1	1	0

Fig. 4.35
Combinational logic synthesis 4

4.3.20 For the 2-input truth table in Fig. 4.35, give the canonical sum of products (SOP) logic diagram.

4.3.21 For the 2-input truth table in Fig. 4.35, give the minterm list.

4.3.22 For the 2-input truth table in Fig. 4.35, give the canonical product of sums (POS) logic expression.

4.3.23 For the 2-input truth table in Fig. 4.35, give the canonical product of sums (POS) logic diagram.

4.3.24 For the 2-input truth table in Fig. 4.35, give the maxterm list.

4.3.25 For the 3-input truth table in Fig. 4.36, give the canonical sum of products (SOP) logic expression.

A	B	C	F
0	0	0	0
0	0	1	1
0	1	0	0
0	1	1	1
1	0	0	1
1	0	1	0
1	1	0	1
1	1	1	0

Fig. 4.36
Combinational logic synthesis 5

4.3.26 For the 3-input truth table in Fig. 4.36, give the canonical sum of products (SOP) logic diagram.

4.3.27 For the 3-input truth table in Fig. 4.36, give the minterm list.

4.3.28 For the 3-input truth table in Fig. 4.36, give the canonical product of sums (POS) logic expression.

4.3.29 For the 3-input truth table in Fig. 4.36, give the canonical product of sums (POS) logic diagram.

4.3.30 For the 3-input truth table in Fig. 4.36, give the maxterm list.

4.3.31 For the 3-input minterm list in Fig. 4.37, give the canonical sum of products (SOP) logic expression.

$$F = \sum_{A,B,C}(2,4,6)$$

Fig. 4.37
Combinational logic synthesis 6

4.3.32 For the 3-input minterm list in Fig. 4.37, give the canonical sum of products (SOP) logic diagram.

4.3.33 For the 3-input minterm list in Fig. 4.37, give the truth table.

4.3.34 For the 3-input minterm list in Fig. 4.37, give the canonical product of sums (POS) logic expression.

4.3.35 For the 3-input minterm list in Fig. 4.37, give the canonical product of sums (POS) logic diagram.

4.3.36 For the 3-input minterm list in Fig. 4.37, give the maxterm list.

4.3.37 For the 3-input maxterm list in Fig. 4.38, give the canonical sum of products (SOP) logic expression.

$$F = \prod_{A,B,C}(2,3,5,6,7)$$

Fig. 4.38
Combinational logic synthesis 7

4.3.38 For the 3-input maxterm list in Fig. 4.38, give the canonical sum of products (SOP) logic diagram.

4.3.39 For the 3-input maxterm list in Fig. 4.38, give the minterm list.

4.3.40 For the 3-input maxterm list in Fig. 4.38, give the canonical product of sums (POS) logic expression.

4.3.41 For the 3-input maxterm list in Fig. 4.38, give the canonical product of sums (POS) logic diagram.

4.3.42 For the 3-input maxterm list in Fig. 4.38, give the truth table.

4.3.43 For the 3-input truth table in Fig. 4.39, give the canonical sum of products (SOP) logic expression.

A	B	C	F
0	0	0	1
0	0	1	1
0	1	0	0
0	1	1	0
1	0	0	1
1	0	1	1
1	1	0	0
1	1	1	1

Fig. 4.39
Combinational logic synthesis 8

4.3.44 For the 3-input truth table in Fig. 4.39, give the canonical sum of products (SOP) logic diagram.

4.3.45 For the 3-input truth table in Fig. 4.39, give the minterm list.

4.3.46 For the 3-input truth table in Fig. 4.39, give the canonical product of sums (POS) logic expression.

4.3.47 For the 3-input truth table in Fig. 4.39, give the canonical product of sums (POS) logic diagram.

4.3.48 For the 3-input truth table in Fig. 4.39, give the maxterm list.

4.3.49 For the 4-input truth table in Fig. 4.40, give the canonical sum of products (SOP) logic expression.

A	B	C	D	F
0	0	0	0	0
0	0	0	1	1
0	0	1	0	0
0	0	1	1	1
0	1	0	0	0
0	1	0	1	0
0	1	1	0	0
0	1	1	1	0
1	0	0	0	0
1	0	0	1	1
1	0	1	0	0
1	0	1	1	1
1	1	0	0	0
1	1	0	1	0
1	1	1	0	0
1	1	1	1	0

Fig. 4.40
Combinational logic synthesis 9

4.3.50 For the 4-input truth table in Fig. 4.40, give the canonical sum of products (SOP) logic diagram.

4.3.51 For the 4-input truth table in Fig. 4.40, give the minterm list.

4.3.52 For the 4-input truth table in Fig. 4.40, give the canonical product of sums (POS) logic expression.

4.3.53 For the 4-input truth table in Fig. 4.40, give the canonical product of sums (POS) logic diagram.

4.3.54 For the 4-input truth table in Fig. 4.40, give the maxterm list.

4.3.55 For the 4-input minterm list in Fig. 4.41, give the canonical sum of products (SOP) logic expression.

$$F = \sum_{A,B,C,D}(4,5,7,12,13,15)$$

Fig. 4.41
Combinational logic synthesis 10

4.3.56 For the 4-input minterm list in Fig. 4.41, give the canonical sum of products (SOP) logic diagram.

4.3.57 For the 4-input minterm list in Fig. 4.41, give the truth table.

4.3.58 For the 4-input minterm list in Fig. 4.41, give the canonical product of sums (POS) logic expression.

4.3.59 For the 4-input minterm list in Fig. 4.41, give the canonical product of sums (POS) logic diagram.

4.3.60 For the 4-input minterm list in Fig. 4.41, give the maxterm list.

4.3.61 For the 4-input maxterm list in Fig. 4.42, give the canonical sum of products (SOP) logic expression.

$$F = \prod_{A,B,C,D}(3,7,11,15)$$

Fig. 4.42
Combinational logic synthesis 11

4.3.62 For the 4-input maxterm list in Fig. 4.42, give the canonical sum of products (SOP) logic diagram.

4.3.63 For the 4-input maxterm list in Fig. 4.42, give the minterm list.

4.3.64 For the 4-input maxterm list in Fig. 4.42, give the canonical product of sums (POS) logic expression.

4.3.65 For the 4-input maxterm list in Fig. 4.42, give the canonical product of sums (POS) logic diagram.

4.3.66 For the 4-input maxterm list in Fig. 4.42, give the truth table.

4.3.67 For the 4-input truth table in Fig. 4.43, give the canonical sum of products (SOP) logic expression.

A	B	C	D	F
0	0	0	0	1
0	0	0	1	0
0	0	1	0	1
0	0	1	1	1
0	1	0	0	1
0	1	0	1	1
0	1	1	0	1
0	1	1	1	1
1	0	0	0	1
1	0	0	1	0
1	0	1	0	1
1	0	1	1	0
1	1	0	0	1
1	1	0	1	0
1	1	1	0	1
1	1	1	1	1

Fig. 4.43
Combinational logic synthesis 12

4.3.68 For the 4-input truth table in Fig. 4.43, give the canonical sum of products (SOP) logic diagram.

4.3.69 For the 4-input truth table in Fig. 4.43, give the minterm list.

4.3.70 For the 4-input truth table in Fig. 4.43, give the canonical product of sums (POS) logic expression.

4.3.71 For the 4-input truth table in Fig. 4.43, give the canonical product of sums (POS) logic diagram.

4.3.72 For the 4-input truth table in Fig. 4.43, give the maxterm list.

Section 4.4: Logic Minimization

4.4.1 For the 2-input truth table in Fig. 4.44, use a K-map to derive a minimized sum of products (SOP) logic expression.

A	B	F
0	0	0
0	1	1
1	0	0
1	1	1

Fig. 4.44
Logic minimization 1

4.4.2 For the 2-input truth table in Fig. 4.44, use a K-map to derive a minimized product of sums (POS) logic expression.

4.4.3 For the 2-input truth table in Fig. 4.45, use a K-map to derive a minimized sum of products (SOP) logic expression.

A	B	F
0	0	0
0	1	1
1	0	1
1	1	1

Fig. 4.45
Logic minimization 2

4.4.4 For the 2-input truth table in Fig. 4.45, use a K-map to derive a minimized product of sums (POS) logic expression.

4.4.5 For the 2-input truth table in Fig. 4.46, use a K-map to derive a minimized sum of products (SOP) logic expression.

A	B	F
0	0	1
0	1	0
1	0	0
1	1	0

Fig. 4.46
Logic minimization 3

4.4.6 For the 2-input truth table in Fig. 4.46, use a K-map to derive a minimized product of sums (POS) logic expression.

4.4.7 For the 2-input truth table in Fig. 4.47, use a K-map to derive a minimized sum of products (SOP) logic expression.

A	B	F
0	0	1
0	1	1
1	0	0
1	1	0

Fig. 4.47
Logic minimization 4

4.4.8 For the 2-input truth table in Fig. 4.47, use a K-map to derive a minimized product of sums (POS) logic expression.

4.4.9 For the 3-input truth table in Fig. 4.48, use a K-map to derive a minimized sum of products (SOP) logic expression.

A	B	C	F
0	0	0	0
0	0	1	1
0	1	0	0
0	1	1	1
1	0	0	1
1	0	1	0
1	1	0	1
1	1	1	0

Fig. 4.48
Logic minimization 5

4.4.10 For the 3-input truth table in Fig. 4.48, use a K-map to derive a minimized product of sums (POS) logic expression.

4.4.11 For the 3-input truth table in Fig. 4.49, use a K-map to derive a minimized sum of products (SOP) logic expression.

A	B	C	F
0	0	0	0
0	0	1	0
0	1	0	1
0	1	1	0
1	0	0	1
1	0	1	0
1	1	0	1
1	1	1	0

Fig. 4.49
Logic minimization 6

4.4.12 For the 3-input truth table in Fig. 4.49, use a K-map to derive a minimized product of sums (POS) logic expression.

4.4.13 For the 3-input truth table in Fig. 4.50, use a K-map to derive a minimized sum of products (SOP) logic expression.

A	B	C	F
0	0	0	1
0	0	1	1
0	1	0	0
0	1	1	0
1	0	0	1
1	0	1	0
1	1	0	0
1	1	1	0

Fig. 4.50
Logic minimization 7

4.4.14 For the 3-input truth table in Fig. 4.50, use a K-map to derive a minimized product of sums (POS) logic expression.

4.4.15 For the 3-input truth table in Fig. 4.51, use a K-map to derive a minimized sum of products (SOP) logic expression.

A	B	C	F
0	0	0	1
0	0	1	1
0	1	0	0
0	1	1	0
1	0	0	1
1	0	1	1
1	1	0	0
1	1	1	1

Fig. 4.51
Logic minimization 8

4.4.16 For the 3-input truth table in Fig. 4.51, use a K-map to derive a minimized product of sums (POS) logic expression.

4.4.17 For the 4-input truth table in Fig. 4.52, use a K-map to derive a minimized sum of products (SOP) logic expression.

A	B	C	D	F
0	0	0	0	0
0	0	0	1	1
0	0	1	0	0
0	0	1	1	1
0	1	0	0	0
0	1	0	1	0
0	1	1	0	0
0	1	1	1	0
1	0	0	0	0
1	0	0	1	1
1	0	1	0	0
1	0	1	1	1
1	1	0	0	0
1	1	0	1	0
1	1	1	0	0
1	1	1	1	0

Fig. 4.52
Logic minimization 9

4.4.18 For the 4-input truth table in Fig. 4.52, use a K-map to derive a minimized product of sums (POS) logic expression.

4.4.19 For the 4-input truth table in Fig. 4.53, use a K-map to derive a minimized sum of products (SOP) logic expression.

A	B	C	D	F
0	0	0	0	0
0	0	0	1	0
0	0	1	0	0
0	0	1	1	0
0	1	0	0	1
0	1	0	1	1
0	1	1	0	0
0	1	1	1	1
1	0	0	0	0
1	0	0	1	0
1	0	1	0	0
1	0	1	1	0
1	1	0	0	1
1	1	0	1	1
1	1	1	0	0
1	1	1	1	1

Fig. 4.53
Combinational logic synthesis 10

4.4.20 For the 4-input truth table in Fig. 4.53, use a K-map to derive a minimized product of sums (POS) logic expression.

4.4.21 For the 4-input truth table in Fig. 4.54, use a K-map to derive a minimized sum of products (SOP) logic expression.

A	B	C	D	F
0	0	0	0	1
0	0	0	1	1
0	0	1	0	1
0	0	1	1	0
0	1	0	0	1
0	1	0	1	1
0	1	1	0	1
0	1	1	1	0
1	0	0	0	1
1	0	0	1	1
1	0	1	0	1
1	0	1	1	0
1	1	0	0	1
1	1	0	1	1
1	1	1	0	1
1	1	1	1	0

Fig. 4.54
Combinational logic synthesis 11

4.4.22 For the 4-input truth table in Fig. 4.54, use a K-map to derive a minimized product of sums (POS) logic expression.

4.4.23 For the 4-input truth table in Fig. 4.55, use a K-map to derive a minimized sum of products (SOP) logic expression.

A	B	C	D	F
0	0	0	0	1
0	0	0	1	0
0	0	1	0	1
0	0	1	1	1
0	1	0	0	1
0	1	0	1	1
0	1	1	0	1
0	1	1	1	1
1	0	0	0	1
1	0	0	1	0
1	0	1	0	1
1	0	1	1	0
1	1	0	0	1
1	1	0	1	0
1	1	1	0	1
1	1	1	1	1

Fig. 4.55
Combinational logic synthesis 12

4.4.24 For the 4-input truth table in Fig. 4.55, use a K-map to derive a minimized product of sums (POS) logic expression.

4.4.25 For the 3-input truth table and K-map in Fig. 4.56, provide the row number(s) of any distinguished one-cells when deriving the minimal SOP logic expression.

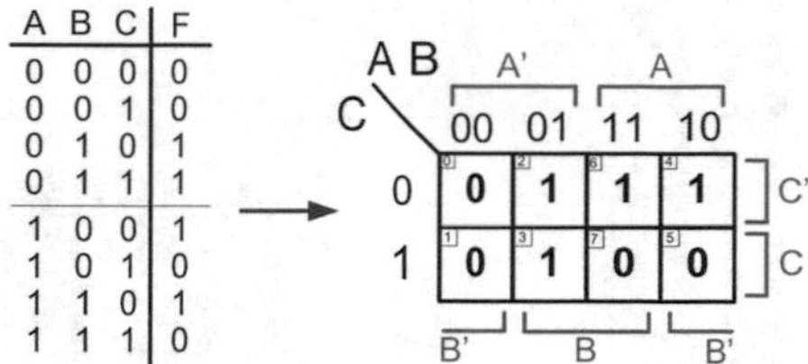

A	B	C	F
0	0	0	0
0	0	1	0
0	1	0	1
0	1	1	1
1	0	0	1
1	0	1	0
1	1	0	1
1	1	1	0

C \ AB	00	01	11	10
0	0	1	1	1
1	0	1	0	0

Fig. 4.56
Combinational logic synthesis 13

4.4.26 For the 3-input truth table and K-map in Fig. 4.56, give the product terms for the essential prime implicants when deriving the minimal SOP logic expression.

4.4.27 For the 3-input truth table and K-map in Fig. 4.56, give the minimal sum of products (SOP) logic expression.

4.4.28 For the 3-input truth table and K-map in Fig. 4.56, give the complete sum of products (SOP) logic expression.

4.4.29 For the 4-input truth table and K-map in Fig. 4.57, provide the row number(s) of any distinguished one-cells when deriving the minimal SOP logic expression.

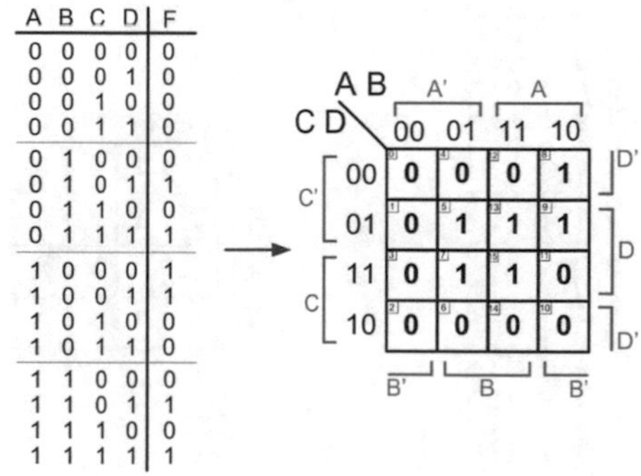

A	B	C	D	F
0	0	0	0	0
0	0	0	1	0
0	0	1	0	0
0	0	1	1	0
0	1	0	0	0
0	1	0	1	1
0	1	1	0	0
0	1	1	1	1
1	0	0	0	1
1	0	0	1	1
1	0	1	0	0
1	0	1	1	0
1	1	0	0	0
1	1	0	1	1
1	1	1	0	0
1	1	1	1	1

CD \ AB	00	01	11	10
00	0	0	0	1
01	0	1	1	1
11	0	1	1	0
10	0	0	0	0

Fig. 4.57
Combinational logic synthesis 14

4.4.30 For the 4-input truth table and K-map in Fig. 4.57, give the product terms for the essential prime implicants when deriving the minimal SOP logic expression.

4.4.31 For the 4-input truth table and K-map in Fig. 4.57, give the minimal sum of products (SOP) logic expression.

4.4.32 For the 4-input truth table and K-map in Fig. 4.57, give the complete sum of products (SOP) logic expression.

4.4.33 For the 4-input truth table and K-map in Fig. 4.58, give the minimal sum of products (SOP) logic expression by exploiting "don't cares."

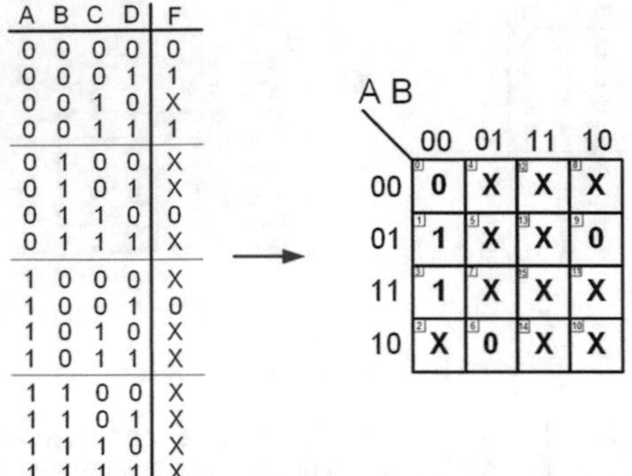

A	B	C	D	F
0	0	0	0	0
0	0	0	1	1
0	0	1	0	X
0	0	1	1	1
0	1	0	0	X
0	1	0	1	X
0	1	1	0	0
0	1	1	1	X
1	0	0	0	X
1	0	0	1	0
1	0	1	0	X
1	0	1	1	X
1	1	0	0	X
1	1	0	1	X
1	1	1	0	X
1	1	1	1	X

CD \ AB	00	01	11	10
00	0	X	X	X
01	1	X	X	0
11	1	X	X	X
10	X	0	X	X

Fig. 4.58
Combinational logic synthesis 15

4.4.34 For the 4-input truth table and K-map in Fig. 4.58, give the minimal product of sums (POS) logic expression by exploiting "don't cares."

4.4.35 For the 4-input truth table and K-map in Fig. 4.59, give the minimal sum of products (SOP) logic expression by exploiting "don't cares."

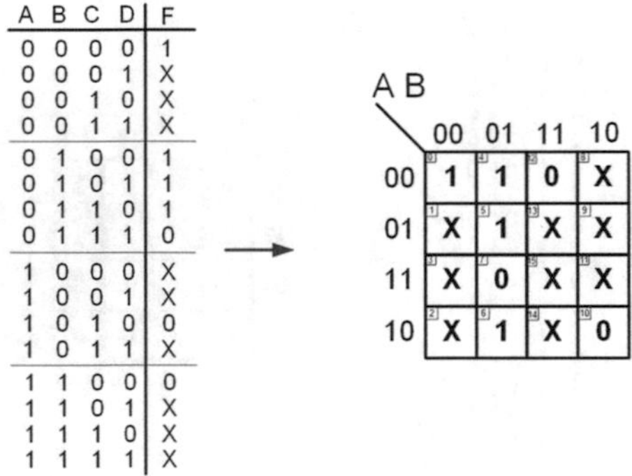

A	B	C	D	F
0	0	0	0	1
0	0	0	1	X
0	0	1	0	X
0	0	1	1	X
0	1	0	0	1
0	1	0	1	1
0	1	1	0	1
0	1	1	1	0
1	0	0	0	X
1	0	0	1	X
1	0	1	0	0
1	0	1	1	X
1	1	0	0	0
1	1	0	1	X
1	1	1	0	X
1	1	1	1	X

CD \ AB	00	01	11	10
00	1	1	0	X
01	X	1	X	X
11	X	0	X	X
10	X	1	X	0

Fig. 4.59
Combinational logic synthesis 16

4.4.36 For the 4-input truth table and K-map in Fig. 4.59, give the minimal product of sums (POS) logic expression by exploiting "don't cares."

Section 4.5: Timing Hazards and Glitches

4.5.1 Describe the situation in which a static-1 timing hazard may occur.

4.5.2 Describe the situation in which a static-0 timing hazard may occur.

4.5.3 In which topology will a static-1 timing hazard occur (SOP, POS, or both)?

4.5.4 In which topology will a static-0 timing hazard occur (SOP, POS, or both)?

4.5.5 For the 3-input truth table and K-map in Fig. 4.56, give the product term that helps eliminate static-1 timing hazards in this circuit.

4.5.6 For the 3-input truth table and K-map in Fig. 4.56, give the sum term that helps eliminate static-0 timing hazards in this circuit.

4.5.7 For the 4-input truth table and K-map in Fig. 4.57, give the product term that helps eliminate static-1 timing hazards in this circuit.

4.5.8 For the 4-input truth table and K-map in Fig. 4.57, give the sum term that helps eliminate static-0 timing hazards in this circuit.

Chapter 5: Verilog (Part 1)

Based on the material presented in Chap. 4, there are a few observations about logic design that are apparent. First, the size of logic circuitry can scale quickly to the point where it is difficult to design by hand. Second, the process of moving from a high-level description of how a circuit works (e.g., a truth table) to a form that is ready to be implemented with real circuitry (e.g., a minimized logic diagram) is straightforward and well-defined. Both of these observations motivate the use of computer-aided design (CAD) tools to accomplish logic design. This chapter introduces hardware description languages (HDLs) as a means to describe digital circuitry using a text-based language. HDLs provide a means to describe large digital systems without the need for schematics, which can become impractical in very large designs. HDLs have evolved to support logic simulation at different levels of abstraction. This provides designers the ability to begin designing and verifying functionality of large systems at a high level of abstraction and postpone the details of the circuit implementation until later in the design cycle. This enables a top-down design approach that is scalable across different logic families. HDLs have also evolved to support automated *synthesis*, which allows the CAD tools to take a functional description of a system (e.g., a truth table) and automatically create the gate-level circuitry to be implemented in real hardware. This allows designers to focus their attention on designing the behavior of a system and not spend as much time performing the formal logic synthesis steps that were presented in Chap. 4. The intent of this chapter is to introduce HDLs and their use in the modern digital design flow. This chapter will cover the basics of designing combinational logic in an HDL and also hierarchical design. The more advanced concepts of HDLs such as sequential logic design, high-level abstraction, and test benches are covered later so that the reader can get started quickly using HDLs to gain experience with the languages and design flow.

There are two dominant hardware description languages in use today. They are VHDL and Verilog. VHDL stands for *very high-speed integrated circuit hardware description language*. Verilog is not an acronym but rather a trade name. The use of these two HDLs is split nearly equally within the digital design industry. Once one language is learned, it is simple to learn the other language, so the choice of the HDL to learn first is somewhat arbitrary. In this text, we will use Verilog to learn the concepts of an HDL. Verilog is more similar to the programming language C and less strict in its type casting than VHDL. Verilog is also widely used in custom integrated circuit design so there is a great deal of documentation and examples readily available online. The goal of this chapter is to provide an understanding of the basic principles of hardware description languages.

Learning Outcomes—After completing this chapter, you will be able to:

5.1 Describe the role of hardware description languages in modern digital design.
5.2 Describe the fundamentals of design abstraction in modern digital design.
5.3 Describe the modern digital design flow based on hardware description languages.
5.4 Describe the fundamental constructs of Verilog.
5.5 Design a Verilog model for a combinational logic circuit using concurrent modeling techniques (continuous signal assignment with logical operators and continuous signal assignment with conditional operators).
5.6 Design a Verilog model for a combinational logic circuit using a structural design approach (gate-level primitives and user-defined primitives).
5.7 Describe the role of a Verilog test bench.

© Springer Nature Switzerland AG 2019
B. J. LaMeres, *Introduction to Logic Circuits & Logic Design with Verilog*,
https://doi.org/10.1007/978-3-030-13605-5_5

5.1 History of Hardware Description Languages

The invention of the integrated circuit is most commonly credited to two individuals who filed patents on different variations of the same basic concept within 6 months of each other in 1959. Jack Kilby filed the first patent on the integrated circuit in February of 1959 titled "Miniaturized Electronic Circuits" while working for *Texas Instruments*. Robert Noyce was the second to file a patent on the integrated circuit in July of 1959 titled "Semiconductor Device and Lead Structure" while at a company he cofounded called *Fairchild Semiconductor.* Kilby went on to win the Nobel Prize in Physics in 2000 for his invention, while Noyce went on to cofound *Intel Corporation* in 1968 with Gordon Moore. In 1971, Intel introduced the first single-chip microprocessor using integrated circuit technology, the *Intel 4004*. This microprocessor IC contained 2300 transistors. This series of inventions launched the semiconductor industry, which was the driving force behind the growth of Silicon Valley and led to 40 years of unprecedented advancement in technology that has impacted every aspect of the modern world.

Gordon Moore, cofounder of Intel, predicted in 1965 that the number of transistors on an integrated circuit would double every 2 years. This prediction, now known as *Moore's law*, has held true since the invention of the integrated circuit. As the number of transistors on an integrated circuit grew, so did the size of the design and the functionality that could be implemented. Once the first microprocessor was invented in 1971, the capability of CAD tools increased rapidly enabling larger designs to be accomplished. These larger designs, including newer microprocessors, enabled the CAD tools to become even more sophisticated and, in turn, yield even larger designs. The rapid expansion of electronic systems based on digital integrated circuits required that different manufacturers needed to produce designs that were compatible with each other. The adoption of logic family standards helped manufacturers ensure their parts would be compatible with other manufacturers at the physical layer (e.g., voltage and current); however, one challenge that was encountered by the industry was a way to document the complex behavior of larger systems. The use of schematics to document large digital designs became too cumbersome and difficult to understand by anyone besides the designer. Word descriptions of the behavior were easier to understand, but even this form of documentation became too voluminous to be effective for the size of designs that were emerging. Simultaneously there was a need to begin simulating the functionality of these large systems prior to fabrication to verify accuracy. Due to the complexity of these systems and the vast potential for design error, it became impractical to verify design accuracy through prototyping.

In 1983, the US Department of Defense (DoD) sponsored a program to create a means to document the behavior of digital systems that could be used across all of its suppliers. This program was motivated by a lack of adequate documentation for the functionality of application-specific integrated circuits (ASICs) that were being supplied to the DoD. This lack of documentation was becoming a critical issue as ASICs would come to the end of their life cycle and need to be replaced. With the lack of a standardized documentation approach, suppliers had difficulty reproducing equivalent parts to those that had become obsolete. The DoD contracted three companies (Texas Instruments, IBM, and Intermetrics) to develop a standardized documentation tool that provided detailed information about both the interface (i.e., inputs and outputs) and the behavior of digital systems. The new tool was to be implemented in a format similar to a programming language. Due to the nature of this type of language-based tool, it was a natural extension of the original project scope to include the ability to *simulate* the behavior of a digital system. The simulation capability was desired to span multiple levels of abstraction to provide maximum flexibility. In 1985, the first version of this tool, called VHDL, was released. In order to gain widespread adoption and ensure consistency of use across the industry, VHDL was turned over to the *Institute of Electrical and Electronic Engineers* (IEEE) for standardization. IEEE is a professional association that defines a broad range of open technology standards. In 1987, IEEE released the first industry standard version of VHDL. The release was titled IEEE 1076-1987. Feedback from the initial version resulted in a major revision of the standard in 1993 titled IEEE 1076-1993. While many minor revisions have been made to the 1993 release, the 1076-1993 standard contains the vast majority of VHDL functionality in use today. The most recent VHDL standard is IEEE 1076-2008.

Also in 1983, the Verilog HDL was developed by *Automated Integrated Design Systems* as a logic simulation language. The development of Verilog took place completely independent from the VHDL project. Automated Integrated Design Systems (renamed *Gateway Design Automation* in 1985) was acquired by CAD tool vendor *Cadence Design Systems* in 1990. In response to the popularity of Verilog's intuitive programming and superior simulation support, and also to stay competitive with the emerging VHDL standard, Cadence made the Verilog HDL open to the public. IEEE once again developed the open standard for this HDL and in 1995 released the Verilog standard titled IEEE 1364-1995. This release has undergone numerous revisions with the most significant occurring in 2001. It is common to refer to the major releases as "Verilog 1995" and Verilog 2001" instead of their official standard numbers.

The development of CAD tools to accomplish automated logic synthesis can be dated back to the 1970s when IBM began developing a series of practical synthesis engines that were used in the design of their mainframe computers; however, the main advancement in logic synthesis came with the founding of a company called *Synopsis* in 1986. Synopsis was the first company to focus on logic synthesis directly from HDLs. This was a major contribution because designers were already using HDLs to describe and simulate their digital systems, and now logic synthesis became integrated in the same design flow. Due to the complexity of synthesizing highly abstract functional descriptions, only lower levels of abstraction that were thoroughly elaborated were initially able to be synthesized. As CAD tool capability evolved, synthesis of higher levels of abstraction became possible, but even today not all functionality that can be described in an HDL can be synthesized.

The history of HDLs, their standardization, and the creation of the associated logic synthesis tools is key to understanding the use and limitations of HDLs. HDLs were originally designed for documentation and behavioral simulation. Logic synthesis tools were developed independently and modified later to work with HDLs. This history provides some background into the most common pitfalls that beginning digital designers encounter, that being that most any type of behavior can be described and simulated in an HDL, but only a subset of well-described functionality can be synthesized. Beginning digital designers are often plagued by issues related to designs that simulate perfectly but that will not synthesize correctly. In this book, an effort is made to introduce Verilog at a level that provides a reasonable amount of abstraction while preserving the ability to be synthesized. Figure 5.1 shows a timeline of some of the major technology milestones that have occurred in the past 150 years in the field of digital logic and HDLs.

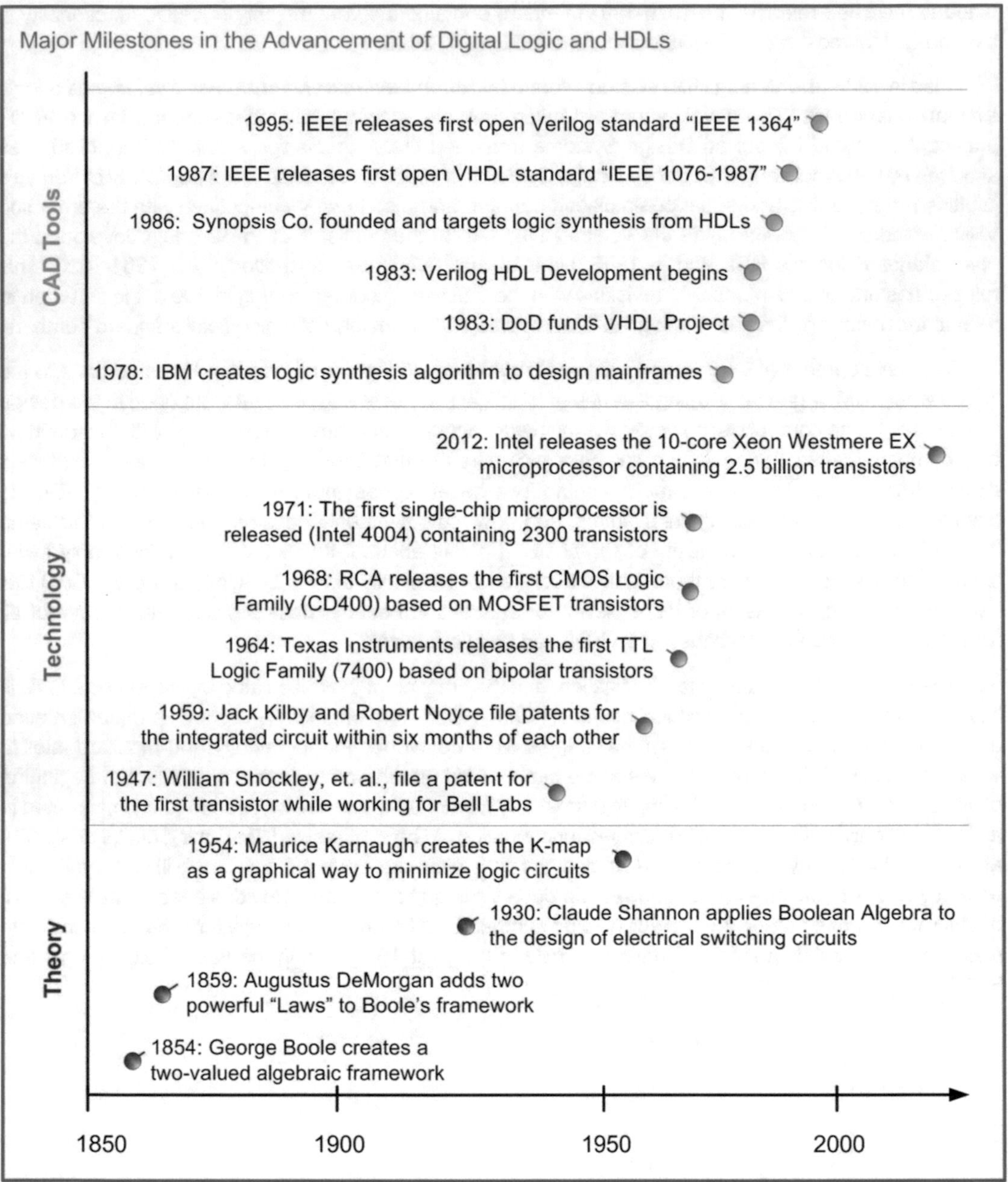

Fig. 5.1
Major milestones in the advancement of digital logic and HDLs

Concept Check

CC5.1 Why does Verilog support modeling techniques that *aren't* synthesizable?

A) There wasn't enough funding available to develop synthesis capability as it all went to the VHDL project.

B) At the time Verilog was created, synthesis was deemed too difficult to implement.

C) To allow Verilog to be used as a generic programming language.

D) Verilog needs to support all steps in the modern digital design flow, some of which are unsynthesizable such as test pattern generation and timing verification.

5.2 HDL Abstraction

HDLs were originally defined to be able to model behavior at multiple levels of abstraction. Abstraction is an important concept in engineering design because it allows us to specify how systems will operate without getting consumed prematurely with implementation details. Also, by removing the details of the lower-level implementation, simulations can be conducted in reasonable amounts of time to model the higher-level functionality. If a full computer system was simulated using detailed models for every MOSFET, it would take an impracticable amount of time to complete. Figure 5.2 shows a graphical depiction of the different layers of abstraction in digital system design.

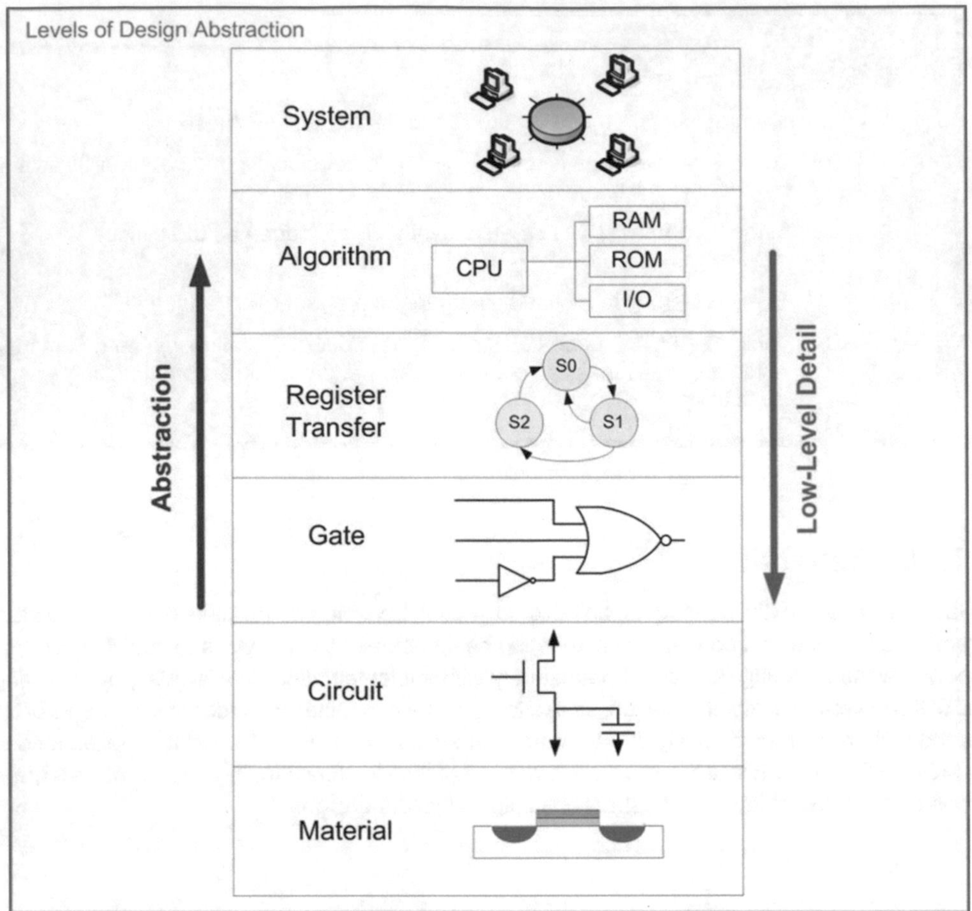

Fig. 5.2
Levels of design abstraction

The highest level of abstraction is the *system level*. At this level, behavior of a system is described by stating a set of broad specifications. An example of a design at this level is a specification such as "the computer system will perform 10 Tera Floating Point Operations per Second (10 TFLOPS) on double precision data and consume no more than 100 Watts of power." Notice that these specifications do not dictate the lower-level details such as the type of logic family or the type of computer architecture to use. One level down from the system level is the *algorithmic level*. At this level, the specifications begin to be broken down into sub-systems, each with an associated behavior that will accomplish a part of the primary task. At this level, the example computer specifications might be broken down into sub-systems such as a central processing unit (CPU) to perform the computation and random-access memory (RAM) to hold the inputs and outputs of the computation. One level down from the algorithmic level is the *register transfer level (RTL)*. At this level, the details of how data is moved between and within sub-systems are described in addition to how the data is manipulated based on system inputs. One level down from the RTL level is the *gate level*. At this level, the design is described using basic gates and registers (or storage elements). The gate level is essentially a schematic (either graphically or text-based) that contains the components and connections that will implement the functionality from the above levels of abstraction. One level down from the gate level is the *circuit level*. The circuit level describes the operation of the basic gates and registers using transistors, wires, and other electrical components such as resistors and capacitors. Finally, the lowest level of design abstraction is the *material level.* This level describes how different materials are combined and shaped in order to implement the transistors, devices, and wires from the circuit level.

HDLs are designed to model behavior at all of these levels with the exception of the material level. While there is some capability to model circuit level behavior such as MOSFETs as ideal switches and pull-up/pull-down resistors, HDLs are not typically used at the circuit level. Another graphical depiction of design abstraction is known as the **Gajski and Kuhn's Y-chart**. A Y-chart depicts abstraction across three different design domains: behavioral, structural, and physical. Each of these design domains contains levels of abstraction (i.e., system, algorithm, RTL, gate, and circuit). An example Y-chart is shown in Fig. 5.3.

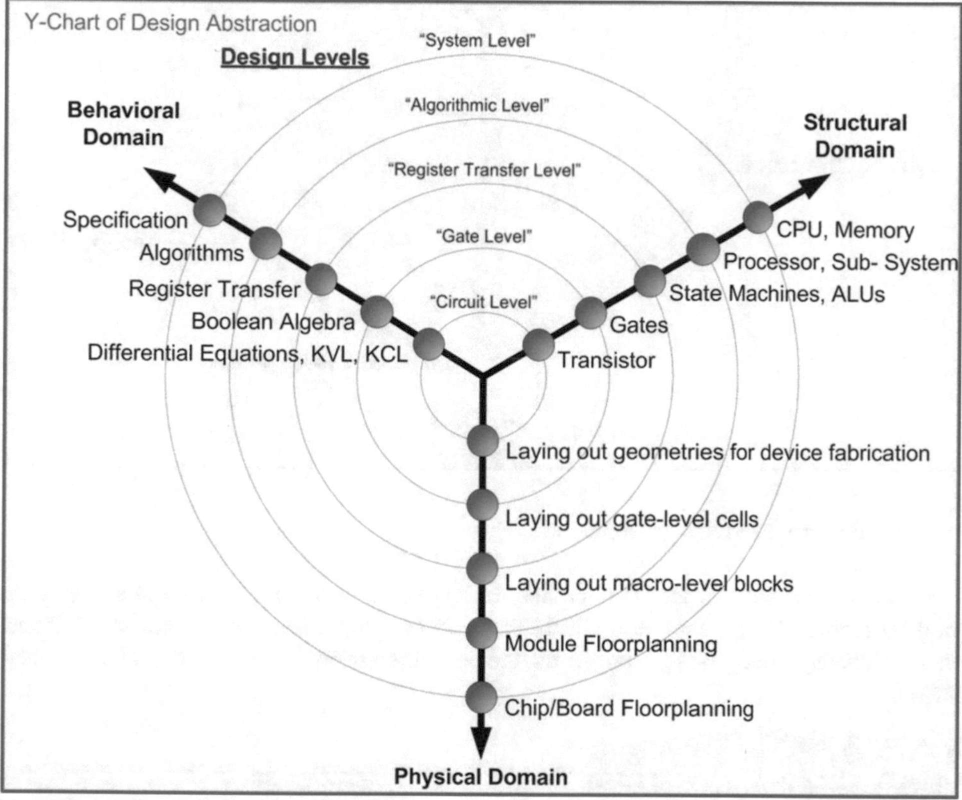

Fig. 5.3
Y-chart of design abstraction

A Y-chart also depicts how the abstraction levels of different design domains are related to each other. A top-down design flow can be visualized in a Y-chart by spiraling inward in a clockwise direction. Moving from the behavioral domain to the structural domain is the process of *synthesis*. Whenever synthesis is performed, the resulting system should be compared with the prior behavioral description. This checking is called *verification*. The process of creating the physical circuitry corresponding to the structural description is called *implementation*. The spiral continues down through the levels of abstraction until the design is implemented at a level that the geometries representing circuit elements (transistors, wires, etc.) are ready to be fabricated in silicon. Figure 5.4 shows the top-down design process depicted as an inward spiral on the Y-chart.

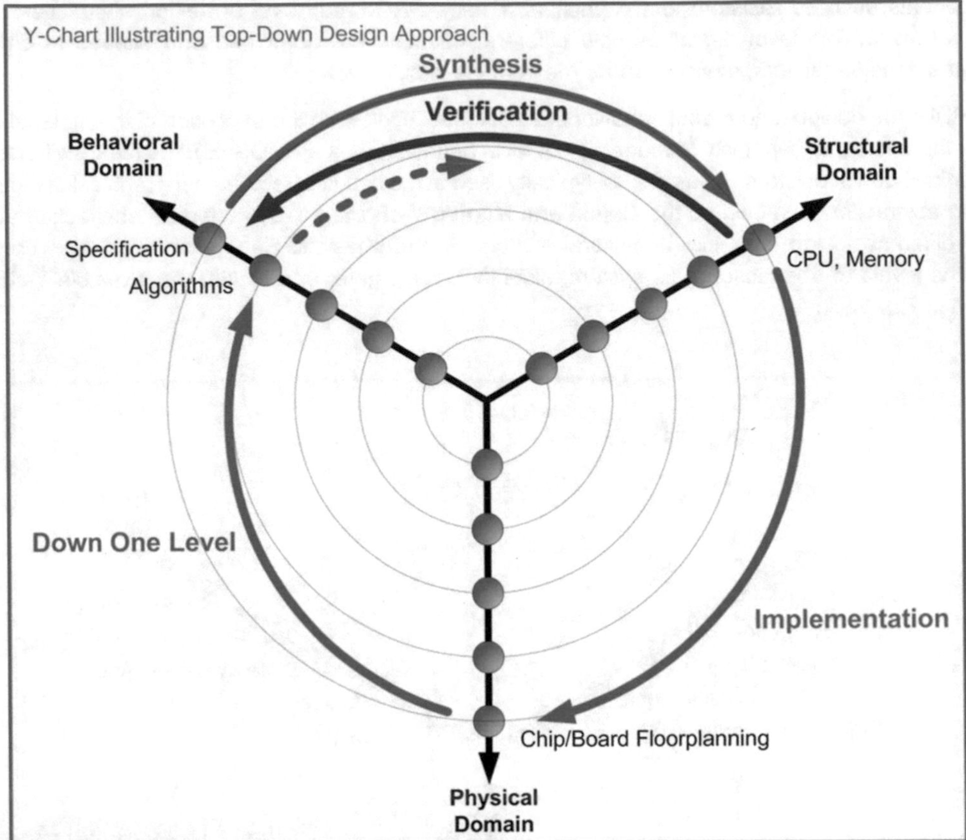

Fig. 5.4
Y-chart illustrating top-down design approach

The Y-chart represents a formal approach for large digital systems. For large systems that are designed by teams of engineers, it is critical that a formal, top-down design process is followed to eliminate potentially costly design errors as the implementation is carried out at lower levels of abstraction.

Concept Check

CC5.2 Why is abstraction an essential part of engineering design?

A) Without abstraction all schematics would be drawn at the transistor level.

B) Abstraction allows computer programs to aid in the design process.

C) Abstraction allows the details of the implementation to be hidden while the higher-level systems are designed. Without abstraction, the details of the implementation would overwhelm the designer.

D) Abstraction allows analog circuit designers to include digital blocks in their systems.

5.3 The Modern Digital Design Flow

When performing a smaller design or the design of fully contained sub-systems, the process can be broken down into individual steps. These steps are shown in Fig. 5.5. This process is given generically and applies to both *classical* and *modern* digital design. The distinction between classical and modern is that modern digital design uses HDLs and automated CAD tools for simulation, synthesis, place and route, and verification.

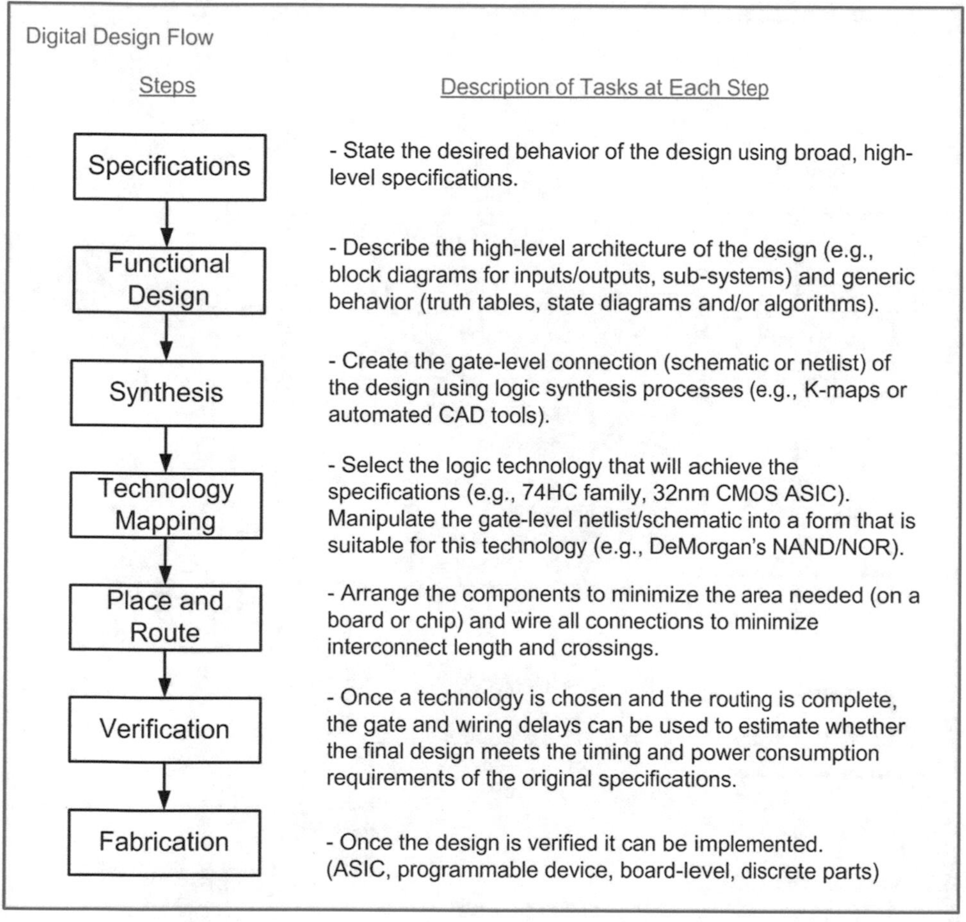

Fig. 5.5
Generic digital design flow

This generic design process flow can be used across classical and modern digital design, although modern digital design allows additional verification at each step using automated CAD tools. Figure 5.6 shows how this flow is used in the classical design approach of a combinational logic circuit.

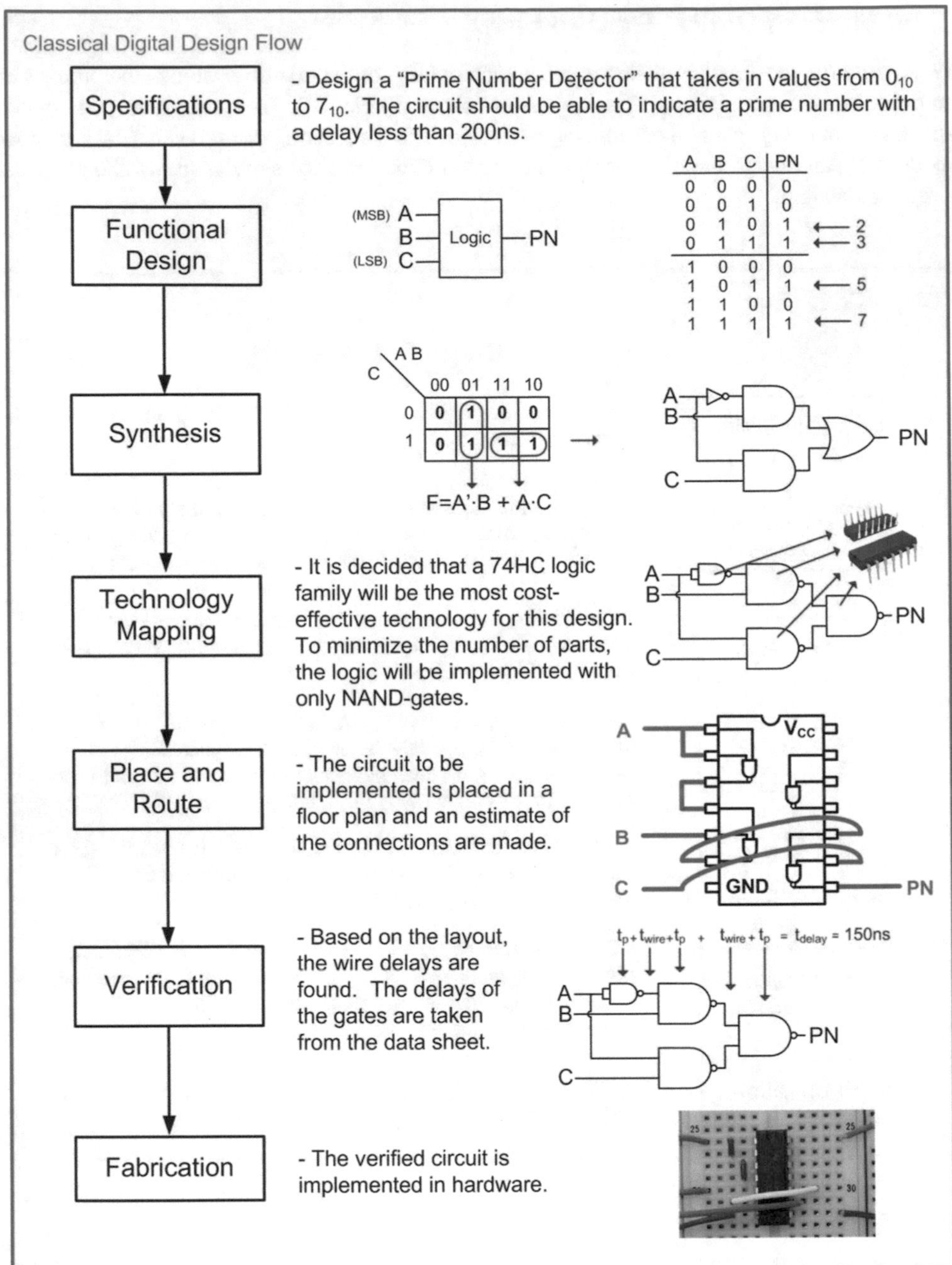

Fig. 5.6
Classical digital design flow

The modern design flow based on HDLs includes the ability to simulate functionality at each step of the process. Functional simulations can be performed on the initial behavioral description of the system. At each step of the design process, the functionality is described in more detail, ultimately moving toward the fabrication step. At each level, the detailed information can be included in the simulation to verify that the functionality is still correct and that the design is still meeting the original specifications. Figure 5.7 shows the modern digital design flow with the inclusion of simulation capability at each step.

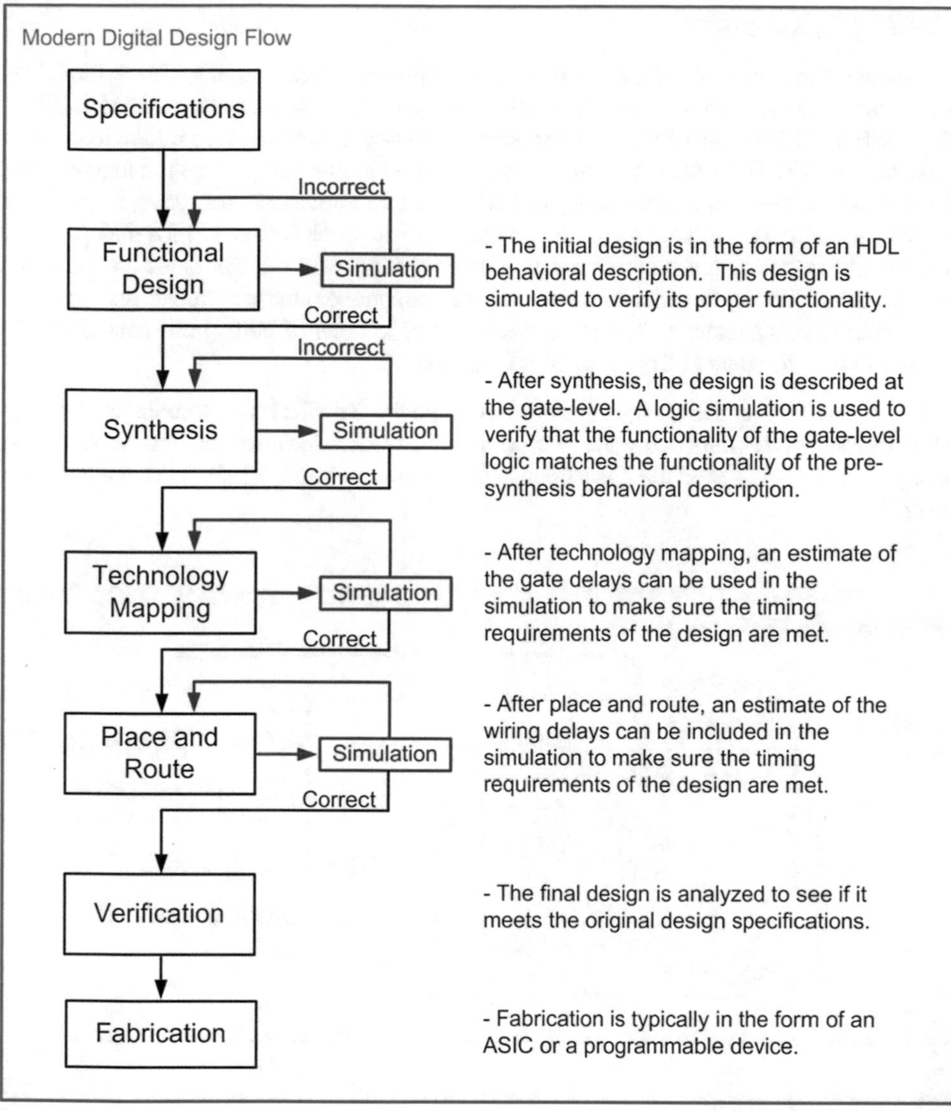

Fig. 5.7
Modern digital design flow

Concept Check

CC5.3 Why did digital designs move from schematic-entry to text-based HDLs?

A) HDL models could be much larger by describing functionality in text similar to traditional programming language.

B) Schematics required sophisticated graphics hardware to display correctly.

C) Schematics symbols became too small as designs became larger.

D) Text was easier to understand by a broader range of engineers.

5.4 Verilog Constructs

Now we begin looking at the details of Verilog. The original Verilog standard (IEEE 1364) has been updated numerous times since its creation in 1995. The most significant update occurred in 2001, which was titled IEEE 1394-2001. In 2005 minor corrections and improvements were added to the standard, which resulted in IEEE 1394-2005. The constructs described in this book reflect the functionality in the IEEE 1394-2005 standard. The functionality of Verilog (e.g., operators, signal types, functions, etc.) is defined within the Verilog standard; thus, it is not necessary to explicitly state that a design is using the IEEE 1394 package because it is inherent in the use of Verilog. This chapter gives an overview of the basic constructs of Verilog in order to model simple combinational logic circuits and begin gaining experience with logic simulations. The more advanced constructs of Verilog are covered in Chap. 8 with examples given throughout Chaps. 9, 10, 11, 12, and 13.

A Verilog design describes a single system in a single file. The file has the suffix *.v. Within the file, the system description is contained within a **module**. The module includes the interface to the system (i.e., the inputs and outputs) and the description of the behavior. Figure 5.8 shows a graphical depiction of a Verilog file.

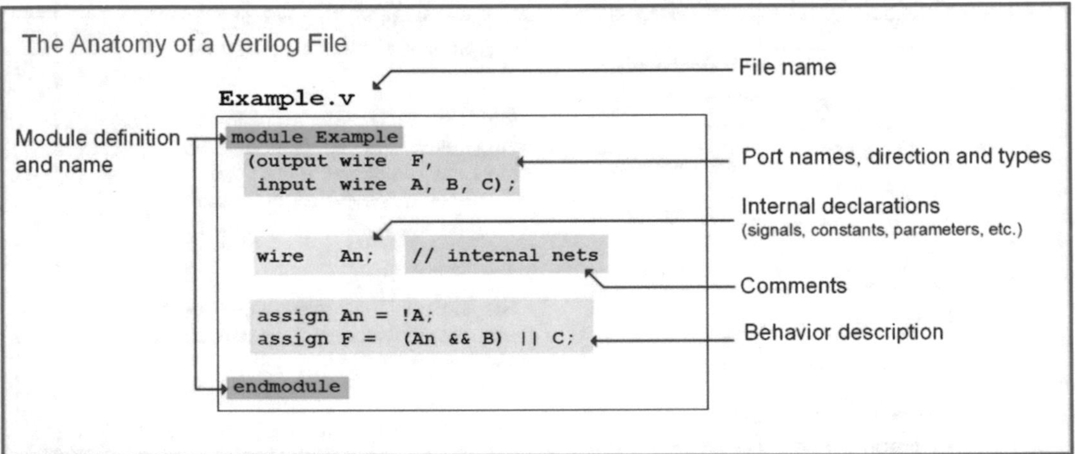

Fig. 5.8
The anatomy of a Verilog file

Verilog is case sensitive. Also, each Verilog assignment, definition, or declaration is terminated with a semicolon (;). As such, line wraps are allowed and do not signify the end of an assignment, definition, or declaration. Line wraps can be used to make Verilog more readable. Comments in Verilog are supported in two ways. The first way is called a *line comment* and is preceded with two slashes (i.e., //). Everything after the slashes is considered a comment until the end of the line. The second comment approach is called a *block comment* and begins with **/*** and ends with a ***/**. Everything between /* and */ is considered a comment. A block comment can span multiple lines. All user-defined names in Verilog must start with an alphabetic letter, not a number. User-defined names are not allowed to be the same as any Verilog keyword. This chapter contains many definitions of syntax in Verilog. The following notations will be used throughout the chapter when introducing new constructs.

bold	= Verilog keyword, use as is, case sensitive.
italics	= User-defined name, case sensitive.
< >	= A required characteristic such as a data type, input/output, etc.

5.4.1 Data Types

In Verilog, every signal, constant, variable, and function must be assigned a *data type*. The IEEE 1394-2005 standard provides a variety of pre-defined data types. Some data types are synthesizable, while others are only for modeling abstract behavior. The following are the most commonly used data types in the Verilog language.

5.4.1.1 Value Set

Verilog supports four basic values that a signal can take on: 0, 1, X, and Z. Most of the pre-defined data types in Verilog store these values. A description of each value supported is given below.

Value	Description
0	A logic zero, or false condition.
1	A logic one, or true condition.
x or **X**	Unknown or uninitialized.
z or **Z**	High impedance, tri-stated, or floating.

In Verilog, these values also have an associated *strength*. The strengths are used to resolve the value of a signal when it is driven by multiple sources. The names, syntax, and relative strengths are given below.

Strength	Description	Strength level
supply1	Supply drive for V_{CC}	7
supply0	Supply drive for V_{SS}, or GND	7
strong1	Strong drive to logic one	6
strong0	Strong drive to logic zero	6
pull1	Medium drive to logic one	5
pull0	Medium drive to logic zero	5
large	Large capacitive	4
weak1	Weak drive to logic one	3
weak0	Weak drive to logic zero	3
medium	Medium capacitive	2
small	Small capacitive	1
highz1	High impedance with weak pull-up to logic one	0
highz0	High impedance with weak pull-down to logic zero	0

When a signal is driven by multiple drivers, it will take on the value of the driver with the highest strength. If the two drivers have the same strength, the value will be *unknown*. If the strength is not specified, it will default to *strong drive*, or level 6.

5.4.1.2 Net Data Types

Every signal within Verilog must be associated with a data type. A *net data type* is one that models an interconnection (aka., a *net*) between components and can take on the values 0, 1, X, and Z. A signal with a net data type must be driven at all times and updates its value when the driver value changes. The most common synthesizable net data type in Verilog is the *wire*. The type wire will be used throughout this text. There are also a variety of other more advanced net data types that model complex digital systems with multiple drivers for the same net. The syntax and description for all Verilog net data types are given below:

Type	Description
wire	A simple connection between components.
wor	Wired-OR. If multiple drivers, their values are OR'd together.
wand	Wired-AND'd. If multiple drivers, their values are AND'd together.
supply0	Used to model the V_{SS}, (GND), power supply (supply strength inherent).
supply1	Used to model the V_{CC} power supply (supply strength inherent).
tri	Identical to **wire**. Used for readability for a net driven by multiple sources.
trior	Identical to **wor**. Used for readability for nets driven by multiple sources.
triand	Identical to **wand**. Used for readability for nets driven by multiple sources.
tri1	Pulls up to logic one when tri-stated.
tri0	Pulls down to logic zero when tri-stated.
trireg	Holds last value when tri-stated (capacitance strength inherent).

Each of these net types can also have an associated *drive strength*. The strength is used in determining the final value of the net when it is connected to multiple drivers.

5.4.1.3 Variable Data Types

Verilog also contains data types that model storage. These are called *variable data types*. A variable data type can take on the values 0, 1, X, and Z but does not have an associated strength. Variable data types will hold the value assigned to them until their next assignment. The syntax and description for the Verilog variable data types are given below.

Type	Description
reg	A variable that models logic storage. Can take on values 0, 1, X, and Z.
integer	A 32-bit, 2's complement variable representing whole numbers between $-2{,}147{,}483{,}648_{10}$ and $+2{,}147{,}483{,}647$.
real	A 64-bit, floating point variable representing real numbers between $-(2.2 \times 10^{-308})_{10}$ and $+(2.2 \times 10^{308})_{10}$.
time	An unsigned, 64-bit variable taking on values from 0_{10} to $+(9.2 \times 10^{18})$.
realtime	Same as **time**. Just used for readability.

5.4.1.4 Vectors

In Verilog, a *vector* is a one-dimensional array of elements. All of the net data types, in addition to the variable type reg, can be used to form vectors. The syntax for defining a vector is as follows:

```
<type> [<MSB_index>:<LSB_index>] vector_name
```

While any range of indices can be used, it is common practice to have the LSB index start at zero.

Example:

```
wire [7:0] Sum;     // This defines an 8-bit vector called "Sum" of type wire. The
                    // MSB is given the index 7 while the LSB is given the index 0.

reg [15:0] Q;       // This defines a 16-bit vector called "Q" of type reg.
```

Individual bits within the vector can be addressed using their index. Groups of bits can be accessed using an index range.

```
Sum[0];                // This is the least significant bit of the vector "Sum" defined
above.
Q[15:8];               // This is the upper 8-bits of the 16-bit vector "Q" defined above.
```

5.4.1.5 Arrays

An *array* is a multidimensional array of elements. This can also be thought of as a "vector of vectors." Vectors within the array all have the same dimensions. To declare an array, the element type and dimensions are defined first followed by the array name and its dimensions. It is common practice to place the start index of the array on the left side of the ":" when defining its dimensions. The syntax for the creation of an array is shown below.

```
<element_type>          [<MSB_index>:<LSB_index>]          array_name
[<array_start_index>:<array_end_index>];
```

Example:

```
reg[7:0] Mem[0:4095];      // Defines an array of 4096, 8-bit vectors of type reg.
integer A[1:100];          // Defines an array of 100 integers.
```

When accessing an array, the name of the array is given first, followed by the index of the element. It is also possible to access an individual bit within an array by adding appending the index of element.

Example:

```
Mem[2];             // This is the 3rd element within the array named "Mem".
                    // This syntax represents an 8-bit vector of type reg.

Mem[2][7];          // This is the MSB of the 3rd element within the array named "Mem".
                    // This syntax represents a single bit of type reg.

A[2];               // This is the 2nd element within the array named "A". Recall
                    //  that A was declared with a starting index of 1.
                    // This syntax represents a 32-bit, signed integer.
```

5.4.1.6 Expressing Numbers Using Different Bases

If a number is simply entered into Verilog without identifying syntax, it is treated as an integer. However, Verilog supports defining numbers in other bases. Verilog also supports an optional bit size and sign of a number. When defining the value of arrays, the "_" can be inserted between numerals to improve readability. The "_" is ignored by the Verilog compiler. Values of numbers can be entered in either upper or lower case (i.e., z or Z, f or F, etc.). The syntax for specifying the base of a number is as follows:

```
<size_in_bits>'<base><value>
```

Note that specifying the size is optional. If it is omitted, the number will default to a 32-bit vector with leading zeros added as necessary. The supported bases are as follows:

Syntax	Description
'b	Unsigned binary
'o	Unsigned octal
'd	Unsigned decimal
'h	Unsigned hexadecimal
'sb	Signed binary
'so	Signed octal
'sd	Signed decimal
'sh	Signed hexadecimal

Example:

```
10               // This is treated as decimal 10, which is a 32-bit signed vector.
4'b1111          // A 4-bit number with the value 1111₂.
8'b1011_0000     // An 8-bit number with the value 10110000₂.
8'hFF            // An 8-bit number with the value 11111111₂.
8'hff            // An 8-bit number with the value 11111111₂.
6'hA             // A 6-bit number with the value 001010₂. Note that leading zeros
                 //   were added to make the value 6-bits.
8'd7             // An 8-bit number with the value 00000111₂.
32'd0            // A 32-bit number with the value 0000_0000₁₆.
'b1111           // A 32-bit number with the value 0000_000F₁₆.
8'bZ             // An 8-bit number with the value ZZZZ_ZZZZ.
```

5.4.1.7 Assigning Between Different Types

Verilog is said to be a weakly typed (or loosely typed) language, meaning that it permits assignments between different data types. This is as opposed to a strongly typed language (such as VHDL) where signal assignments are only permitted between like types. The reason Verilog permits assignment between different types is because it treats all of its types as just group of bits. When assigning between different types, Verilog will automatically truncate or add leading bits as necessary to make the assignment work. The following examples illustrate how Verilog handles a few assignments between different types. Assume that a variable called ABC_TB has been declared as type reg[2:0].

Example:

```
ABC_TB = 2'b00;  // ABC_TB will be assigned 3'b000. A leading bit is automatically added.
ABC_TB = 5;      // ABC_TB will be assigned 3'b101. The integer is truncated to 3-bits.
ABC_TB = 8;      // ABC_TB will be assigned 3'b000. The integer is truncated to 3-bits.
```

5.4.2 The Module

All systems in Verilog are encapsulated inside of a **module**. Modules can include instantiations of lower-level modules in order to support hierarchical designs. The keywords **module** and **endmodule** signify the beginning and end of the system description. When working on large designs, it is common practice to place each module in its own file with the same name.

```
module module_name (port_list);                                  // Pre Verilog-2001
  // port_definitions
  // module_items
endmodule
```

or

```
module module_name (port_list and port_definitions); // Verilog-2001 and after
  // module_items
endmodule
```

5.4.2.1 Port Definitions

The first item within a module is its definition of the inputs and outputs, or ports. Each port needs to have a user-defined name, a direction, and a type. The user-defined port names are case sensitive and must begin an alphabetic character. The port directions are declared to be one of the three types: **input**, **output**, and **inout**. A port can take on any of the previously described data types, but only wires, registers, and integers are synthesizable. Port names with the same type and direction can be listed on the same line separated by commas.

There are two different port definition styles supported in Verilog. Prior to the Verilog-2001 release, the port names were listed within parentheses after the module name. Then within the module, the directionality and type of the ports were listed. Starting with the Verilog-2001 release, the port directions and types could be included alongside the port names within the parenthesis after the module name. This approach mimicked more of an ANSCI-C approach to passing inputs/outputs to a system. In this text, the newer approach to port definition will be used. Example 5.1 shows multiple approaches for defining a module and its ports.

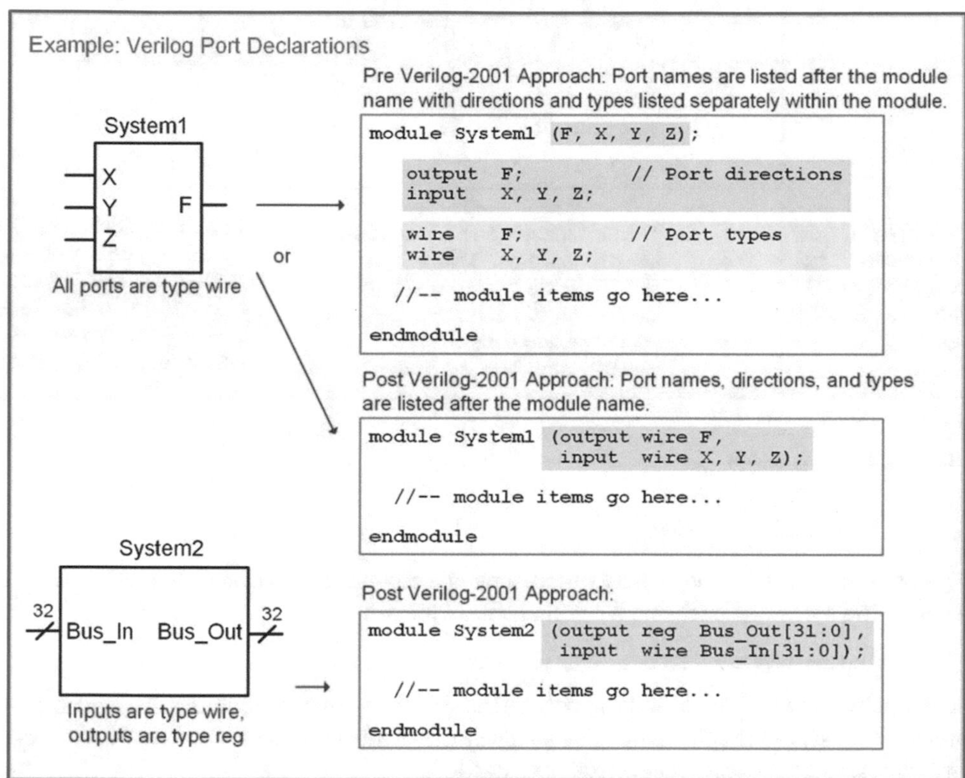

Example 5.1
Declaring Verilog module ports

5.4.2.2 Signal Declarations

A signal that is used for internal connections within a system is declared within the module before its first use. Each signal must be declared by listing its type followed by a user-defined name. Signal names of like type can be declared on the same line separated with a comma. All of the legal data types described above can be used for signals; however, only types net, reg, and integer will synthesize directly. The syntax for a signal declaration is as follows:

```
<type> name;
```

Example:

```
wire  node1;          // declare a signal named "node1" of type wire
reg   Q2, Q1, Q0;     // declare three signals named "Q2", "Q1", and "Q0", all of type
reg
wire  [63:0] bus1;    // declare a 64-bit vector named "bus1" with all bits of type wire
integer i,j;          // declare two integers called "i" and "j"
```

Verilog supports a hierarchical design approach; thus, signal names can be the same within a sub-system as those at a higher level without conflict. Figure 5.9 shows an example of legal signal naming in a hierarchical design.

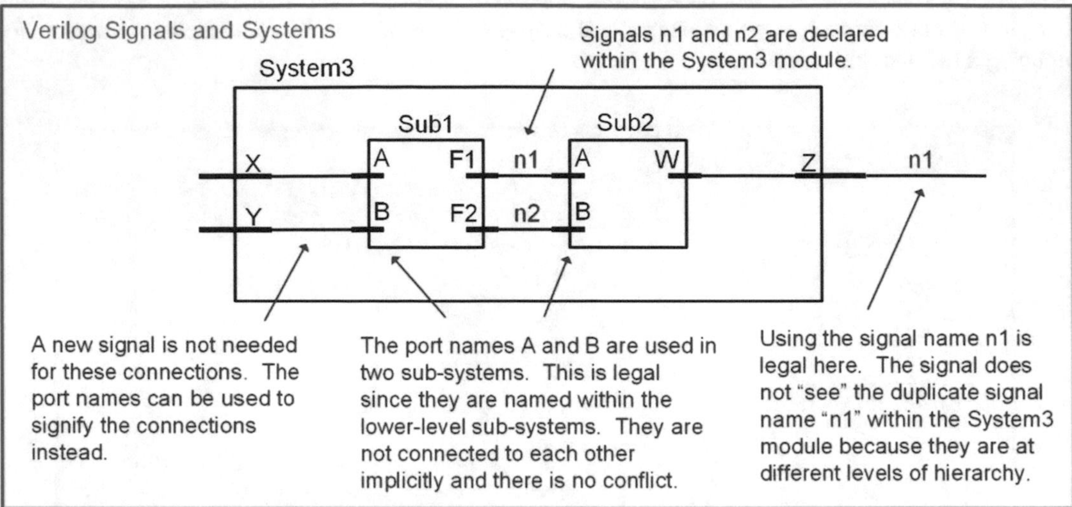

Fig. 5.9
Verilog signals and systems

5.4.2.3 Parameter Declarations

A parameter, or constant, is useful for representing a quantity that will be used multiple times in the architecture. The syntax for declaring a parameter is as follows:

```
parameter <type> constant_name = <value>;
```

Note that the type is optional and can only be **integer**, **time**, **real**, or **realtime**. If a type is provided, the parameter will have the same properties as a variable of the same time. If the type is excluded, the parameter will take on the type of the value assigned to it.

Example:

```
parameter BUS_WIDTH = 64;
parameter NICKEL    = 8'b0000_0101;
```

Once declared, the constant name can be used throughout the module. The following example illustrates how we can use a constant to define the size of a vector. Notice that since we defined the constant to be the actual width of the vector (i.e., 32-bits), we need to subtract one from its value when defining the indices (i.e., [31:0]).

Example:

```
wire [BUS_WIDTH-1:0] BUS_A;              // It is acceptable to add a "space" for
readability
```

5.4.2.4 Compiler Directives

A compiler directive provides additional information to the simulation tool on how to interpret the Verilog model. A compiler directive is placed before the module definition and is preceded with a backtick (i.e., `). Note that this is not an apostrophe. A few of the most commonly used compiler directives are as follows:

Syntax	Description
`timescale <unit>,<precision>	Defines the timescale of the delay unit and its smallest precision.
`include <filename>	Includes additional files in the compilation.
`define <macroname> <value>	Declares a global constant.

Example:

```
'timescale 1ns/1ps    // Declares the unit of time is 1 ns with a precision of 1ps.
                      // The precision is the smallest amount that the time can
                      // take on. For example, with this directive the number
                      // 0.001 would be interpreted as 0.001 ns, or 1 ps.
                      // However, the number 0.0001 would be interpreted as 0 since
                      // it is smaller than the minimum precision value.
```

5.4.3 Verilog Operators

There are a variety of pre-defined operators in the Verilog standard. It is important to note that operators are defined to work on specific data types and that not all operators are synthesizable.

5.4.3.1 Assignment Operator

Verilog uses the equal sign (**=**) to denote an assignment. The left-hand side (LHS) of the assignment is the target signal. The right-hand side (RHS) contains the input arguments and can contain both signals, constants, and operators.

Example:

```
F1 = A;      // F1 is assigned the signal A
F2 = 8'hAA;  // F2 is an 8-bit vector and is assigned the value 10101010_2
```

5.4.3.2 Bitwise Logical Operators

Bitwise operators perform logic functions on individual bits. The inputs to the operation are single bits and the output is a single bit. In the case where the inputs are vectors, each bit in the first vector is operated on by the bit in the same position from the second vector. If the vectors are not the same length, the shorter vector is padded with leading zeros to make both lengths equal. Verilog contains the following bitwise operators:

Syntax	Operation
~	Negation
&	AND
|	OR
^	XOR
~^ or **^~**	XNOR
<<	Logical shift left (fill empty LSB location with zero)
>>	Logical shift right (fill empty MSB location with zero)

Example:

```
~X          // invert each bit in X
X & Y       // AND each bit of X with each bit of Y
X | Y       // OR each bit of X with each bit of Y
X ^ Y       // XOR each bit of X with each bit of Y
X ~^ Y      // XNOR each bit of X with each bit of Y
X << 3      // Shift X left 3 times and fill with zeros
Y >> 2      // Shift Y right 2 times and fill with zeros
```

5.4.3.3 Reduction Logic Operators

A *reduction* operator is one that uses each bit of a vector as individual inputs into a logic operation and produces a single bit output. Verilog contains the following reduction logic operators.

Syntax	Operation
&	AND all bits in the vector together (1-bit result)
~&	NAND all bits in the vector together (1-bit result)
|	OR all bits in the vector together (1-bit result)
~|	NOR all bits in the vector together (1-bit result)
^	XOR all bits in the vector together (1-bit result)
~^ or **^~**	XNOR all bits in the vector together (1-bit result)

Example:

```
&X          // AND all bits in vector X together
~&X         // NAND all bits in vector X together
|X          // OR all bits in vector X together
~|X         // NOR all bits in vector X together
^X          // XOR all bits in vector X together
~^X         // XNOR all bits in vector X together
```

5.4.3.4 Boolean Logic Operators

A Boolean logic operator is one that returns a value of TRUE (1) or FALSE (0) based on a logic operation of the input operations. These operations are used in decision statements.

Syntax	Operation
!	Negation
&&	AND
||	OR

Example:

```
!X          // TRUE if all values in X are 0, FALSE otherwise
X && Y      // TRUE if the bitwise AND of X and Y results in all ones, FALSE otherwise
X || Y      // TRUE if the bitwise OR of X and Y results in all ones, FALSE otherwise
```

5.4.3.5 Relational Operators

A relational operator is one that returns a value of TRUE (1) or FALSE (0) based on a comparison of two inputs.

Syntax	Description
==	Equality
!=	Inequality
<	Less than
>	Greater than
<=	Less than or equal
>=	Greater than or equal

Example:

```
X == Y     // TRUE if X is equal to Y, FALSE otherwise
X != Y     // TRUE if X is not equal to Y, FALSE otherwise
X < Y      // TRUE if X is less than Y, FALSE otherwise
X > Y      // TRUE if X is greater than Y, FALSE otherwise
X <= Y     // TRUE if X is less than or equal to Y, FALSE otherwise
X >= Y     // TRUE if X is greater than or equal to Y, FALSE otherwise
```

5.4.3.6 Conditional Operators

Verilog contains a conditional operator that can be used to provide a more intuitive approach to modeling logic statements. The keyword for the conditional operator is **?** with the following syntax:

```
<target_net> = <Boolean_condition> ? <true_assignment> : <false_assignment>;
```

This operator specifies a Boolean condition in which if evaluated TRUE, the *true_assignment* will be assigned to the target. If the Boolean condition is evaluated FALSE, the *false_assignment* portion of the operator will be assigned to the target. The values in this assignment can be signals or logic values. The Boolean condition can be any combination of the Boolean operators described above. Nested conditional operators can also be implemented by inserting subsequent conditional operators in place of the *false_value*.

Example:

```
F = (A == 1'b0) ? 1'b1 : 1'b0;                // If A is a zero, F=1, otherwise F=0.
                                                   This models an inverter.

F = (sel == 1'b0) ? A : B;                    // If sel is a zero, F=A, otherwise F=B.
                                                   This models a selectable switch.

F = ((A == 1'b0) && (B == 1'b0)) ? 1'b'0 :    // Nested conditional statements.
    ((A == 1'b0) && (B == 1'b1)) ? 1'b'1 :    //    This models an XOR gate.
    ((A == 1'b1) && (B == 1'b0)) ? 1'b'1 :
    ((A == 1'b1) && (B == 1'b1)) ? 1'b'0;

F = ( !C && (!A || B) ) ? 1'b1 : 1'b0;        // This models the logic expression
                                              //    F = C'·(A'+B).
```

5.4.3.7 Concatenation Operator

In Verilog, the curly brackets (i.e., **{}**) are used to concatenate multiple signals. The target of this operation must be the same size of the sum of the sizes of the input arguments.

Example:

```
Bus1[7:0] = {Bus2[7:4], Bus3[3:0]}; // Assuming Bus1, Bus2, and Bus3 are all 8-bit
                                    //  vectors, this operation takes the upper 4-bits
of
                                    // Bus2, concatenates them with the lower 4-bits of
                                    // Bus3, and assigns the 8-bit combination to Bus1.

BusC = {BusA, BusB};                // If BusA and BusB are 4-bits, then BusC
                                    //  must be 8-bits.

BusC[7:0] = {4'b0000, BusA};        // This pads the 4-bit vector BusA with 4x leading
                                    //  zeros and assigns to the 8-bit vector BusC.
```

5.4.3.8 Replication Operator

Verilog provides the ability to concatenate a vector with itself through the *replication operator*. This operator uses double curly brackets (i.e., **{{}}**) and an integer indicating the number of replications to be performed. The replication syntax is as follows:

```
{<number_of_replications>{<vector_name_to_be_replicated>}}
```

Example:

```
BusX = {4{Bus1}};              // This is equivalent to: BusX = {Bus1, Bus1, Bus1, Bus1};
BusY = {2{A,B}};               // This is equivalent to: BusY = {A, B, A, B};
BusZ = {Bus1, {2{Bus2}}};      // This is equivalent to: BusZ = {Bus1, Bus2, Bus2};
```

5.4.3.9 Numerical Operators

Verilog also provides a set of numerical operators as follows:

Syntax	Operation
+	Addition
–	Subtraction (when placed between arguments)
–	2's complement negation (when placed in front of an argument)
*	Multiplication
/	Division
%	Modulus
**	Raise to the power
<<<	Shift to the left, fill with zeros
<<<	Shift to the right, fill with sign bit

Example:

```
X + Y        // Add X to Y
X - Y        // Subtract Y from X
-X           // Take the two's complement negation of X
X * Y        // Multiply X by Y
X / Y        // Divide X by Y
X % Y        // Modulus X/Y
X ** Y       // Raise X to the power of Y
X <<< 3      // Shift X left 3 times, fill with zeros
X >>> 2      // Shift X right 2 times, fill with sign bit
```

Verilog will allow the use of these operators on arguments of different sizes, types, and signs. The rules of the operations are as follows:

- If two vectors are of different sizes, the smaller vector is expanded to the size of the larger vector.
  - If the smaller vector is unsigned, it is padded with zeros.
  - If the smaller vector is signed, it is padded with the sign bit.
- If one of the arguments is real, then the arithmetic will take place using real numbers.
- If one of the arguments is unsigned, then all arguments will be treated as unsigned.

5.4.3.10 Operator Precedence

The following is the order of precedence of the Verilog operators:

<table>
<tr><th>Operators</th><th>Precedence</th><th>Notes</th></tr>
<tr><td>! ~ + -</td><td>Highest</td><td>Bitwise/unary</td></tr>
<tr><td>{} {{}}</td><td></td><td>Concatenation/replication</td></tr>
<tr><td>()</td><td>↓</td><td>No operation, just parenthesis</td></tr>
<tr><td>**</td><td></td><td>Power</td></tr>
<tr><td>* / %</td><td></td><td>Binary multiply/divide/modulo</td></tr>
<tr><td>+ -</td><td>↓</td><td>Binary addition/subtraction</td></tr>
<tr><td><< >> <<< >>></td><td></td><td>Shift operators</td></tr>
<tr><td>< <= > >=</td><td></td><td>Greater/less than comparisons</td></tr>
<tr><td>== !=</td><td>↓</td><td>Equality/inequality comparisons</td></tr>
<tr><td>& ~&</td><td></td><td>AND/NAND operators</td></tr>
<tr><td>^ ~^</td><td></td><td>XOR/XNOR operators</td></tr>
<tr><td>| ~|</td><td>↓</td><td>OR/NOR operators</td></tr>
<tr><td>&&</td><td></td><td>Boolean AND</td></tr>
<tr><td>||</td><td></td><td>Boolean OR</td></tr>
<tr><td>?:</td><td>Lowest</td><td>Conditional operator</td></tr>
</table>

CONCEPT CHECK

CC5.4(a) What revision of Verilog added the ability to list the port names, types, and directions just once after the module name?

A) Verilog-1995.

B) Verilog-2001.

C) Verilog-2005.

D) SystemVerilog.

CC5.4(b) What is the difference between types wire and reg?

A) They are the same.

B) The type wire is a simple interconnection while reg will hold the value of its last assignment.

C) The type wire is for scalars while the type reg is for vectors.

D) Only wire is synthesizable.

5.5 Modeling Concurrent Functionality in Verilog

It is important to remember that Verilog is a hardware description language, not a programming language. In a programming language, the lines of code are executed sequentially as they appear in the source file. In Verilog, the lines of code represent the behavior of real hardware. Thus, the assignments are executed concurrently unless specifically noted otherwise.

5.5.1 Continuous Assignment

Verilog uses the keyword **assign** to denote a continuous signal assignment. After this keyword, an assignment is made using the = symbol. The left-hand side (LHS) of the assignment is the target signal and must be a net type. The right-hand side (RHS) contains the input arguments and can contain nets, regs, constants, and operators. A continuous assignment models combinational logic. Any change to the RHS of the expression will result in an update to the LHS target net. The net being assigned to must be declared prior to the first continuous assignment. Multiple continuous assignments can be made to the same net. When this happens, the assignment containing signals with the highest drive strength will take priority.

Example:

```
assign F1 = A;        // F1 is updated anytime A changes, where A is a signal
assign F2 = 1'b0;     // F2 is assigned the value 0
assign F3 = 4'hAA;    // F3 is an 8-bit vector and is assigned the value 10101010₂
```

Each individual assignment will be executed concurrently and synthesized as separate logic circuits. Consider the following example.

Example:

```
assign X = A;
assign Y = B;
assign Z = C;
```

When simulated, these three lines of Verilog will make three separate signal assignments at the exact same time. This is different from a programming language that will first assign A to X, then B to Y, and finally C to Z. In Verilog this functionality is identical to three separate wires. This description will be directly synthesized into three separate wires.

Below is another example of how continuous signal assignments in Verilog differ from a sequentially executed programming language.

Example:

```
assign A = B;
assign B = C;
```

In a Verilog simulation, the signal assignments of C to B and B to A will take place at the same time. This means during synthesis, the signal B will be eliminated from the design since this functionality describes two wires in series. Automated synthesis tools will eliminate this unnecessary signal name. This is not the same functionality that would result if this example was implemented as a sequentially executed computer program. A computer program would execute the assignment of B to A first and then assign the value of C to B second. In this way, B represents a storage element that is passed to A before it is updated with C.

5.5.2 Continuous Assignment with Logical Operators

Each of the logical operators described in Sect. 5.4.3.2 can be used in conjunction with concurrent signal assignments to create individual combinational logic circuits. Example 5.2 shows how to design a Verilog model of a combinational logic circuit using this approach.

Example: Modeling Combinational Logic using Continuous Assignment with Logical Operators

Implement the following truth table using <u>continuous assignment with logical operators</u>.

A	B	C	F
0	0	0	1
0	0	1	0
0	1	0	1
0	1	1	0
1	0	0	0
1	0	1	0
1	1	0	1
1	1	1	0

Let's call the module *SystemX*. First, let's declare the ports. The module will have three inputs (A, B, C) and one output (F). We'll use the type wire for all inputs/outputs so that this will synthesize directly into real circuitry.

SystemX.v

A
B
C
F

Now we can design the behavior. We will create a canonical sum of products logic expression for this truth table using minterms.

$$F = \Sigma_{A,B,C}(0,2,6) = A'\cdot B'\cdot C' + A'\cdot B\cdot C' + A\cdot B\cdot C'$$

Drawing out the logic diagram will help us understand which internal signals need to be declared for the interim connections. Since there is a need for the complement of each of the inputs, the first set of logic will be three inverters. We'll need to create three wires to hold the inverted versions of the inputs. Let's call them An, Bn and Cn. We'll also need three wires to hold the outputs of the AND gates. Let's call them m0, m2 and m6. Using these internal wires, the port names, and logical operators, we can describe the behavior of the logic expression above.

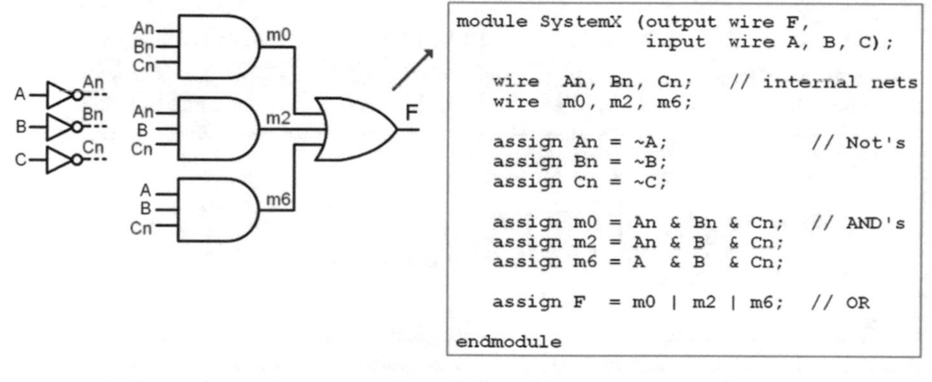


```
module SystemX (output wire F,
                input  wire A, B, C);

   wire  An, Bn, Cn;     // internal nets
   wire  m0, m2, m6;

   assign An = ~A;                // Not's
   assign Bn = ~B;
   assign Cn = ~C;

   assign m0 = An & Bn & Cn;     // AND's
   assign m2 = An & B  & Cn;
   assign m6 = A  & B  & Cn;

   assign F  = m0 | m2 | m6;     // OR

endmodule
```

Example 5.2
Modeling combinational logic using continuous assignment with logical operators

5.5.3 Continuous Assignment with Conditional Operators

Logical operators are good for describing the behavior of small circuits; however, in the prior example, we still needed to create the canonical sum of products logic expression by hand before describing the functionality in Verilog. The true power of an HDL is when the behavior of the system can be described fully without requiring any hand design. The conditional operator allows us to describe a continuous assignment using Boolean conditions that affect the values of the result. In this approach, we use the conditional operator (**?**) in conjunction with the continuous assignment keyword **assign**. Example 5.3 shows how to design a Verilog model of a combinational logic circuit using continuous assignment with conditional operators. Note that this example uses the same truth table as in Example 5.2 to illustrate a comparison between approaches.

Example: Modeling Combinational Logic using Continuous Assignment with Conditional Operators (1)

Implement the following truth table using a continuous assignment with conditional operators.

A	B	C	F
0	0	0	1
0	0	1	0
0	1	0	1
0	1	1	0
1	0	0	0
1	0	1	0
1	1	0	1
1	1	1	0

We can implement the entire truth table in its current form by nesting conditional operators to explicitly list out each possible input code and its corresponding output as follows:

```
module SystemX (output wire F,
                input  wire A, B, C);

   assign F = ((A == 1'b0) && (B == 1'b0) && (C == 1'b0)) ? 1'b1 :
              ((A == 1'b0) && (B == 1'b0) && (C == 1'b1)) ? 1'b0 :
              ((A == 1'b0) && (B == 1'b1) && (C == 1'b0)) ? 1'b1 :
              ((A == 1'b0) && (B == 1'b1) && (C == 1'b1)) ? 1'b0 :
              ((A == 1'b1) && (B == 1'b0) && (C == 1'b0)) ? 1'b0 :
              ((A == 1'b1) && (B == 1'b0) && (C == 1'b1)) ? 1'b0 :
              ((A == 1'b1) && (B == 1'b1) && (C == 1'b0)) ? 1'b1 :
              ((A == 1'b1) && (B == 1'b1) && (C == 1'b1)) ? 1'b0 :
               1'b0;

endmodule
```

We can reduce the length of this model by only explicitly listing the input conditions for when the output is TRUE and allowing the final FALSE value to cover all other inputs.

```
module SystemX (output wire F,
                input  wire A, B, C);

   assign F = ((A == 1'b0) && (B == 1'b0) && (C == 1'b0)) ? 1'b1 :
              ((A == 1'b0) && (B == 1'b1) && (C == 1'b0)) ? 1'b1 :
              ((A == 1'b1) && (B == 1'b1) && (C == 1'b0)) ? 1'b1 :
               1'b0;

endmodule
```

Example 5.3
Modeling combinational logic using continuous assignment with conditional operators (1)

In the prior example, the conditional operator was based on a truth table. Conditional operators can also be used to model logic expressions. Example 5.4 shows how to design a Verilog model of a combinational logic circuit when the logic expression is already known. Note that this example again uses the same truth table as in Examples 5.2 and 5.3 to illustrate a comparison between approaches.

Example: Modeling Combinational Logic using Continuous Assignment with Conditional Operators (2)

Implement the following truth table using a continuous assignment with conditional operators.

In this example, a K-map was used to find a minimized logic expression of:

$$F = C' \cdot (A' + B)$$

A	B	C	F
0	0	0	1
0	0	1	0
0	1	0	1
0	1	1	0
1	0	0	0
1	0	1	0
1	1	0	1
1	1	1	0

We can implement the conditional operator using input variables and Boolean operators to directly model the logic expression.

```
module SystemX (output wire F,
                input  wire A, B, C);

   assign F = ( !C && (!A || B) ) ? 1'b1 : 1'b0;

endmodule
```

Example 5.4
Modeling combinational logic using continuous assignment with conditional operators (2)

5.5.4 Continuous Assignment with Delay

Verilog provides the ability to model gate delays when using a continuous assignment. The **#** is used to indicate a delayed assignment. For combinational logic circuits, the delay can be specified for all transitions, for rising and falling transitions separately, and for rising, falling, and transitions to the value *off* separately. A transition to *off* refers to a transition to Z. If only one delay parameter is specified, it is used to model all delays. If two delay parameters are specified, the first parameter is used for the rise time delay, while the second is used to model the fall time delay. If three parameters are specified, the third parameter is used to model the transition to off. Parentheses are optional but recommended when using multiple delay parameters.

```
assign#(<del_all>)                      <target_net>=<RHS_nets, operators, etc...>;
assign#(<del_rise, del_fall>)           <target_net>=<RHS_nets, operators, etc...>;
assign#(<del_rise, del_fall, del_off>) <target_net>=<RHS_nets, operators, etc...>;
```

Example:

```
assign #1       F = A;  // Delay of 1 on all transitions.
assign #(2,3)   F = A;  // Delay of 2 for rising transitions and 3 for falling.
assign #(2,3,4) F =  A; // Delay of 2 for rising, 3 for falling, and 4 for off
transitions.
```

When using delay, it is typical to include the **`timescale** directive to provide the units of the delay being specified. Example 5.5 shows a graphical depiction of using delay with continuous assignments when modeling combinational logic circuits.

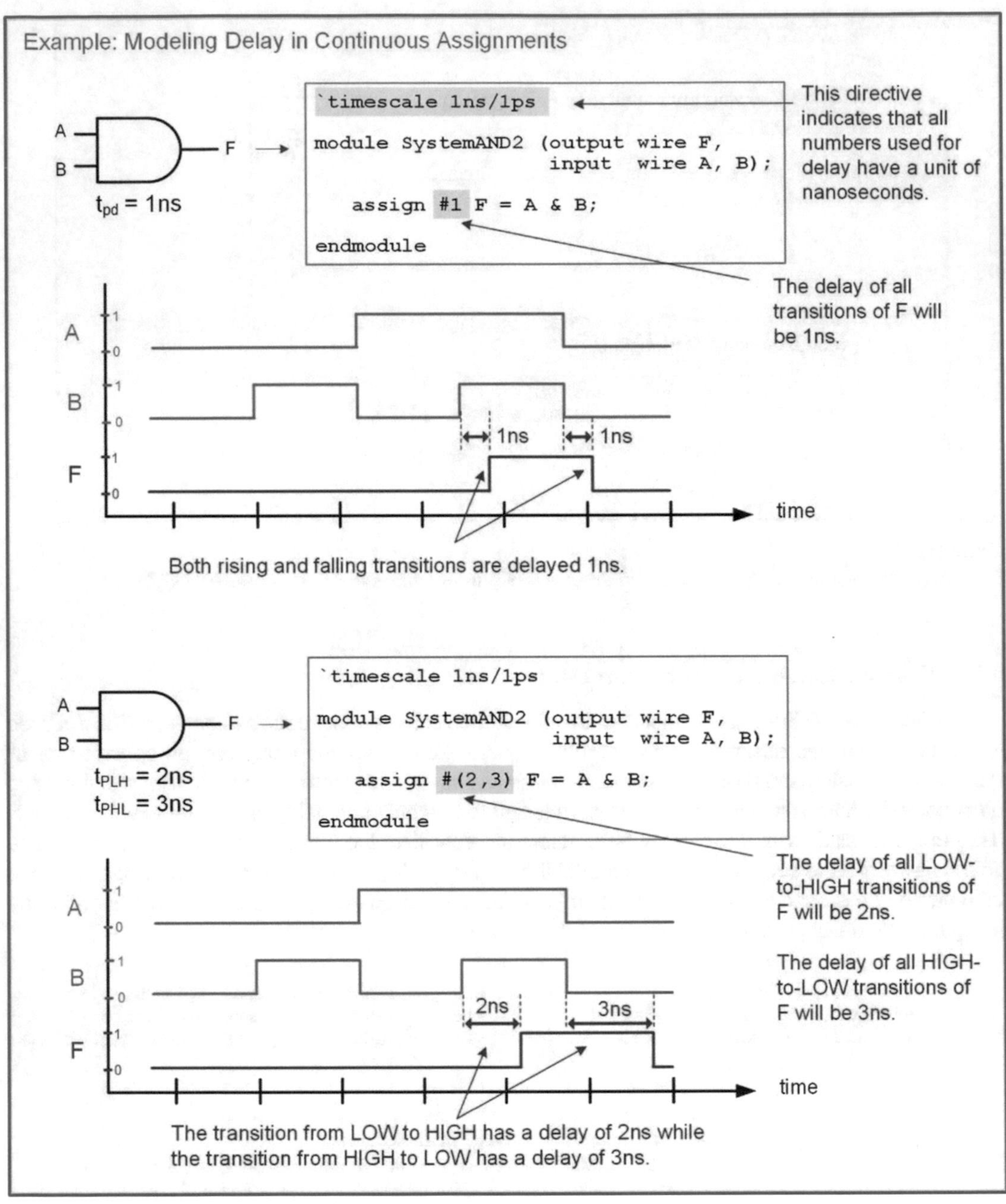

Example 5.5
Modeling delay in continuous assignments

Verilog also provides a mechanism to model a range of delays that are selected by a switch set in the CAD compiler. There are three delays categories that can be specified: *minimum*, *typical*, and *maximum*. The delays are separated by a ":". The following is the syntax of how to use the delay range capability.

```
assign #(<min>:<typ>:<max>) <target_net> = <RHS_nets, operators, etc...>;
```

Example:

```
assign #(1:2:3)                 F=A; // Specifying a range of delays for all transitions.
assign #(1:1:2, 2:2:3)          F=A; // Specifying a range of delays for rising/falling.
assign #(1:1:2, 2:2:3, 4:4:5)   F=A; // Specifying a range of delays for each transition.
```

The delay modeling capability in continuous assignment is designed to model the behavior of real combinational logic with respect to short duration pulses. When a pulse is shorter than the delay of the combinational logic gate, the pulse is ignored. Ignoring brief input pulses on the input accurately models the behavior of on-chip gates. When the input pulse is faster than the delay of the gate, the output of the gate does not have time to respond. As a result, there will not be a logic change on the output. This is called *inertial delay* modeling and is the default behavior when using continuous assignments. Example 5.6 shows a graphical depiction of inertial delay behavior in Verilog.

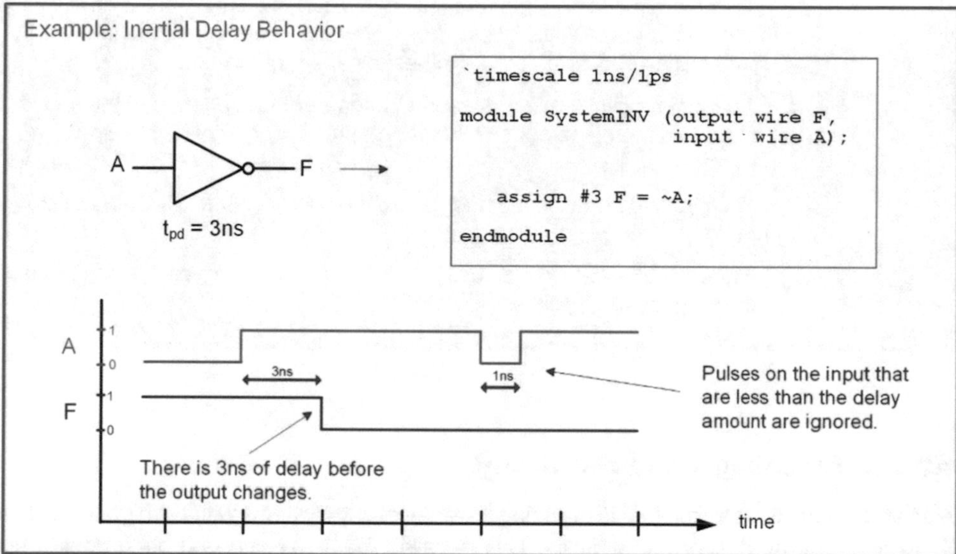

Example 5.6
Inertial delay modeling when using continuous assignment

CONCEPT CHECK

CC5.5(a) Why is concurrency such an important concept in HDLs?

A) Concurrency is a feature of HDLs that can't be modeled using schematics.

B) Concurrency allows automated synthesis to be performed.

C) Concurrency allows logic simulators to display useful system information.

D) Concurrency is necessary to model real systems that operate in parallel.

CC5.5(b) Why does modeling combinational logic in its canonical form with continuous assignment with logical operators defeat the purpose of the modern digital design flow?

A) It requires the designer to first create the circuit using the classical digital design approach and then enter it into the HDL in a form that is essentially a text-based netlist. This doesn't take advantage of the abstraction capabilities and automated synthesis in the modern flow.

B) It cannot be synthesized because the order of precedence of the logical operators in Verilog doesn't match the precedence defined in Boolean algebra.

C) The circuit is in its simplest form so there is no work for the synthesizer to do.

D) It doesn't allow an *else* clause to cover the outputs for any remaining input codes not explicitly listed.

5.6 Structural Design and Hierarchy

Structural design in Verilog refers to including lower-level sub-systems within a higher-level module in order to produce the desired functionality. This is called *hierarchy* and is a good design practice because it enables design partitioning. A purely structural design will not contain any behavioral constructs in the module such as signal assignments but instead just contain the instantiation and interconnections of other sub-systems. A sub-system in Verilog is simply another module that is called by a higher-level module. Each lower-level module that is called is executed concurrently by the calling module.

5.6.1 Lower-Level Module Instantiation

The term *instantiation* refers to the *use* or *inclusion* of a lower-level module within a system. In Verilog, the syntax for instantiating a lower-level module is as follows.

```
module_name <instance_identifier> (port mapping...);
```

The first portion of the instantiation is the module name that is being called. This must match the lower-level module name exactly, including case. The second portion of the instantiation is an optional instance identifier. An instance identifier is useful when instantiating multiple instances of the same lower-level module. The final portion of the instantiation is the port mapping. There are two techniques to connect signals to the ports of the lower-level module, *explicit* and *positional*.

5.6.1.1 Explicit Port Mapping

In explicit port mapping, the names of the ports of the lower-level sub-system are provided along with the signals they are being connected to. The lower-level port name is preceded with a period (.), while the signal it is being connected is enclosed within parentheses. The port connections can be listed in any order since the details of the connection (i.e., port name to signal name) are explicit. Each connection is separated by a comma. The syntax for explicit port mapping is as follows:

```
module_name<instance identifier> (.port_name1(signal1), .port_name2(signal2), etc.);
```

Example 5.7 shows how to design a Verilog model of a hierarchical system that consists of two lower-level modules.

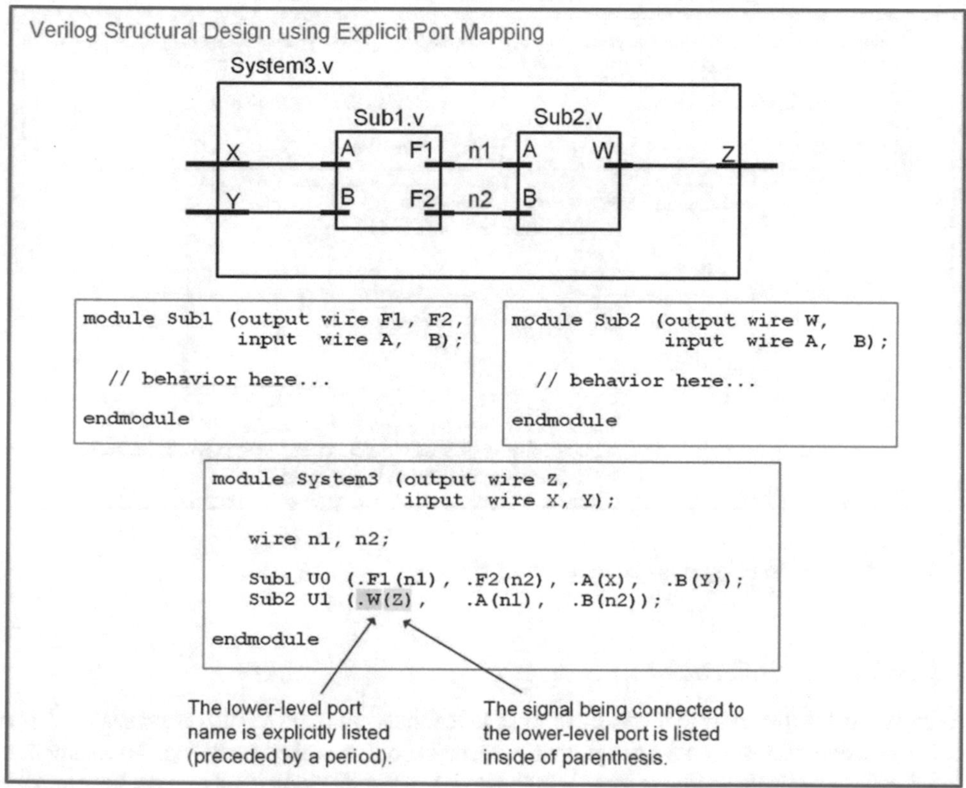

Example 5.7
Verilog structural design using explicit port mapping

5.6.1.2 Positional Port Mapping

In positional port mapping, the names of the ports of the lower-level modules are not explicitly listed. Instead, the signals to be connected to the lower-level system are listed in the same order in which the ports were defined in the sub-system. Each signal name is separated by a comma. This approach requires less text to describe the connection but can also lead to misconnections due to inadvertent mistakes in the signal order. The syntax for positional port mapping is as follows:

```
module_name : <instance_identifier> (signal1, signal2, etc.);
```

Example 5.8 shows how to create the same structural Verilog model as in Example 5.7, but using positional port mapping instead.

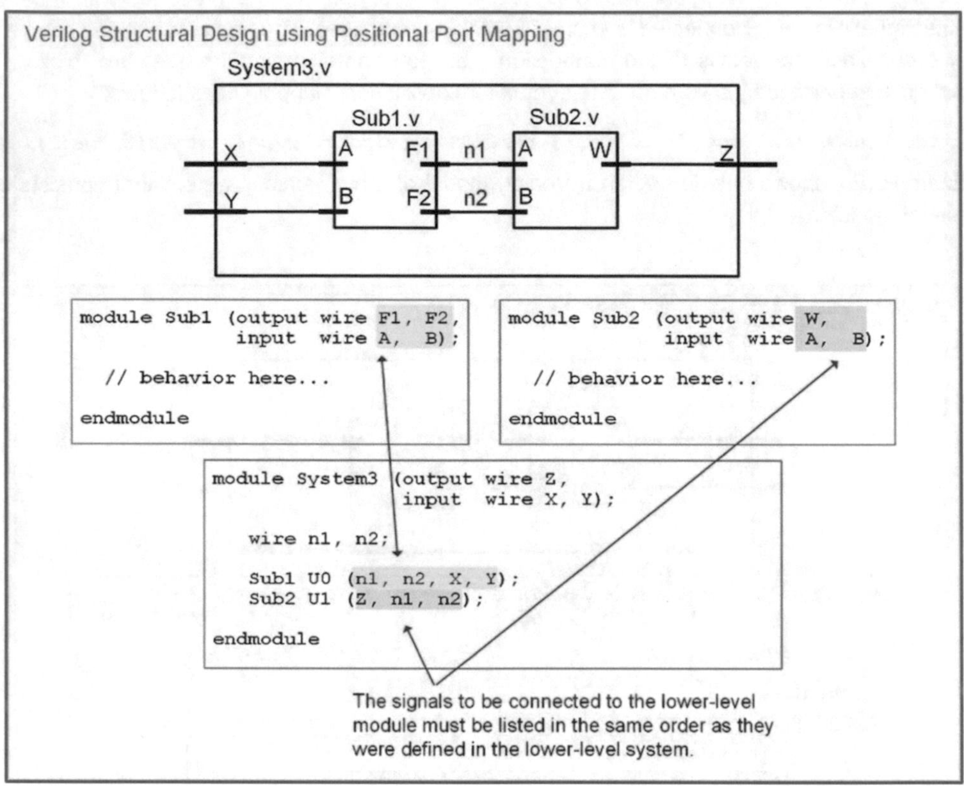

Example 5.8
Verilog structural design using positional port mapping

5.6.2 Gate-Level Primitives

Verilog provides the ability to model basic logic functionality through the use of *primitives*. A primitive is a logic operation that is simple enough that it doesn't require explicit modeling. An example of this behavior can be a basic logic gate or even a truth table. Verilog provides a set of *gate-level primitives* to model simple logic operations. These gate-level primitives are **not()**, **and()**, **nand()**, **or()**, **nor()**, **xor()**, and **xnor()**. Each of these primitives is instantiated as lower-level sub-systems with positional port mapping. The port order for each primitive has the output listed first followed by the input(s). The output and each of the inputs are scalars. Gate-level primitives do not need to be explicitly created as they are provided as part of the Verilog standard. One of the benefits of using gate-level primitives is that the number of inputs is easily scaled as each primitive can accommodate an increasing number of inputs automatically. Furthermore, modeling using this approach essentially provides a gate-level netlist, so it represents a very low-level, detailed gate-level implementation that is ready for technology mapping. Example 5.9 shows how to use gate-level primitives to model the behavior of a combinational logic circuit.

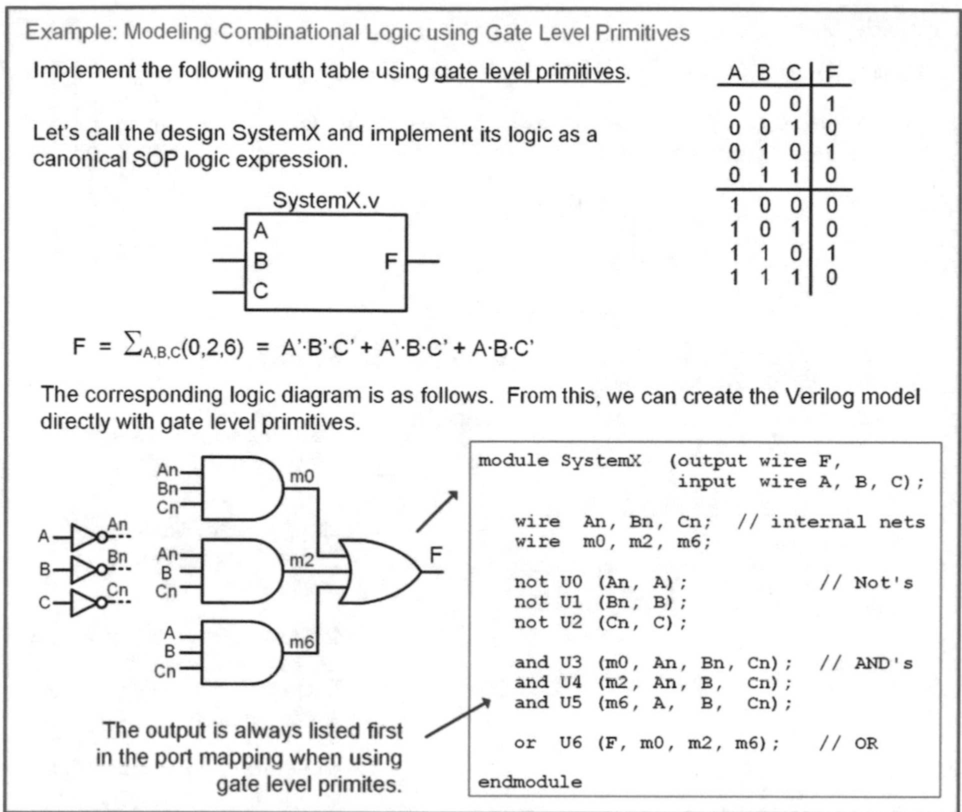

Example 5.9
Modeling combinational logic circuits using gate-level primitives

5.6.3 User-Defined Primitives

A **user-defined primitive** (UDP) is a system that describes the behavior of a low-level component using a logic table. This is very useful for creating combinational logic functionality that will be used numerous times. UDPs are also useful for large truth tables where it is more convenient to list the functionality in table form. UDPs are lower-level sub-systems that are intended to be instantiated in higher-level modules just like gate-level primitives, with the exception that the UPD needs to be created in its own file. The syntax for a UDP is as follows:

```
primitive primitive_name (output output_name,
                          input  input_name1, input_name2, ...);
  table
    in1_val in2_val ... : out_val;
    in1_val in2_val ... : out_val;
    :
  endtable
endprimitive
```

A UDP must list its output(s) first in the port definition. It also does not require types to be defined for the ports. For combinational logic UDPs, all ports are assumed to be of type wire. Example 5.10 shows how to design a user-defined primitive to implement a combinational logic circuit.

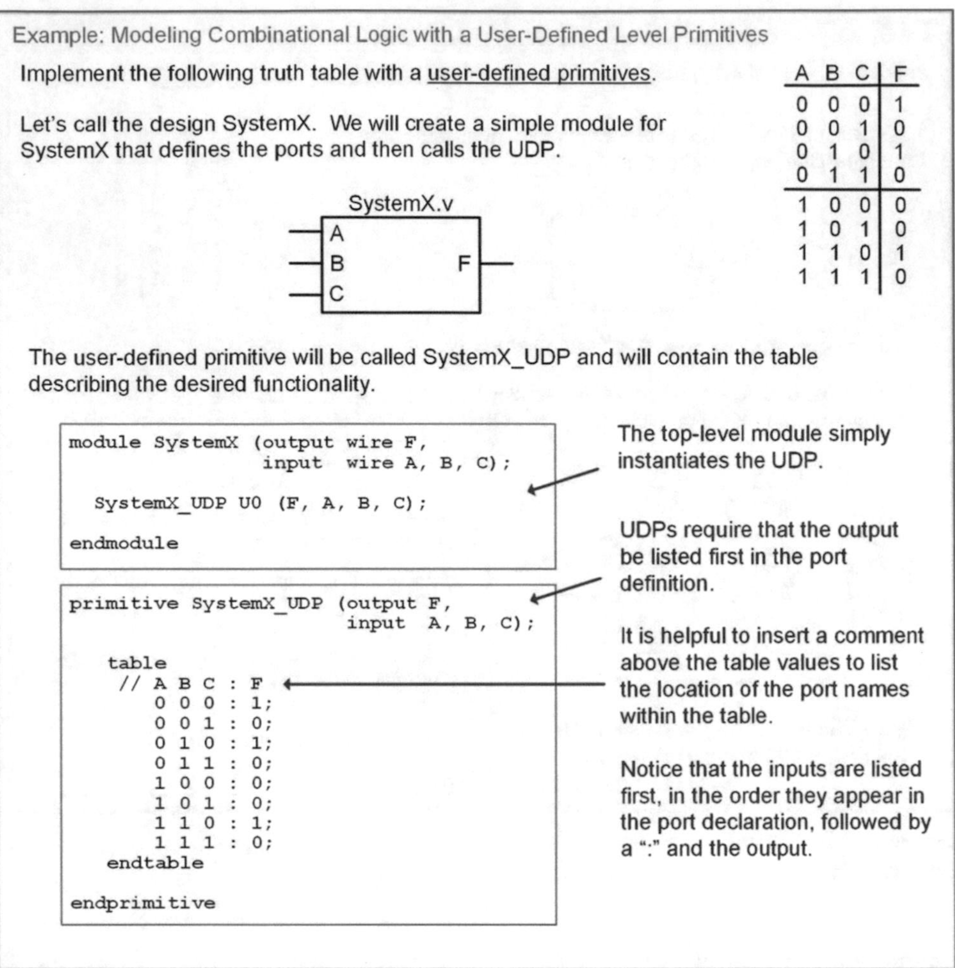

Example 5.10
Modeling combinational logic circuits with a user-defined primitive

5.6.4 Adding Delay to Primitives

Delay can be added to primitives using the same approach as described in Sect. 5.5.4. The delay is inserted after the primitive name but before the instance name.

Example:

```
not #2 U0 (An, A);                  // Gate level primitive for an inverter with delay of 2.
and #3 U3 (m0, An, Bn, Cn);         // Gate level primitive for an AND gate with delay of 3.
SystemX_UDP #1 U0 (F, A, B, C); // UDP with a delay of 1.
```

Concept Check

CC5.6 Does the use of lower-level sub-modules model concurrent functionality? Why?

A) No. Since the lower-level behavior of the module being instantiated may contain nonconcurrent behavior, it is not known what functionality will be modeled.

B) Yes. The modules are treated like independent sub-systems whose behavior runs in parallel just as if separate parts were placed in a design.

5.7 Overview of Simulation Test Benches

One of the essential components of the modern digital design flow is verifying functionality through simulation. This simulation takes place at many levels of abstraction. For a system to be tested, there needs to be a mechanism to generate input patterns to drive the system and then observe the outputs to verify correct operation. The mechanism to do this in Verilog is called a *test bench*. A test bench is a file in Verilog that has no inputs or outputs. The test bench instantiates the system to be tested as a lower-level module. The test bench generates the input conditions and drives them into the input ports of the system being tested. Verilog contains numerous methods to generate stimulus patterns. Since a test bench will not be synthesized, very abstract behavioral modeling can be used to generate the inputs. The output of the system can be viewed as a waveform in a simulation tool. Verilog also has the ability to check the outputs against expected results and notify the user if differences occur. Figure 5.10 gives an overview of how test benches are used in Verilog. The techniques to generate the stimulus patterns are covered in Chap. 8.

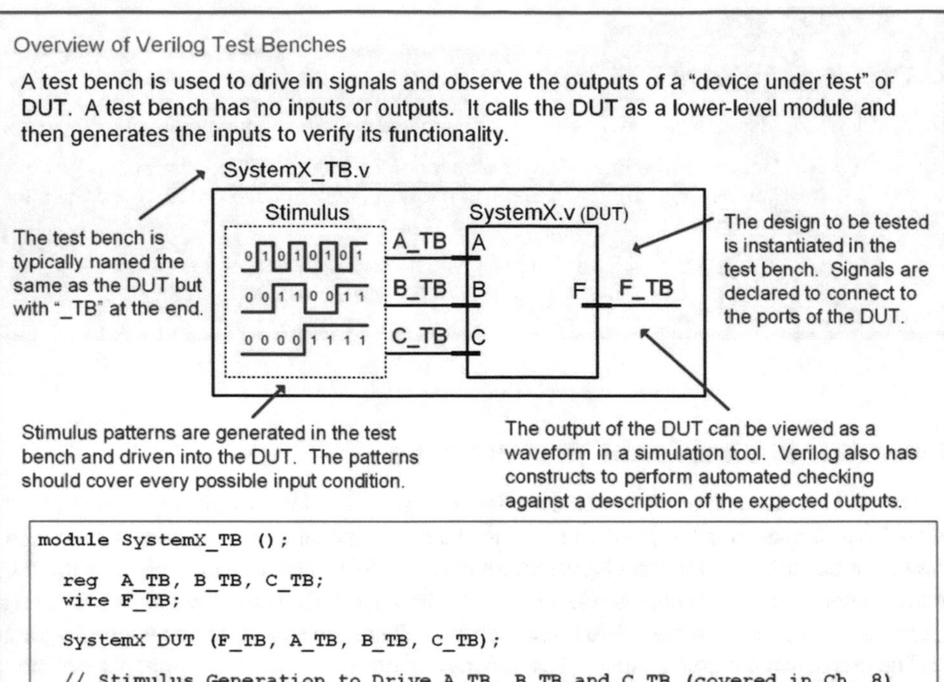


```
module SystemX_TB ();

  reg  A_TB, B_TB, C_TB;
  wire F_TB;

  SystemX DUT (F_TB, A_TB, B_TB, C_TB);

  // Stimulus Generation to Drive A_TB, B_TB and C_TB (covered in Ch. 8)

  // Automated Output Checking & Reporting for F_TB  (covered in Ch. 8)

endmodule
```

Fig. 5.10
Overview of Verilog test benches

Concept Check

CC5.7 How can the output of a DUT be verified when it is connected to a signal that does not go anywhere?

A) It can't. The output must be routed to an output port on the test bench.

B) The values of any dangling signal are automatically written to a text file.

C) It is viewed in the logic simulator as either a waveform or text listing.

D) It can't. A signal that does not go anywhere will cause an error when the Verilog file is compiled.

Summary

- The modern digital design flow relies on computer- aided engineering (CAE) and computer-aided design (CAD) tools to manage the size and complexity of today's digital designs.
- Hardware description languages (HDLs) allow the functionality of digital systems to be entered using text. VHDL and Verilog are the two most common HDLs in use today.
- In the 1980s, two major HDLs emerged, VHDL and Verilog. VHDL was sponsored by the Department of Defense, while Verilog was driven by the commercial industry. Both were later standardized by IEEE.
- The ability to automatically synthesize a logic circuit from a Verilog behavioral description became possible approximately 10 years after the original definition of Verilog. As such, only a subset of the behavioral modeling techniques in Verilog can be automatically synthesized.
- HDLs can model digital systems at different levels of design abstraction. These include the *system*, *algorithmic*, *RTL*, *gate*, and *circuit* levels. Designing at a higher level of abstraction allows more complex systems to be modeled without worrying about the details of the implementation.
- In a Verilog source file, all functionality is contained within a module. The first portion of the module is the port definition. The second portion contains declarations of internal signals/constants/ parameters. The third portion contains the description of the behavior.
- A *port* is an input or output to a system that is defined as part of the initial module statement. A *signal*, or *net*, is an internal connection within the system that is declared inside of the module. A signal is not visible outside of the system.
- Instantiating other modules from within a higher-level module is how Verilog implements hierarchy. A lower-level module can be instantiated as many times as desired. An instance identifier is useful in keeping track of each instantiation. The ports of the component can be connected using either *explicit* or *positional port mapping*.
- *Concurrency* is the term that describes operations being performed in parallel. This allows real-world system behavior to be modeled.
- Verilog provides the *continuous assignment* operator to support modeling concurrent systems. Complex logic circuits can be implemented by using continuous assignment with *logical operators* or *conditional operators*.
- Verilog sub-systems are also treated as concurrent sub-systems.
- Delay can be modeled in Verilog for all transitions, or for individual transitions (rise, fall, off). A range of delays can also be provided (min:typ:max). Delay can be added to continuous assignments and sub-system instantiations.
- Gate-level primitives are provided in Verilog to implement basic logic functions (not, and, nand, or, nor, xor, xnor). These primitives are instantiated just like any other lower-level sub-system.
- User-defined primitives are supported in Verilog that allow the functionality of a circuit to be described in table form.
- A *simulation test bench* is a Verilog file that drives stimulus into a device under test (DUT). Test benches do not have inputs or outputs and are not synthesizable.

Exercise Problems

Section 5.1: History of HDLs

5.1.1 What was the original purpose of Verilog?

5.1.2 Can all of the functionality that can be described in Verilog be simulated?

5.1.3 Can all of the functionality that can be described in Verilog be synthesized?

Section 5.2: HDL Abstraction

5.2.1 Give the level of design abstraction that the following statement relates to: *if there is ever an error in the system, it should return to the reset state.*

5.2.2 Give the level of design abstraction that the following statement relates to: *once the design is implemented in a sum of products form, DeMorgan's Theorem will be used to convert it to a NAND-gate only implementation.*

5.2.3 Give the level of design abstraction that the following statement relates to: *the design will be broken down into two sub-systems, one that will handle data collection and the other that will control data flow.*

5.2.4 Give the level of design abstraction that the following statement relates to: *the interconnect on the IC should be changed from aluminum to copper to achieve the performance needed in this design.*

5.2.5 Give the level of design abstraction that the following statement relates to: *the MOSFETs need to be able to drive at least 8 other loads in this design.*

5.2.6 Give the level of design abstraction that the following statement relates to: *this system will contain 1 host computer and support up to 1000 client computers.*

5.2.7 Give the design domain that the following activity relates to: *drawing the physical layout of the CPU will require 6 months of engineering time.*

5.2.8 Give the design domain that the following activity relates to: *the CPU will be connected to four banks of memory.*

5.2.9 Give the design domain that the following activity relates to: *the fan-in specifications for this logic family require excessive logic circuitry to be used.*

5.2.10 Give the design domain that the following activity relates to: *the performance specifications for this system require 1 TFLOP at <5 W.*

Section 5.3: The Modern Digital Design Flow

5.3.1 Which step in the modern digital design flow does the following statement relate to: *a CAD tool will convert the behavioral model into a gate-level description of functionality.*

5.3.2 Which step in the modern digital design flow does the following statement relate to: *after realistic gate and wiring delays are determined, one last simulation should be performed to make sure the design meets the original timing requirements.*

5.3.3 Which step in the modern digital design flow does the following statement relate to: *if the memory is distributed around the perimeter of the CPU, the wiring density will be minimized.*

5.3.4 Which step in the modern digital design flow does the following statement relate to: *the design meets all requirements so now I'm building the hardware that will be shipped.*

5.3.5 Which step in the modern digital design flow does the following statement relate to: *the system will be broken down into three sub-systems with the following behaviors.*

5.3.6 Which step in the modern digital design flow does the following statement relate to: *this system needs to have 10 Gbytes of memory.*

5.3.7 Which step in the modern digital design flow does the following statement relate to: *to meet the power requirements, the gates will be implemented in the 74HC logic family.*

Section 5.4: Verilog Constructs

5.4.1 What is the name of the main design unit in Verilog?

5.4.2 What portion of the Verilog module describes the inputs and outputs.

5.4.3 What step is necessary if a system requires internal connections?

5.4.4 What are all the possible values that a Verilog net type can take on?

5.4.5 What is the highest strength that a value can take on in Verilog.

5.4.6 What is the range of decimal numbers that can be represented using the type *integer* in Verilog?

5.4.7 What is the width of the vector defined using the type *[63:0] wire*?

5.4.8 What is the syntax for indexing the most significant bit in the type *[31:0] wire*? Assume the vector is named *example*.

5.4.9 What is the syntax for indexing the least significant bit in the type *[31:0] wire*? Assume the vector is named *example*.

5.4.10 What is the difference between a *wire* and *reg* type?

5.4.11 How many bits is the type *integer* by default?

5.4.12 How many bits is the type *real* by default?

Section 5.5: Modeling Concurrent Functionality in Verilog

5.5.1 Design a Verilog model to implement the behavior described by the 3-input minterm list shown in Fig. 5.11. Use continuous assignment with logical operators. Declare your module and ports to match the block diagram provided. Use the type wire for your ports.

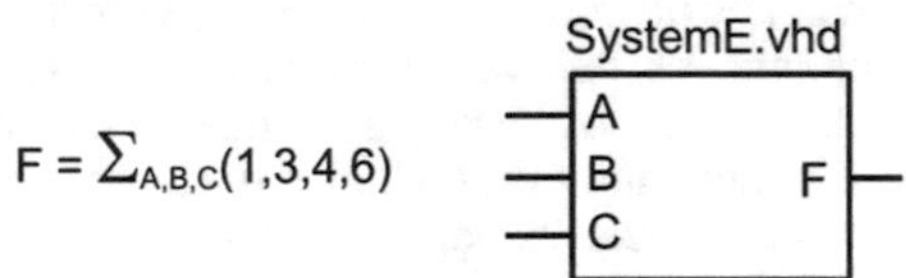

Fig. 5.11
System E functionality

5.5.2 Design a Verilog model to implement the behavior described by the 3-input minterm list shown in Fig. 5.11. Use continuous assignment with conditional operators. Declare your module and ports to match the block diagram provided. Use the type wire for your ports.

5.5.3 Design a Verilog model to implement the behavior described by the 3-input maxterm list shown in Fig. 5.12. Use continuous assignment with logical operators. Declare your module and ports to match the block diagram provided. Use the type wire for your ports.

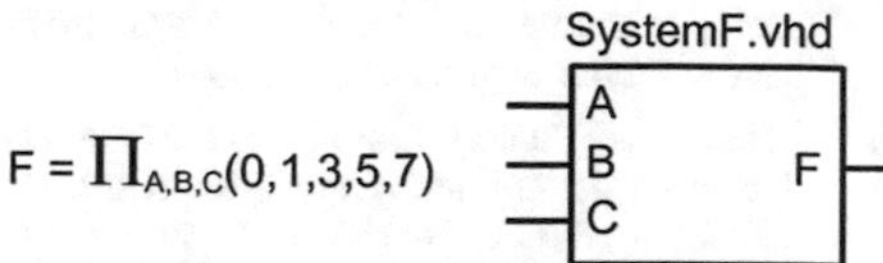

Fig. 5.12
System F functionality

5.5.4 Design a Verilog model to implement the behavior described by the 3-input maxterm list shown in Fig. 5.12. Use continuous assignment with conditional operators. Declare your module and ports to match the block diagram provided. Use the type wire for your ports.

5.5.5 Design a Verilog model to implement the behavior described by the 3-input truth table shown in Fig. 5.13. Use continuous assignment with logical operators. Declare your module and ports to match the block diagram provided. Use the type wire for your ports.

A	B	C	F
0	0	0	1
0	0	1	1
0	1	0	0
0	1	1	0
1	0	0	1
1	0	1	0
1	1	0	0
1	1	1	0

SystemG.vhd

A — B — C — F

Fig. 5.13
System G functionality

5.5.6 Design a Verilog model to implement the behavior described by the 3-input truth table shown in Fig. 5.13. Use continuous assignment with conditional operators. Declare your module and ports to match the block diagram provided. Use the type wire for your ports.

5.5.7 Design a Verilog model to implement the behavior described by the 4-input minterm list shown in Fig. 5.14. Use continuous assignment and logical operators. Declare your module and ports to match the block diagram provided. Use the type wire for your ports.

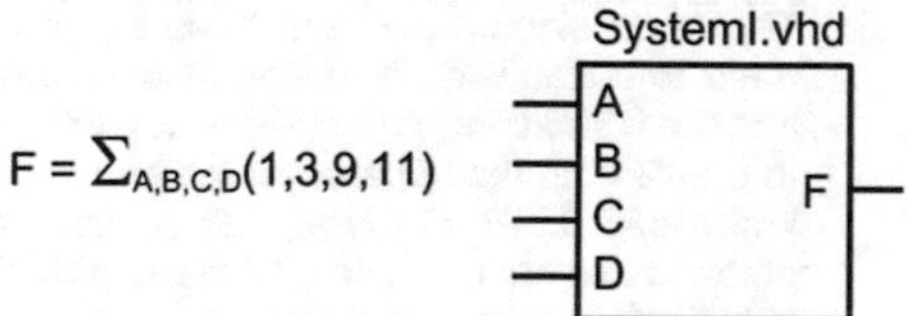

Fig. 5.14
System I functionality

5.5.8 Design a Verilog model to implement the behavior described by the 4-input minterm list shown in Fig. 5.14. Use continuous assignment and conditional operators. Declare your module and ports to match the block diagram provided. Use the type wire for your ports.

5.5.9 Design a Verilog model to implement the behavior described by the 4-input maxterm list shown in Fig. 5.15. Use continuous assignment and logical operators. Declare your module and ports to match the block diagram provided. Use the type wire for your ports.

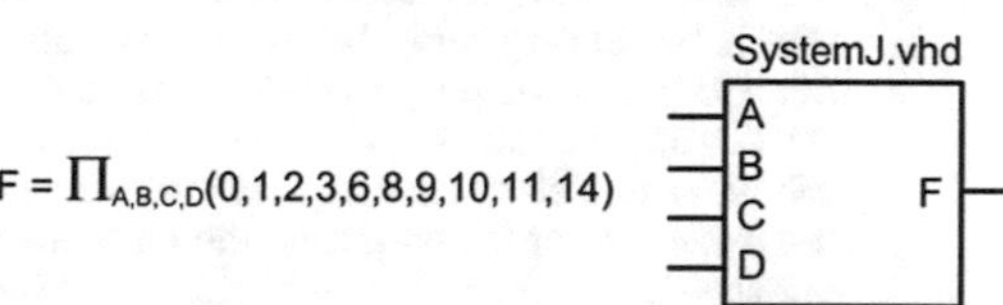

Fig. 5.15
System J functionality

5.5.10 Design a Verilog model to implement the behavior described by the 4-input maxterm list shown in Fig. 5.15. Use continuous assignment and conditional operators. Declare your module and ports to match the block diagram provided. Use the type wire for your ports.

5.5.11 Design a Verilog model to implement the behavior described by the 4-input truth table shown in Fig. 5.16. Use continuous assignment and logical operators. Declare your module and ports to match the block diagram provided. Use the type wire for your ports.

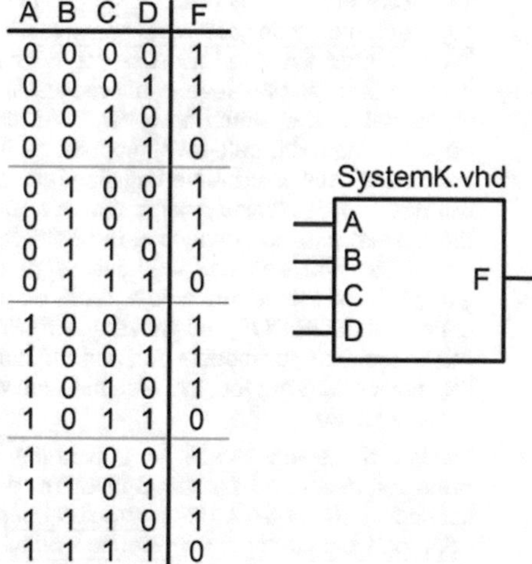

A	B	C	D	F
0	0	0	0	1
0	0	0	1	1
0	0	1	0	1
0	0	1	1	0
0	1	0	0	1
0	1	0	1	1
0	1	1	0	1
0	1	1	1	0
1	0	0	0	1
1	0	0	1	1
1	0	1	0	1
1	0	1	1	0
1	1	0	0	1
1	1	0	1	1
1	1	1	0	1
1	1	1	1	0

Fig. 5.16
System K functionality

5.5.12 Design a Verilog model to implement the behavior described by the 4-input truth table shown in Fig. 5.16. Use continuous assignment and conditional operators. Declare your module and ports to match the block diagram provided. Use the type wire for your ports.

Section 5.6: Structural Design in Verilog

5.6.1 Design a Verilog model to implement the behavior described by the 3-input minterm list shown in Fig. 5.11. Use a structural design approach based on gate- level primitives. This is considered *structural* because you will need to instantiate the gate-level primitives just like a traditional sub-system; however, you don't need to create the gate-level modules as they are already built into the Verilog standard. You will need to determine a logic expression for the system prior to connecting the gate- level primitives. You can use whatever approach you prefer to create the logic expression (i.e., canonical SOP/POS, minimized SOP/POS, etc.). Declare your module and ports to match the block diagram provided. Use the type wire for your ports.

5.6.2 Design a Verilog model to implement the behavior described by the 3-input minterm list shown in Fig. 5.11. Use a structural design approach based on a user-defined primitive. This is considered *structural* because you will need to instantiate the user-defined primitive just like a traditional sub-system. You will need to create both the upper-level module and the lower-level UDP. Declare your module and ports to match the block diagram provided. Use the type wire for your ports.

5.6.3 Design a Verilog model to implement the behavior described by the 3-input maxterm list shown in Fig. 5.12. Use a structural design approach based on gate- level primitives. This is considered *structural* because you will need to instantiate the gate-level primitives just like a traditional sub-system; however, you don't need to create the gate-level modules as they are already built into the Verilog standard. You will need to determine a logic expression for the system prior to connecting the gate- level primitives. You can use whatever approach you prefer to create the logic expression (i.e., canonical SOP/POS, minimized SOP/POS, etc.). Declare your module and ports to match the block diagram provided. Use the type wire for your ports.

5.6.4 Design a Verilog model to implement the behavior described by the 3-input maxterm list shown in Fig. 5.12. Use a structural design approach based on a user-defined primitive. This is considered *structural* because you will need to instantiate the user-defined primitive just like a traditional sub-system. You will need to create both the upper-level module and the lower-level UDP. Declare your module and ports to match the block diagram provided. Use the type wire for your ports.

5.6.5 Design a Verilog model to implement the behavior described by the 3-input truth table shown in Fig. 5.13. Use a structural design approach based on gate-level primitives. This is considered *structural* because you will need to instantiate the gate-level primitives just like a traditional sub-system; however, you don't need to create the gate-level modules as they are already built into the Verilog standard. You will need to determine a logic expression for the system prior to connecting the gate- level primitives. You can use whatever approach you prefer to create the logic expression (i.e., canonical SOP/POS, minimized SOP/POS, etc.). Declare your module and ports to match the block diagram provided. Use the type wire for your ports.

5.6.6 Design a Verilog model to implement the behavior described by the 3-input truth table shown in Fig. 5.13. Use a structural design approach based on a user-defined primitive. This is considered *structural* because you will need to instantiate the user-defined primitive just like a traditional sub-system. You will need to create both the upper-level module and the lower-level UDP. Declare your module and ports to match the block diagram provided. Use the type wire for your ports.

5.6.7 Design a Verilog model to implement the behavior described by the 4-input minterm list shown in Fig. 5.14. Use a structural design approach based on gate- level primitives. This is considered *structural* because you will need to instantiate the gate-level primitives just like a traditional sub-system; however, you don't need to create the gate-level modules as they are already built into the Verilog standard. You will need to determine a logic expression for the system prior to connecting the gate- level primitives. You can use whatever approach you prefer to create the logic expression (i.e., canonical SOP/POS, minimized SOP/POS, etc.). Declare your module and ports to match the block diagram provided. Use the type wire for your ports.

5.6.8 Design a Verilog model to implement the behavior described by the 4-input minterm list shown in Fig. 5.14. Use a structural design approach based on a user- defined primitive. This is considered *structural* because you will need to instantiate the user-defined primitive just like a traditional sub-system. You will need to create both the upper-level module and the lower-level UDP. Declare your module and ports to match the block diagram provided. Use the type wire for your ports.

5.6.9 Design a Verilog model to implement the behavior described by the 4-input maxterm list shown in Fig. 5.15. Use a structural design approach based on gate- level primitives. This is considered *structural* because you will need to instantiate the gate-level primitives just like a traditional sub-system; however, you don't need to create the gate-level modules as they are already built into the Verilog standard. You will need to determine a logic expression for the system prior to connecting the gate- level primitives. You can use whatever approach you prefer to create the logic expression (i.e., canonical SOP/POS, minimized SOP/POS, etc.). Declare your module and ports to match the block diagram provided. Use the type wire for your ports.

5.6.10 Design a Verilog model to implement the behavior described by the 4-input maxterm list shown in Fig. 5.15. Use a structural design approach based on a user- defined primitive. This is considered *structural* because you will need to instantiate the user-defined primitive just like a traditional sub-system. You will need to create both the upper-level module and the lower-level UDP. Declare your module and ports to match the block diagram provided. Use the type wire for your ports.

5.6.11 Design a Verilog model to implement the behavior described by the 4-input truth table shown in Fig. 5.16. Use a structural design approach based on gate-level primitives. This is considered *structural* because you will need to instantiate the gate-level primitives just like a traditional sub-system; however, you don't need to create the gate-level modules as they are already built into the Verilog standard. You will need to determine a logic expression for the system prior to connecting the gate- level primitives. You can use whatever approach you prefer to create the logic expression (i.e., canonical SOP/POS, minimized SOP/POS, etc.). Declare your module and ports to match the block diagram provided. Use the type wire for your ports.

5.6.12 Design a Verilog model to implement the behavior described by the 4-input truth table shown in Fig. 5.16. Use a structural design approach based on a user-defined primitive. This is considered *structural* because you will need to instantiate the user-defined primitive just like a traditional sub-system. You will need to create both the upper-level module and the lower-level UDP. Declare your module and ports to match the block diagram provided. Use the type wire for your ports.

Section 5.7: Overview of Simulation Test Benches

5.7.1 What is the purpose of a test bench?

5.7.2 Does a test bench have input and output ports?

5.7.3 Can a test bench be simulated?

5.7.4 Can a test bench be synthesized?

Chapter 6: MSI Logic

This chapter introduces a group of combinational logic building blocks that are commonly used in digital design. As we move into systems that are larger than individual gates, there are naming conventions that are used to describe the size of the logic. Table 6.1 gives these naming conventions. In this chapter, we will look at *medium- scale integrated circuit* (MSI) logic. Each of these building blocks can be implemented using the combinational logic design steps covered in Chaps. 4 and 5. The goal of this chapter is to provide an understanding of the basic principles of MSI logic.

Commonly Used Names to Describe The Size of Digital Logic

Name	Example	# of Transistors
SSI - Small Scale Integrated Circuits	Individual Gates (NAND, INV)	10's
MSI - Medium Scale Integrated Circuits	Decoders, Multiplexers	100's
LSI – Large Scale Integrated Circuits	Arithmetic Circuits, RAM	1k – 10k
VLSI – Very Large Scale Integrated Circuits	Microprocessors	100k – 1M

While there are names for logic sizes above 1M transistor such as ULSI (Ultra), the term "VLSI" is now used to describe all integrated circuits that are so large they require CAD tools for their design, synthesis and implementation.

Table 6.1
Naming convention for the size of digital systems

Learning Outcomes—After completing this chapter, you will be able to:

6.1 Design a decoder circuit using both the classical digital design approach and the modern HDL-based approach.
6.2 Design an encoder circuit using both the classical digital design approach and the modern HDL-based approach.
6.3 Design a multiplexer circuit using both the classical digital design approach and the modern HDL-based approach.
6.4 Design a demultiplexer circuit using both the classical digital design approach and the modern HDL-based approach.

6.1 Decoders

A decoder is a circuit that takes in a binary *code* and has outputs that are asserted for specific values of that code. The code can be of any type or size (e.g., unsigned, two's complement, etc.). Each output will assert for only specific input codes. Since combinational logic circuits only produce a single output, this means that within a decoder, there will be a separate combinational logic circuit for each output.

6.1.1 Example: One-Hot Decoder

A one-hot decoder is a circuit that has n inputs and 2^n outputs. Each output will assert for one and only one input code. Since there are 2^n outputs, there will always be one and only one output asserted at

© Springer Nature Switzerland AG 2019

B. J. LaMeres, *Introduction to Logic Circuits & Logic Design with Verilog*,
https://doi.org/10.1007/978-3-030-13605-5_6

any given time. Example 6.1 shows the process of designing a 2-to-4 one-hot decoder by hand (i.e., using the classical digital design approach).

Example: 2-to-4 One-Hot Decoder - Logic Synthesis by Hand

The block diagram and truth table for this system are as follows:

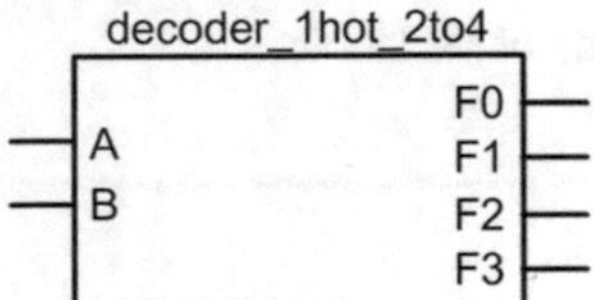

A	B	F3	F2	F1	F0
0	0	0	0	0	1
0	1	0	0	1	0
1	0	0	1	0	0
1	1	1	0	0	0

Each output asserts for a specific input code. This is where the term "one-hot" comes from. Each output is only "hot" for one input code.

When designing this circuit, each output needs to have its own separate combinational logic circuit. This is the same as if there were four separate truth tables. This design could be implemented using 4x, 2-input K-maps to form the logic expressions for these outputs; however, by inspection a minterm list for each output will be the most minimal circuit.

$$F0 = \Sigma_{A,B}(0) = A' \cdot B' \qquad F2 = \Sigma_{A,B}(2) = A \cdot B'$$

$$F1 = \Sigma_{A,B}(1) = A' \cdot B \qquad F3 = \Sigma_{A,B}(3) = A \cdot B$$

When implementing the final decoder, the input inversions for A and B can be shared across all of the AND gates.

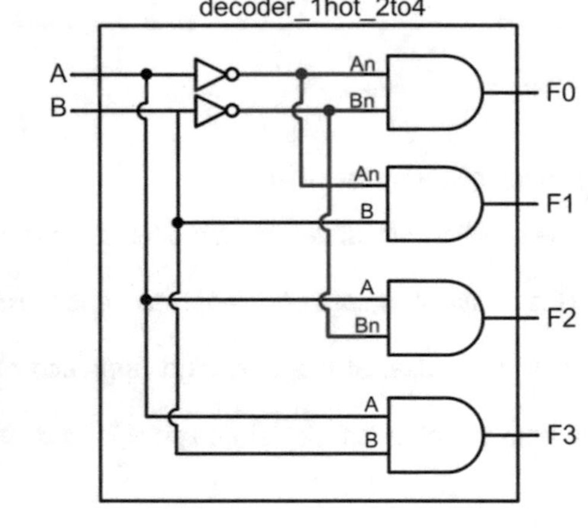

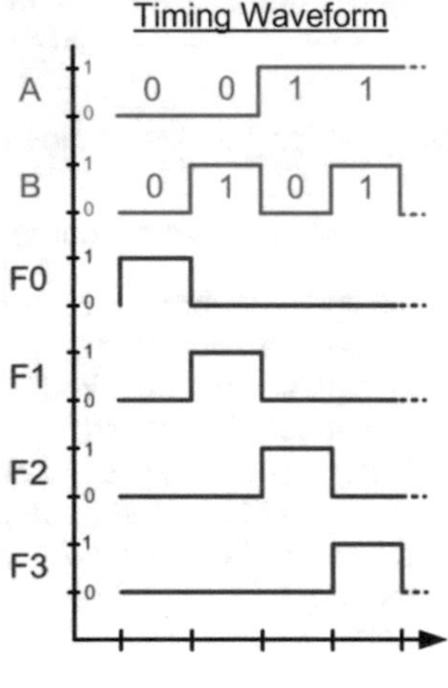

Example 6.1
2-to-4 one-hot decoder: logic synthesis by hand

As decoders get larger, it is necessary to use hardware description languages to model their behavior. Example 6.2 shows how to model a 3-to-8 one-hot decoder in Verilog with continuous assignment and logic operators.

Example: 3-to-8 One-Hot Decoder – Verilog Modeling using Logical Operators

The block diagram and truth table for this system are as follows:

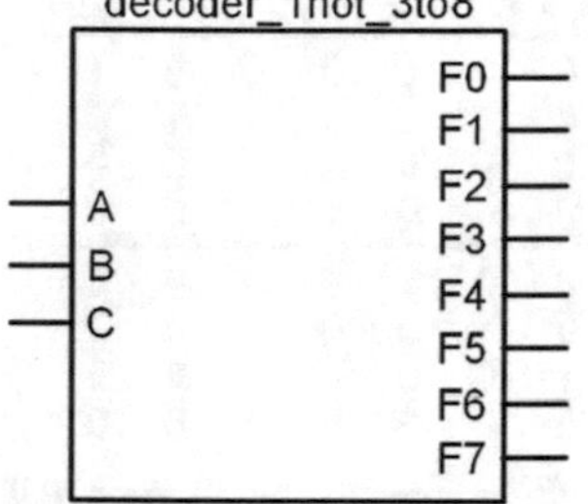

A	B	C	F7	F6	F5	F4	F3	F2	F1	F0
0	0	0	0	0	0	0	0	0	0	1
0	0	1	0	0	0	0	0	0	1	0
0	1	0	0	0	0	0	0	1	0	0
0	1	1	0	0	0	0	1	0	0	0
1	0	0	0	0	0	1	0	0	0	0
1	0	1	0	0	1	0	0	0	0	0
1	1	0	0	1	0	0	0	0	0	0
1	1	1	1	0	0	0	0	0	0	0

To implement this in Verilog using logical operators, we must first determine the logic that will be used in the continuous assignment. Again, since each logic function only has one input code corresponding to an output of '1', the minterm can be used to implement the logic.

$F0 = \sum_{A,B,C}(0) = A'\cdot B'\cdot C'$ $F4 = \sum_{A,B,C}(4) = A\cdot B'\cdot C'$

$F1 = \sum_{A,B,C}(1) = A'\cdot B'\cdot C$ $F5 = \sum_{A,B,C}(5) = A\cdot B'\cdot C$

$F2 = \sum_{A,B,C}(2) = A'\cdot B\cdot C'$ $F6 = \sum_{A,B,C}(6) = A\cdot B\cdot C'$

$F3 = \sum_{A,B,C}(3) = A'\cdot B\cdot C$ $F7 = \sum_{A,B,C}(7) = A\cdot B\cdot C$

In Verilog, each of the outputs requires a separate continuous assignment.

```
module decoder_1hot_3to8
  (output wire F0, F1, F2, F3, F4, F5, F6, F7,
   input  wire A, B, C);

  assign F0 = ~A & ~B & ~C;
  assign F1 = ~A & ~B &  C;
  assign F2 = ~A &  B & ~C;
  assign F3 = ~A &  B &  C;
  assign F4 =  A & ~B & ~C;
  assign F5 =  A & ~B &  C;
  assign F6 =  A &  B & ~C;
  assign F7 =  A &  B &  C;

endmodule
```

Example 6.2
3-to-8 one-hot decoder: Verilog modeling using logical operators

This description can be further simplified by using vector notation for the ports and describing the functionality using conditional operators. Example 6.3 shows how to model the 3-to-8 one-hot decoder in Verilog using continuous assignment with conditional operators.

Example: 3-to-8 One-Hot Decoder – Verilog Modeling using Conditional Operators

The block diagram and truth table for this system are as follows. Notice that the input and output ports now use vectors in order to create a more compact description.

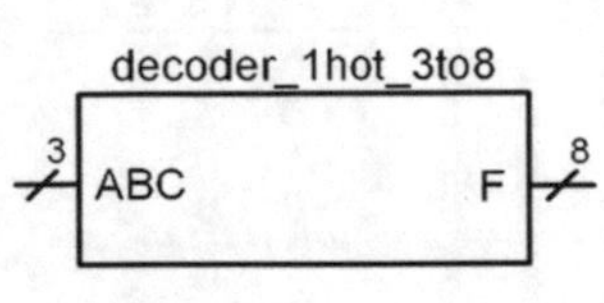

ABC	F(7)	F(6)	F(5)	F(4)	F(3)	F(2)	F(1)	F(0)
"000"	0	0	0	0	0	0	0	1
"001"	0	0	0	0	0	0	1	0
"010"	0	0	0	0	0	1	0	0
"011"	0	0	0	0	1	0	0	0
"100"	0	0	0	1	0	0	0	0
"101"	0	0	1	0	0	0	0	0
"110"	0	1	0	0	0	0	0	0
"111"	1	0	0	0	0	0	0	0

The following shows a technique to model the decoder using continuous assignment with conditional operators. Note that the output will be "unknown" (X) if the input code is not one of the eight possible binary input values.

```
module decoder_1hot_3to8 (output wire [7:0] F,
                          input  wire [2:0] ABC);

   assign F = (ABC == 3'b000) ? 8'b0000_0001 :
              (ABC == 3'b001) ? 8'b0000_0010 :
              (ABC == 3'b010) ? 8'b0000_0100 :
              (ABC == 3'b011) ? 8'b0000_1000 :
              (ABC == 3'b100) ? 8'b0001_0000 :
              (ABC == 3'b101) ? 8'b0010_0000 :
              (ABC == 3'b110) ? 8'b0100_0000 :
              (ABC == 3'b111) ? 8'b1000_0000 :
              8'bXXXX_XXXX;

endmodule
```

Example 6.3
3-to-8 one-hot decoder: Verilog modeling using conditional operators

6.1.2 Example: 7-Segment Display Decoder

A 7-segment display decoder is a circuit used to drive character displays that are commonly found in applications such as digital clocks and household appliances. A character display is made up of seven individual LEDs, typically labeled a–g. The input to the decoder is the binary equivalent of the decimal or Hex character that is to be displayed. The output of the decoder is the arrangement of LEDs that will form the character. Decoders with 2-inputs can drive characters "0" to "3." Decoders with 3-inputs can drive characters "0" to "7." Decoders with 4-inputs can drive characters "0" to "F" with the case of the Hex characters being "A, b, c or C, d, E and F."

Let's look at an example of how to design a 3-input, 7-segment decoder by hand. The first step in the process is to create the truth table for the outputs that will drive the LEDs in the display. We'll call these outputs F_a, F_b, …, F_g. Example 6.4 shows how to construct the truth table for the 7-segment display decoder. In this table, a logic 1 corresponds to the LED being ON.

Example: 7-Segment Display Decoder - Truth Table

LED Labels

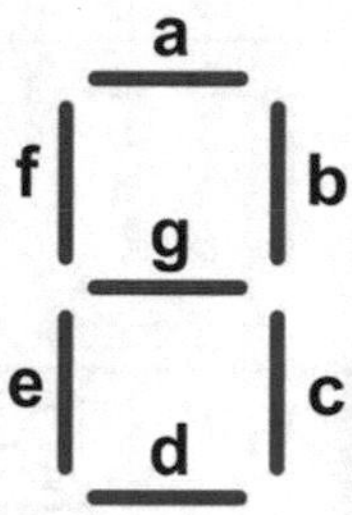

a

f b

g

e c

d

A	B	C		F_a	F_b	F_c	F_d	F_e	F_f	F_g
0	0	0		1	1	1	1	1	1	0
0	0	1		0	1	1	0	0	0	0
0	1	0		1	1	0	1	1	0	1
0	1	1		1	1	1	1	0	0	1
1	0	0		0	1	1	0	0	1	1
1	0	1		1	0	1	1	0	1	1
1	1	0		1	0	1	1	1	1	1
1	1	1		1	1	1	0	0	0	0

Example 6.4
7-Segment display decoder: truth table

If we wish to design this decoder by hand, we need to create seven separate combinational logic circuits. Each of the outputs (F_a – F_g) can be put into a 3-input K-map to find the minimized logic expression. Example 6.5 shows the design of the decoder from the truth table in Example 6.4 by hand.

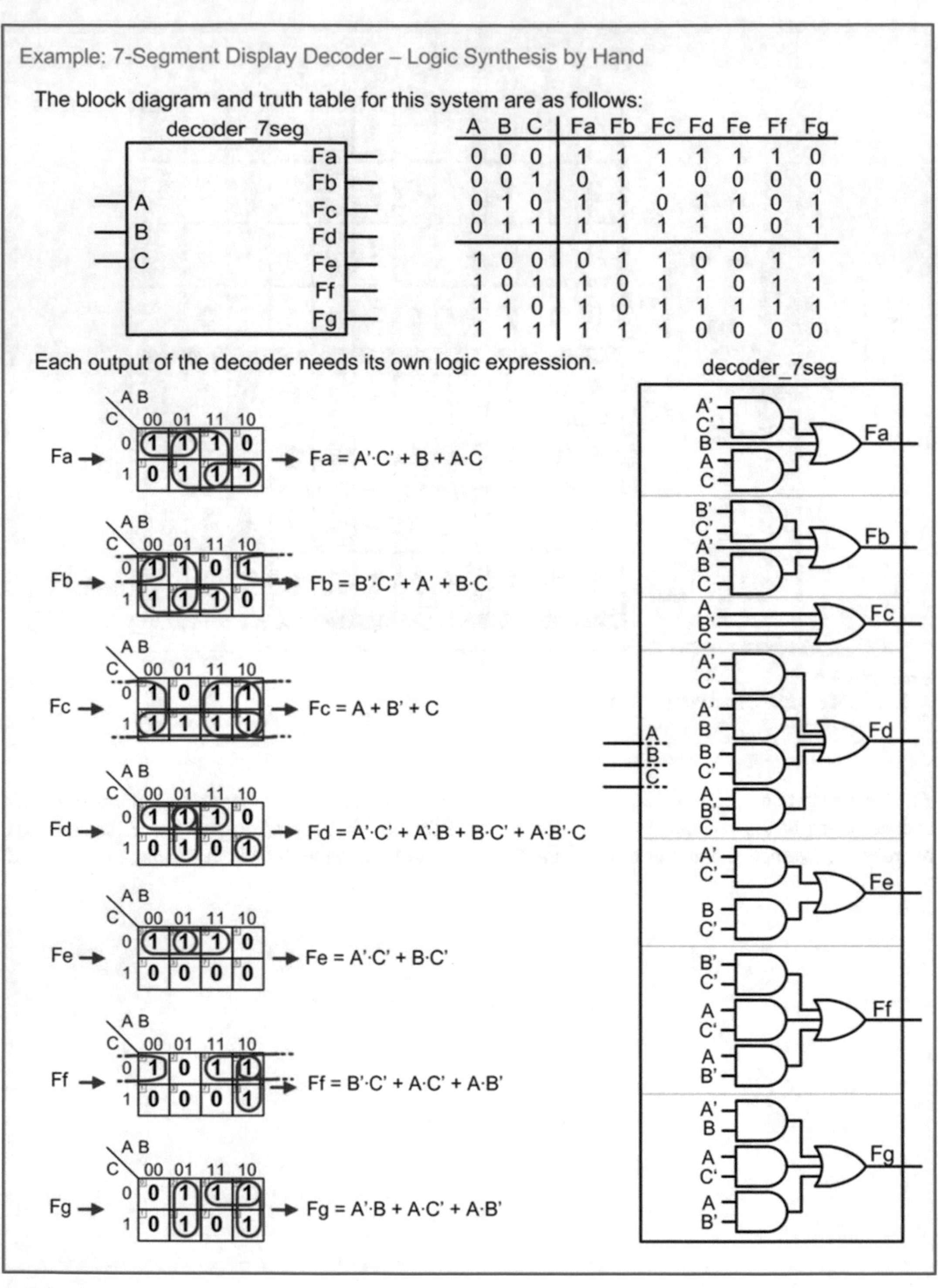

A	B	C	Fa	Fb	Fc	Fd	Fe	Ff	Fg
0	0	0	1	1	1	1	1	1	0
0	0	1	0	1	1	0	0	0	0
0	1	0	1	1	0	1	1	0	1
0	1	1	1	1	1	1	0	0	1
1	0	0	0	1	1	0	0	1	1
1	0	1	1	0	1	1	0	1	1
1	1	0	1	0	1	1	1	1	1
1	1	1	1	1	1	0	0	0	0

Example 6.5
7-Segment display decoder: logic synthesis by hand

This same functionality can be implemented in Verilog using concurrent modeling techniques. Example 6.6 shows how to model the 7-segment decoder in Verilog using continuous assignment with logic operators.

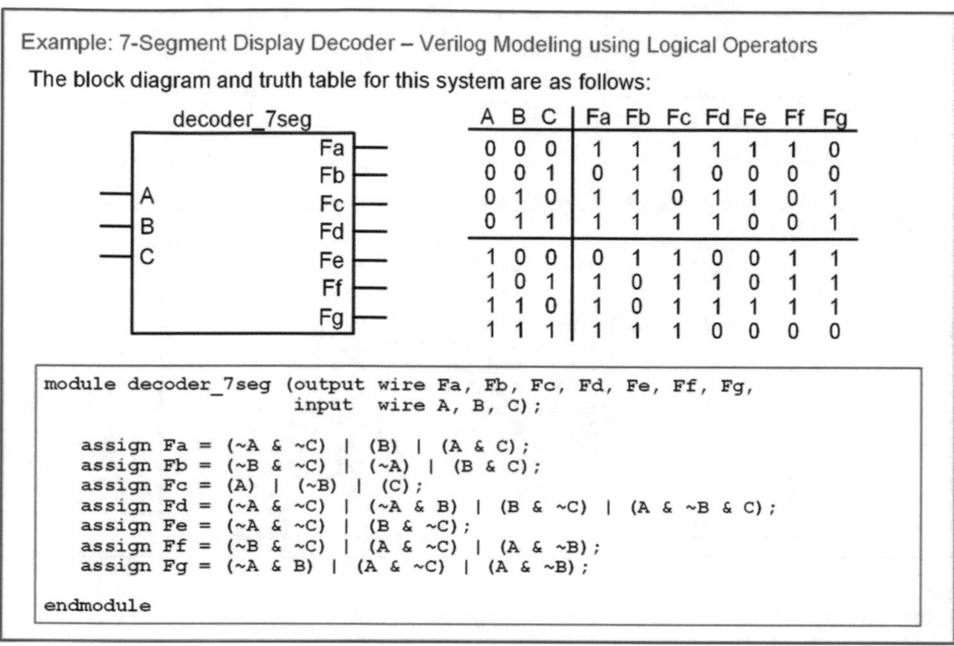
Example: 7-Segment Display Decoder – Verilog Modeling using Logical Operators

The block diagram and truth table for this system are as follows:

A	B	C	Fa	Fb	Fc	Fd	Fe	Ff	Fg
0	0	0	1	1	1	1	1	1	0
0	0	1	0	1	1	0	0	0	0
0	1	0	1	1	0	1	1	0	1
0	1	1	1	1	1	1	0	0	1
1	0	0	0	1	1	0	0	1	1
1	0	1	1	0	1	1	0	1	1
1	1	0	1	0	1	1	1	1	1
1	1	1	1	1	1	0	0	0	0

```
module decoder_7seg (output wire Fa, Fb, Fc, Fd, Fe, Ff, Fg,
                     input  wire A, B, C);

   assign Fa = (~A & ~C) | (B) | (A & C);
   assign Fb = (~B & ~C) | (~A) | (B & C);
   assign Fc = (A) | (~B) | (C);
   assign Fd = (~A & ~C) | (~A & B) | (B & ~C) | (A & ~B & C);
   assign Fe = (~A & ~C) | (B & ~C);
   assign Ff = (~B & ~C) | (A & ~C) | (A & ~B);
   assign Fg = (~A & B) | (A & ~C) | (A & ~B);

endmodule
```

Example 6.6
7-Segment display decoder: Verilog modeling using logical operators

Again, a more compact description of the decoder can be accomplished if the ports are described as vectors and a conditional operator is used. Example 6.7 shows how to model the 7-segment decoder in Verilog using continuous assignment with conditional operators.

Example: 7-Segment Decoder – Verilog Modeling using Conditional Operators

The block diagram and truth table for this system are as follows:

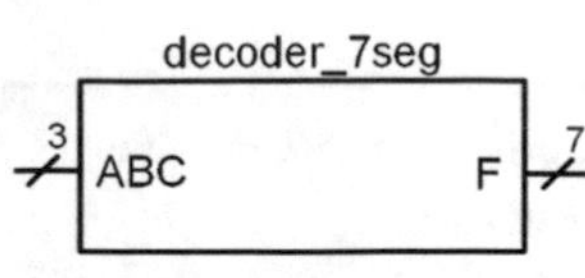

ABC	a F(6)	b F(5)	c F(4)	d F(3)	e F(2)	f F(1)	g F(0)
"000"	1	1	1	1	1	1	0
"001"	0	1	1	0	0	0	0
"010"	1	1	0	1	1	0	1
"011"	1	1	1	1	0	0	1
"100"	0	1	1	0	0	1	1
"101"	1	0	1	1	0	1	1
"110"	1	0	1	1	1	1	1
"111"	1	1	1	0	0	0	0

The following shows a technique to model the decoder using continuous assignment with conditional operators.

```
module decoder_7seg (output wire [6:0] F,
                     input  wire [2:0] ABC);

   assign F = (ABC == 3'b000) ? 7'b111_1110 :
              (ABC == 3'b001) ? 7'b011_0000 :
              (ABC == 3'b010) ? 7'b110_1101 :
              (ABC == 3'b011) ? 7'b111_1001 :
              (ABC == 3'b100) ? 7'b011_0011 :
              (ABC == 3'b101) ? 7'b101_1011 :
              (ABC == 3'b110) ? 7'b101_1111 :
              (ABC == 3'b111) ? 7'b111_0000 :
              8'bXXXX_XXXX;

endmodule
```

Example 6.7
7-Segment display decoder: Verilog modeling using conditional operators

Concept Check

CC6.1 In a decoder, a logic expression is created for each output. Once all of the output logic expressions are found, how can the decoder logic be further minimized?

A) By using K-maps to find the output logic expressions.

B) By buffering the inputs so that they can drive a large number of other gates.

C) By identifying any logic terms that are used in multiple locations (inversions, product terms, and sum terms) and sharing the interim results among multiple circuits in the decoder.

D) By ignoring fan-out.

6.2 Encoders

An encoder works in the opposite manner as a decoder. An assertion on a specific input port corresponds to a unique code on the output port.

6.2.1 Example: One-Hot Binary Encoder

A one-hot binary encoder has n outputs and 2^n inputs. The output will be an n-bit, binary code which corresponds to an assertion on one and only one of the inputs. Example 6.8 shows the process of designing a 4-to-2 binary encoder by hand (i.e., using the classical digital design approach).

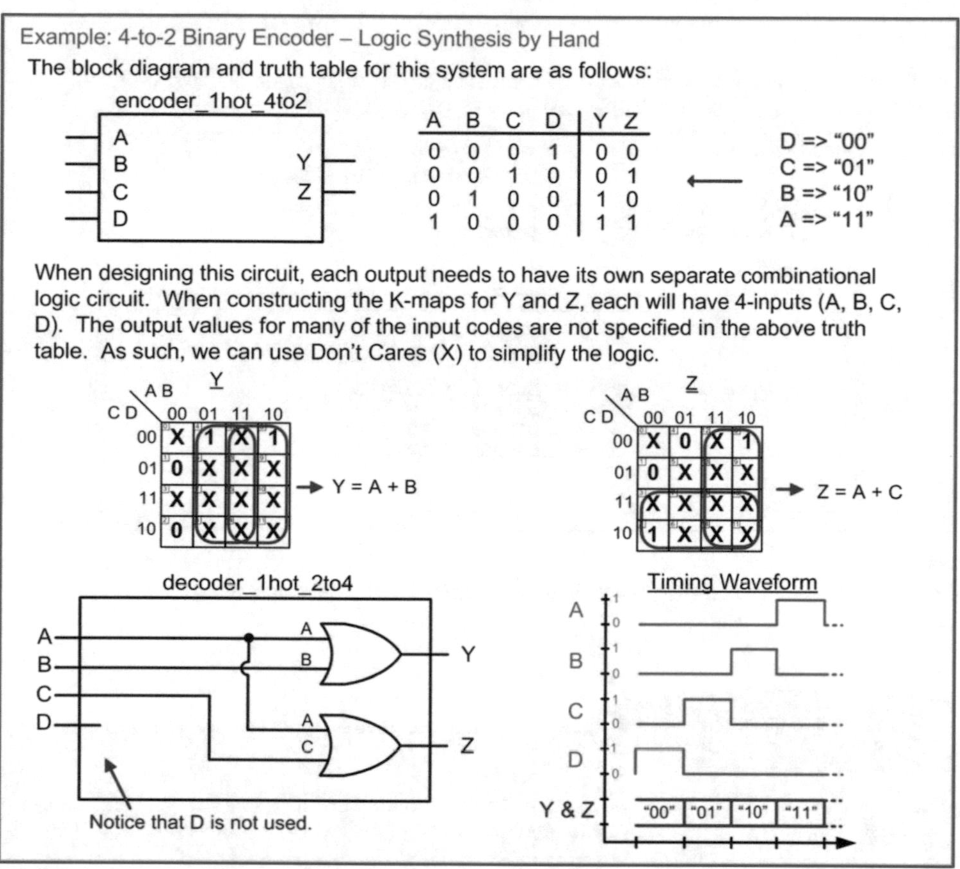

Example 6.8
4-to-2 binary encoder: logic synthesis by hand

In Verilog, an encoder can be implemented using continuous assignment with either logical or conditional operators. Example 6.9 shows how to model the encoder in Verilog using these techniques.

Example: 4-to-2 Binary Encoder – Verilog Modeling using Continuous Assignment

The block diagram and truth table for this system are as follows:

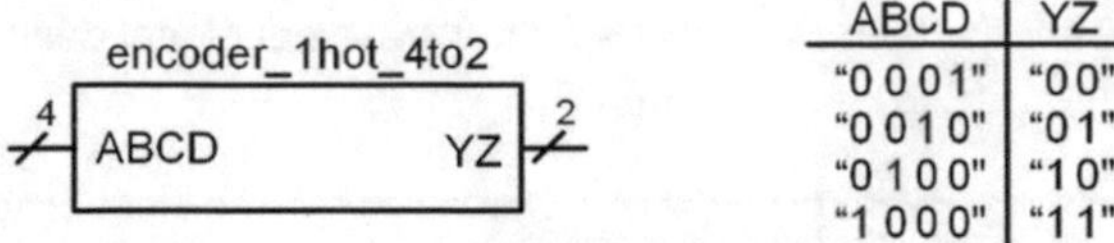

ABCD	YZ
"0001"	"00"
"0010"	"01"
"0100"	"10"
"1000"	"11"

The following are two different ways to implement the behavior of the encoder with continuous assignment: (1) with logical operators; and (2) with conditional operators.

(1)

```
module encoder_1hot_4to2 (output wire [1:0] YZ,
                          input  wire [3:0] ABCD);

   assign YZ[1] = ABCD[3] | ABCD[2];
   assign YZ[0] = ABCD[3] | ABCD[1];

endmodule
```

(2)

```
module encoder_1hot_4to2 (output wire [1:0] YZ,
                          input  wire [3:0] ABCD);

   assign YZ = (ABCD == 4'b0001) ? 2'b00 :
               (ABCD == 4'b0010) ? 2'b01 :
               (ABCD == 4'b0100) ? 2'b10 :
               (ABCD == 4'b1000) ? 2'b11 :
               2'bXX;
endmodule
```

Example 6.9
4-to-2 binary encoder: Verilog modeling using logical and conditional operators

Concept Check

CC6.2 If it is desired to have the outputs of an encoder produce 0's for all input codes not defined in the truth table, can "don't cares" be used when deriving the minimized logic expressions? Why?

A) No. Don't cares aren't used in encoders.

B) Yes. Don't cares can always be used in K-maps.

C) Yes. All that needs to be done is to treat each X as a 0 when forming the most minimal prime implicant.

D) No. Each cell in the K-map corresponding to an undefined input code needs to contain a 0 so don't cares are not applicable.

6.3 Multiplexers

A multiplexer is a circuit that passes one of its multiple inputs to a single output based on a select input. This can be thought of as a digital switch. The multiplexer has n select lines, 2^n inputs, and one output. Example 6.10 shows the process of designing a 2-to-1 multiplexer by hand (i.e., using the classical digital design approach).

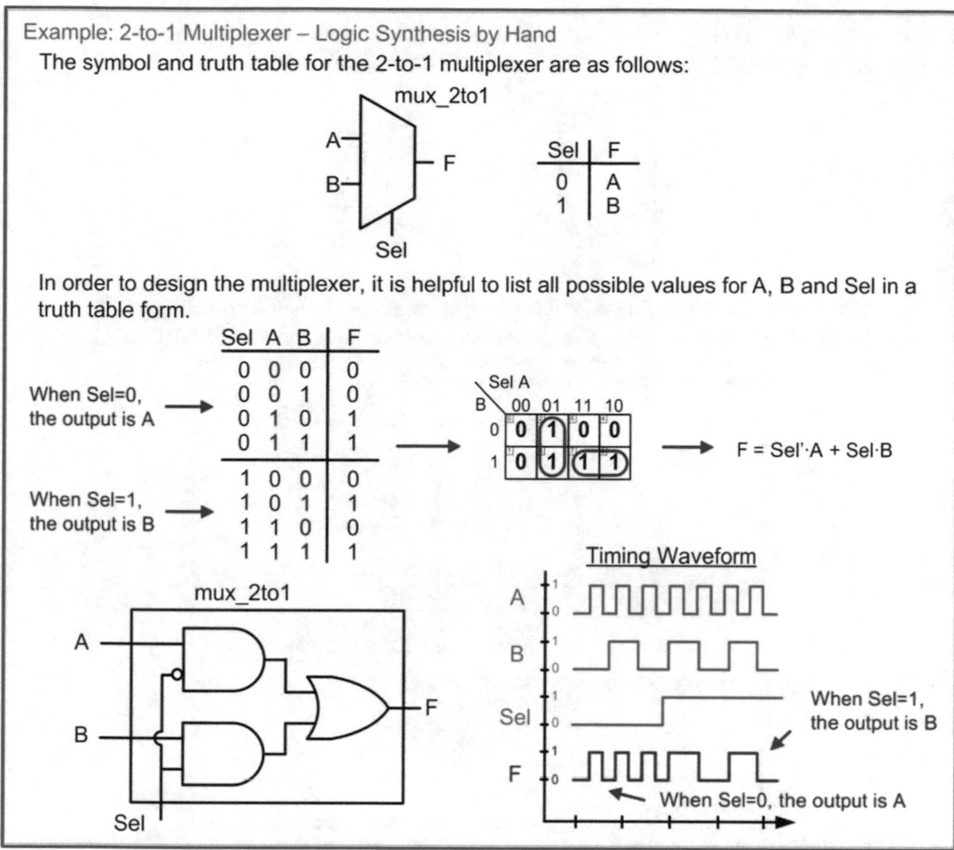

Example 6.10
2-to-1 multiplexer: logic synthesis by hand

In Verilog, a multiplexer can be implemented using continuous assignment with either logical or conditional operators. Example 6.11 shows how to model the multiplexer in Verilog using these techniques.

Example: 4-to-1 Multiplexer – Verilog Modeling using Continuous Assignment

The symbol and truth table for the 4-to-1 multiplexer are as follows:

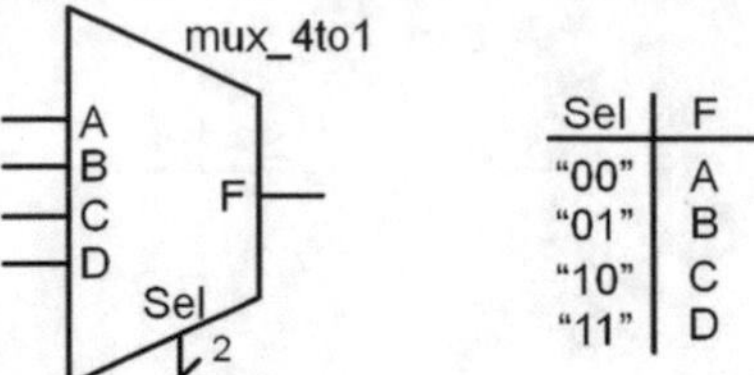

Sel	F
"00"	A
"01"	B
"10"	C
"11"	D

The following are two different ways to implement the behavior of the multiplexer with continuous assignment: (1) with logical operators; and (2) with conditional operators.

(1)

```
module mux_4to1 (output wire F,
                 input  wire A, B, C, D,
                 input  wire [1:0] Sel);

   assign F = (A & ~Sel[1] & ~Sel[0]) |
              (B & ~Sel[1] &  Sel[0]) |
              (C &  Sel[1] & ~Sel[0]) |
              (D &  Sel[1] &  Sel[0]);

endmodule
```

(2)

```
module mux_4to1 (output wire F,
                 input  wire A, B, C, D,
                 input  wire [1:0] Sel);

   assign F = (Sel == 2'b00) ? A :
              (Sel == 2'b01) ? B :
              (Sel == 2'b10) ? C :
              (Sel == 2'b11) ? D :
              1'bX;

endmodule
```

Example 6.11
4-to-1 multiplexer: Verilog modeling using logical and conditional operators

Concept Check

CC6.3 How are the product terms in a multiplexer based on the identity theorem?

A) Only the select product term will pass its input to the final sum term. Since all of the unselected product terms output 0, the input will be passed through the sum term because anything OR'd with a 0 is itself.

B) The select lines are complemented such that they activate only one OR gate.

C) The select line inputs will produce 1's on the inputs of the selected product term. This allows the input signal to pass through the selected AND gate because anything AND'd with a 1 is itself.

D) The select line inputs will produce 0's on the inputs of the selected sum term. This allows the input signal to pass through the selected OR gate because anything OR'd with a 0 is itself.

6.4 Demultiplexers

A demultiplexer works in a complementary fashion to a multiplexer. A demultiplexer has one input that is routed to one of its multiple outputs. The output that is active is dictated by a select input. A demux has *n* select lines that chooses to route the input to one of its 2^n outputs. When an output is not selected, it outputs a logic 0. Example 6.12 shows the process of designing a 1-to-2 demultiplexer by hand (i.e., using the classical digital design approach).

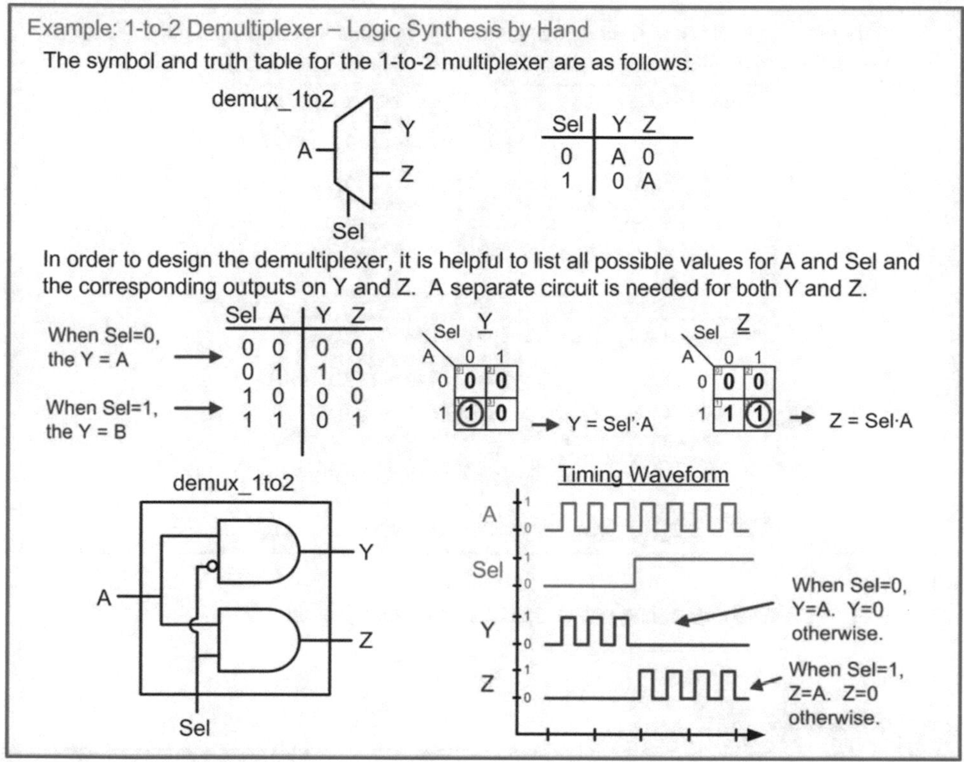

Example 6.12
1-to-2 demultiplexer: logic synthesis by hand

In Verilog, a demultiplexer can be implemented using continuous assignment with either logical or conditional operators. Example 6.13 shows how to model the demultiplexer in Verilog using these techniques.

Example: 1-to-4 Demultiplexer – Verilog Modeling using Continuous Assignment

The symbol and truth table for the 1-to-4 demultiplexer are as follows:

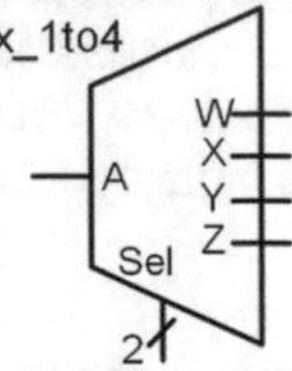

Sel	W	X	Y	Z
"00"	A	0	0	0
"01"	0	A	0	0
"10"	0	0	A	0
"11"	0	0	0	A

The following are two different ways to implement the behavior of the demultiplexer with continuous assignment: (1) with logical operators; and (2) with conditional operators.

(1)

```
module demux_1to4 (output wire W, X, Y, Z,
                   input  wire A,
                   input  wire [1:0] Sel);

   assign W = (A & ~Sel[1] & ~Sel[0]);
   assign X = (A & ~Sel[1] &  Sel[0]);
   assign Y = (A &  Sel[1] & ~Sel[0]);
   assign Z = (A &  Sel[1] &  Sel[0]);

endmodule
```

(2)

```
module demux_1to4 (output wire W, X, Y, Z,
                   input  wire A,
                   input  wire [1:0] Sel);

   assign W = (Sel == 2'b00) ? A : 1'b0;
   assign X = (Sel == 2'b01) ? A : 1'b0;
   assign Y = (Sel == 2'b10) ? A : 1'b0;
   assign Z = (Sel == 2'b11) ? A : 1'b0;

endmodule
```

Example 6.13
1-to-4 demultiplexer: Verilog modeling using logical and conditional operators

Concept Check

CC6.4 How many select lines are needed in a 1-to-64 demultiplexer?

A) 1 B) 4 C) 6 D) 64

Summary

- The term medium-scale integrated circuit (MSI) logic refers to a set of basic combinational logic circuits that implement simple, commonly used functions such as decoders, encoders, multiplexers, and demultiplexers. MSI logic can also include operations such as comparators and simple arithmetic circuits.
- While an MSI logic circuit may have multiple outputs, each output requires its own unique logic expression that is based on the system inputs.
- A decoder is a system that has a greater number of outputs than inputs. The behavior of each output is based on each unique input code.

- An encoder is a system that has a greater number of inputs than outputs. A compressed output code is produced based on which input(s) lines are asserted.
- A multiplexer is a system that has one output and multiple inputs. At any given time, one and only one input is routed to the output based on the value on a set of *select lines*. For *n* select lines, a multiplexer can support 2^n inputs.
- A demultiplexer is a system that has one input and multiple outputs. The input is routed to one of the outputs depending on the value on a set of select lines. For n select lines, a demultiplexer can support 2^n outputs.
- HDLs are particularly useful for describing MSI logic due to their abstract modeling capability. Through the use of Boolean conditions and vector assignments, the behavior of MSI logic can be modeled in a compact and intuitive manner.

Exercise Problems

Section 6.1: Decoders

6.1.1 Design a 4-to-16 one-hot decoder by hand. The block diagram and truth table for the decoder are given in Fig. 6.1. Give the minimized logic expressions for each output (i.e., F_0, F_1, …, F_{15}) and the full logic diagram for the system.

4-to-16 One-Hot Decoder

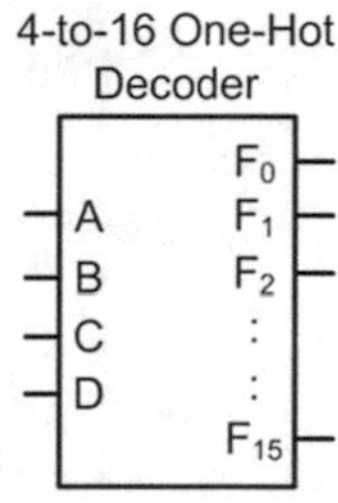

A B C D	F_{15}	F_{14}	F_{13}	F_{12}	F_{11}	F_{10}	F_9	F_8	F_7	F_6	F_5	F_4	F_3	F_2	F_1	F_0
0 0 0 0	0	0	0	0	0	0	0	0	0	0	0	0	0	0	0	1
0 0 0 1	0	0	0	0	0	0	0	0	0	0	0	0	0	0	1	0
0 0 1 0	0	0	0	0	0	0	0	0	0	0	0	0	0	1	0	0
0 0 1 1	0	0	0	0	0	0	0	0	0	0	0	0	1	0	0	0
0 1 0 0	0	0	0	0	0	0	0	0	0	0	0	1	0	0	0	0
0 1 0 1	0	0	0	0	0	0	0	0	0	0	1	0	0	0	0	0
0 1 1 0	0	0	0	0	0	0	0	0	0	1	0	0	0	0	0	0
0 1 1 1	0	0	0	0	0	0	0	0	1	0	0	0	0	0	0	0
1 0 0 0	0	0	0	0	0	0	0	1	0	0	0	0	0	0	0	0
1 0 0 1	0	0	0	0	0	0	1	0	0	0	0	0	0	0	0	0
1 0 1 0	0	0	0	0	0	1	0	0	0	0	0	0	0	0	0	0
1 0 1 1	0	0	0	0	1	0	0	0	0	0	0	0	0	0	0	0
1 1 0 0	0	0	0	1	0	0	0	0	0	0	0	0	0	0	0	0
1 1 0 1	0	0	1	0	0	0	0	0	0	0	0	0	0	0	0	0
1 1 1 0	0	1	0	0	0	0	0	0	0	0	0	0	0	0	0	0
1 1 1 1	1	0	0	0	0	0	0	0	0	0	0	0	0	0	0	0

Fig. 6.1
4-to-16 one-hot decoder functionality

6.1.2 Design a Verilog model for a 4-to-16 one-hot decoder using continuous assignment and gate-level primitives. Use the module port definition given in Fig. 6.2.

```
module decoder_1hot_4to16
  (output wire [15:0] F,
   input  wire [3:0]  ABCD);
                      :
```

Fig. 6.2
4-to-16 one-hot decoder module definition

6.1.3 Design a Verilog model for a 4-to-16 one-hot decoder using continuous assignment and logical operators. Use the module port definition given in Fig. 6.2.

6.1.4 Design a Verilog model for a 4-to-16 one-hot decoder using continuous assignment and conditional operators. Use the module port definition given in Fig. 6.2.

6.1.5 Design a 4-input, 7-segment HEX character decoder by hand. The system has four inputs called A, B, C, and D. The system has seven outputs called F_a, F_b, F_c, F_d, F_e, F_f, and F_g. These outputs drive the individual LEDs within the display. A logic 1 on an output corresponds to the LED being ON. The display will show the HEX characters 0–9, A, b, c, d, E, and F corresponding to the 4-bit input code on A. A template for creating the truth tables for this system is provided in Fig. 6.3. Provide the minimized logic expressions for each of the seven outputs and the overall logic diagram for the decoder.

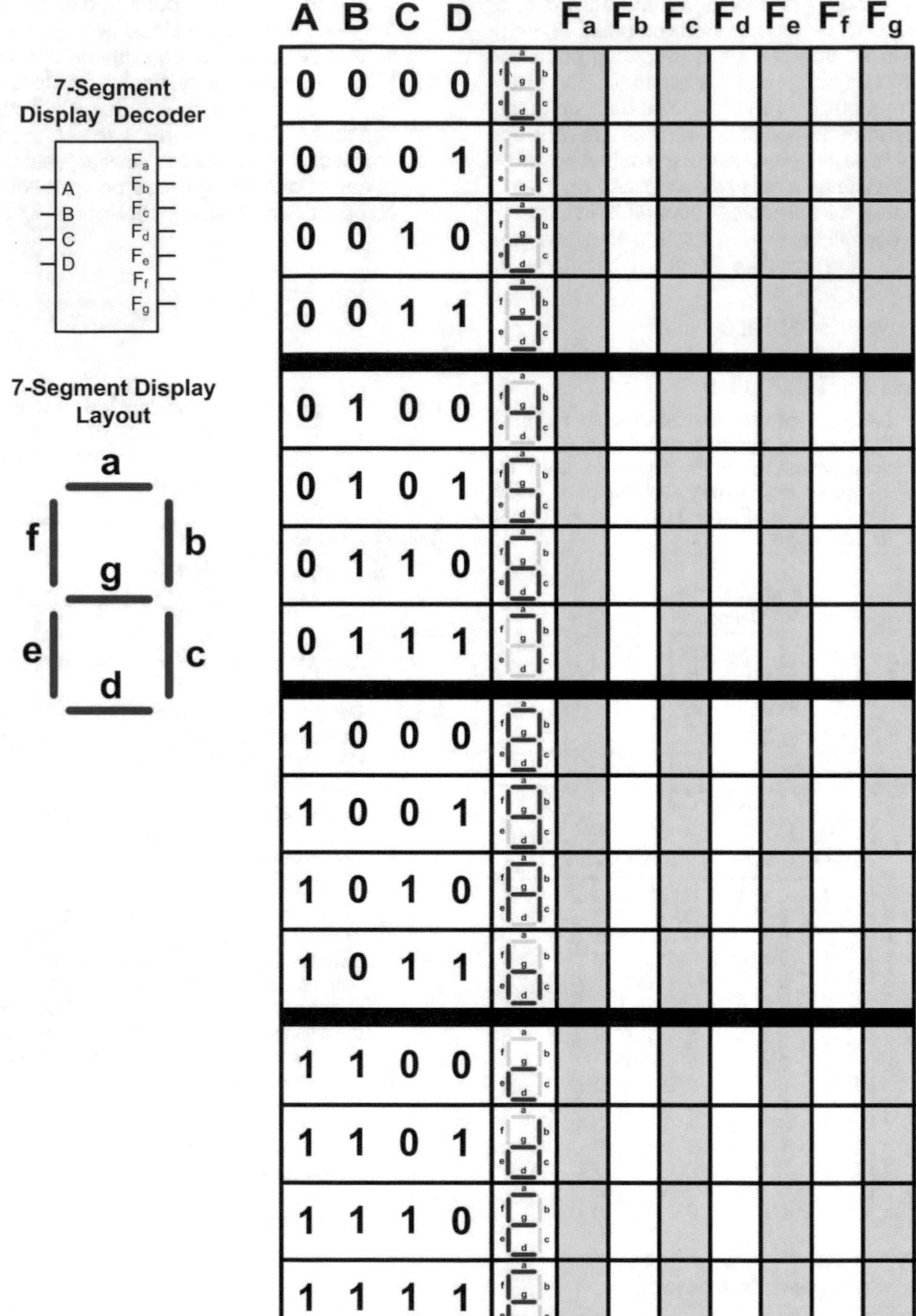

A	B	C	D		F_a	F_b	F_c	F_d	F_e	F_f	F_g
0	0	0	0								
0	0	0	1								
0	0	1	0								
0	0	1	1								
0	1	0	0								
0	1	0	1								
0	1	1	0								
0	1	1	1								
1	0	0	0								
1	0	0	1								
1	0	1	0								
1	0	1	1								
1	1	0	0								
1	1	0	1								
1	1	1	0								
1	1	1	1								

Fig. 6.3
7-segment display decoder truth table

6.1.6 Design a Verilog model for a 4-input, 7-segment HEX character decoder using continuous assignment and logical operators. Use the module port definition given in Fig. 6.4 for your design. The system has a 4-bit input vector called ABCD and a 7-bit output vector called F. The individual scalars within the output vector (i.e., F[6:0]) correspond to the character display segments a, b, c, d, e, f, and g, respectively. A logic 1 on an output corresponds to the LED being ON. The display will show the HEX characters 0–9, A, b, c, d, E, and F corresponding to the 4-bit input code on A. A template for creating the truth table is provided in. The signals in this table correspond to the ports in this problem as follows: $F_a = F(6)$, $F_b = F(5)$, $F_c = F(4)$, $F_d = F(3)$, $F_e = F(2)$, $F_f = F(1)$, and $F_g = F(0)$.

```
module decoder_7seg_4in
  (output wire [6:0] F,
   input  wire [3:0] ABCD);
                  :
```

Fig. 6.4
7-segment display decoder module definition

6.1.7 Design a Verilog model for a 4-input, 7-segment HEX character decoder using continuous assignment and conditional operators. Use the module port definition given in Fig. 6.4 for your design. The system has a 4-bit input vector called ABCD and a 7-bit output vector called F. The individual scalars within the output vector (i.e., F[6:0]) correspond to the character display segments a, b, c, d, e, f, and g, respectively. A logic 1 on an output corresponds to the LED being ON. The display will show the HEX characters 0–9, A, b, c, d, E, and F corresponding to the 4-bit input code on A. A template for creating the truth table is provided in. The signals in this table correspond to the ports in this problem as follows: $F_a = F(6)$, $F_b = F(5)$, $F_c = F(4)$, $F_d = F(3)$, $F_e = F(2)$, $F_f = F(1)$, and $F_g = F(0)$.

Section 6.2: Encoders

6.2.1 Design an 8-to-3 binary encoder by hand. The block diagram and truth table for the encoder are given in Fig. 6.5. Give the logic expressions for each output and the full logic diagram for the system.

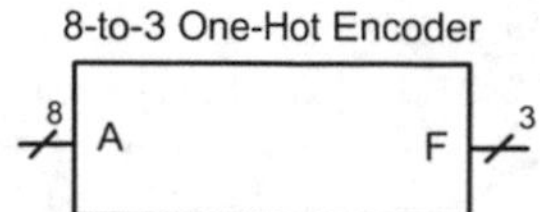

A(7)	A(6)	A(5)	A(4)	A(3)	A(2)	A(1)	A(0)	F(2)	F(1)	F(0)
0	0	0	0	0	0	0	1	0	0	0
0	0	0	0	0	0	1	0	0	0	1
0	0	0	0	0	1	0	0	0	1	0
0	0	0	0	1	0	0	0	0	1	1
0	0	0	1	0	0	0	0	1	0	0
0	0	1	0	0	0	0	0	1	0	1
0	1	0	0	0	0	0	0	1	1	0
1	0	0	0	0	0	0	0	1	1	1

Fig. 6.5
8-to-3 one-hot encoder functionality

6.2.2 Design a Verilog model for an 8-to-3 binary encoder using continuous assignment and gate-level primitives. Use the module port definition given in Fig. 6.6.

```
module encoder_8to3_binary
  (output wire [2:0] F,
   input  wire [7:0] A);
                   :
```

Fig. 6.6
8-to-3 one-hot encoder module functionality

6.2.3 Design a Verilog model for an 8-to-3 binary encoder using continuous assignment and logical operators. Use the module port definition given in Fig. 6.6.

6.2.4 Design a Verilog model for an 8-to-3 binary encoder using continuous assignment and conditional operators. Use the module port definition given in Fig. 6.6.

Section 6.3: Multiplexers

6.3.1 Design an 8-to-1 multiplexer by hand. The block diagram and truth table for the multiplexer are given in Fig. 6.7. Give the minimized logic expressions for the output and the full logic diagram for the system.

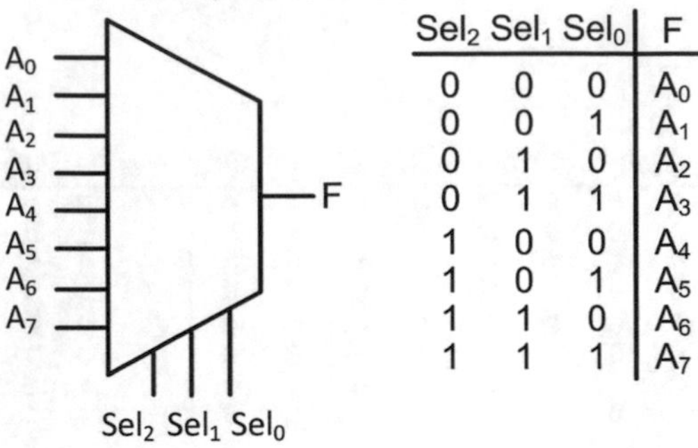

Sel_2	Sel_1	Sel_0	F
0	0	0	A_0
0	0	1	A_1
0	1	0	A_2
0	1	1	A_3
1	0	0	A_4
1	0	1	A_5
1	1	0	A_6
1	1	1	A_7

Fig. 6.7
8-to-1 multiplexer functionality

6.3.2 Design a Verilog model for an 8-to-1 multiplexer using continuous assignment and gate-level primitives. Use the module port definition given in Fig. 6.8.

```
module mux_8to1
  (output wire F,
   input  wire [7:0] A,
   input  wire [2:0] Sel);
                     :
```

Fig. 6.8
8-to-1 multiplexer module definition

6.3.3 Design a Verilog model for an 8-to-1 multiplexer using continuous assignment and logical operators. Use the module port definition given in Fig. 6.8.

6.3.4 Design a Verilog model for an 8-to-1 multiplexer using continuous assignment and conditional operators. Use the module port definition given in Fig. 6.8.

Section 6.4: Demultiplexers

6.4.1 Design a 1-to-8 demultiplexer by hand. The block diagram and truth table for the demultiplexer are given in Fig. 6.9. Give the minimized logic expressions for each output and the full logic diagram for the system.

1-to-8 Demultiplexer

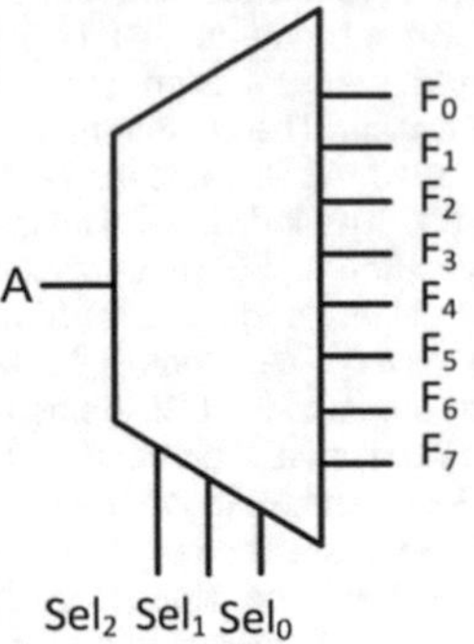

Sel_2	Sel_1	Sel_0	F_7	F_6	F_5	F_4	F_3	F_2	F_1	F_0
0	0	0	0	0	0	0	0	0	0	A
0	0	1	0	0	0	0	0	0	A	0
0	1	0	0	0	0	0	0	A	0	0
0	1	1	0	0	0	0	A	0	0	0
1	0	0	0	0	0	A	0	0	0	0
1	0	1	0	0	A	0	0	0	0	0
1	1	0	0	A	0	0	0	0	0	0
1	1	1	A	0	0	0	0	0	0	0

Fig. 6.9
1-to-8 demultiplexer functionality

6.4.2 Design a Verilog model for a 1-to-8 demultiplexer using continuous assignment and gate-level primitives. Use the module port definition given in Fig. 6.10 for your design.

```
module demux_1to8
  (output wire [7:0] F,
   input  wire A,
   input  wire [2:0] Sel);
                     :
```

Fig. 6.10
1-to-8 demultiplexer module definition

6.4.3 Design a Verilog model for a 1-to-8 demultiplexer using continuous assignment and logical operators. Use the module port definition given in Fig. 6.10 for your design.

6.4.4 Design a Verilog model for a 1-to-8 demultiplexer using continuous assignment and conditional operators. Use the module port definition given in Fig. 6.10 for your design.

Chapter 7: Sequential Logic Design

In this chapter we begin looking at sequential logic design. Sequential logic design differs from combinational logic design in that the outputs of the circuit depend not only on the current values of the inputs but also on the *past* values of the inputs. This is different from the combinational logic design where the output of the circuitry depends only on the current values of the inputs. The ability of a sequential logic circuit to base its outputs on both the current and past inputs allows more sophisticated and intelligent systems to be created. We begin by looking at sequential logic storage devices, which are used to hold the past values of a system. This is followed by an investigation of timing considerations of sequential logic circuits. We then look at some useful circuits that can be created using only sequential logic storage devices. Finally, we look at one of the most important logic circuits in digital systems, the finite-state machine. The goal of this chapter is to provide an understanding of the basic operation of sequential logic circuits.

Learning Outcomes—After completing this chapter, you will be able to:

7.1 Describe the operation of a sequential logic storage device.
7.2 Describe sequential logic timing considerations.
7.3 Design a variety of common circuits based on sequential storage devices (toggle flops, ripple counters, switch debouncers, and shift registers).
7.4 Design a finite-state machine using the classical digital design approach.
7.5 Design a counter using the classical digital design approach and using an HDL-based, structural approach.
7.6 Describe the finite-state machine reset condition.
7.7 Analyze a finite-state machine to determine its functional operation and maximum clock frequency.

7.1 Sequential Logic Storage Devices

7.1.1 The Cross-Coupled Inverter Pair

The first thing that is needed in sequential logic is a storage device. The fundamental storage device in sequential logic is based on a positive feedback configuration. Consider the circuit in Fig. 7.1. This circuit configuration is called the *cross-coupled inverter pair.* In this circuit if the input of U1 starts with a value of 1, it will produce an output of Q = 0. This output is fed back to the input of U2, thus producing an output of Qn = 1. Qn is fed back to the original input of U1, thus reinforcing the initial condition. This circuit will *hold*, or *store,* a logic 0 without being driven by any other inputs. This circuit operates in a complementary manner when the initial value of U1 is a 0. With this input condition, the circuit will store a logic 1 without being driven by any other inputs.

© Springer Nature Switzerland AG 2019

B. J. LaMeres, *Introduction to Logic Circuits & Logic Design with Verilog*,
https://doi.org/10.1007/978-3-030-13605-5_7

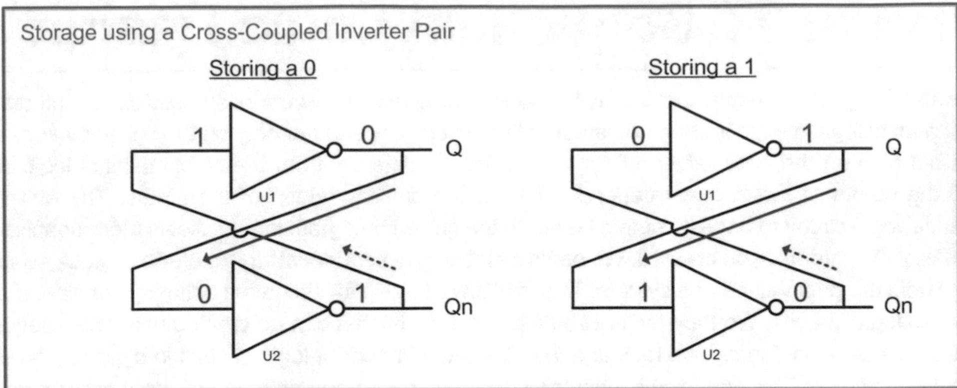

Fig. 7.1
Storage using a cross-coupled inverter pair

7.1.2 Metastability

The cross-coupled inverter pair in Fig. 7.1 exhibits what is called *metastable* behavior due to its positive feedback configuration. Metastability refers to when a system can exist in a state of equilibrium when undisturbed but can be moved to a different, more stable state of equilibrium when sufficiently disturbed. Systems that exhibit *high levels* of metastability have an equilibrium state that is highly unstable, meaning that if disturbed even slightly the system will move rapidly to a more stable point of equilibrium. The cross-coupled inverter pair is a highly metastable system. This system actually contains three equilibrium states. The first is when the input of U1 is exactly between a logic 0 and logic 1 (i.e., $V_{CC}/2$). In this state, the output of U1 is also exactly $V_{CC}/2$. This voltage is fed back to the input of U2, thus producing an output of exactly $V_{CC}/2$ on U2. This in turn is fed back to the original input on U1 reinforcing the initial state. Despite this system being at equilibrium in this condition, this state is highly unstable. With minimal disturbance to any of the nodes within the system, it will move rapidly to one of the two more stable states. The two stable states for this system are when Q = 0 or when Q = 1 (see Fig. 7.1). Once the transition begins between the unstable equilibrium state toward one of the two more stable states, the positive feedback in the system continually reinforces the transition until the system reaches its final state. In electrical systems, this initial disturbance is caused by the presence of *noise*, or unwanted voltage in the system. Noise can come from many sources including random thermal motion of charge carriers in the semiconductor materials, electromagnetic energy, or naturally occurring ionizing particles. Noise is present in every electrical system so the cross-coupled inverter pair will never be able to stay in the unstable equilibrium state where all nodes are at $V_{CC}/2$.

The cross-coupled inverter pair has two stable states; thus it is called a *bistable* element. In order to understand the bistable behavior of this circuit, let's look at its behavior when the initial input value on U1 is set directly between a logic 0 and logic 1 (i.e., $V_{CC}/2$) and how a small amount of noise will cause the system to move toward a stable state. Recall that an inverter is designed to have an output that quickly transitions between a logic LOW and HIGH in order to minimize the time spent in the uncertainty region. This is accomplished by designing the inverter to have what is called *gain*. Gain can be thought of as a multiplying factor that is applied to the input of the circuit when producing the output (i.e., $V_{out} = \text{gain} \cdot V_{in}$). The gain for an inverter will be negative since the output moves in the opposite direction of the input. The inverter is designed to have a very high gain such that even the smallest change on the input when in the transition region will result in a large change on the output. Consider the behavior of this circuit shown in Fig. 7.2. In this example, let's represent the gain of the inverter as $-g$ and see how the system responds when a small positive voltage noise (V_n) is added to the $V_{CC}/2$ input on U1.

Examining Metastability – Moving Toward the State Q=0.

Let's consider how this circuit responds when its initial value at the input to U1 is directly in between a 0 and a 1 (e.g., $V_{CC}/2$).

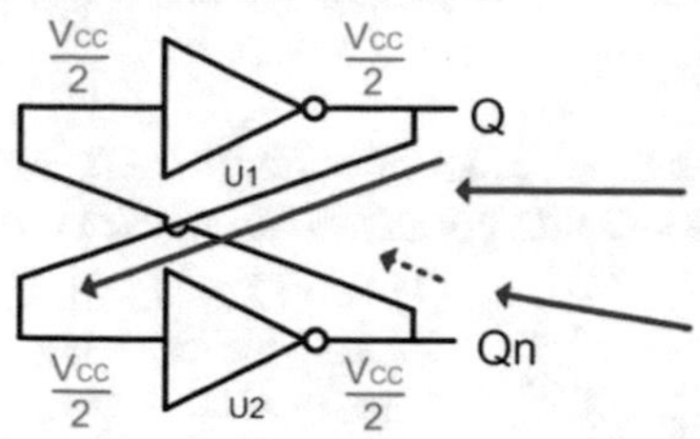

The input to U1 is $V_{CC}/2$, which creates an output of $V_{CC}/2$.

The output of U1 is fed to the input of U2, again producing an output of $V_{CC}/2$ on U2.

The output of U2 is fed to the input of U1, thus reinforcing the original value of $V_{CC}/2$. We can say that the circuit is in an *equilibrium state*.

Now let's consider how this circuit responds when a small amount of positive noise (V_n) is added to the input of U1 when it is at $V_{CC}/2$. The $V_{CC}/2$ component is not shown for simplicity.

(1) A small amount of noise is added to $V_{CC}/2$ at the input of U1. This pushes it slightly toward a logic 1.

V_n $-g$ $-g \cdot V_n$ Q   U1

(2) This noise is amplified by the inverter with a negative gain, pushing it slightly toward a logic 0.

(3) The amplified noise is fed to the input of U2.

$-g \cdot V_n$ $-g$ $+g^2 \cdot V_n$ Qn   U2

(4) The noise is amplified again, thus creating an even larger, positive voltage that is fed back to the original input of U1.

(5) When the noise is fed back to the input of U1, it pushes it even more toward a logic 1.

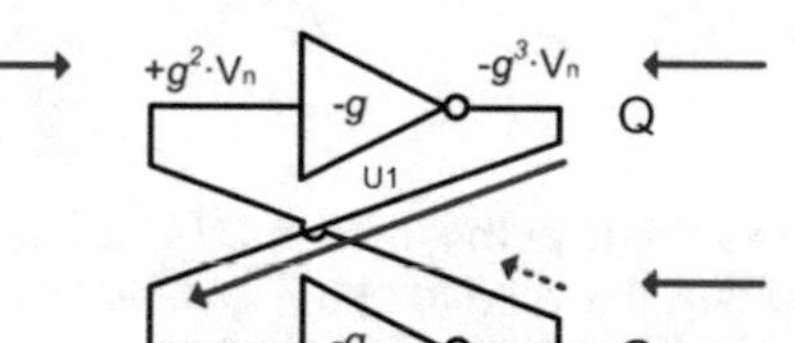

(6) The noise is amplified further, pushing the output even more toward a logic 0.

(7) The amplified noise is fed to the input of U2.

(8) The noise is amplified again, thus creating an even larger, positive voltage that is fed back to the original input of U1.

This process continues until the voltage at the input of U1 reaches V_{CC} and cannot be increased further. Simultaneously, the voltage at the input to U2 is decreased until it reaches GND and cannot be decreased further. At that point, the system is at a stable state and will store Q=0.

The system reaches stability once the input of U1 cannot be increased any further.

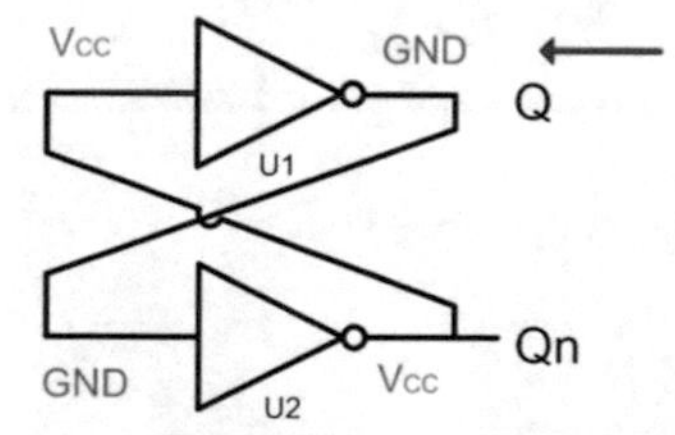

In this stable state, the system is holding, or storing a value of Q=0.

Fig. 7.2
Examining metastability moving toward the state Q = 0

Figure 7.3 shows how the system responds when a small negative voltage noise ($-V_n$) is added to the $V_{CC}/2$ input on U1.

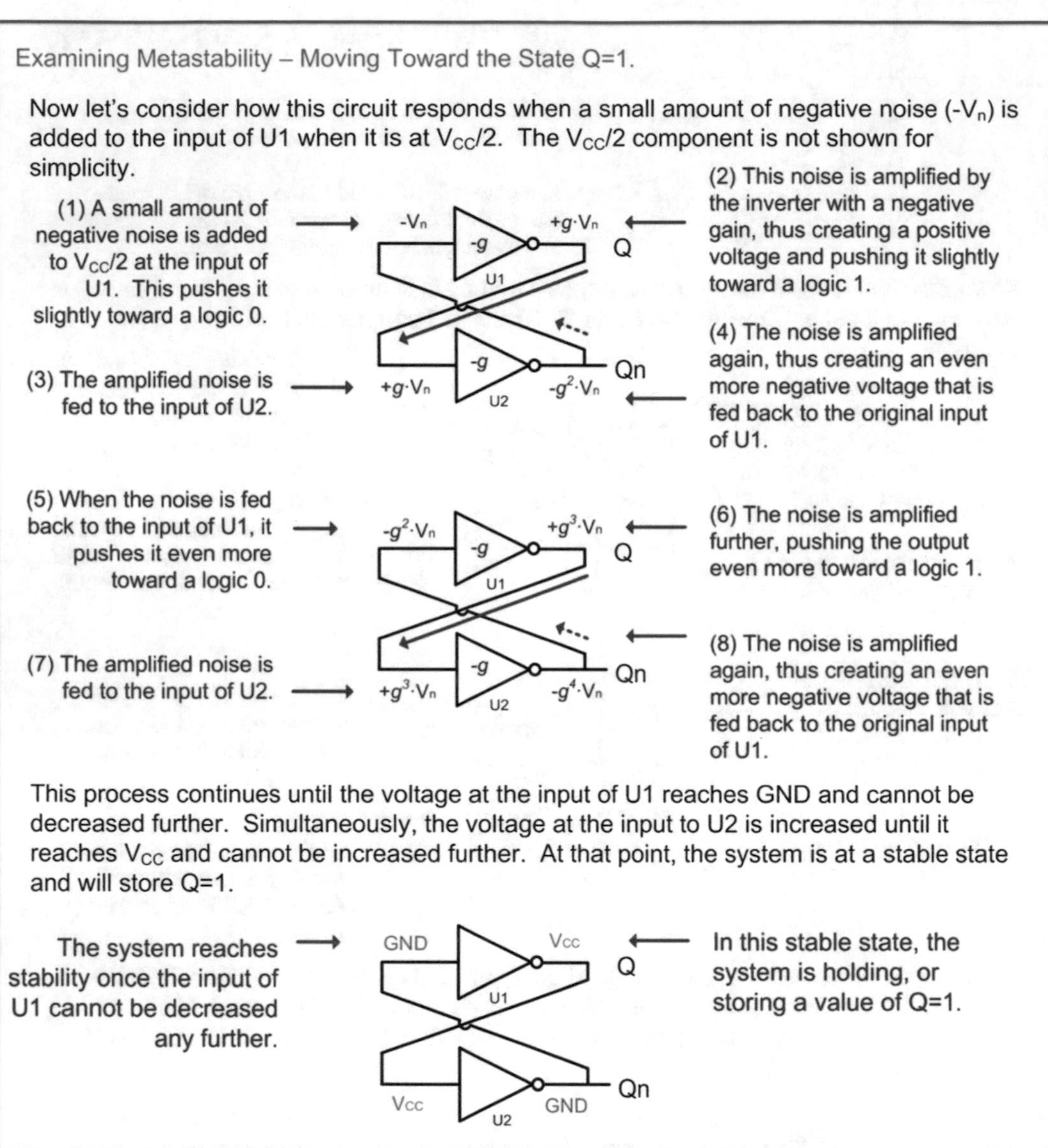

Fig. 7.3
Examining metastability moving toward the state Q = 1

7.1.3 The SR Latch

While the cross-coupled inverter pair is the fundamental storage concept for sequential logic, there is no mechanism to set the initial value of Q. All that is guaranteed is that the circuit will store a value in one of the two stable states (Q = 0 or Q = 1). The *SR Latch* provides a means to control the initial values in this positive feedback configuration by replacing the inverters with NOR gates. In this circuit, S stands for *set* and indicates when the output is forced to a logic 1 (Q = 1), and R stands for *reset* and indicates when the output is forced to a logic 0 (Q = 0). When both S = 0 and R = 0, the SR Latch is put into a *store* mode, and it will hold the last value of Q. In all of these input conditions, Qn is the complement of Q. Consider the behavior of the SR Latch during its store state shown in Fig. 7.4.

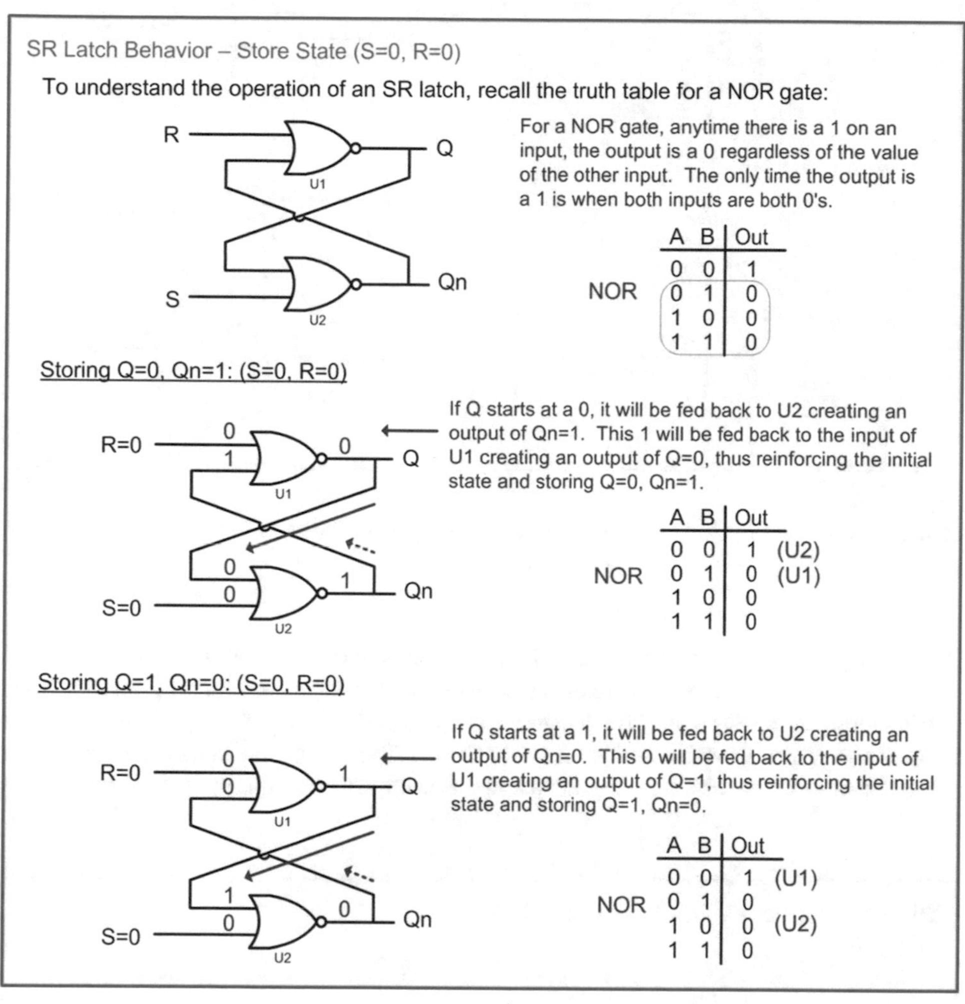

Fig. 7.4
SR Latch behavior: store state (S = 0, R = 0)

The SR Latch has two input conditions that will force the outputs to known values. The first condition is called the *set* state. In this state, the inputs are configured as S = 1 and R = 0. This input condition will force the outputs to Q = 1 (e.g., setting Q) and Qn = 0. The second input condition is called the *reset* state. In this state the inputs are configured as S = 0 and R = 1. This input condition will force the outputs to Q = 0 (i.e., resetting Q) and Qn = 1. Consider the behavior of the SR Latch during its set and reset states shown in Fig. 7.5.

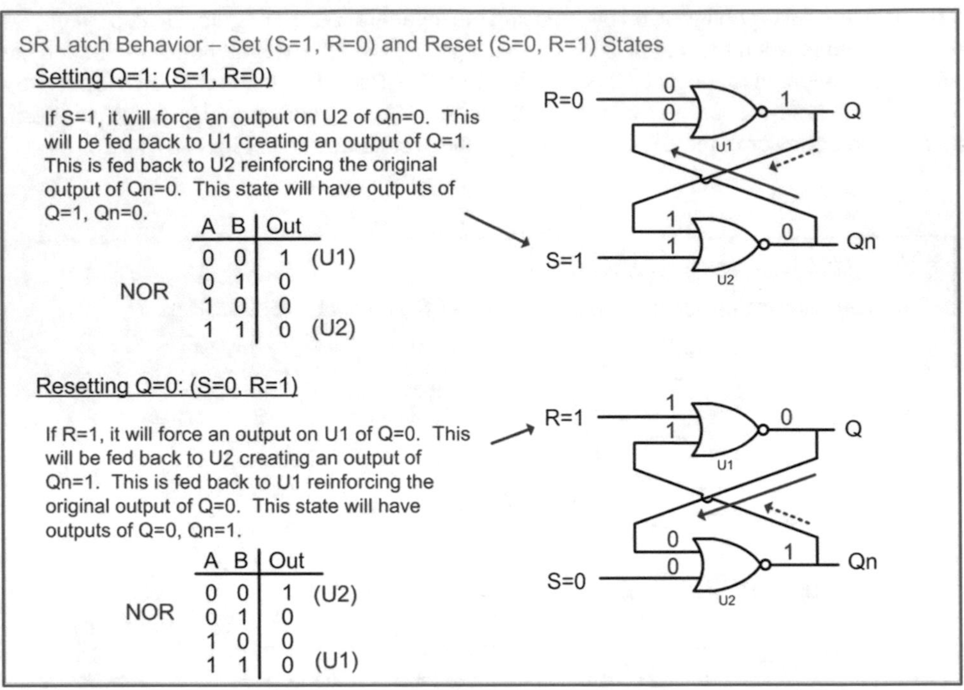

Fig. 7.5
SR Latch behavior: set (S = 1, R = 0) and reset (S = 0, R = 1) states

The final input condition for the SR Latch leads to potential metastability and should be avoided. When S = 1 and R = 1, the outputs of the SR Latch will both go to logic 0's. The problem with this state is that if the inputs subsequently change to the store state (S = 0, R = 0), the outputs will go metastable and then settle in one of the two stable states (Q = 0 or Q = 1). The reason this state is avoided is because the final resting state of the SR Latch is random and unknown. Consider this operation shown in Fig. 7.6.

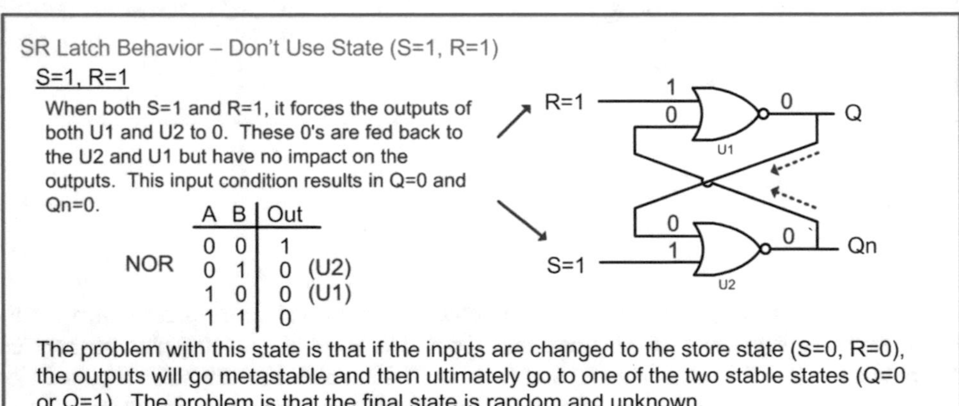

Fig. 7.6
SR Latch behavior: don't use state (S = 1 and R = 1)

Figure 7.7 shows the final truth table for the SR Latch.

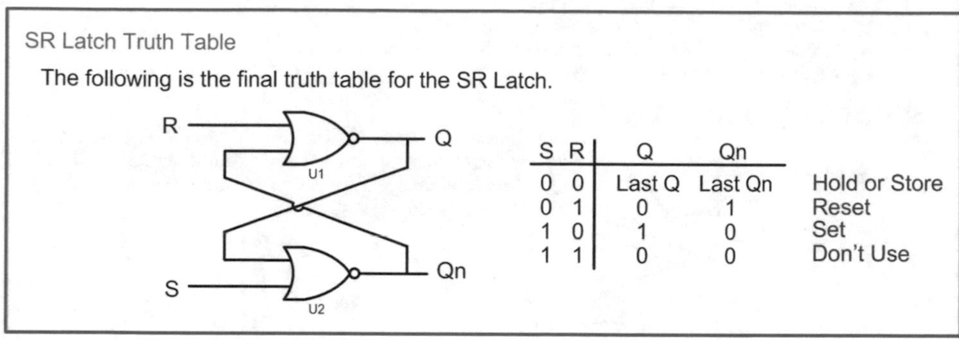

S	R	Q	Qn	
0	0	Last Q	Last Qn	Hold or Store
0	1	0	1	Reset
1	0	1	0	Set
1	1	0	0	Don't Use

Fig. 7.7
SR Latch truth table

The SR Latch has some drawbacks when it comes to implementation with real circuitry. First, it takes two independent inputs to control the outputs. Second, the state where S = 1 and R = 1 causes problems when real propagation delays are considered through the gates. Since it is impossible to match the delays exactly between U1 and U2, the SR Latch may occasionally enter this state and experience momentary metastable behavior. In order to address these issues, a number of improvements can be made to this circuit to create two of the most commonly used storage devices in sequential logic, the *D-Latch* and the *D-flip-flop*. In order to understand the operation of these storage devices, two incremental modifications are made to the SR Latch. The first is called the *S'R' Latch*, and the second is the *SR Latch with enable*. These two circuits are rarely implemented and are only explained to understand how the SR Latch is modified to create a D-Latch and ultimately a D-flip-flop.

7.1.4 The S'R' Latch

The S'R' Latch operates in a similar manner as the SR Latch with the exception that the input codes corresponding to the store, set, and reset states are complemented. To accomplish this complementary behavior, the S'R' Latch is implemented with NAND gates configured in a positive feedback configuration. In this configuration, the S'R' Latch will store the last output when S' = 1, R' = 1. It will set the output (Q = 1) when S' = 0, R' = 1. Finally, it will reset the output (Q = 0) when S' = 1, R' = 0. Consider the behavior of the S'R' Latch during its store state shown in Fig. 7.8.

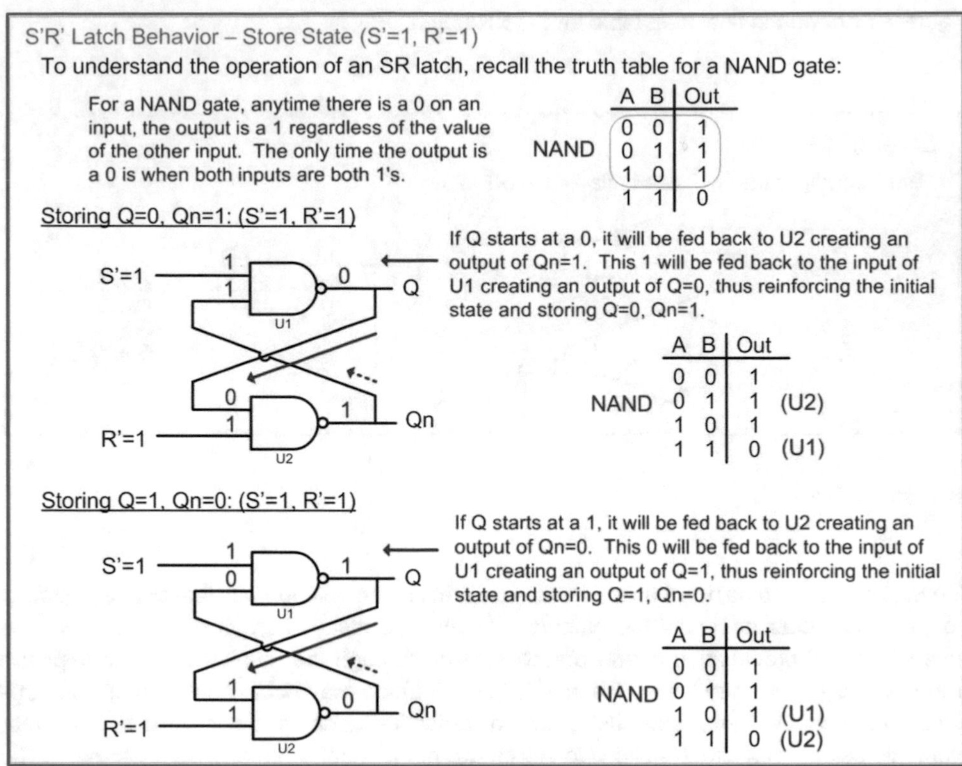

Fig. 7.8
S'R' Latch behavior: store state (S′ = 1, R' = 1)

Just as with the SR Latch, the S'R' Latch has two input configurations to control the values of the outputs. Consider the behavior of the S'R' Latch during its set and reset states shown in Fig. 7.9.

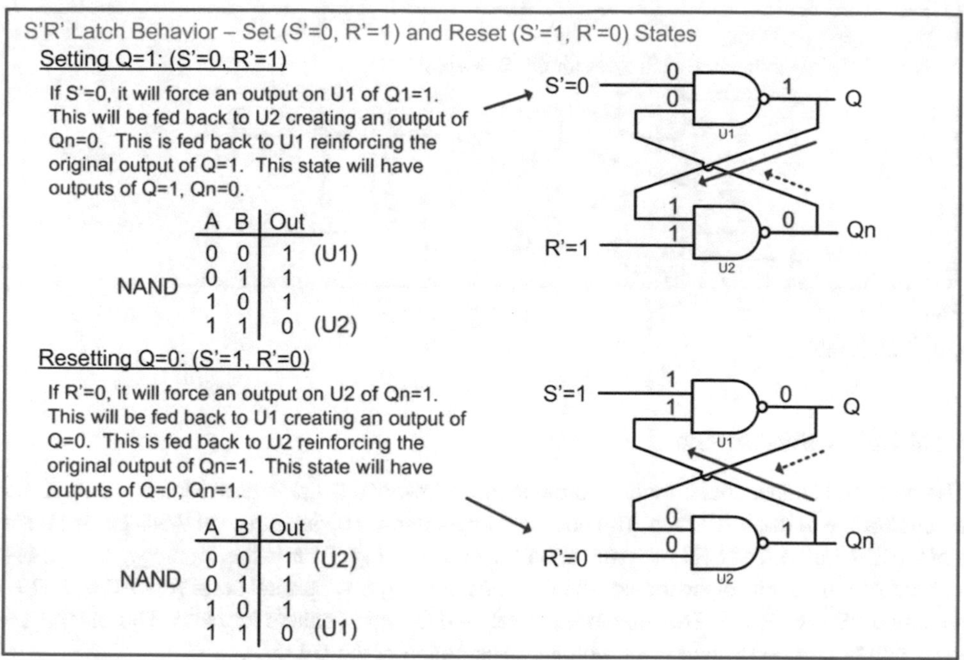

Fig. 7.9
S'R' Latch behavior: set (S' = 0, R' = 1) and reset (S' = 1, R' = 0) states

And finally, just as with the SR Latch, the S'R' Latch has a state that leads to potential metastability and should be avoided. Consider the operation of the S'R' Latch when the inputs are configured as S' = 0 and R' = 0 shown in Fig. 7.10.

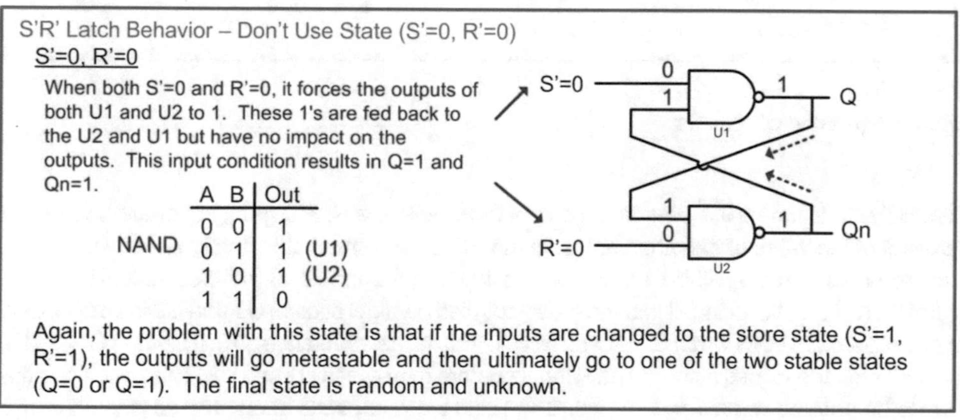

Fig. 7.10
S'R' Latch behavior: don't use state (S' = 0 and R' = 0)

The final truth table for the S'R' Latch is given in Fig. 7.11.

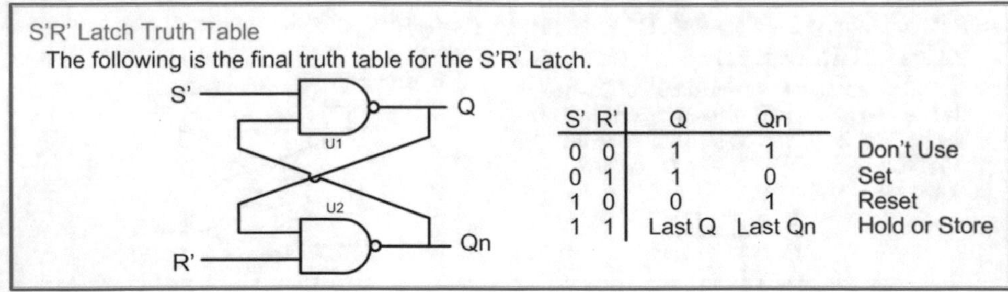

Fig. 7.11
S'R' Latch truth table

7.1.5 SR Latch with Enable

The next modification that is made in order to move toward a D-Latch and ultimately a D-flip-flop is to add an *enable* line to the S'R' Latch. The enable is implemented by adding two NAND gates on the input stage of the S'R' Latch. The SR Latch with enable is shown in Fig. 7.12. In this topology, the use of NAND gates changes the polarity of the inputs, so this circuit once again has a set state where S = 1, R = 0 and a reset state of S = 0, R = 1. The enable line is labeled *C*, which stands for *clock*. The rationale for this will be demonstrated upon moving through the explanation of the D-Latch.

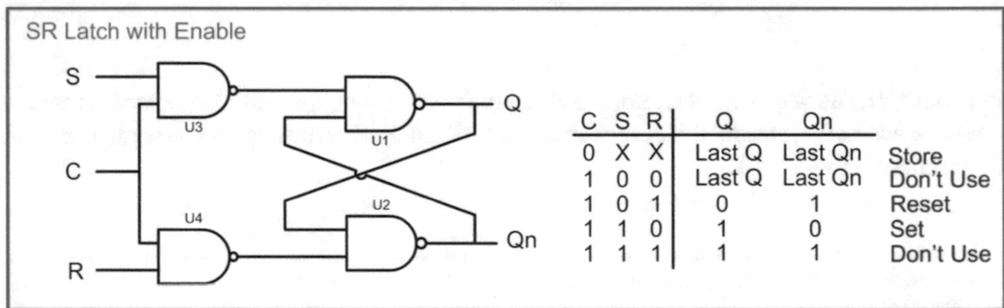

Fig. 7.12
SR Latch with enable schematic

Recall that any time a 0 is present on one of the inputs to a NAND gate, the output will always be a 1 regardless of the value of the other inputs. In the SR Latch with enable configuration, any time C = 0, the outputs of U3 and U4 will be 1's and will be fed into the inputs of the cross-coupled NAND gate configuration (U1 and U2). Recall that the cross-coupled configuration of U1 and U2 is an S'R' Latch and will be put into a store state when S' = 1 and R' = 1. This is the *store state* (C = 0). When C = 1, it has the effect of inverting the values of the S and R inputs before they reach U1 and U2. This condition allows the *set state* to be entered when S = 1, R = 0, C = 1 and the *reset state* to be entered when S = 0, R = 1, C = 1. Consider this operation in Fig. 7.13.

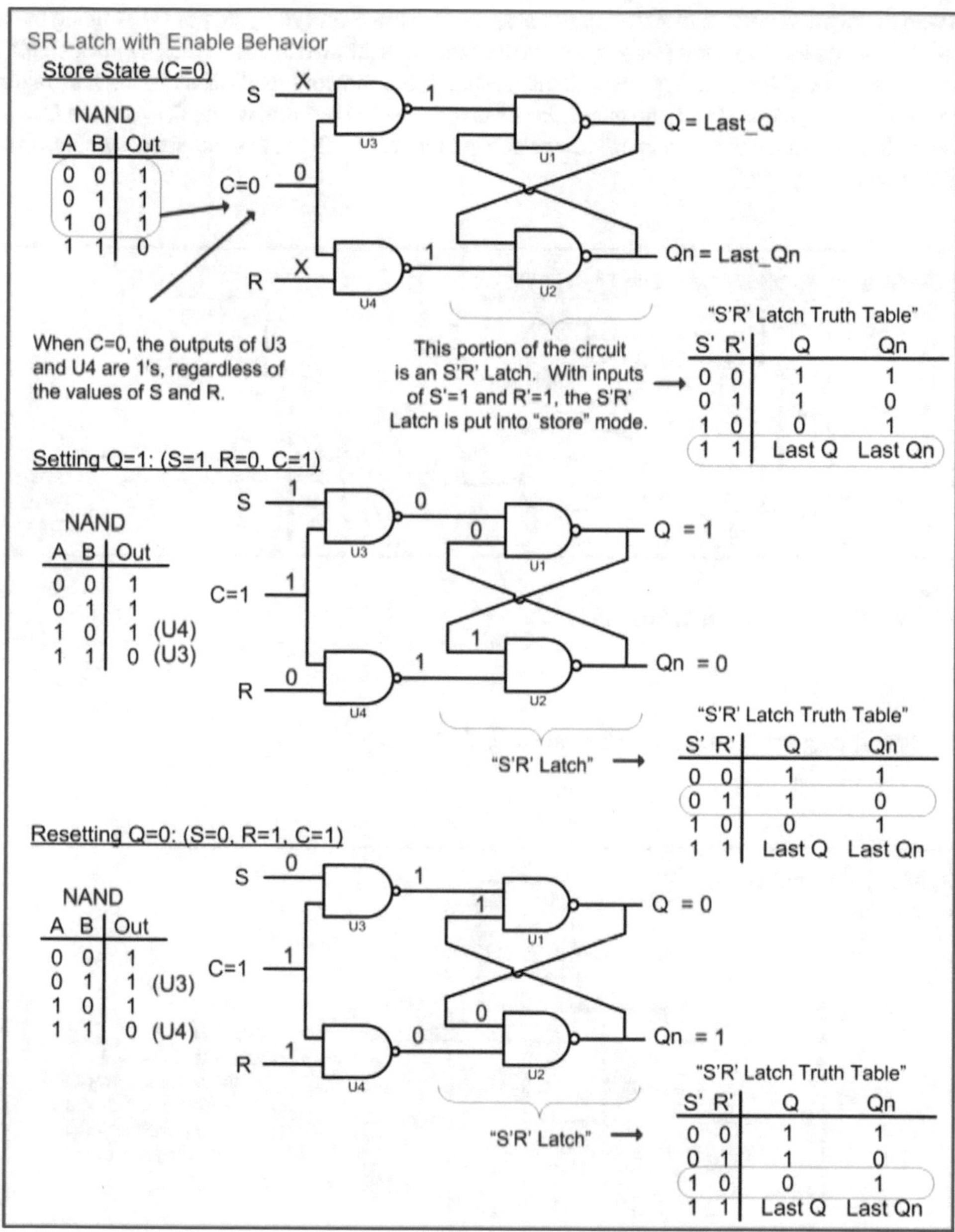

Fig. 7.13
SR Latch with enable behavior: store, set, and reset

Again, there is a potential metastable state when S = 1, R = 1 and C = 1 that should be avoided. There is also a second store state when S = 0, R = 0, and C = 1 that is not used because storage is to be dictated by the C input.

7.1.6 The D-Latch

The SR Latch with enable can be modified to create a new storage device called a D-Latch. Instead of having two separate input lines to control the outputs of the latch, the R input of the latch is instead driven with an inverted version of the S input. This prevents the S and R inputs from ever being the same value and removes the two "Don't Use" states in the truth table shown in Fig. 7.12. The new, single input is renamed D to stand for *data*. This new circuit still has the behavior that it will store the last value of Q and Qn when C = 0. When C = 1, the output will be Q = 1 when D = 1 and will be Q = 0 when D = 0. The behavior of the output when C = 1 is called *tracking* the input. The D-Latch schematic, symbol, and truth table are given in Fig. 7.14.

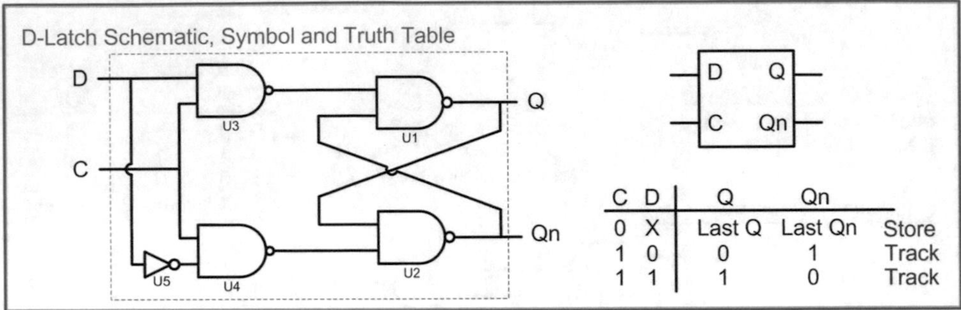

Fig. 7.14
D-Latch schematic, symbol, and truth table

The timing diagram for the D-Latch is shown in Fig. 7.15.

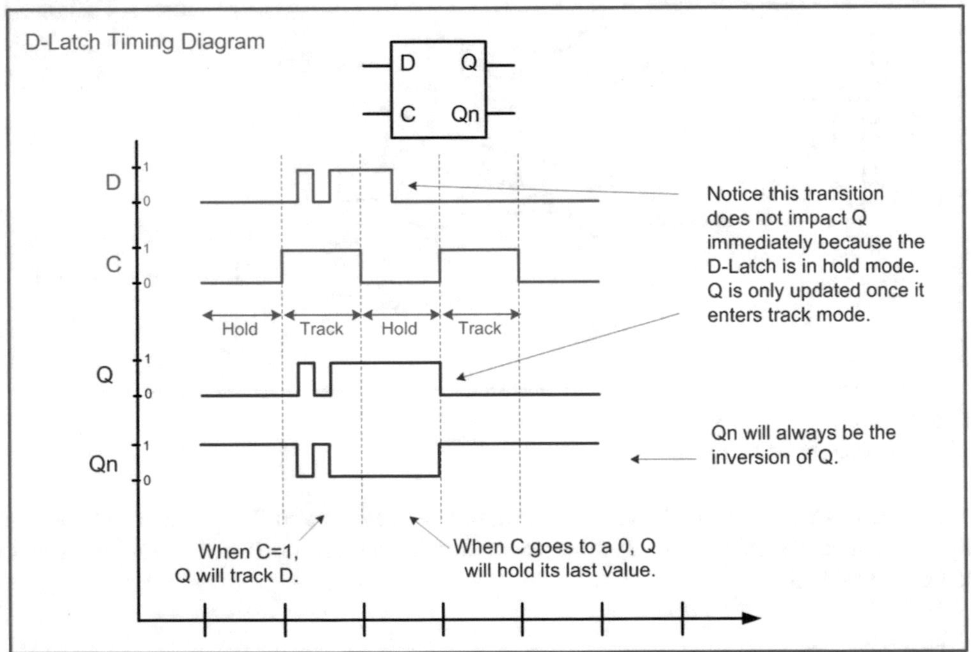

Fig. 7.15
D-Latch timing diagram

7.1.7 The D-Flip-Flop

The final and most widely used storage device in sequential logic is the *D-flip-flop*. The D-flip-flop is similar in behavior to the D-Latch with the exception that the store mode is triggered by a transition, or edge on the clock signal instead of a level. This allows the D-flip-flop to implement higher frequency systems since the outputs are updated in a shorter amount of time. The schematic, symbol, and truth table are given in Fig. 7.16 for a rising edge-triggered D-flip-flop. To indicate that the device is edge sensitive, the input for the clock is designated with a ">." The U3 inverter in this schematic creates the rising edge behavior. If U3 is omitted, this circuit would be a negative edge-triggered D-flip-flop.

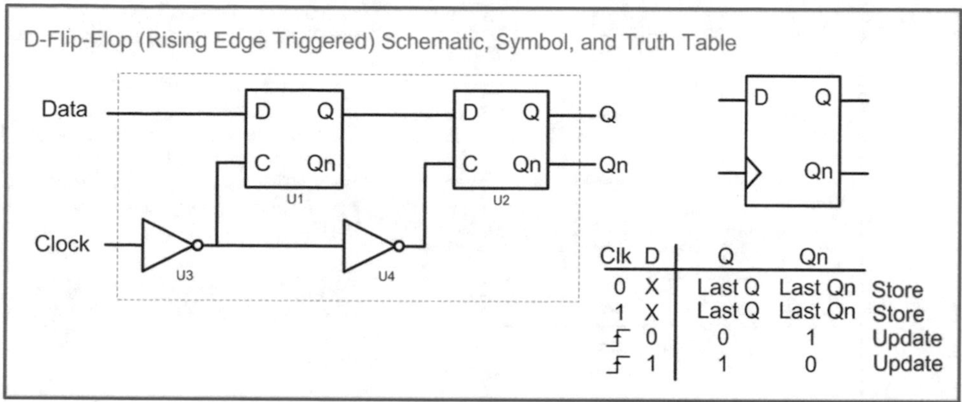

Clk	D	Q	Qn	
0	X	Last Q	Last Qn	Store
1	X	Last Q	Last Qn	Store
↑	0	0	1	Update
↑	1	1	0	Update

Fig. 7.16
D-flip-flop (rising edge triggered) schematic, symbol, and truth table

The D-flip-flop schematic shown above is called a *master/slave* configuration because of how the data is passed through the two D-Latches (U1 and U2). Due to the U4 inverter, the two D-Latches will always be in complementary modes. When U1 is in hold mode, U2 will be in track mode and vice versa. When the clock signal transitions HIGH, U1 will store the last value of data. During the time when the clock is HIGH, U2 will enter track mode and pass this value to Q. In this way, the data is latched into the storage device on the rising edge of the clock and is present on Q. This is the *master* operation of the device because U1, or the first D-Latch, is holding the value, and the second D-Latch (the *slave*) is simply passing this value to the output Q. When the clock transitions LOW, U2 will store the output of U1. Since there is a finite delay through U1, the U2 D-Latch is able to store the value before U1 fully enters track mode. U2 will drive Q for the duration of the time that the clock is LOW. This is the *slave* operation of the device because U2, or the second D-Latch, is holding the value. During the time the clock is LOW, U1 is in track mode, which passes the input data to the middle of the D-flip-flop preparing for the next rising edge of the clock. The master/slave configuration creates a behavior where the Q output of the D-flip-flop is only updated with the value of D on a rising edge of the clock. At all other times, Q holds the last value of D. An example timing diagram for the operation of a rising edge D-flip-flop is given in Fig. 7.17.

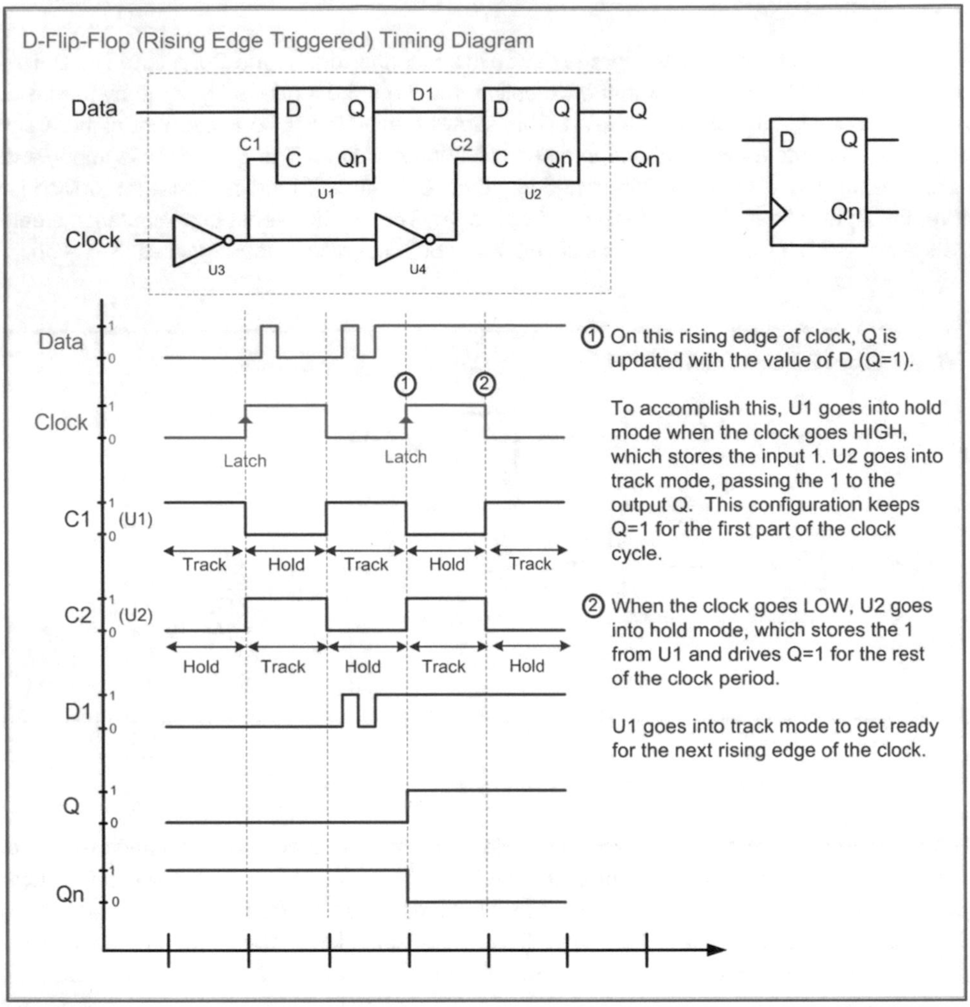

Fig. 7.17
D-flip-flop (rising edge triggered) timing diagram

D-flip-flops often have additional signals that will set the initial conditions of the outputs that are separate from the clock. A *reset* input is used to force the outputs to Q = 0, Qn = 1. A *preset* input is used to force the outputs to Q = 1, Qn = 0. In most modern D-flip-flops, these inputs are active LOW, meaning that the line is asserted when the input is a 0. Active LOW inputs are indicated by placing an inversion bubble on the input pin of the symbol. These lines are typically *asynchronous*, meaning that when they are asserted, action is immediately taken to alter the outputs. This is different from a *synchronous* input in which action is only taken on the edge of the clock. Figure 7.18 shows the symbols and truth tables for two D-flip-flop variants, one with an active LOW reset and another with both an active LOW reset and active LOW preset.

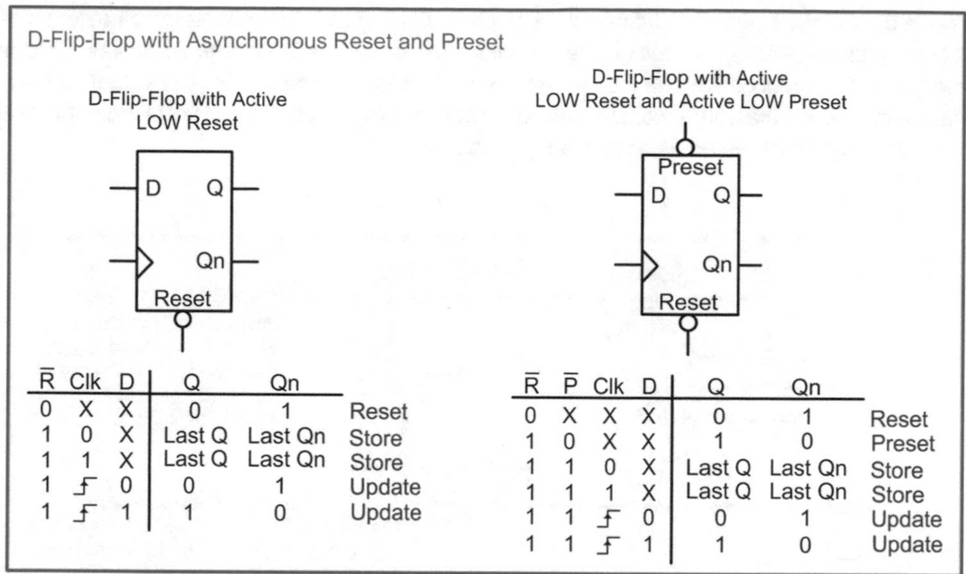

$\overline{R}$	Clk	D	Q	Qn	
0	X	X	0	1	Reset
1	0	X	Last Q	Last Qn	Store
1	1	X	Last Q	Last Qn	Store
1	↑	0	0	1	Update
1	↑	1	1	0	Update

$\overline{R}$	$\overline{P}$	Clk	D	Q	Qn	
0	X	X	X	0	1	Reset
1	0	X	X	1	0	Preset
1	1	0	X	Last Q	Last Qn	Store
1	1	1	X	Last Q	Last Qn	Store
1	1	↑	0	0	1	Update
1	1	↑	1	1	0	Update

Fig. 7.18
D-flip-flop with asynchronous reset and preset

D-flip-flops can also be created with an *enable* line. An enable line controls whether or not the output is updated. Enable lines are synchronous, meaning that when they are asserted, the outputs will be updated on the rising edge of the clock. When de-asserted, the outputs are not updated. This behavior in effect ignores the clock input when de-asserted. Figure 7.19 shows the symbol and truth table for a D-flip-flop with a synchronous enable.

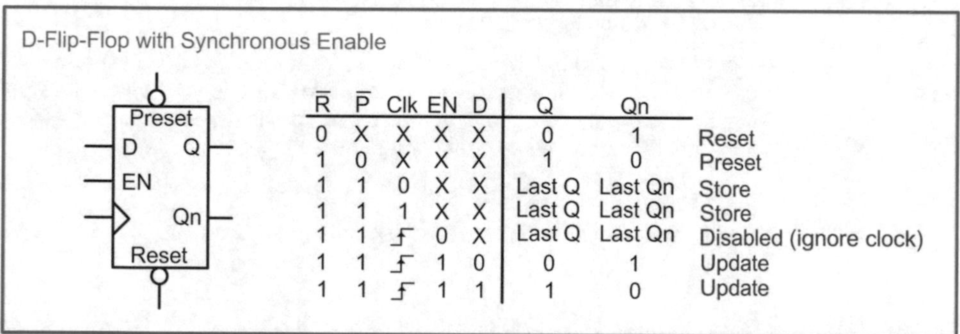

$\overline{R}$	$\overline{P}$	Clk	EN	D	Q	Qn	
0	X	X	X	X	0	1	Reset
1	0	X	X	X	1	0	Preset
1	1	0	X	X	Last Q	Last Qn	Store
1	1	1	X	X	Last Q	Last Qn	Store
1	1	↑	0	X	Last Q	Last Qn	Disabled (ignore clock)
1	1	↑	1	0	0	1	Update
1	1	↑	1	1	1	0	Update

Fig. 7.19
D-flip-flop with synchronous enable

The behavior of the D-flip-flop allows us to design systems that are *synchronous* to a clock signal. A clock signal is a periodic square wave that dictates when events occur in a digital system. A synchronous system based on D-flip-flops will allow the outputs of its storage devices to be updated upon a rising edge of the clock. This is advantageous because when the Q outputs are storing values, they can be used as inputs for combinational logic circuits. Since combinational logic circuits contain a certain amount of propagation delay before the final output is calculated, the D-flip-flop can hold the inputs at a steady value while the output is generated. Since the input on a D-flip-flop is ignored during all other times, the output of a combinational logic circuit can be fed back as an input to a D-flip-flop. This gives a system the ability to generate outputs based on the current values of inputs in addition to past values of the inputs that are being held on the outputs of D-flip-flops. This is the definition of sequential logic. An example synchronous, sequential system is shown in Fig. 7.20.

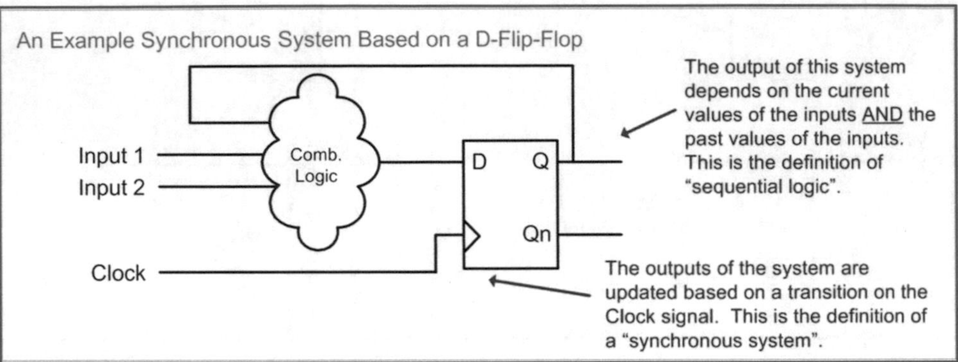

Fig. 7.20
An example synchronous system based on a D-flip-flop

Concept Check

CC7.1(a) What will always cause a digital storage device to come out of metastability and settle in one of its two stable states? Why?

A) The power supply. The power supply provides the necessary current for the device to overcome metastability.

B) Electrical noise. Noise will always push the storage device toward one state or another. Once the storage device starts moving toward one of its stable states, the positive feedback of the storage device will reinforce the transition until the output eventually comes to rest in a stable state.

C) A reset. A reset will put the device into a known stable state.

D) A rising edge of clock. The clock also puts the device into a known stable state.

CC7.1(b) What was the purpose of replacing the inverters in the cross-coupled inverter pair with NOR gates to form the SR Latch?

A) NOR gates are easier to implement in CMOS.

B) To provide the additional output Qn.

C) To provide more drive strength for storing.

D) To provide inputs to explicitly set the value being stored.

7.2 Sequential Logic Timing Considerations

There are a variety of timing specifications that need to be met in order to successfully design circuits using sequential storage devices. The first specification is called the *setup time* (t_{setup} or t_s). The setup time specifies how long the data input needs to be at a steady state *before* the clock event. The second specification is called the *hold time* (t_{hold} or t_h). The hold time specifies how long the data input needs to be at a steady state *after* the clock event. If these specifications are violated (i.e., the input transitions too close to the clock transition), the storage device will not be able to determine whether the input was a 1 or 0 and will go metastable. The time a storage device will remain metastable is a deterministic value and is specified by the part manufacturer (t_{meta}). In general, metastability should be avoided; however, knowing the maximum duration of metastability for a storage device allows us to design circuits to overcome potential metastable conditions. During the time the device is metastable, the output will have random behavior. It may go to a steady state 1, a steady state 0, or toggle between a 0 and 1 uncontrollably. Once the device comes out of metastability, it will come to rest in one of its two stable states (Q = 0 or Q = 1). The final resting state is random and unknown. Another specification for sequential storage devices is the delay from the time a clock transition occurs to the point that the data is present on the Q output. This specification is called the *clock-to-Q* delay and is given the notation t_{CQ}. These specifications are shown in Fig. 7.21.

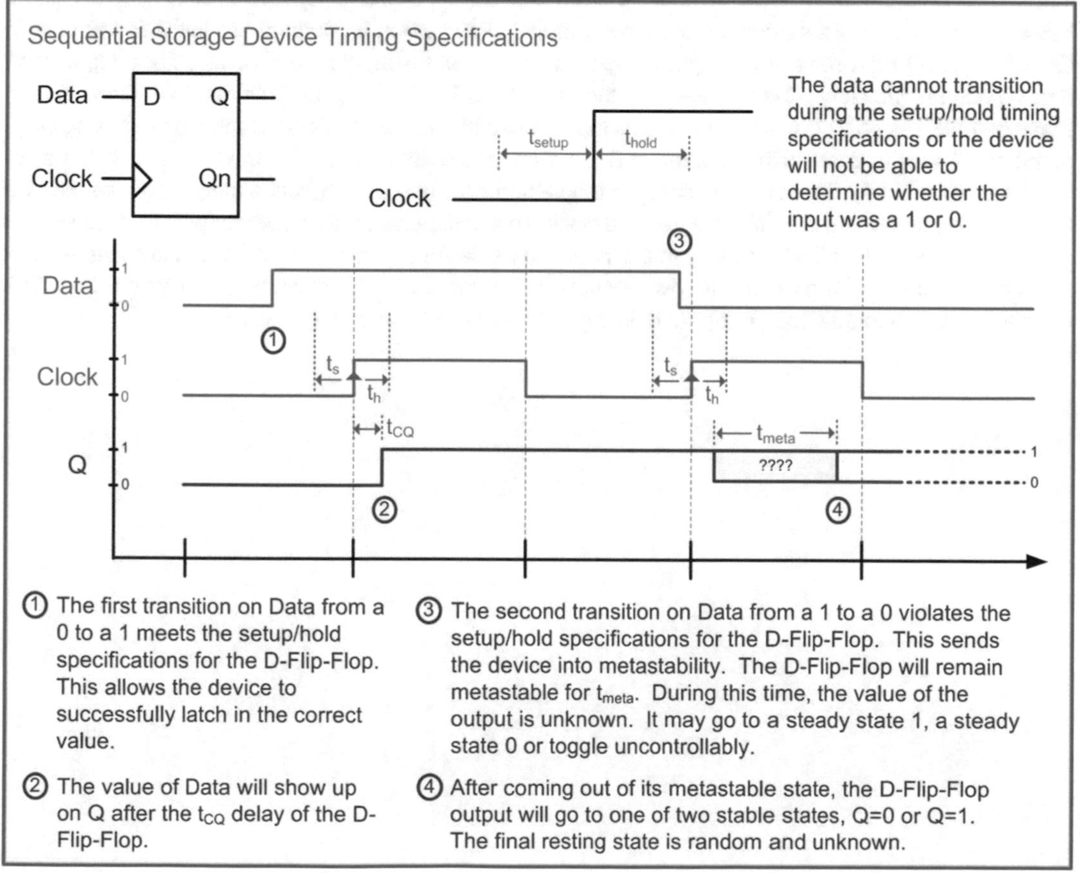

Fig. 7.21
Sequential storage device timing specifications

Concept Check

CC7.2 Which D-flop-flop timing specification requires all of combinational logic circuits in the system to settle on their final output before a triggering clock edge can occur?

A) t_{setup} B) t_{hold} C) t_{CQ} D) t_{meta}

7.3 Common Circuits Based on Sequential Storage Devices

Sequential logic storage devices give us the ability to create sophisticated circuits that can make decisions based on the current and past values of the inputs; however, there are a variety of simple yet useful circuits that can be created with only these storage devices. This section will introduce a few of these circuits.

7.3.1 Toggle Flop Clock Divider

A *toggle flop* is a circuit that contains a D-flip-flop configured with its Qn output wired back to its D input. This configuration is also commonly referred to as a *T-flip-flop* or *T-flop*. In this circuit, the only input is the clock signal. Let's examine the behavior of this circuit when its outputs are initialized to Q = 0, Qn = 1. Since Qn is wired to the D input, a logic 1 is present on the input before the first clock edge. Upon a rising edge of the clock, Q is updated with the value of D. This puts the outputs at Q = 1, Qn = 0. With these outputs, now a logic 0 is present on the input before the next clock edge. Upon the next rising edge of the clock, Q is updated with the value of D. This time the outputs go to Q = 0, Qn = 1. This behavior continues indefinitely. The circuit is called a toggle flop because the outputs simply toggle between a 0 and 1 every time there is a rising edge of the clock. This configuration produces outputs that are square waves with exactly half the frequency of the incoming clock. As a result, this circuit is also called a *clock divider*. This circuit can be given its own symbol with a label of "T" indicating it is a toggle flop. The configuration of a toggle flop (T-flop) and timing diagram are shown in Fig. 7.22.

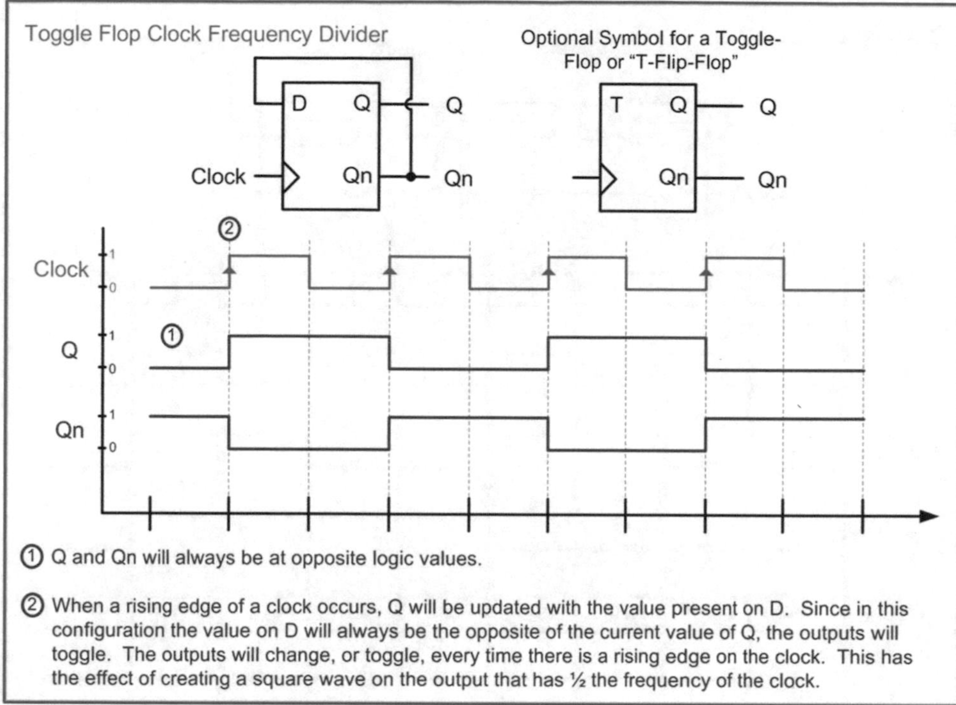

Fig. 7.22
Toggle flop clock frequency divider

7.3.2 Ripple Counter

The toggle flop configuration can be used to create a simple binary counter called a *ripple counter.* In this configuration, the Qn output of a toggle flop is used as the clock for a subsequent toggle flop. Since the output of the first toggle flop is a square wave that is 1/2 the frequency of the incoming clock, this configuration will produce an output on the second toggle flop that is 1/4 the frequency of the incoming clock. This is by nature the behavior of a binary counter. The output of this counter is present on the Q pins of each toggle flop. Toggle flops are added until the desired width of the counter is achieved with each toggle flop representing one bit of the counter. Since each toggle flop produces the clock for the subsequent latch, the clock is said to *ripple* through the circuit, hence the name ripple counter. A 3-bit ripple counter is shown in Fig. 7.23.

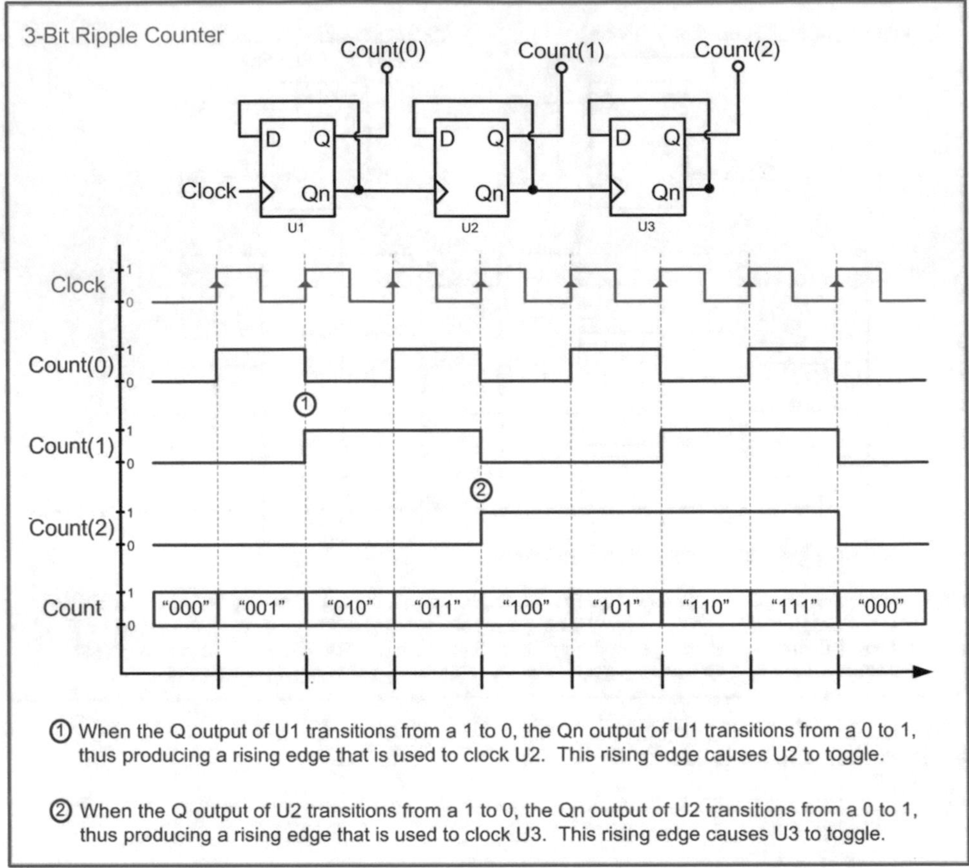

Fig. 7.23
3-Bit ripple counter

7.3.3 Switch Debouncing

Another useful circuit based on sequential storage devices is a switch debouncer. Mechanical switches have a well-known issue of not producing clean logic transitions on their outputs when pressed. This becomes problematic when using a switch to create an input for a digital device because it will cause unwanted logic level transitions on the output of the gate. In the case of a clock input, this unwanted transition can cause a storage device to unintentionally latch incorrect data.

The primary cause of these unclean logic transitions is due to the physical vibrations of the metal contacts when they collide with each other during a button press or switch actuation. Within a mechanical switch, there is typically one contact that is fixed and another that is designed to move when the button is pressed. The contact that is designed to move can be thought of as a beam that is fixed on one side and free on the other. As the free side of the beam moves toward the fixed contact in order to close the circuit, it will collide and then vibrate just as a tuning fork does when struck. The vibration will eventually diminish, and the contact will come to rest, thus making a clean electrical connection; however, during the vibration period, the moving contact will *bounce* up and down on the destination contact. This bouncing causes the switch to open and close multiple times before coming to rest in the closed position. This phenomenon is accurately referred to as *switch bounce*. Switch bounce is present in all mechanical switches and gets progressively worse as the switches are used more and more.

Figure 7.24 shows some of the common types of switches found in digital systems. The term *pole* is used to describe the number of separate circuits controlled by the switch. The term *throw* is used to describe the number of separate closed positions the switch can be in.

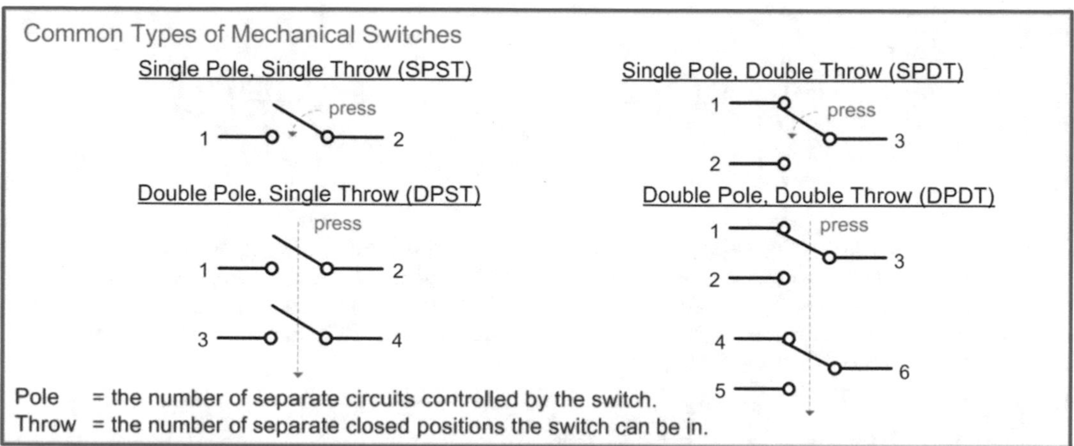

Fig. 7.24
Common types of mechanical switches

Let's look at switch bounce when using a SPST switch to provide an input to a logic gate. A SPST requires a resistor and can be configured to provide either a logic HIGH or LOW when in the open position and the opposite logic level when in the closed position. The example configuration in Fig. 7.25 provides a logic LOW when in the open position and a logic HIGH when in the closed position. In the open position, the input to the gate (SW) is pulled to GND to create a logic LOW. In the closed position, the input to the gate is pulled to V_{CC} to create a logic HIGH. A resistor is necessary to prevent a short circuit between V_{CC} and GND when the switch is closed. Since the input current specification for a logic gate is very small, the voltage developed across the resistor due to the gate input current is negligible. This means that the resistor can be inserted in the pull-down network without developing a noticeable voltage. When the switch closes, the free-moving contact will bounce off of the destination contact numerous times before settling in the closed position. During the time while the switch is bouncing, the switch will repeatedly toggle between the open (HIGH) and closed (LOW) positions.

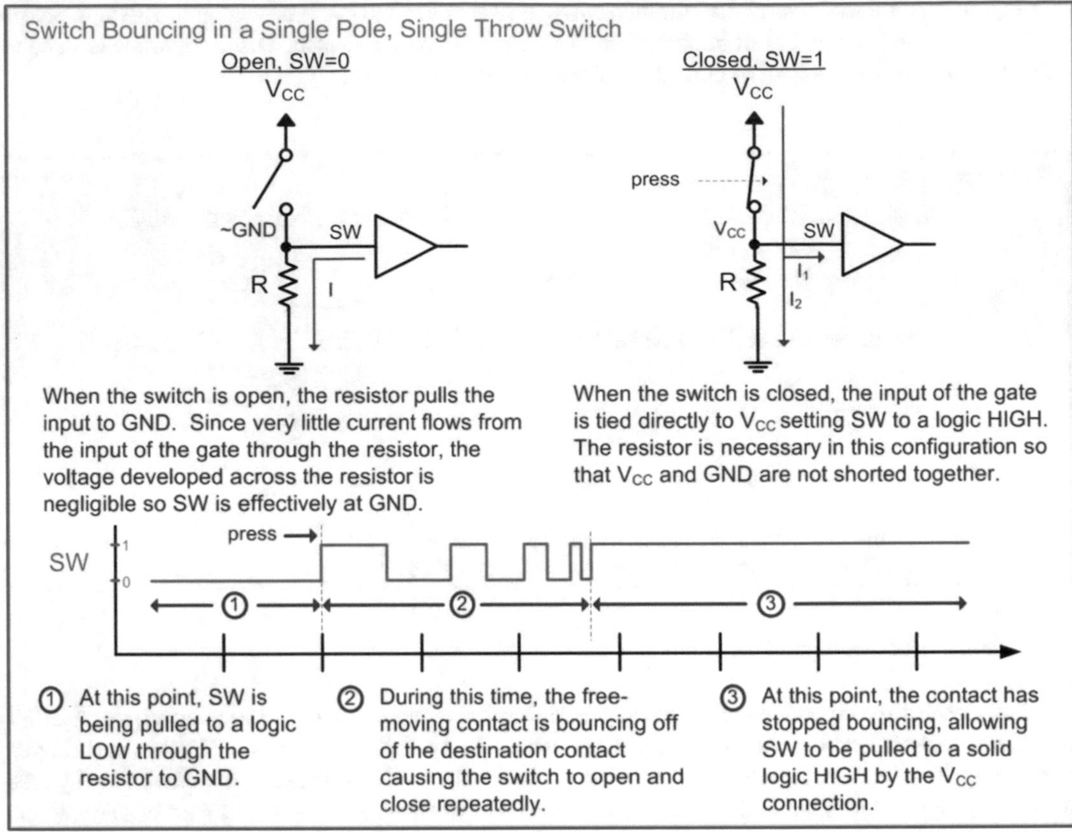

Fig. 7.25
Switch bouncing in a single pole, single throw switch

A possible solution to eliminate this switch bounce is to instead use a SPDT switch in conjunction with a sequential storage device. Before looking at this solution, we need to examine an additional condition introduced by the SPDT switch. The SPDT switch has what is known as *break-before-make* behavior. The term *break* is used to describe when a switch is open, while the term *make* is used to describe when the switch is closed. When a SPDT switch is pressed, the input will be floating during the time when the free-moving contact is transitioning toward the destination contact. During this time, the output of the switch is unknown and can cause unwanted logic transitions if it is being used to drive the input of a logic gate.

Let's look at switch bounce when using a SPDT switch without additional circuitry to handle bouncing. A SPDT has two positions that the free-moving contact can make a connection to (i.e., double throw). When using this switch to drive a logic level into a gate, one position is configured as a logic HIGH and the other a logic LOW. Consider the SPDT switch configuration in Fig. 7.26. Position 1 of the SPDT switch is connected to GND, while position 2 is connected to V_{CC}. When unpressed the switch is in position 1. When pressed, the free-moving contact will transition from position 1 to 2. During the transition, the free-moving contact is floating. This creates a condition where the input to the gate (SW) is unknown. This floating input will cause unpredictable behavior on the output of the gate. Upon reaching position 2, the free-moving contact will bounce off of the destination contact. This will cause the input of the logic gate to toggle between a logic HIGH and floating repeatedly until the free-moving contact comes to rest in position 2.

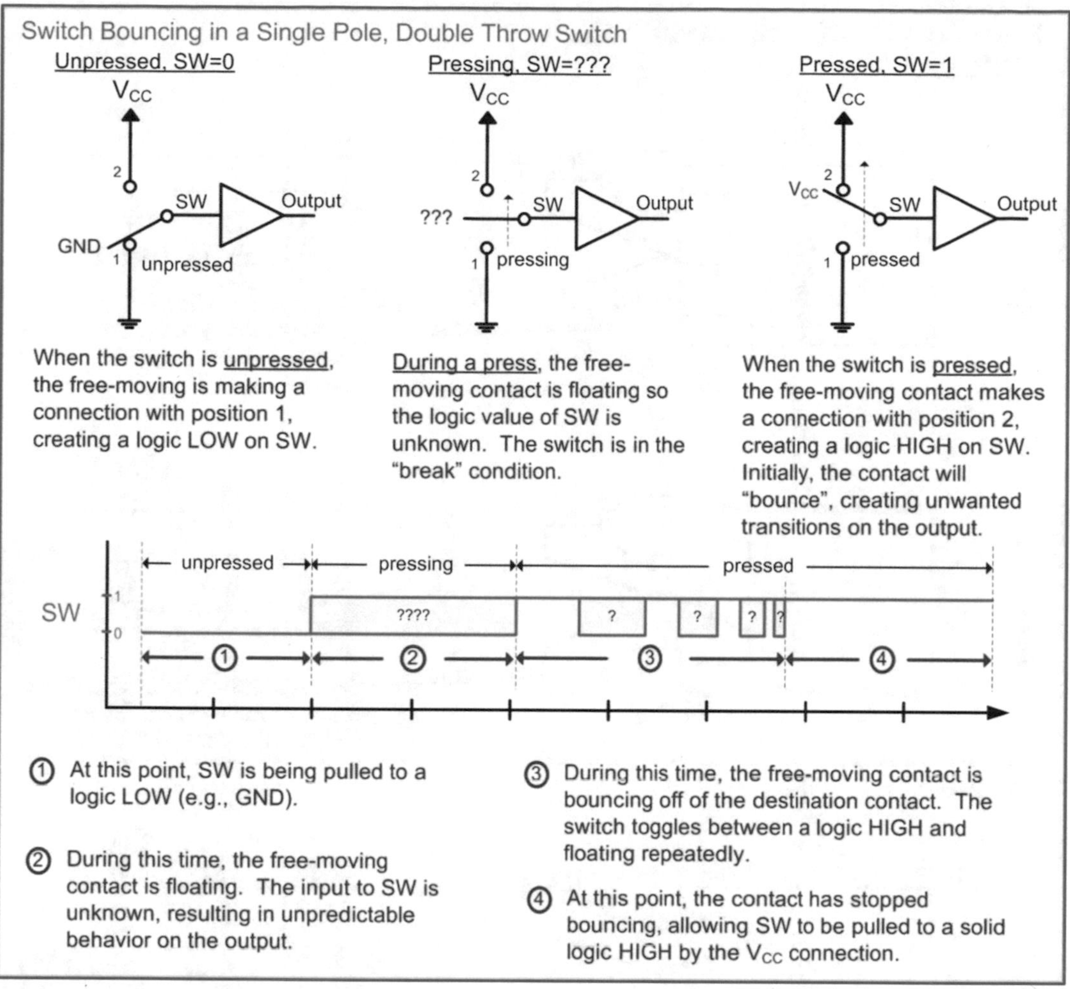

Fig. 7.26
Switch bouncing in a single pole, double throw switch

The SPDT switch is ideal for use with an S'R' Latch in order to produce a clean logic transition. This is because during the *break* portion of the transition, an S'R' Latch can be used to hold the last value of the switch. This is unique to the SPDT configuration. The SPST switch in comparison does not have the *break* characteristic; rather it always drives a logic level in both of its possible positions. Consider the debounce circuit for a SPDT switch in Fig. 7.27. This circuit is based on an S'R' Latch with two pull-up resistors. Since the S'R' Latch is created using NAND gates, this circuit is commonly called a *NAND-debounce* circuit. In the unpressed configuration, the switch drives S' = 0, and the R2 pull-up resistor drives R' = 1. This creates a logic 0 on the output of the circuit (Qn = 0). During a switch press, the free-moving contact is floating; thus it is not driving in a logic level into the S'R' Latch. Instead, both pull-up resistors pull S' and R' to 1's. This puts the latch into its hold mode, and the output will remain at a logic 0 (Qn = 0). Once the free-moving contact reaches the destination contact, the switch will drive R' = 0. Since at this point the R1 pull-up is driving S' = 1, the latch outputs a logic 1 (Qn = 1). When the free-moving contact bounces off of the destination contact, it will put the latch back into the hold mode; however, this time the last value that will be held is Qn = 1. As the switch continues to bounce, the latch will move between the Qn = 1 and Qn = "Last Qn" states, both of which produce an output of 1. In this way, the SPDT switch in conjunction with the S'R' Latch produces a clean 0 to 1 logic transition despite the break-before-make behavior of the switch and the contact bounce.

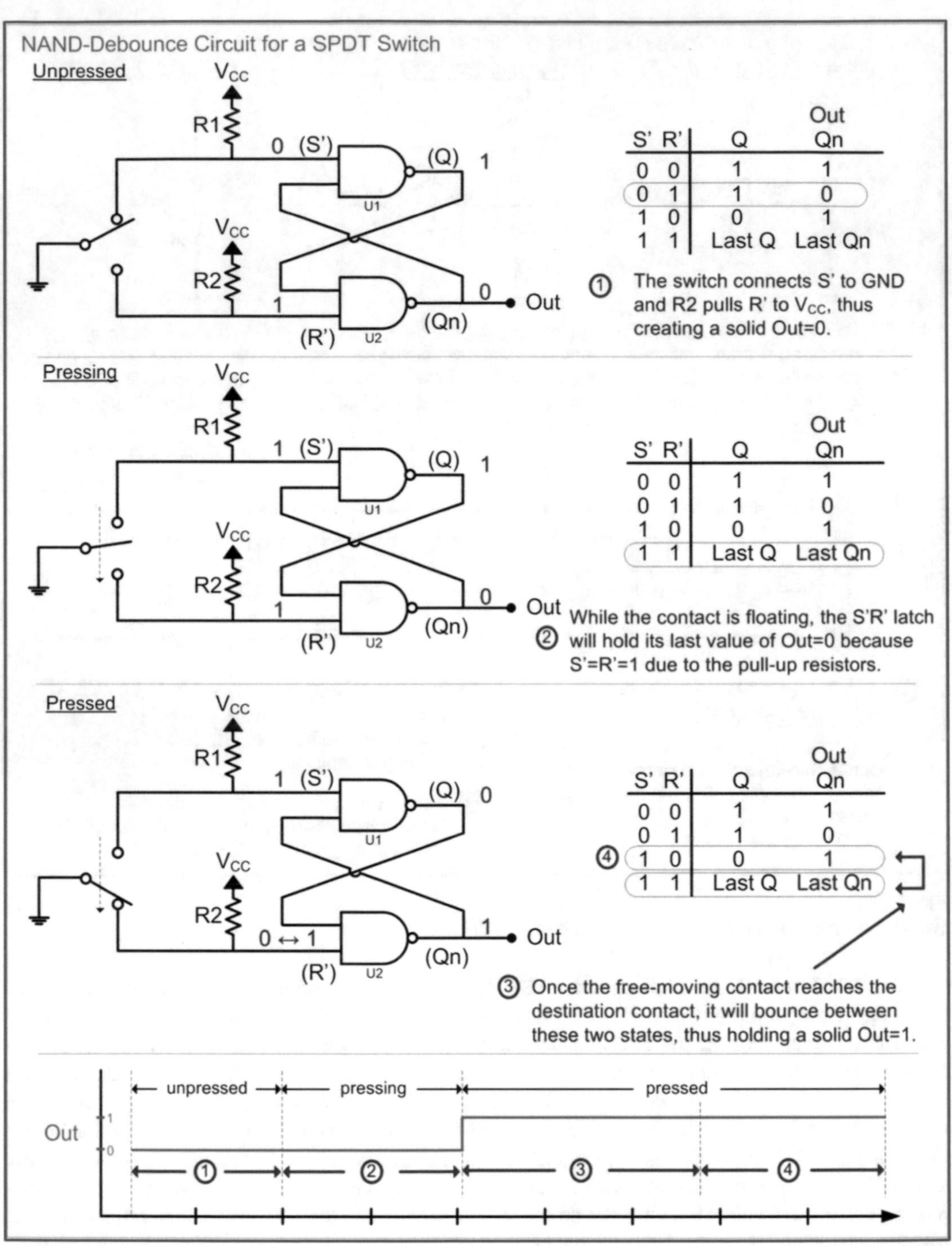

Fig. 7.27
NAND-debounce circuit for a SPDT switch

7.3.4 Shift Registers

A *shift register* is a chain of D-flip-flops that each is connected to a common clock. The output of the first D-flip-flop is connected to the input of the second D-flip-flop. The output of the second D-flip-flop is connected to the input of the third D-flip-flop and so on. When data is present on the input to the first D-flip-flop, it will be latched upon the first rising edge of the clock. On the second rising edge of the clock, the same data will be latched into the second D-flip-flop. This continues on each rising edge of the clock until the data has been *shifted* entirely through the chain of D-flip-flops. Shift registers are commonly used to convert a serial string of data into a parallel format. If an n-bit, serial sequence of information is clocked into the shift register, after n clocks the data will be held on each of the D-flip-flop outputs. At this moment, the n-bits can be read as a parallel value. Consider the shift register configuration shown in Fig. 7.28.

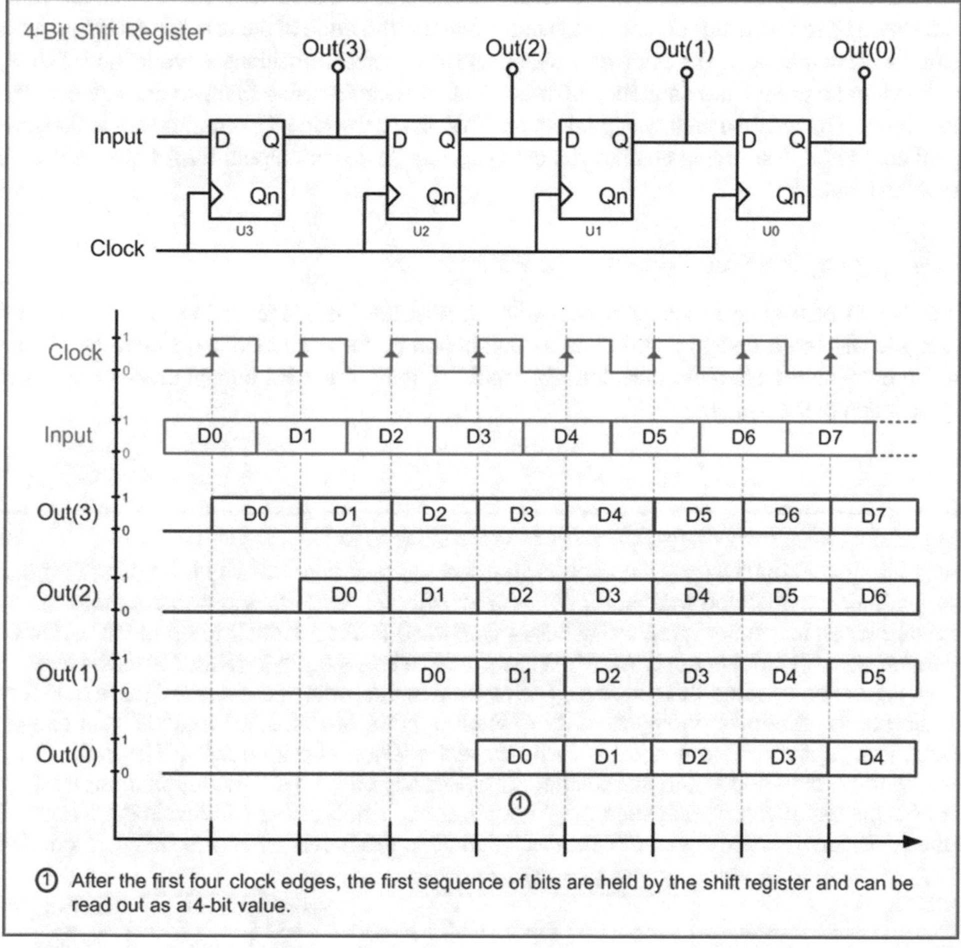

Fig. 7.28
4-Bit shift register

CONCEPT CHECK

CC7.3 Which D-flip-flop timing specification is most responsible for the ripple delay in a ripple counter?

A) t_{setup} B) t_{hold} C) t_{CQ} D) t_{meta}

7.4 Finite-State Machines

Now we turn our attention to one of the most powerful sequential logic circuits, the finite-state machine (FSM). A FSM, or *state machine*, is a circuit that contains a predefined number of states (i.e., a finite number of states). The machine can exist in one and only one state at a time. The circuit *transitions* between states based on a triggering event, most commonly the edge of a clock, in addition to the values of any inputs of the machine. The number of states and all possible transitions is predefined. Through the use of states and a predefined sequence of transitions, the circuit is able to make decisions on the next state to transition based on a history of past states. This allows the circuit to create outputs that are more intelligent compared to a simple combinational logic circuit that has outputs based only on the current values of the inputs.

7.4.1 Describing the Functionality of a FSM

The design of a state machine begins with an abstract word description of the desired circuit behavior. We will use a design example of a push-button motor controller to describe all of the steps involved in creating a finite-state machine. Example 7.1 starts the FSM design process by stating the word description of the system.

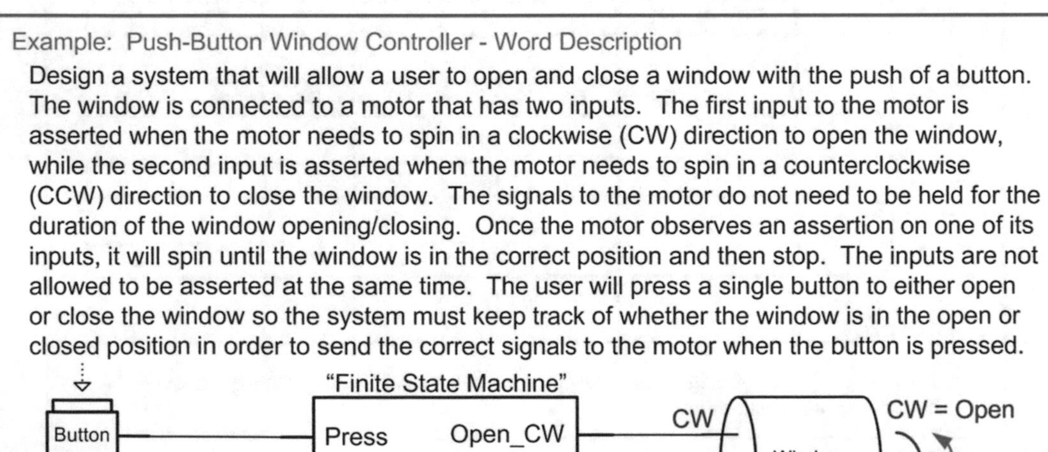

Example 7.1
Push-button window controller: word description

7.4.1.1 State Diagrams

A state diagram is a graphical way to describe the functionality of a finite-state machine. A state diagram is a form of a directed graph, in which each state (or vertex) within the system is denoted as a circle and given a descriptive name. The names are written inside of the circles. The transitions between states are denoted using arrows with the input conditions causing the transitions written next to them. Transitions (or edges) can move to different states upon particular input conditions or remain in the same state. For a state machine implemented using sequential logic storage, an evaluation of when to transition states is triggered every time the storage devices update their outputs. For example, if the system was implemented using rising edge-triggered D-flip-flops, then an evaluation would occur on every rising edge of the clock.

There are two different types of output conditions for a state machine. The first is when the output only depends on the current state of the machine. This type of system is called a *Moore machine*. In this case, the outputs of the system are written inside of the state circles. This indicates the output value that will be generated for each specific state. The second output condition is when the outputs depend on both the current state and the system inputs. This type of system is called a *Mealy machine*. In this case, the outputs of the system are written next to the state transitions corresponding to the appropriate input values. Outputs in a state diagram are typically written inside of parentheses. Example 7.2 shows the construction of the state diagram for our push-button window controller design.

Example: Push-Button Window Controller - State Diagram

1) Defining the States - For this design, we will define two finite states. The first state is when the window is in the closed position. Let's call this state "w_closed". The second state is when the window is in the open position. Let's call this state "w_open". Each of these two states will be represented in the state diagram as circles. The names of the states are written inside of the circles.

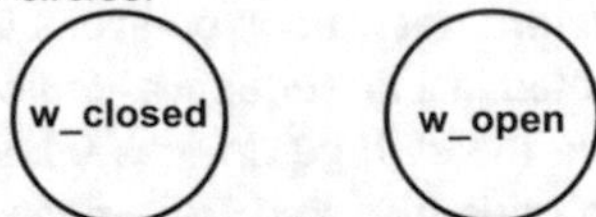

2) Defining the Transitions - We now describe the transitions between states using arrows and labeling the arrows with the input conditions that trigger each transition. For this design, when the machine is in the "w_closed" state, a button press (Press=1) will cause a transition to the "w_open" state. When the button is not pressed, the machine will remain in the "w_closed" state (Press=0). When the machine is in the "w_open" state, a button press (Press=1) will cause a transition to the "w_closed" state, while the button not being pressed (Press=0) will keep the machine in the "w_open" state.

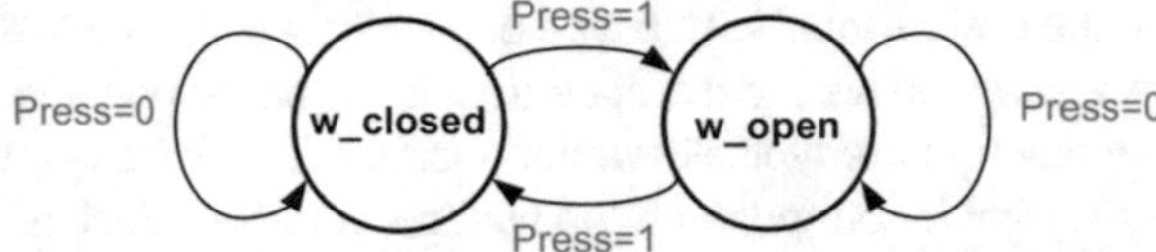

3) Defining the Outputs – We now describe the outputs of the system. For this design, the system will output the appropriate motor control signals upon a button press. This means that the outputs depend on both the current state and the current inputs. This is by definition a *Mealy Machine*. As such, the outputs are listed next to the state transitions. By listing the outputs in this location, both the current state and the input values producing the outputs are indicated. When this machine is in either the w_closed or w_open states and the button is NOT pressed, the outputs Open_CW and Close_CCW are both 0's. When the machine is in w_closed state and the button is pressed, the Open_CW output is asserted to rotate the motor clockwise and open the window. When the machine is in w_open state and the button is pressed, the Close_CCW output is asserted to rotate the motor counterclockwise and close the window. The final state diagram for this system is shown below.

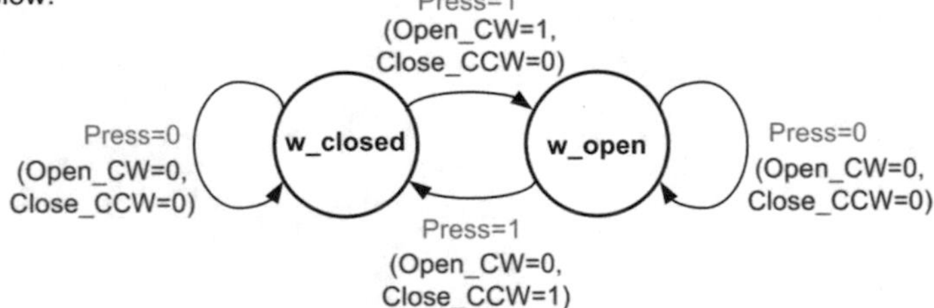

Example 7.2
Push-button window controller: state diagram

7.4.1.2 State Transition Tables

The state diagram can now be described in a table format that is similar to a truth table. This puts the state machine behavior in a form that makes logic synthesis straightforward. The table contains the same information as in the state diagram. The state that the machine exists in is called the *current state*. For each current state that the machine can reside in, every possible input condition is listed along with the destination state of each transition. The destination state for a transition is called the *next state*. Also listed in the table are the outputs corresponding to each current state and, in the case of a Mealy machine, the output corresponding to each input condition. Example 7.3 shows the construction of the state transition table for the push-button window controller design. This information is identical to the state diagram given in Example 7.2.

Example: Push-Button Window Controller - State Transition Table

A state transition table contains the same information as the state diagram but in a tabular format. This format is similar to a truth table and makes logic synthesis straight forward. Each state and input condition is listed in the table along with the corresponding next state and outputs.

	(Input)		(Outputs)	
Current State	Press	Next State	Open_CW	Close_CCW
w_closed	0	w_closed	0	0
w_closed	1	w_open	1	0
w_open	0	w_open	0	0
w_open	1	w_closed	0	1

Example 7.3
Push-button window controller: state transition table

7.4.2 Logic Synthesis for a FSM

Once the behavior of the state machine has been described, it can be directly synthesized. There are three main components of a state machine: the state memory, the next state logic, and the output logic. Figure 7.29 shows a block diagram of a state machine highlighting these three components. The *next state logic* block is a group of combinational logic that produces the next state signals based on the current state and any system inputs. The *state memory* holds the current state of the system. The current state is updated with next state on every rising edge of the clock, which is indicated with the ">" symbol within the block. This behavior is created using D-flip-flops where the current state is held on the Q outputs of the D-flip-flops, while the next state is present on the D inputs of the D-flip-flops. In this way, every rising edge of the clock will trigger an evaluation of which state to move to next. This decision is based on the current state and the current inputs. The *output logic* block is a group of combinational logic that creates the outputs of the system. This block always uses the current state as an input and, depending on the type of machine (Mealy vs. Moore), uses the system inputs. It is useful to keep this block diagram in mind when synthesizing finite-state machines as it will aid in keeping the individual design steps separate and clear.

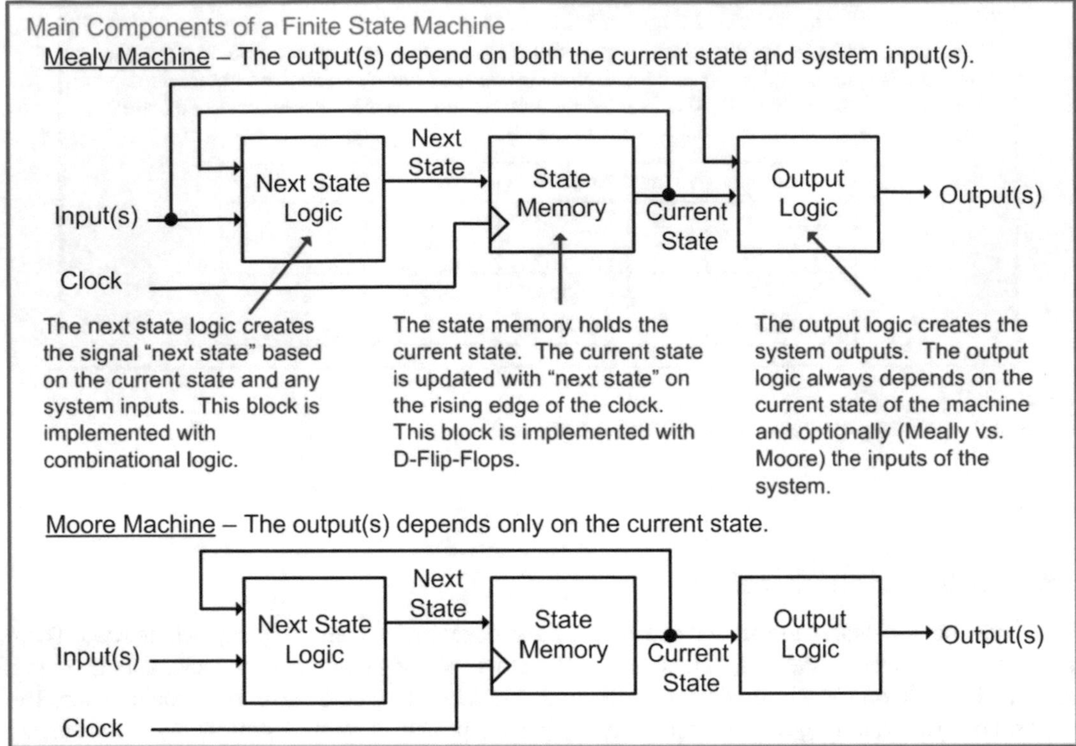

Fig. 7.29
Main components of a finite-state machine

7.4.2.1 State Memory

The state memory is the circuitry that will hold the current state of the machine. Upon a rising edge of a clock, it will update the current state with the next state. At all other times, the next state input is ignored. This gives time for the next state logic circuitry to compute the results for the next state. This behavior is identical to that of a D-flip-flop; thus the state memory is simply one or more D-flip-flops. The number of D-flip-flops required depends on how the states are *encoded*. *State encoding* is the process of assigning a binary value to the descriptive names of the states from the state diagram and state transition tables. Once the descriptive names have been converted into representative codes using 1's and 0's, the states can be implemented in real circuitry. The assignment of codes is arbitrary and can be selected in order to minimize the circuitry needed in the machine.

There are three main styles of state encoding. The first is straight ***binary encoding***. In this approach the state codes are simply a set of binary counts (i.e., 00, 01, 10, 11…). The binary counts are assigned starting at the beginning of the state diagram and incrementally assigned toward the end. This type of encoding has the advantage that it is very efficient in minimizing the number of D-flip-flops needed for the state memory. With n D-flip-flops, 2^n states can be encoded. When a large number of states are required, the number of D-flip-flops can be calculated using the rules of logarithmic math. Example 7.4 shows how to solve for the number of bits needed in the binary state code based on the number of states in the machine.

Solving For the Number of Bits Needed for Binary State Encoding

Problem: You are designing a state machine that has 41 unique states and are going to encode the states in binary. How many D-Flip-Flops do you need?

Solution: Each D-Flip-Flops will hold one bit of the state code. If the state memory has n-bits, it can encode 2^n states using binary encoding. We can use logarithms in order to solve for the n in the exponent.

$$2^n = (\text{\# of states})$$

$$\log(2^n) = \log(\text{\# of states})$$

$$n \cdot \log(2) = \log(\text{\# of states})$$

$$n = \frac{\log(\text{\# of states})}{\log(2)}$$

$$n = \frac{\log(41)}{\log(2)}$$

$$n = 5.36$$

Rounding up to the next whole number means that we need 6 bits, or 6-D-Flip-Flops to encode 41 states in binary.

Check: To check this, let's plug 6 back into the original expression. If we have 6 bits, we can encode 2^6 states, or 64 states. This is enough to encode our 41 states. If we had 1 less bit (e.g., 5), we could only encode up to $2^5=32$ states, so we require 6 bits for this state encoding. Note that not all of the possible binary values are used as state codes.

Example 7.4
Solving for the number of bits needed for binary state encoding

The second type of state encoding is called ***gray code encoding***. A gray code is one in which the value of a code differs by only one bit from any of its neighbors, (i.e., 00, 01, 11, 10. . .). A gray code is useful for reducing the number of bit transitions on the state codes when the machine has a transition sequence that is linear. Reducing the number of bit transitions can reduce the amount of power consumption and noise generated by the circuit. When the state transitions of a machine are highly nonlinear, a gray code encoding approach does not provide any benefit. Gray code is also an efficient coding approach. With *n* D-flip-flops, 2^n states can be encoded just as in binary encoding. Figure 7.30 shows the process of creating n-bit, gray code patterns.

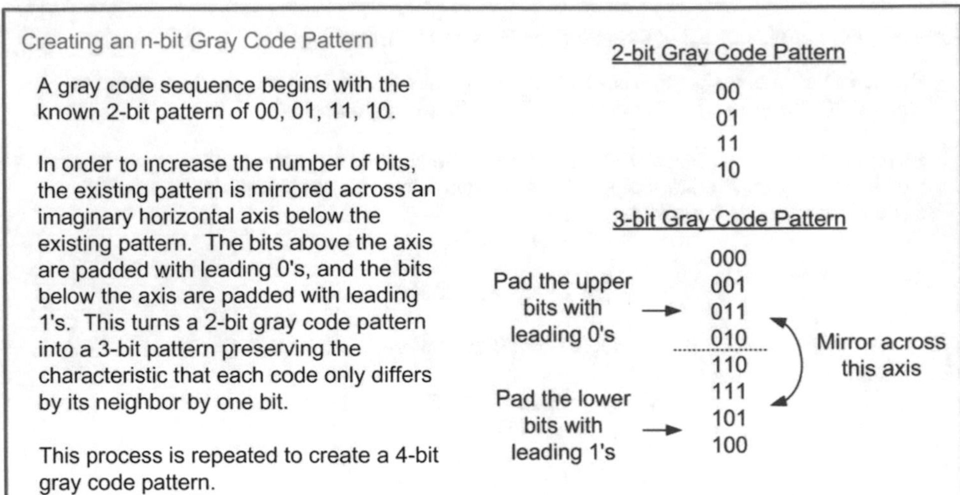

Fig. 7.30
Creating an n-bit gray code pattern

The third common technique to encode states is using ***one-hot encoding***. In this approach, a separate D-flip-flop is asserted for each state in the machine. For an *n-state* machine, this encoding approach requires *n* D-flip-flops. For example, if a machine had three states, the one-hot state codes would be "001", "010," and "100." This approach has the advantage that the next state logic circuitry is very simple; further, there is less chance that the different propagation delays through the next state logic will cause an inadvertent state to be entered. This approach is not as efficient as binary and gray code in terms of minimizing the number of D-flip-flops because it requires one D-flip-flop for each state; however, in modern digital integrated circuits that have abundant D-flip-flops, one-hot encoding is commonly used.

Figure 7.31 shows the differences between these three state encoding approaches.

Comparison of Different State Encoding Approaches

A state machine has eight unique states named S0, S1, … S7. The following is an example of how these states can be encoded using binary, gray code and one-hot.

State Name	Binary	Gray Code	One-Hot
S0	000	000	00000001
S1	001	001	00000010
S2	010	011	00000100
S3	011	010	00001000
S4	100	110	00010000
S5	101	111	00100000
S6	110	101	01000000
S7	111	100	10000000

Fig. 7.31
Comparison of different state encoding approaches

Once the codes have been assigned to the state names, each of the bits within the code must be given a unique signal name. The signal names are necessary because the individual bits within the state code are going to be implemented with real circuitry, so each signal name will correspond to an actual node in the logic diagram. These individual signal names are called ***state variables***. Unique variable names are needed for both the current state and next state signals. The current state variables are driven by the Q outputs of the D-flip-flops holding the state codes. The next state variables are driven by the next state logic circuitry and are connected to the D inputs of the D-flip-flops. State variable names are commonly chosen that are descriptive both in terms of their purpose and connection location. For example, current state variables are often given the names Q, Q_cur, or Q_current to indicate that they come from the Q outputs of the D-flip-flops. Next state variables are given names such as Q*, Q_nxt, or Q_next to indicate that they are the *next* value of Q and are connected to the D input of the D-flip-flops. Once state codes and state variable names are assigned, the state transition table is updated with the detailed information.

Returning to our push-button window controller example, let's encode our states in straight binary and use the state variable names of Q_cur and Q_nxt. Example 7.5 shows the process of state encoding and the new state transition table.

Example: Push-Button Window Controller - State Encoding

This state machine contains two states, w_closed and w_open. The following are the three possible ways these states could be encoded.

State Name	Binary	Gray Code	One-Hot
w_closed	0	0	01
w_open	1	1	10

Since this machine is so small, there is no difference between the binary and gray code approaches. Both of these techniques will require one D-Flip-Flop to hold the state code. The one-hot approach will require two D-Flip-Flops. Let's choose binary state encoding for this example. Let's use the state variable names Q_cur and Q_nxt.

Once the state codes and state variables are chosen, the state transition table is updated with the new detailed information about the design.

Current State		Input	Next State		Outputs	
	Q_cur	Press		Q_nxt	Open_CW	Close_CCW
w_closed	0	0	w_closed	0	0	0
w_closed	0	1	w_open	1	1	0
w_open	1	0	w_open	1	0	0
w_open	1	1	w_closed	0	0	1

Example 7.5
Push-button window controller: state encoding

7.4.2.2 Next State Logic

The next step in the state machine design is to synthesize the next state logic. The next state logic will compute the values of the next state variables based on the current state and the system inputs. Recall that a combinational logic function drives one and only one output bit. This means that every bit within the next state code needs to have a dedicated combinational logic circuit. The state transition table contains all of the necessary information to synthesize the next state logic including the exact output values of each next state variable for each and every input combination of state code and system input(s).

In our push-button window controller example, we only need to create one combinational logic circuit because there is only one next state variable (Q_nxt). The inputs to the combinational logic circuit are Q_cur and Press. Notice that the state transition table was created such that the order of the input values are listed in a binary count just as in a formal truth table formation. This makes synthesizing the combinational logic circuit straightforward. Example 7.6 shows the steps to synthesize the next state logic for this the push-button window controller.

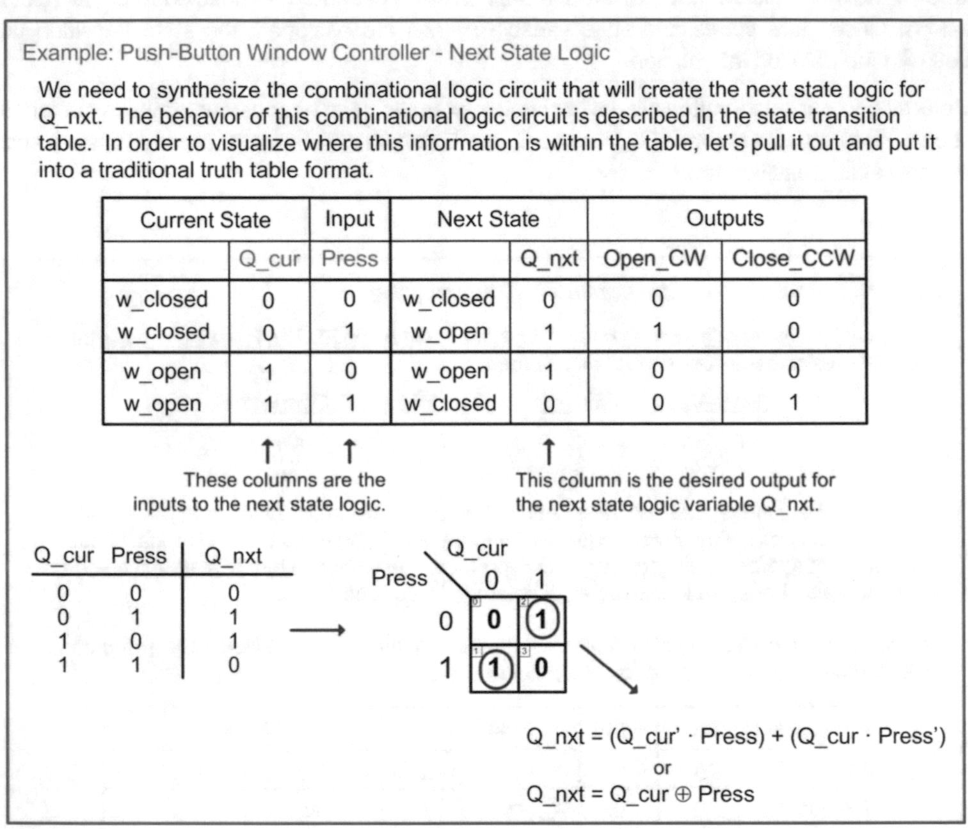

Current State		Input	Next State		Outputs	
	Q_cur	Press		Q_nxt	Open_CW	Close_CCW
w_closed	0	0	w_closed	0	0	0
w_closed	0	1	w_open	1	1	0
w_open	1	0	w_open	1	0	0
w_open	1	1	w_closed	0	0	1

Q_cur	Press	Q_nxt
0	0	0
0	1	1
1	0	1
1	1	0

Example 7.6
Push-button window controller: next state logic

7.4.2.3 Output Logic

The next step in the state machine design is to synthesize the output logic. The output logic will compute the values of the system outputs based on the current state and, in the case of a Mealy machine, the system inputs. Each of the output signals will require a dedicated combinational logic circuit. Again, the state transition table contains all of the necessary information to synthesize the output logic.

In our push-button window controller example, we need to create one circuit to compute the output "Open_CW" and one circuit to compute the output "Close_CCW." In this example, the inputs to these circuits are the current state (Q_cur) and the system input (Press). Example 7.7 shows the steps to synthesize the output logic for the push-button window controller.

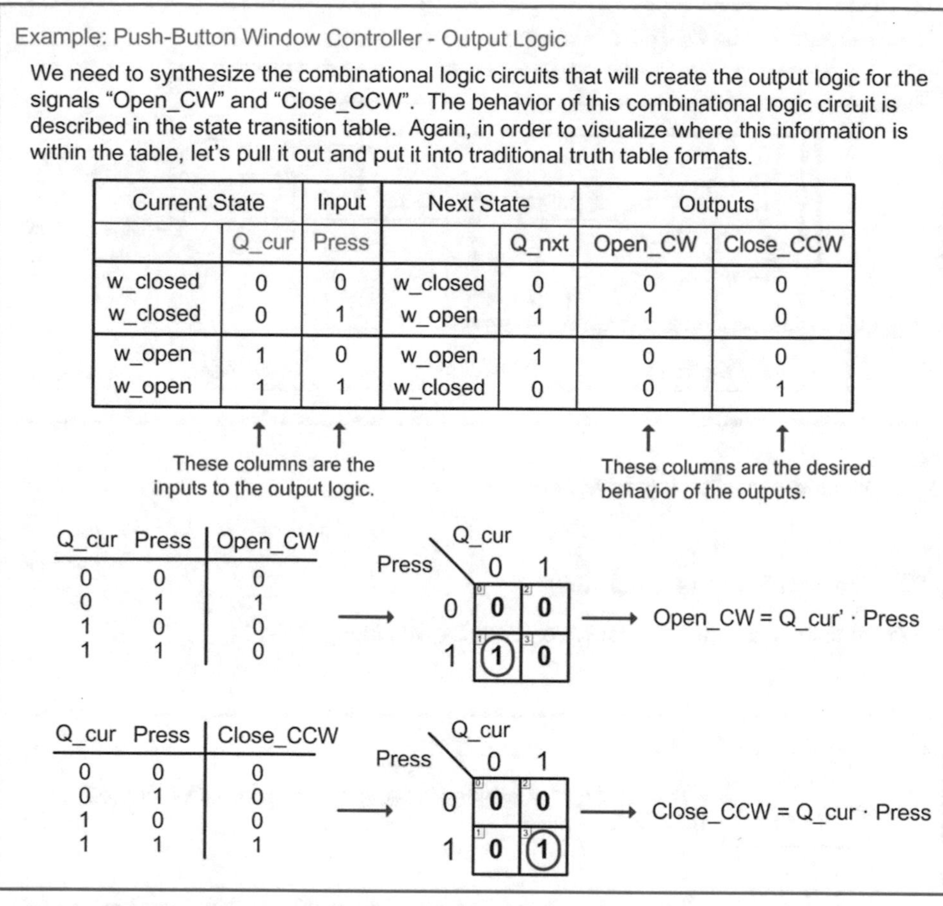

Example: Push-Button Window Controller - Output Logic

We need to synthesize the combinational logic circuits that will create the output logic for the signals "Open_CW" and "Close_CCW". The behavior of this combinational logic circuit is described in the state transition table. Again, in order to visualize where this information is within the table, let's pull it out and put it into traditional truth table formats.

Current State		Input	Next State		Outputs	
	Q_cur	Press		Q_nxt	Open_CW	Close_CCW
w_closed	0	0	w_closed	0	0	0
w_closed	0	1	w_open	1	1	0
w_open	1	0	w_open	1	0	0
w_open	1	1	w_closed	0	0	1

Q_cur	Press	Open_CW
0	0	0
0	1	1
1	0	0
1	1	0

Q_cur	Press	Close_CCW
0	0	0
0	1	0
1	0	0
1	1	1

Example 7.7
Push-button window controller: output logic

7.4.2.4 The Final Logic Diagram

The final step of the design of the state machine is to create the logic diagram. It is useful to recall the block diagram for a state machine from Fig. 7.29. A logic diagram begins by entering the state memory. Recall that the state memory consists of D-flip-flops that hold the current state code. One D-flip-flop is needed for every current state variable. When entering the D-flip-flops, it is useful to label them with the current state variable they will be holding. The next part of the logic diagram is the next state logic. Each of the combinational logic circuits that compute the next state variables should be drawn to the left of D-flip-flop holding the corresponding current state variable. The output of each next state logic circuit is connected to the D input of the corresponding D-flip-flop. Finally, the output logic is entered with the inputs to the logic coming from the current state and potentially from the system inputs.

Example 7.8 shows the process for creating the final logic diagram for our push-button window controller. Notice that the state memory is implemented with one D-flip-flop since there is only 1-bit in the current state code (Q_cur). The next state logic is a combinational logic circuit that computes Q_nxt based on the values of Q_cur and Press. Finally, the output logic consists of two separate combinational logic circuits to compute the system outputs Open_CW and Close_CCW based on Q_cur and Press. In this diagram the Qn output of the D-flip-flop could have been used for the inverted versions of Q_cur; however, inversion bubbles were used instead in order to make the diagram more readable.

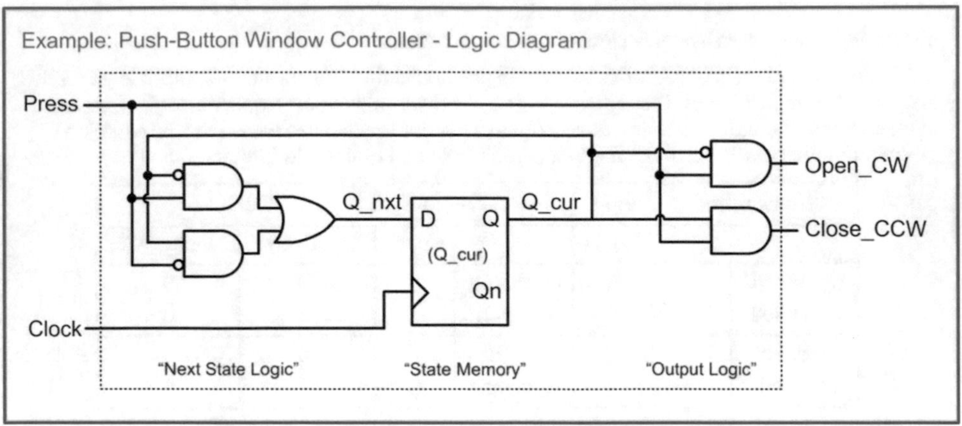

Example 7.8
Push-button window controller: logic diagram

7.4.3 FSM Design Process Overview

The entire finite-state machine design process is given in Fig. 7.32.

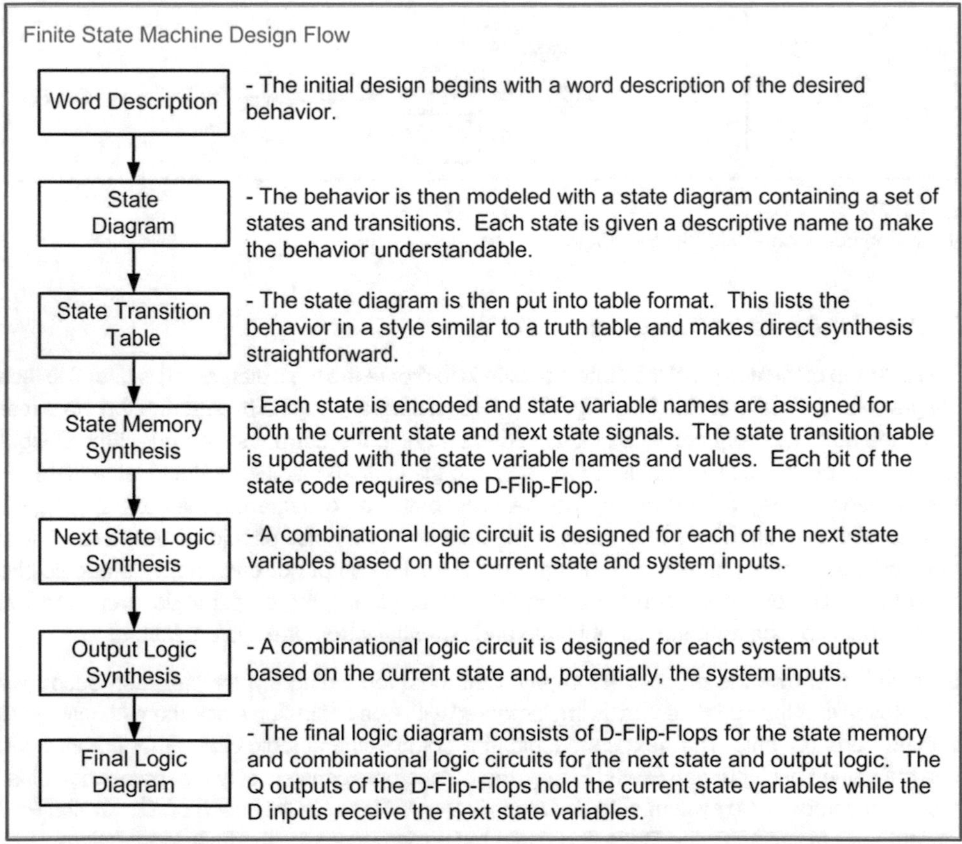

Fig. 7.32
Finite-state machine design flow

7.4.4 FSM Design Examples

7.4.4.1 Serial Bit Sequence Detector

Let's consider the design of a 3-bit serial sequence detector. Example 7.9 provides the word description, state diagram, and state transition table for this finite-state machine.

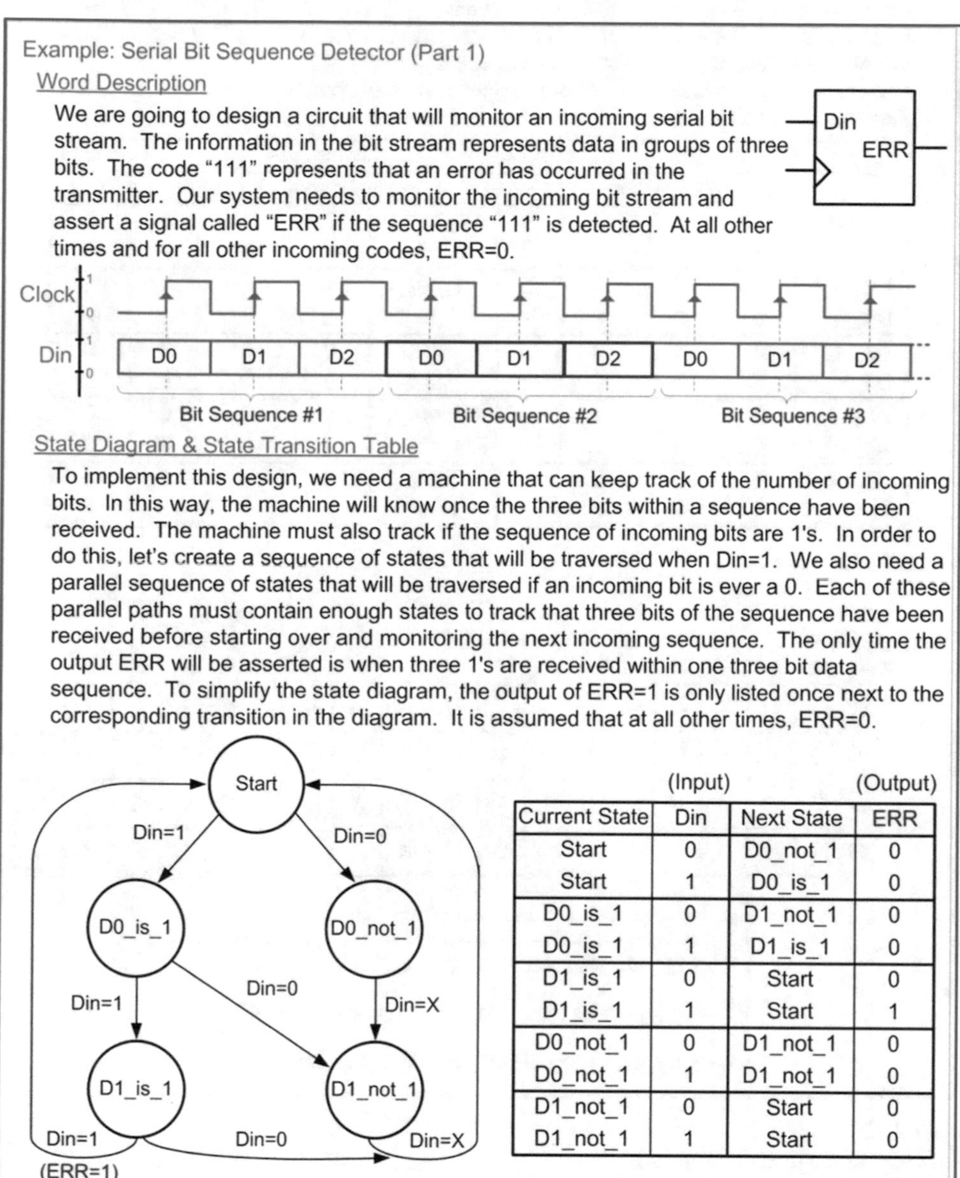

Example: Serial Bit Sequence Detector (Part 1)

Word Description

We are going to design a circuit that will monitor an incoming serial bit stream. The information in the bit stream represents data in groups of three bits. The code "111" represents that an error has occurred in the transmitter. Our system needs to monitor the incoming bit stream and assert a signal called "ERR" if the sequence "111" is detected. At all other times and for all other incoming codes, ERR=0.

State Diagram & State Transition Table

To implement this design, we need a machine that can keep track of the number of incoming bits. In this way, the machine will know once the three bits within a sequence have been received. The machine must also track if the sequence of incoming bits are 1's. In order to do this, let's create a sequence of states that will be traversed when Din=1. We also need a parallel sequence of states that will be traversed if an incoming bit is ever a 0. Each of these parallel paths must contain enough states to track that three bits of the sequence have been received before starting over and monitoring the next incoming sequence. The only time the output ERR will be asserted is when three 1's are received within one three bit data sequence. To simplify the state diagram, the output of ERR=1 is only listed once next to the corresponding transition in the diagram. It is assumed that at all other times, ERR=0.

Current State	(Input) Din	Next State	(Output) ERR
Start	0	D0_not_1	0
Start	1	D0_is_1	0
D0_is_1	0	D1_not_1	0
D0_is_1	1	D1_is_1	0
D1_is_1	0	Start	0
D1_is_1	1	Start	1
D0_not_1	0	D1_not_1	0
D0_not_1	1	D1_not_1	0
D1_not_1	0	Start	0
D1_not_1	1	Start	0

Example 7.9
Serial bit sequence detector (part 1)

Example 7.10 provides the state encoding and next state logic synthesis for the 3-bit serial bit sequence detector.

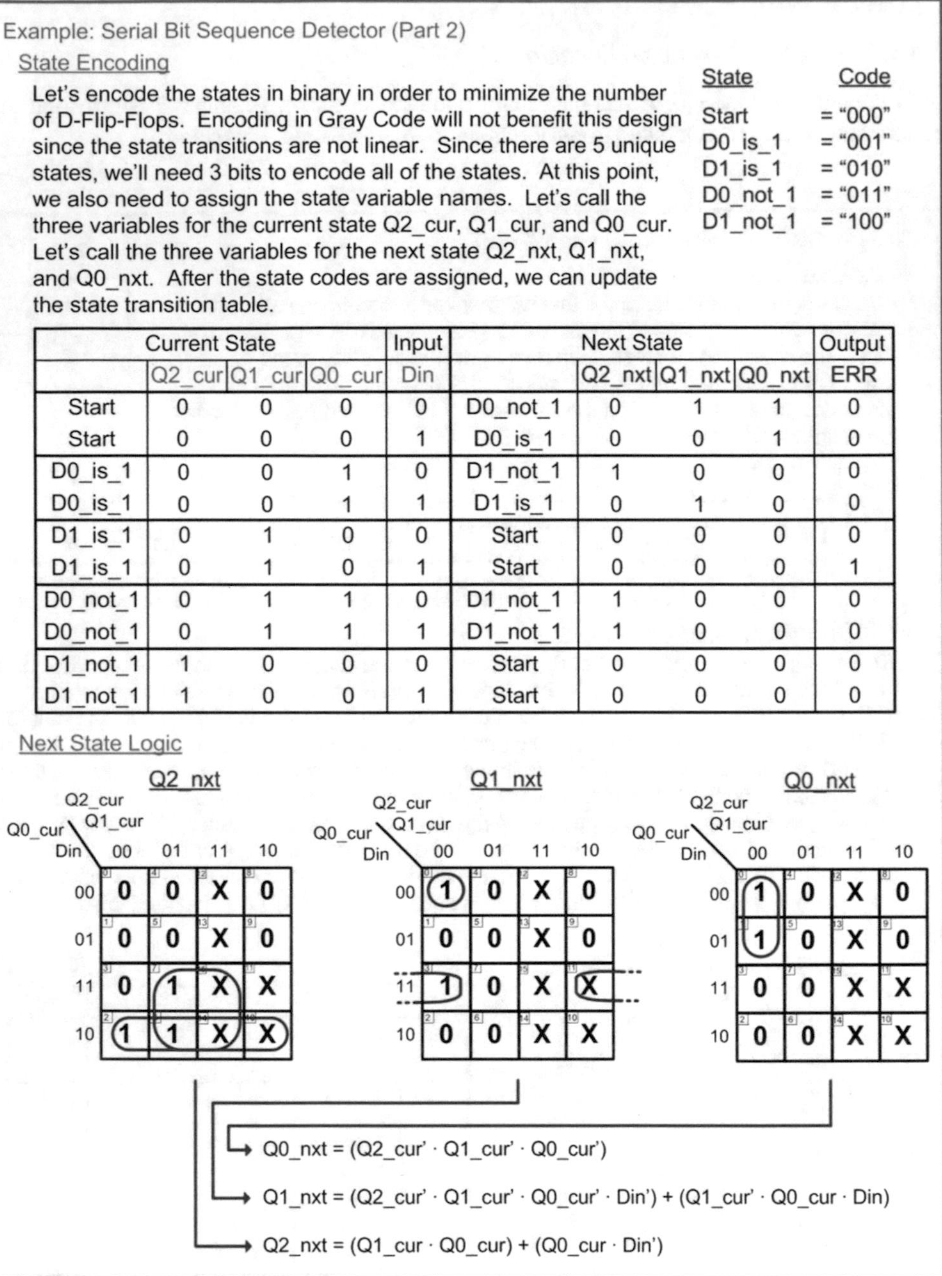

Example: Serial Bit Sequence Detector (Part 2)

State Encoding

Let's encode the states in binary in order to minimize the number of D-Flip-Flops. Encoding in Gray Code will not benefit this design since the state transitions are not linear. Since there are 5 unique states, we'll need 3 bits to encode all of the states. At this point, we also need to assign the state variable names. Let's call the three variables for the current state Q2_cur, Q1_cur, and Q0_cur. Let's call the three variables for the next state Q2_nxt, Q1_nxt, and Q0_nxt. After the state codes are assigned, we can update the state transition table.

State	Code
Start	= "000"
D0_is_1	= "001"
D1_is_1	= "010"
D0_not_1	= "011"
D1_not_1	= "100"

	Current State			Input	Next State				Output
	Q2_cur	Q1_cur	Q0_cur	Din		Q2_nxt	Q1_nxt	Q0_nxt	ERR
Start	0	0	0	0	D0_not_1	0	1	1	0
Start	0	0	0	1	D0_is_1	0	0	1	0
D0_is_1	0	0	1	0	D1_not_1	1	0	0	0
D0_is_1	0	0	1	1	D1_is_1	0	1	0	0
D1_is_1	0	1	0	0	Start	0	0	0	0
D1_is_1	0	1	0	1	Start	0	0	0	1
D0_not_1	0	1	1	0	D1_not_1	1	0	0	0
D0_not_1	0	1	1	1	D1_not_1	1	0	0	0
D1_not_1	1	0	0	0	Start	0	0	0	0
D1_not_1	1	0	0	1	Start	0	0	0	0

Next State Logic

Q0_nxt = (Q2_cur' · Q1_cur' · Q0_cur')

Q1_nxt = (Q2_cur' · Q1_cur' · Q0_cur' · Din') + (Q1_cur' · Q0_cur · Din)

Q2_nxt = (Q1_cur · Q0_cur) + (Q0_cur · Din')

Example 7.10
Serial bit sequence detector (part 2)

Example 7.11 shows the output logic synthesis and final logic diagram for the 3-bit serial bit sequence detector.

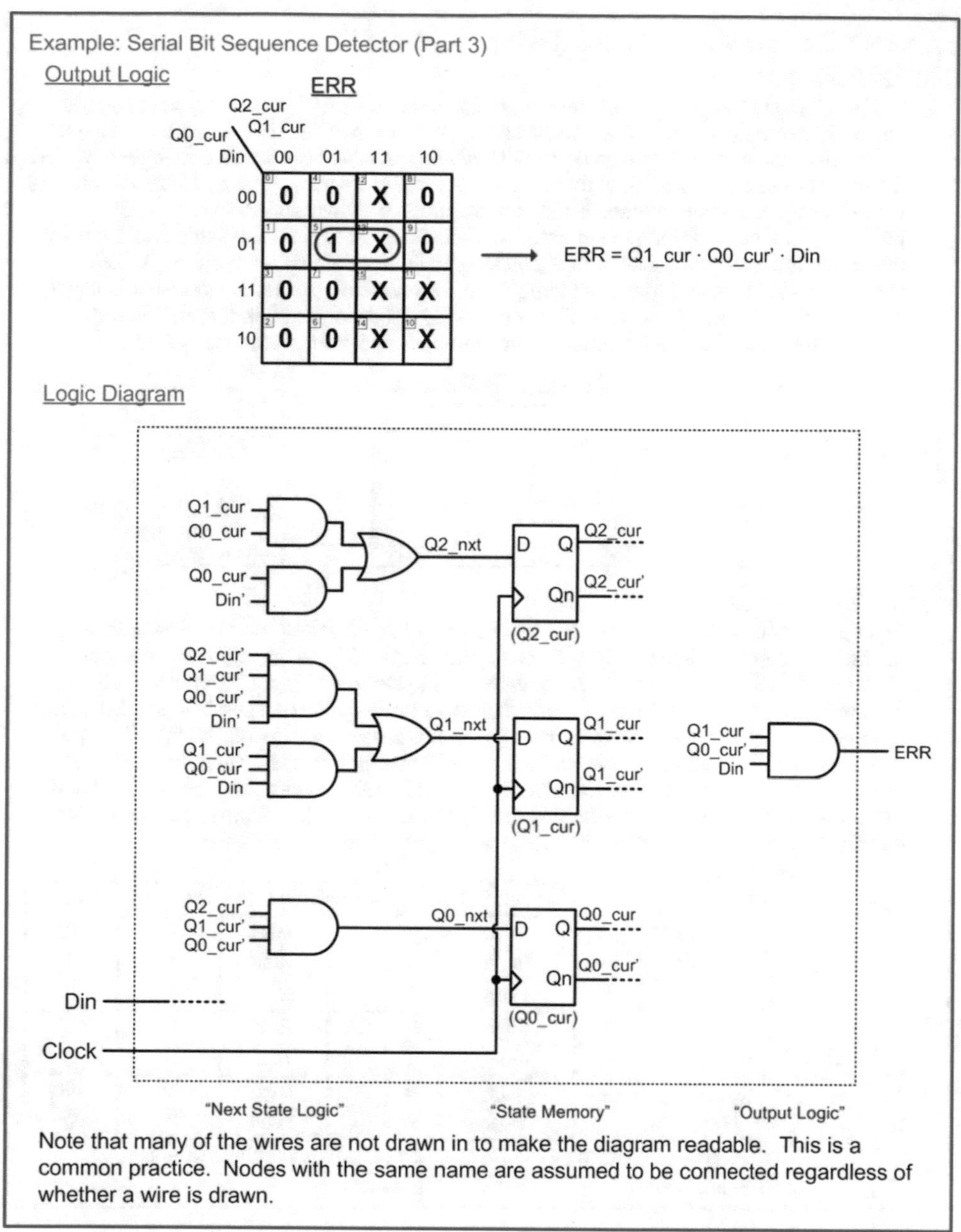

Example 7.11
Serial bit sequence detector (part 3)

7.4.4.2 Vending Machine Controller

Let's now look at the design of a simple vending machine controller. Example 7.12 provides the word description, state diagram, and state transition table for this finite-state machine.

Example: Vending Machine Controller (Part 1)

Word Description

We are going to design a simple vending machine controller. The vending machine will sell bottles of water for 75¢. Customers can enter either a dollar bill or quarters. Once a sufficient amount of money is entered, the vending machine will dispense a bottle of water. If the user entered a dollar it will return one quarter in change. A "Money Receiver" detects when money has been entered. The receiver sends two logic signals to our circuit indicating whether a dollar bill or quarter was received. A "Bottle Dispenser" system holds the water bottles and will release one bottle when its input signal is asserted. A "Coin Return" system holds quarters for change and will release one quarter when its input signal is asserted. The money receiver will reject money if a dollar and quarter are entered simultaneously or if a dollar is entered once the user has started entering quarters.

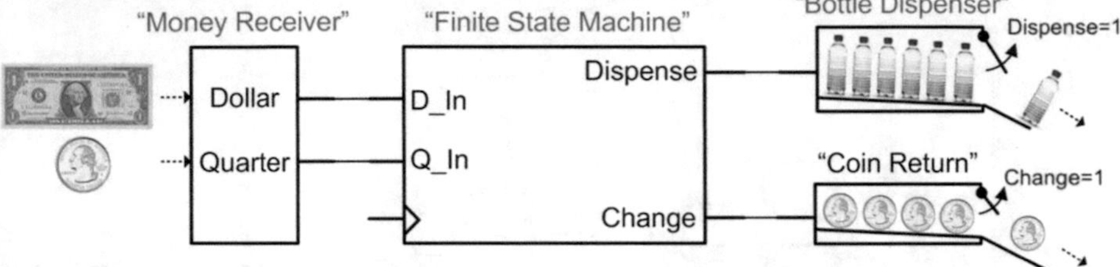

State Diagram and State Transition Table

To implement this state machine, we will need an initial state that the machine will wait in until a customer enters money (Wait). If a dollar is entered, the machine will assert the "Dispense" signal to release a bottle of water and assert the "Change" signal to give one quarter in change. We do not need an additional state for the condition of when a dollar is entered because the machine will simply assert the output signals and return to the Wait state. When the customer pays with quarters, our machine needs to keep track of how many quarters have been received. We'll need two interim states that keep track of how many quarters have been entered (25¢ and 50¢). Once the third quarter has been entered, our machine will assert the "Dispense" signal and return to the Wait state.

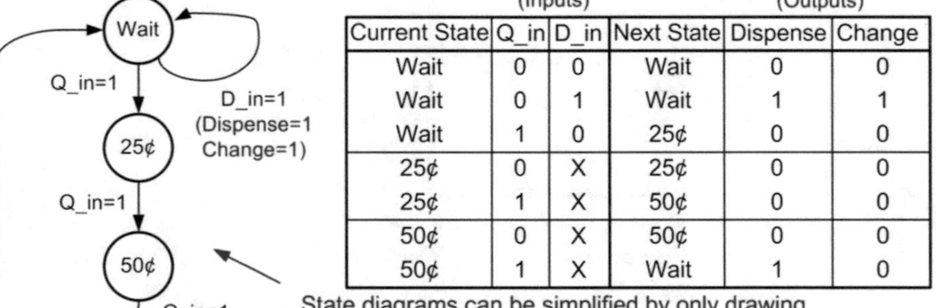

	(Inputs)			(Outputs)	
Current State	Q_in	D_in	Next State	Dispense	Change
Wait	0	0	Wait	0	0
Wait	0	1	Wait	1	1
Wait	1	0	25¢	0	0
25¢	0	X	25¢	0	0
25¢	1	X	50¢	0	0
50¢	0	X	50¢	0	0
50¢	1	X	Wait	1	0

State diagrams can be simplified by only drawing transitions when a signal is asserted.

Example 7.12
Vending machine controller (part 1)

Example 7.13 provides the state encoding and next state logic synthesis for the simple vending machine controller.

Example: Vending Machine Controller (Part 2)

State Encoding

Let's encode the states in binary and name the current state variables Q1_cur and Q0_cur and the next state variables Q1_nxt and Q0_nxt. In this table we list out all possible values the current state and the inputs to make the table more complete.

State	Code
Wait	= "00"
25¢	= "01"
50¢	= "10"

Current State			Input		Next State			Outputs	
	Q1_cur	Q0_cur	Q_in	D_in		Q1_nxt	Q0_nxt	Dispense	Change
Wait	0	0	0	0	Wait	0	0	0	0
Wait	0	0	0	1	Wait	0	0	1	1
Wait	0	0	1	0	25¢	0	1	0	0
Wait	0	0	1	1	Wait	0	0	0	0
25¢	0	1	0	0	25¢	0	1	0	0
25¢	0	1	0	1	25¢	0	1	0	0
25¢	0	1	1	0	50¢	1	0	0	0
25¢	0	1	1	1	25¢	0	1	0	0
50¢	1	0	0	0	50¢	1	0	0	0
50¢	1	0	0	1	50¢	1	0	0	0
50¢	1	0	1	0	Wait	0	0	1	0
50¢	1	0	1	1	50¢	1	0	0	0

Next State Logic

The next state logic for this counter depends on both the current state variables and the system input Up. We can again take advantage of don't cares for the unused state code to minimize the logic.

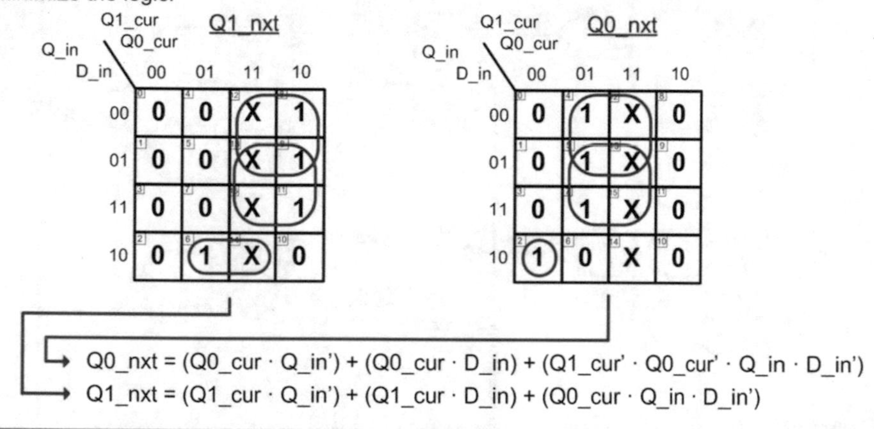

Q0_nxt = (Q0_cur · Q_in') + (Q0_cur · D_in) + (Q1_cur' · Q0_cur' · Q_in · D_in')

Q1_nxt = (Q1_cur · Q_in') + (Q1_cur · D_in) + (Q0_cur · Q_in · D_in')

Example 7.13
Vending machine controller (part 2)

Example 7.14 shows the output logic synthesis and final logic diagram for the vending machine controller.

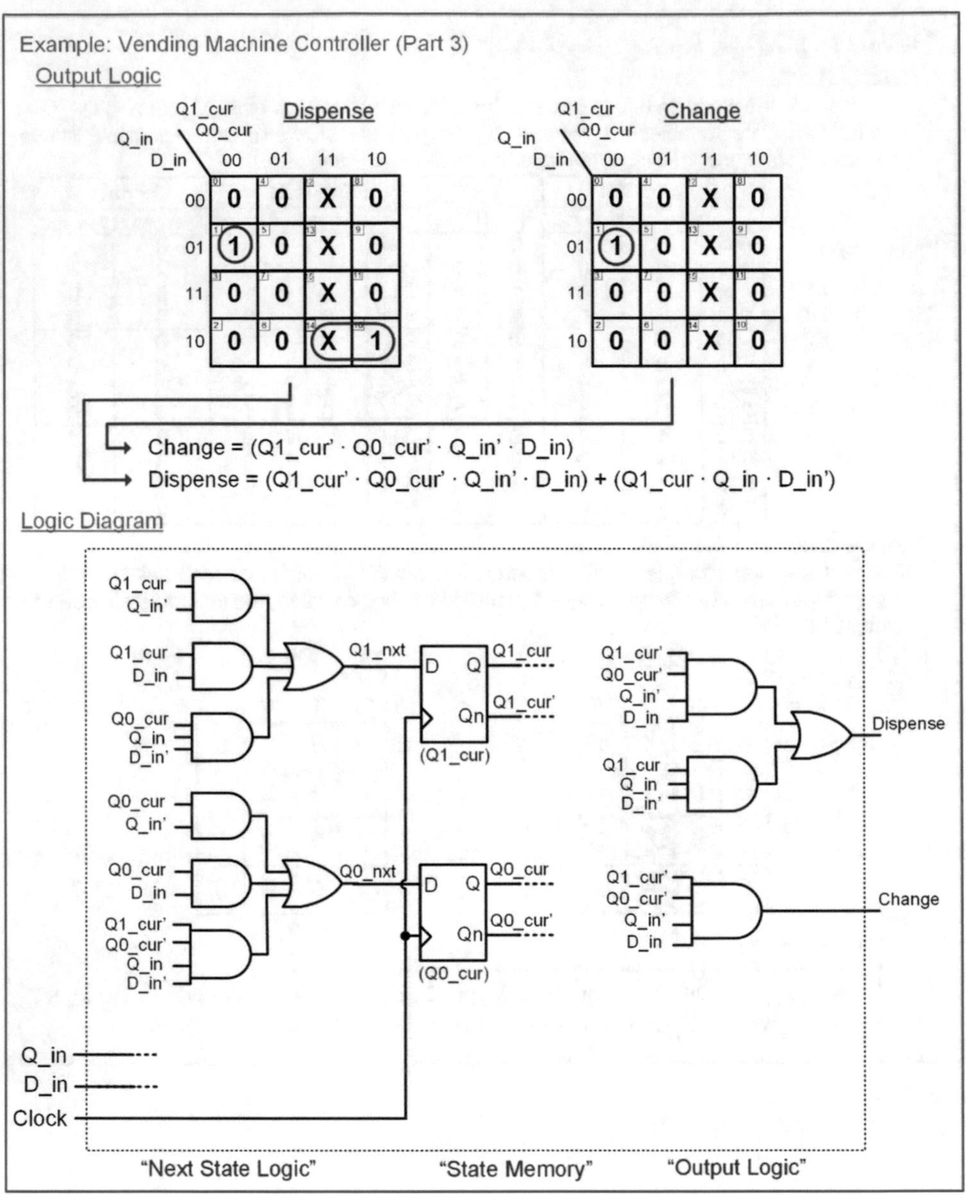

Example 7.14
Vending machine controller (part 3)

Concept Check

CC7.4(a) What allows a finite-state machine to make more intelligent decisions about the system outputs compared to combinational logic alone?

A) A finite-state machine has knowledge about the past inputs.

B) The D-flip-flops allow the outputs to be generated more rapidly.

C) The next state and output logic allows the finite-state machine to be more complex and implement larger truth tables.

D) A synchronous system is always more intelligent.

CC7.4(b) When designing a finite-state machine, many of the details of the implementation can be abstracted. At what design step do the details of the implementation start being considered?

A) The state diagram step.

B) The state transition table step.

C) The state memory synthesis step.

D) The word description.

CC7.4(c) What impact does adding an additional state have on the implementation of the state memory logic in a finite-state machine?

A) It adds an additional D-flip-flop.

B) It adds a new state code that must be supported.

C) It adds more combinational logic to the logic diagram.

D) It reduces the speed that the machine can run at.

CC7.4(d) Which of the following statements about the next state logic is FALSE?

A) It is always combinational logic.

B) It always uses the current state as one of its inputs.

C) Its outputs are connected to the D inputs of the D-flip-flops in the state memory.

D) It uses the results of the output logic as part of its inputs.

CC7.4(e) Why does the output logic stage of a finite-state machine always use the current state as one of its inputs?

A) If it didn't, it would simply be a separate combinational logic circuit and not be part of the finite-state machine.

B) To make better decisions about what the system outputs should be.

C) Because the next state logic is located too far away.

D) Because the current state is produced on every triggering clock edge.

CC7.4(f) What impact does asserting a reset have on a finite-state machine?

A) It will cause the output logic to produce all zeros.

B) It will cause the next state logic to produce all zeros.

C) It will set the current state code to all zeros.

D) It will start the system clock.

7.5 Counters

A *counter* is a special type of finite-state machine. A counter will traverse the states within a state diagram in a linear fashion continually circling around all states. This behavior allows a special type of output topology called *state-encoded outputs*. Since each state in the counter represents a unique counter output, the states can be encoded with the associated counter output value. In this way, the current state code of the machine can be used as the output of the entire system.

7.5.1 2-Bit Binary Up Counter

Let's consider the design of a 2-bit binary up counter. Example 7.15 provides the word description, state diagram, state transition table, and state encoding for this counter.

Example: 2-Bit Binary Up Counter (Part 1)

Word Description

We are going to design a 2-bit binary up counter. The counter will increment by 1 on every rising edge of the clock ("00", "01", "10", "11"). When the counter reaches "11", it will start over counting at "00". The output of the counter is called CNT.

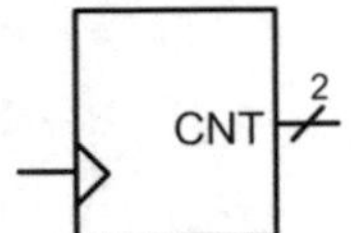

State Diagram & State Transition Table

The state diagram for this counter is below. Notice that there are no inputs to the state machine. Also notice that the machine transitions in a linear pattern through the states and continually repeats the sequence of states. The outputs of this machine depend only on the current state so they are written inside of the state circles. This is a Moore machine.

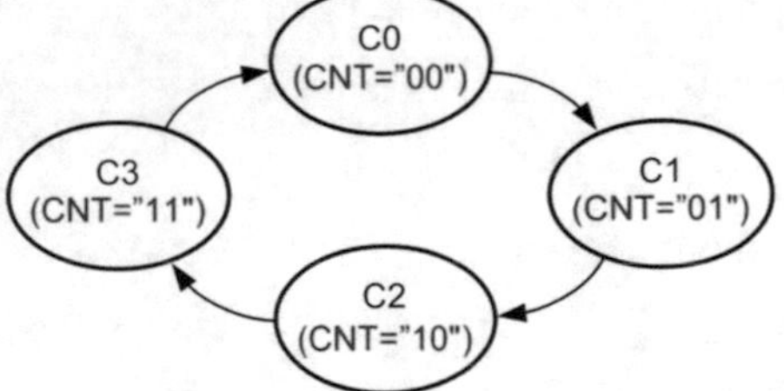

Current State	Next State	(Output) CNT
C0	C1	"00"
C1	C2	"01"
C2	C3	"10"
C3	C0	"11"

State Encoding

When implementing this counter, we can use "state-encoded outputs". This means that we choose the state codes so that they match the desired output at each state. This allows the machine to simply use the current state variables for the system outputs. Let's name the current state variables Q1_cur and Q0_cur and the next state variables Q1_nxt and Q0_nxt. The state code assignments and updated state transition table are below.

State	Code
C0	= "00"
C1	= "01"
C2	= "10"
C3	= "11"

Current State			Next State			Outputs
	Q1_cur	Q0_cur		Q1_nxt	Q0_nxt	CNT
C0	0	0	C1	0	1	"00"
C1	0	1	C2	1	0	"01"
C2	1	0	C3	1	1	"10"
C3	1	1	C0	0	0	"11"

Example 7.15
2-Bit binary up counter (part 1)

Example 7.16 shows the next state and output logic synthesis, the final logic diagram, and resultant representative timing diagram for the 2-bit binary up counter.

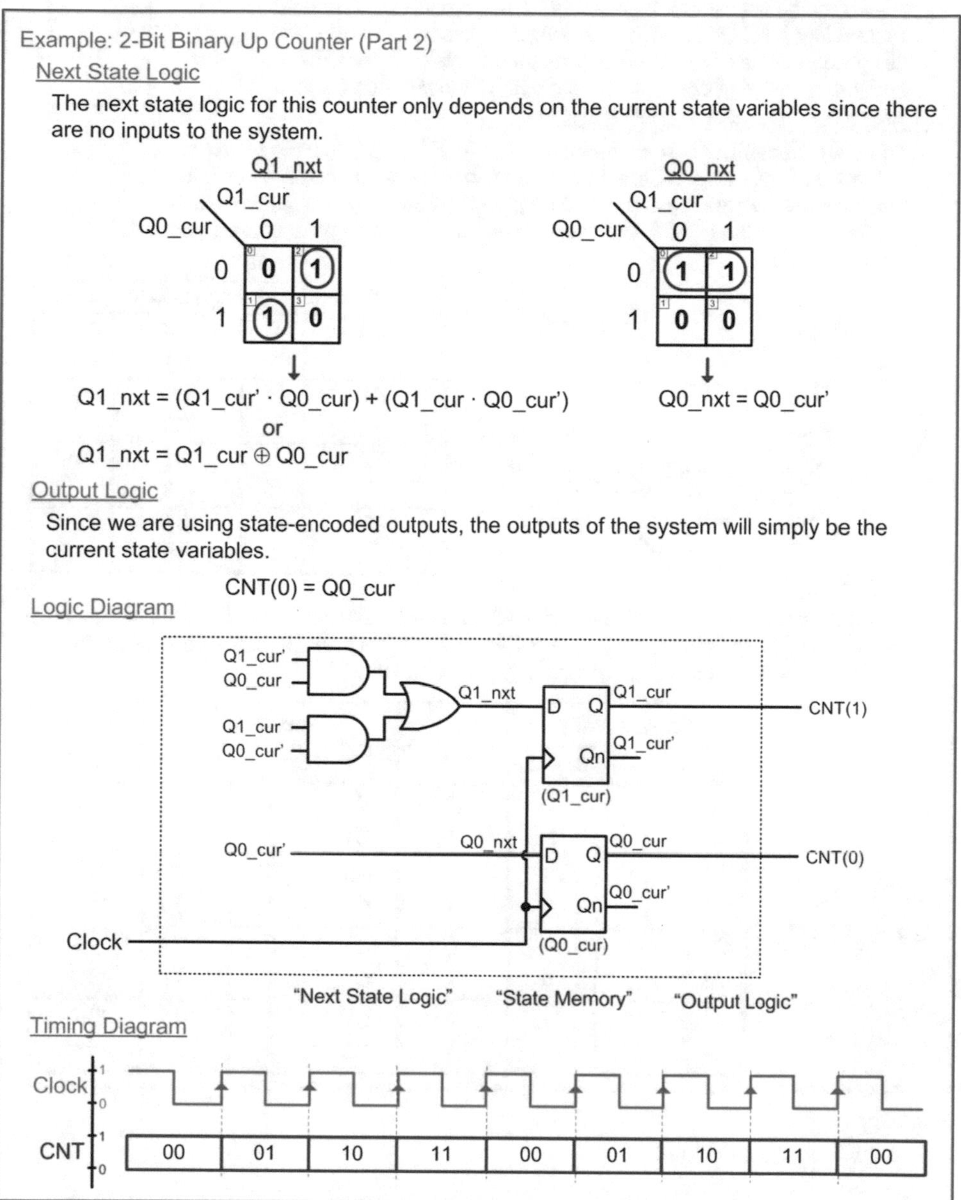

Example 7.16
2-Bit binary up counter (part 2)

7.5.2 2-Bit Binary Up/Down Counter

Let's now consider a 2-bit binary up/down counter. In this type of counter, there is an input that dictates whether the counter increments or decrements. This counter can still be implemented as a Moore machine and use state-encoded outputs. Example 7.17 provides the word description, state diagram, state transition table, and state encoding for this counter.

Example: 2-Bit Binary Up/Down Counter (Part 1)

Word Description

We are going to design a 2-bit binary up/down counter. When the system input "Up" is asserted, the counter will increment by 1 on every rising edge of the clock. When Up=0, the counter will decrement by 1 on every rising edge of the clock. The output of the counter is called CNT.

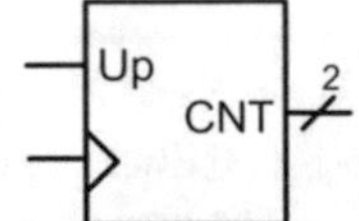

State Diagram & State Transition Table

The state diagram for this counter is below. In this diagram, if the input Up=1, the machine will traverse the states in order to create an incrementing count. If the input Up=0, the machine will traverse the states in the opposite order. The outputs of this machine again only depend on the current state so they are written inside of the state circles. This is a Moore machine.

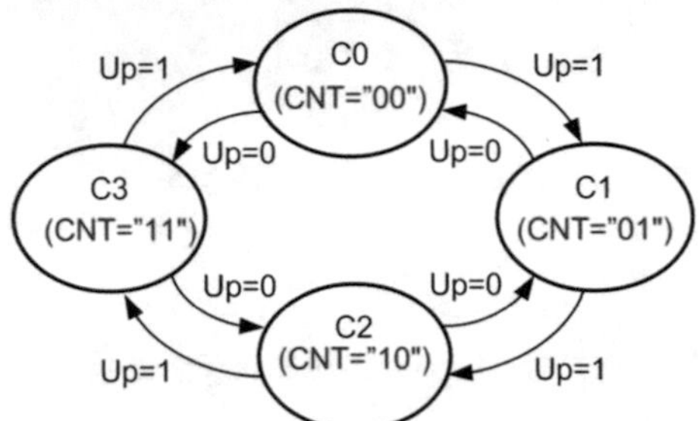

Current State	Up (Input)	Next State	CNT (Output)
C0	0	C3	"00"
	1	C1	
C1	0	C0	"01"
	1	C2	
C2	0	C1	"10"
	1	C3	
C3	0	C2	"11"
	1	C0	

State Encoding

Again, this counter will use "state-encoded outputs". Let's name the current state variables Q1_cur and Q0_cur and the next state variables Q1_nxt and Q0_nxt. The state code assignments and updated state transition table are below.

State	Code
C0	= "00"
C1	= "01"
C2	= "10"
C3	= "11"

Current State			Input	Next State			Outputs
	Q1_cur	Q0_cur	Up		Q1_nxt	Q0_nxt	CNT
C0	0	0	0	C3	1	1	"00"
C0	0	0	1	C1	0	1	"00"
C1	0	1	0	C0	0	0	"01"
C1	0	1	1	C2	1	0	"01"
C2	1	0	0	C1	0	1	"10"
C2	1	0	1	C3	1	1	"10"
C3	1	1	0	C2	1	0	"11"
C3	1	1	1	C0	0	0	"11"

Example 7.17
2-Bit binary up/down counter (part 1)

Example 7.18 shows the next state and output logic synthesis, the final logic diagram, and resultant representative timing diagram for the 2-bit binary up/down counter.

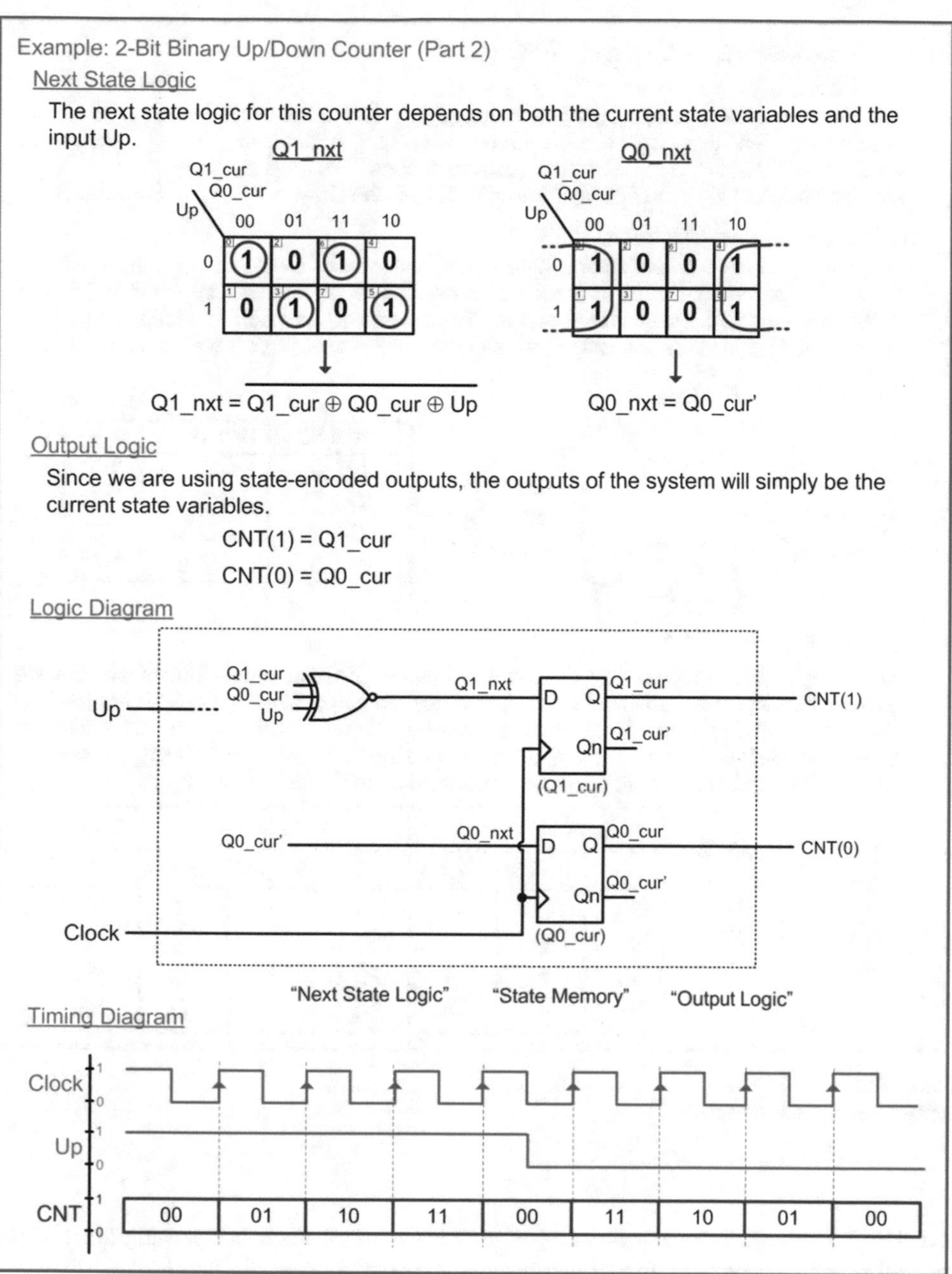

Example 7.18
2-Bit binary up/down counter (part 2)

7.5.3 2-Bit Gray Code Up Counter

A gray code counter is one in which the output only differs by one bit from its prior value. This type of counter can be implemented using state-encoded outputs by simply encoding the states in gray code. Let's consider the design of a 2-bit gray code up counter. Example 7.19 provides the word description, state diagram, state transition table, and state encoding for this counter.

Example: 2-Bit Gray Code Up Counter (Part 1)

Word Description

We are going to design a 2-bit gray code up counter. The counter will output an incrementing gray code pattern on every rising edge of the clock ("00", "01", "11", "10). When the counter reaches "11", it will start over counting at "00". The output of the counter is called Gray.

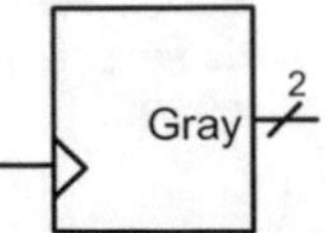

State Diagram & State Transition Table

The state diagram for this counter is below. Notice that there are no inputs to the state machine. Also notice that the machine transitions in a linear pattern through the states and continually repeats the sequence of states. The outputs of this machine depend only on the current state, so they are written inside of the state circles. This is a Moore machine.

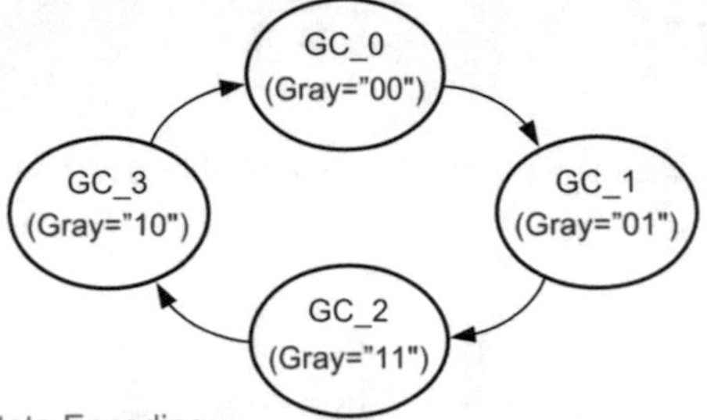

Current State	Next State	Gray (Output)
GC_0	GC_1	"00"
GC_1	GC_2	"01"
GC_2	GC_3	"11"
GC_3	GC_0	"10"

State Encoding

When implementing this counter, we can use "state-encoded outputs". This means that we choose the state codes so that they match the desired output at each state. This allows the machine to simply use the current state variables for the system outputs. Let's name the current state variables Q1_cur and Q0_cur and the next state variables Q1_nxt and Q0_nxt. The state code assignments and updated state transition table are below.

State	Code
GC_0	= "00"
GC_1	= "01"
GC_2	= "11"
GC_3	= "10"

Current State			Next State			Outputs
	Q1_cur	Q0_cur		Q1_nxt	Q0_nxt	Gray
GC_0	0	0	GC_1	0	1	"00"
GC_1	0	1	GC_2	1	1	"01"
GC_2	1	1	GC_3	1	0	"11"
GC_3	1	0	GC_0	0	0	"10"

Example 7.19
2-Bit gray code up counter (part 1)

Example 7.20 shows the next state and output logic synthesis, the final logic diagram, and resultant representative timing diagram for the 2-bit gray code up counter.

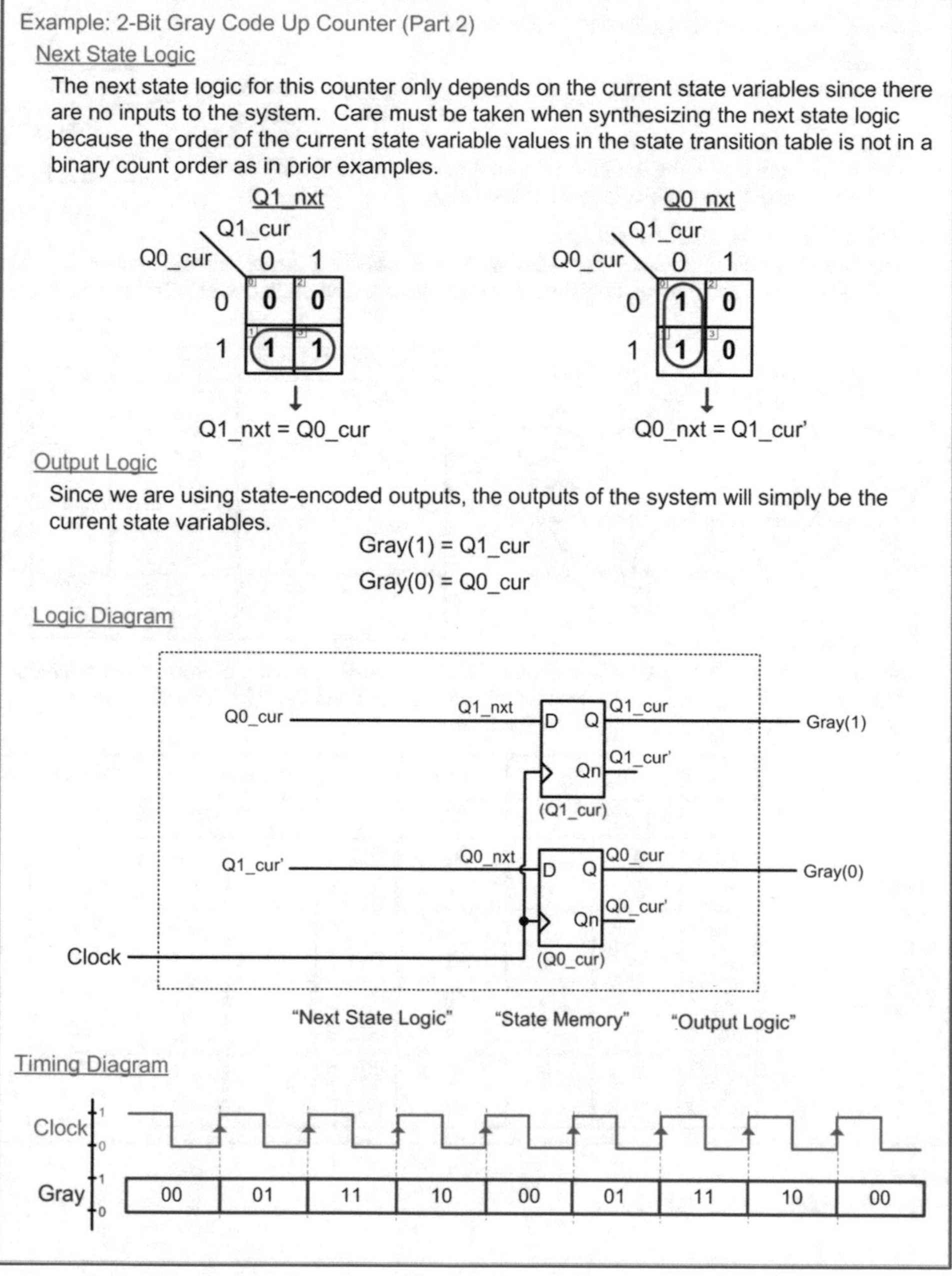

Example 7.20
2-Bit gray code up counter (part 2)

7.5.4 2-Bit Gray Code Up/Down Counter

Let's now consider a 2-bit gray code up/down counter. In this type of counter, there is an input that dictates whether the counter increments or decrements. This counter can still be implemented as a Moore machine and use state-encoded outputs. Example 7.21 provides the word description, state diagram, state transition table, and state encoding for this counter.

Example: 2-Bit Gray Code Up/Down Counter (Part 1)

Word Description

We are going to design a 2-bit gray code up/down counter. When the system input "Up" is asserted, the counter will output an incrementing gray code pattern on every rising edge of the clock ("00", "01", "11", "10"). When the input Up=0, the counter will output a decrementing gray code pattern. The output of the counter is called Gray.

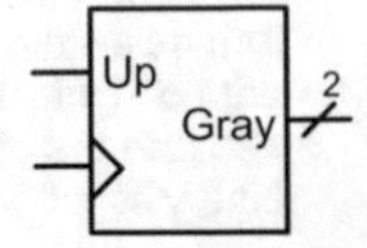

State Diagram & State Transition Table

The state diagram for this counter is below. The outputs of this machine again only depend on the current state, so they are written inside of the state circles. This is a Moore machine.

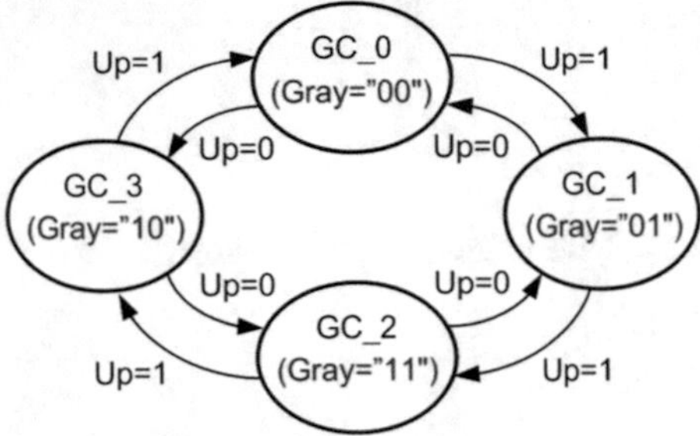

Current State	Up (Input)	Next State	Gray (Output)
GC_0	0	GC_3	"00"
	1	GC_1	
GC_1	0	GC_0	"01"
	1	GC_2	
GC_2	0	GC_1	"11"
	1	GC_3	
GC_3	0	GC_2	"10"
	1	GC_0	

State Encoding

Again, this counter will use "state-encoded outputs". Let's name the current state variables Q1_cur and Q0_cur and the next state variables Q1_nxt and Q0_nxt. The state code assignments and updated state transition table are below.

State	Code
GC_0	= "00"
GC_1	= "01"
GC_2	= "11"
GC_3	= "10"

Current State			Input	Next State			Outputs
	Q1_cur	Q0_cur	Up		Q1_nxt	Q0_nxt	Gray
GC_0	0	0	0	GC_3	1	0	"00"
GC_0	0	0	1	GC_1	0	1	"00"
GC_1	0	1	0	GC_0	0	0	"01"
GC_1	0	1	1	GC_2	1	1	"01"
GC_2	1	1	0	GC_1	0	1	"11"
GC_2	1	1	1	GC_3	1	0	"11"
GC_3	1	0	0	GC_2	1	1	"10"
GC_3	1	0	1	GC_0	0	0	"10"

Example 7.21
2-Bit gray code up/down counter (part 1)

Example 7.22 shows the next state and output logic synthesis, the final logic diagram, and resultant representative timing diagram for the 2-bit gray code up/down counter.

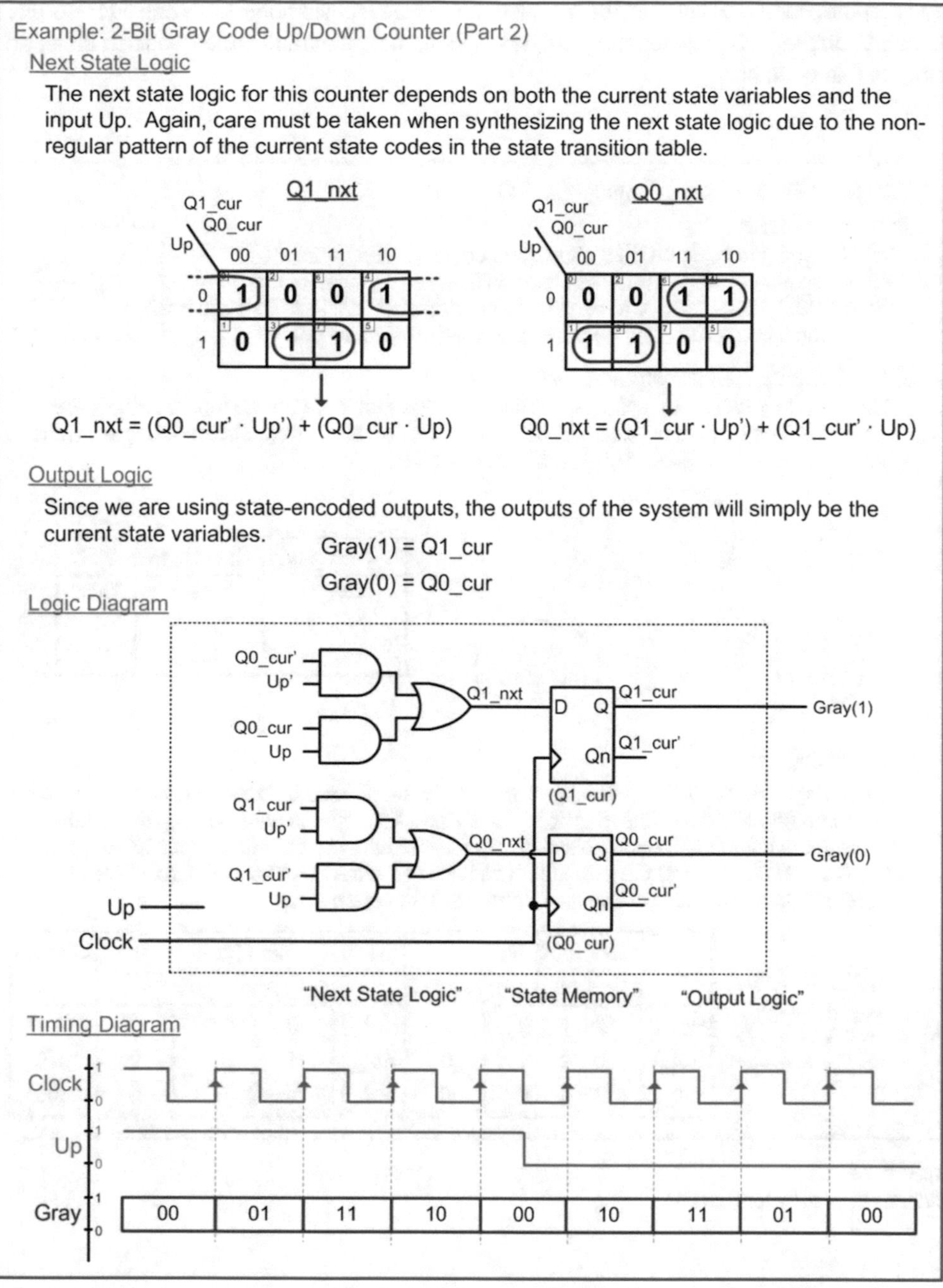

Example 7.22
2-Bit gray code up/down counter (part 2)

7.5.5 3-Bit One-Hot Up Counter

A one-hot counter creates an output in which one and only one bit is asserted at a time. In an *up counter* configuration, the assertion is made on the least significant bit first, followed by the next higher significant bit, and so on (i.e., 001, 010, 100, 001…). A one-hot counter can be created using state-encoded outputs. For a *n*-bit counter, the machine will require *n* D-flip-flops. Let's consider a 3-bit one-hot up counter. Example 7.23 provides the word description, state diagram, state transition table, and state encoding for this counter.

Example: 3-Bit One-Hot Up Counter (Part 1)

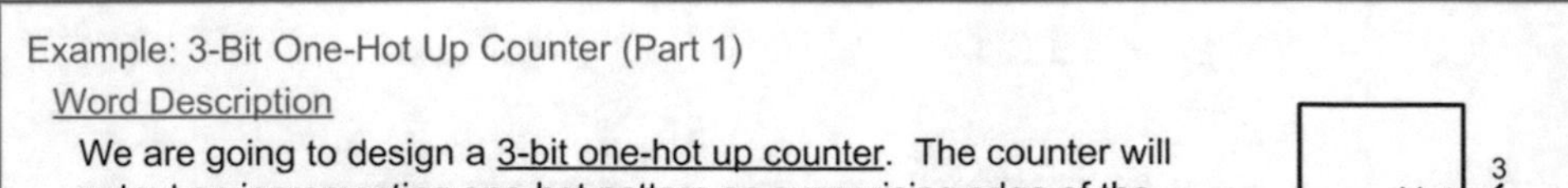

Word Description

We are going to design a 3-bit one-hot up counter. The counter will output an incrementing one-hot pattern on every rising edge of the clock ("001", "010", "100"). When the counter reaches "100", it will start over counting at "001". The output of the counter is called Hot.

State Diagram & State Transition Table

The state diagram for this counter is below. Notice that there are no inputs to the state machine. The outputs of this machine depend only on the current state so they are written inside of the state circles. This is a Moore machine.

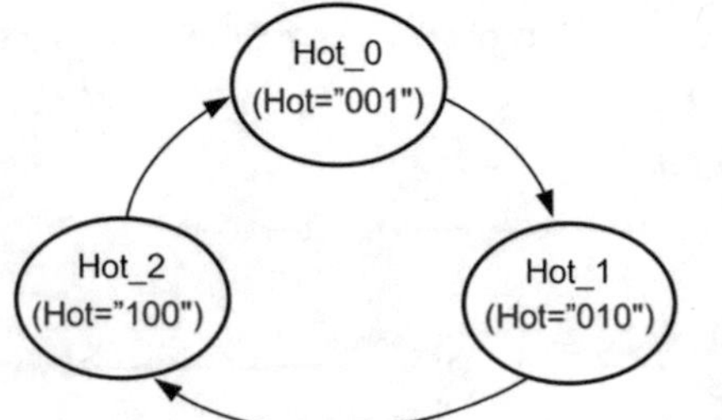

Current State	Next State	Hot (Output)
Hot_0	Hot_1	"001"
Hot_1	Hot_2	"010"
Hot_2	Hot_0	"100"

State Encoding

When implementing this counter, we can use "state-encoded outputs". Using one-hot state encoding requires three bits to encode the states. This means we'll need three variables for both the current state and next state. Let's name the current state variables Q2_cur, Q1_cur and Q0_cur and the next state variables Q2_nxt, Q1_nxt and Q0_nxt. The state code assignments and updated state transition table are below.

State	Code
Hot_0	= "001"
Hot_1	= "010"
Hot_2	= "100"

Current State				Next State				Outputs
	Q2_cur	Q1_cur	Q0_cur		Q2_nxt	Q1_nxt	Q0_nxt	Hot
Hot_0	0	0	1	Hot_1	0	1	0	"001"
Hot_1	0	1	0	Hot_2	1	0	0	"010"
Hot_2	1	0	0	Hot_0	0	0	1	"100"

Example 7.23
3-Bit one-hot up counter (part 1)

Example 7.24 shows the next state and output logic synthesis, the final logic diagram, and resultant representative timing diagram for the 3-bit one-hot up counter.

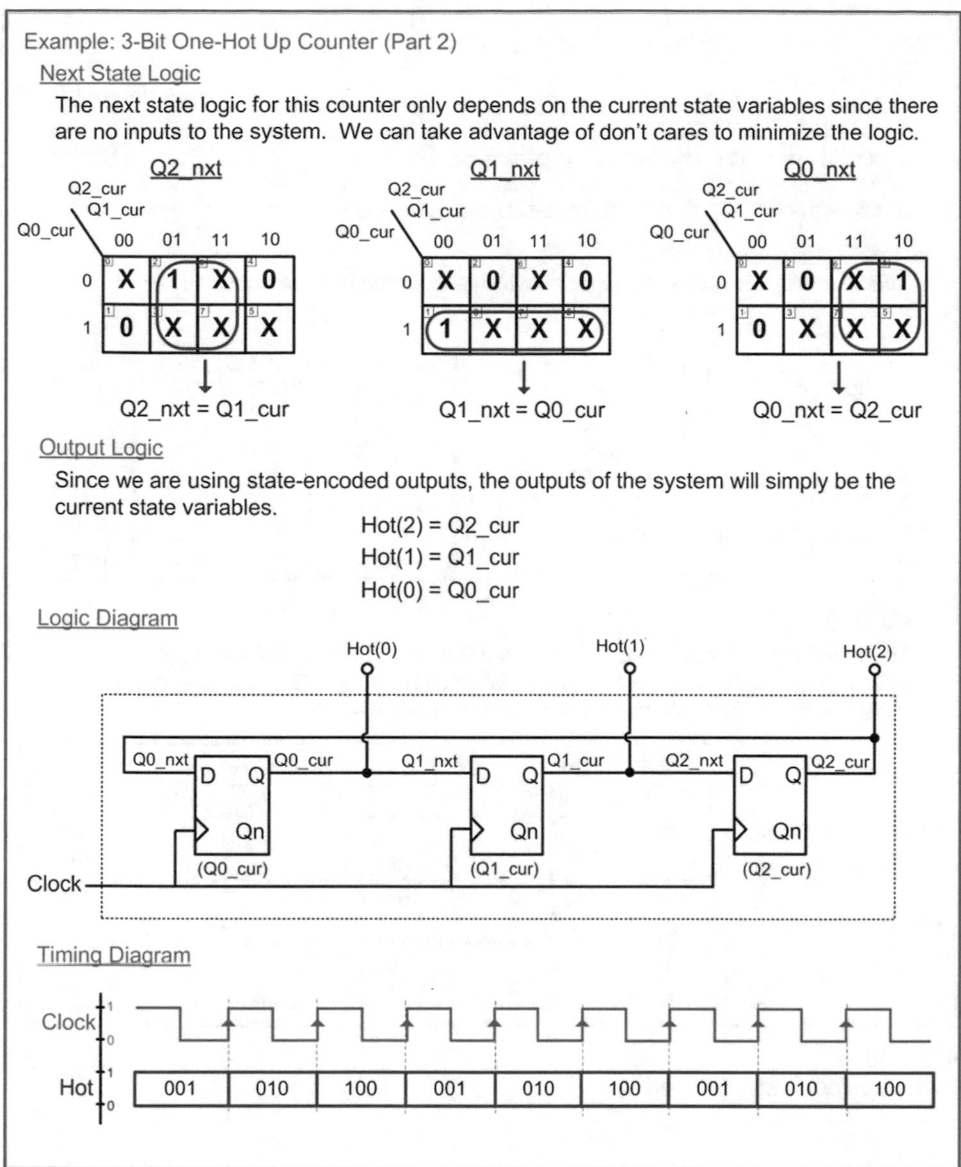

Example 7.24
3-Bit one-hot up counter (part 2)

7.5.6 3-Bit One-Hot Up/Down Counter

Let's now consider a 3-bit one-hot up/down counter. In this type of counter, there is an input that dictates whether the counter increments or decrements. This counter can still be implemented as a Moore machine and use state-encoded outputs. Example 7.25 provides the word description, state diagram, state transition table, and state encoding for this counter.

Example: 3-Bit One-Hot Up/Down Counter (Part 1)

Word Description

We are going to design a 3-bit one-hot up/down counter. When the system input "Up" is asserted, the counter will output an incrementing one-hot pattern on every rising edge of the clock ("001", "010", "100"). When the input Up=0, the counter will output a decrementing one-hot pattern ("100", "010", "001"). The output of the counter is called Hot.

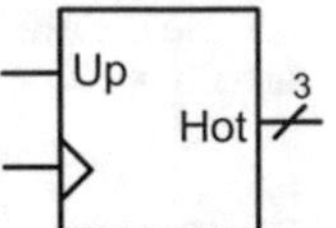

State Diagram & State Transition Table

The state diagram and state transition table for this counter are below.

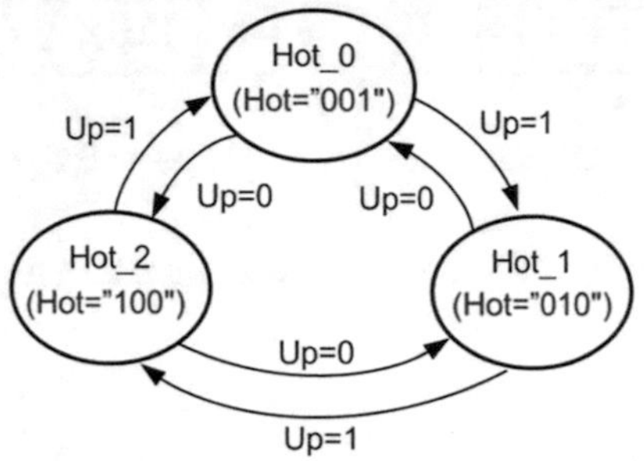

Current State	Up (Input)	Next State	Hot (Output)
Hot_0	0	Hot_2	"001"
Hot_0	1	Hot_1	"001"
Hot_1	0	Hot_0	"010"
Hot_1	1	Hot_2	"010"
Hot_2	0	Hot_1	"100"
Hot_2	1	Hot_0	"100"

State Encoding

Let's use "state-encoded outputs" and name the current state variables Q2_cur, Q1_cur and Q0_cur and the next state variables Q2_nxt, Q1_nxt and Q0_nxt. The state code assignments and updated state transition table are below.

State	Code
Hot_0	= "001"
Hot_1	= "010"
Hot_2	= "100"

Current State				Input	Next State				Outputs
	Q2_cur	Q1_cur	Q0_cur	Up		Q2_nxt	Q1_nxt	Q0_nxt	Hot
Hot_0	0	0	1	0	Hot_2	1	0	0	"001"
Hot_0	0	0	1	1	Hot_1	0	1	0	"001"
Hot_1	0	1	0	0	Hot_0	0	0	1	"010"
Hot_1	0	1	0	1	Hot_2	1	0	0	"010"
Hot_2	1	0	0	0	Hot_1	0	1	0	"100"
Hot_2	1	0	0	1	Hot_0	0	0	1	"100"

Example 7.25
3-Bit one-hot up/down counter (part 1)

Example 7.26 shows the next state and output logic synthesis for the 3-bit one-hot up/down counter.

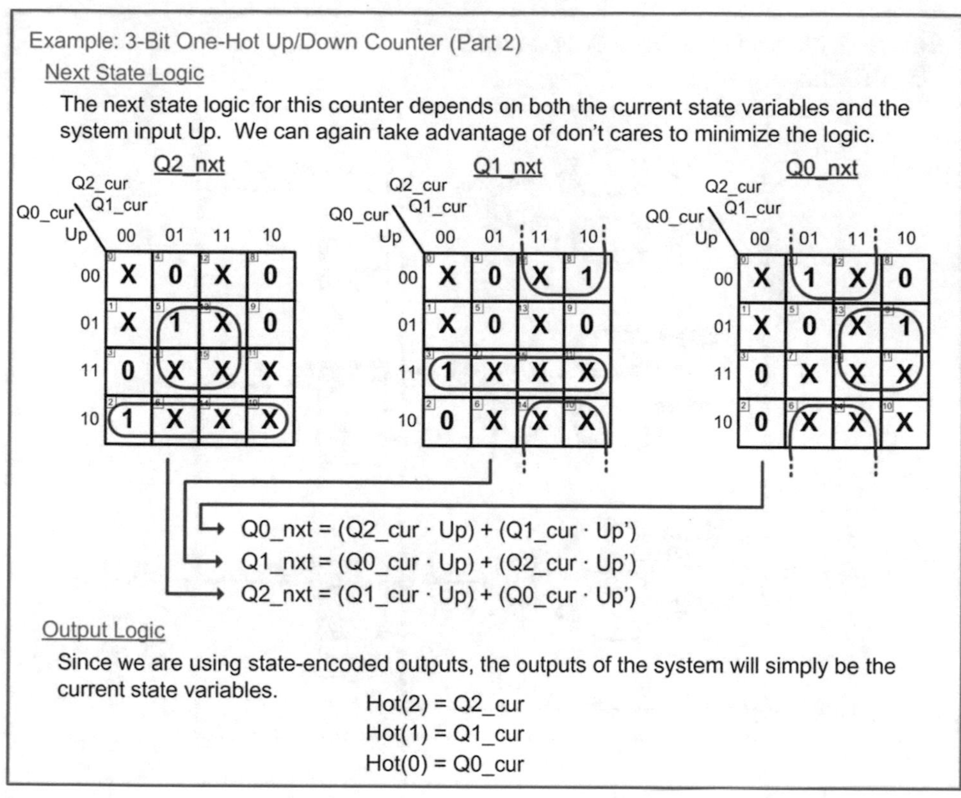

Example 7.26
3-Bit one-hot up/down counter (part 2)

Finally, Example 7.27 shows the logic diagram and resultant representative timing diagram for the counter.

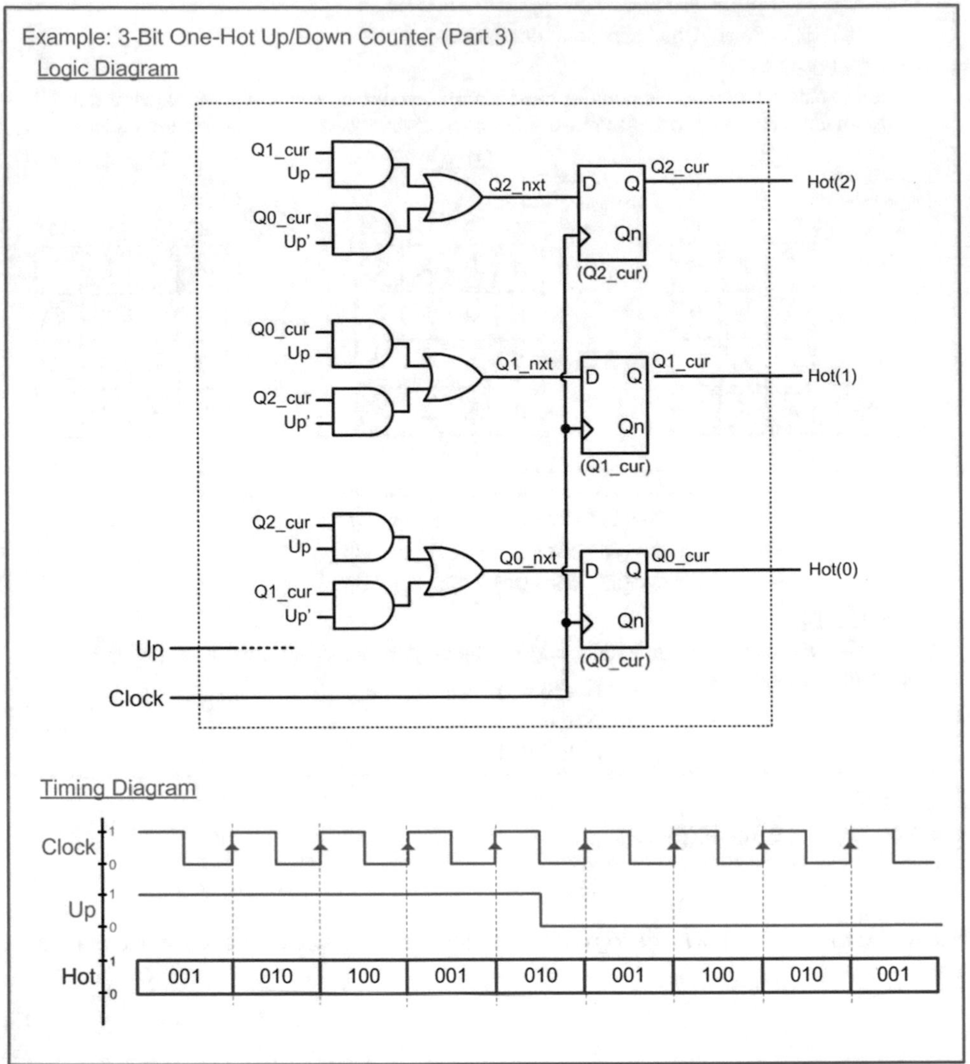

Example 7.27
3-Bit one-hot up/down counter (part 3)

Concept Check

CC7.5 What characteristic of a counter makes it a special case of a finite-state machine?

A) The state transitions are mostly linear, which reduces the implementation complexity.

B) The outputs are always a gray code.

C) The next state logic circuitry is typically just sum terms.

D) There is never a situation where a counter could be a Mealy machine.

7.6 Finite-State Machine's Reset Condition

The one-hot counter designs in Examples 7.23 and 7.25 where the first FSM examples that had an initial state that was not encoded with all 0's. Notice that all of the other FSM examples had initial states with state codes comprised of all 0's (e.g., w_closed = 0, S0 = "00", C0 = "00", GC_0 = "00", etc.). When the initial state is encoded with all 0's, the FSM can be put into this state by asserting the reset line of all of the D-flip-flops in the state memory. By asserting the reset line, the Q outputs of all of the D-flip-flops are forced to 0's. This sets the initial current state value to whatever state is encoded with all 0's. The initial state of a machine is often referred to as the *reset state*. The circuitry to initialize state machines is often omitted from the logic diagram as it is assumed that the necessary circuitry will exist in order to put the state machine into the reset state. If the reset state is encoded with all 0's, then the reset line can be used alone; however, if the reset state code contains 1's, then both the reset *and* preset lines must be used to put the machine into the reset state upon start up. Let's look at the behavior of the one-hot up counter again. Figure 7.33 shows how using the reset lines of the D-flip-flops alone will cause the circuit to operate incorrectly. Instead, a combination of the reset and preset lines must be used to get the one-hot counter into its initial state of Hot_0 = "001".

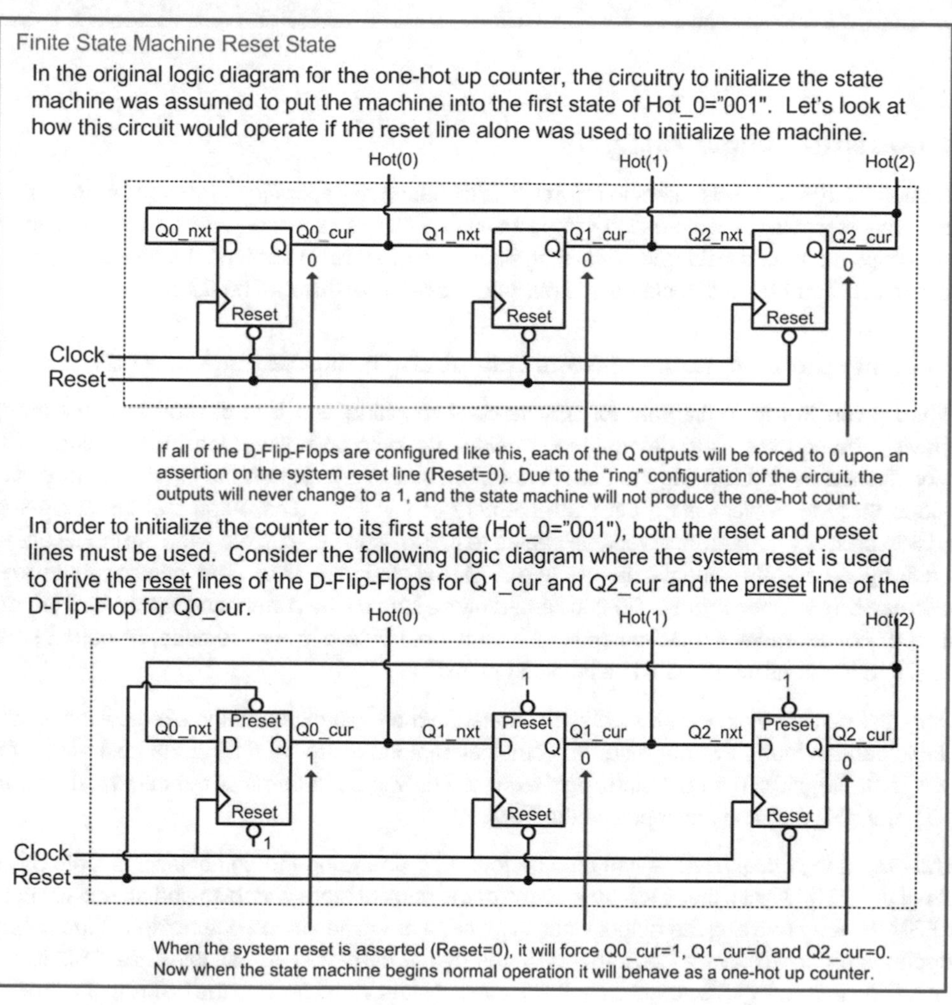

Fig. 7.33
Finite-state machine reset state

Resets are most often asynchronous so that they can immediately alter the state of the FSM. If a reset was implemented in a synchronous manner and there was a clock failure, the system could not be reset since there would be no more subsequent clock edges that would recognize that the reset line was asserted. An asynchronous reset allows the system to be fully restarted even in the event of a clock failure.

CONCEPT CHECK

CC7.6 What is the downside of using D-flip-flops that do not have preset capability in a finite-state machine?

A) The finite-state machine will run slower.

B) The next state logic will be more complex.

C) The output logic will not be able to support both Mealy- and Moore-type machine architectures.

D) The start-up state can never have a 1 in its state code.

7.7 Sequential Logic Analysis

Sequential logic analysis refers to the act of deciphering the operation of a circuit from its final logic diagram. This is similar to combinational logic analysis with the exception that the storage capability of the D-flip-flops must be considered. This analysis is also used to understand the timing of a sequential logic circuit and can be used to predict the maximum clock rate that can be used.

7.7.1 Finding the State Equations and Output Logic Expressions of a FSM

When given the logic diagram for a finite-state machine and it is desired to reverse-engineer its behavior, the first step is to determine the next state logic and output logic expressions. This can be accomplished by first labeling the current and next state variables on the inputs and outputs of the D-flip-flops that are implementing the state memory of the FSM. The outputs of the D-flip-flops are labeled with arbitrary current state variable names (e.g., Q1_cur, Q0_cur, etc.) and the inputs are labeled with arbitrary next state variable names (e.g., Q1_nxt, Q0_nxt, etc.). The numbering of the state variables can be assigned to the D-flip-flops arbitrarily as long as the current and next state bit numbering is matched. For example, if a D-flip-flop is labeled to hold bit 0 of the state code, its output should be labeled Q**0**_cur and its input should be labeled Q**0**_nxt.

Once the current state variable nets are labeled in the logic diagram, the expressions for the next state logic can be found by analyzing the combinational logic circuity driving the next state variables (e.g., Q1_nxt, Q0_nxt). The next state logic expressions will be in terms of the current state variables (e.g., Q1_cur, Q0_cur) and any inputs to the FSM.

The output logic expressions can also be found by analyzing the combinational logic driving the outputs of the FSM. Again, these will be in terms of the current state variables and potentially the inputs to the FSM. When analyzing the output logic, the type of machine can be determined. If the output logic only depends on combinational logic that is driven by the current state variables, the FSM is a Moore machine. If the output logic depends on both the current state variables and the FSM inputs, the FSM is a Mealy machine. An example of this analysis approach is given in Example 7.28.

Example: Determining the Next State Logic and Output Logic Expressions of a FSM

Given: The following finite state machine logic diagram.

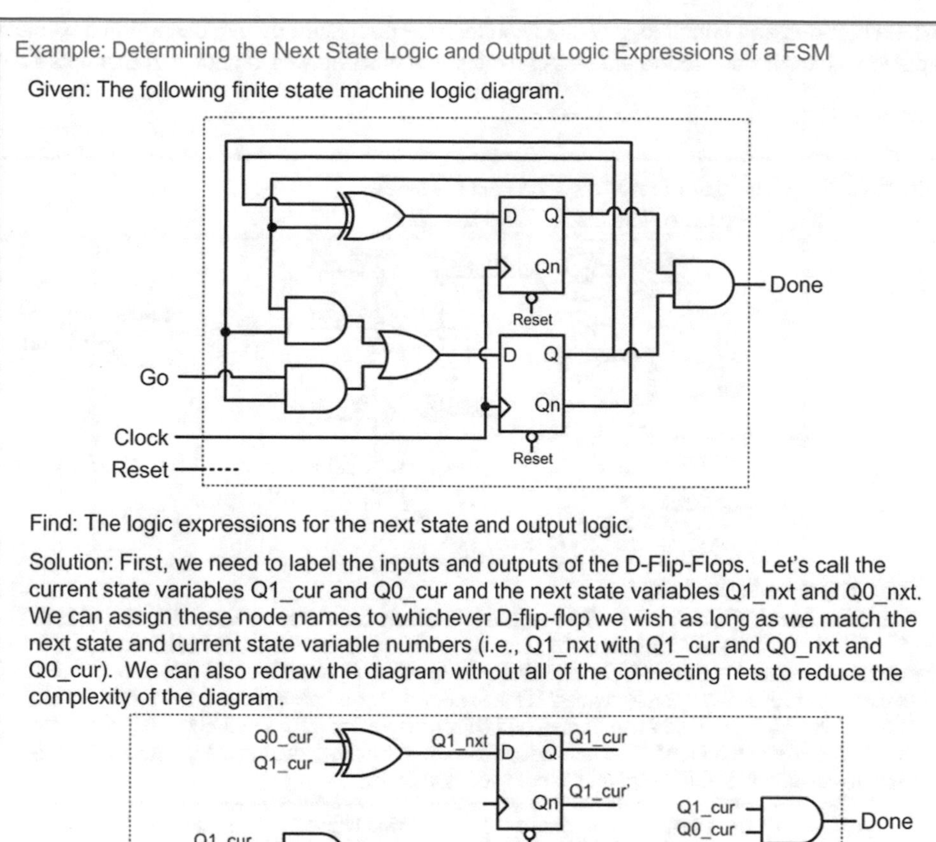

Find: The logic expressions for the next state and output logic.

Solution: First, we need to label the inputs and outputs of the D-Flip-Flops. Let's call the current state variables Q1_cur and Q0_cur and the next state variables Q1_nxt and Q0_nxt. We can assign these node names to whichever D-flip-flop we wish as long as we match the next state and current state variable numbers (i.e., Q1_nxt with Q1_cur and Q0_nxt and Q0_cur). We can also redraw the diagram without all of the connecting nets to reduce the complexity of the diagram.

From this drawing the next state logic and output logic expressions can be found directly.

Q1_nxt = Q0_cur ⊕ Q1_cur
Q0_nxt = (Q1_cur · Q0_cur') + (Go · Q0_cur')

Done = Q1_cur · Q0_cur

Example 7.28
Determining the next state logic and output logic expression of a FSM

7.7.2 Finding the State Transition Table of a FSM

Once the next state logic and output logic expressions are known, the state transition table can be created. It is useful to assign more descriptive names to all possible state codes in the FSM. The number of unique states possible depends on how many D-flip-flops are used in the state memory of the FSM. For example, if the FSM uses two D-flip-flops, there are four unique state codes (i.e., 00, 01, 10, 11). We can assign descriptive names such as S0 = 00, S1 = 01, S2 = 10, and S3 = 11. When first creating the transition table, we assign labels and list each possible state code. If a particular code is not used, it can be removed from the transition table at the end of the analysis. The state code that the machine will start in can be found by analyzing its reset and preset connections. This code is typically listed first in the table. The transition table is then populated with all possible combinations of current states and inputs.

The next state codes and output logic values can then be populated by evaluating the next state logic and output logic expressions found earlier. An example of this analysis is shown in Example 7.29.

Example: Determining the State Transition Table of a FSM

Given: The following finite state machine logic diagram.

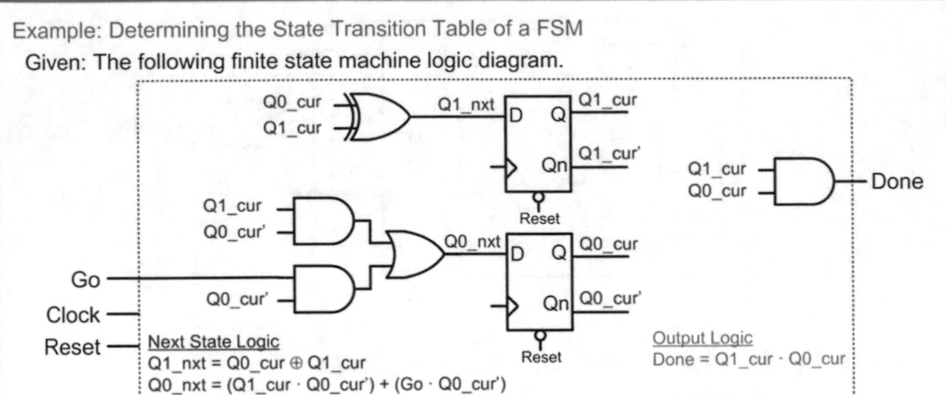

Find: The state transition table.

Solution: Since there are two D-Flip-Flops in this circuit there can be four unique state codes (00, 01, 10, and 11). We notice that the reset condition for this FSM will initialize this machine to state 00. We will insert this state code in the table as the first state. Also, we can assign four arbitrary state names to these codes. Let's use S0=00, S1=01, S2=10, and S3=11. We can list these state names and current state codes in the table along with every possible value of the input. The last step is to simply evaluate the logic expressions for the next state variables and the output to complete the table.

Current State			Input	Next State			Outputs
	Q1_cur	Q0_cur	Go		Q1_nxt	Q0_nxt	Done
S0	0	0	0	S0	0	0	0
S0	0	0	1	S1	0	1	0
S1	0	1	0	S2	1	0	0
S1	0	1	1	S2	1	0	0
S2	1	0	0	S3	1	1	0
S2	1	0	1	S3	1	1	0
S3	1	1	0	S0	0	0	1
S3	1	1	1	S0	0	0	1

The current state codes and Go are used as the inputs into the next state logic and output logic expressions.

These values are calculated using the next state logic and output logic expressions. The state names for the next states are added last.

Example 7.29
Determining the state transition table of a FSM

7.7.3 Finding the State Diagram of a FSM

Once the state transition table is found, creating the state diagram becomes possible. We start the diagram with the state corresponding to the reset state. We then draw how the FSM transitions between each of its possible states based on the inputs to the machine and list the corresponding outputs. An example of this analysis is shown in Example 7.30.

Example: Determining the State Diagram of a FSM

Given: The following state transition table that has been created from a FSM logic diagram.

Current State			Input	Next State			Outputs
	Q1_cur	Q0_cur	Go		Q1_nxt	Q0_nxt	Done
S0	0	0	0	S0	0	0	0
S0	0	0	1	S1	0	1	0
S1	0	1	0	S2	1	0	0
S1	0	1	1	S2	1	0	0
S2	1	0	0	S3	1	1	0
S2	1	0	1	S3	1	1	0
S3	1	1	0	S0	0	0	1
S3	1	1	1	S0	0	0	1

Find: The state diagram.

Solution: The reset condition for this FSM is S0=00 based on the way that the resets of the D-Flip-Flops were connected in the prior logic diagram. This allows us to begin drawing the state diagram starting in S0. From this state we simply list the next state based on the input Go. We notice that the machine will stay in S0 when Go=0 and will transition to S1 when Go=1. We then notice that the machine transitions from S1-to-S2, from S2-to-S3, and from S3-to-S0 regardless of the input value. We can draw these transitions with the input condition Go=X.

For the output Done, we notice that it only depends on the current state, thus this is a Moore machine. For this type of machine we can write the output value within the state bubbles. The final state diagram is as follows.

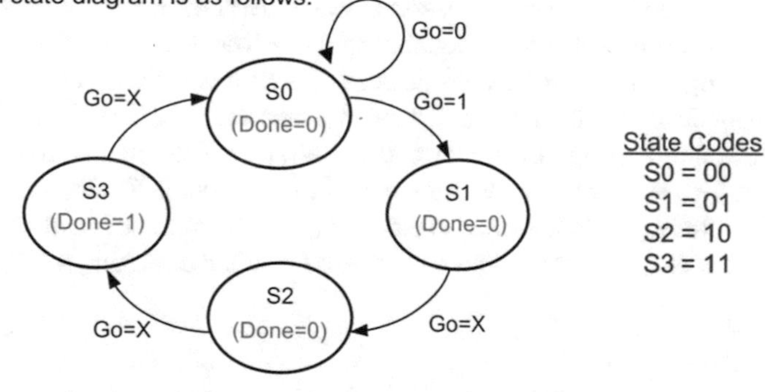

State Codes
S0 = 00
S1 = 01
S2 = 10
S3 = 11

Example 7.30
Determining the state diagram of a FSM

7.7.4 Determining the Maximum Clock Frequency of a FSM

The maximum clock frequency is often one of the banner specifications for a digital system. The clock frequency of a FSM depends on a variety of timing specifications within the sequential circuit including the setup and hold time of the D-flip-flop, the clock-to-Q delay of the D-flip-flop, the combinational logic delay driving the input of the D-flip-flop, the delay of the interconnect that wires the circuit together, and the desired margin for the circuit. The basic concept of analyzing the timing of FSM is to determine how long we must wait after a rising (assuming a rising edge-triggered D-flip-flop) clock edge occurs until the subsequent rising clock edge can occur. The amount of time that must be allowed between rising clock edges depends on how much delay exists in the system. A sufficient amount of time must exist between clock edges to allow the logic computations to settle so that on the next clock edge, the D-flip-flops can latch in a new value on their inputs.

Let's examine all of the sources of delay in a FSM. Let's begin by assuming that all logic values are at a stable value and we experience a rising clock edge. The value present on the D input of the D-flip-flop is latched into the storage device and will appear on the Q output after one clock-to-Q delay of the device (t_{CQ}). Once the new value is produced on the output of the D-flip-flop, it is then used by a variety of combinational logic circuits to produce the next state codes and the outputs of the FSM. The next state code computation is typically longer than the output computation so let's examine that path. The new value on Q propagates through the combinational logic circuitry and produces the next state code at the D input of the D-flip-flop. The delay to produce this next state code includes wiring delay in addition to gate delay. When analyzing the delay of the combinational logic circuitry (t_{cmb}) and the delay of the interconnect (t_{int}), the worst-case path is always considered. Once the new logic value is produced by the next state logic circuitry, it must remain stable for a certain amount of time in order to meet the D-flip-flop's setup specification (t_{setup}). Once this specification is met, the D-flip-flop *could* be clocked with the next clock edge; however, this represents a scenario without any *margin* in the timing. This means that if anything in the system caused the delay to increase even slightly, the D-flip-flop could go metastable. To avoid this situation, margin is included in the delay (t_{margin}). This provides some padding so that the system can reliably operate. A margin of 10% is typical in digital systems. The time that must exist between rising clock edges is then simply the sum of all of these sources of delay ($t_{CQ} + t_{cmb} + t_{int} + t_{setup} + t_{margin}$). Since the time between rising clock edges is defined as the period of the signal (T), this value is also the definition of the period of the fastest clock. Since the frequency of a signal is simply $f = 1/T$, the maximum clock frequency for the FSM is the reciprocal of the sum of the delay.

One specification that is not discussed in the above description is the hold time of the D-flip-flop (t_{hold}). The hold specification is the amount of time that the input to the D-flip-flop must remain constant after the clock edge. In modern storage devices, this time is typically very small and considerably less than the t_{CQ} specification. If the hold specification is less than t_{CQ}, it can be ignored because the output of the D-flip-flop will not change until after one t_{CQ} anyway. This means that the hold requirements are inherently met. This is the situation with the majority of modern D-flip-flops. In the rare case that the hold time is greater than t_{CQ}, then it is used in place of t_{CQ} in the summation of delays. Figure 7.34 gives the summary of the maximum clock frequency calculation when analyzing a FSM.

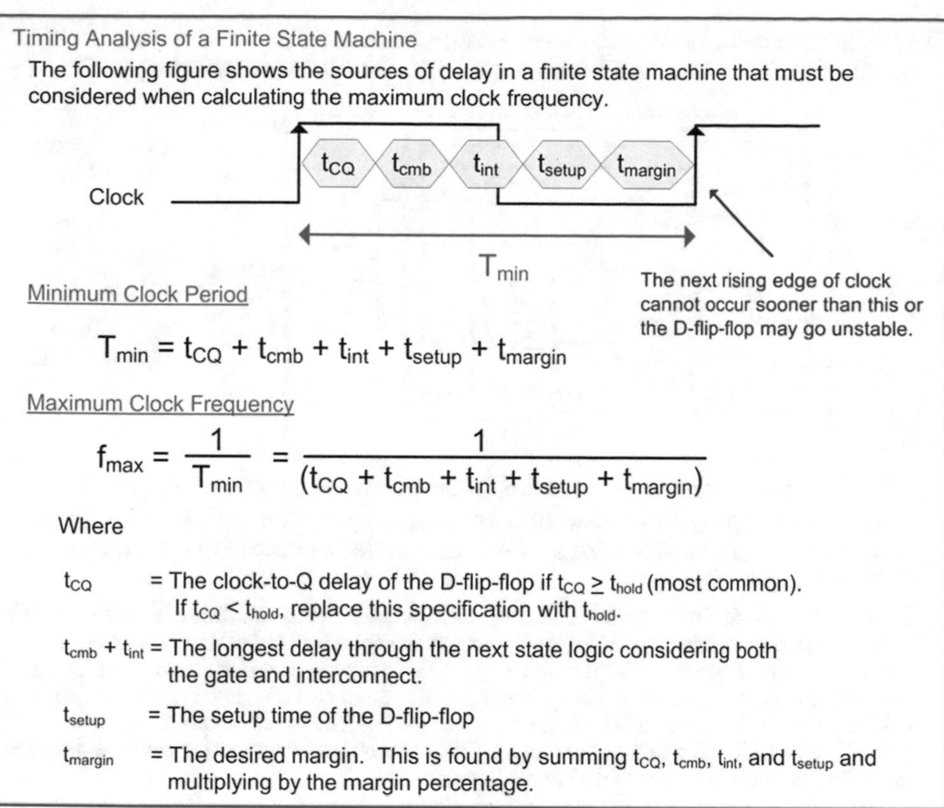

Fig. 7.34
Timing analysis of a finite-state machine

Let's take a look at an example of how to use this analysis. Example 7.31 shows this analysis for the FSM analyzed in prior sections but this time considering the delay specifications of each device.

Example: Determining the Maximum Clock Frequency of a FSM

Given: The following finite state machine logic diagram with the associated delays.

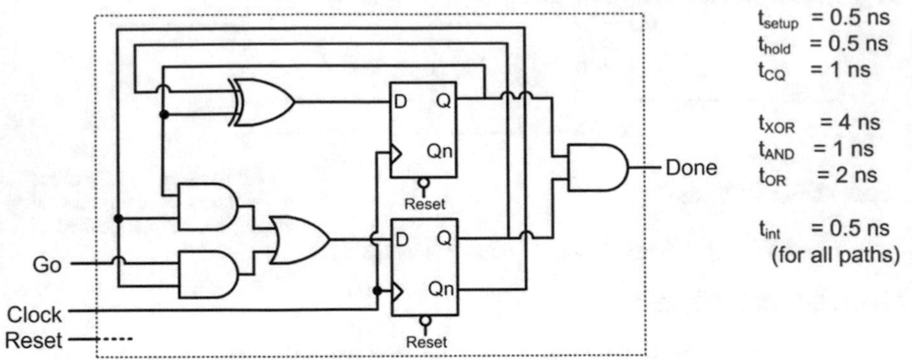

t_{setup} = 0.5 ns
t_{hold} = 0.5 ns
t_{CQ} = 1 ns

t_{XOR} = 4 ns
t_{AND} = 1 ns
t_{OR} = 2 ns

t_{int} = 0.5 ns
(for all paths)

Find: The maximum clock frequency this FSM can operate at with a timing margin of 10%.

Solution: First, we need to decide whether to use t_{CQ} or t_{hold} in our delay calculation. In this example, $t_{CQ} > t_{hold}$ so we will use t_{CQ}. When $t_{CQ} \geq t_{hold}$ the hold specification of the D-flip-flop is inherently met.

Next, we need to find the longest combinational logic and interconnect path. Since it is given that all interconnect paths are identical at 0.5ns, we simply need to find the longest gate delay path. There are three paths in this FSM. The first is the next state logic circuit using the XOR gate with 4ns of delay (t_{XOR}). The second is the next state logic expression using the SOP form with a delay of 3ns ($t_{AND} + t_{OR}$ =1ns + 2ns). The third path is through the output logic circuit with a delay of 1ns (t_{AND}). The longest combinational logic path is through the XOR gate so in our calculation we will use t_{cmb}=4ns.

Next, we need to calculate the exact value of the 10% margin required. The margin is found by summing all other real delays in the signal path and multiplying by the margin percentage. For this example:

$$t_{margin} = (t_{CQ} + t_{cmb} + t_{int} + t_{setup})\cdot(0.1)$$
$$t_{margin} = (1ns + 4ns + 0.5ns + 0.5ns)\cdot(0.1)$$
$$t_{margin} = 0.6ns$$

Now we can plug all of our delays directly into the equation for the maximum clock frequency:

$$f_{max} = \frac{1}{(t_{CQ} + t_{cmb} + t_{int} + t_{setup} + t_{margin})} = \frac{1}{(1ns + 4ns + 0.5ns + 0.5ns + 0.6ns)}$$

$$f_{max} = 151\ MHz$$

Example 7.31
Determining the maximum clock frequency of a FSM

Concept Check

CC7.7 What is the risk of running the clock above its maximum allowable frequency in a finite-state machine?

A) The power consumption may drop below the recommended level.

B) The setup and hold specifications of the D-flip-flops may be violated, which may put the machine into an unwanted state.

C) The states may transition too quickly to be usable.

D) The crystal generating the clock may become unstable.

Summary

- Sequential logic refers to a circuit that bases its outputs on both the present and past values of the inputs. Past values are held in sequential logic storage device.
- All sequential logic storage devices are based on a cross-coupled feedback loop. The positive feedback loop formed in this configuration will hold either a 1 or a 0. This is known as a bistable device.
- If the inputs of the feedback loop in a sequential logic storage device are driven to exactly between a 1 and a 0 (i.e., $V_{cc}/2$) and then released, the device will go *metastable*. Metastability refers to the behavior where the device will ultimately be pushed toward one of the two stable states in the system, typically by electrical noise. Once the device begins moving toward one of the stable states, the positive feedback will reinforce the transition until it reaches the stable state. The stable state that the device will move toward is random and unknown.
- Cross-coupled inverters are the most basic form of the positive feedback loop configuration. To give the ability to drive the outputs of the storage device to known values, the inverters are replaced with NOR gates to form the SR Latch. A variety of other modifications can be made to the loop configuration to ultimately produce a D-Latch and D-flip-flop.
- A D-flip-flop will update its Q output with the value on its D input on every triggering edge of a clock. The amount of time that it takes for the Q output to update after a triggering clock edge is called the "t-clock-to-Q" (t_{CQ}) specification.
- The setup and hold times of a D-flip-flop describe how long before (t_{setup}) and after (t_{hold}) the triggering clock edge that the data on the D input of the device must be stable. If the D input transitions too close to the triggering clock edge (i.e., violating a setup or hold specification), then the device will go metastable, and the ultimate value on Q is unknown.
- A synchronous system is one in which all logic transitions occur based on a single timing event. The timing event is typically the triggering edge of a clock.
- There are a variety of common circuits that can be accomplished using just sequential storage devices. Examples of these circuits include switch debouncing, toggle flops, ripple counters, and shift registers.
- A finite-state machine (FSM) is a system that produces outputs based on the current value of the inputs and a history of past inputs. The history of inputs is recorded as *states* that the machine has been in. As the machine responds to new inputs, it transitions between states. This allows a finite-state machine to make more sophisticated decisions about what outputs to produce by knowing its history.
- A state diagram is a graphical way to describe the behavior of a FSM. States are represented using circles, and transitions are represented using arrows. Outputs are listed either inside of the state circle or next to the transition arrow.
- A state transition table contains the same information as a state diagram but in tabular format. This allows the system to be more easily synthesized because the information is in a form similar to a truth table.
- The first step in FSM synthesis is creating the *state memory*. The state memory consists of a set of D-flip-flops that hold the current state of the FSM. Each state in the FSM must be assigned a binary code. The type of encoding is arbitrary; however, there are certain encoding types that are commonly used such as binary, gray code, and one-hot. Once the codes are assigned, state variables need to be defined for each bit position for both the current state and the next state codes. The state variables for the current state represent the Q outputs of the D-flip-flops, which hold the current state code. The state variables for the next state code represent the D inputs of the D-flip-flops. A D-flip-flop is needed for each bit in the state code. On the triggering edge of a clock, the current state will be updated with the next state code.
- The second step in FSM synthesis is creating the *next state logic*. The next state logic is combinational logic circuitry that produces the next state codes based on the current state variables and any system inputs. The next state logic drives the D inputs of the D-flip-flops in the state memory.
- The third step in FSM synthesis is creating the *output logic*. The output logic is combinational logic circuitry that produces the system outputs based on the current state and, potentially, the system inputs.
- The output logic always depends on the current state of a FSM. If the output logic also depends on the system inputs, the machine

is a *Mealy* machine. If the output logic does not depend on the system inputs, the machine is a *Moore* machine.

- A counter is a special type of finite-state machine in which the states are traversed linearly. The linear progression of states allows the next state logic to be simplified. The complexity of the output logic in a counter can also be reduced by encoding the states with the desired counter output for that state. This technique, known as *state-encoded outputs,* allows the system outputs to simply be the current state of the FSM.
- The *reset state* of a FSM is the state that the machine will go to when it begins operation. The state code for the reset state must be configured using the reset and/or preset lines of the D-flip-flops. If only reset lines are used on the D-flip-flops, the reset state must be encoded using only zeros.
- Given the logic diagram for a state machine, the logic expression for the next state memory and the output logic can be determined by analyzing the combinational logic driving the D inputs of the state memory (i.e., the next state logic) and the combinational logic driving the system outputs (i.e., the output logic).
- Given the logic diagram for a state diagram, the state diagram can be determined by first finding the logic expressions for the next state and output logic. The number of D-flip-flops in the logic diagram can then be used to calculate the possible number of state codes that the machine has. The state codes are then used to calculate the next state logic and output values. From this information, a state transition table can be created and, in turn, the state diagram.
- The maximum frequency of a FSM is found by summing all sources of time delay that must be accounted for before the next triggering edge of the clock can occur. These sources include t_{CQ}, the worst-case combinational logic path, the worst-case interconnect delay path, the setup/hold time of the D-flip-flops, and any margin that is to be included. The sum of these timing delays represents the smallest period (T) that the clock can have. This is then converted to frequency.
- If the t_{CQ} time is greater than or equal to the hold time, the hold time can be ignored in the maximum frequency calculation. This is because the outputs of the D-flip-flops are inherently *held* while the D-flip-flops are producing the next output value. The time it takes to change the outputs after a triggering clock edge is defined as t_{CQ}. This means as long as $t_{CQ} \geq t_{hold}$, the hold time specification is inherently met since the logic driving the next state codes uses the Q outputs of the D-flip-flops.

Exercise Problems

For some of the following exercise problems, you will be asked to design a Verilog model and perform a functional simulation. You will be provided with a test bench for each of these problems. The details of how to create your own Verilog test bench are provided later in Chap. 8. For some of the following exercise problems, you will be asked to use D-flip-flops as part of a Verilog design. You will be provided with the model of the D-flip-flop and can declare it as a sub-system in your design. The Verilog module definition for a D-flip-flop is given in Fig. 7.35. Keep in mind that this D-flip-flop has an active LOW reset. This means that when the reset line is pulled to a 0, the outputs will go to Q = 0, Qn = 1. When the reset line is LOW, the incoming clock is ignored. Once the reset line goes HIGH, the D-flip-flop resumes normal behavior. The details of how to create your own model of a D-flip-flop are provided later in Chap. 8.

Rising Edge Triggered D-Flip-Flop
with Active LOW Reset

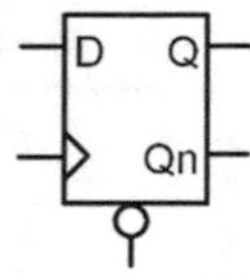

dflipflop.v

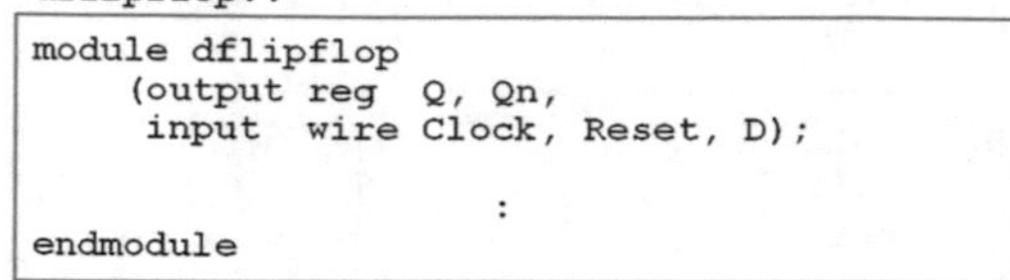

```
module dflipflop
    (output reg  Q, Qn,
     input  wire Clock, Reset, D);

                    :
endmodule
```

Fig. 7.35
D-flip-flop module definition

Section 7.1: Sequential Logic Storage Devices

7.1.1 What does the term *metastability* refer to in a sequential storage device?

7.1.2 What does the term *bistable* refer to in a sequential storage device?

7.1.3 You are given a cross-coupled inverter pair in which all nodes are set to $V_{cc}/2$. Why will this configuration always move to a more stable state?

7.1.4 An SR Latch essentially implements the same cross-coupled feedback loop to store information as in a cross-coupled inverter pair. What is the purpose of using NOR gates instead of inverters in the SR Latch configuration?

7.1.5 Why isn't the input condition S = R = 1 used in an SR Latch?

7.1.6 How will the output Q behave in an SR Latch if the inputs continuously switch between S = 0, R = 1 and S = 1, R = 1 every 10 ns?

7.1.7 How do D-flip-flops enable synchronous systems?

7.1.8 What signal in the D-flip-flop in Fig. 7.35 has the highest priority?

7.1.9 For the timing diagram shown in Fig. 7.36, draw the outputs Q and Qn for a rising edge-triggered D-flip-flop with active LOW.

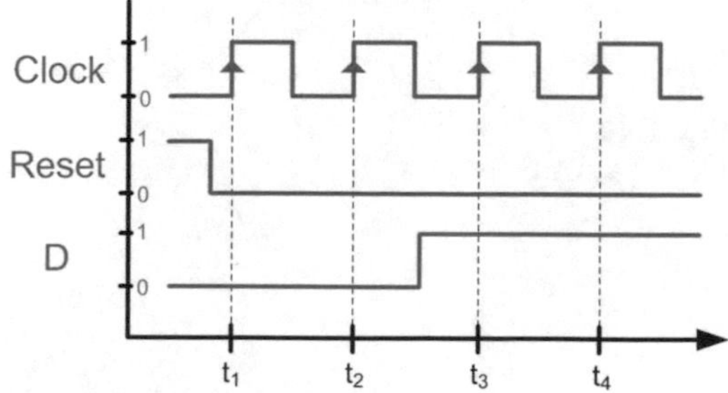

Fig. 7.36
D-flip-flop timing exercise 1

7.1.10 For the timing diagram shown in Fig. 7.37, draw the outputs Q and Qn for a rising edge triggered D-flip-flop with active LOW.

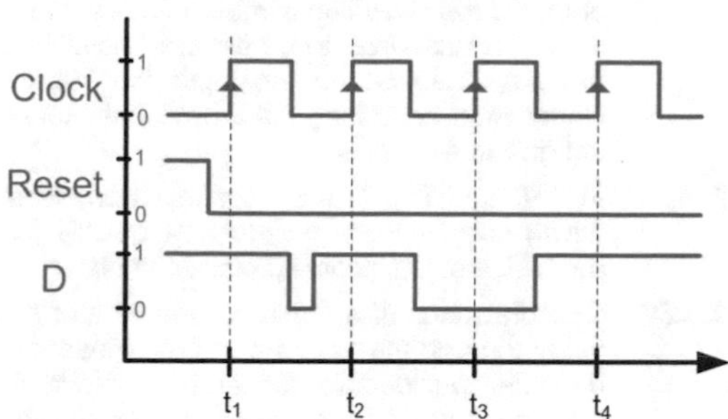

Fig. 7.37
D-flip-flop timing exercise 2

7.1.11 For the timing diagram shown in Fig. 7.38, draw the outputs Q and Qn for a rising edge triggered D-flip-flop with active LOW.

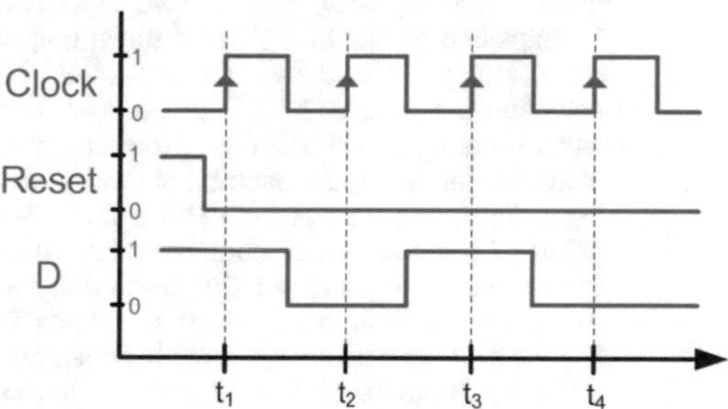

Fig. 7.38
D-flip-flop timing exercise 3

Section 7.2: Sequential Logic Timing Considerations

7.2.1 What timing specification is violated in a D-flip-flop when the data is not held long enough *before* the triggering clock edge occurs?

7.2.2 What timing specification is violated in a D-flip-flop when the data is not held long enough *after* the triggering clock edge occurs?

7.2.3 What is the timing specification for a D-flip-flop that describes how long after the triggering clock edge occurs that the new data will be present on the Q output?

7.2.4 What is the timing specification for a D-flip-flop that describes how long after the device goes metastable that the outputs will settle to known states.

7.2.5 If the Q output of a D-flip-flop is driving the D input of another D-flip-flop from the same logic family, can the hold time be ignored if it is less than the clock-to-Q delay? Provide an explanation as to why or why not.

Section 7.3: Common Circuits Based on Sequential Storage Devices

7.3.1 In a toggle flop (T-flop) configuration, the Qn output of the D-flip-flop is routed back to the D input. This can lead to a hold time violation if the output arrives on the input too quickly. Under what condition(s) is a hold time violate not an issue?

7.3.2 In a toggle flop (T-flop) configuration, what timing specifications dictate how quickly the next edge of the incoming clock can occur?

7.3.3 One drawback of a ripple counter is that the delay through the cascade of D-flip-flops can become considerable for large counters. At what point does the delay of a ripple counter prevent it from being useful?

7.3.4 A common use of a ripple counter is in the creation of a *2^n programmable clock divider*. In a ripple counter, bit(0) has a frequency that is exactly 1/2 of the incoming clock, bit(1) has a frequency that is exactly 1/4 of the incoming clock, bit(2) has a frequency that is exactly 1/8 of the incoming clock, etc. This behavior can be exploited to create a divided down output clock that is divided by multiples of 2^n by selecting a particular bit of the counter. The typical configuration of this programmable clock divider is to route each bit of the counter to an input of a multiplexer. The select lines going to the multiplexer choose which bit of the counter are used as the divided down clock output. This architecture is shown in Fig. 7.39. Design a Verilog model to implement the programmable clock divider shown in this figure. Use the module port definition provided in this figure for your design. Use a 4-bit ripple counter to produce four divided versions of the clock (1/2, 1/4, 1/8, and 1/16). Your system will take in two select lines that will choose which version of the clock is to be routed to the output. Instantiate the D-flip-flop model provided to implement the ripple counter. Implement the 4-to-1 multiplexer using continuous assignment. The multiplexer does not need to be its own sub-system.

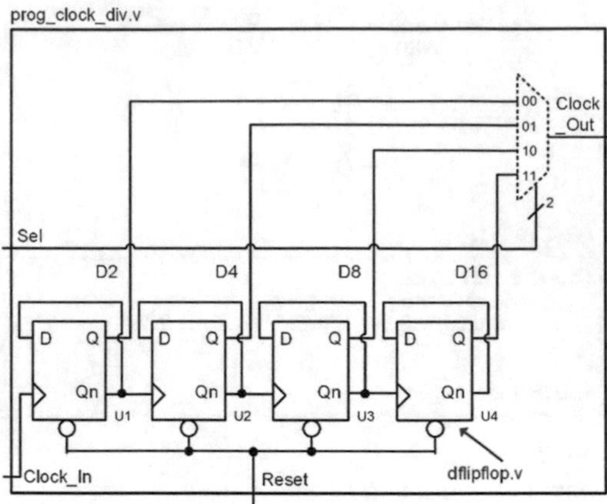


```
module prog_clock_div
    (output wire Clock_Out,
     input  wire Clock_In, Reset,
     input  wire Sel[1:0]);
                :
endmodule
```

Fig. 7.39
Programmable clock module description

7.3.5 What phenomenon causes switch bounce in a SPST switch?

7.3.6 What two phenomena causes switch bounce in a SPDT switch?

Section 7.4: Finite-State Machines

7.4.1 For the state diagram in Fig. 7.40, how many D-Flip-Flops will this machine take if the states are encoded in binary?

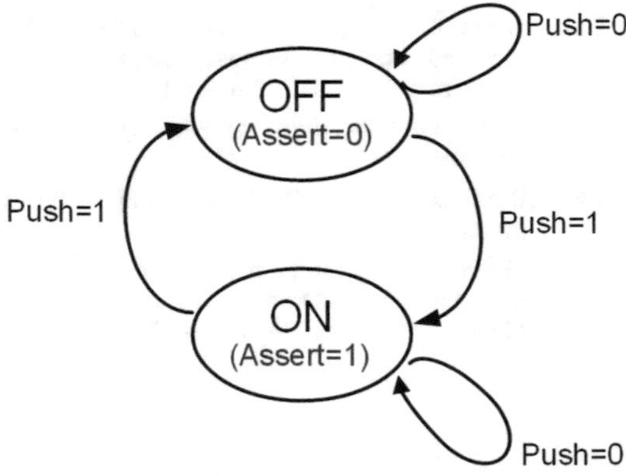

Fig. 7.40
FSM 0 state diagram

7.4.2 For the state diagram in Fig. 7.40, how many D-flip-flops will this machine take if the states are encoded in gray code?

7.4.3 For the state diagram in Fig. 7.40, how many D-flip-flops will this machine take if the states are encoded in one-hot?

7.4.4 For the state diagram in Fig. 7.40, is this a Mealy or Moore machine?

The next set of questions are about the design of a finite-state machine by hand to implement the behavior described by the state diagram in Fig. 7.40. For this design, you will name the current state variable Q0_cur and name the next state variable Q0_nxt. You will also use the following state codes:

OFF = '0'
ON = '1'

7.4.5 For the state diagram in Fig. 7.40, what is the next state logic expression for Q0_nxt?

7.4.6 For the state diagram in Fig. 7.40, what is the output logic expression for Assert?

7.4.7 For the state diagram in Fig. 7.40, provide the final logic diagram for this machine.

7.4.8 Design a Verilog model to implement the behavior described by the state diagram in Fig. 7.40. Use the module port definition provided in Fig. 7.41 for your design. Name the current state variables Q1_cur and Q0_cur, and name the next state variables Q1_nxt and Q0_nxt. Instantiate the D-flip-flop model provided to implement your state memory. Use continuous assignment with logical operators for the implementation of your next state and output logic.

fsm0.v

```
module fsm0
    (output wire Assert,
     input  wire Clock, Reset,
     input  wire Push);
                      :
endmodule
```

Fig. 7.41
FSM 0 module definition

7.4.9 Design a Verilog model to implement the behavior described by the state diagram in Fig. 7.40. Use the module port definition provided in Fig. 7.41 for your design. Name the current state variables Q1_cur and Q0_cur and name the next state variables Q1_nxt and Q0_nxt. Instantiate the D-flip-flop model provided to implement your state memory. Use continuous assignment with conditional operators for the implementation of your next state and output logic.

7.4.10 Design a Verilog model to implement the behavior described by the state diagram in Fig. 7.40. Use the module port definition provided in Fig. 7.41 for your design. Name the current state variables Q1_cur and Q0_cur, and name the next state variables Q1_nxt and Q0_nxt. Instantiate the D-flip-flop model provided to implement your state memory. Use user-defined primitives for the implementation of your next state and output logic.

7.4.11 For the state diagram in Fig. 7.42, how many D-flip-flops will this machine take if the states are encoded in binary?

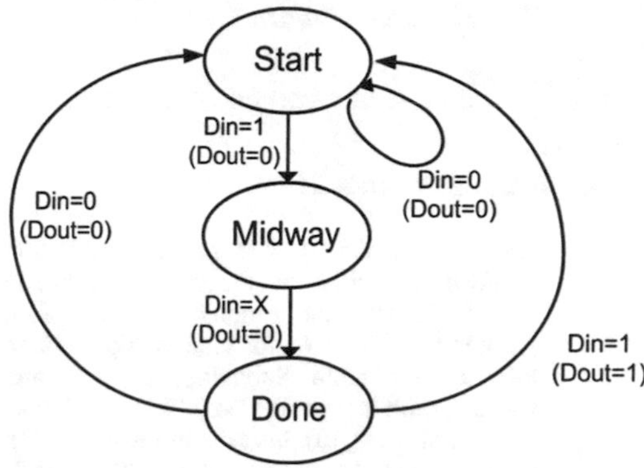

Fig. 7.42
FSM 1 state diagram

7.4.12 For the state diagram in Fig. 7.42, how many D-flip-flops will this machine take if the states are encoded in gray code?

7.4.13 For the state diagram in Fig. 7.42, how many D-flip-flops will this machine take if the states are encoded in one-hot?

7.4.14 For the state diagram in Fig. 7.42, is this a Mealy or Moore machine?

The next set of questions are about the design of a finite-state machine by hand to implement the behavior described by the state diagram in Fig. 7.42. For this design, you will name the current state variables Q1_cur and Q0_cur and name the next state variables Q1_nxt and Q0_nxt. You will also use the following state codes:

Start = "00"
Midway = "01"
Done = "10"

7.4.15 For the state diagram in Fig. 7.42, what is the next state logic expression for Q1_nxt?

7.4.16 For the state diagram in Fig. 7.42, what is the next state logic expression for Q0_nxt?

7.4.17 For the state diagram in Fig. 7.42, what is the output logic expression for Dout?

7.4.18 For the state diagram in Fig. 7.42, provide the final logic diagram for this machine.

7.4.19 Design a Verilog model to implement the behavior described by the state diagram in Fig. 7.42. Use the module port definition provided in Fig. 7.43 for your design. Name the current state variables Q1_cur and Q0_cur, and name the next state variables

Q1_nxt and Q0_nxt. Instantiate the D-flip-flop model provided to implement your state memory. Use continuous assignment with logical operators for the implementation of your next state and output logic.

fsm1.v

```
module fsm1
    (output wire Dout,
     input  wire Clock, Reset,
     input  wire Din);
                    :
endmodule
```

Fig. 7.43
FSM 1 module description

7.4.20 Design a Verilog model to implement the behavior described by the state diagram in Fig. 7.42. Use the module port definition provided in Fig. 7.43 for your design. Name the current state variables Q1_cur and Q0_cur, and name the next state variables Q1_nxt and Q0_nxt. Instantiate the D-flip-flop model provided to implement your state memory. Use continuous assignment with conditional operators for the implementation of your next state and output logic.

7.4.21 Design a Verilog model to implement the behavior described by the state diagram in Fig. 7.42. Use the module port definition provided in Fig. 7.43 for your design. Name the current state variables Q1_cur and Q0_cur, and name the next state variables Q1_nxt and Q0_nxt. Instantiate the D-flip-flop model provided to implement your state memory. Use user-defined primitives for the implementation of your next state and output logic.

7.4.22 For the state diagram in Fig. 7.44, how many D-flip-flops will this machine take if the states are encoded in binary?

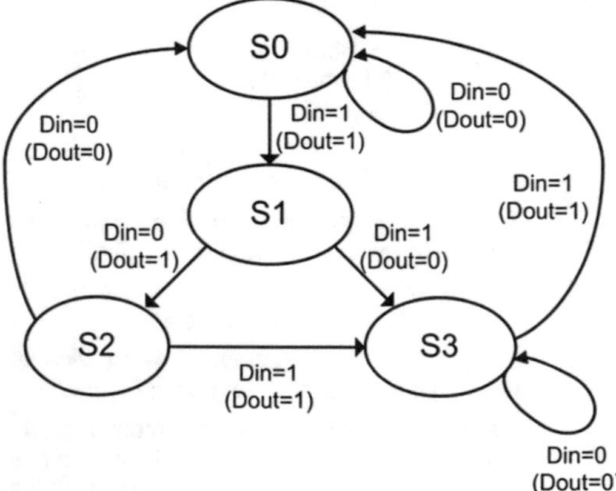

Fig. 7.44
FSM 2 state diagram

7.4.23 For the state diagram in Fig. 7.44, how many D-flip-flops will this machine take if the states are encoded in gray code?

7.4.24 For the state diagram in Fig. 7.44, how many D-flip-flops will this machine take if the states are encoded in one-hot?

7.4.25 For the state diagram in Fig. 7.44, is this a Mealy or Moore machine?

The next set of questions are about the design of a finite-state machine by hand to implement the behavior described by the state diagram in Fig. 7.44. For this design, you will name the current state variables Q1_cur and Q0_cur and name the next state variables Q1_nxt and Q0_nxt. You will also use the following state codes:

S0 = "00"
S1 = "01"
S2 = "10"
S3 = "11"

7.4.26 For the state diagram in Fig. 7.44, what is the next state logic expression for Q1_nxt?

7.4.27 For the state diagram in Fig. 7.44, what is the next state logic expression for Q0_nxt?

7.4.28 For the state diagram in Fig. 7.44, what is the output logic expression for Dout?

7.4.29 For the state diagram in Fig. 7.44, provide the final logic diagram for this machine.

7.4.30 Design a Verilog model to implement the behavior described by the state diagram in Fig. 7.44. Use the module port definition provided in Fig. 7.45 for your design. Name the current state variables Q1_cur and Q0_cur, and name the next state variables Q1_nxt and Q0_nxt. Instantiate the D-flip-flop model provided to implement your state memory. Use continuous assignment with logical operators for the implementation of your next state and output logic.

fsm2.v

```
module fsm2
    (output wire Dout,
     input  wire Clock, Reset,
     input  wire Din);
                    :
endmodule
```

Fig. 7.45
FSM 2 module description

7.4.31 Design a Verilog model to implement the behavior described by the state diagram in Fig. 7.44. Use the module port definition provided in Fig. 7.45 for your design. Name the current state variables Q1_cur and Q0_cur, and name the next state variables Q1_nxt and Q0_nxt. Instantiate the D-flip-flop model provided to implement your state memory. Use continuous assignment with logical operators for the implementation of your next state and output logic.

7.4.32 Design a Verilog model to implement the behavior described by the state diagram in Fig. 7.44. Use the module port definition provided in Fig. 7.45 for your design. Name the current state variables Q1_cur and Q0_cur, and name the next state variables Q1_nxt and Q0_nxt. Instantiate the D-flip-flop model provided to implement your state memory. Use user-defined primitives for the implementation of your next state and output logic.

7.4.33 For the state diagram in Fig. 7.46, how many D-flip-flops will this machine take if the states are encoded in binary?

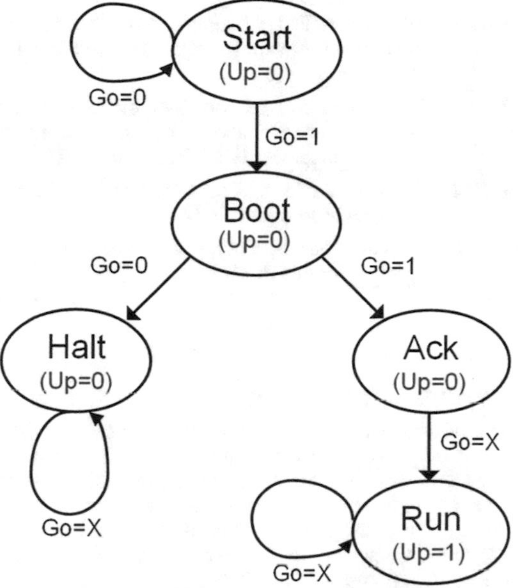

Fig. 7.46
FSM 3 state diagram

7.4.34 For the state diagram in Fig. 7.46, how many D-flip-flops will this machine take if the states are encoded in gray code?

7.4.35 For the state diagram in Fig. 7.46, how many D-flip-flops will this machine take if the states are encoded in one-hot?

7.4.36 For the state diagram in Fig. 7.46, is this a Mealy or Moore machine?

The next set of questions are about the design of a finite-state machine by hand to implement the behavior described by the state diagram in Fig. 7.46. For this design, you will name the current state variables Q2_cur, Q1_cur, and Q0_cur and name the next state variables Q2_nxt, Q1_nxt, and Q0_nxt. You will also use the following state codes:

Start = "000"
Boot = "001"
Halt = "010"
Ack = "011"
Run = "100"

7.4.37 For the state diagram in Fig. 7.46, what is the next state logic expression for Q2_nxt?

7.4.38 For the state diagram in Fig. 7.46, what is the next state logic expression for Q1_nxt?

7.4.39 For the state diagram in Fig. 7.46, what is the next state logic expression for Q0_nxt?

7.4.40 For the state diagram in Fig. 7.46, what is the output logic expression for Up?

7.4.41 For the state diagram in Fig. 7.46, provide the final logic diagram for this machine.

7.4.42 Design a Verilog model to implement the behavior described by the state diagram in Fig. 7.46. Use the module port definition provided in Fig. 7.47 for your design. Name the current state variables Q2_cur, Q1_cur, and Q0_cur, and name the next state variables Q2_nxt, Q1_nxt, and Q0_nxt. Instantiate the D-flip-flop model provided to implement your state memory. Use continuous assignment with logical operators for the implementation of your next state and output logic.

fsm3.v

```
module fsm3
    (output wire Up,
     input  wire Clock, Reset,
     input  wire Go);
                        :
endmodule
```

Fig. 7.47
FSM 3 module description

7.4.43 Design a Verilog model to implement the behavior described by the state diagram in Fig. 7.46. Use the module port definition provided in Fig. 7.47 for your design. Name the current state variables Q2_cur, Q1_cur, and Q0_cur, and name the next state variables Q2_nxt, Q1_nxt, and Q0_nxt. Instantiate the D-flip-flop model provided to implement your state memory. Use continuous assignment with logical operators for the implementation of your next state and output logic.

7.4.44 Design a Verilog model to implement the behavior described by the state diagram in Fig. 7.46. Use the module port definition provided in Fig. 7.47 for your design. Name the current state variables Q2_cur, Q1_cur, and Q0_cur, and name the next state variables Q2_nxt, Q1_nxt, and Q0_nxt. Instantiate the D-flip-flop model provided to implement your state memory. Use user-defined primitives for the implementation of your next state and output logic.

The next set of questions are about the design of a 4-bit serial bit sequence detector by hand similar to the one described in Example 7.9. The input to your state detector is called *DIN*, and the output is called *FOUND*. Your detector will assert FOUND anytime there is a 4-bit sequence of "0101." For all other input sequences, the output is not asserted.

7.4.45 For your 4-bit serial bit sequence detector, provide the state diagram for this FSM.

7.4.46 For your 4-bit serial bit sequence detector, how many D-flip-flops does it take to implement the state memory for this FSM if you encode your states in binary?

7.4.47 For your 4-bit serial bit sequence detector, provide the state transition table for this FSM.

7.4.48 For your 4-bit serial bit sequence detector, synthesize and provide the combinational logic expressions for the next state logic.

7.4.49 For your 4-bit serial bit sequence detector, synthesize and provide the combinational logic expression for the output logic for FOUND.

7.4.50 For your 4-bit serial bit sequence detector, is this FSM a Mealy or Moore machine?

7.4.51 For your 4-bit serial bit sequence detector, provide the logic diagram for this FSM.

The next set of questions are about the design of a 20-cent vending machine controller by hand similar to the one described in Example 7.12. Your controller will take in nickels and dimes and dispense a product anytime the customer has entered 20 cents. Your FSM has two inputs, *Nin* and *Din*. Nin is asserted whenever the customer enters a nickel, while Din is asserted anytime the customer enters a dime. Your FSM has two outputs, *dispense* and *change*. Dispense is asserted anytime the customer has entered at least 20 cents and change is asserted anytime the customer has entered more than 20 cents and needs a nickel in change.

7.4.52 For your 20-cent vending machine controller, provide the state diagram for this FSM.

7.4.53 For your 20-cent vending machine controller, how many D-flip-flops does it take to implement the state memory for this FSM if you encode your states in binary?

7.4.54 For your 20-cent vending machine controller, provide the state transition table for this FSM.

7.4.55 For your 20-cent vending machine controller, synthesize and provide the combinational logic expressions for the next state logic.

7.4.56 For your 20-cent vending machine controller, synthesize and provide the combinational logic expression for the output logic for Dispense and Change.

7.4.57 For your 20-cent vending machine controller, is this FSM a Mealy or Moore machine?

7.4.58 For your 20-cent vending machine controller, provide the logic diagram for this FSM.

The next set of questions are about the design of a finite-state machine by hand that controls a traffic light at the intersection of a busy highway and a seldom used side road. You will be designing the control signals for just the red, yellow, and green lights facing the highway. Under normal conditions, the highway has a green light. The side road has a car detector that indicates when a car pulls up by asserting a signal called *CAR*. When CAR is asserted, you will change the highway traffic light from green to yellow. Once yellow, you will always go to red. Once in the red position, a built-in timer will begin a countdown and provide your controller a signal called *TIMEOUT* when 15 seconds has passed. Once TIMEOUT is asserted, you will change the highway traffic light back to green. Your system will have three outputs *GRN*, *YLW*, and *RED* that control when the highway facing traffic lights are on (1 = ON, 0 = OFF).

7.4.59 For your traffic light controller, provide the state diagram for this FSM.

7.4.60 For your traffic light controller, how many D-flip-flops does it take to implement the state memory for this FSM if you encode your states in binary?

7.4.61 For your traffic light controller, provide the state transition table for this FSM.

7.4.62 For your traffic light controller, synthesize and provide the combinational logic expressions for the next state logic.

7.4.63 For your traffic light controller, synthesize and provide the combinational logic expression for the output logic for GRN, YLW, and RED.

7.4.64 For your traffic light controller, is this FSM a Mealy or Moore machine?

7.4.65 For your traffic light controller, provide the logic diagram for this FSM.

Section 7.5: Counters

The next set of questions are about the design a 3-bit binary up counter by hand. This state machine will need eight states and require three bits for the state variable codes. Name the current state variables Q2_cur, Q1_cur, and Q0_cur and the next state variables Q2_nxt, Q1_nxt, and Q0_nxt. The output of your counter will be a 3-bit vector called *Count*.

7.5.1 For your 3-bit binary up counter, what is the next state logic expression for Q2_nxt?

7.5.2 For your 3-bit binary up counter, what is the next state logic expression for Q1_nxt?

7.5.3 For your 3-bit binary up counter, what is the next state logic expression for Q0_nxt?

7.5.4 For your 3-bit binary up counter, what is the output logic expression for Count(2)?

7.5.5 For your 3-bit binary up counter, what is the output logic expression for Count(1)?

7.5.6 For your 3-bit binary up counter, what is the output logic expression for Count(0)?

7.5.7 For your 3-bit binary up counter, provide the logic diagram.

7.5.8 Design a Verilog model for a 3-bit binary up counter. Instantiate the D-flip-flop model provided to implement your state memory. Use whatever concurrent modeling approach you wish to model the next state and output logic. Use the module port definition provided in Fig. 7.48 for your design.

counter_3bit_binary_up.v

```
module counter_3bit_binary_up
    (output wire Count[2:0],
     input  wire Clock, Reset);
                           :
endmodule
```

Fig. 7.48
3-Bit binary up counter module definition

The next set of questions are about the design a 3-bit binary up/down counter by hand. The counter will have an input called *Up* that will dictate the direction of the counter. When Up = 1, the counter should increment and when Up = 0 it should decrement. This state machine will need eight states and require three bits for the state variable codes. Name the current state variables Q2_cur, Q1_cur, and Q0_cur and the next state variables Q2_nxt, Q1_nxt, and Q0_nxt. The output of your counter will be a 3-bit vector called *Count*.

7.5.9 For your 3-bit binary up/down counter, what is the next state logic expression for Q2_nxt?

7.5.10 For your 3-bit binary up/down counter, what is the next state logic expression for Q1_nxt?

7.5.11 For your 3-bit binary up/down counter, what is the next state logic expression for Q0_nxt?

7.5.12 For your 3-bit binary up/down counter, what is the output logic expression for Count(2)?

7.5.13 For your 3-bit binary up/down counter, what is the output logic expression for Count(1)?

7.5.14 For your 3-bit binary up/down counter, what is the output logic expression for Count(0)?

7.5.15 For your 3-bit binary up/down counter, provide the logic diagram.

7.5.16 Design a Verilog model for a 3-bit binary up/down counter. Instantiate the D-flip-flop model provided to implement your state memory. Use whatever concurrent modeling approach you wish to model the next state and output logic. Use the module port definition provided in Fig. 7.49 for your design.

counter_3bit_binary_up_down.v

```
module counter_3bit_binary_up_down
    (output wire Count[2:0],
     input  wire Clock, Reset,
     input  wire Up);
                           :
endmodule
```

Fig. 7.49
3-Bit binary up/down counter module definition

The next set of questions are about the design a 3-bit gray code up counter by hand. This counter will create a pattern of 000→001→011→010→110→111→101→100 and then repeat. This state machine will need eight states and require three bits for the state variable codes. Name the current state variables Q2_cur, Q1_cur, and Q0_cur and the next state variables Q2_nxt, Q1_nxt, and Q0_nxt. The output of your counter will be a 3-bit vector called *Count*.

7.5.17 For your 3-bit gray code up counter, what is the next state logic expression for Q2_nxt?

7.5.18 For your 3-bit gray code up counter, what is the next state logic expression for Q1_nxt?

7.5.19 For your 3-bit gray code up counter, what is the next state logic expression for Q0_nxt?

7.5.20 For your 3-bit gray code up counter, what is the output logic expression for Count(2)?

7.5.21 For your 3-bit gray code up counter, what is the output logic expression for Count(1)?

7.5.22 For your 3-bit gray code up counter, what is the output logic expression for Count(0)?

7.5.23 For your 3-bit gray code up counter, provide the logic diagram.

7.5.24 Design a Verilog model for a 3-bit gray code up counter. Instantiate the D-flip-flop model provided to implement your state memory. Use whatever concurrent modeling approach you wish to model the next state and output logic. Use the module port definition provided in Fig. 7.50 for your design.

counter_3bit_graycode_up.v

```
module counter_3bit_graycode_up
    (output wire Count[2:0],
     input  wire Clock, Reset);
                           :
endmodule
```

Fig. 7.50
3-Bit gray code up counter module definition

The next set of questions are about the design a 3-bit gray code up/down counter by hand. The counter will have an input called *Up* that will dictate the direction of the counter. When Up = 1, the counter should increment, and when Up = 0, it should decrement. When incrementing up, the counter will create the pattern 000→001→011→010→110→111→101→100 and then repeat. This state machine will need eight states and require three bits for the state variable codes. Name the current state variables Q2_cur, Q1_cur, and Q0_cur and the next state variables Q2_nxt, Q1_nxt, and Q0_nxt. The output of your counter will be a 3-bit vector called *Count*.

7.5.25 For your 3-bit gray code up/down counter, what is the next state logic expression for Q2_nxt?

7.5.26 For your 3-bit gray code up/down counter, what is the next state logic expression for Q1_nxt?

7.5.27 For your 3-bit gray code up/down counter, what is the next state logic expression for Q0_nxt?

7.5.28 For your 3-bit gray code up/down counter, what is the output logic expression for Count(2)?

7.5.29 For your 3-bit gray code up/down counter, what is the output logic expression for Count(1)?

7.5.30 For your 3-bit gray code up/down counter, what is the output logic expression for Count(0)?

7.5.31 For your 3-bit gray code up/down counter, provide the logic diagram.

7.5.32 Design a Verilog model for a 3-bit gray code up/down counter. Instantiate the D-flip-flop model provided to implement your state memory. Use whatever concurrent modeling approach you wish to model the next state and output logic. Use the module port definition provided in Fig. 7.51 for your design.

counter_3bit_graycode_up_down.v

```
module counter_3bit_graycode_up_down
    (output wire Count[2:0],
     input  wire Clock, Reset,
     input  wire Up);
                  :
endmodule
```

Fig. 7.51
3-Bit gray code up/down counter module definition

The next set of questions are about the design a 4-bit one-hot up counter by hand. This counter will create a pattern of 0001→0010→0100→1000 and then repeat. This FSM should use state-encoded outputs to simplify the output logic. This state machine will need four states and require four bits for the state variable codes. Name the current state variables Q3_cur, Q2_cur, Q1_cur, and Q0_cur and the next state variables Q3_nxt, Q2_nxt, Q1_nxt, and Q0_nxt. The output of your counter will be a 4-bit vector called *Count*.

7.5.33 For your 4-bit one-hot up counter, what is the next state logic expression for Q3_nxt?

7.5.34 For your 4-bit one-hot up counter, what is the next state logic expression for Q2_nxt?

7.5.35 For your 4-bit one-hot up counter, what is the next state logic expression for Q1_nxt?

7.5.36 For your 4-bit one-hot up counter, what is the next state logic expression for Q0_nxt?

7.5.37 For your 4-bit one-hot up counter, what is the output logic expression for Count(3)?

7.5.38 For your 4-bit one-hot up counter, what is the output logic expression for Count(2)?

7.5.39 For your 4-bit one-hot up counter, what is the output logic expression for Count(1)?

7.5.40 For your 4-bit one-hot up counter, what is the output logic expression for Count(0)?

7.5.41 For your 4-bit one-hot up counter, provide the logic diagram.

7.5.42 Design a Verilog model for a 4-bit one-hot up counter. Instantiate the D-flip-flop model provided to implement your state memory. Use whatever concurrent signal assignment modeling approach you wish to model the next state and output logic. Use the Verilog module definition provided in Fig. 7.52 for your design.

counter_4bit_onehot_up.v

```
module counter_4bit_onehot_up
    (output wire Count[3:0],
     input  wire Clock, Reset);
                     :
endmodule
```

Fig. 7.52
4-Bit one-hot up counter module description

The next set of questions are about the design of a binary counter that goes from 0000_2 (0_{10}) to 1001_2 (9_{10}) by hand. This counter will create a pattern of 0000→0001→0010→0011→0100→0101→0110→0111→1000→1001 and then repeat. This type of counter is known as a *binary-coded decimal* (BCD) counter. This FSM should use state-encoded outputs to simplify the output logic. This state machine will need ten states and require four bits for the state variable codes. Name the current state variables Q3_cur, Q2_cur, Q1_cur, and Q0_cur and the next state variables Q3_nxt, Q2_nxt, Q1_nxt, and Q0_nxt. The output of your counter will be a 4-bit vector called *Count.*

7.5.43 For your BCD counter, what is the next state logic expression for Q3_nxt?

7.5.44 For your BCD counter, what is the next state logic expression for Q2_nxt?

7.5.45 For your BCD counter, what is the next state logic expression for Q1_nxt?

7.5.46 For your BCD counter, what is the next state logic expression for Q0_nxt?

7.5.47 For your BCD counter, what is the output logic expression for Count(3)?

7.5.48 For your BCD counter, what is the output logic expression for Count(2)?

7.5.49 For your BCD counter, what is the output logic expression for Count(1)?

7.5.50 For your BCD counter, what is the output logic expression for Count(0)?

7.5.51 For your BCD counter, provide the logic diagram.

Section 7.6: Finite-State Machine's Reset Condition

7.6.1 Are resets typically synchronous or asynchronous?

7.6.2 Why is it necessary to have a reset/preset condition in a finite-state machine?

7.6.3 How does the reset/preset condition correspond to the behavior described in the state diagram?

7.6.4 When is it necessary to also use the preset line (s) of a D-flip-flop instead of just the reset line (s) when implementing the state memory of a finite-state machine?

7.6.5 If a finite-state machine has eight unique states that are encoded in binary and all D-flip-flops used for the state memory use their reset lines, what is the state code that the machine will go to upon reset?

Section 7.7: Sequential Logic Analysis

7.7.1 For the finite state machine logic diagram in Fig. 7.53, give the next state logic expression for *Q_nxt*.

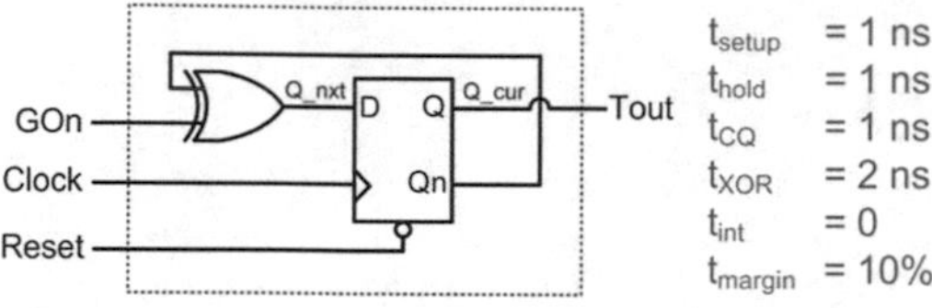

Fig. 7.53
Sequential logic analysis 1

7.7.2 For the finite-state machine logic diagram in Fig. 7.53, give the output logic expression for *Tout*.

7.7.3 For the finite-state machine logic diagram in Fig. 7.53, give the state transition table.

7.7.4 For the finite-state machine logic diagram in Fig. 7.53, give the state diagram.

7.7.5 For the finite-state machine logic diagram in Fig. 7.53, give the maximum clock frequency.

7.7.6 For the finite-state machine logic diagram in Fig. 7.54, give the next state logic expression for *Q_nxt*.

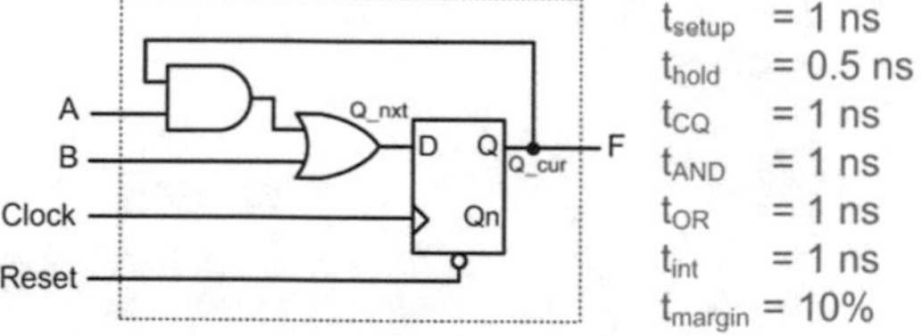

Fig. 7.54
Sequential logic analysis 2

7.7.7 For the finite-state machine logic diagram in Fig. 7.54, give the output logic expression for *F*.

7.7.8 For the finite-state machine logic diagram in Fig. 7.54, give the state transition table.

7.7.9 For the finite-state machine logic diagram in Fig. 7.54, give the state diagram.

7.7.10 For the finite-state machine logic diagram in Fig. 7.54, give the maximum clock frequency.

7.7.11 For the finite-state machine logic diagram in Fig. 7.55, give the next state logic expressions for *Q1_nxt* and *Q0_nxt*.

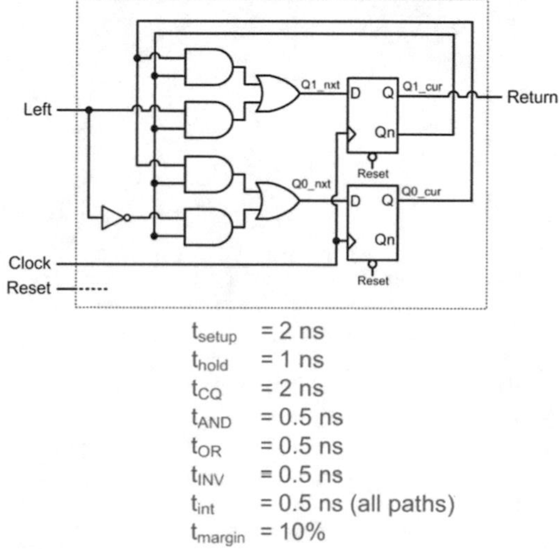

Fig. 7.55
Sequential logic analysis 3

7.7.12 For the finite-state machine logic diagram in Fig. 7.55, give the output logic expression for *Return*.

7.7.13 For the finite-state machine logic diagram in Fig. 7.55, give the state transition table.

7.7.14 For the finite-state machine logic diagram in Fig. 7.55, give the state diagram.

7.7.15 For the finite-state machine logic diagram in Fig. 7.55, give the maximum clock frequency.

Chapter 8: Verilog (Part 2)

In Chap. 5, Verilog was presented as a way to describe the behavior of concurrent systems. The modeling techniques presented were appropriate for combinational logic because these types of circuits have outputs dependent only on the current values of their inputs. This means a model that continuously performs signal assignments provides an accurate model of this circuit behavior. In Chap. 7, sequential logic storage devices were presented that did not continuously update their outputs based on the instantaneous values of their inputs. Instead, sequential storage devices only update their outputs based upon an event, most often the edge of a clock signal. The modeling techniques presented in Chap. 5 are unable to accurately describe this type of behavior. In this chapter, we describe the Verilog constructs to model signal assignments that are triggered by an event in order to accurately model sequential logic. We can then use these techniques to describe more complex sequential logic circuits such as finite state machines and register transfer level systems. This chapter will also present how to create test benches and look at more advanced features that are commonly used in Verilog to model modern systems. The goal of this chapter is to give an understanding of the full capability of hardware description languages.

Learning Outcomes—After completing this chapter, you will be able to:

8.1 Describe the behavior of Verilog procedural assignment and how they are used to model sequential logic circuits.
8.2 Model combinational logic circuits using a Verilog procedural assignment and conditional programming constructs.
8.3 Describe the functionality of common Verilog system tasks.
8.4 Design a Verilog test bench to verify the functional operation of a system.

8.1 Procedural Assignment

Verilog uses *procedural assignment* to model signal assignments that are based on an event. An *event* is most commonly a transition of a signal. This provides the ability to model sequential logic circuits such as D-flip-flops and finite state machines by triggering assignments off of a clock edge. Procedural assignments can only drive variable data types (i.e., reg, integer, real, and time); thus, they are ideal for modeling storage devices. Procedural signal assignments can be evaluated in the order they are listed; thus, they are able to model sequential assignments.

A procedural assignment can also be used to model combinational logic circuits by making signal assignments when any of the inputs to the model change. Despite the left-hand side of the assignment not being able to be of type wire in procedural assignment, modern synthesizers will recognize properly designed combinational logic models and produce the correct circuit implementation. Procedural assignment also supports standard programming constructs such as if-else decisions, case statements, and loops. This makes procedural assignment a powerful modeling approach in Verilog and is the most common technique for designing digital systems and creating test benches.

8.1.1 Procedural Blocks

All procedural signal assignments must be enclosed within a procedural *block*. Verilog has two types of procedural blocks, *initial* and *always*.

© Springer Nature Switzerland AG 2019
B. J. LaMeres, *Introduction to Logic Circuits & Logic Design with Verilog*,
https://doi.org/10.1007/978-3-030-13605-5_8

8.1.1.1 Initial Blocks

An initial block will execute all of the statements embedded within it one time at the beginning of the simulation. An initial block is not used to model synthesizable behavior. It is instead used within test benches to either set the initial values of repetitive signals or to model the behavior of a signal that only has a single set of transitions. The following is the syntax for an initial block.

```
initial
  begin                  // an optional ": name" can be added after the begin keyword
    signal_assignment_1
    signal_assignment_2
              :
   end
```

Let's look at a simple model of how an initial block is used to model the reset line in a test bench. In the following example, the signal "Reset_TB" is being driven into a DUT. At the beginning of the simulation, the initial value of Reset_TB is set to a logic zero. The second assignment will take place after a delay of 15 time units. The second assignment statement sets Reset_TB to a logic one. The assignments in this example are evaluated in sequence in the order they are listed due to the delay operator. Since the initial block executes only once, Reset_TB will stay at the value of its last assignment for the remainder of the simulation.

Example:

```
initial
 begin
     Reset_TB = 1'b0;
  #15 Reset_TB = 1'b1;
 end
```

8.1.1.2 Always Blocks

An *always* block will execute forever or for the duration of the simulation. An always block can be used to model synthesizable circuits in addition to non-synthesizable behavior in test benches. The following is the syntax for an always block.

```
always
  begin
        signal_assignment_1
        signal_assignment_2
                :
   end
```

Let's look at a simple model of how an always block can be used to model a clock line in a test bench. In the following example, the value of the signal Clock_TB will continuously change its logic value every 10 time units.

Example:

```
always
   begin
     #10 Clock_TB = ~Clock_TB;
   end
```

By itself, the above always block will not work because when the simulation begins, Clock_TB does not have an initial value so the simulator will not know what the value of Clock_TB is at time zero. It will also not know what the output of the negation operation (~) will be at time unit 10. The following example shows the correct way of modeling a clock signal using a combination of initial and always blocks. Verilog allows assignments to the same variable from multiple procedural blocks, so the following example is valid. Note that when the simulation begins, Clock_TB is assigned a logic zero. This provides a known value for the signal at time zero and also allows the always block negation to have a deterministic value. The example below will create a clock signal that will toggle every 10 time units.

Example:

```
initial
  begin
    Clock_TB = 1'b0;
  end

always
  begin
    #10 Clock_TB = ~Clock_TB;
  end
```

8.1.1.3 Sensitivity Lists

A *sensitivity list* is used in conjunction with a procedural block to trigger when the assignments within the block are executed. The symbol @ is used to indicate a sensitivity list. Signals can then be listed within parenthesis after the @ symbol that will trigger the procedural block. The following is the base syntax for a sensitivity list.

```
always @ (signal1, signal2)
 begin
  signal_assignment_1
  signal_assignment_2
      :
 end
```

In this syntax, any transition on any of the signals listed within the parenthesis will cause the always block to trigger and all of its assignments to take place one time. After the always block ends, it will await the next signal transition in the sensitivity list to trigger again. The following example shows how to model a simple 3-input AND gate. In this example, any transition on inputs A, B, or C will cause the block to trigger and the assignment to F to occur.

Example:

```
always @ (A, B, C)
  begin
    F = A & B & C;
  end
```

Verilog also supports keywords to limit triggering of the block to only rising edge or falling edge transitions. The keywords are **posedge** and **negedge**. The following is the base syntax for an edge sensitive block. In this syntax, only rising edge transitions on signal1 or falling edge transitions on signal2 will cause the block to trigger.

```
always @ (posedge signal1, negedge signal2)
  begin
    signal_assignment_1
    signal_assignment_2
              :
  end
```

Sensitivity lists can also contain Boolean operators to more explicitly describe behavior. The following syntax is identical to the syntax above.

```
always @ (posedge signal1 or negedge signal2)
  begin
    signal_assignment_1
    signal_assignment_2
              :
  end
```

The ability to model edge sensitivity allows us to model sequential circuits. The following example shows how to model a simple D-flip-flop.

Example:

```
always @ (posedge Clock)
 begin
   Q = D;       // Note: This model does not include a reset.
 end
```

In Verilog-2001, the syntax to support sensitivity lists that will trigger based on any signal listed on the right-hand side of any assignment within the block was added. This syntax is **@***. The following example how to use this modeling approach to model a 3-input AND gate.

Example:

```
always @*
  begin
    F = A & B & C;
  end
```

8.1.2 Procedural Statements

There are two kinds of signal assignments that can be used within a procedural block, **blocking** and **non-blocking**.

8.1.2.1 Blocking Assignments

A *blocking assignment* is denoted with the = symbol, and the evaluation and assignment of each statement take place immediately. Each assignment within the block is executed in parallel. When this behavior is coupled with a sensitivity list that contains all of the inputs to the system, this approach can model synthesizable combinational logic circuits. This approach provides the same functionality as continuous assignments outside of a procedural block. The reason that designers use blocking assignments instead of continuous assignment is that more advanced programming constructs are supported within Verilog procedural blocks. These will be covered in the next section. Example 8.1 shows how to use blocking assignments within a procedural block to model a combinational logic circuit.

Example: Using Blocking Assignments to Model Combinational Logic

In this model, each of the inputs A, B, and C are listed in the sensitivity list so that the procedural block is triggered on any input transition. When using blocking assignments, the assignments inside of the block are evaluated and executed immediately. These two behaviors allow us to model combinational logic.

```
module BlockingEx1 (output reg  S,
                    input  wire A, B, C);

  reg  n1;

  always @ (A, B, C)
   begin
     n1 = A ^ B;      // statement 1
     S  = n1 ^ C;     // statement 2
   end

endmodule
```

Resulting Circuit

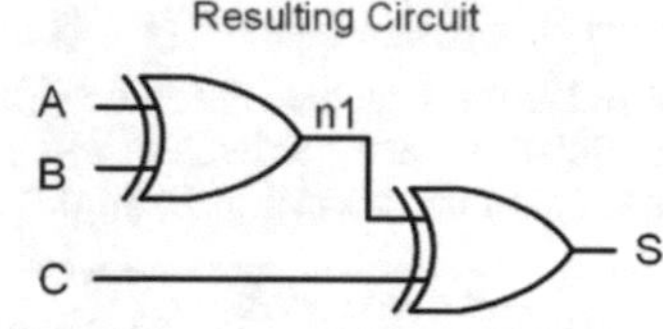

Both statement 1 and statement 2 are treated as separate circuits that execute concurrently when using blocking assignments.

Example 8.1
Using blocking assignments to model combinational logic

8.1.2.2 Non-blocking Assignments

A *non-blocking assignment* is denoted with the <= symbol. When using non-blocking assignments, the assignment to the target signal is deferred until the end of the procedural block. This allows the assignments to be executed in the order they are listed in the block without cascading interim assignments through the list. When this behavior is coupled with triggering the block off of a clock signal, this approach can model synthesizable sequential logic circuits. Example 8.2 shows an example of using non-blocking assignments to model a sequential logic circuit.

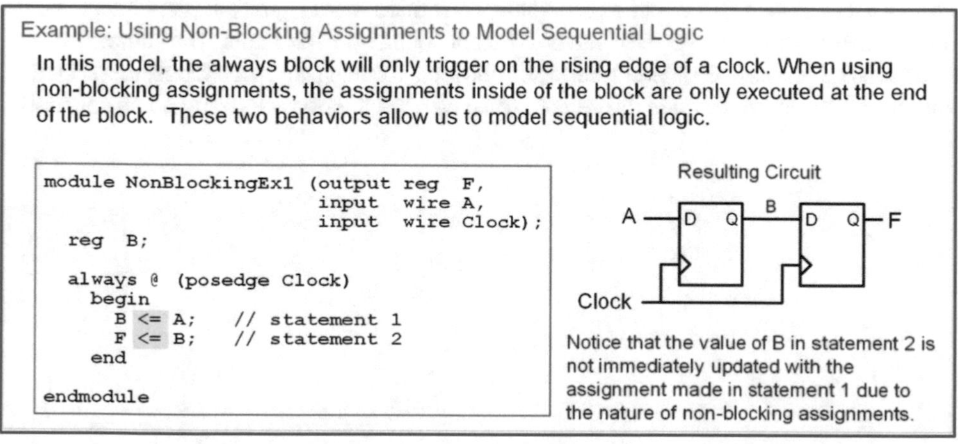

Example: Using Non-Blocking Assignments to Model Sequential Logic

In this model, the always block will only trigger on the rising edge of a clock. When using non-blocking assignments, the assignments inside of the block are only executed at the end of the block. These two behaviors allow us to model sequential logic.

```
module NonBlockingEx1 (output reg  F,
                       input  wire A,
                       input  wire Clock);
  reg  B;

  always @ (posedge Clock)
    begin
      B <= A;    // statement 1
      F <= B;    // statement 2
    end

endmodule
```


Notice that the value of B in statement 2 is not immediately updated with the assignment made in statement 1 due to the nature of non-blocking assignments.

Example 8.2
Using non-blocking assignments to model sequential logic

The difference between blocking and non-blocking assignments is subtle and is often one of the most difficult concepts to grasp when first learning Verilog. One source of confusion comes from the fact that blocking and non-blocking assignments *can* produce the same results when they either contains a single assignment or a list of assignments that don't have any signal interdependencies. A *signal interdependency* refers to when a signal that is the target of an assignment (i.e., on the LHS of an assignment) is used as an argument (i.e., on the RHS of an assignment) in subsequent statements. Example 8.3 shows two models that produce the same results regardless of whether a blocking or non-blocking assignment is used.

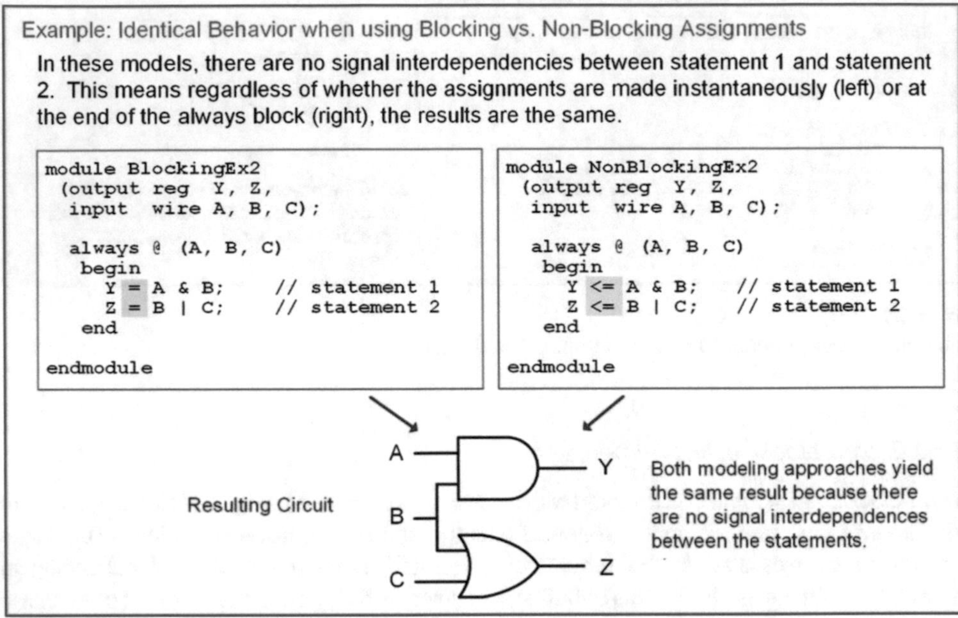

Example 8.3
Identical behavior when using blocking vs. non-blocking assignments

When a list of statements within a procedural block *does* have signal interdependencies, blocking and non-blocking assignments will have different behavior. Example 8.4 shows how signal interdependencies will cause different behavior between blocking and non-blocking assignments. In this example, all inputs are listed in the sensitivity list with the intent of modeling combinational logic.

Example: Different Behavior when using Blocking vs. Non-Blocking Assignments (1)

In these examples, there is a signal interdependency and all of the inputs are listed in the sensitivity list. By listing all of the inputs in the sensitivity list, the intent is to model combinational logic such that the outputs update anytime there is a change on the inputs.

CORRECT

```
module BlockingEx3
 (output reg  S,
  input  wire A, B, C);

  reg  n1;

  always @ (A, B, C)
   begin
     n1 = A ^ B;    // statement 1
     S  = n1 ^ C;   // statement 2
   end
```

```
module NonBlockingEx3
 (output reg  S,
  input  wire A, B, C);

  reg  n1;

  always @ (A, B, C)
   begin
     n1 <= A ^ B;  // statement 1
     S  <= n1 ^ C; // statement 2
   end
```

In both cases, statement 1 (the assignment to **n1**) will produce the same result. This is because the assignment only depends on the inputs A and B. Since A and B are listed in the sensitivity list, any change on them will trigger the block and their current value will be used in the assignment to n1.

However, statement 2 (the assignment to **S**) contains a signal interdependency and will NOT produce the same result in both cases.

Blocking Assignment Case (=): Blocking assignments take place immediately. This means that the assignment to n1 in statement 1 takes place immediately and the updated value of n1 is used in statement 2. When used in conjunction with listing all of the inputs in the sensitivity list, this approach successfully models **combinational logic**.

Non-Blocking Assignment Case (<=): Non-blocking assignments take place at the end of the procedural block. This means that the assignment to n1 in statement 1 does not take place before the assignment in statement 2. The value of n1 used in statement 2 will be the value of n1 when the block is triggered, not the new value of n1 assigned within the block. Said another way, the value of n1 used in statement 2 will be the value of n1 from the "prior" time the block triggered, or the "last" value of n1. This does **not model combinational logic**.

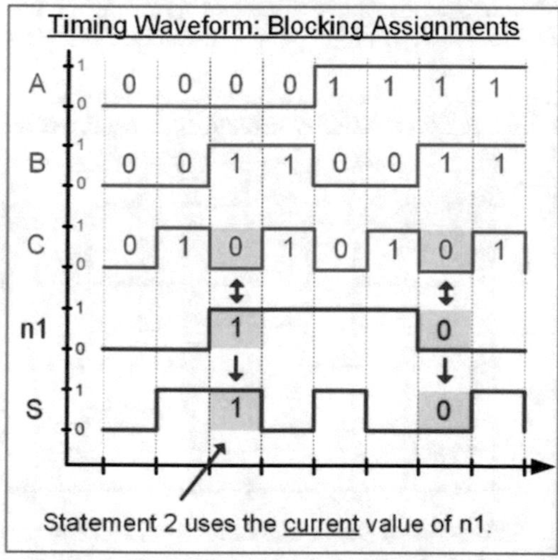

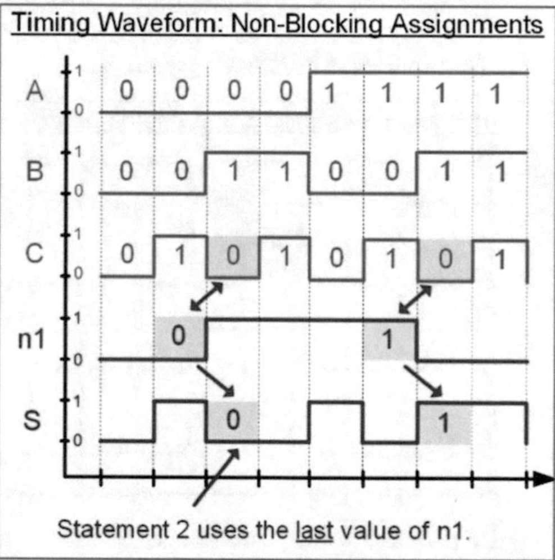

Example 8.4
Different behavior when using blocking vs. non-blocking assignments (1)

Example 8.5 shows another case where signal interdependencies will cause different behavior between blocking and non-blocking assignments. In this example, the procedural block is triggered by the rising edge of a clock signal with the intent of modeling two stages of sequential logic.

Example: Different Behavior when using Blocking vs. Non-Blocking Assignments (2)

In these examples, there is a signal interdependency and the sensitivity list is triggered by the edge of a clock. The intent of this model is to create a 2-stage sequential logic circuit such that the outputs only update when there is a rising edge of the clock.

```
module BlockingEx4
 (output reg  F,
  input  wire A,
  input  wire Clock);

  reg  B;

  always @  (posedge Clock)
    begin
      B = A;    // statement 1
      F = B;    // statement 2
    end

endmodule
```

CORRECT

```
module NonBlockingEx4
 (output reg  F,
  input  wire A,
  input  wire Clock);

  reg  B;

  always @  (posedge Clock)
    begin
      B <= A;    // statement 1
      F <= B;    // statement 2
    end

endmodule
```

In both cases, the procedural block will trigger on the rising edge of a clock. Also in each case, the assignment in statement 1 will produce the same result.

However, statement 2 has a signal dependency and will NOT produce the same result in both cases.

Blocking Assignment Case (=): Blocking assignments take place immediately. This means that the assignment of A to B and then B to F will result in simply F = A. This will still result in sequential logic, but the output F will be updated with the input A on every rising edge of clock.

Non-Blocking Assignment Case (<=): Non-blocking assignments take place at the end of the procedural block. This means that an input on A will be stored to B on rising edge of clock, but not to F on the same edge. Instead, the subsequent clock edge will cause the result stored in B to be stored to F. This will result in sequential logic that has two edge triggered storage elements instead of just one as in the blocking example.

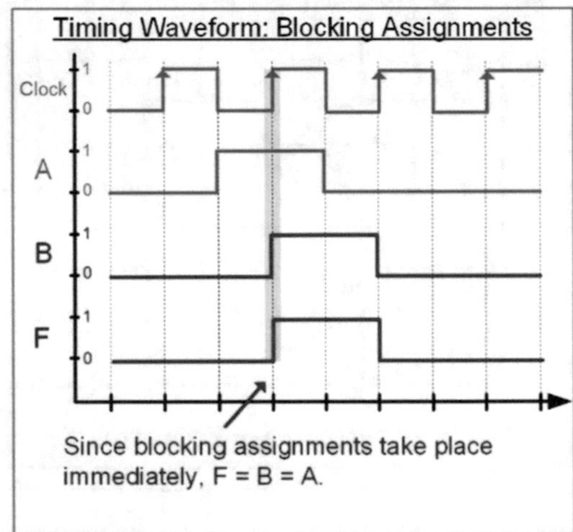

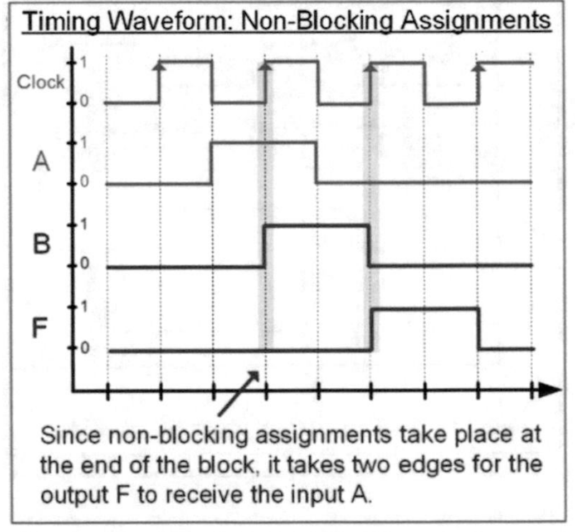

Example 8.5
Different behavior when using blocking vs. non-blocking assignments (2)

While the behavior of these procedural assignments can be confusing, there are two design guidelines that can make creating accurate, synthesizable models straightforward. They are:

1. When modeling **combinational logic**, use *blocking assignments* and *list every input in the sensitivity list.*
2. When modeling **sequential logic**, use *non-blocking assignments* and *only list the clock and reset lines* (if applicable) in the sensitivity list.

8.1.3 Statement Groups

A statement group refers to how the statements in a block are processed. Verilog supports two types of statement groups: **begin/end** and **fork/join**. When using begin/end, all statements enclosed within the group will be evaluated in the order they are listed. When using a fork/join, all statements enclosed within the group will be evaluated in parallel. When there is only one statement within procedural block, a statement group is not needed. For multiple statements in a procedural block, a statement group is required. Statement groups can contain an optional name that is appended after the first keyword preceded by a ":." Example 8.6 shows a graphical depiction of the difference between begin/end and fork/join groups. Note that this example also shows the syntax for naming the statement groups.

Example: Behavior of Statement Groups begin/end vs. fork/join

When using the statement group begin/end, all statements are evaluated in sequence.
When using the statement group fork/join, all statements are executed in parallel.

```
module StatementGroupEx1 ();

  reg [7:0] S_TB;

  initial
    begin: Ex1              // group name
           S_TB = 8'h00;
      #10  S_TB = 8'h55;
      #15  S_TB = 8'hAA;
    end

endmodule
```

```
module StatementGroupEx2 ();

  reg [7:0] S_TB;

  initial
    fork: Ex2               // group name
           S_TB = 8'h00;
      #10  S_TB = 8'h55;
      #15  S_TB = 8'hAA;
    join

endmodule
```

In the begin/end example, the statements are executed in order. This treats the delays as a sequence. At time 10, the signal S_TB is assigned 8'h55. Then 15 time units **later**, it is assigned 8'hAA. The assignment of 8'hAA takes place at absolute time 25.

In the fork/join example, the statements are executed in parallel. This treats the delays as taking place in absolute time. At absolute time unit 10, the signal S_TB is assigned 8'h55. At absolute time unit 15, the signal S_TB is assigned 8'hAA.

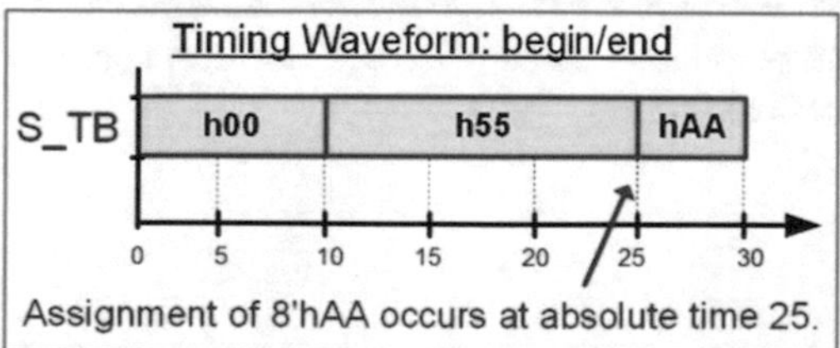

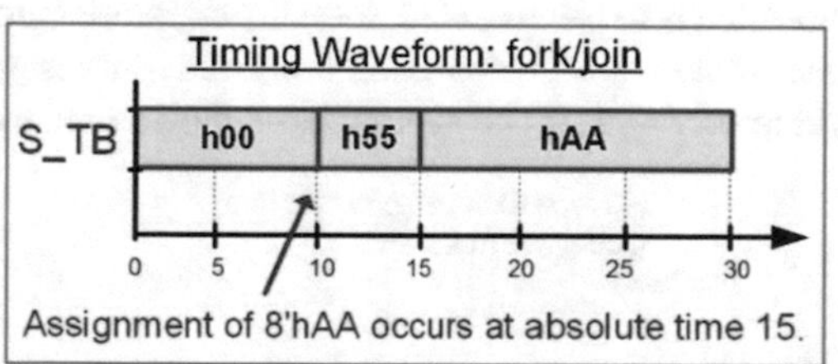

Example 8.6
Behavior of statement groups begin/end vs. fork/join

8.1.4 Local Variables

Local variables can be declared within a procedural block. The statement group must be named, and the variables will not be visible outside of the block. Variables can only be of variable type.

Example:

```
initial
 begin: stim_block    // it is required to name the block when declaring local variables
  integer i;          // local variables can only be of variable type
  i=2;
 end
```

Concept Check

CC8.1 If a model of a combinational logic circuit excludes one of its inputs from the sensitivity list, what is the implied behavior?

A) A storage element because the output will be held at its last value when the unlisted input transitions.

B) An infinite loop.

C) A don't care will be used to form the minimal logic expression.

D) Not applicable because this syntax will not compile.

8.2 Conditional Programming Constructs

One of the more powerful features that procedural blocks provide in Verilog is the ability to use conditional programming constructs such as if-else decisions, case statements, and loops. These constructs are only available within a procedural block and can be used to model both combinational and sequential logic.

8.2.1 if-else Statements

An *if-else* statement provides a way to make conditional signal assignments based on Boolean conditions. The **if** portion of statement is followed by a Boolean condition that if evaluated TRUE will cause the signal assignment listed after it to be performed. If the Boolean condition is evaluated FALSE, the statements listed after the **else** portion are executed. If multiple statements are to be executed in either the if or else portion, then the statement group keywords begin/end need to be used. If only one statement is to be executed, then the statement group keywords are not needed. The else portion of the statement is not required, and if omitted, no assignment will take place when the Boolean condition is evaluated FALSE. The syntax for an if-else statement is as follows:

```
if (<boolean_condition>)
   true_statement
else
   false_statement
```

The syntax for an if-else statement with multiple true/false statements is as follows:

```
if (<boolean_condition>)
  begin
   true_statement_1
   true_statement_2
  end
else
  begin
   false_statement_1
   false_statement_2
  end
```

If more than one Boolean condition is required, additional if-else statements can be embedded within the *else* clause of the preceding if statement. The following shows an example of if-else statements implementing two Boolean conditions.

```
if (<boolean_condition_1>)
   true_statement_1
else if (<boolean_condition_2>)
   true_statement_2
else
   false_statement
```

Let's look at using an if-else statement to describe the behavior of a combinational logic circuit. Recall that a combinational logic circuit is one in which the output depends on the instantaneous values of the inputs. This behavior can be modeled by placing all of the inputs to the circuit in the sensitivity list of an always block and using blocking assignments. Using this approach, a change on any of the inputs in the sensitivity list will trigger the block, and the assignments will take place immediately. Example 8.7 shows how to model a 3-input combinational logic circuit using if-else statements within a procedural always block.

Example: Using If-Else Statements to Model Combinational Logic

Implement the following truth table using an if-else statement within a procedural block.

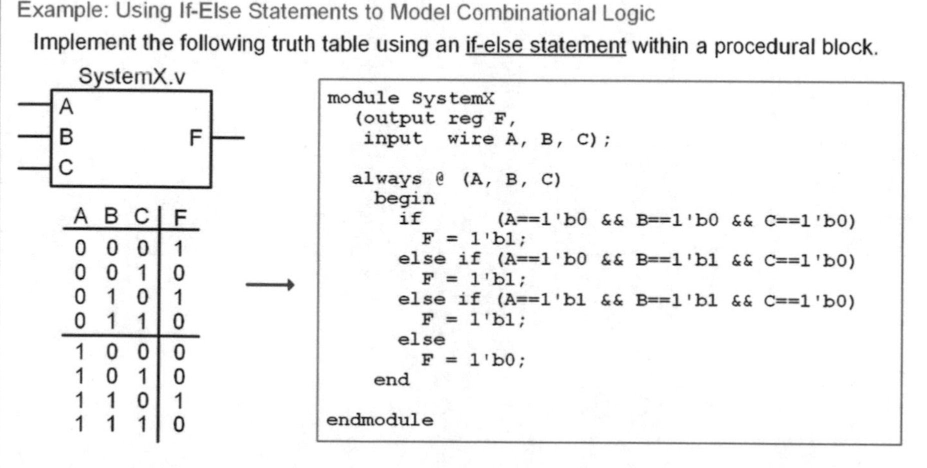

A	B	C	F
0	0	0	1
0	0	1	0
0	1	0	1
0	1	1	0
1	0	0	0
1	0	1	0
1	1	0	1
1	1	1	0

```
module SystemX
  (output reg F,
   input  wire A, B, C);

  always @ (A, B, C)
    begin
      if       (A==1'b0 && B==1'b0 && C==1'b0)
        F = 1'b1;
      else if (A==1'b0 && B==1'b1 && C==1'b0)
        F = 1'b1;
      else if (A==1'b1 && B==1'b1 && C==1'b0)
        F = 1'b1;
      else
        F = 1'b0;
    end

endmodule
```

When modeling combinational logic using a procedural assignment, all of the inputs to the circuit must be listed in the sensitivity list and blocking assignments are used.

In this model, three nested if-else statements are used to explicitly describe the input conditions corresponding to an output of a one. For all other input codes, the else clause is used to drive the output to a zero.

Example 8.7
Using if-else statements to model combinational logic

8.2.2 case Statements

A *case* statement is another technique to model signal assignments based on Boolean conditions. As with the if-else statement, a case statement can only be used inside of a procedural block. The statement begins with the keyword **case** followed by the input signal name that assignments will be based off of enclosed within parenthesis. The case statement can be based on multiple input signal names by concatenating the signals within the parenthesis. Then a series of input codes followed by the corresponding assignment is listed. The keyword **default** can be used to provide the desired signal assignment for any input codes not explicitly listed. When multiple input conditions have the same assignment statement, they can be listed on the same line comma-delimited to save space. The keyword **endcase** is used to denote the end of the case statement. The following is the syntax for a case statement.

```
case (<input_name>)
   input_val_1 : statement_1
   input_val_2 : statement_2
                 :
   input_val_n : statement_n
   default     : default_statement
endcase
```

Example 8.8 shows how to model a 3-input combinational logic circuit using a case statement within a procedural block. Note in this example the inputs are scalars, so they must be concatenated so that the input values can be listed as 3-bit vectors. In this example, there are three versions of the model provided. The first explicitly lists out all binary input codes. This approach is more readable because it mirrors a truth table form. The second approach only lists the input codes corresponding to an output of one and uses the default clause to handle all other input codes. The third approach shows how to list multiple input codes with the same assignment on the same line using a comma-delimited series.

Example: Using Case Statements to Model Combinational Logic

Implement the following truth table using a <u>case statement</u> within a procedural block.

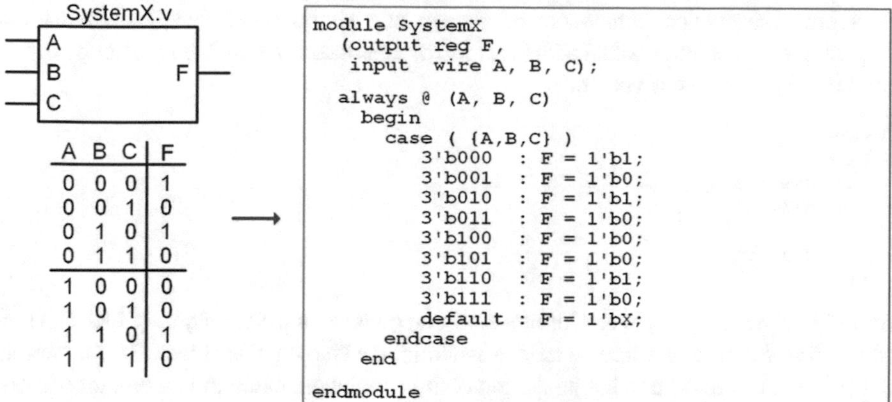

A	B	C	F
0	0	0	1
0	0	1	0
0	1	0	1
0	1	1	0
1	0	0	0
1	0	1	0
1	1	0	1
1	1	1	0

```
module SystemX
  (output reg F,
   input  wire A, B, C);

  always @ (A, B, C)
    begin
      case ( {A,B,C} )
        3'b000   : F = 1'b1;
        3'b001   : F = 1'b0;
        3'b010   : F = 1'b1;
        3'b011   : F = 1'b0;
        3'b100   : F = 1'b0;
        3'b101   : F = 1'b0;
        3'b110   : F = 1'b1;
        3'b111   : F = 1'b0;
        default  : F = 1'bX;
      endcase
    end

endmodule
```

In this model, each binary input code is explicitly listed to create a model that mirrors the truth table format. Note that since the inputs are scalars, they must be concatenated so that 3-bit vectors can be listed as the input conditions. A default condition is needed to provide the output assignment for input codes containing X or Z.

Below are two alternative approaches of using a case statement that are more compact.

```
case ( {A,B,C} )
   3'b000   : F = 1'b1;
   3'b010   : F = 1'b1;
   3'b110   : F = 1'b1;
   default  : F = 1'b0;
endcase
```

In this approach, only the input codes corresponding to an output of one are explicitly listed. The default clause is used to handle all other input codes corresponding to an output of zero.

```
case ( {A,B,C} )
    3'b000, 3'b010, 3'b111   : F = 1'b1;
    default                  : F = 1'b0;
endcase
```

In this model, the input codes corresponding to the output assignment of one are listed on the same line, comma-delimited.

Example 8.8
Using case statements to model combinational logic

If-else statements can be embedded within a case statement, and, conversely, case statements can be embedded within an if-else statement.

8.2.3 casez and casex Statements

Verilog provides two additional case statements that support don't cares in the input conditions. The **casez** statement allows the symbols **?** and **Z** to represent a don't care. The **casex** statement extends the casez statement by also interpreting **X** as a don't care. Care should be taken when using the casez and casex statement as it is easy to create unintended logic when using don't cares in the input codes.

8.2.4 forever Loops

A *loop* within Verilog provides a mechanism to perform repetitive assignments infinitely. This is useful in test benches for creating stimulus such as clocks or other periodic signals. We have already covered a looping construct in the form of an always block. An always block provides a loop with a starting condition. Verilog provides additional looping constructs to model more sophisticated behavior. All looping constructs must reside with a procedural block.

The simplest looping construct is the **forever** loop. As with other conditional programming constructs, if multiple statements are associated with the forever loop, they must be enclosed within a statement group. If only one statement is used, the statement group is not needed. A forever loop within an initial block provides identical behavior as an always loop without a sensitivity loop. It is important to provide a time step event or delay within a forever loop or it will cause a simulation to hang. The following is the syntax for a forever loop in Verilog.

```
forever
  begin
    statement_1
    statement_2
         :
    statement_n
  end
```

Consider the following example of a forever loop that generates a clock signal (CLK) with a period of 10 time units. In this example, the forever loop is embedded within an initial block. This allows the initial value of CLK to be set to zero upon the beginning of the simulation. Once the forever loop is entered, it will execute indefinitely. Notice that since there is only one statement after the forever keyword, a statement group (i.e., begin/end) is not needed.

Example:

```
initial
  begin
    CLK = 0;

    forever
      #10 CLK = ~CLK;

  end
```

8.2.5 while Loops

A **while** loop provides a looping structure with a Boolean condition that controls its execution. The loop will only execute as long as the Boolean condition is evaluated true. The following is the syntax for a Verilog while loop.

```
while (<boolean_condition>)
  begin
   statement_1
   statement_2
        :
   statement_n
  end
```

Let's implement the previous example of a loop that generates a clock signal (CLK) with a period of 10 time units as long as EN = 1. The TRUE Boolean condition for the while loop is EN = 1. When EN = 0, the while loop will be skipped. When the loop becomes inactive, CLK will hold its last assigned value.

Example:

```
initial
  begin
    CLK = 0;

    while (EN == 1)
       #10 CLK = ~CLK;

  end
```

8.2.6 repeat Loops

A **repeat** loop provides a looping structure that will execute a fixed number of times. The following is the syntax for a Verilog repeat loop.

```
repeat (<number_of_loops>)
  begin
    statement_1
    statement_2
        :
    statement_n
  end
```

Let's implement the previous example of a loop that generates a clock signal (CLK) with a period of 10 time units, except this time we'll use a repeat loop to only produce ten clock transitions, or five full periods of CLK.

Example:

```
initial
  begin
    CLK = 0;
    repeat (10)
      #10 CLK = ~CLK;
  end
```

8.2.7 for loops

A **for** loop provides the ability to create a loop that can automatically update an internal variable. A *loop variable* within a for loop is altered each time through the loop according to a *step assignment*. The starting value of the loop variable is provided using an *initial assignment*. The loop will execute as long as a Boolean condition associated with the loop variable is TRUE. The following is the syntax for a Verilog for loop:

```
for (<initial_assignment>; <Boolean_condition>; <step_assignment>)
  begin
    statement_1
    statement_2
        :
    statement_n
  end
```

The following is an example of creating a simple counter using the loop variable. The loop variable *i* was declared as an integer prior to this block. The signal count is also of type integer. The loop variable will start at 0 and increment by 1 each time through the loop. The loop will execute as long as i < 15 or 16 times total. For loops allow the loop variable to be used in signal assignments within the block.

Example:

```
initial
  begin
    for (i=0; i<15; i=i+1)
      #10 Count = i;
  end
```

8.2.8 disable

Verilog provides the ability to stop a loop using the keyword **disable**. The disable function only works on named statement groups. The disable function is typically used after a certain fixed amount of time or within a conditional construct such as an if-else or case statement that is triggered by a control signal.

Consider the following forever loop example that will generate a clock signal (CLK) but only when an enable (EN) is asserted. When EN = 0, the loop will disable, and the simulation will end.

Example:

```
initial
  begin
     CLK = 0;
     forever
        begin: loop_ex
          if (EN == 1)
            #10 CLK = ~CLK;
          else
            disable loop_ex; // The group name to be disabled comes after the keyword
        end
  end
```

Concept Check

CC8.2 When using an if-else statement to model a combinational logic circuit, is using the *else* clause the same as using *don't cares* when minimizing a logic expression with a K-map?

A) Yes. The else clause allows the synthesizer to assign whatever output values are necessary in order to create the most minimal circuit.

B) No. The else clause explicitly states the output values for all input codes not listed in the if portion of the statement. This is the same as filling in the truth table with specific values for all input codes covered by the else clause, and the synthesizer will create the logic expression accordingly.

8.3 System Tasks

A system task in Verilog is one that is used to insert additional functionality into a model that is not associated with real circuitry. There are three main groups of system tasks in Verilog: (1) text output, (2) file input/output, and (3) simulation control. All system tasks begin with a **$** and are only used during simulation. These tasks are ignored by synthesizers, so they can be included in real circuit models. All system tasks must reside within procedural blocks.

8.3.1 Text Output

Text output system tasks are used to print strings and variable values to the console or transcript of a simulation tool. The syntax follows ANSI C where double quotes (“”) are used to denote the text string to be printed. Standard text can be entered within the string in addition to variables. Variable can be printed in two ways. The first is to simply list the variable in the system task function outside of the double quotes. In this usage, the default format to be printed will be decimal unless a task is used with a different default format. The second way to print a variable is within a text string. In this usage, a unique code is inserted into the string indicating the format of how to print the value. After the string, a comma-separated list of the variable name(s) is listed that corresponds positionally to the codes within the string. The following are the most commonly used text output system tasks:

Task	Description
$display()	Print text string when statement is encountered *and* append a newline.
$displayb()	Same as $display, but default format of any arguments is binary.
$displayo()	Same as $display, but default format of any arguments is octal.
$displayh()	Same as $display, but default format of any arguments is hexadecimal.
$write()	Same as $display, but the string is printed *without* a newline.
$writeb()	Same as $write, but default format of any arguments is binary.
$writeo()	Same as $write, but default format of any arguments is octal.
$writeh()	Same as $write, but default format of any arguments is hexadecimal.
$strobe()	Same as $display, but printing occurs *after* all simulation events are executed.
$strobeb()	Same as $strobe, but default format of any arguments is binary.
$strobeo()	Same as $strobe, but default format of any arguments is octal.
$strobeh()	Same as $strobe, but default format of any arguments is hexadecimal.
$monitor()	Same as $display, but printing occurs when the value of an argument *changes*.
$monitorb()	Same as $monitor, but default format of any arguments is binary.
$monitoro()	Same as $monitor, but default format of any arguments is octal.
$monitorh()	Same as $monitor, but default format of any arguments is hexadecimal.
$monitoron	Begin tracking argument changes in subsequent $monitor tasks.
$monitoroff	Stop tracking argument changes in subsequent $monitor tasks.

The following is a list of the most common text formatting codes for printing variables within a string.

Code	Format
%b	Binary values
%o	Octal values
%d	Decimal values
%h	Hexadecimal values
%f	Real values using decimal form
%e	Real values using exponential form
%t	Time values
%s	Character strings
%m	Hierarchical name of scope (no argument required when printing)
%l	Configuration library binding (no argument required when printing)

The format letters in these codes are not case sensitive (i.e., %d and %D are equivalent). Each of these formatting codes can also contain information about truncation of leading and trailing digits. Rounding will take place when numbers are truncated. The formatting syntax is as follows:

```
%<number_of_leading_digits>.<number_of_trailing_digits><format_code_letter>
```

There are also a set of string formatting and character escapes that are supported for use with the text output system tasks.

Code	Description
\n	Print a new line.
\t	Print a tab.
\”	Print a quote (“).
\cr	Print a backslash (\).
%%	Print a percent sign (%).

The following is a set of examples using common text output system tasks. For these examples, assume two variables have been declared and initialized as follows: A = 3 (integer) and B = 45.6789 (real). Recall that Verilog uses 32-bit codes to represent type integer and real.

Example:

```
$display("Hello World");    // Will print: Hello World
$display("A = %b", A);      // This will print: A = 00000000000000000000000000000011
$display("A = %o", A);      // This will print: A = 00000000003
$display("A = %d", A);      // This will print: A = 3
$display("A = %h", A);      // This will print: A = 00000003
$display("A = %4.0b", A);   // This will print: A = 0011

$display("B = %f", B);      // This will print: B = 45.678900
$display("B = %2.0f", B);   // This will print: B = 46
$display("B = %2.1f", B);   // This will print: B = 45.7
$display("B = %2.2f", B);   // This will print: B = 45.68

$display("B = %e", B);      // This will print: B = 4.567890e+001
$display("B = %1.0e", B);   // This will print: B = 5e+001
$display("B = %1.1e", B);   // This will print: B = 4.6e+001
$display("B = %2.2e", B);   // This will print: B = 4.57e+001

$write("A is ", A, "\n");   // This will print: A is 3
$writeb("A is ", A, "\n");  // This will print: A is 00000000000000000000000000000011
$writeo("A is ", A, "\n");  // Will print: A is 00000000003
$writeh("A is ", A, "\n");  // Will print: A is 00000003
```

8.3.2 File Input/Output

File I/O system tasks allow a Verilog module to create and/or access data files is the same way files are handled in ANSI C. This is useful when the results of a simulation are large and need to be stored in a file as opposed to viewing in a waveform or transcript window. This is also useful when complex stimulus vectors are to be read from an external file and driven into a device under test. Verilog supports the following file I/O system task functions:

Task	Description
$fopen()	Opens a file and returns a unique file descriptor.
$fclose()	Closes the file associated with the descriptor.
$fdisplay()	Same as $display but statements are directed to the file descriptor.
$fwrite()	Same as $write but statements are directed to the file descriptor.
$fstrobe()	Same as $strobe but statements are directed to the file descriptor.
$fmonitor()	Same as $monitor but statements are directed to the file descriptor.
$readmemb()	Read binary data from file and insert into previously defined memory array.
$readmemh()	Read hexadecimal data from file and insert into previously defined memory array

The **$fopen()** function will either create and open or open an existing file. Each file that is opened is given a unique integer called a *file descriptor* that is used to identify the file in other I/O functions. The integer must be declared prior to the first use of $fopen. A file name argument is required and provided within double quotes. By default, the file is opened for writing. If the file name doesn't exist, it will be created. If the file name does exist, it will be overwritten. An optional *file_type* can be provided that gives specific action for the file opening including opening an existing file and appending to a file. The following are the supported codes for $fopen():

$fopen types	Description
"**r**" or "**rb**"	Open file for reading.
"**w**" or "**wb**"	Create for writing.
"**a**" or "**ab**"	Open for writing and append to the end of file.
"**r+**" or "**r + b**" or "**rb+**"	Open for update, reading or writing file.
"**w+**" or "**w + b**" or "**wb+**"	Create for update.
"**a+**" or "**a + b**" or "**ab+**"	Open or create for update, append to the end of file.

Once a file is open, data can be written to it using the **$fdisplay()**, **$fwrite()**, **$fstrobe()**, and **$fmonitor()** tasks. These functions require two arguments. The first argument is the file descriptor and the second is the information to be written. The information follows the same syntax as the I/O system tasks. The following example shows how to create a file and write data to it. This example will create a new file called "Data_out.txt" and write two lines of text to it with the values of variables A and B.

Example:

```
integer A = 3;
real  B = 45.6789;
integer FILE_1;

initial
  begin
    FILE_1 = $fopen("Data_out.txt", "w");
    $fdisplay(FILE_1, "A is %d", A);
    $fdisplay(FILE_1, "B is %f", B);
    $fclose(FILE_1);
  end
```

When reading data from a file, the functions **$readmemb()** and **$readmemh()** can be used. These tasks require that a storage array be declared that the contents of the file can be read into. These tasks have two arguments, the first being the name of the file and the second being the name of the storage array to store the file contents into. The following example shows how to read the contents of a file into a storage array called "memory." Assume the file contains eight lines, each containing a 3-bit vector. The vectors start at 000 and increment to 111, and each symbol will be interpreted as binary using the $readmemb() task. The storage array "memory" is declared to be an 8x3 array of type reg. The $readmemb() task will insert each line of the file into each 3-bit vector location within "memory." To illustrate how the data is stored, this example also contains a second procedural block that will print the contents of the storage element to the transcript.

Example:

```
reg[2:0] memory[7:0];

initial
  begin: Read_Block
    $readmemb("Data_in.txt", memory);
  end
initial
  begin: Print_Block
    $display("printing memory %b", memory[0]);   // This will print "000"
    $display("printing memory %b", memory[1]);   // This will print "001"
    $display("printing memory %b", memory[2]);   // This will print "010"
    $display("printing memory %b", memory[3]);   // This will print "011"
    $display("printing memory %b", memory[4]);   // This will print "100"
    $display("printing memory %b", memory[5]);   // This will print "101"
    $display("printing memory %b", memory[6]);   // This will print "110"
    $display("printing memory %b", memory[7]);   // This will print "111"
  end
```

8.3.3 Simulation Control and Monitoring

Verilog also provides a set of simulation control and monitoring tasks. The following are the most commonly used tasks in this group.

Task	Description	
$finish()	Finishes simulation and exits.	
$stop()	Halts the simulation and enters an interactive debug mode.	
$time()	Returns the current simulation time as a 64-bit vector.	
$stime()	Returns the current simulation time as a 32-bit integer.	
$realtime()	Returns the current simulation time as a 32-bit real number.	
$timeformat()	Controls the format used by the %t code in print statements.	
	The arguments are: (<unit>, <precision>, <suffix>, <min_field_width>)	
	where:	
	<unit>	0 = 1 s
		−1 = 100 ms
		−2 = 10 ms
		−3 = 1 ms
		−4 = 100 us
		−5 = 10 us
		−6 = 1 us
		−7 = 100 ns
		−8 = 10 ns
		−9 = 1 ns
		−10 = 100 ps
		−11 = 10 ps
		−12 = 1 ps
		−13 = 100 fs
		−14 = 10 fs
		−15 = 1 fs
	<precision> = The number of decimal points to display.	
	<suffix> = A string to be appended to time to indicate units.	
	<min_field_width> = The minimum number of characters to display.	

The following shows an example of how these tasks can be used.

Example:

```
initial
  begin

   $timeformat (-9, 2, "ns", 10);
   $display("Stimulus starting at time: %t", $time);

   #10    A_TB=0; B_TB=0; C_TB=0;
   #10    A_TB=0; B_TB=0; C_TB=1;
   #10    A_TB=0; B_TB=1; C_TB=0;
   #10    A_TB=0; B_TB=1; C_TB=1;
   #10    A_TB=1; B_TB=0; C_TB=0;
   #10    A_TB=1; B_TB=0; C_TB=1;
   #10    A_TB=1; B_TB=1; C_TB=0;
   #10    A_TB=1; B_TB=1; C_TB=1;

   $display("Simulation stopping at time: %t", $time);
  end
```

This example will result in the following statements printed to the simulator transcript:

```
Stimulus starting at time:    0.00ns
Simulation stopping at time: 80.00ns
```

CONCEPT CHECK

CC8.3 How can Verilog system tasks be included in synthesizable circuit models when they provide inherently unsynthesizable functionality?

A) They can't. System tasks can only be used in test benches.

B) The "$" symbol tells the CAD tool that the task can be ignored during synthesis.

C) The designer must only use system tasks that model sequential logic.

D) The designer must only use system tasks that model combinational logic.

8.4 Test Benches

The functional verification of Verilog designs is accomplished through simulation using a *test bench*. A test bench is a Verilog model that instantiates the system to be tested as a sub-system, generates the input patterns to drive into the sub-system, and observes the outputs. The system being tested is often called a *device under test (DUT)* or *unit under test (UUT)*. Test benches are only used for simulation, so they can use abstract modeling techniques that are unsynthesizable to generate the stimulus patterns. Verilog conditional programming constructions and system tasks can also be used to report on the status of a test and also automatically check that the outputs are correct.

8.4.1 Common Stimulus Generation Techniques

When creating stimulus for combinational logic circuits, it is common to use a procedural block to generate all possible input patterns to drive the DUT and especially any transitions that may cause timing errors. Example 8.9 shows a test bench for a combinational logic circuit where an initial block contains a series of delayed assignments to provide the stimulus to the DUT. This block creates every possible input pattern, delayed by a fixed amount. Note that the initial block will only execute once. If the patterns were desired to repeat indefinitely, an always block without a sensitivity list could be used instead.

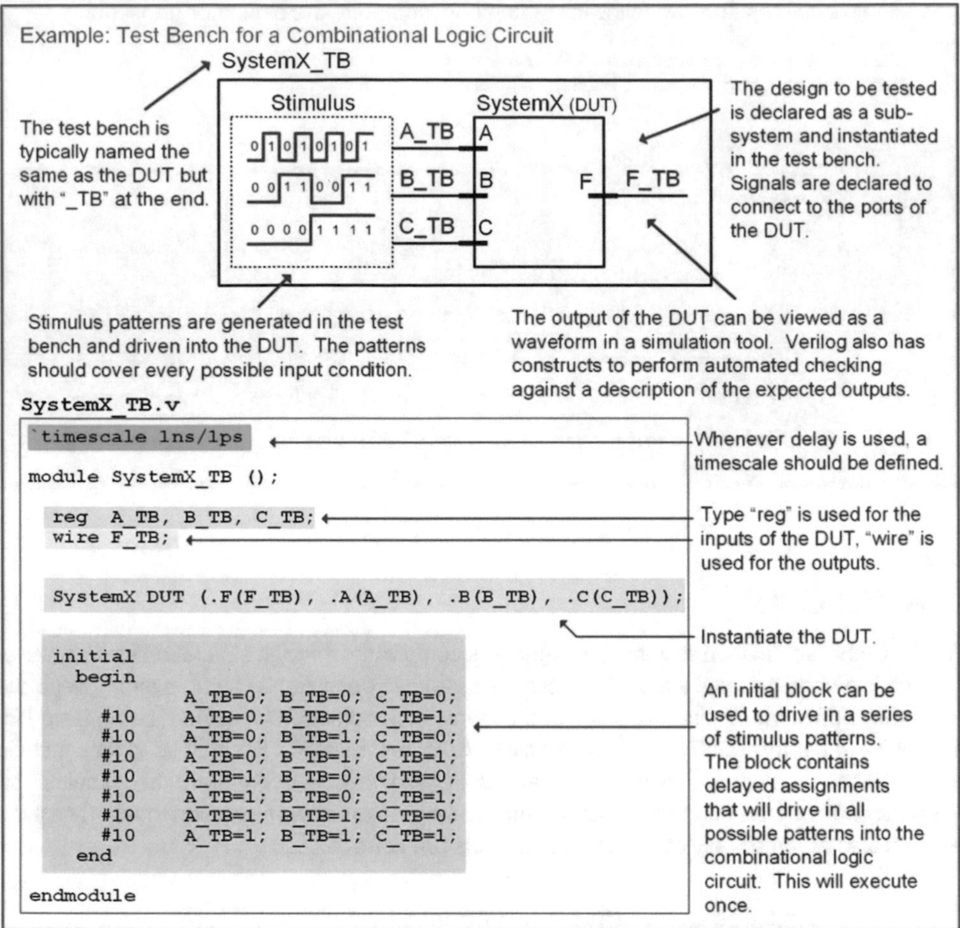


```
`timescale 1ns/1ps

module SystemX_TB ();

  reg  A_TB, B_TB, C_TB;
  wire F_TB;

  SystemX DUT (.F(F_TB), .A(A_TB), .B(B_TB), .C(C_TB));

  initial
    begin
              A_TB=0; B_TB=0; C_TB=0;
      #10     A_TB=0; B_TB=0; C_TB=1;
      #10     A_TB=0; B_TB=1; C_TB=0;
      #10     A_TB=0; B_TB=1; C_TB=1;
      #10     A_TB=1; B_TB=0; C_TB=0;
      #10     A_TB=1; B_TB=0; C_TB=1;
      #10     A_TB=1; B_TB=1; C_TB=0;
      #10     A_TB=1; B_TB=1; C_TB=1;
    end

endmodule
```

Example 8.9
Test bench for a combinational logic circuit

Multiple procedural blocks can be used within a Verilog module to provide parallel functionality. Using both initial and always blocks allows the test bench to drive both repetitive and aperiodic signals. Initial and always blocks can also be used to drive the same signal in order to provide a starting value and a repetitive pattern. Example 8.10 shows a test bench for a rising edge triggered D-flip-flop with an asynchronous, active LOW reset in which multiple procedural blocks are used to generate the stimulus patterns for the DUT.

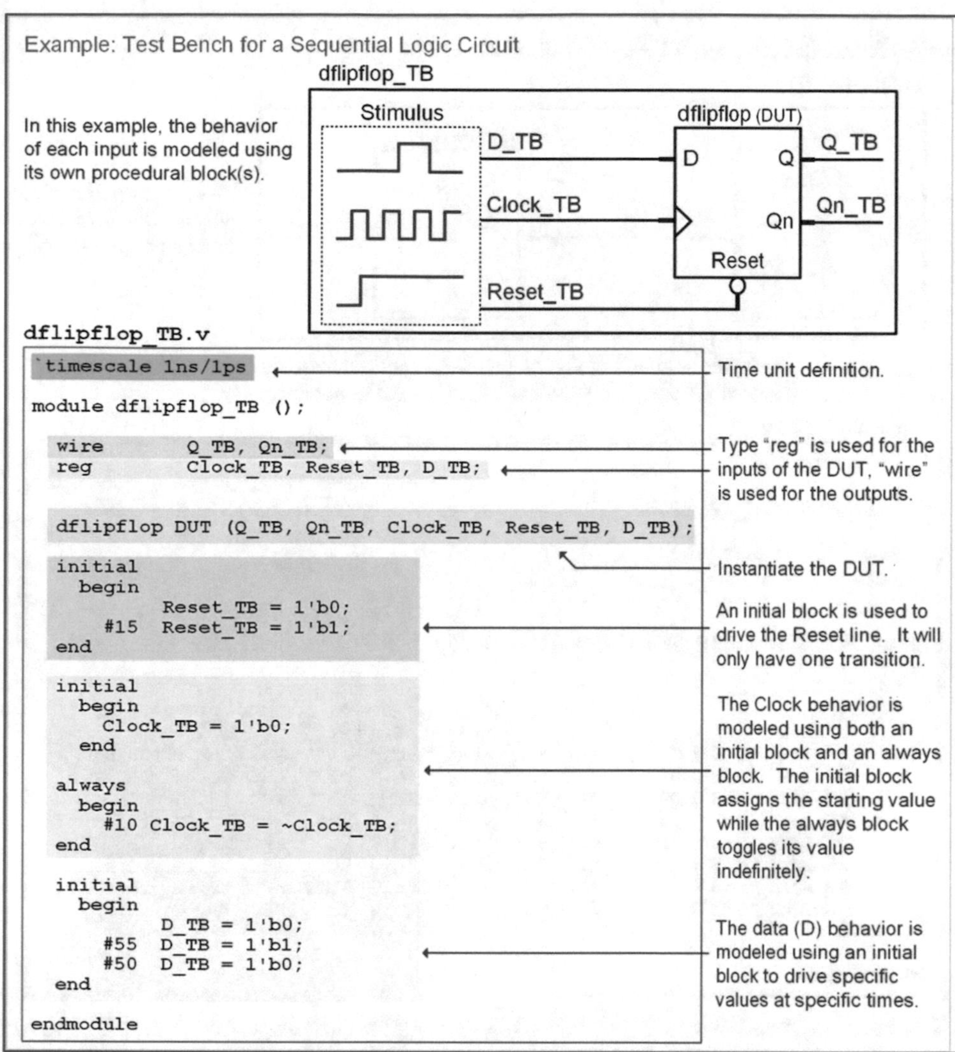

Example 8.10
Test bench for a sequential logic circuit

8.4.2 Printing Results to the Simulator Transcript

In the past test bench examples, the input and output values are observed using either the waveform or listing tool within the simulator tool. It is also useful to print the values of the simulation to a transcript window to track the simulation as each statement is processed. Messages can be printed that show the status of the simulation in addition to the inputs and outputs of the DUT using the text output system tasks. Example 8.11 shows a test bench that prints the inputs and output to the transcript of the simulation tool. Note that the test bench must wait some amount of delay before evaluating the output, even if the DUT does not contain any delay.

Example: Printing Test Bench Results to the Transcript

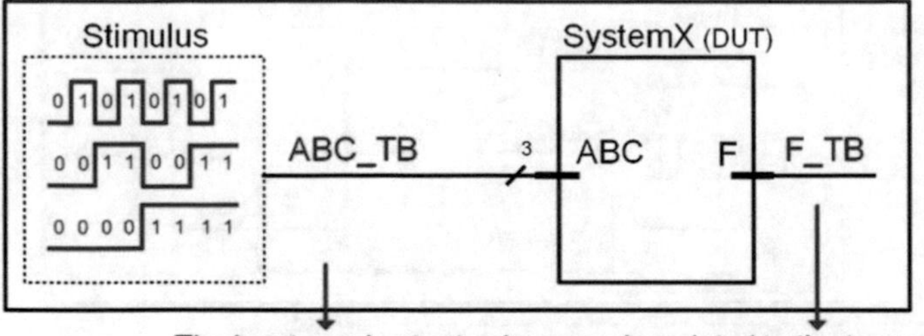

The test bench needs to wait for the DUT to respond before printing the output.

The inputs and output values can be printed to the transcript using either $display(), $write(), $strobe(), or $monitor().

SystemX_TB.v

```
`timescale 1ns/1ps

module SystemX_TB ();

  reg   [2:0] ABC_TB;
  wire        F_TB;


 SystemX DUT (.F(F_TB), .ABC(ABC_TB));

  initial
    begin
                                $display("ABC | F");
         ABC_TB=3'b000;   #1    $display("%b | %b", ABC_TB, F_TB);
      #9 ABC_TB=3'b001;   #1    $display("%b | %b", ABC_TB, F_TB);
      #9 ABC_TB=3'b010;   #1    $display("%b | %b", ABC_TB, F_TB);
      #9 ABC_TB=3'b011;   #1    $display("%b | %b", ABC_TB, F_TB);
      #9 ABC_TB=3'b100;   #1    $display("%b | %b", ABC_TB, F_TB);
      #9 ABC_TB=3'b101;   #1    $display("%b | %b", ABC_TB, F_TB);
      #9 ABC_TB=3'b110;   #1    $display("%b | %b", ABC_TB, F_TB);
      #9 ABC_TB=3'b111;   #1    $display("%b | %b", ABC_TB, F_TB);
    end

endmodule
```

Even if the DUT model does not contain delay, the test bench needs to delay before evaluating the output.

The above test bench will print out the following message to the transcript of the simulator tool.

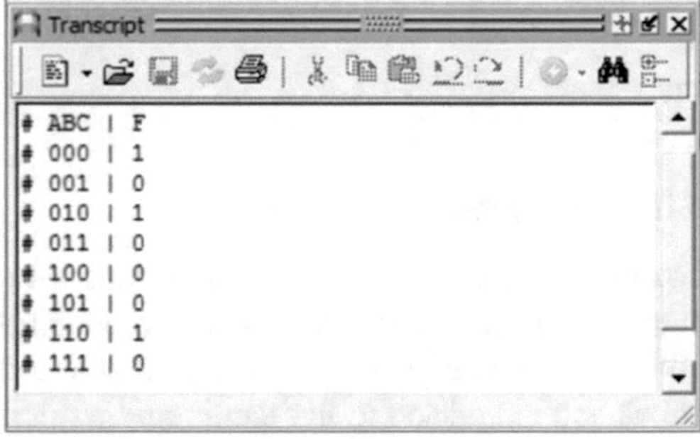

Example 8.11
Printing test bench results to the transcript

8.4.3 Automatic Result Checking

Test benches can also perform automated checking of the results using the conditional programming constructs described earlier in this chapter. Example 8.12 shows an example of a test bench that uses if-else statements to check the output of the DUT and print a PASS/FAIL message to the transcript.

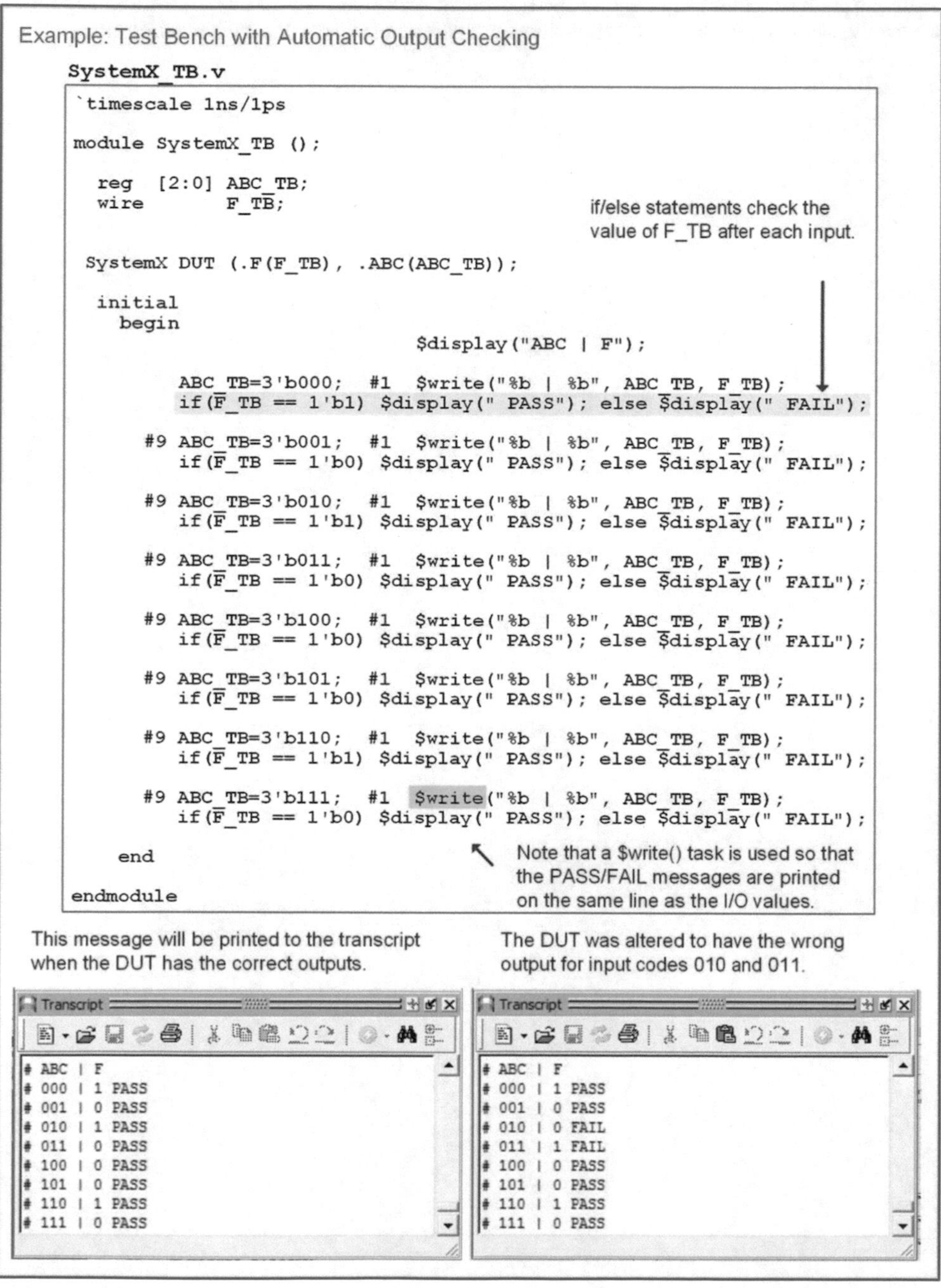


```
`timescale 1ns/1ps

module SystemX_TB ();

  reg   [2:0] ABC_TB;
  wire        F_TB;

 SystemX DUT (.F(F_TB), .ABC(ABC_TB));

  initial
    begin
                               $display("ABC | F");

          ABC_TB=3'b000;  #1   $write("%b | %b", ABC_TB, F_TB);
          if(F_TB == 1'b1) $display(" PASS"); else $display(" FAIL");

       #9 ABC_TB=3'b001;  #1   $write("%b | %b", ABC_TB, F_TB);
          if(F_TB == 1'b0) $display(" PASS"); else $display(" FAIL");

       #9 ABC_TB=3'b010;  #1   $write("%b | %b", ABC_TB, F_TB);
          if(F_TB == 1'b1) $display(" PASS"); else $display(" FAIL");

       #9 ABC_TB=3'b011;  #1   $write("%b | %b", ABC_TB, F_TB);
          if(F_TB == 1'b0) $display(" PASS"); else $display(" FAIL");

       #9 ABC_TB=3'b100;  #1   $write("%b | %b", ABC_TB, F_TB);
          if(F_TB == 1'b0) $display(" PASS"); else $display(" FAIL");

       #9 ABC_TB=3'b101;  #1   $write("%b | %b", ABC_TB, F_TB);
          if(F_TB == 1'b0) $display(" PASS"); else $display(" FAIL");

       #9 ABC_TB=3'b110;  #1   $write("%b | %b", ABC_TB, F_TB);
          if(F_TB == 1'b1) $display(" PASS"); else $display(" FAIL");

       #9 ABC_TB=3'b111;  #1   $write("%b | %b", ABC_TB, F_TB);
          if(F_TB == 1'b0) $display(" PASS"); else $display(" FAIL");

    end

endmodule
```

Example 8.12
Test bench with automatic output checking

8.4.4 Using Loops to Generate Stimulus

When creating stimulus that follow regular patterns such as counting, loops can be an effective way to produce the input vectors. A for loop is especially useful for generating exhaustive stimulus patterns for combinational logic circuits. An integer loop variable can increment within the for loop and then be assigned to the DUT inputs as type reg. Recall that in Verilog, when an integer is assigned to a variable of type reg, it is truncated to matched the size of the reg. This allows a binary count to be created for an input stimulus pattern by using an integer loop variable that increments within a for loop. Example 8.13 shows how the stimulus for a combinational logic circuit can be produced with a for loop.

Example: Using a Loop to Generate Stimulus in a Test Bench

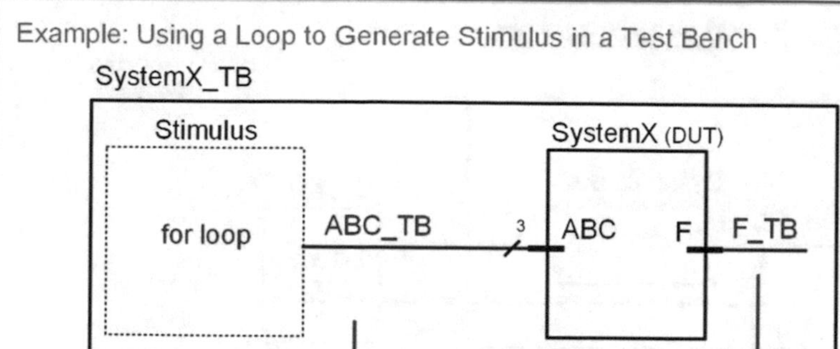

The inputs and output values will be printed to the transcript using $display().

SystemX_TB.v

```
`timescale 1ns/1ps

module SystemX_TB ();

  reg   [2:0] ABC_TB;
  wire        F_TB;
  integer     i;

  SystemX DUT (.F(F_TB), .ABC(ABC_TB));

  initial
    begin

      for (i=0; i<8; i=i+1)
        begin
          ABC_TB = i;
          #10 $display("i=%d, ABC=%b, F=%b", i, ABC_TB, F_TB);
        end

    end

endmodulee
```

When using a for loop, the loop variable must be declared.

Each time through the loop, *i* will increment by one. It will count from 0 to 7.

The bottom 3-bits of the integer *i* are used in the assignment to ABC_TB.

The above test bench will print out the following message to the transcript of the simulator tool.

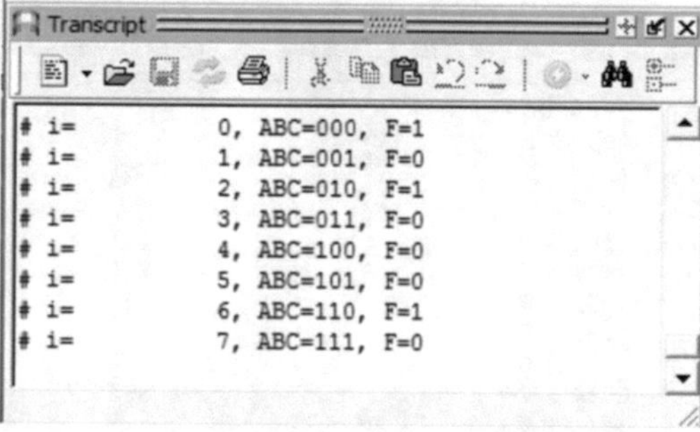

Example 8.13
Using a loop to generate stimulus in a test bench

8.4.5 Using External Files in Test Benches

There are often cases where the results of a test bench need to be written to an external file, either because they are too verbose or because there needs to be a stored record. Verilog allows writing to external files via the file I/O system tasks (i.e., $fdisplay(), $fwrite(), $fstrong(), and $fmonitor()). Example 8.14 shows a test bench in which the input vectors and the output of the DUT are written to an external file using the $fdisplay() system task.

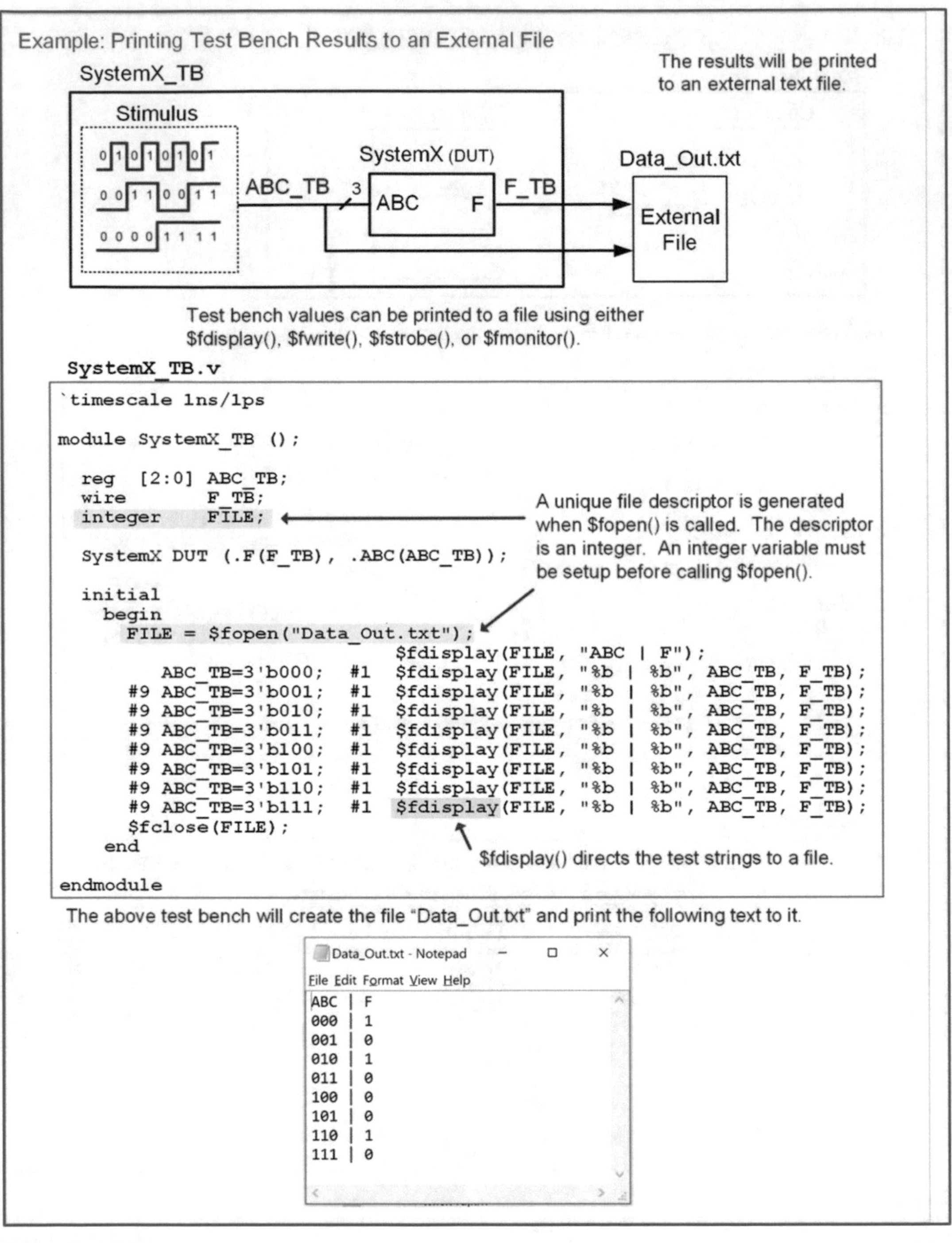


```
`timescale 1ns/1ps

module SystemX_TB ();

  reg   [2:0] ABC_TB;
  wire        F_TB;
  integer     FILE;

  SystemX DUT (.F(F_TB), .ABC(ABC_TB));

  initial
    begin
      FILE = $fopen("Data_Out.txt");
                                  $fdisplay(FILE, "ABC | F");
         ABC_TB=3'b000;   #1      $fdisplay(FILE, "%b | %b", ABC_TB, F_TB);
      #9 ABC_TB=3'b001;   #1      $fdisplay(FILE, "%b | %b", ABC_TB, F_TB);
      #9 ABC_TB=3'b010;   #1      $fdisplay(FILE, "%b | %b", ABC_TB, F_TB);
      #9 ABC_TB=3'b011;   #1      $fdisplay(FILE, "%b | %b", ABC_TB, F_TB);
      #9 ABC_TB=3'b100;   #1      $fdisplay(FILE, "%b | %b", ABC_TB, F_TB);
      #9 ABC_TB=3'b101;   #1      $fdisplay(FILE, "%b | %b", ABC_TB, F_TB);
      #9 ABC_TB=3'b110;   #1      $fdisplay(FILE, "%b | %b", ABC_TB, F_TB);
      #9 ABC_TB=3'b111;   #1      $fdisplay(FILE, "%b | %b", ABC_TB, F_TB);
      $fclose(FILE);
    end

endmodule
```

```
ABC | F
000 | 1
001 | 0
010 | 1
011 | 0
100 | 0
101 | 0
110 | 1
111 | 0
```

Example 8.14
Printing test bench results to an external file

It is often the case that the input vectors are either too large to enter manually or were created by a separate program. In either case, a useful technique in test benches is to read input vectors from an external file. Example 8.15 shows an example where the input stimulus vectors for a DUT are read from an external file using the $readmemb() system task.

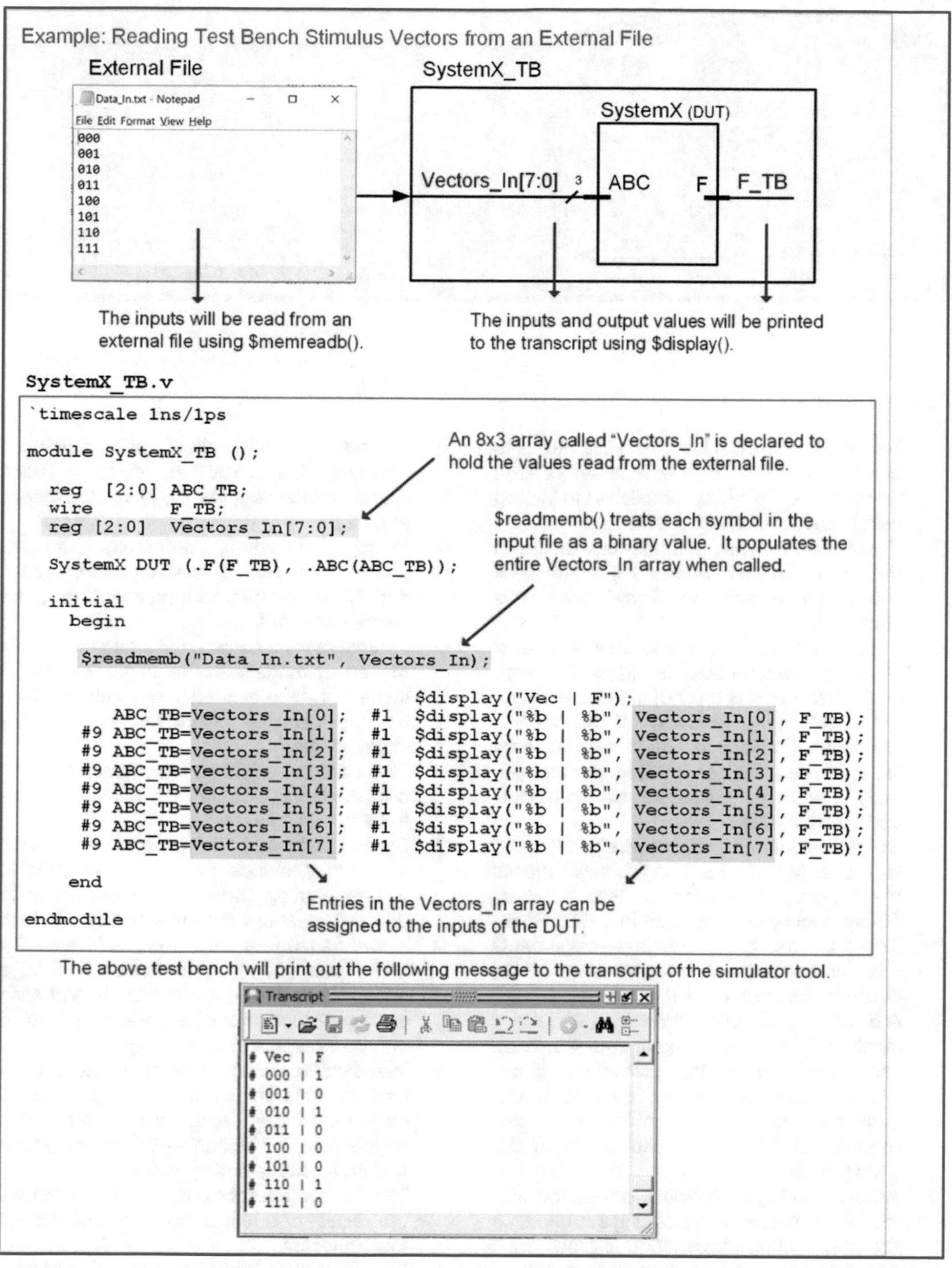

Example 8.15
Reading test bench stimulus vectors from an external file

Concept Check

CC8.4 Could a test bench ever use always blocks and sensitivity lists exclusively to create its stimulus? Why or why not?

A) Yes. The signal assignments will simply be made when the block ends.

B) No. Since a sensitivity list triggers when there is a change on one or more of the signals listed, the blocks in the test bench would never trigger because there is no method to make the initial signal transition.

Summary

- To model sequential logic, an HDL needs to be able to trigger signal assignments based on an event. This is accomplished in Verilog using *procedural assignment*.
- There are two types of procedural blocks in Verilog, *initial* and *always*. An initial block executes one time. An always block runs continually.
- A *sensitivity* list is a way to control when a Verilog procedural block is triggered. A sensitivity list contains a list of signals. If any of the signals in the sensitivity list transitions, it will cause the block to trigger. If a sensitivity list is omitted, the block will trigger immediately. Sensitivity lists are most commonly used with always blocks.
- Sensitivity lists and always blocks are used to model synthesizable logic. Initial blocks are typically only used in test benches. Always blocks are also used in test benches.
- There are two types of signal assignments that can be used within a procedural block, blocking, and non-blocking.
- A blocking assignment is denoted with the = symbol. All blocking assignments are made immediately within the procedural block. Blocking assignments are used to model combinational logic. Combinational logic models list all input to the circuit in the sensitivity list.
- A non-blocking assignment is denoted with the <= symbol. All non-blocking assignments are made when the procedural block ends and are evaluated in the order they appeared in the block. Blocking assignments are used to model sequential logic. Sequential logic models list only the clock and reset in the sensitivity list.
- Variables can be defined within a procedural block as long as the block is named.
- Procedural blocks allow more advanced modeling constructs in Verilog. These include *if-else statements*, *case statements*, *and loops*.
- Verilog provides numerous looping constructs including *forever, while*, *repeat*, and *for.* Loops can be terminated using the *disable* keyword.
- *System tasks* provide additional functionality to Verilog models. Tasks begin with the $ symbol and are omitted from synthesis. System tasks can be included in synthesizable logic models.
- There are three groups of system tasks: text output, file input/output, and simulation control and monitoring.
- System tasks that perform printing functions can output strings in addition to variable values. Verilog provides a mechanism to print the variable values in a variety of format.
- A *test bench* is a way to simulate a device under test (DUT) by instantiating it as a sub-system, driving in stimulus, and observing the outputs. Test benches do not have inputs or outputs and are unsynthesizable.
- Test benches for combinational logic typically exercise the DUT under an exhaustive set of stimulus vectors. These include all possible logic inputs in addition to critical transitions that could cause timing errors.
- Text I/O system tasks provide a way to print the results of a test bench to the simulation tool transcript.
- File I/O system tasks provide a way to print the results of a test bench to an external file and also to read in stimulus vectors from an external file.
- Conditional programming constructs can be used within a test bench to perform automatic checking of the outputs of a DUT within a test bench.

- Loops can be used in test benches to automatically generate stimulus patterns. A for loop is a convenient technique to produce a counting pattern.
- Assignment from an integer to a reg in a for loop is allowed. The binary value of the integer is truncated to fit the size of the reg vector.

Exercise Problems

Section 8.1: Procedural Assignment

8.1.1 When using a sensitivity list with a procedural block, what will cause the block to *trigger*?

8.1.2 When a sensitivity list is not used with a procedural block, when will the block trigger?

8.1.3 When are statements executed when using blocking assignments?

8.1.4 When are statements executed when using non-blocking assignments?

8.1.5 When is it possible to exclude statement groups from a procedural block?

8.1.6 What is the difference between a begin/end and fork/join group when each contain multiple statements?

8.1.7 What is the difference between a begin/end and fork/join group when each contain only a single statement?

8.1.8 What type of procedural assignment is used when modeling combinational logic?

8.1.9 What type of procedural assignment is used when modeling sequential logic?

8.1.10 What signals should be listed in the sensitivity list when modeling combinational logic?

8.1.11 What signals should be listed in the sensitivity list when modeling sequential logic?

Section 8.2: Conditional Programming Constructs

8.2.1 Design a Verilog model to implement the behavior described by the 4-input truth table in Fig. 8.1. Use procedural assignment and an if-else statement. Declare the module to match the block diagram provided. Use the type wire for the inputs and type reg for the output. Hint: Notice that there are far more input codes producing F = 0 than producing F = 1. Can you use this to your advantage to make your if-else statement simpler?

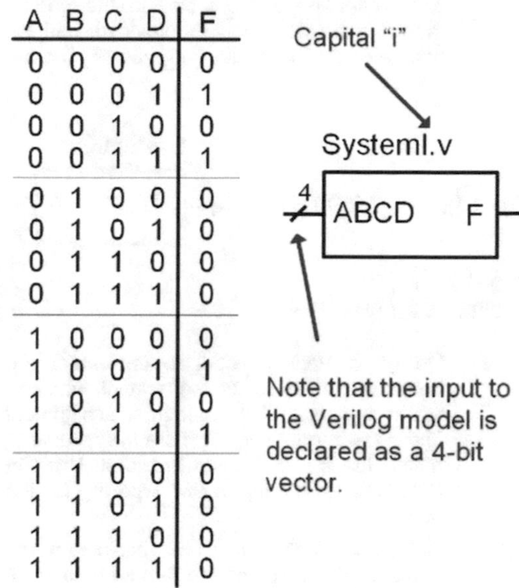

A	B	C	D	F
0	0	0	0	0
0	0	0	1	1
0	0	1	0	0
0	0	1	1	1
0	1	0	0	0
0	1	0	1	0
0	1	1	0	0
0	1	1	1	0
1	0	0	0	0
1	0	0	1	1
1	0	1	0	0
1	0	1	1	1
1	1	0	0	0
1	1	0	1	0
1	1	1	0	0
1	1	1	1	0

Fig. 8.1
System I functionality

8.2.2 Design a Verilog model to implement the behavior described by the 4-input truth table in Fig. 8.1. Use procedural assignment and a case statement. Declare the module to match the block diagram provided. Use the type wire for the inputs and type reg for the output.

8.2.3 Design a Verilog model to implement the behavior described by the 4-input minterm list in Fig. 8.2. Use procedural assignment and an if-else statement. Declare the module to match the block diagram provided. Use the type wire for the inputs and type reg for the output.

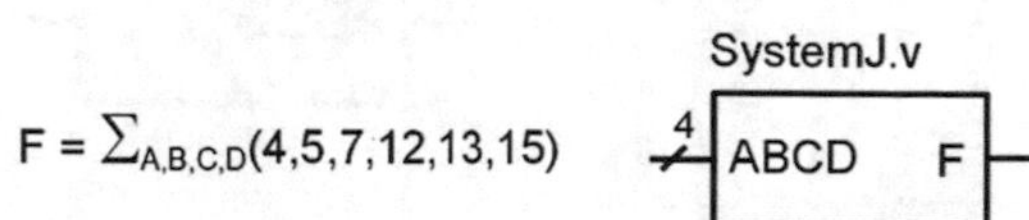

$$F = \Sigma_{A,B,C,D}(4,5,7,12,13,15)$$

Fig. 8.2
System J functionality

8.2.4 Design a Verilog model to implement the behavior described by the 4-input minterm list in Fig. 8.2. Use procedural assignment and a case statement. Declare the module to match the block diagram provided. Use the type wire for the inputs and type reg for the output.

8.2.5 Design a Verilog model to implement the behavior described by the 4-input maxterm list in Fig. 8.3. Use procedural assignment and an if-then statement. Declare the module to match the block diagram provided. Use the type wire for the inputs and type reg for the output.

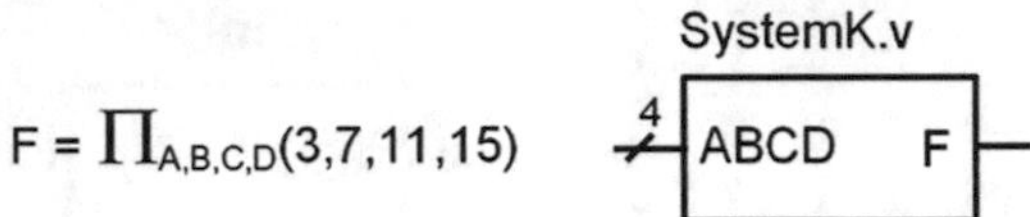

Fig. 8.3
System K functionality

8.2.6 Design a Verilog model to implement the behavior described by the 4-input maxterm list in Fig. 8.3. Use procedural assignment and a case statement. Declare the module to match the block diagram provided. Use the type wire for the inputs and type reg for the output.

8.2.7 Design a Verilog model to implement the behavior described by the 4-input truth table in Fig. 8.4. Use procedural assignment and an if-else statement. Declare the module to match the block diagram provided. Use the type wire for the inputs and type reg for the output. Hint: Notice that there are far more input codes producing F = 1 than producing F = 0. Can you use this to your advantage to make your if-else statement simpler?

A	B	C	D	F
0	0	0	0	1
0	0	0	1	0
0	0	1	0	1
0	0	1	1	1
0	1	0	0	1
0	1	0	1	1
0	1	1	0	1
0	1	1	1	1
1	0	0	0	1
1	0	0	1	0
1	0	1	0	1
1	0	1	1	0
1	1	0	0	1
1	1	0	1	0
1	1	1	0	1
1	1	1	1	1

SystemL.v

4 ABCD F

Fig. 8.4
System L functionality

8.2.8 Design a Verilog model to implement the behavior described by the 4-input truth table in Fig. 8.4. Use procedural assignment and a case statement. Declare the module to match the block diagram provided. Use the type wire for the inputs and type reg for the output.

8.2.9 Figure 8.5 shows the topology of a *4-bit shift register* when implemented structurally using D-Flip-Flops. Design a Verilog model to describe this functionality using a single procedural block and non-blocking assignments instead of instantiating D-Flip-Flops. The figure also provides the block diagram for the module port definition. Use the type wire for the inputs and type reg for the outputs.

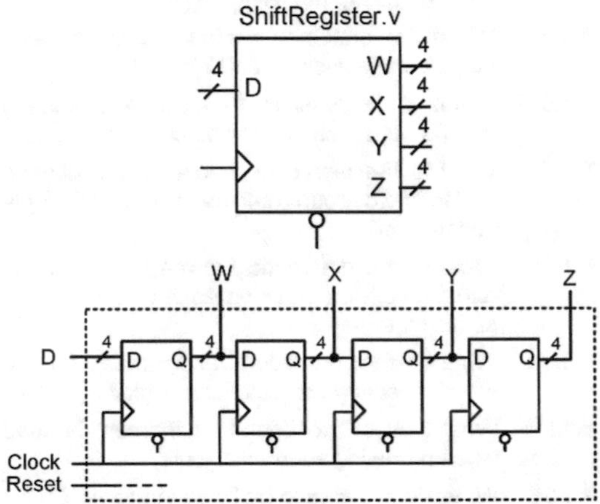

Fig. 8.5
4-bit shift register functionality

8.2.10 Design a Verilog model for a counter using a for loop with an output type of integer. Figure 8.6 shows the block diagram for the module definition. The counter should increment from 0 to 31 and then start over. Use delay in your loop to update the counter value every 10 ns. Consider using the loop variable of the for loop to generate your counter value.

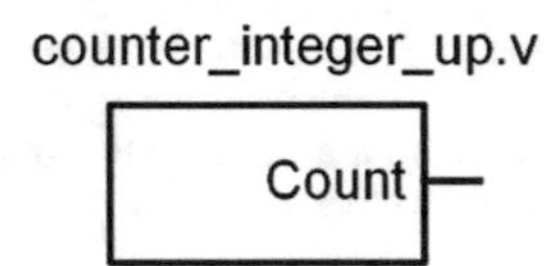

Fig. 8.6
Integer counter block diagram

8.2.11 Design a Verilog model for a counter using a for loop with an output type of reg[4:0]. Figure 8.7 shows the block diagram for the module definition. The counter should increment from 00000_2 to 11111_2 and then start over. Use delay in your loop to update the counter value every 10 ns. Consider using the loop variable of the for loop to generate an integer version of your count value and then assign it to the output variable of type reg[4:0].

counter_5bit_binary_up.v

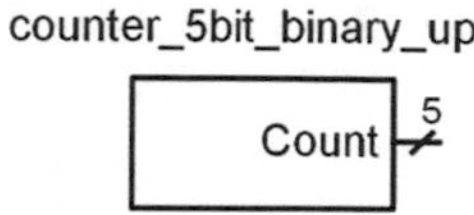

Fig. 8.7
5-Bit binary counter block diagram

Section 8.3: System Tasks

8.3.1 Are system tasks synthesizable? Why or why not?

8.3.2 What is the difference between the tasks $display() and $write()?

8.3.3 What is the difference between the tasks $display() and $monitor()?

8.3.4 What is the data type returned by the task $fopen()?

Section 8.4: Test Benches

8.4.1 Design a Verilog test bench to verify the functional operation of the system in Fig. 8.1. Your test bench should drive in every possible input code for the vector ABCD (i.e., "0000", "0001", "0010", . . ., "1111"). Have your test bench change the input pattern every 10 ns using delay within your procedural block.

8.4.2 Design a Verilog test bench to verify the functional operation of the system in Fig. 8.1 with automatic checking. Your test bench should drive in every possible input code for the vector ABCD (i.e., "0000", "0001", "0010", . . ., "1111"). Have your test bench change the input pattern every 10 ns using delay within your procedural block. Use conditional statements to check whether the output of the DUT is correct. For each input vector, print a message using $display() that indicates the current input vector being tested, the resulting output of your DUT, and whether the DUT output is correct.

8.4.3 Design a Verilog test bench to verify the functional operation of the system in Fig. 8.2. Your test bench should drive in every possible input code for the vector ABCD (i.e., "0000", "0001", "0010", . . ., "1111"). Have your test bench change the input pattern every 10 ns using delay within your procedural block.

8.4.4 Design a Verilog test bench to verify the functional operation of the system in Fig. 8.2 with automatic checking. Your test bench should drive in every possible input code for the vector ABCD (i.e., "0000", "0001", "0010", . . ., "1111"). Have your test bench change the input pattern every 10 ns using delay within your procedural block. Use conditional statements to check whether the output of the DUT is correct. For each input vector, print a message using $display() that indicates the current input vector being tested, the resulting output of your DUT, and whether the DUT output is correct.

8.4.5 Design a Verilog test bench to verify the functional operation of the system in Fig. 8.3. Your test bench should drive in every possible input code for the vector ABCD (i.e., "0000", "0001", "0010", . . ., "1111"). Have your test bench change the input pattern every 10 ns using delay within your procedural block.

8.4.6 Design a Verilog test bench to verify the functional operation of the system in Fig. 8.3 with automatic checking. Your test bench should drive in every possible input code for the vector ABCD (i.e., "0000", "0001", "0010", . . ., "1111"). Have your test bench change the input pattern every 10 ns using delay within your procedural block. Use conditional statements to check whether the output of the DUT is correct. For each input vector, print a message using $display() that indicates the current input vector being tested, the resulting output of your DUT, and whether the DUT output is correct.

8.4.7 Design a Verilog test bench to verify the functional operation of the system in Fig. 8.4. Your test bench should drive in every possible input code for the vector ABCD (i.e., "0000", "0001", "0010", . . ., "1111"). Have your test bench change the input pattern every 10 ns using delay within your procedural block.

8.4.8 Design a Verilog test bench to verify the functional operation of the system in Fig. 8.4 with automatic checking. Your test bench should drive in every possible input code for the vector ABCD (i.e., "0000", "0001", "0010", . . ., "1111"). Have your test bench change the input pattern every 10 ns using delay within your procedural block. Use conditional statements to check whether the output of the DUT is correct. For each input vector, print a message using $display() that indicates the current input vector being tested, the resulting output of your DUT, and whether the DUT output is correct.

8.4.9 Design a Verilog test bench to verify the functional operation of the system in Fig. 8.4. Your test bench should drive in every possible input code for the vector ABCD (i.e., "0000", "0001", "0010", . . ., "1111"). Have your test bench change the input pattern every 10 ns using delay within your procedural block. Print the results to an external file named "output_vectors.txt" using $fdisplay().

8.4.10 Design a Verilog test bench that reads in test vectors from an external file to verify the functional operation of the system in Fig. 8.4. Create an input text file called "input_vectors.txt" that contains each input code for the vector ABCD (i.e., "0000", "0001", "0010", . . ., "1111"), each on a separate line in the file. Your test bench should read in the vectors using $readmemb(), drive each code into the DUT, and print the results to the transcript using $display().

Chapter 9: Behavioral Modeling of Sequential Logic

In this chapter, we will look at modeling sequential logic using the more sophisticated behavioral modeling techniques presented in Chap. 8. We will begin by looking at modeling sequential storage devices. Next, we will look at the behavioral modeling of finite-state machines. Finally, we will look at register transfer level, or RTL modeling. The goal of this chapter is to provide an understanding of how hardware description languages can be used to create behavioral models of synchronous digital systems.

Learning Outcomes—After completing this chapter, you will be able to:

9.1 Design a Verilog behavioral model for a sequential logic storage device.
9.2 Describe the process for creating a Verilog behavioral model for a finite-state machine.
9.3 Design a Verilog behavioral model for a finite-state machine.
9.4 Design a Verilog behavioral model for a counter.
9.5 Design a Verilog register transfer level (RTL) model of a synchronous digital system.

9.1 Modeling Sequential Storage Devices in Verilog

9.1.1 D-Latch

Let's begin with the model of a simple D-Latch. Since the outputs of this sequential storage device are not updated continuously, its behavior is modeled using a procedural assignment. Since we want to create a synthesizable model of sequential logic, non-blocking assignments are used. In the sensitivity list, we need to include the C input since it controls when the D-Latch is in track or store mode. We also need to include the D input in the sensitivity list because during the track mode, the output Q will be assigned the value of D, so any change on D needs to trigger the procedural assignments. The use of an if-else statement is used to model the behavior during track mode (C = 1). Since the behavior is not explicitly stated for when C = 0, the outputs will hold their last value, which allows us to simply omit the else portion of the if statement to complete the model. Example 9.1 shows the behavioral model for a D-Latch.

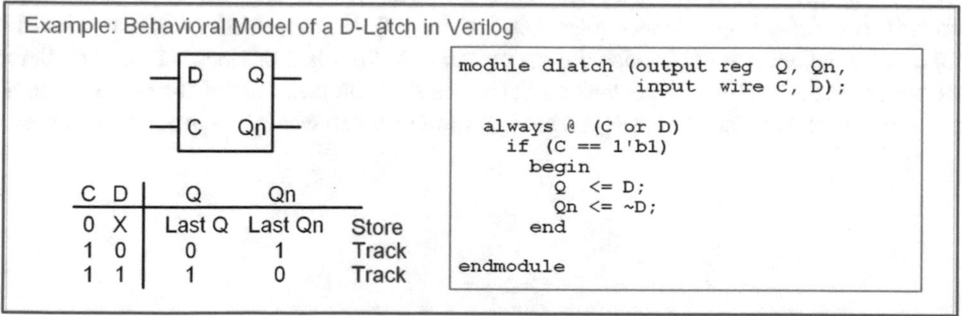

C	D	Q	Qn	
0	X	Last Q	Last Qn	Store
1	0	0	1	Track
1	1	1	0	Track

```
module dlatch (output reg  Q, Qn,
               input  wire C, D);

  always @ (C or D)
    if (C == 1'b1)
      begin
        Q  <= D;
        Qn <= ~D;
      end

endmodule
```

Example 9.1
Behavioral model of a D-latch in verilog

© Springer Nature Switzerland AG 2019
B. J. LaMeres, *Introduction to Logic Circuits & Logic Design with Verilog*,
https://doi.org/10.1007/978-3-030-13605-5_9

9.1.2 D-Flip-Flop

The rising edge behavior of a D-flip-flop is modeled using a (posedge Clock) Boolean condition in the sensitivity list of a procedural block. Example 9.2 shows the behavioral model for a rising edge-triggered D-flip-flop with both Q and Qn outputs.

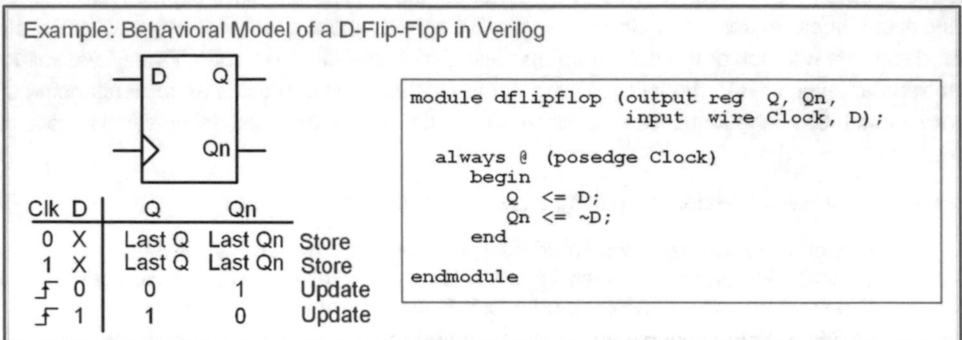

Clk	D	Q	Qn	
0	X	Last Q	Last Qn	Store
1	X	Last Q	Last Qn	Store
↑	0	0	1	Update
↑	1	1	0	Update

```
module dflipflop (output reg  Q, Qn,
                  input  wire Clock, D);

  always @ (posedge Clock)
    begin
      Q  <= D;
      Qn <= ~D;
    end

endmodule
```

Example 9.2
Behavioral model of a D-flip-flop in verilog

9.1.3 D-Flip-Flop with Asynchronous Reset

D-flip-flops typically have a reset line to initialize their outputs to known states (e.g., Q = 0, Qn = 1). Resets are asynchronous, meaning whenever they are asserted, assignments to the outputs take place immediately. If a reset was *synchronous*, the outputs would only update on the next rising edge of the clock. This behavior is undesirable because if there is a system failure, there is no guarantee that a clock edge will ever occur. Thus, the reset may never take place. Asynchronous resets are more desirable not only to put the D-flip-flops into a known state at start-up, but also to recover from a system failure that may have impacted the clock signal. In order to model this asynchronous behavior, the reset signal is included in the sensitivity list. This allows both clock and the reset transitions to trigger the procedural block. The edge sensitivity of the reset can be specified using posedge (active HIGH) or negedge (active LOW). Within the block an if-else statement is used to determine whether the reset has been asserted or a rising edge of the clock has occurred. The if-else statement first checks whether the reset input has been asserted since it has the highest priority. If it has, it makes the appropriate assignments to the outputs (Q = 0, Qn = 1). If the reset has not been asserted, the *else* clause is executed, which corresponds to a rising edge of clock (Q <= D, Qn <= ~D). No other assignments are listed in the block; thus, the outputs are only updated on a transition of the reset or clock. At all other times, the outputs remain at their current value, thus modeling the store behavior of the D-flip-flop. Example 9.3 shows the behavioral model for a rising edge-triggered D-flip-flop with an asynchronous, active LOW reset.

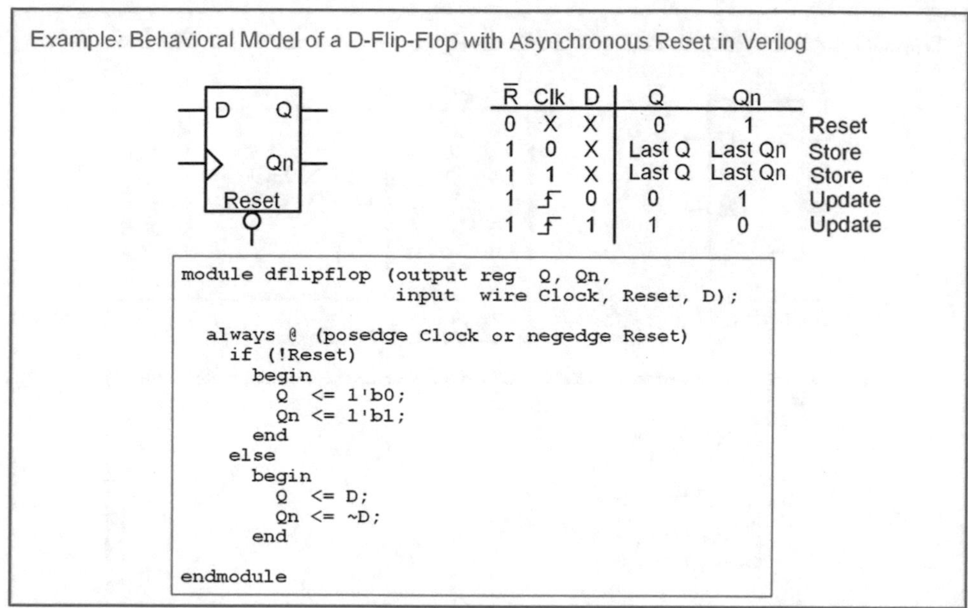

$\overline{R}$	Clk	D	Q	Qn	
0	X	X	0	1	Reset
1	0	X	Last Q	Last Qn	Store
1	1	X	Last Q	Last Qn	Store
1	↑	0	0	1	Update
1	↑	1	1	0	Update

```
module dflipflop (output reg  Q, Qn,
                  input  wire Clock, Reset, D);

  always @ (posedge Clock or negedge Reset)
    if (!Reset)
      begin
        Q  <= 1'b0;
        Qn <= 1'b1;
      end
    else
      begin
        Q  <= D;
        Qn <= ~D;
      end

endmodule
```

Example 9.3
Behavioral model of a D-flip-flop with asynchronous reset in verilog

9.1.4 D-Flip-Flop with Asynchronous Reset and Preset

A D-flip-flop with both an asynchronous reset and asynchronous preset is handled in a similar manner as the D-flip-flop in the prior section. The preset input is included in the sensitivity list in order to trigger the block whenever a transition occurs on either the clock, reset, or preset inputs. The edge sensitivity keywords are used to dictated whether the preset is active HIGH or LOW. Nested if-else statements are used to first check whether a reset has occurred; then whether a preset has occurred; and finally, whether a rising edge of the clock has occurred. Example 9.4 shows the model for a rising edge-triggered D-flip-flop with asynchronous, active LOW reset and preset.

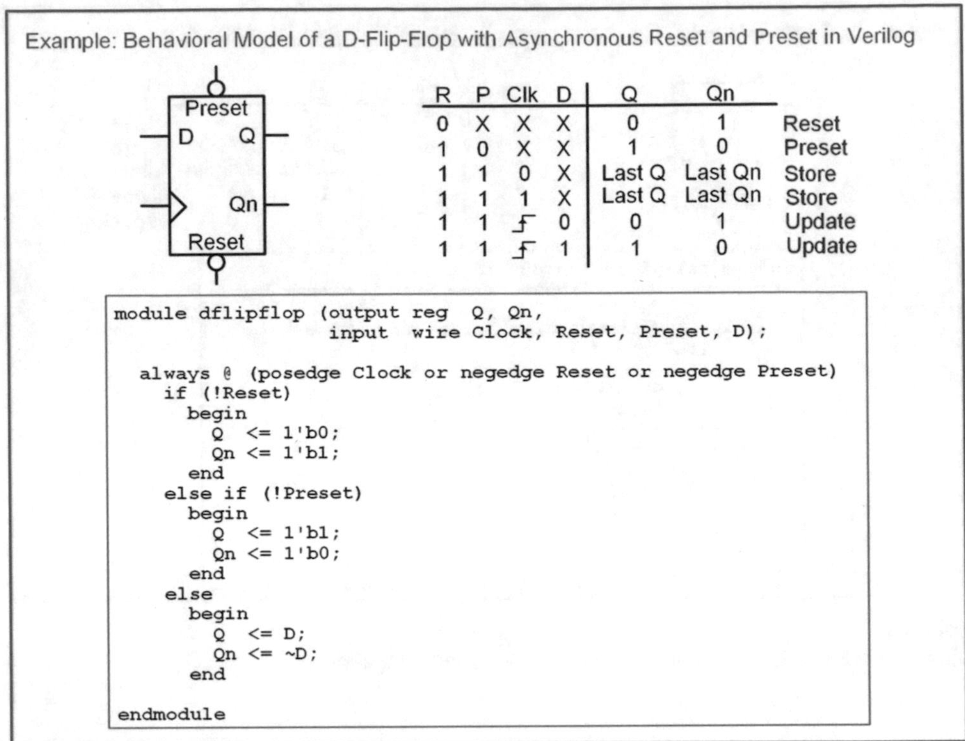

Example: Behavioral Model of a D-Flip-Flop with Asynchronous Reset and Preset in Verilog

R	P	Clk	D	Q	Qn	
0	X	X	X	0	1	Reset
1	0	X	X	1	0	Preset
1	1	0	X	Last Q	Last Qn	Store
1	1	1	X	Last Q	Last Qn	Store
1	1	↑	0	0	1	Update
1	1	↑	1	1	0	Update

```
module dflipflop (output reg  Q, Qn,
                  input  wire Clock, Reset, Preset, D);

  always @ (posedge Clock or negedge Reset or negedge Preset)
    if (!Reset)
      begin
        Q  <= 1'b0;
        Qn <= 1'b1;
      end
    else if (!Preset)
      begin
        Q  <= 1'b1;
        Qn <= 1'b0;
      end
    else
      begin
        Q  <= D;
        Qn <= ~D;
      end

endmodule
```

Example 9.4
Behavioral model of a D-flip-flop with asynchronous reset and preset in verilog

9.1.5 D-Flip-Flop with Synchronous Enable

An enable input is also a common feature of modern D-flip-flops. Enable inputs are synchronous, meaning that when they are asserted, action is only taken on the rising edge of the clock. This means that the enable input is not included in the sensitivity list of the always block. Since enable is only considered when there is a rising edge of the clock, the logic for the enable is handled in a nested if-else statement that is included in the section that models the behavior for when a rising edge of clock is detected. Example 9.5 shows the model for a D-flip-flop with a synchronous enable (EN) input. When EN = 1, the D-flip-flop is enabled, and assignments are made to the outputs only on the rising edge of the clock. When EN = 0, the D-flip-flop is disabled and assignments to the outputs are not made. When disabled, the D-flip-flop effectively ignores rising edges on the clock and the outputs remain at their last values.

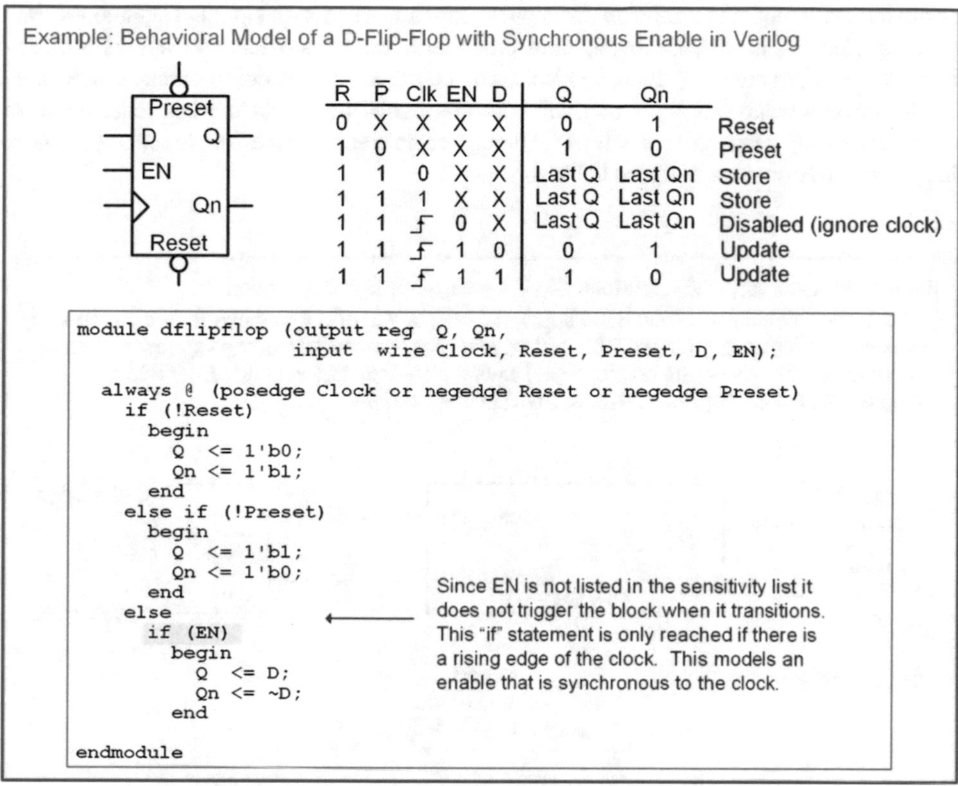

$\overline{R}$	$\overline{P}$	Clk	EN	D	Q	Qn	
0	X	X	X	X	0	1	Reset
1	0	X	X	X	1	0	Preset
1	1	0	X	X	Last Q	Last Qn	Store
1	1	1	X	X	Last Q	Last Qn	Store
1	1	↑	0	X	Last Q	Last Qn	Disabled (ignore clock)
1	1	↑	1	0	0	1	Update
1	1	↑	1	1	1	0	Update

```
module dflipflop (output reg  Q, Qn,
                  input  wire Clock, Reset, Preset, D, EN);

  always @ (posedge Clock or negedge Reset or negedge Preset)
    if (!Reset)
      begin
        Q  <= 1'b0;
        Qn <= 1'b1;
      end
    else if (!Preset)
      begin
        Q  <= 1'b1;
        Qn <= 1'b0;
      end
    else
      if (EN)
        begin
          Q  <= D;
          Qn <= ~D;
        end

endmodule
```


Example 9.5
Behavioral model of a D-flip-flop with synchronous enable in verilog

Concept Check

CC9.1 Why is the D input not listed in the sensitivity list of a D-flip-flop?

A) To simplify the behavioral model.

B) To avoid a setup time violation if D transitions too closely to the clock.

C) Because a rising edge of clock is needed to make the assignment.

D) Because the outputs of the D-flip-flop are not updated when D changes.

9.2 Modeling Finite-State Machines in Verilog

Finite-state machines can be easily modeled using the behavioral constructs from Chap. 8. The most common modeling practice for FSMs is to declare two signals of type reg that are called *current_state* and *next_state*. Then a parameter is declared for each descriptive state name in the state diagram. A parameter also requires a value, so the state encoding can be accomplished during the parameter declaration. Once the signals and parameters are created, all of the procedural assignments in the state machine model can use the descriptive state names in their signal assignments. Within the Verilog state machine model, three separate procedural blocks are used to describe each of the functional blocks, *state memory*, *next state logic*, and *output logic*. In order to examine how to model a finite-state machine using this approach, let's use the push-button window controller example from Chap. 7. Example 9.6 gives the overview of the design objectives for this example and the state diagram describing the behavior to be modeled in Verilog.

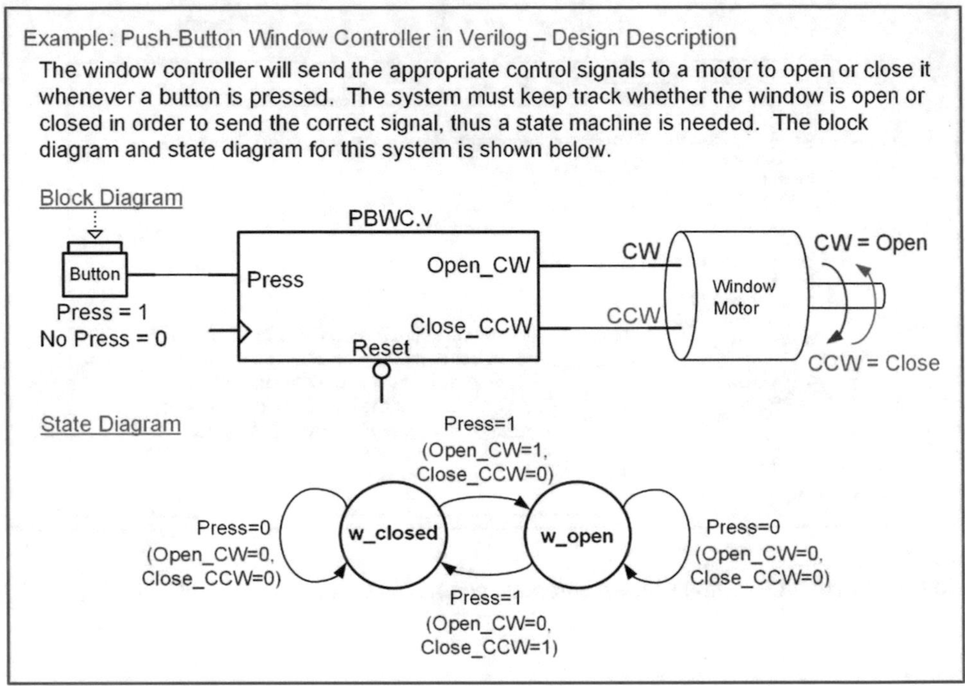

Example 9.6
Push-button window controller in verilog: design description

Let's begin by defining the ports of the module. The system has an input called *Press* and two outputs called *Open_CW* and *Close_CCW*. The system also has clock and reset inputs. We will design the system to update on the rising edge of the clock and have an asynchronous, active LOW, reset. Example 9.7 shows the port definitions for this example. Note that outputs are declared as type reg, while inputs are declared as type wire.

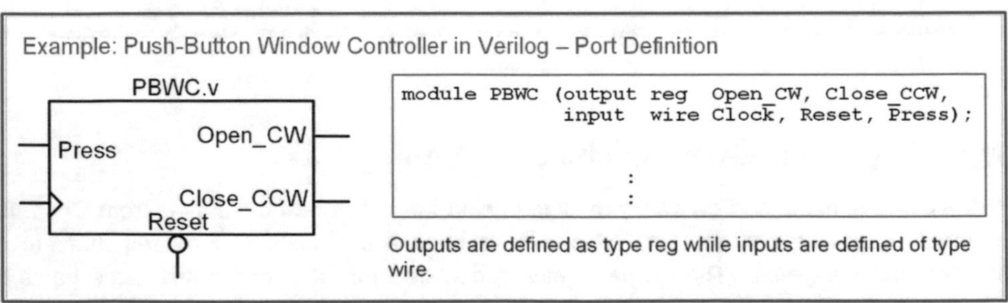

Example 9.7
Push-button window controller in verilog: port definition

9.2.1 Modeling the States

Now we begin designing the finite-state machine in Verilog using behavioral modeling constructs. The first step is to create two signals that will be used for the state variables. In this text we will always name these signals *current_state* and *next_state*. The signal current_state will represent the outputs of the D-flip-flops forming the state memory and will hold the current state code. The signal next_state will represent the D inputs to the D-flip-flops forming the state memory and will receive the value from the next state logic circuitry. Since the FSM will be modeled using procedural assignment, both of these signals will be declared of type reg. The width of the reg vector depends on the number of states in the machine and the encoding technique chosen. The next step is to declare parameters for each of the descriptive state names in the state diagram. The state encoding must be decided at this point. The following syntax shows how to declare the current_state and next_state signals and the parameters. Note that since this machine only has two states, the width of these signals is only 1-bit.

```
reg         current_state, next_state;
parameter   w_closed = 1'b0,
            w_open   = 1'b1;
```

9.2.2 The State Memory Block

Now that we have variables and parameters for the states of the FSM, we can create the model for the state memory. State memory is modeled using its own procedural block. This block models the behavior of the D-flip-flops in the FSM that are holding the current state on their Q outputs. Each time there is a rising edge of the clock, the current state is updated with the next state value present on the D inputs of the D-flip-flops. This block must also model the reset condition. For this example, we will have the state machine go to the *w_closed* state when Reset is asserted. At all other times, the block will simply update current_state with next_state on every rising edge of the clock. The block model is very similar to the model of a D-flip-flop. This is as expected since this block will synthesize into one or more D-flip-flops to hold the current state. The sensitivity list contains only Clock and Reset and assignments are only made to the signal current_state. The following syntax shows how to model the state memory of this FSM example.

```
always @ (posedge Clock or negedge Reset)
  begin: STATE_MEMORY
    if (!Reset)
       current_state <= w_closed;
    else
       current_state <= next_state;
  end
```

9.2.3 The Next State Logic Block

Now we model the next state logic of the FSM using a second procedural block. Recall that the next state logic is combinational logic; thus, we need to include all of the input signals that the circuit considers in the next state calculation in the sensitivity list. The current_state signal will always be included in the sensitivity list of the next state logic block in addition to any inputs to the system. For this example, the system has one other input called *Press*. This block makes assignments to the next_state signal. It is common to use a case statement to separate out the assignments that occur at each state. At each state within the case statement, an if-else statement is used to model the assignments for different input conditions on Press. The following syntax shows how to model the next state logic of this FSM example. Notice that we include a *default* clause in the case statement to ensure that the state machine has a path back to the reset state in the case of an unexpected fault.

```
always @ (current_state or Press)
  begin: NEXT_STATE_LOGIC
    case (current_state)
      w_closed : if (Press == 1'b1) next_state = w_open; else next_state = w_closed;
      w_open   : if (Press == 1'b1) next_state = w_closed; else next_state = w_open;
      default  : next_state = w_closed;
    endcase
  end
```

9.2.4 The Output Logic Block

Now we model the output logic of the FSM using a third procedural block. Recall that output logic is combinational logic; thus, we need to include all of the input signals that this circuit considers in the output assignments. The current_state will always be included in the sensitivity list. If the FSM is a Mealy machine, then the system inputs will also be included in the sensitivity list. If the machine is a Moore machine, then only the current_state will be present in the sensitivity list. For this example, the FSM is a Mealy machine, so the input Press needs to be included in the sensitivity list. Note that this block only makes assignments to the outputs of the machine (Open_CW and Close_CCW). The following syntax shows how to model the output logic of this FSM example. Again, we include a *default* clause to ensure that the state machine has explicit output behavior in the case of a fault.

```
always @ (current_state or Press)
   begin: OUTPUT_LOGIC
     case (current_state)
       w_closed : if (Press == 1'b1)
                    begin
                      Open_CW  = 1'b1;
                      Close_CCW = 1'b0;
                    end
                  else
                    begin
                      Open_CW  = 1'b0;
                      Close_CCW = 1'b0;
                    end
       w_open   : if (Press == 1'b1)
                    begin
                      Open_CW  = 1'b0;
                      Close_CCW = 1'b1;
                    end
                  else
                    begin
                      Open_CW  = 1'b0;
                      Close_CCW = 1'b0;
                    end
       default : begin
                      Open_CW  = 1'b0;
                      Close_CCW = 1'b0;
                    end
   endcase
end
```

Putting this all together yields a behavioral model for the FSM that can be simulated and synthesized. Example 9.8 shows the entire model for this example.

Example: Push-Button Window Controller in Verilog – Full Model

```
module PBWC (output reg  Open_CW, Close_CCW,
             input  wire Clock, Reset, Press);

  reg       current_state, next_state;
  parameter w_closed = 1'b0,
            w_open   = 1'b1;

  always @ (posedge Clock or negedge Reset)
    begin: STATE_MEMORY
      if (!Reset)
        current_state <= w_closed;
      else
        current_state <= next_state;
    end

  always @ (current_state or Press)
    begin: NEXT_STATE_LOGIC
      case (current_state)
        w_closed : if (Press == 1'b1)
                      next_state = w_open;
                   else
                      next_state = w_closed;
        w_open   : if (Press == 1'b1)
                      next_state = w_closed;
                   else
                      next_state = w_open;
        default  : next_state = w_closed;
      endcase
    end

  always @ (current_state or Press)
     begin: OUTPUT_LOGIC
       case (current_state)
         w_closed : if (Press == 1'b1)
                      begin
                        Open_CW   = 1'b1;
                        Close_CCW = 1'b0;
                      end
                    else
                      begin
                        Open_CW   = 1'b0;
                        Close_CCW = 1'b0;
                      end
         w_open   : if (Press == 1'b1)
                      begin
                        Open_CW   = 1'b0;
                        Close_CCW = 1'b1;
                      end
                    else
                      begin
                        Open_CW   = 1'b0;
                        Close_CCW = 1'b0;
                      end
         default : begin
                        Open_CW   = 1'b0;
                        Close_CCW = 1'b0;
                   end
       endcase
   end

endmodule
```

Declaration of state variables and state encoding.

State memory block. This is sequential logic so non-blocking assignments are used. This block only makes assignments to the signal "current_state".

Next state logic block. This is combinational logic so blocking assignments are used. This block only makes assignments to the signal "next_state".

Output logic block. This is combinational logic so blocking assignments are used. Since this is a Mealy machine, the current state and input are listed in the sensitivity list. This block only makes assignments to the outputs "Open_CW" and "Close_CCW".

Example 9.8
Push-button window controller in verilog: full model

Example 9.9 shows the simulation waveform for this state machine. This functional simulation was performed using ModelSim-Altera Starter Edition 10.1d. A macro file was used to display the current and next state variables using their parameter names instead of their state codes. This allows the functionality of the FSM to be more easily observed. This approach will be used for the rest of the FSM examples in this book.

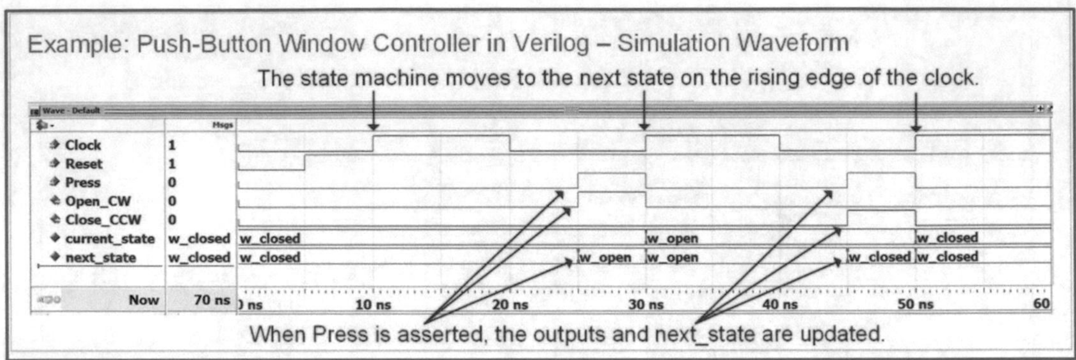

Example 9.9
Push-button window controller in verilog: simulation waveform

9.2.5 Changing the State Encoding Approach

In the prior example, we only had two states and they were encoded as w_closed = 1'b0; w_open_1'b1. This encoding technique is considered *binary*; however, a *gray code* approach would yield the same codes since the width of the variables were only one bit. The way that state variables and state codes are assigned in Verilog makes is straightforward to change the state codes. The only consideration that must be made is expanding the size of the current_state and next_state variables to accommodate the new state codes. The following example shows how the state encoding would look if a *one-hot* approach was used (w_closed = 2'b01; w_open_2'b10). Note that the state variables now must be two bits wide. This means the state variables need to be declared as type reg[1:0]. Example 9.10 shows the resulting simulation waveforms. The simulation waveform shows the value of the state codes instead of the state names.

```
reg [1:0]  current_state, next_state;
parameter  w_closed = 2'b01,
           w_open   = 2'b10;
```

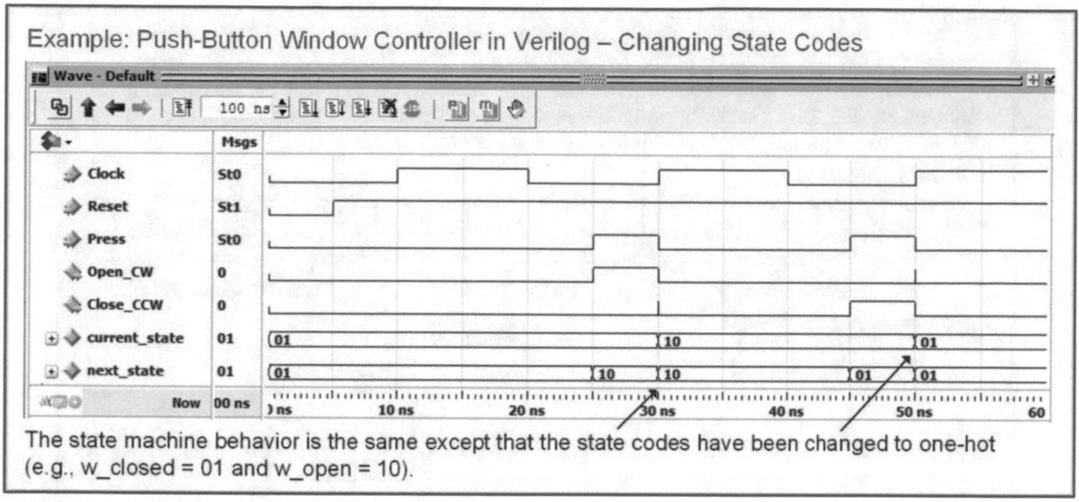

Example 9.10
Push-button window controller in verilog: changing state codes

CONCEPT CHECK

CC9.2 Why is it always a good design approach to model a generic finite-state machine using three processes?

A) For readability.

B) So that it is easy to identify whether the machine is a Mealy or Moore.

C) So that the state memory process can be reused in other FSMs.

D) Because each of the three sub-systems of a FSM has unique inputs and outputs that should be handled using dedicated processes.

9.3 FSM Design Examples in Verilog

This section presents a set of example finite-state machine designs using the behavioral modeling constructs of Verilog. These examples are the same state machines that were presented in Chap. 7.

9.3.1 Serial Bit Sequence Detector in Verilog

Let's look at the design of the serial bit sequence detector finite-state machine from Chap. 7 using the behavioral modeling constructs of Verilog. Example 9.11 shows the design description and port definition for this state machine.

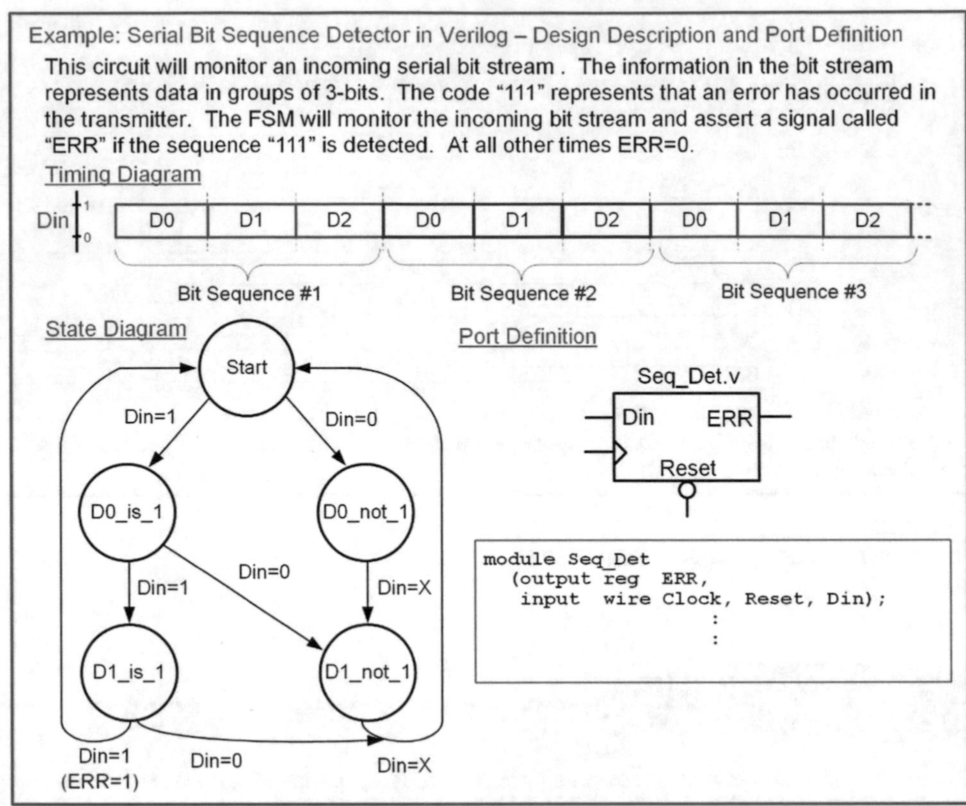

Example 9.11
Serial bit sequence detector in verilog: design description and port definition

Example 9.12 shows the full model for the serial bit sequence detector. Notice that the states are encoded in binary, which requires three bits for the variables current_state and next_state.

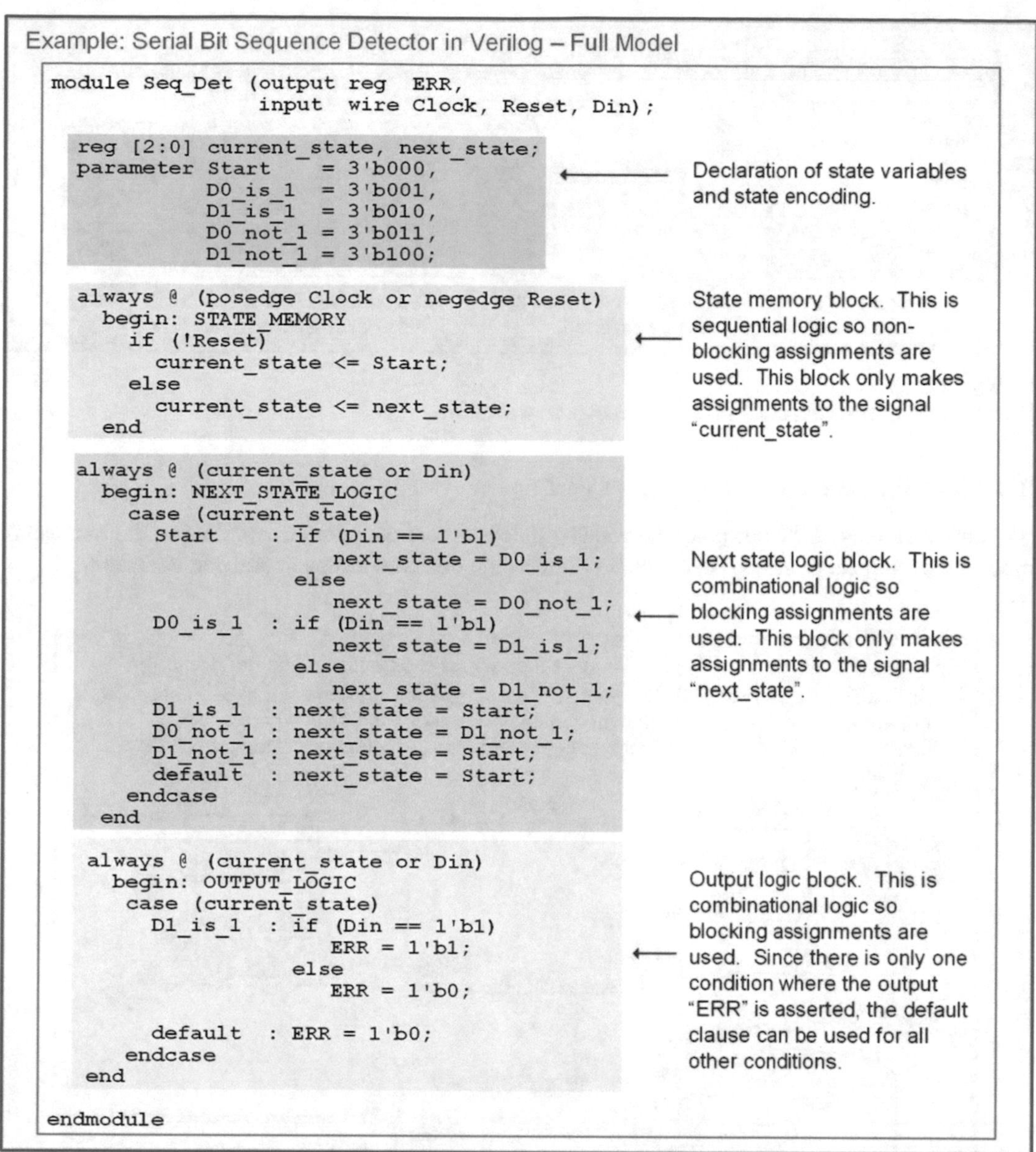

Example 9.12
Serial bit sequence detector in verilog: full model

Example 9.13 shows the functional simulation waveform for this design.

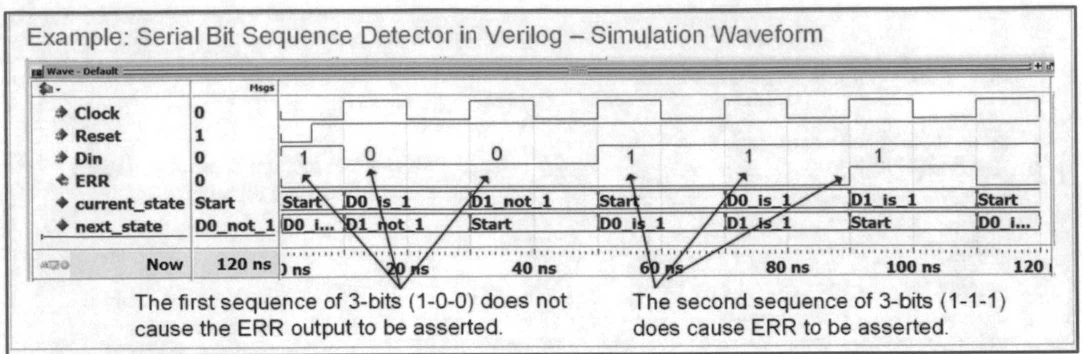

Example 9.13
Serial bit sequence detector in verilog: simulation waveform

9.3.2 Vending Machine Controller in Verilog

Let's now look at the design of the vending machine controller from Chap. 7 using the behavioral modeling constructs of Verilog. Example 9.14 shows the design description and port definition.

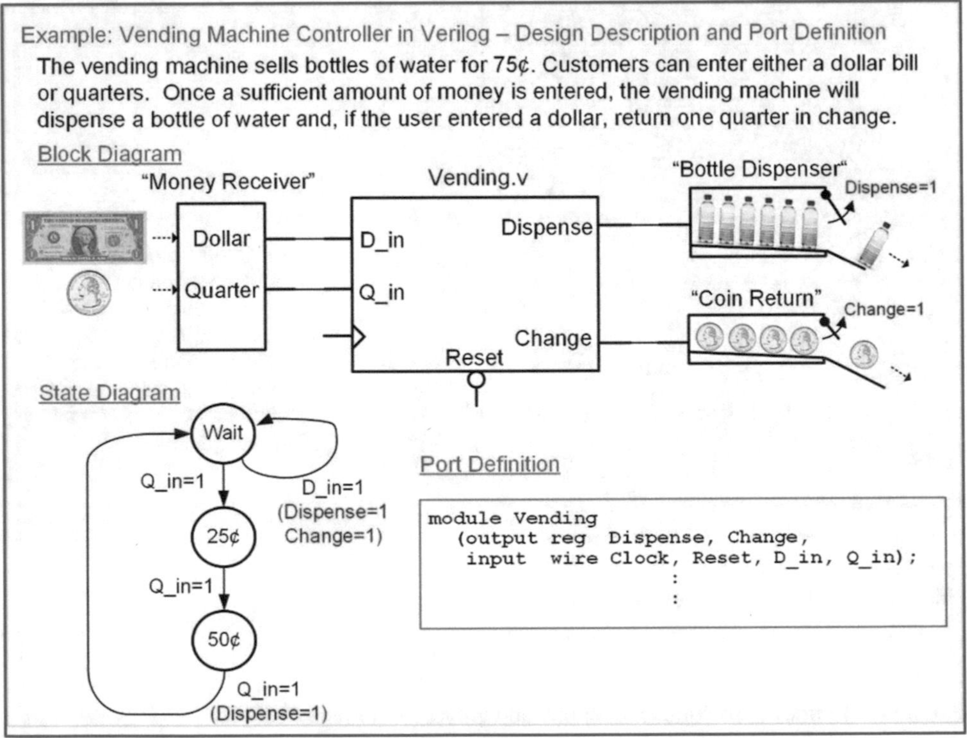

Example 9.14
Vending machine controller in verilog: design description and port definition

Example 9.15 shows the full model for the vending machine controller. In this model, the descriptive state names Wait, 25¢, and 50¢ cannot be used directly. This is because Verilog user-defined names cannot begin with a number. Instead, the letter "s" is placed in front of the state names in order to make them legal Verilog names (i.e., sWait, s25, s50).

Example: Vending Machine Controller in Verilog – Full Model

```
module Vending (output reg  Dispense, Change,
                input  wire Clock, Reset, D_in, Q_in);

  reg [1:0] current_state, next_state;
  parameter sWait = 2'b00, s25 = 2'b01, s50 = 2'b10;

  always @ (posedge Clock or negedge Reset)
    begin: STATE_MEMORY
      if (!Reset)
        current_state <= sWait;
      else
        current_state <= next_state;
    end

  always @ (current_state or D_in or Q_in)
    begin: NEXT_STATE_LOGIC
      case (current_state)
        sWait    : if (Q_in == 1'b1)
                      next_state = s25;
                   else
                      next_state = sWait;
        s25      : if (Q_in == 1'b1)
                      next_state = s50;
                   else
                      next_state = s25;
        s50      : if (Q_in == 1'b1)
                      next_state = sWait;
                   else
                      next_state = s50;
        default  : next_state = sWait;
      endcase
    end

  always @ (current_state or D_in or Q_in)
    begin: OUTPUT_LOGIC
      case (current_state)
        sWait    : if (D_in == 1'b1)
                     begin
                       Dispense = 1'b1; Change = 1'b1;
                     end
                   else
                     begin
                       Dispense = 1'b0; Change = 1'b0;
                     end
        s25      : begin
                     Dispense = 1'b0; Change = 1'b0;
                   end
        s50      : if (Q_in == 1'b1)
                     begin
                       Dispense = 1'b1; Change = 1'b0;
                     end
                   else
                     begin
                       Dispense = 1'b0; Change = 1'b0;
                     end
        default  : begin
                     Dispense = 1'b0; Change = 1'b0;
                   end
      endcase
    end
endmodule
```

State variables and state encoding.

State memory block.

Next state logic block.

Output logic block.

Example 9.15
Vending machine controller in verilog: full model

Example 9.16 shows the resulting simulation waveform for this design.

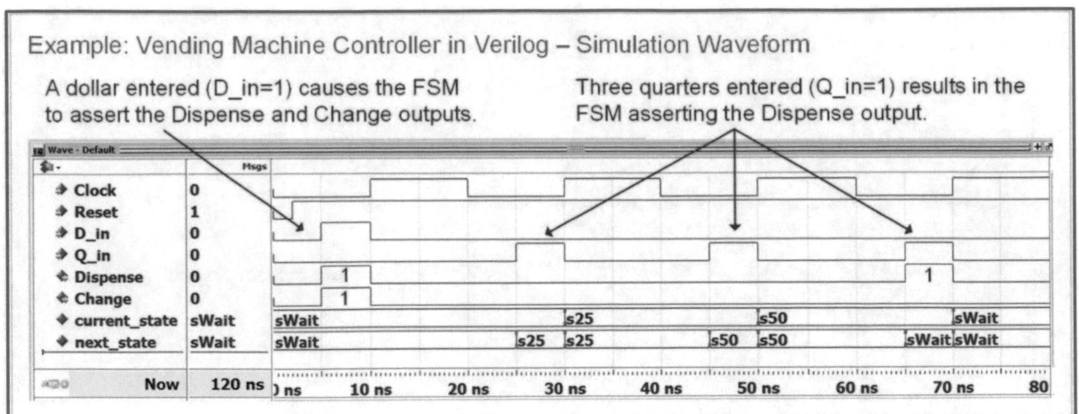

Example 9.16
Vending machine controller in verilog: simulation waveform

9.3.3 2-Bit, Binary Up/Down Counter in Verilog

Let's now look at how a simple counter can be implemented using the three-block behavioral modeling approach in Verilog. Example 9.17 shows the design description and port definition for the 2-bit, binary up/down counter FSM from Chap. 7.

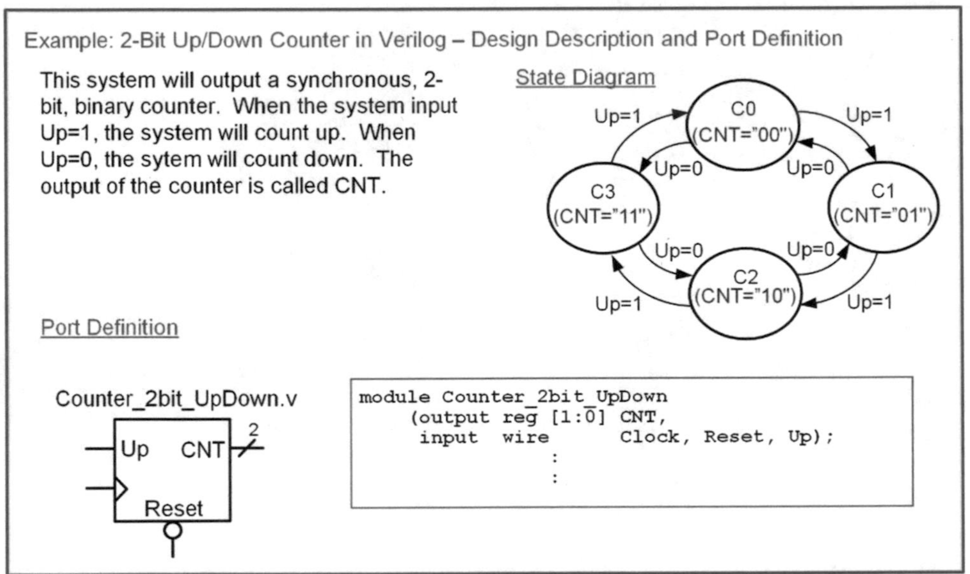

Example 9.17
2-bit up/down counter in verilog: design description and port definition

Example 9.18 shows the full model for the 2-bit up/down counter using the three-block modeling approach. Since a counter's outputs only depend on the current state, counters are Moore machines. This simplifies the output logic block since it only needs to contain the current state in its sensitivity list.

Example: 2-Bit Up/Down Counter in Verilog – Full Model (Three Block Approach)

```
module Counter_2bit_UpDown (output reg [1:0] CNT,
                             input  wire      Clock, Reset, Up);

  reg [1:0] current_state, next_state;
  parameter C0 = 2'b00,
            C1 = 2'b01,
            C2 = 2'b10,
            C3 = 2'b11;

  always @ (posedge Clock or negedge Reset)
    begin: STATE_MEMORY
      if (!Reset)
        current_state <= C0;
      else
        current_state <= next_state;
    end

  always @ (current_state or Up)
    begin: NEXT_STATE_LOGIC
      case (current_state)
        C0      : if (Up == 1'b1) next_state = C1; else next_state = C3;
        C1      : if (Up == 1'b1) next_state = C2; else next_state = C0;
        C2      : if (Up == 1'b1) next_state = C3; else next_state = C1;
        C3      : if (Up == 1'b1) next_state = C0; else next_state = C2;
        default : next_state = C0;
      endcase
    end

   always @ (current_state)
     begin: OUTPUT_LOGIC
      case (current_state)
        C0      : CNT = 2'b00;
        C1      : CNT = 2'b01;
        C2      : CNT = 2'b10;
        C3      : CNT = 2'b11;
        default : CNT = 2'b00;
      endcase
   end

endmodule
```

← State variables and state encoding.

← State memory block.

↙ Next state logic block.

← Output logic block. Note that since this is a Moore machine only the current state is listed in the sensitivity list.

Example 9.18
2-bit up/down counter in verilog: full model (three-block approach)

Example 9.19 shows the resulting simulation waveform for this counter finite-state machine.

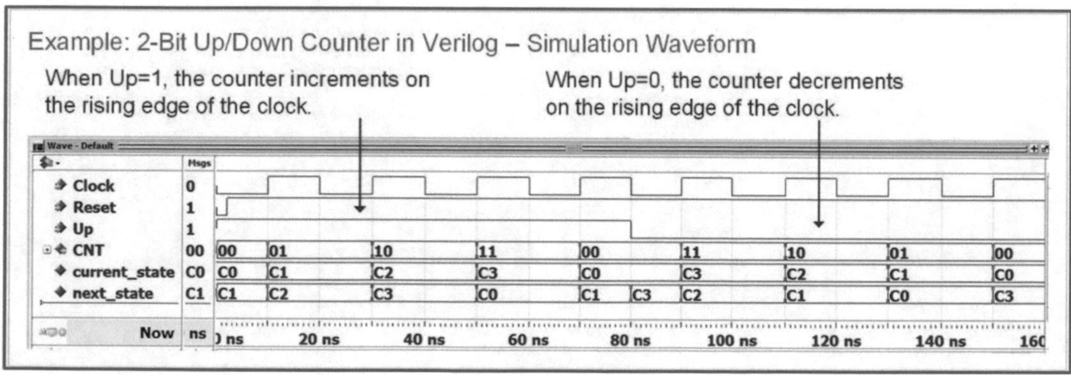

Example 9.19
2-bit up/down counter in verilog: simulation waveform

Concept Check

CC9.3 The procedural block for the state memory is nearly identical for all finite-state machines with one exception. What is it?

A) The sensitivity list may need to include a preset signal.

B) Sometimes it is modeled using an SR latch storage approach instead of with D-flip-flop behavior.

C) The name of the reset state will be different.

D) The current_state and next_state signals are often swapped.

9.4 Modeling Counters in Verilog

Counters are a special case of finite-state machines because they move linearly through their discrete states (either forward or backward) and typically are implemented with state-encoded outputs. Due to this simplified structure and widespread use in digital systems, Verilog allows counters to be modeled using a single-procedural block with arithmetic operators (i.e., + and –). This enables a more compact model and allows much wider counters to be implemented in a practical manner.

9.4.1 Counters in Verilog Using a Single-Procedural Block

Let's look at how we can model a 4-bit, binary up counter with an output called *CNT*. We want to model this counter using the "+" operator to avoid having to explicitly define a state code for each state as in the three-block modeling approach to FSMs. The "+" operator works on the type reg, so the counting behavior can simply be modeled using CNT <= CNT + 1. The procedural block also needs to handle the reset condition. Both the Clock and Reset signals are listed in the sensitivity list. Within the block, an if-else statement is used to handle both the reset and increment behaviors. Example 9.20 shows the Verilog model and simulation waveform for this counter. When the counter reaches its maximum value of "1111," it rolls over to "0000" and continues counting because it is declared to only contain 4-bits.

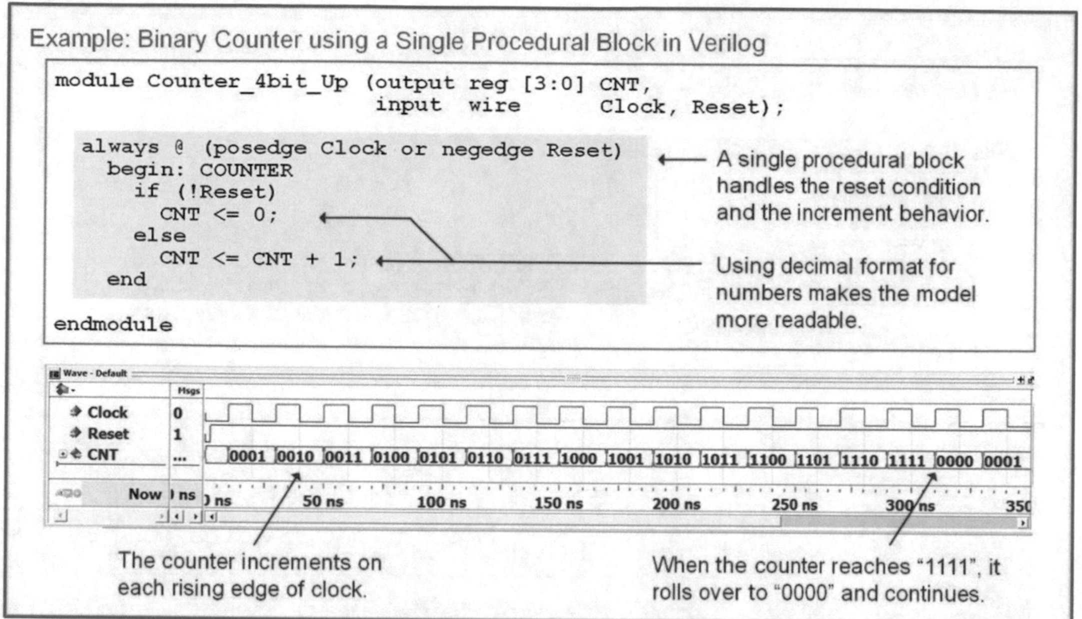

Example 9.20
Binary counter using a single-procedural block in verilog

9.4.2 Counters with Range Checking

When a counter needs to have a maximum range that is different from the maximum binary value of the count vector (i.e., $<2^n$-1), then the procedural block needs to contain *range checking* logic. This can be modeled by inserting a nested if-else statement beneath of the else clause that handles the behavior for when the counter receives a rising clock edge. This nested if-else first checks whether the count has reached its maximum value. If it has, it is reset back to it minimum value. If it hasn't, the counter is incremented as usual. Example 9.21 shows the Verilog model and simulation waveform for a counter with a minimum count value of 0_{10} and a maximum count value of 10_{10}. This counter still requires 4-bits to be able to encode 10_{10}.

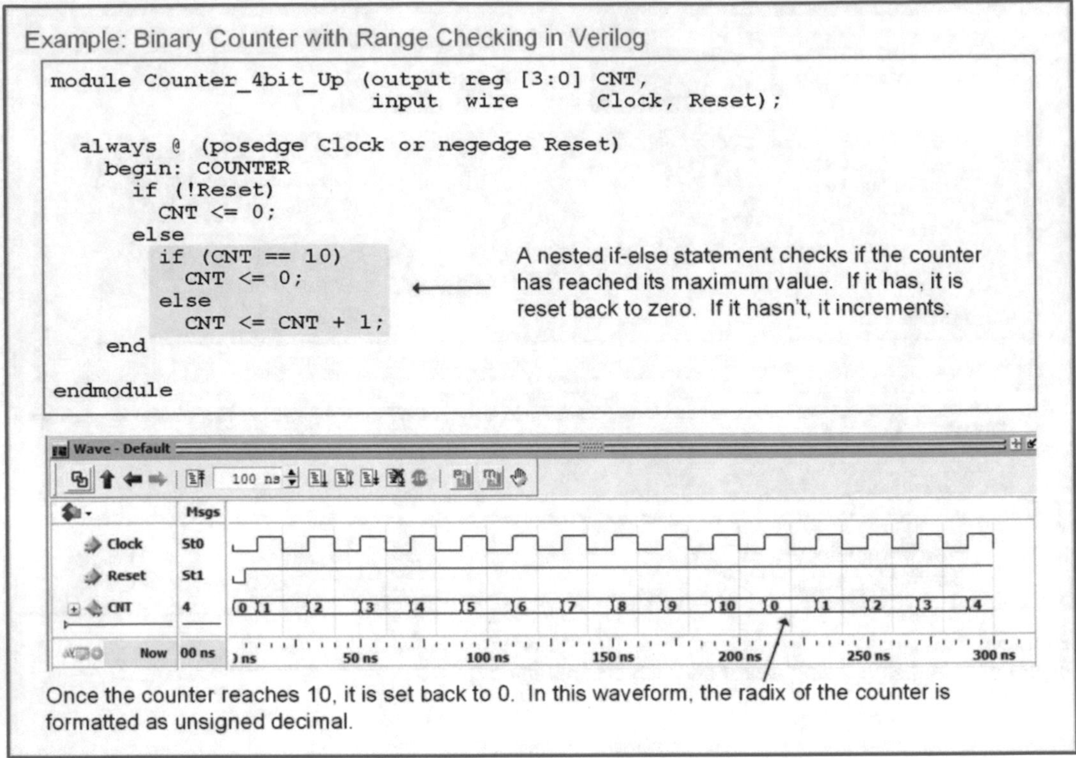

Example 9.21
Binary counter with range checking in verilog

9.4.3 Counters with Enables in Verilog

Including an *enable* in a counter is a common technique to prevent the counter from running continuously. When the enable is asserted, the counter will increment on the rising edge of the clock as usual. When the enable is de-asserted, the counter will simply hold its last value. Enable lines are synchronous, meaning that they are only evaluated on the rising edge of the clock. As such, they are modeled using a nested if-else statement within the main if-else statement checking for a rising edge of the clock. Example 9.22 shows an example model for a 4-bit counter with enable.

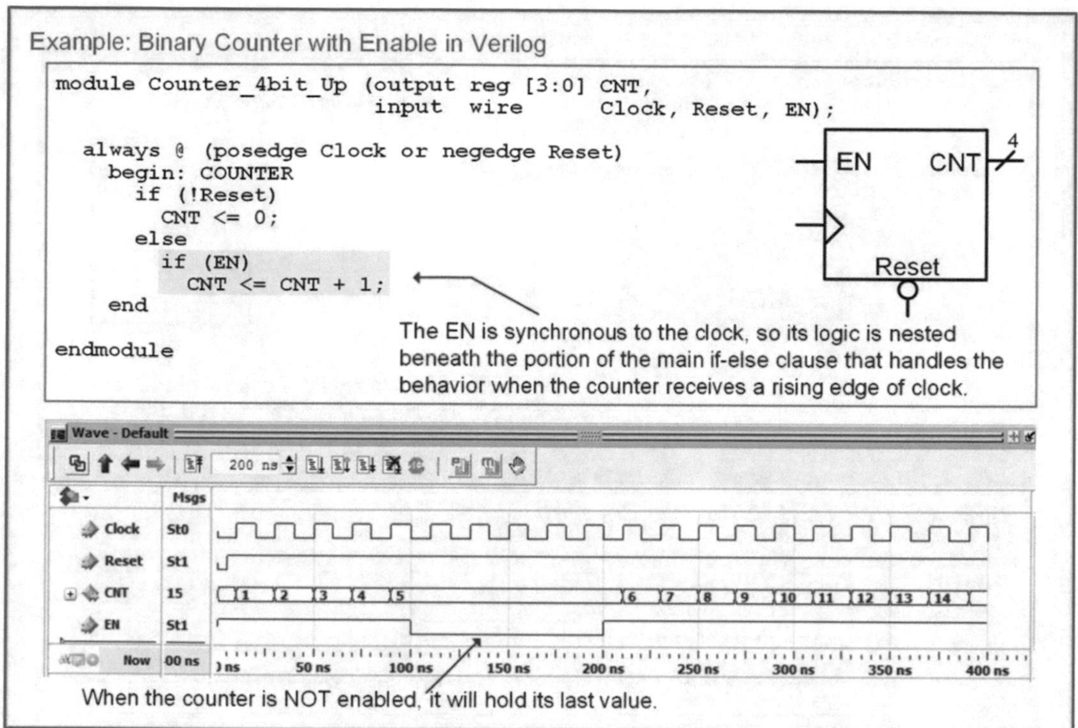

Example 9.22
Binary counter with enable in verilog

9.4.4 Counters with Loads

A counter with a *load* has the ability to set the counter to a specified value. The specified value is provided on an input port (i.e., CNT_in) with the same width as the counter output (CNT). A synchronous load input signal (i.e., Load) is used to indicate when the counter should set its value to the value present on CNT_in. Example 9.23 shows an example model for a 4-bit counter with load capability.

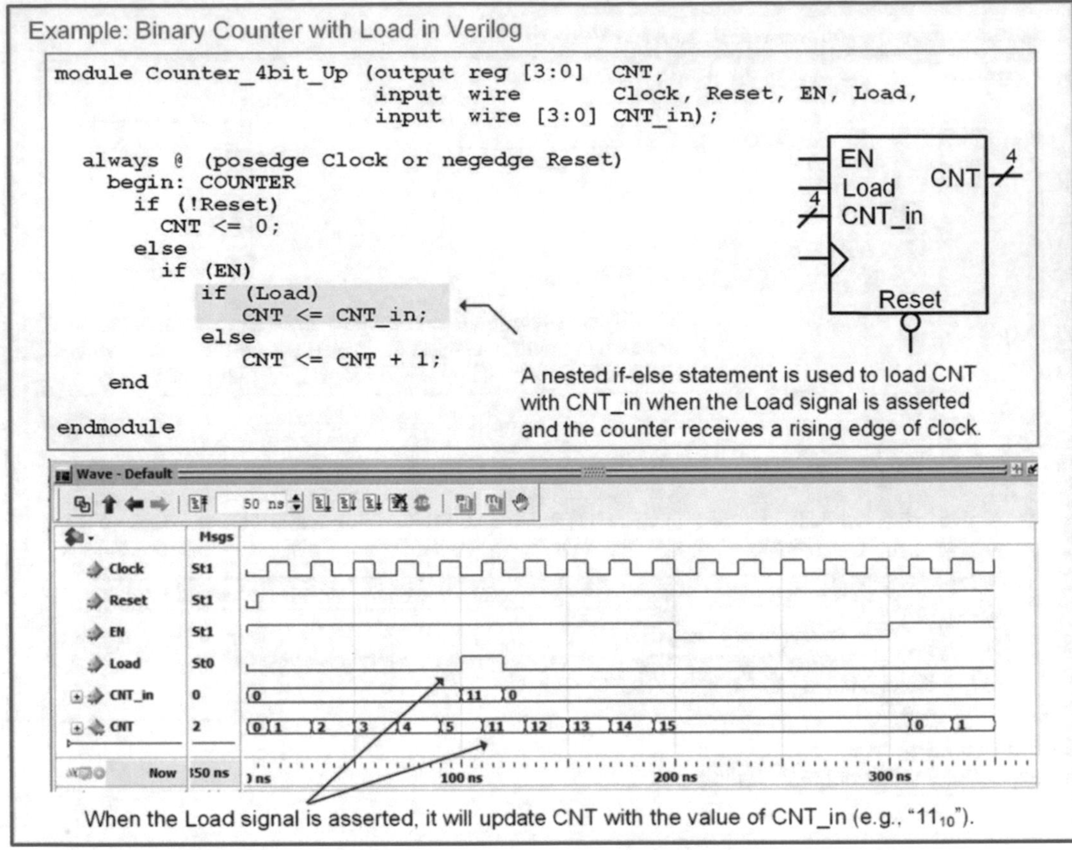

Example 9.23
Binary counter with load in verilog

Concept Check

CC9.4 If a counter is modeled using only one procedural block in Verilog, is it still a finite-state machine? Why or why not?

A) Yes. It is just a special case of a FSM that can easily be modeled using one block. Synthesizers will recognize the single block model as a FSM.

B) No. Using only one block will synthesize into combinational logic. Without the ability to store a state, it is not a finite-state machine.

9.5 RTL Modeling

Register transfer level modeling refers to a level of design abstraction in which vector data is moved and operated on in a synchronous manner. This design methodology is widely used in data path modeling and computer system design.

9.5.1 Modeling Registers in Verilog

The term *register* describes a group of D-Flip-Flops running off of the same clock, reset, and enable inputs. Data is moved in and out of the bank of D-flip-flops as a vector. Logic operations can be made on the vectors and are latched into the register on a clock edge. A register is a higher level of abstraction that allows vector data to be stored without getting into the details of the lower-level implementation of the D-flip-flops and combinational logic. Example 9.24 shows an RTL model of an 8-bit, synchronous register. This circuit has an active LOW, asynchronous reset that will cause the 8-bit output *Reg_Out* to go to 0 when it is asserted. When the reset is not asserted, the output will be updated with the 8-bit input *Reg_In* if the system is enabled (EN = 1), and there is a rising edge on the clock. If the register is disabled (EN = 0), the input clock is ignored. At all other times, the output holds its last value.

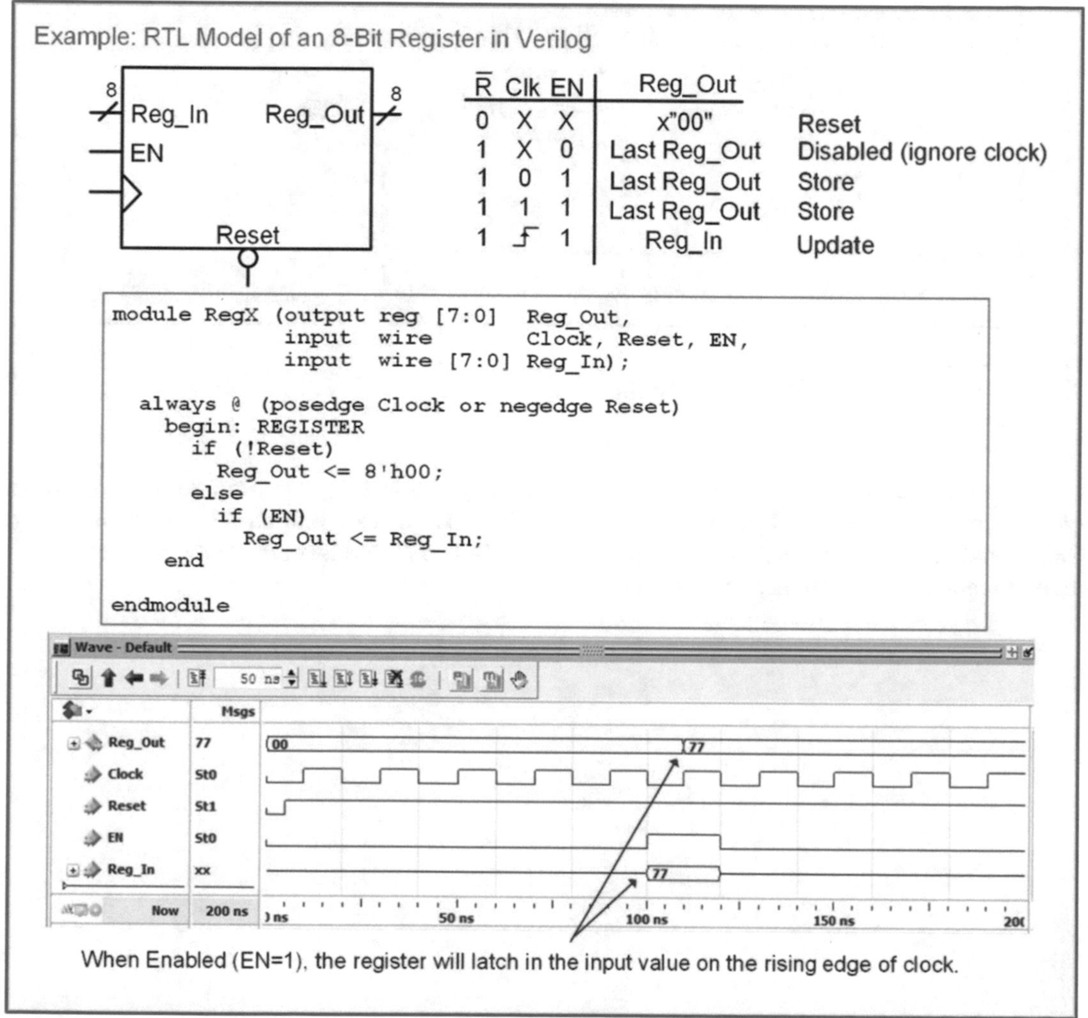

Example 9.24
RTL model of an 8-bit register in verilog

9.5.2 Registers as Agents on a Data Bus

One of the powerful topologies that registers can easily model is a multi-drop bus. In this topology, multiple registers are connected to a data bus as receivers, or *agents*. Each agent has an enable line that controls when it latches information from the data bus into its storage elements. This topology is synchronous, meaning that each agent and the driver of the data bus is connected to the same clock signal. Each agent has a dedicated, synchronous enable line that is provided by a system controller elsewhere in the design. Example 9.25 shows this multi-drop bus topology. In this example system, three registers (A, B, and C) are connected to a data bus as receivers. Each register is connected to the same clock and reset signals. Each register has its own dedicated enable line (A_EN, B_EN, and C_EN).

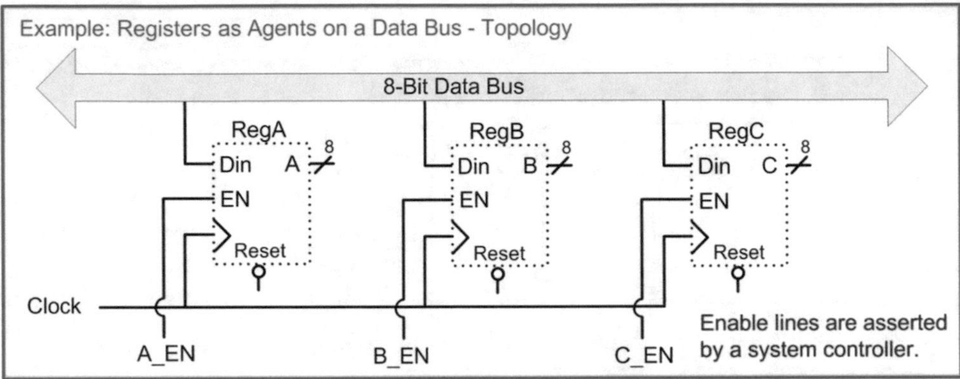

Example 9.25
Registers as agents on a data bus: system topology

This topology can be modeled using RTL abstraction by treating each register as a separate procedural block. Example 9.26 shows how to describe this topology with an RTL model in Verilog. Notice that the three procedural blocks modeling the A, B, and C registers are nearly identical to each other except for the signal names they use.

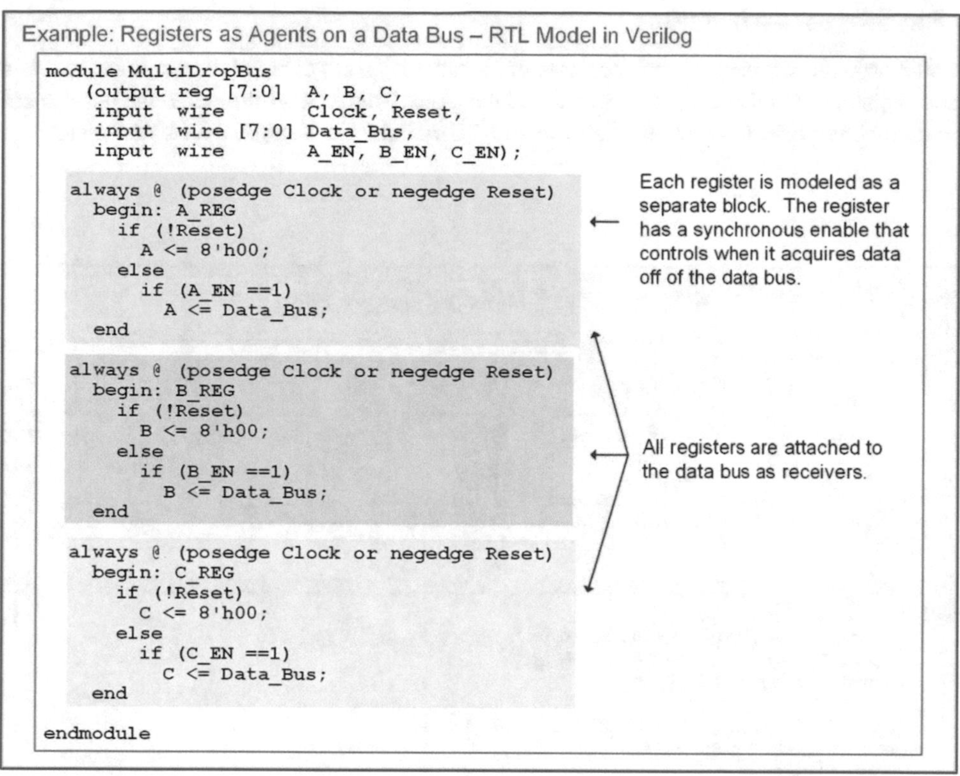

Example: Registers as Agents on a Data Bus – RTL Model in Verilog

```
module MultiDropBus
   (output reg [7:0]  A, B, C,
    input  wire       Clock, Reset,
    input  wire [7:0] Data_Bus,
    input  wire       A_EN, B_EN, C_EN);

  always @ (posedge Clock or negedge Reset)
    begin: A_REG
      if (!Reset)
        A <= 8'h00;
      else
        if (A_EN ==1)
          A <= Data_Bus;
    end

  always @ (posedge Clock or negedge Reset)
    begin: B_REG
      if (!Reset)
        B <= 8'h00;
      else
        if (B_EN ==1)
          B <= Data_Bus;
    end

  always @ (posedge Clock or negedge Reset)
    begin: C_REG
      if (!Reset)
        C <= 8'h00;
      else
        if (C_EN ==1)
          C <= Data_Bus;
    end

endmodule
```


Example 9.26
Registers as agents on a data bus: RTL model in verilog

Example 9.27 shows the resulting simulation waveform for this system. Each register is updated with the value on the data bus whenever its dedicated enable line is asserted.

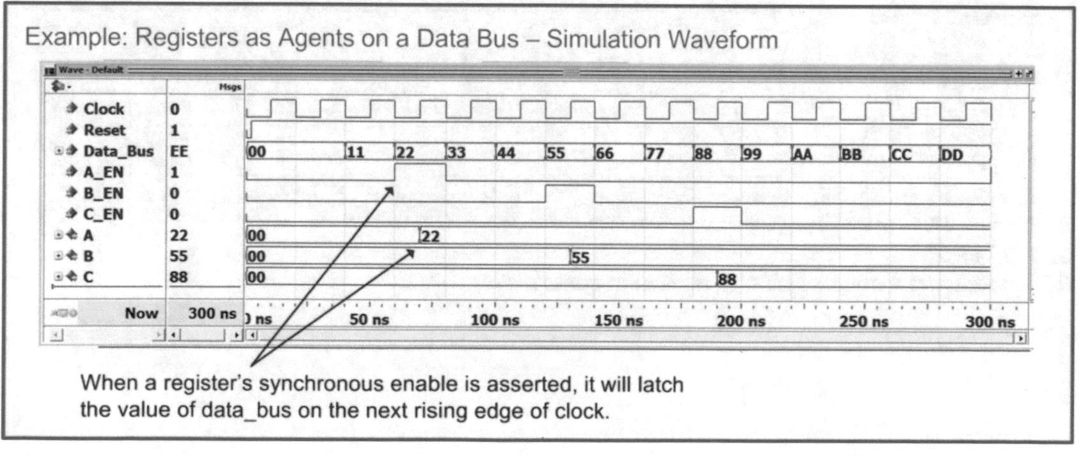

Example 9.27
Registers as agents on a data bus: simulation waveform

9.5.3 Shift Registers in Verilog

A shift register is a circuit which consists of multiple registers connected in series. Data is shifted from one register to another on the rising edge of the clock. This type of circuit is often used in serial-to-parallel data converters. Example 9.28 shows an RTL model for a 4-stage, 8-bit shift register.

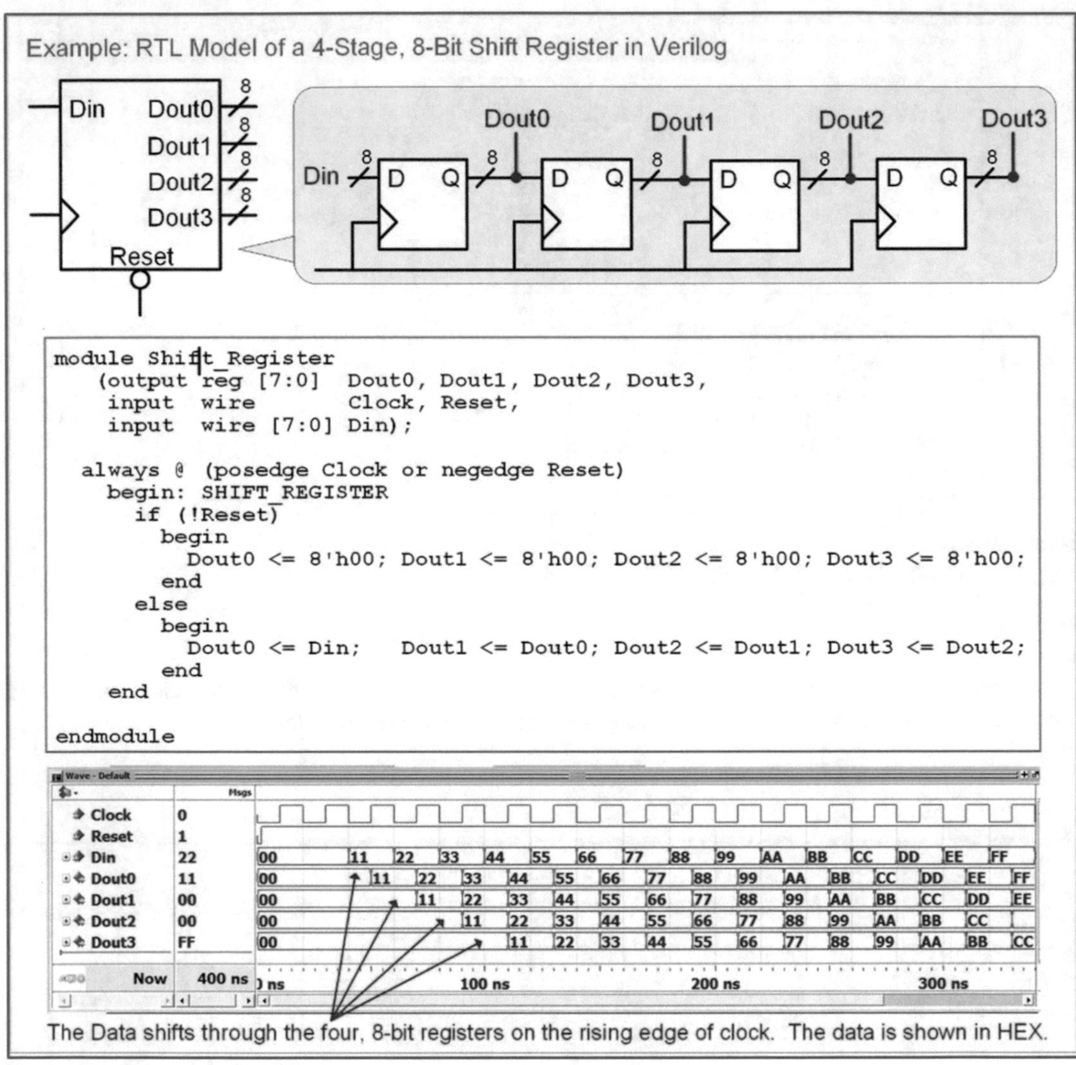


```
module Shift_Register
   (output reg [7:0]  Dout0, Dout1, Dout2, Dout3,
    input  wire       Clock, Reset,
    input  wire [7:0] Din);

  always @ (posedge Clock or negedge Reset)
    begin: SHIFT_REGISTER
      if (!Reset)
        begin
          Dout0 <= 8'h00; Dout1 <= 8'h00; Dout2 <= 8'h00; Dout3 <= 8'h00;
        end
      else
        begin
          Dout0 <= Din;   Dout1 <= Dout0; Dout2 <= Dout1; Dout3 <= Dout2;
        end
    end

endmodule
```

Example 9.28
RTL model of a 4-stage, 8-bit shift register in verilog

Concept Check

CC9.5 Does RTL modeling synthesize as combinational logic, sequential logic, or both? Why?

A) Combinational logic. Since only one process is used for each register, it will be synthesized using basic gates.

B) Sequential logic. Since the sensitivity list contains clock and reset, it will synthesize into only D-flip-flops.

C) Both. The model has a sensitivity list containing clock and reset and uses an if-else statement indicative of a D-flip-flop. This will synthesize a D-flip-flop to hold the value for each bit in the register. In addition, the ability to manipulate the inputs into the register (using either logical operators, arithmetic operators, or choosing different signals to latch) will synthesize into combinational logic in front of the D input to each D-flip-flop.

Summary

- A synchronous system is modeled with a procedural block and a sensitivity list. The clock and reset signals are always listed by themselves in the sensitivity list. Within the block is an if-else statement. The *if* clause of the statement handles the asynchronous reset condition, while the *else* clause handles the synchronous signal assignments.
- Edge sensitivity is modeled within a procedural block using the *(posedge Clock or negedge reset)* syntax in the sensitivity lists.
- Most D-flip-flops and registers contain a synchronous *enable* line. This is modeled using a nested if-else statement within the main procedural block's if-else statement. The nested if-else goes beneath the clause for the synchronous signal assignments.
- Generic finite-state machines are modeled using three separate procedural blocks that describe the behavior of the next state logic, the state memory, and the output logic. Separate blocks are used because each of the three functions in a FSM is dependent on different input signals.
- In Verilog, descriptive state names can be created for a FSM using parameters. Two signals are first declared called *current_state* and *next_state* of type reg. Then a parameter is defined for each unique state in the machine with the state name and desired state code. Throughout the rest of the model, the unique state names can be used as both assignments to current_state/ next_state and as inputs in case and if-else statements. This approach allows the model to be designed using readable syntax while providing a synthesizable design.
- Counters are a special type of finite-state machine that can be modeled using a single-procedural block. Only the clock and reset signals are listed in the sensitivity list of the counter block.
- Registers are modeled in Verilog in a similar manner to a D-flip-flop with a synchronous enable. The only difference is that the inputs and outputs are vectors.
- Register transfer level, or RTL, modeling provides a higher level of abstraction for moving and manipulating vectors of data in a synchronous manner.

Exercise Problems

Section 9.1: Modeling Sequential Storage Devices in Verilog

9.1.1 How does a Verilog model for a D-flip-flop handle treating reset as the highest priority input?

9.1.2 For a Verilog model of a D-flip-flop with a synchronous enable (EN), why isn't EN listed in the sensitivity list?

9.1.3 For a Verilog model of a D-flip-flop with a synchronous enable (EN), what is the impact of listing EN in the sensitivity list?

9.1.4 For a Verilog model of a D-flip-flop with a synchronous enable (EN), why is the behavior of the enable modeled using a nested if-else statement under the else clause handling the logic for the clock edge input?

Section 9.2: Modeling Finite-State Machines in Verilog

9.2.1 What is the advantage of using *parameters* for the state when modeling a finite-state machine?

9.2.2 What is the advantage of having to assign the state codes during the parameter declaration for the state names when modeling a finite-state machine?

9.2.3 When using the three-procedural block behavioral modeling approach for finite-state machines, does the next state logic block model combinational or sequential logic?

9.2.4 When using the three-procedural block behavioral modeling approach for finite-state machines, does the state memory block model combinational or sequential logic?

9.2.5 When using the three-procedural block behavioral modeling approach for finite-state machines, does the output logic block model combinational or sequential logic?

9.2.6 When using the three-procedural block behavioral modeling approach for finite-state machines, what inputs are listed in the sensitivity list of the next state logic block?

9.2.7 When using the three-procedural block behavioral modeling approach for finite-state machines, what inputs are listed in the sensitivity list of the state memory block?

9.2.8 When using the three-procedural block behavioral modeling approach for finite-state machines, what inputs are listed in the sensitivity list of the output logic block?

9.2.9 When using the three-procedural block behavioral modeling approach for finite-state machines, how can the signals listed in the sensitivity list of the output logic block immediately indicate whether the FSM is a Mealy or a Moore machine?

9.2.10 Why is it not a good design approach to combine the next state logic and output logic behavior into a single-procedural block?

Section 9.3: FSM Design Examples in Verilog

9.3.1 Design a Verilog behavioral model to implement the finite-state machine described by the state diagram in Fig. 9.1. Use the port definition provided in this figure for your design. Use the three-block approach to modeling FSMs described in this chapter for your design. Model the state variables using parameters and encode the states in binary using the following state codes: Start = "00," Midway = "01," Done = "10."

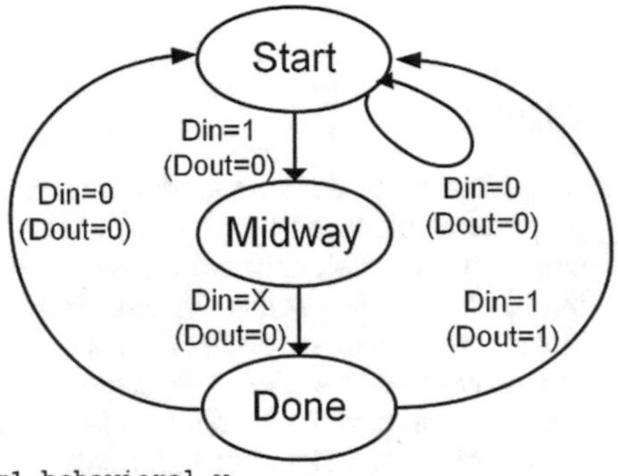

fsm1_behavioral.v

```
module fsm1_behavioral
   (output reg  Dout,
    input  wire Clock, Reset, Din);
                 :
```

Fig. 9.1
FSM 1 state diagram and module definition

9.3.2 Design a Verilog behavioral model to implement the finite-state machine described by the state diagram in Fig. 9.1. Use the port definition provided in this figure for your design. Use the three-block approach to modeling FSMs described in this chapter for your design. Model the state variables using parameters, and encode the states in one-hot using the following state codes: Start = "001," Midway = "010," Done = "100."

9.3.3 Design a Verilog behavioral model to implement the finite-state machine described by the state diagram in Fig. 9.2. Use the port definition provided in this figure for your design. Use the three-block approach to modeling FSMs described in this chapter for your design. Model the state variables using parameters and encode the states in binary using the following state codes: S0 = "00," S1 = "01," S2 = "10," and S3 = "11."

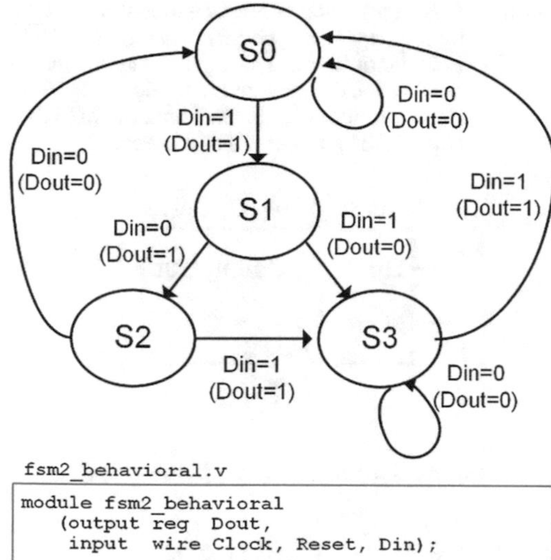

fsm2_behavioral.v

```
module fsm2_behavioral
   (output reg  Dout,
    input  wire Clock, Reset, Din);
             :
```

Fig. 9.2
FSM 2 state diagram and module definition

9.3.4 Design a Verilog behavioral model to implement the finite-state machine described by the state diagram in Fig. 9.2. Use the port definition provided in this figure for your design. Use the three-block approach to modeling FSMs described in this chapter for your design. Model the state variables using parameters and encode the states in one-hot using the following state codes: S0 = "0001," S1 = "0010," S2 = "0100," and S3 = "1000,"

9.3.5 Design a Verilog behavioral model for a 4-bit serial bit sequence detector similar to Example 9.11. Use the port definition provided in Fig. 9.3. Use the three-block approach to modeling FSMs described in this chapter for your design. The input to your sequence detector is called *DIN* and the output is called *FOUND*. Your detector will assert FOUND anytime there is a 4-bit sequence of "0101." Model the states in this machine with parameters. Choose any state encoding approach you wish.

Seq_Det_behavioral.v

```
module Seq_Det_behavioral
   (output reg  FOUND,
    input  wire Clock, Reset,
    input  wire DIN);
               :
```

Fig. 9.3
Sequence detector module definition

9.3.6 Design a Verilog behavioral model for a 20-cent vending machine controller similar to Example 9.14. Use the port definition provided in Fig. 9.4. Use the three-block approach to modeling FSMs described in this chapter for your design. Your controller will take in nickels and dimes and dispense a product anytime the customer has entered 20 cents. Your FSM has two inputs, *Nin* and *Din*. Nin is asserted whenever the customer enters a nickel, while Din is asserted anytime the customer enters a dime. Your FSM has two outputs, *Dispense* and *Change*. Dispense is asserted anytime the customer has entered at least 20 cents and Change is asserted anytime the customer has entered more than 20 cents and needs a nickel in change. Model the states in this machine with parameters. Choose any state encoding approach you wish.

Vending_behavioral.v

```
module Vending_behavioral
   (output reg  Dispense, Change,
    input  wire Clock, Reset,
    input  wire Nin, Din);
                 :
```

Fig. 9.4
Vending machine module definition

9.3.7 Design a Verilog behavioral model for a finite-state machine for a traffic light controller. Use the port definition provided in Fig. 9.5. This is the same problem description as in exercise 7.4.15. This time, you will implement the functionality using the behavioral modeling techniques presented in this chapter. Your FSM will control a traffic light at the intersection of a busy highway and a seldom used side road. You will be designing the control signals for just the red, yellow, and green lights facing the highway. Under normal conditions, the highway has a green light. The side road has car detector that indicates when car pulls up by asserting a signal called *CAR*. When CAR is asserted, you will change the highway traffic light from green to yellow and then from yellow to red. Once in the red position, a built-in timer will begin a countdown and provide your controller a signal called *TIMEOUT* when 15 seconds has passed. Once TIMEOUT is asserted, you will change the highway traffic light back to green. Your system will have three outputs *GRN*, *YLW*, and *RED*, which control when the highway facing traffic lights are on (1 = ON, 0 = OFF). Model the states in this machine with parameters. Choose any state encoding approach you wish.

tlc_behavioral.v

```
module tlc_behavioral
   (output reg  GRN, YLW, RED,
    input  wire Clock, Reset,
    input  wire CAR, TIMEOUT);
                :
```

Fig. 9.5
Traffic light controller module definition

Section 9.4: Modeling Counters in Verilog

9.4.1 Design a Verilog behavioral model for a 16-bit, binary up counter using a single-procedural block. The block diagram for the port definition is shown in Fig. 9.6.

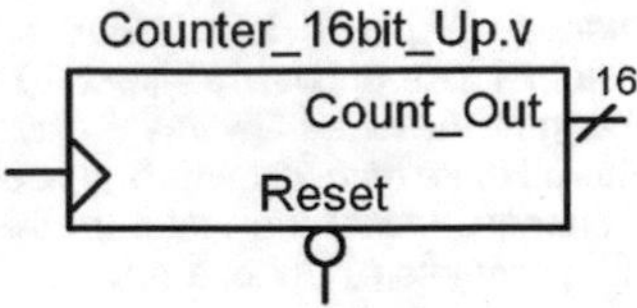

Fig. 9.6
16-bit binary up counter block diagram

9.4.2 Design a Verilog behavioral model for a 16-bit, binary up counter with range checking using a single-procedural block. The block diagram for the port definition is shown in Fig. 9.6. Your counter should count up to 60,000 and then start over at 0.

9.4.3 Design a Verilog behavioral model for a 16-bit, binary up counter with enable using a single-procedural block. The block diagram for the port definition is shown in Fig. 9.7.

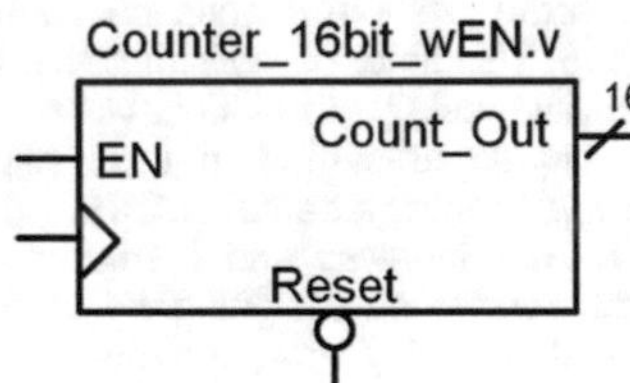

Fig. 9.7
16-bit binary up counter with enable block diagram

9.4.4 Design a Verilog behavioral model for a 16-bit, binary up counter with enable and load using a single-procedural block. The block diagram for the port definition is shown in Fig. 9.8.

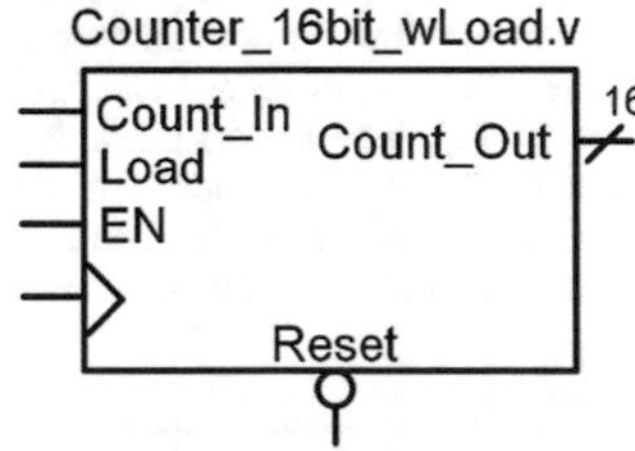

Fig. 9.8
16-bit binary up counter with load block diagram

9.4.5 Design a Verilog behavioral model for a 16-bit, binary up/down counter using a single-procedural block. The block diagram for the port definition is shown in Fig. 9.9. When Up = 1, the counter will increment. When Up = 0, the counter will decrement.

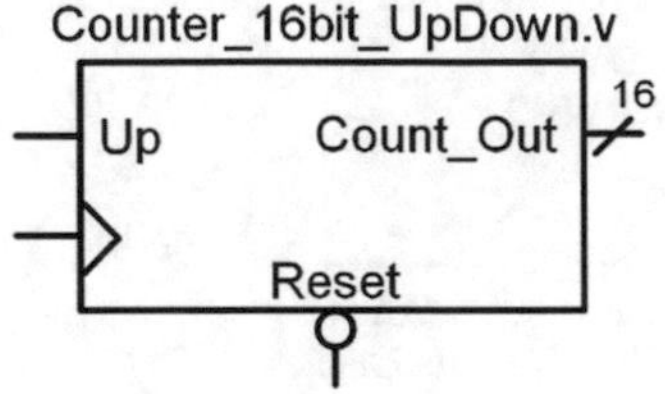

Fig. 9.9
16-bit binary up/down counter block diagram

Section 9.5: RTL Modeling

9.5.1 In *register transfer level* modeling, how does the width of the register relate to the number of D-flip-flops that will be synthesized?

9.5.2 In *register transfer level* modeling, how is the synchronous data movement managed if all registers are using the same clock?

9.5.3 Design a Verilog RTL model of a 32-bit, synchronous register. The block diagram for the port definition is shown in Fig. 9.10. The register has a synchronous enable. The register should be modeled using a single-procedural block.

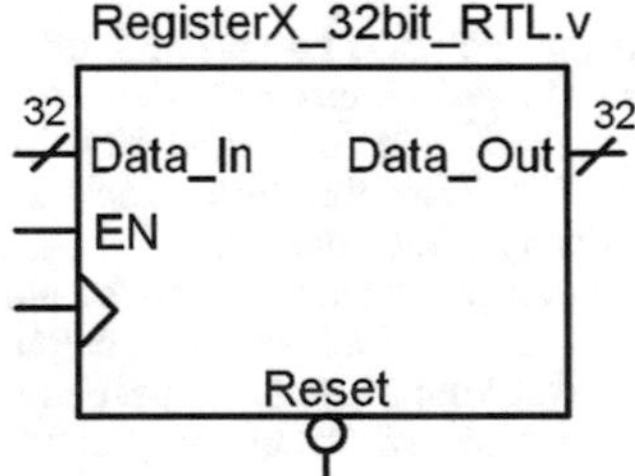

Fig. 9.10
32-bit register block diagram

9.5.4 Design a Verilog RTL model of an 8-stage, 16-bit shift register. The block diagram for the port definition is shown in Fig. 9.11. Each stage of the shift register will be provided as an output of the system (A, B, C, D, E, F, G, and H). The shift register should be modeled using a single-procedural block.

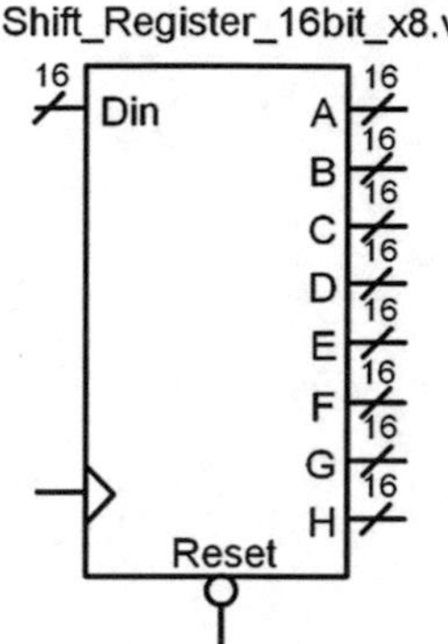

Fig. 9.11
16-bit shift register block diagram

9.5.5 Design a Verilog RTL model of the multi-drop bus topology in Fig. 9.12. Each of the 16-bit registers (RegA, RegB, RegC, and RegD) will latch the contents of the 16-bit data bus if their enable line is asserted. Each register should be modeled using an individual procedural block.

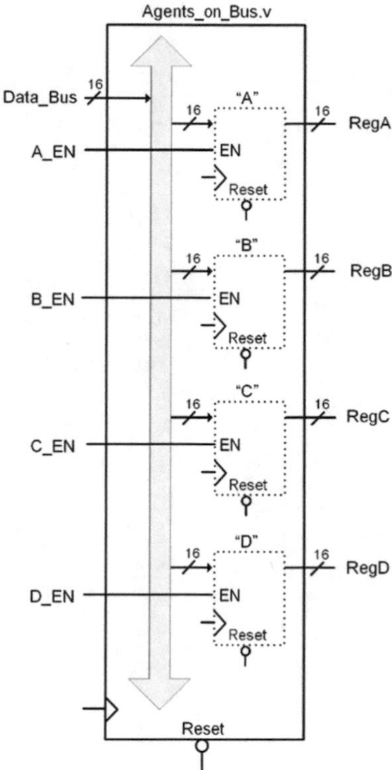

Fig. 9.12
Agents on a bus block diagram

Chapter 10: Memory

This chapter introduces the basic concepts, terminology, and roles of memory in digital systems. The material presented here will not delve into the details of the device physics or low-level theory of operation. Instead, the intent of this chapter is to give a general overview of memory technology and its use in computer systems in addition to how to model memory in Verilog. The goal of this chapter is to give an understanding of the basic principles of semiconductor-based memory systems.

Learning Outcomes—After completing this chapter, you will be able to:

10.1 Describe the basic architecture and terminology for semiconductor-based memory systems.
10.2 Describe the basic architecture of non-volatile memory systems.
10.3 Describe the basic architecture of volatile memory systems.
10.4 Design a Verilog behavioral model of a memory system.

10.1 Memory Architecture and Terminology

The term *memory* is used to describe a system with the ability to store digital information. The term *semiconductor memory* refers to systems that are implemented using integrated circuit technology. These types of systems store the digital information using transistors, fuses, and/or capacitors on a single semiconductor substrate. Memory can also be implemented using technology other than semiconductors. Disk drives store information by altering the polarity of magnetic fields on a circular substrate. The two magnetic polarities (north and south) are used to represent different logic values (i.e., 0 or 1). Optical disks use lasers to burn pits into reflective substrates. The binary information is represented by light either being reflected (no pit) or not reflected (pit present). Semiconductor memory does not have any moving parts, so it is called *solid-state memory* and can hold more information per unit area than disk memory. Regardless of the technology used to store the binary data, all memory has common attributes and terminology that are discussed in this chapter.

10.1.1 Memory Map Model

The information stored in memory is called the **data**. When information is placed into memory, it is called a **write**. When information is retrieved from memory, it is called a **read**. In order to access data in memory, an **address** is used. While data can be accessed as individual bits, in order to reduce the number of address locations needed, data is typically grouped into *N-bit words*. If a memory system has $N = 8$, this means that 8-bits of data are stored at each address. The number of address locations is described using the variable *M*. The overall size of the memory is typically stated by saying "M×N." For example, if we had a 16×8 memory system, that means that there are 16 address locations, each capable of storing a byte of data. This memory would have a **capacity** of $16 \times 8 = 128$ bits. Since the address is implemented as a binary code, the number of lines in the address bus (n) will dictate the number of address locations that the memory system will have ($M = 2^n$). Figure 10.1 shows a graphical depiction of how data resides in memory. This type of graphic is called a *memory map model*.

© Springer Nature Switzerland AG 2019

B. J. LaMeres, *Introduction to Logic Circuits & Logic Design with Verilog*,
https://doi.org/10.1007/978-3-030-13605-5_10

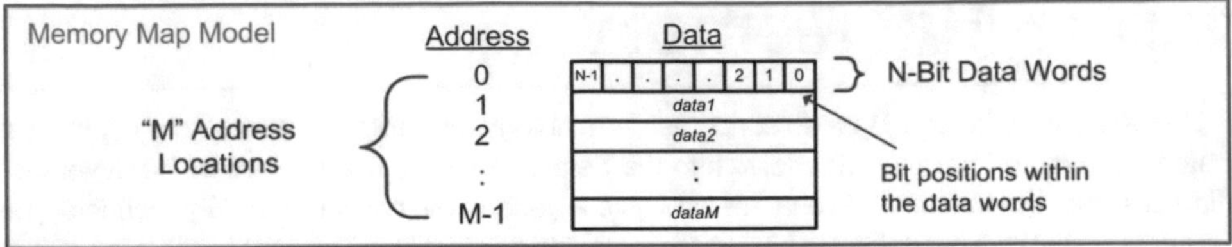

Fig. 10.1
Memory map model

10.1.2 Volatile Versus Non-volatile Memory

Memory is classified into two categories depending on whether it can store information when power is removed or not. The term **non-volatile** is used to describe memory that *holds* information when the power is removed, while the term **volatile** is used to describe memory that loses its information when power is removed. Historically, volatile memory is able to run at faster speeds compared to non-volatile memory, so it is used as the primary storage mechanism while a digital system is running. Non-volatile memory is necessary in order to hold critical operation information for a digital system such as start-up instructions, operations systems, and applications.

10.1.3 Read-Only Versus Read/Write Memory

Memory can also be classified into two categories with respect to how data is accessed. **Read-only memory (ROM)** is a device that cannot be written to during normal operation. This type of memory is useful for holding critical system information or programs that should not be altered while the system is running. **Read/write** memory refers to memory that can be read and written to during normal operation and is used to hold temporary data and variables.

10.1.4 Random Access Versus Sequential Access

Random-access memory (RAM) describes memory in which any location in the system can be accessed at any time. The opposite of this is **sequential access** memory, in which not all address locations are immediately available. An example of a sequential access memory system is a tape drive. In order to access the desired address in this system, the tape spool must be spun until the address is in a position that can be observed. Most semiconductor memory in modern systems is random access. The terms RAM and ROM have been adopted, somewhat inaccurately, to also describe groups of memory with particular behavior. While the term ROM technically describes a system that cannot be written to, it has taken on the additional association of being the term to describe non-volatile memory. While the term RAM technically describes how data is accessed, it has taken on the additional association of being the term to describe volatile memory. When describing modern memory systems, the terms RAM and ROM are used most commonly to describe the characteristics of the memory being used; however, modern memory systems can be both read/write and non-volatile, and the majority of memory is random access.

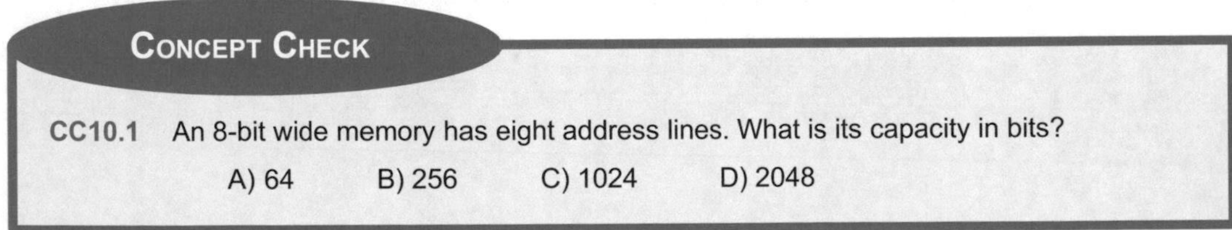

Concept Check

CC10.1 An 8-bit wide memory has eight address lines. What is its capacity in bits?

A) 64 B) 256 C) 1024 D) 2048

10.2 Non-volatile Memory Technology

10.2.1 ROM Architecture

This section describes some of the most common non-volatile memory technologies used to store digital information. An address decoder is used to access individual data words within the memory system. The address decoder asserts one and only one *word line* (WL) for each unique binary address that is present on its input. This operation is identical to a binary-to-one-hot decoder. For an n-bit address, the decoder can access 2^n, or M words in memory. The word lines historically run horizontally across the memory array; thus they are often called *row lines*, and the word line decoder is often called the *row decoder. Bit lines* (BL) run perpendicular to the word lines in order to provide individual bit storage access at the intersection of the bit and word lines. These lines typically run vertically through the memory array; thus they are often called *column lines*. The output of the memory system (i.e., Data_Out) is obtained by providing an address and then reading the word from buffered versions of the bit lines. When a system provides individual bit access to a row, or access to multiple data words sharing a row line, a column decoder is used to route the appropriate bit line(s) to the data out port.

In a traditional ROM array, each bit line contains a pull-up network to V_{CC}. This provides the ability to store a logic 1 at all locations within the array. If a logic 0 is desired at a particular location, an NMOS pull-down transistor is inserted. The gate of the NMOS is connected to the appropriate word line, and the drain of the NMOS is connected to the bit line. When reading, the word line is asserted and turns on the NMOS transistor. This pulls the bit line to GND and produces a logic 0 on the output. When the NMOS transistor is excluded, the bit line remains at a logic 1 due to the pull-up network. Figure 10.2 shows the basic architecture of a ROM.

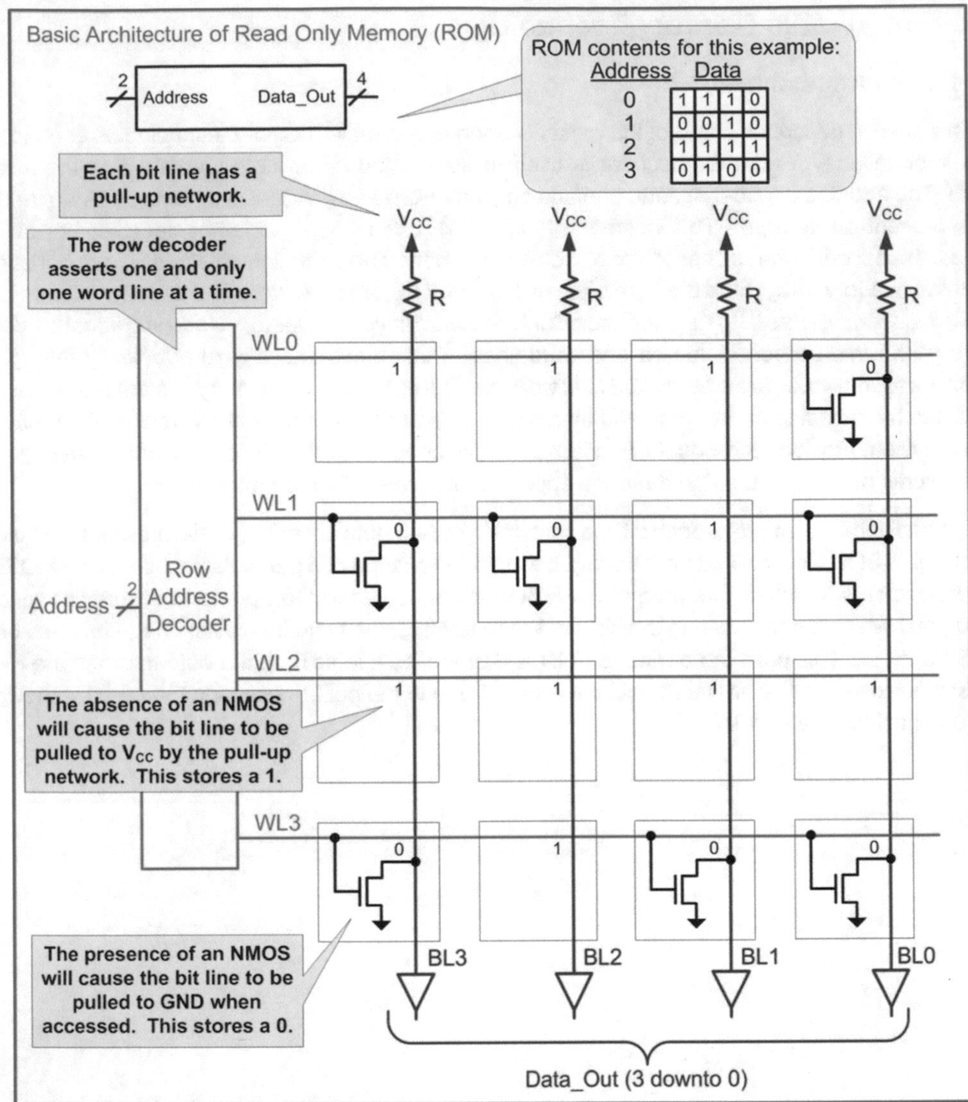

Fig. 10.2
Basic architecture of read-only memory (ROM)

Figure 10.3 shows the operation of a ROM when information is being read.

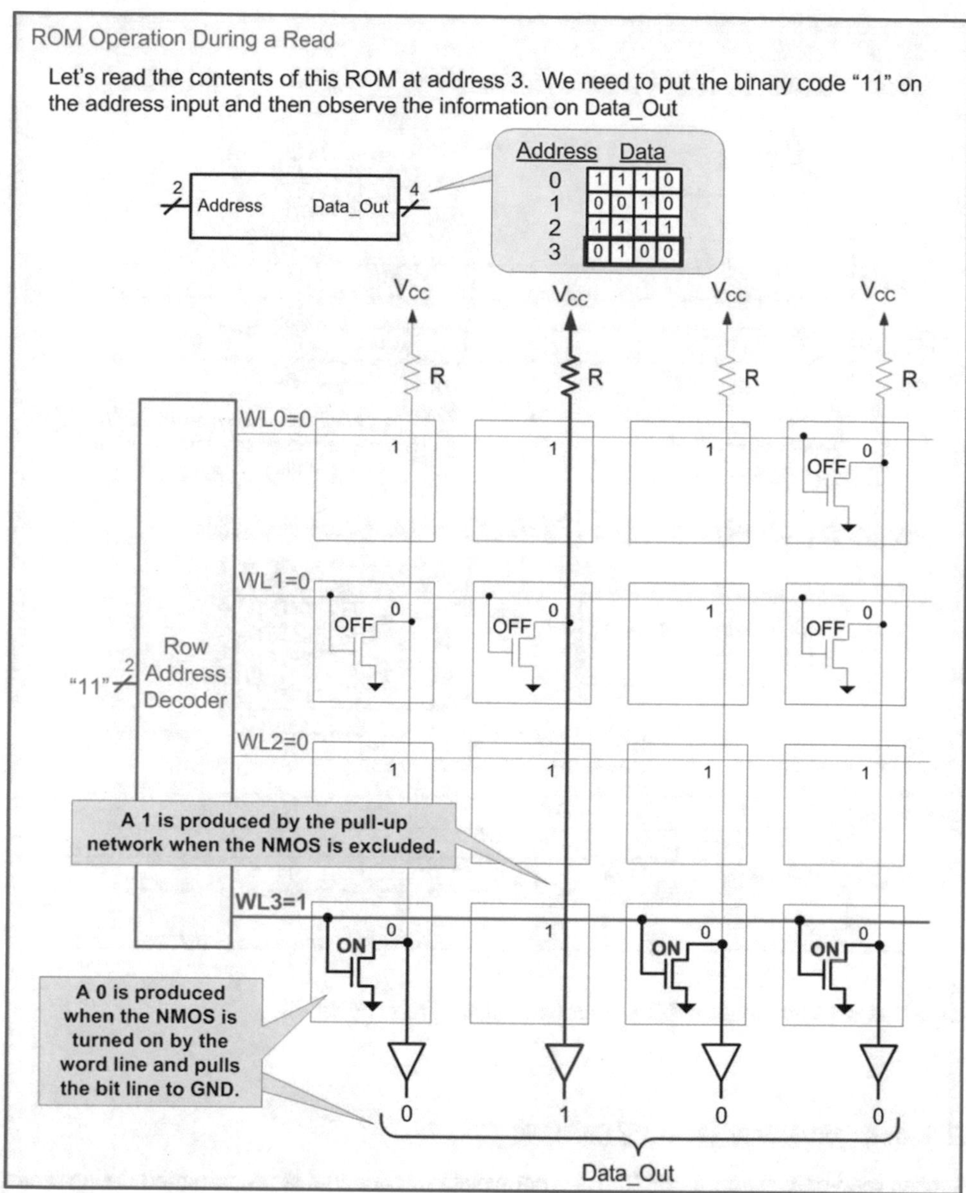

Fig. 10.3
ROM operation during a read

Memory can be designed to be either asynchronous or synchronous. **Asynchronous memory** updates its data outputs immediately upon receiving an address. **Synchronous memory** only updates its data outputs on the rising edge of a clock. The term *latency* is used to describe the delay between when a signal is sent to the memory (either the address in an asynchronous system or the clock in a synchronous system) and when the data is available. Figure 10.4 shows a comparison of the timing diagrams between asynchronous and synchronous ROM systems during a read cycle.

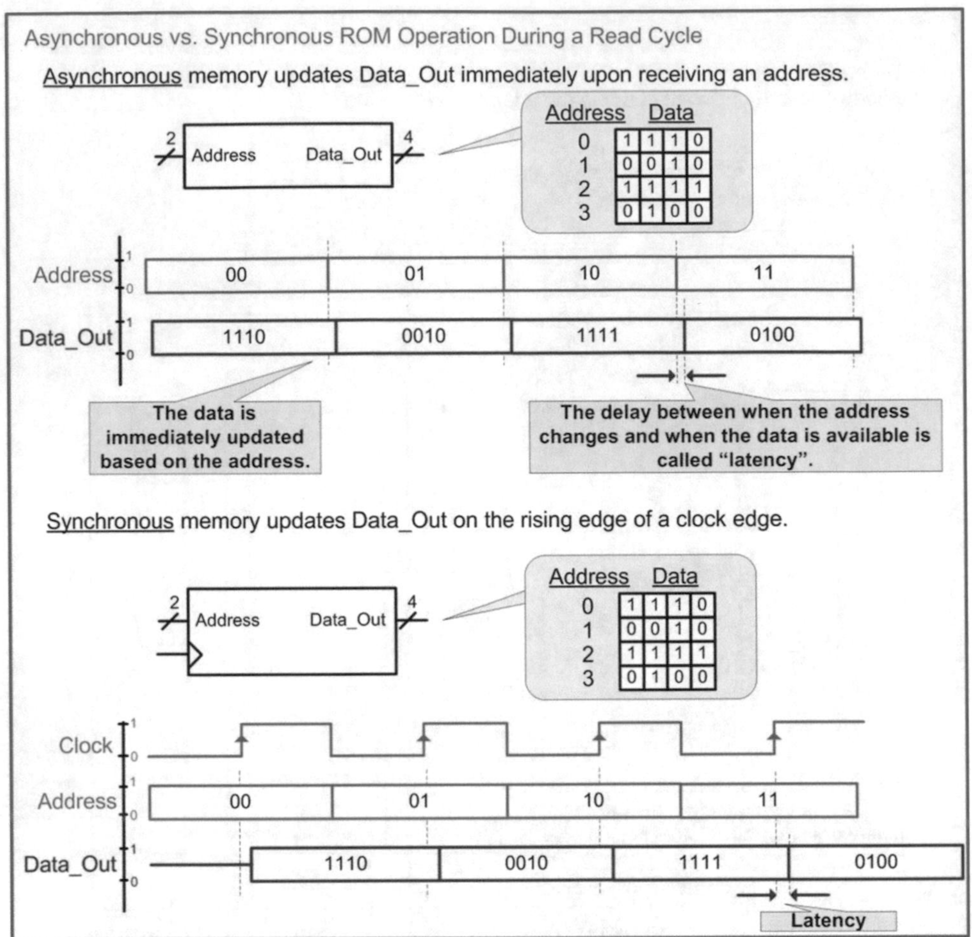

Fig. 10.4
Asynchronous vs. synchronous ROM operation during a read cycle

10.2.2 Mask Read-Only Memory (MROM)

A mask read-only memory (MROM) is a non-volatile device that is programmed during fabrication. The term *mask* refers to a transparent plate that contains patterns to create the features of the devices on an integrated circuit using a process called photolithography. An MROM is fabricated with all of the features necessary for the memory device with the exception of the final connections between the NMOS transistors and the word and bit lines. This allows the majority of the device to be created prior to knowing what the final information to be stored is. Once the desired information to be stored is provided by the customer, the fabrication process is completed by adding connections between certain NMOS transistors and the word/bit lines in order to create logic 0's. Figure 10.5 shows an overview of the MROM programming process.

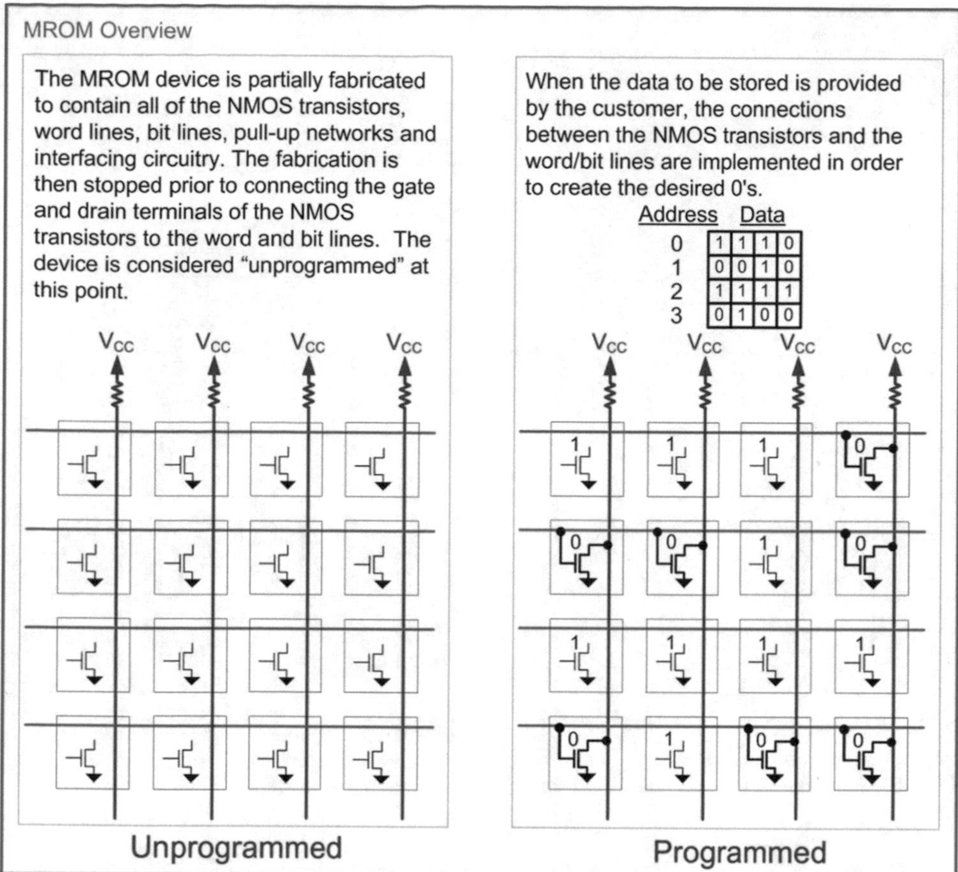

Fig. 10.5
MROM overview

10.2.3 Programmable Read-Only Memory (PROM)

A programmable read-only memory (PROM) is created in a similar manner as an MROM except that the programming is accomplished post-fabrication through the use of fuses or anti-fuses. A **fuse** is an electrical connection that is normally conductive. When a certain amount of current is passed through the fuse, it will melt, or *blow*, and create an open circuit. The amount of current necessary to open the fuse is much larger than the current the fuse would conduct during normal operation. An **anti-fuse** operates in the opposite manner as a fuse. An anti-fuse is normally an open circuit. When a certain amount of current is forced into the anti-fuse, the insulating material breaks down and creates a conduction path. This turns the anti-fuse from an open circuit into a wire. Again, the amount of current necessary to close the anti-fuse is much larger than the current the anti-fuse would experience during normal operation. A PROM uses fuses or anti-fuses in order to connect/disconnect the NMOS transistors in the ROM array to the word/bit lines. A PROM programmer is used to *burn* the fuses or anti-fuses. A PROM can only be programmed once in this manner; thus it is a read-only memory and non-volatile. A PROM has the advantage that programming can take place quickly as opposed to an MROM that must be programmed through device fabrication. Figure 10.6 shows an example PROM device based on fuses.

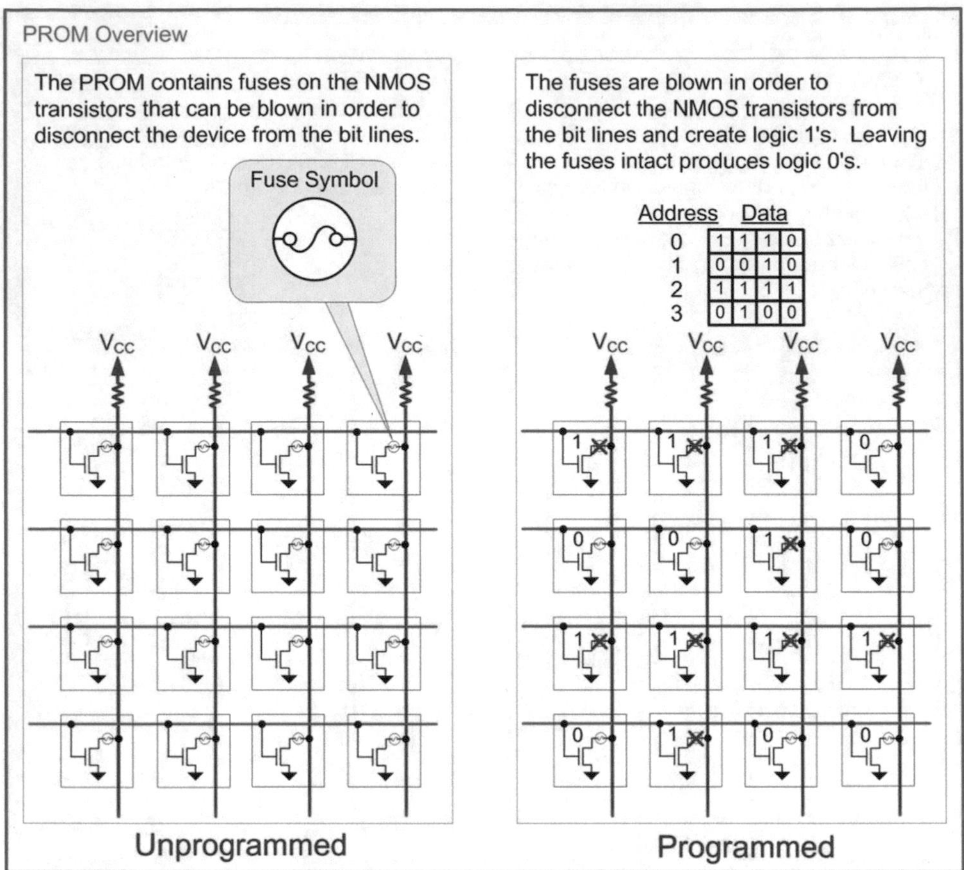

Fig. 10.6
PROM overview

10.2.4 Erasable Programmable Read-Only Memory (EPROM)

As an improvement to the one-time programming characteristic of PROMs, an electrically programmable ROM with the ability to be erased with ultraviolet (UV) light was created. The erasable programmable read-only memory (EPROM) is based on a **floating-gate transistor**. In a floating-gate transistor, an additional metal-oxide structure is added to the gate of an NMOS. This has the effect of increasing the threshold voltage. The geometry of the second metal-oxide is designed such that the threshold voltage is high enough that normal CMOS logic levels are not able to turn the transistor on (i.e., $V_{T1} \geq V_{CC}$). This threshold can be changed by applying a large electric field across the two metal structures in the gate. This causes charge to tunnel into the secondary oxide, ultimately changing it into a conductor. This phenomenon is called Fowler-Nordheim tunneling. The new threshold voltage is low enough that normal CMOS logic levels are not able to turn the transistors off (i.e., $V_{T2} \leq$ GND). This process is how the device is *programmed*. This process is accomplished using a dedicated programmer; thus the EPROM must be removed from its system to program. Figure 10.7 shows an overview of a floating-gate transistor and how it is programmed.

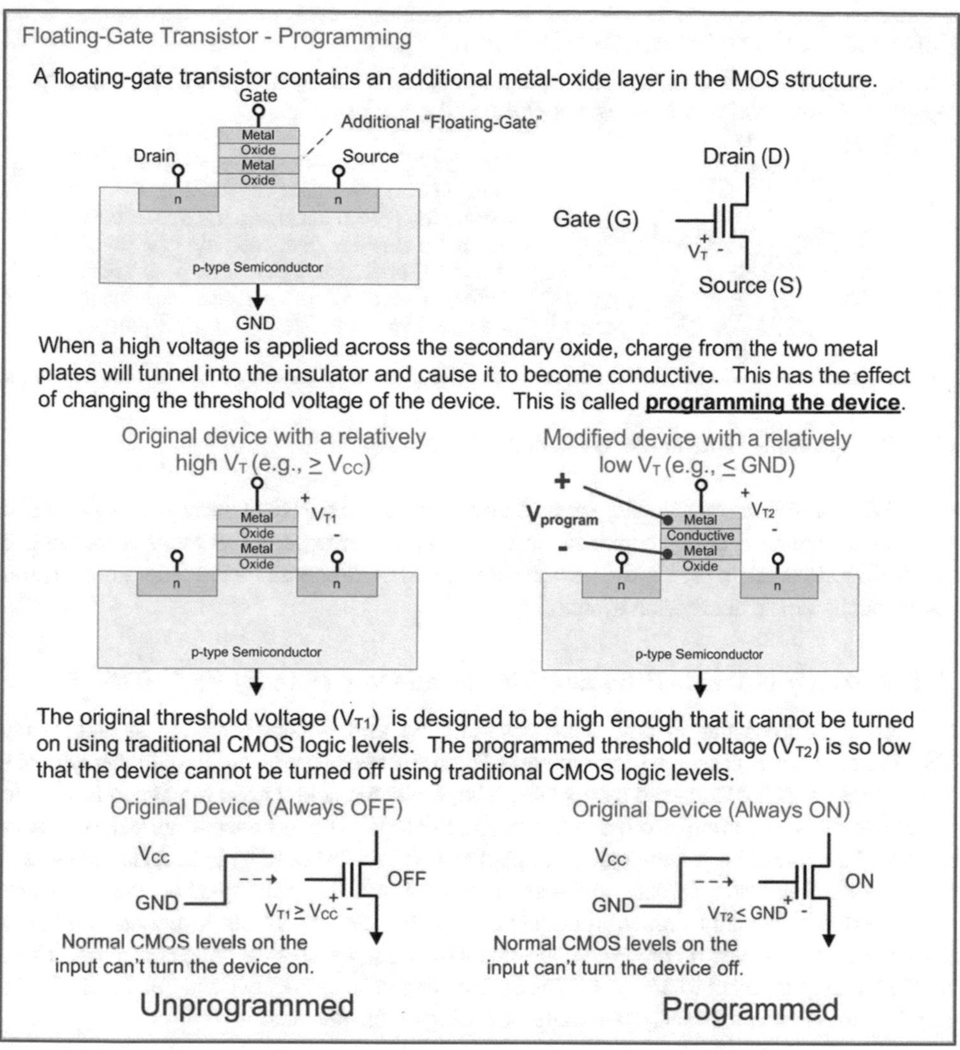

Fig. 10.7
Floating-gate transistor – Programming

In order to change the floating-gate transistor back into its normal state, the device is exposed to a strong ultraviolet light source. When the UV light strikes the trapped charge in the secondary oxide, it transfers enough energy to the charge particles that they can move back into the metal plates in the gate. This, in effect, *erases* the device and restores it back to a state with a high threshold voltage. EPROMs contain a transparent window on the top of their package that allows the UV light to strike the devices. The EPROM must be removed from its system to perform the erase procedure. When the UV light erase procedure is performed, every device in the memory array is erased. EPROMs are a significant improvement over PROMs because they can be programmed multiple times; however, the programming and erase procedures are manually intensive and require an external programmer and external eraser. Figure 10.8 shows the erase procedure for a floating-gate transistor using UV light.

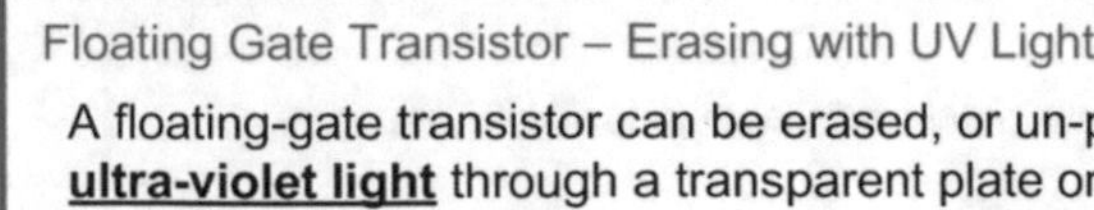

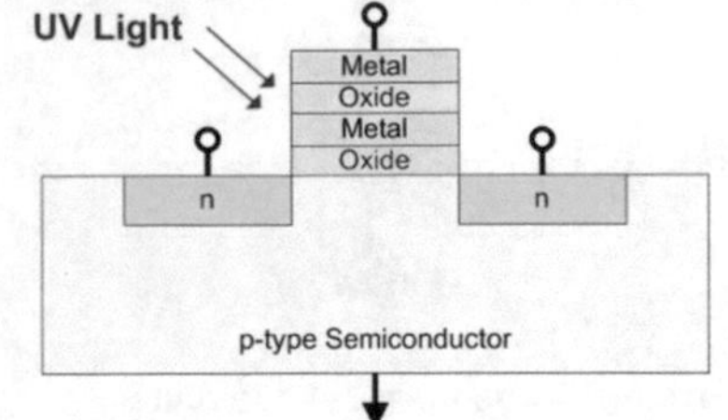

Fig. 10.8
Floating-gate transistor – Erasing with UV light

An EPROM array is created in the exact same manner as in a PROM array with the exception that additional programming circuitry is placed on the IC and a transparent window is included on the package to facilitate erasing. An EPROM is non-volatile and read-only since the programming procedure takes place outside of its destination system.

10.2.5 Electrically Erasable Programmable Read-Only Memory (EEPROM)

In order to address the inconvenient programming and erasing procedures associated with EPROMs, the electrically erasable programmable ROM (EEPROM) was created. In this type of circuit, the floating-gate transistor is erased by applying a large electric field across the secondary oxide. This electric field provides the energy to move the trapped charge from the secondary oxide back into the metal plates of the gate. The advantage of this approach is that the circuitry to provide the large electric field can be generated using circuitry on the same substrate as the memory array, thus eliminating the need for an external UV light eraser. In addition, since the circuitry exists to generate large on-chip voltages, the device can also be programmed without the need for an external programmer. This allows an EEPROM to be programmed and erased while it resides in its target environment. Figure 10.9 shows the procedure for erasing a floating-gate transistor using an electric field.

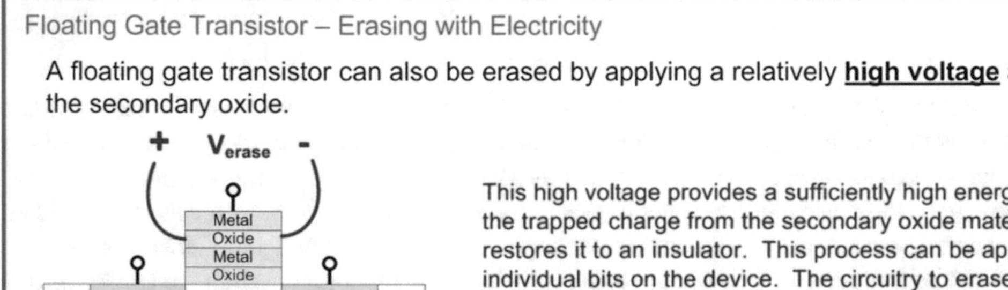

Fig. 10.9
Floating-gate transistor – Erasing with electricity

Early EEPROMs were very slow and had a limited number of program/erase cycles; thus they were classified into the category of non-volatile, read-only memory. Modern floating-gate transistors are now capable of access times on scale with other volatile memory systems; thus they have evolved into one of the few non-volatile, read/write memory technologies used in computer systems today.

10.2.6 FLASH Memory

One of the early drawbacks of EEPROM was that the circuitry that provided the capability to program and erase individual bits also added to the size of each individual storage element. FLASH EEPROM was a technology that attempted to improve the density of floating-gate memory by programming and erasing in large groups of data, known as *blocks*. This allowed the individual storage cells to shrink and provided higher-density memory parts. This new architecture was called *NAND FLASH* and provided faster write and erase times coupled with higher-density storage elements. The limitation of NAND FLASH was that reading and writing could only be accomplished in a block-by-block basis. This characteristic precluded the use of NAND FLASH for run-time variables and data storage but was well suited for streaming applications such as audio/video and program loading. As NAND FLASH technology advanced, the block size began to shrink, and software adapted to accommodate the block-by-block data access. This expanded the applications that NAND FLASH could be deployed in. Today, NAND FLASH memory is used in nearly all portable devices (e.g., smartphones, tablets, etc.), and its use in solid-state hard drives is on pace to replace hard disk drives and optical disks as the primary non-volatile storage medium in modern computers.

In order to provide individual word access, NOR FLASH was introduced. In NOR FLASH, circuitry is added to provide individual access to data words. This architecture provided faster read times than NAND FLASH, but the additional circuitry causes the write and erase times to be slower and the individual storage cell size to be larger. Due to NAND FLASH having faster write times and higher density, it is seeing broader scale adoption compared to NOR FLASH despite only being able to access information in blocks. NOR FLASH is considered random-access memory, while NAND FLASH is typically not; however, as the block size of NAND FLASH is continually reduced, its use for variable storage is becoming more attractive. All FLASH memory is non-volatile and read/write.

Concept Check

CC10.2 Which of the following is suitable for implementation in a read-only memory?

A) Variables that a computer program needs to continuously update

B) Information captured by a digital camera

C) A computer program on a spacecraft

D) Incoming digitized sound from a microphone

10.3 Volatile Memory Technology

This section describes some common volatile memory technologies used to store digital information.

10.3.1 Static Random-Access Memory (SRAM)

Static random-access memory (SRAM) is a semiconductor technology that stores information using a cross-coupled inverter feedback loop. Figure 10.10 shows the schematic for the basic SRAM storage cell. In this configuration, two access transistors (M1 and M2) are used to read and write from the storage cell. The cell has two complementary ports called bit line (BL) and bit line' (BLn). Due to the inverting functionality of the feedback loop, these two ports will always be the complement of each other. This behavior is advantageous because the two lines can be compared to each other to determine the data value. This allows the voltage levels used in the cell to be lowered while still being able to detect the stored data value. Word lines are used to control the access transistors. This storage element takes six CMOS transistors to implement and is often called a 6T configuration. The advantage of this memory cell is that it has very fast performance compared to other sub-systems because of its underlying technology being simple CMOS transistors. SRAM cells are commonly implemented on the same IC substrate as the rest of the system, thus allowing a fully integrated system to be realized. SRAM cells are used for cache memory in computer systems.

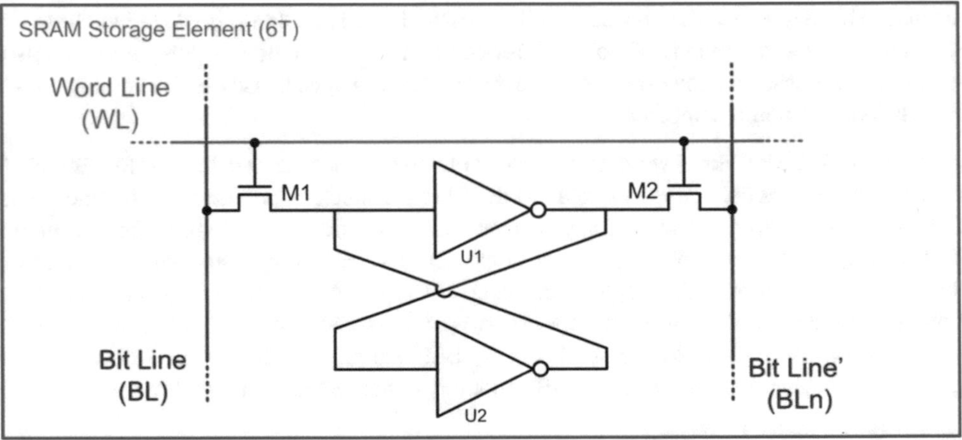

Fig. 10.10
SRAM storage element (6T)

To build an SRAM memory system, cells are arranged in an array pattern. Figure 10.11 shows a 4×4 SRAM array topology. In this configuration, word lines are shared horizontally across the array in order to provide addressing capability. An address decoder is used to convert the binary-encoded address into the appropriate word line assertions. N storage cells are attached to the word line to provide the desired data word width. Bit lines are shared vertically across the array in order to provide data access (either read or write). A data line controller handles whether data is read from or written to the cells based on an external write enable (WE) signal. When WE is asserted (WE = 1), data will be written to the cells. When WE is de-asserted (WE = 0), data will be read from the cells. The data line controller also handles determining the correct logic value read from the cells by comparing BL to BLn. As more cells are added to the bit lines, the signal magnitude being driven by the storage cells diminishes due to the additional loading of the other cells. This is where having complementary data signals (BL and BLn) is advantageous because this effectively doubles the magnitude of the storage cell outputs. The comparison of BL to BLn is handled using a *differential amplifier* that produces a full logic level output even when the incoming signals are very small.

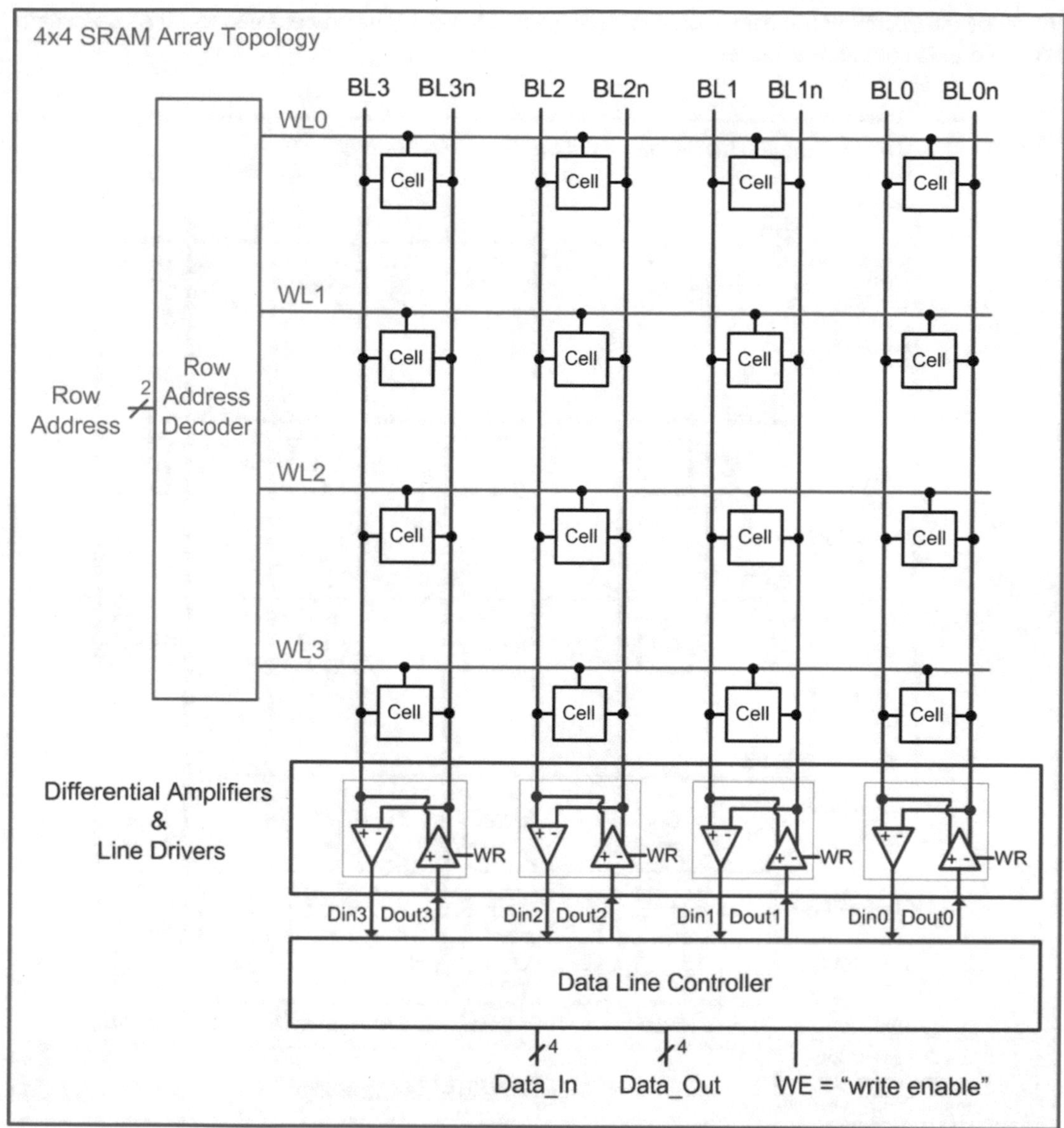

Fig. 10.11
4×4 SRAM array topology

SRAM is volatile memory because when the power is removed, the cross-coupled inverters are not able to drive the feedback loop and the data is lost. SRAM is also read/write memory because the storage cells can be easily read from or written to during normal operation.

Let's look at the operation of the SRAM array when writing the 4-bit word "0111" to address "01." Figure 10.12 shows a graphical depiction of this operation. In this write cycle, the row address decoder observes the address input "01" and asserts WL1. Asserting this word line enables all of the access transistors (i.e., M1 and M2 in Fig. 10.10) of the storage cells in this row. The line drivers are designed to have a stronger drive strength than the inverters in the storage cells so that they can override their values during a write. The information "0111" is present on the Data_In bus, and the write enable control line is asserted (WE = 1) to indicate a write. The data line controller passes the information to be stored to the line drivers, which in turn converts each input into complementary signals and drives the bit lines. This overrides the information in each storage cell connected to WL1. The address decoder then de-asserts WL1 and the information is stored.

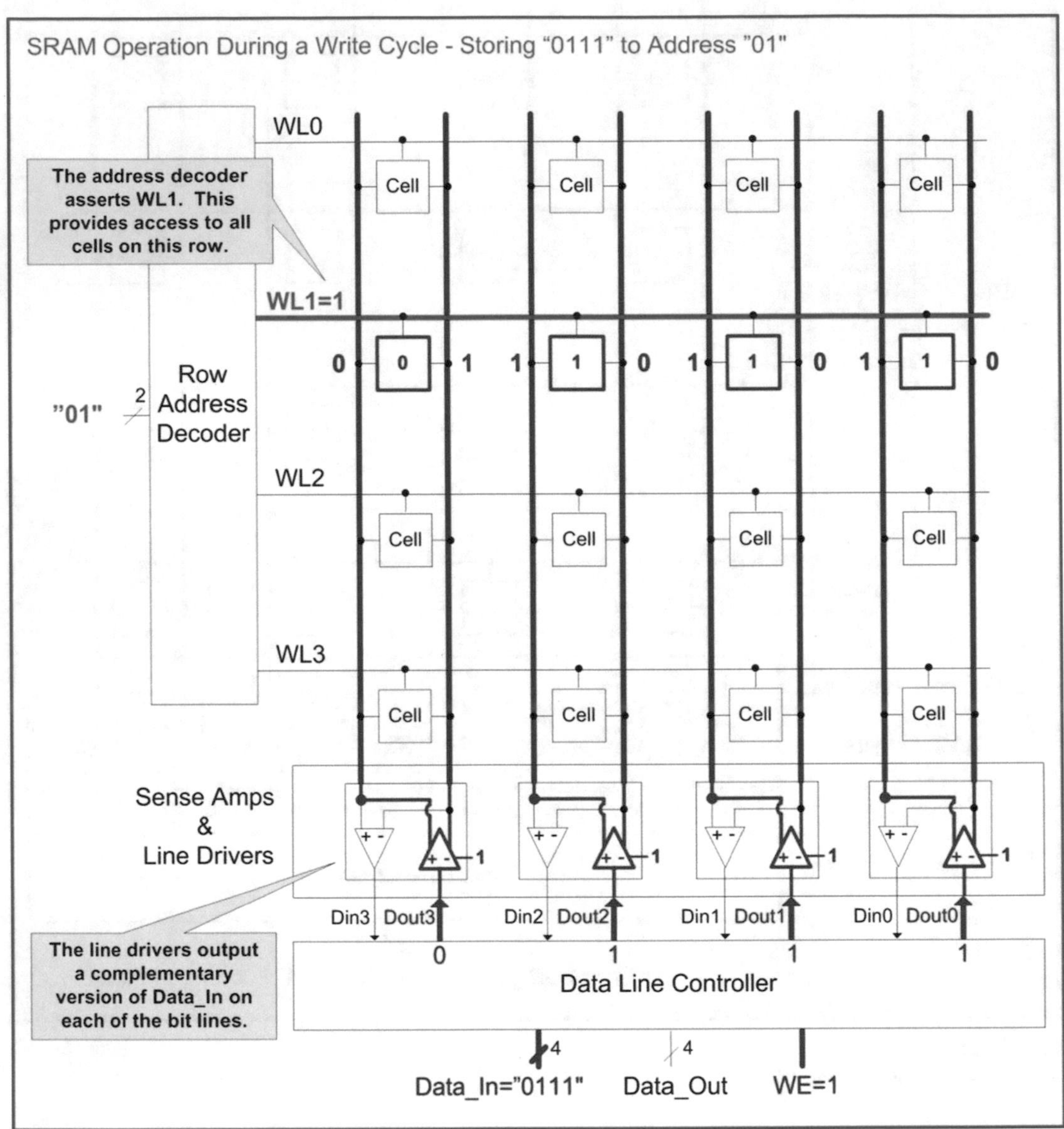

Fig. 10.12
SRAM operation during a write cycle – Storing "0111" to Address "01"

Now let's look at the operation of the SRAM array when reading a 4-bit word from address "10." Let's assume that this row was storing the value "1010." Figure 10.13 shows a graphical depiction of this operation. In this read cycle, the row address decoder asserts WL2, which allows the SRAM cells to drive their respective bit lines. Note that each cell drives a complementary version of its stored value. The input control line is de-asserted (WE = 0), which indicates that the sense amplifiers will read the BL and BLn lines in order to determine the full logic value stored in each cell. This logic value is then routed to the Data_Out port of the array. In an SRAM array, reading from the cell does not impact the contents of the cell. Once the read is complete, WL2 is de-asserted and the read cycle is complete.

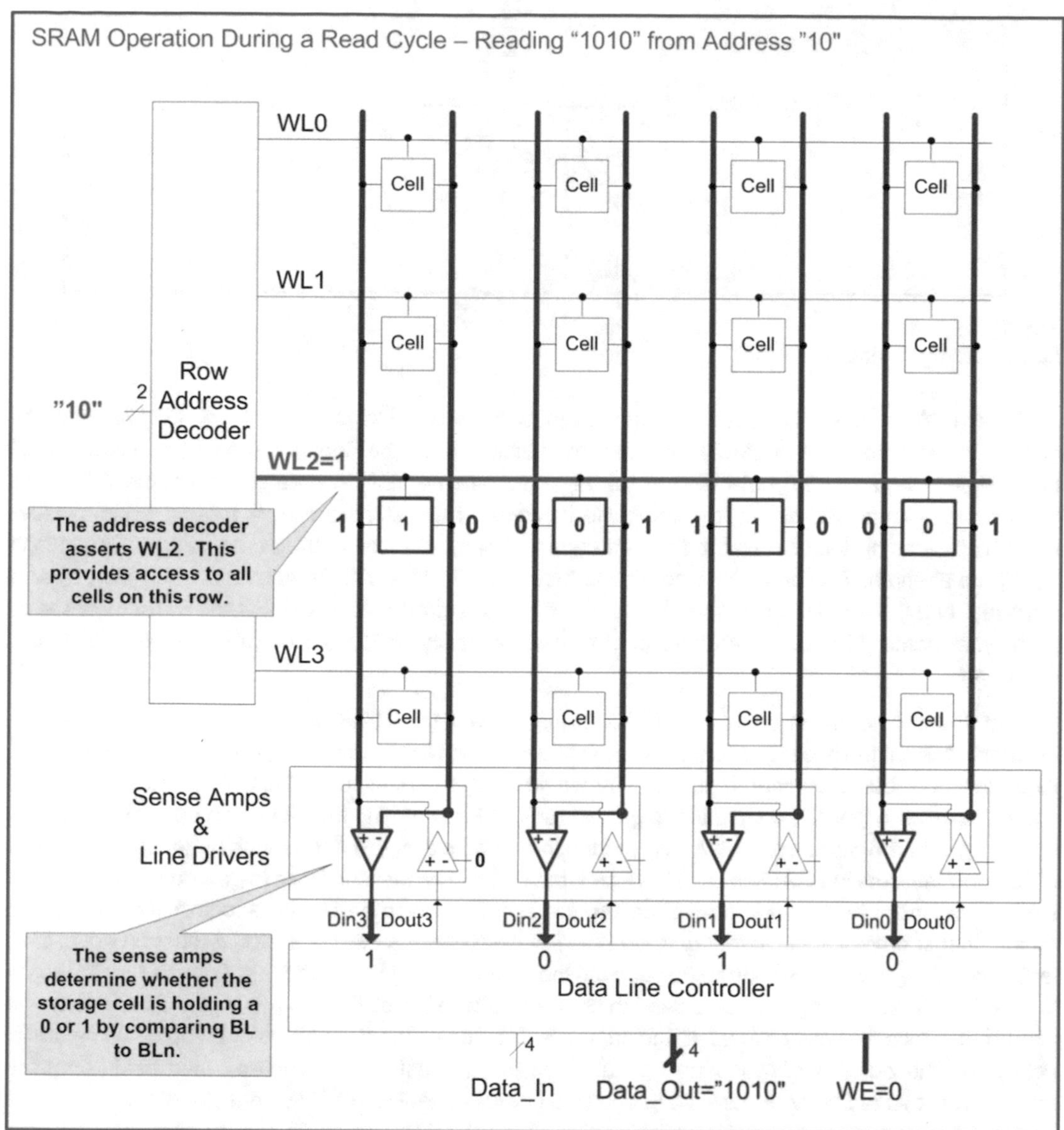

Fig. 10.13
SRAM operation during a read cycle – Reading "0101" from Address "10"

10.3.2 Dynamic Random-Access Memory (DRAM)

Dynamic random-access memory (DRAM) is a semiconductor technology that stores information using a capacitor. A capacitor is a fundamental electrical device that stores charge. Figure 10.14 shows the schematic for the basic DRAM storage cell. The capacitor is accessed through a transistor (M1). Since this storage element takes one transistor and one capacitor, it is often referred to as a 1T1C configuration. Just as in SRAM memory, word lines are used to access the storage elements. The term *digit line* is used to describe the vertical connection to the storage cells. DRAM has an advantage over SRAM in that the storage element requires less area to implement. This allows DRAM memory to have much higher density compared to SRAM.

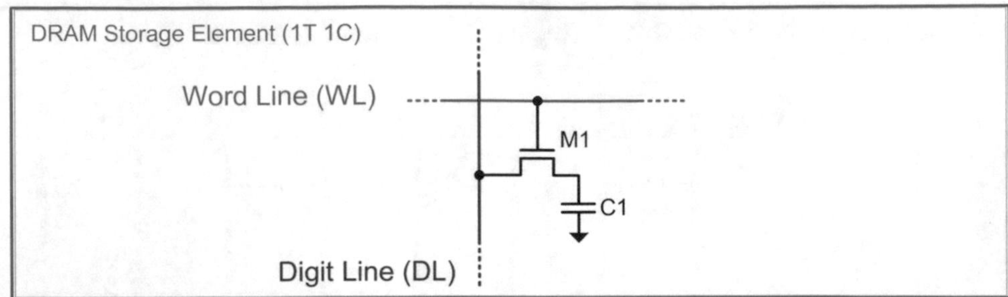

Fig. 10.14
DRAM storage element (1T 1C)

There are a variety of considerations that must be accounted for when using DRAM. First, the charge in the capacitor will slowly dissipate over time due to the capacitors being non-ideal. If left unchecked, eventually the data held in the capacitor will be lost. In order to overcome this issue, DRAM has a dedicated circuit to *refresh* the contents of the storage cell. A refresh cycle involves periodically reading the value stored on the capacitor and then writing the same value back again at full signal strength. This behavior also means that that DRAM is volatile because when the power is removed, and the refresh cycle cannot be performed, the stored data is lost. DRAM is also considered read/write memory because the storage cells can be easily read from or written to during normal operation.

Another consideration when using DRAM is that the voltage of the word line must be larger than V_{CC} in order to turn on the access transistor. In order to turn on an NMOS transistor, the gate terminal must be larger than the source terminal by at least a threshold voltage (V_T). In traditional CMOS circuit design, the source terminal is typically connected to ground (0v). This means the transistor can be easily turned on by driving the gate with a logic 1 (i.e., V_{CC}) since this creates a V_{GS} voltage much larger than V_T. This is not always the case in DRAM. In DRAM, the source terminal is not connected to ground but rather to the storage capacitor. In the worst-case situation, the capacitor could be storing a logic 1 (i.e., V_{CC}). This means that in order for the word line to be able to turn on the access transistor, it must be equal to or larger than ($V_{CC} + V_T$). This is an issue because the highest voltage that the DRAM device has access to is V_{CC}. In DRAM, a *charge pump* is used to create a voltage larger than $V_{CC} + V_T$ that is driven on the word lines. Once this voltage is used, the charge is lost, so the line must be pumped up again before its next use. The process of "pumping up" takes time that must be considered when calculating the maximum speed of DRAM. Figure 10.15 shows a graphical depiction of this consideration.

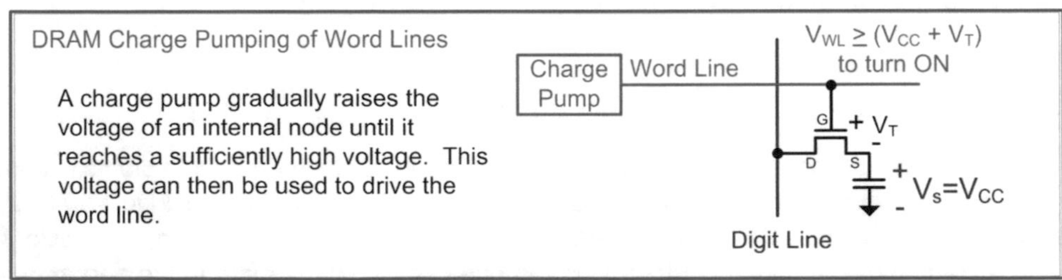

Fig. 10.15
DRAM charge pumping of word lines

Another consideration when using DRAM is how the charge in the capacitor develops into an actual voltage on the digital line when the access transistor is closed. Consider the simple 4×4 array of DRAM cells shown in Fig. 10.16. In this topology, the DRAM cells are accessed using the same approach as in the SRAM array from Fig. 10.11.

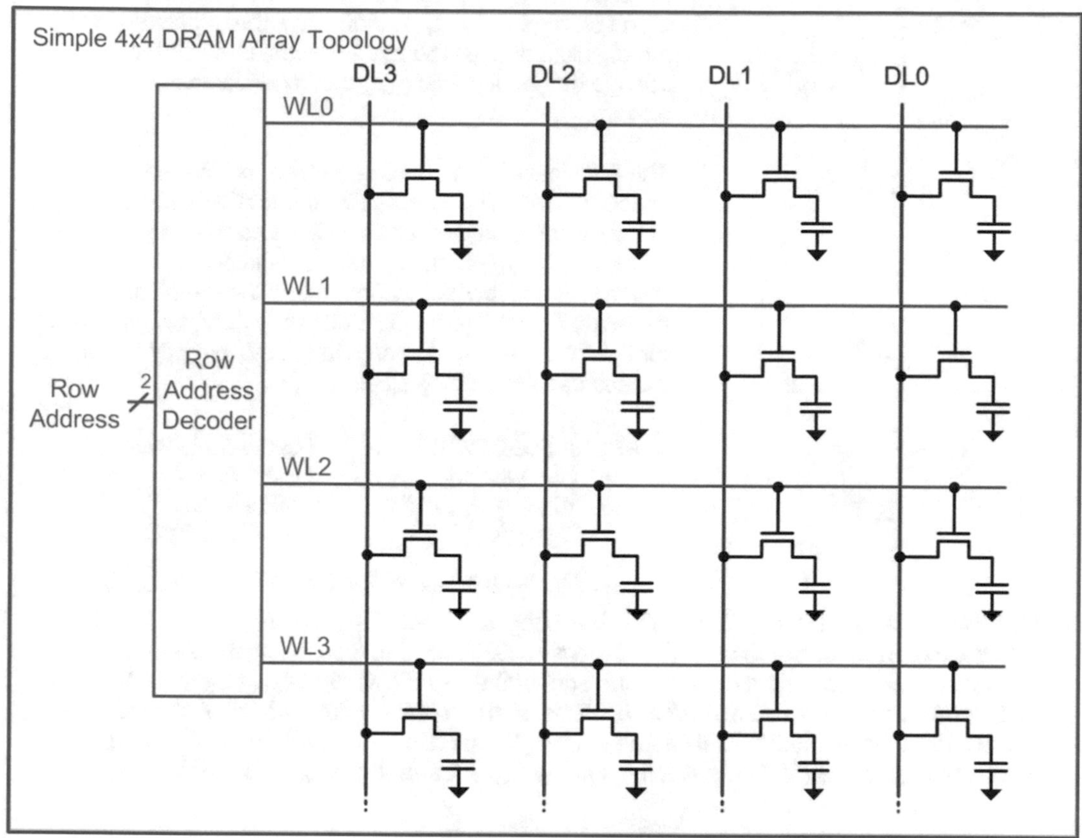

Fig. 10.16
Simple 4×4 DRAM array topology

One of the limitations of this simple configuration is that the charge stored in the capacitors cannot develop a full voltage level across the digit line when the access transistor is closed. This is because the digit line itself has capacitance that impacts how much voltage will be developed. In practice, the capacitance of the digit line (C_{DL}) is much larger than the capacitance of the storage cell (C_S) due to having significantly more area and being connected to numerous other storage cells. This becomes an issue because when the storage capacitor is connected to the digit line, the resulting voltage on the digit line (V_{DL}) is much less than the original voltage on the storage cell (V_S). This behavior is known as *charge sharing* because when the access transistor is closed, the charge on both capacitors is distributed across both devices and results in a final voltage that depends on the initial charge in the system and the values of the two capacitors. Example 10.1 shows an example of how to calculate the final digit line voltage when the storage cell is connected.

Example: Calculating the Final Digit Line Voltage in a DRAM Based on Charge Sharing

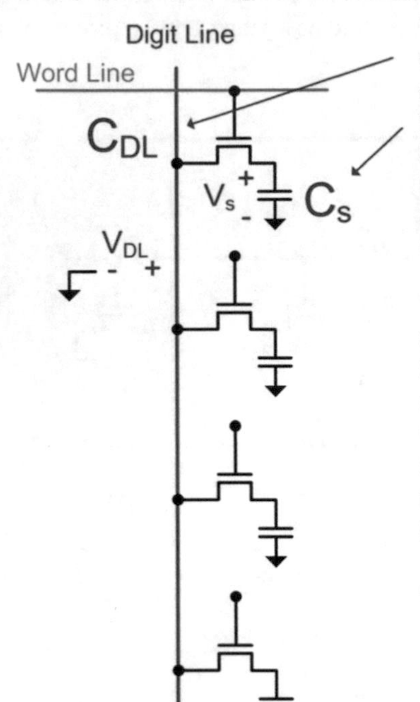

To illustrate how charge sharing limits the voltage that is developed on the digit line, let's consider a simple example where the cell is storing a logic 1 (V_S=+3.3v) and the digit line is initially set to V_{DL}=1.65v. The capacitance of the storage cell is C_S=10 pF while the capacitance of the digit line is C_{DL}=150 pF. We want to solve for the voltage on the digit line after the access transistor is closed.

The principle that guides this problem is "charge conservation". This means that the total amount of charge in the system can neither be created nor destroyed. The amount of charge in the system is dictated by the initial voltage across the capacitors. Since the definition of capacitance is "Charge per Volt", or C=Q/V, we can solve for the total amount of charge in the system prior to the access transistor being closed.

Q_{init} →	Q in the storage cell	+	Q on the digital line
	$C_S = Q_S / V_S$		$C_{DL} = Q_{DL} / V_{DL}$
	$10\ pF = Q_S / 3.3\ v$		$150\ pF = Q_{DL} / 1.65\ v$
	$Q_S = 33\ pC$		$Q_{DL} = 247.5\ pC$

$$Q_{init} = Q_s + Q_{DL} = 33\ pC + 247.5\ pC = \underline{280.5\ pC}$$

Once the access transistor closes, the two voltages (V_S and V_{DL}) are connected together and are forced to the same voltage (V_S=V_{DL}=V_{final}). Also after the access transistor closes, the final capacitance of the system is the sum of the two capacitors (C_{final}=C_S+C_{DL}=160 pF) since capacitors in parallel are additive. Using charge conservation, the initial charge in the system is equivalent to the final charge in the system (Q_{final}=Q_{init}=280.5 pC). From these values we can calculate the final voltage in the system after the access transistor closes.

$$C_{final} = Q_{final} / V_{final}$$
$$160\ pF = 280.5\ pC / V_{final}$$
$$V_{final}=1.75\ v$$

This means that when storage cell is connected to the digit line, it only moves the voltage by 0.1v (1.76v-1.65v), or 100mv. This is a problem because this voltage difference is not sufficient to be detected using a standard logic gate.

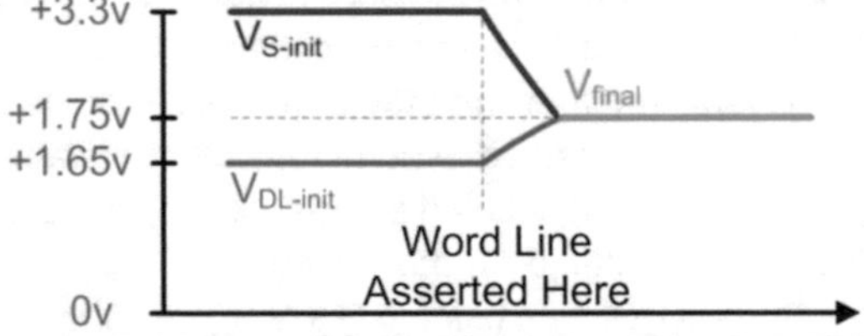

Example 10.1
Calculating the final digit line voltage in a DRAM based on charge sharing

The issue with the charge sharing behavior of a DRAM cell is that the final voltage on the word line is not large enough to be detected by a standard logic gate or latch. In order to overcome this issue, modern DRAM arrays use complementary storage cells and sense amplifiers. The complementary cells store the original data and its complement. Two digit lines (DL and DLn) are used to read the contents of the storage cells. DL and DLn are initially pre-charged to exactly $V_{CC}/2$. When the access transistors are closed, the storage cells will share their charge with the digit lines and move them slightly away from $V_{CC}/2$ in different directions. This allows twice the voltage difference to be developed during a read. A sense amplifier is then used to boost this small voltage difference into a full logic level that can be read by a standard logic gate or latch. Figure 10.17 shows the modern DRAM array topology based on complementary storage cells.

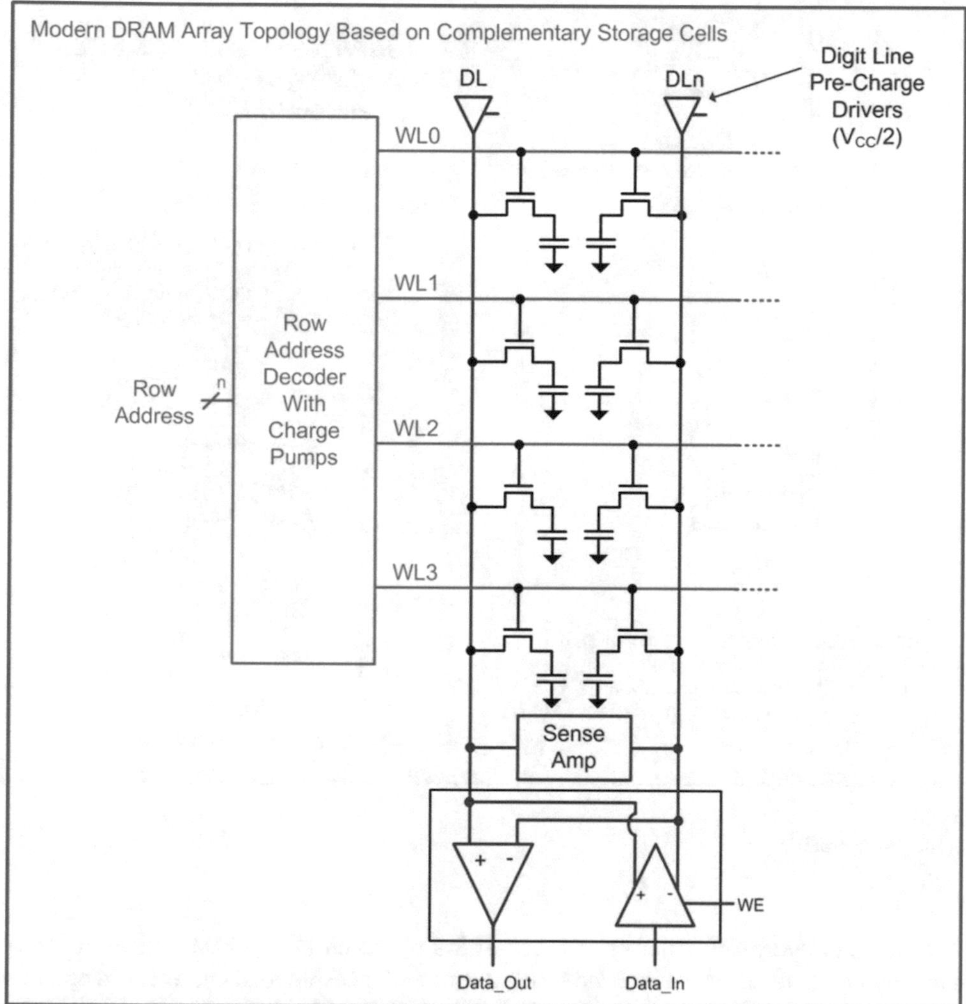

Fig. 10.17
Modern DRAM array topology based on complementary storage cells

The sense amplifier is designed to boost small voltage deviations from $V_{CC}/2$ on DL and DLn to full logic levels. The sense amplifier sits in-between DL and DLn and has two complementary networks, the N-sense amplifier and the P-sense amplifier. The N-sense amplifier is used to pull a signal that is below $V_{CC}/2$ (either DL or DLn) down to GND. A control signal (N-Latch or NLATn) is used to turn on this network. The P-sense amplifier is used to pull a signal that is above $V_{CC}/2$ (either DL or DLn) up to V_{CC}. A control signal (Active Pull-Up or ACT) is used to turn on this network. The two networks are activated in a sequence with the N-sense network activating first. Figure 10.18 shows an overview of the operation of a DRAM sense amplifier.

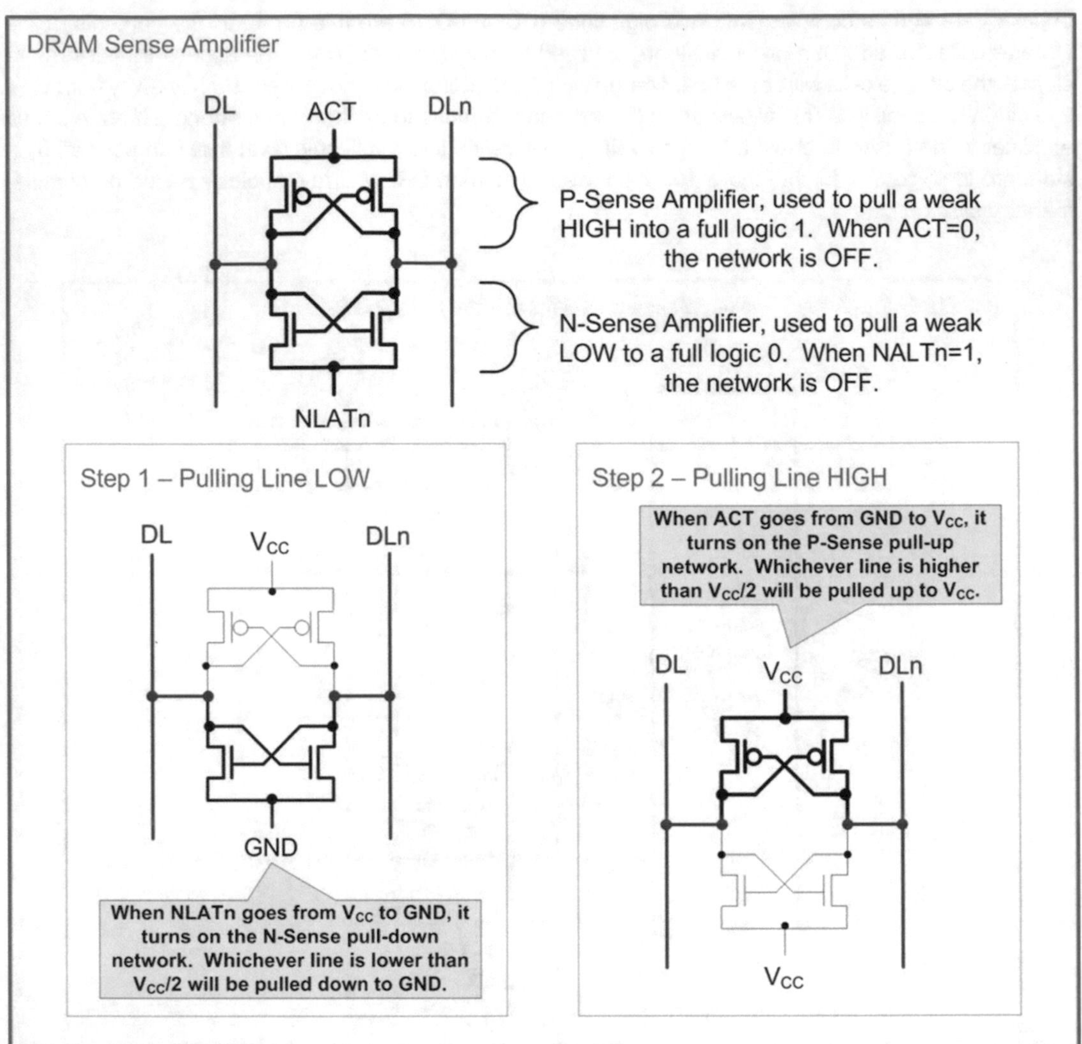

Fig. 10.18
DRAM sense amplifier

Let's now put everything together and look at the operation of a DRAM system during a read operation. Figure 10.19 shows a simplified timing diagram of a DRAM read cycle. This diagram shows the critical signals and their values when reading a logic 1. Notice that there is a sequence of steps that must be accomplished before the information in the storage cells can be retrieved.

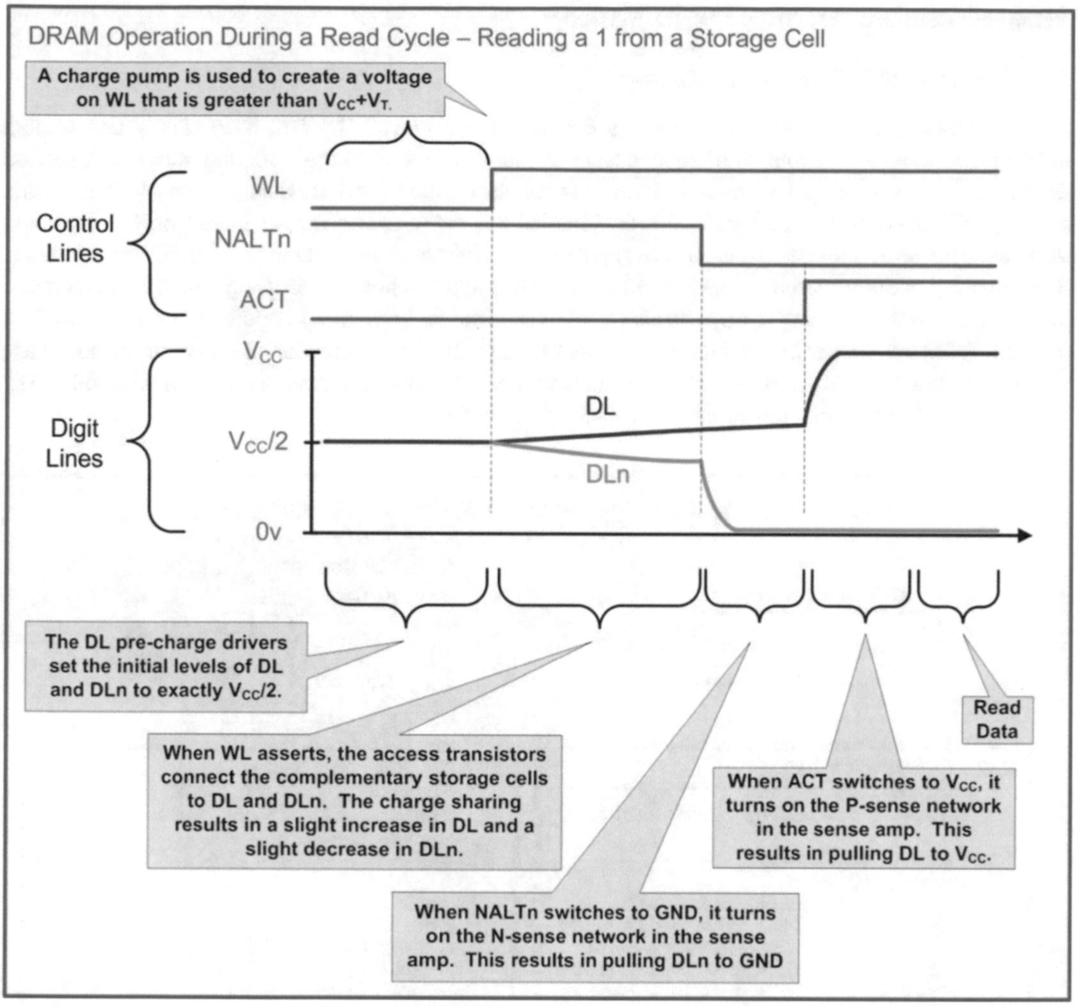

DRAM Operation During a Read Cycle – Reading a 1 from a Storage Cell

A charge pump is used to create a voltage on WL that is greater than $V_{CC}+V_T$.

Control Lines

WL

NALTn

ACT

Digit Lines

V_{CC}

$V_{CC}/2$

0v

DL

DLn

The DL pre-charge drivers set the initial levels of DL and DLn to exactly $V_{CC}/2$.

When WL asserts, the access transistors connect the complementary storage cells to DL and DLn. The charge sharing results in a slight increase in DL and a slight decrease in DLn.

When NALTn switches to GND, it turns on the N-sense network in the sense amp. This results in pulling DLn to GND

When ACT switches to V_{CC}, it turns on the P-sense network in the sense amp. This results in pulling DL to V_{CC}.

Read Data

Fig. 10.19
DRAM operation during a read cycle – Reading a 1 from a storage cell

A DRAM write operation is accomplished by opening the access transistors to the complementary storage cells using WL, disabling the pre-charge drivers and then writing full logic level signals to the storage cells using the Data_In line driver.

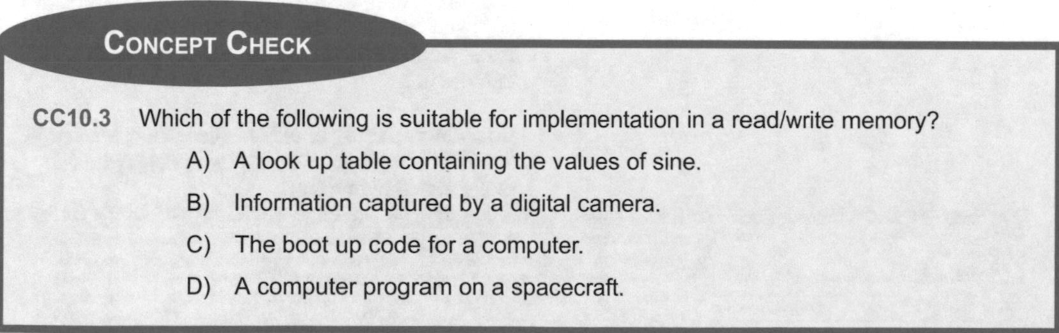

Concept Check

CC10.3 Which of the following is suitable for implementation in a read/write memory?

A) A look up table containing the values of sine.

B) Information captured by a digital camera.

C) The boot up code for a computer.

D) A computer program on a spacecraft.

10.4 Modeling Memory with Verilog

10.4.1 Read-Only Memory in Verilog

A read-only memory in Verilog can be defined in two ways. The first is to simply use a case statement to define the contents of each location in memory based on the incoming address. A second approach is to declare an array and then initialize its contents. When using an array, a separate procedural block handles assigning the contents of the array to the output based on the incoming address. The array can be initialized using either an initial block or through the file I/O system tasks $readmemb() or $readmemh(). Example 10.2 shows two approaches for modeling a 4×4 ROM memory. In this example the memory is asynchronous, meaning that as soon as the address changes, the data from the ROM will appear immediately. To model this asynchronous behavior, the procedural blocks are sensitive to the incoming address. In the simulation, each possible address is provided (i.e., "00," "01," "10," and "11) to verify that the ROM was initialized correctly.

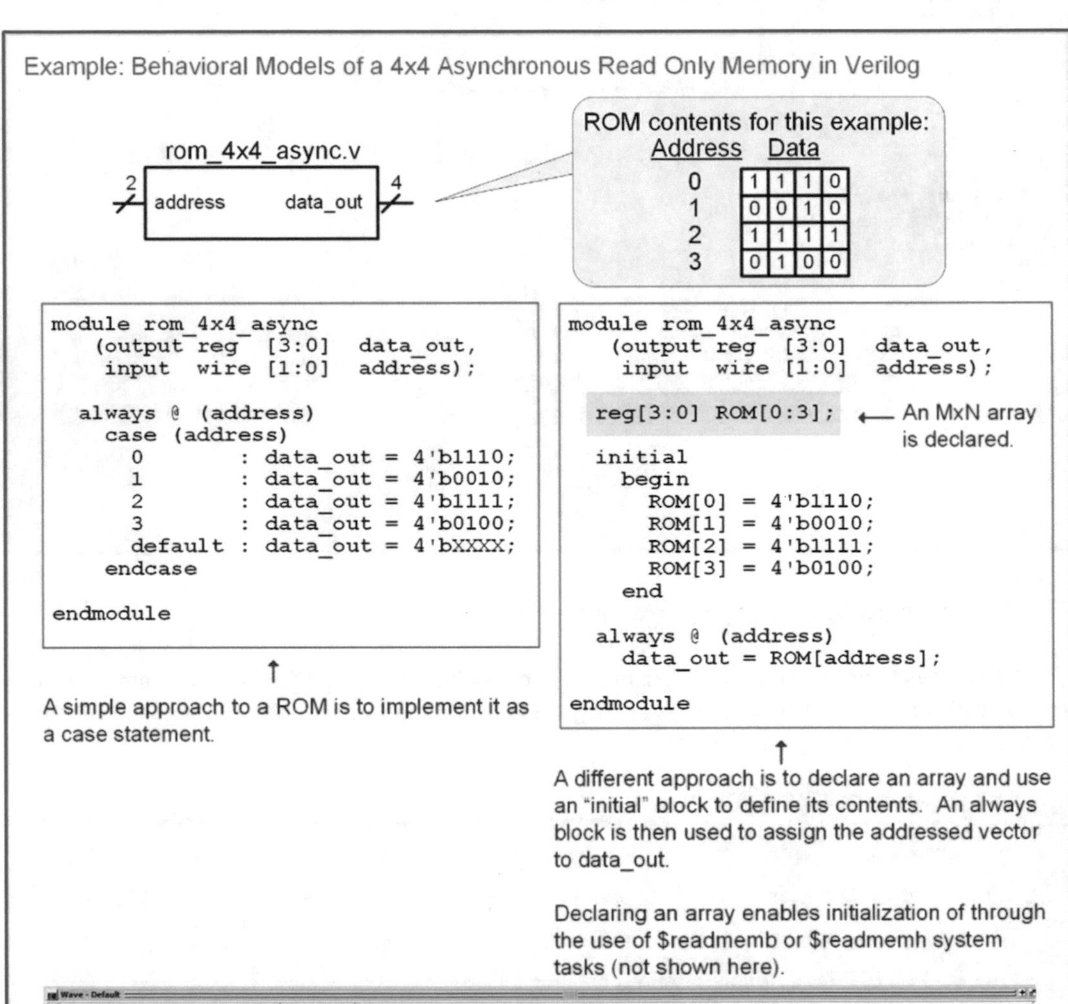

Example 10.2
Behavioral models of a 4×4 asynchronous read-only memory in Verilog

A synchronous ROM can be created in a similar manner as in the asynchronous approach. The only difference is that in a synchronous ROM, a clock edge is used to trigger the procedural block that updates data_out. A sensitivity list is used that contains the clock to trigger the assignment. Example 10.3 shows two Verilog models for a synchronous ROM. Notice that prior to the first clock edge, the simulator does not know what to assign to data_out, so it lists the value as unknown (X).

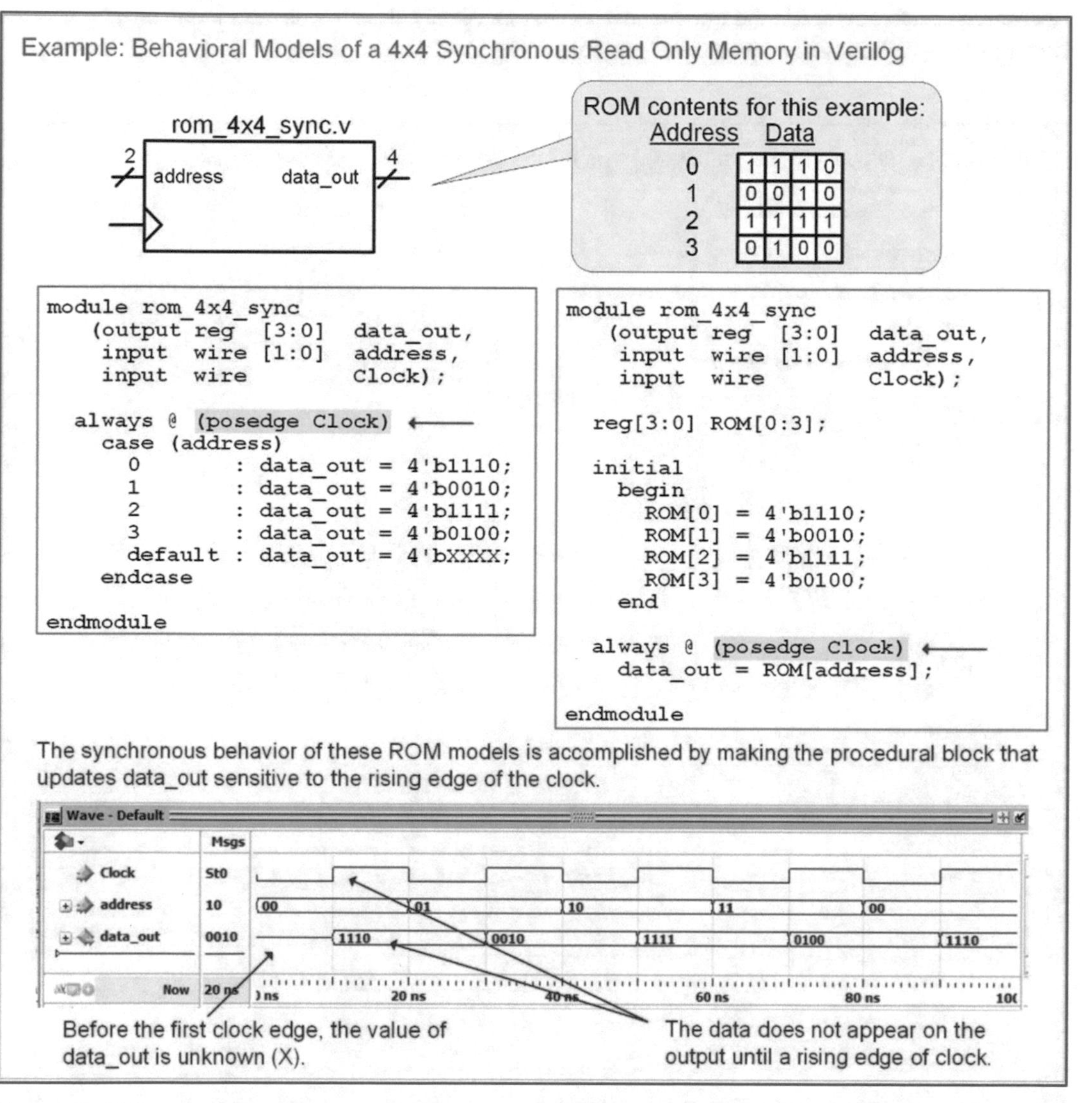

Example 10.3
Behavioral models of a 4×4 synchronous read-only memory in Verilog

10.4.2 Read/Write Memory in Verilog

In a simple read/write memory model, there is an output port that provides data when reading (data_out) and an input port that receives data when writing (data_in). Within the module, an array signal is declared with elements of type reg. To write to the array, signal assignments are made from the data_in port to the element within the array corresponding to the incoming address. To read from the array, the data_out port is assigned the element within the array corresponding to the incoming address. A *write enable* (WE) signal tells the system when to write to the array (WE = 1) or when to read from the array (WE = 0). In an asynchronous R/W memory, data is immediately written to the array when WE = 1 and data is immediately read from the array when WE = 0. This is modeled using a procedural block with a sensitivity list containing every input to the system. Example 10.4 shows an asynchronous R/W 4×4 memory system and functional simulation results. In the simulation, each address is initially read from to verify that it does not contain data. The data_out port produces unknown (X) for the initial set of read operations. Each address in the array is then written to. Finally, the array is read from verifying that the data that was written can be successfully retrieved.

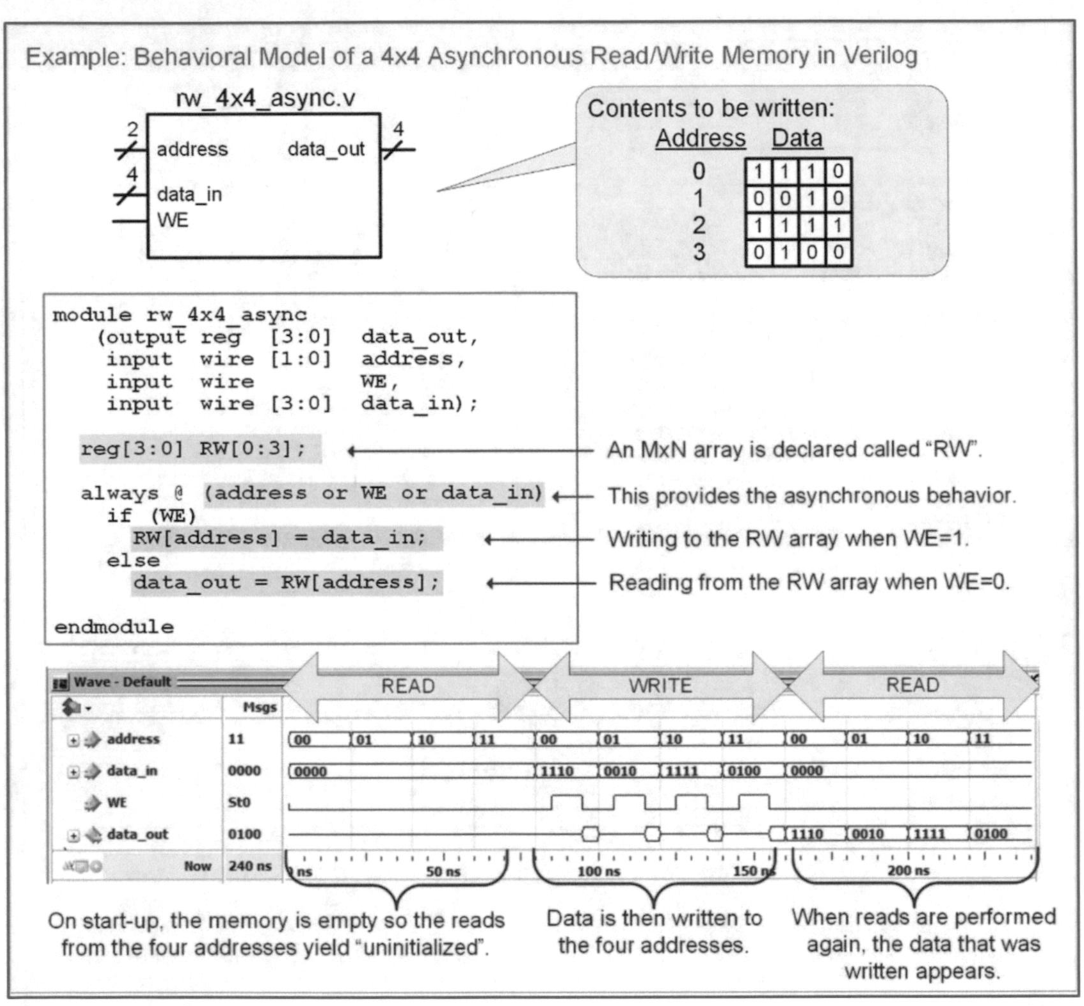

Example 10.4
Behavioral model of a 4×4 asynchronous read/write memory in Verilog

A synchronous read/write memory is made in a similar manner with the exception that a clock is used to trigger the procedural block managing the signal assignments. In this case, the WE signal acts as a synchronous control signal indicating whether assignments are read from or written to the RW array. Example 10.5 shows the Verilog model for a synchronous read/write memory and the simulation waveform showing both read and write cycles.

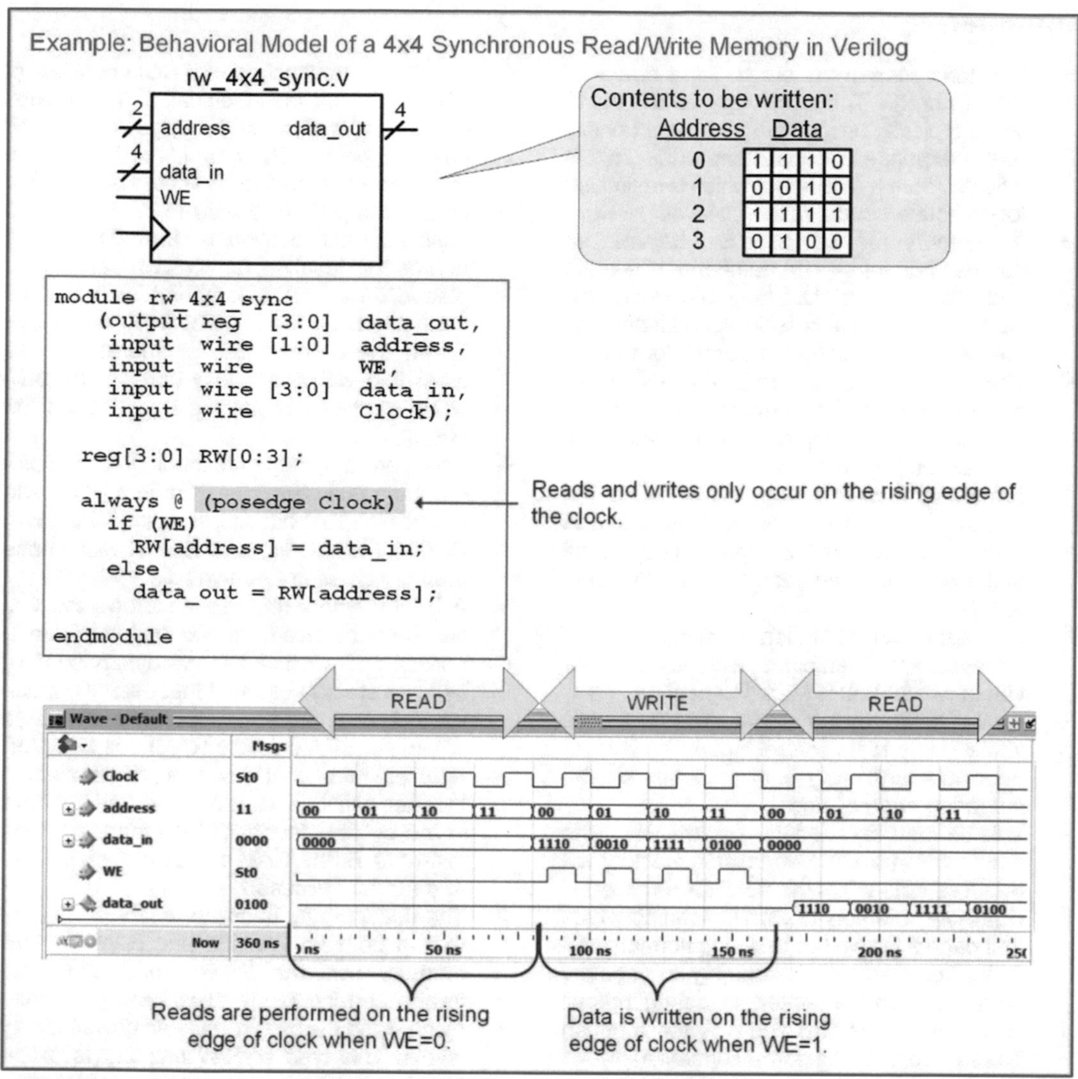

Example 10.5
Behavioral model of a 4×4 synchronous read/write memory in Verilog

Concept Check

CC10.4 Explain the advantage of modeling memory in Verilog without going into the details of the storage cell operation.

A) It allows the details of the storage cell to be abstracted from the functional operation of the memory system.

B) It is too difficult to model the analog behavior of the storage cell.

C) There are too many cells to model so the simulation would take too long.

D) It lets both ROM and R/W memory to be modeled in a similar manner.

Summary

- The term memory refers to large arrays of digital storage. The technology used in memory is typically optimized for storage density at the expense of control capability. This is different from a D-flip-flop, which is optimized for complete control at the bit level.
- A memory device always contains an address bus input. The number of bits in the address bus dictates how many storage locations can be accessed. An n-bit address bus can access 2^n (or M) storage locations.
- The width of each storage location (N) allows the density of the memory array to be increased by reading and writing vectors of data instead of individual bits.
- A memory map is a graphical depiction of a memory array. A memory map is useful to give an overview of the capacity of the array and how different address ranges of the array are used.
- A read is an operation in which data is retrieved from memory. A write is an operation in which data is stored to memory.
- An asynchronous memory array responds immediately to its control inputs. A synchronous memory array only responds on the triggering edge of clock.
- Volatile memory will lose its data when the power is removed. Non-volatile memory will retain its data when the power is removed.
- Read-only memory (ROM) is a memory type that cannot be written to during normal operation. Read/write (R/W) memory is a memory type that can be written to during normal operation. Both ROM and R/W memory can be read from during normal operation.
- Random-access memory (RAM) is a memory type in which any location in memory can be accessed at any time. In sequential access memory, the data can only be retrieved in a linear sequence. This means that in sequential memory the data cannot be accessed arbitrarily.
- The basic architecture of a ROM consists of intersecting bit lines (vertical) and word lines (horizontal) that contain storage cells at their crossing points. The data is read out of the ROM array using the bit lines. Each bit line contains a pull-up resistor to initially store a logic 1 at each location. If a logic 0 is desired at a certain location, a pull-down transistor is placed on a particular bit line with its gate connected to the appropriate word line. When the storage cell is addressed, the word line will assert and turn on the pull-down transistor producing a logic 0 on the output.
- There are a variety of technologies to implement the pull-down transistor in a ROM. Different ROM architectures include MROMs, PROMs, EPROMs, and EEPROMs. These memory types are non-volatile.
- A R/W memory requires a storage cell that can be both read from and written to during normal operation. A DRAM (dynamic RAM) cell is a storage element that uses a capacitor to hold charge corresponding to a logic value. An SRAM (static RAM) cell is a storage element that uses a cross-coupled inverter pair to hold the value being stored in the positive feedback loop formed by the inverters. Both DRAM and SRAM are volatile and random access.
- The floating-gate transistor enables memory that is both non-volatile and R/W. Modern memory systems based on floating-gate transistor technology allow writing to take place using the existing system power supply levels. This type of R/W memory is called FLASH. In FLASH memory, the information is read out in blocks; thus it is not technically random access.
- Memory can be modeled in Verilog using an array data type consisting of elements of type reg.

Exercise Problems

Section 10.1: Memory Architecture and Terminology

10.1.1 For a 512k × 32 memory system, how many unique address locations are there? Give the exact number.

10.1.2 For a 512k × 32 memory system, what is the data width at each address location?

10.1.3 For a 512k × 32 memory system, what is the *capacity* in bits?

10.1.4 For a 512k × 32-bit memory system, what is the *capacity* in bytes?

10.1.5 For a 512k × 32 memory system, how wide does the incoming address bus need to be in order to access every unique address location?

10.1.6 Name the type of memory with the following characteristic: *when power is removed, the data is lost.*

10.1.7 Name the type of memory with the following characteristic: *when power is removed, the memory still holds its information.*

10.1.8 Name the type of memory with the following characteristic: *it can only be read from during normal operation.*

10.1.9 Name the type of memory with the following characteristic: *during normal operation, it can be read and written to.*

10.1.10 Name the type of memory with the following characteristic: *data can be accessed from any address location at any time.*

10.1.11 Name the type of memory with the following characteristic: *data can only be accessed in consecutive order, thus not every location of memory is available instantaneously.*

Section 10.2: Non-volatile Memory Technology

10.2.1 Name the type of memory with the following characteristic: *this memory is non-volatile, read/write, and only provides data access in blocks.*

10.2.2 Name the type of memory with the following characteristic: *this memory uses a floating-gate transistor, can be erased with electricity, and provides individual bit access.*

10.2.3 Name the type of memory with the following characteristic: *this memory is non-volatile, read/write, and provides word-level data access.*

10.2.4 Name the type of memory with the following characteristic: *this memory uses a floating-gate transistor that is erased with UV light.*

10.2.5 Name the type of memory with the following characteristic: *this memory is programmed by blowing fuses or anti-fuses.*

10.2.6 Name the type of memory with the following characteristic: *this memory is partially fabricated prior to knowing the information to be stored.*

Section 10.3: Volatile Memory Technology

10.3.1 How many transistors does it take to implement an SRAM cell?

10.3.2 Why doesn't an SRAM cell require a refresh cycle?

10.3.3 Design a Verilog model for the SRAM system shown in Fig. 10.20. Your storage cell should be designed such that its contents can be overwritten by the line driver. Consider using signal strengths for this behavior (e.g., strong1 will overwrite a weak0). You will need to create a system for the differential line driver with enable. This driver will need to contain a high impedance state when disabled. Both your line driver (Din) and receiver (Dout) are differential. These systems can be modeled using simple if-else statements. Create a test bench for your system that will write a 0 to the cell, then read it back to verify the 0 was stored, and then repeat the write/read cycles for a 1.

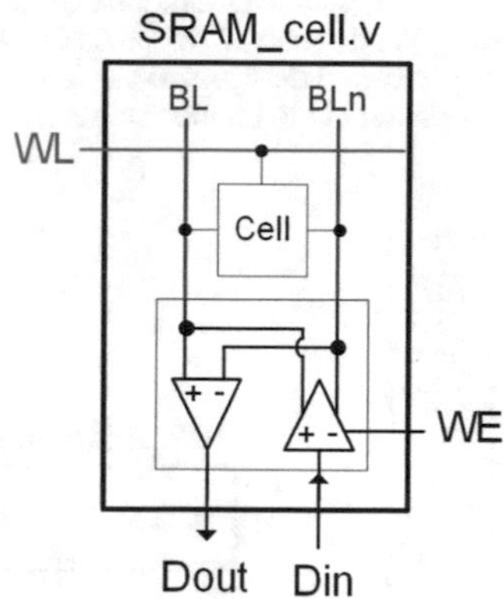

Fig. 10.20
SRAM cell block diagram

10.3.4 Why is a DRAM cell referred to as a 1 T 1C configuration?

10.3.5 Why is a charge pump necessary on the word lines of a DRAM array?

10.3.6 Why does a DRAM cell require a refresh cycle?

10.3.7 For the DRAM storage cell shown in Fig. 10.21, solve for the final voltage on the digit line after the access transistor (M1) closes if initially $V_S = V_{CC}$ (**i.e., the cell is storing a 1**). In this system, $C_S = 5$ pF, $C_{DL} = 10$ pF, and $V_{CC} = +3.4$v. Prior to the access transistor closing, the digit line is pre-charged to $V_{CC}/2$.

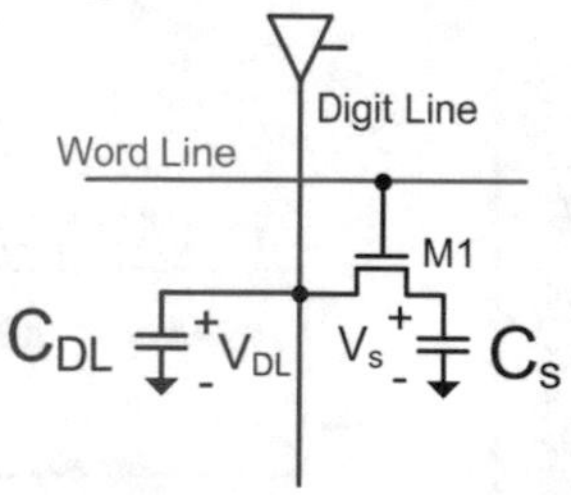

Fig. 10.21
DRAM charge sharing exercise

10.3.8 For the DRAM storage cell shown in Fig. 10.21, solve for the final voltage on the digit line after the access transistor (M1) closes if initially $V_S = \text{GND}$ (**i.e., the cell is storing a 0**). In this system, $C_S = 5$ pF, $C_{DL} = 10$ pF, and $V_{CC} = +3.4$v. Prior to the access transistor closing, the digit line is pre-charged to $V_{CC}/2$.

Section 10.4: Modeling Memory with Verilog

10.4.1 Design a Verilog model for the 16×8, asynchronous, read-only memory system shown in Fig. 10.22. The system should contain the information provided in the memory map. Create a test bench to simulate your model by reading from each of the 16 unique addresses and observing data_out to verify it contains the information in the memory map.

Address	Data
0	x"00"
1	x"11"
2	x"22"
3	x"33"
4	x"44"
5	x"55"
6	x"66"
7	x"77"
8	x"88"
9	x"99"
10	x"AA"
11	x"BB"
12	x"CC"
13	x"DD"
14	x"EE"
15	x"FF"

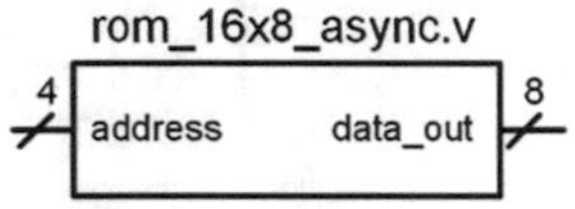

rom_16x8_async.v
4 address
data_out 8

Fig. 10.22
16×8 asynchronous ROM block diagram

10.4.2 Design a Verilog model for the 16×8, synchronous, read-only memory system shown in Fig. 10.23. The system should contain the information provided in the memory map. Create a test bench to simulate your model by reading from each of the 16 unique addresses and observing data_out to verify it contains the information in the memory map.

Address	Data
0	x"FF"
1	x"EE"
2	x"DD"
3	x"CC"
4	x"BB"
5	x"AA"
6	x"99"
7	x"88"
8	x"77"
9	x"66"
10	x"55"
11	x"44"
12	x"33"
13	x"22"
14	x"11"
15	x"00"

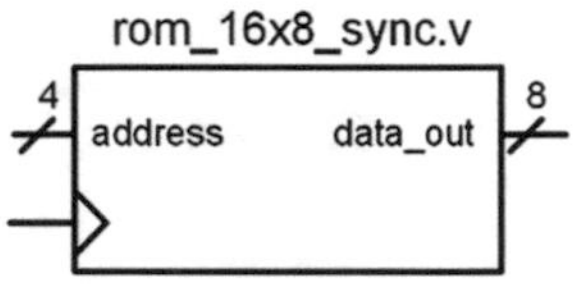

rom_16x8_sync.v
4 address
data_out 8

Fig. 10.23
16×8 synchronous ROM block diagram

10.4.3 Design a Verilog model for the 16×8, asynchronous, read/write memory system shown in Fig. 10.24. Create a test bench to simulate your model. Your test bench should first read from all of the address locations to verify they are uninitialized. Next, your test bench should write unique information to each of the address locations. Finally, your test bench should read from each address location to verify that the information that was written was stored and can be successfully retrieved.

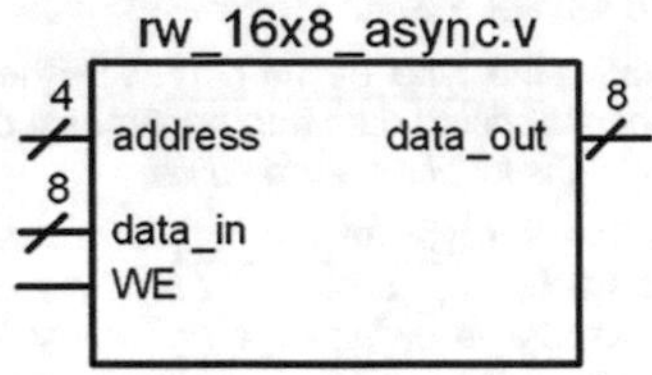

Fig. 10.24
16×8 asynchronous R/W memory block diagram

10.4.4 Design a Verilog model for the 16×8, synchronous, read/write memory system shown in Fig. 10.25. Create a test bench to simulate your model. Your test bench should first read from all of the address locations to verify they are uninitialized. Next, your test bench should write unique information to each of the address locations. Finally, your test bench should read from each address location to verify that the information that was written was stored and can be successfully retrieved.

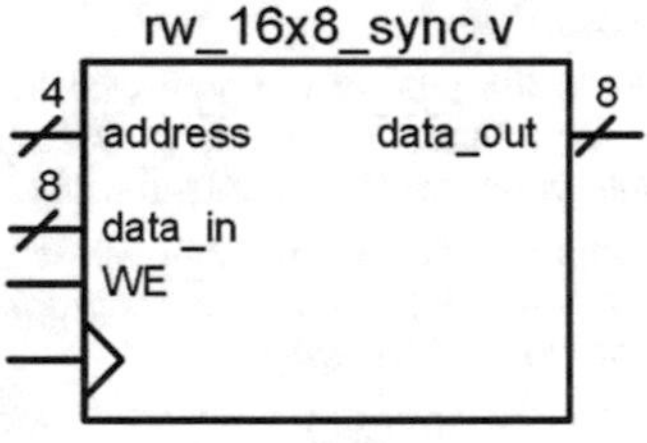

Fig. 10.25
16×8 synchronous R/W memory block diagram

Chapter 11: Programmable Logic

This chapter provides an overview of programmable logic devices (PLDs). The term PLD is used as a generic description for any circuit that can be programmed to implement digital logic. The technology and architectures of PLDs have advanced over time. A historical perspective is given on how the first programmable devices evolved into the programmable technologies that are prevalent today. The goal of this chapter is to provide a basic understanding of the principles of programmable logic devices.

Learning Outcomes—After completing this chapter, you will be able to:

11.1 Describe the basic architecture and evolution of programmable logic devices.
11.2 Describe the basic architecture of field-programmable gate arrays (FPGAs).

11.1 Programmable Arrays

11.1.1 Programmable Logic Array (PLA)

One of the first commercial PLDs developed using modern integrated circuit technology was the **programmable logic array (PLA)**. In 1970, Texas Instrument introduced the PLA with an architecture that supported the implementation of arbitrary, sum of products logic expressions. The PLA was fabricated with a dense array of AND gates, called an *AND plane*, and a dense array of OR gates, called an *OR plane*. Inputs to the PLA each had an inverter in order to provide the original variable and its complement. Arbitrary SOP logic expressions could be implemented by creating connections between the inputs, the AND plane, and the OR plane. The original PLAs were fabricated with all of the necessary features except the final connections to implement the SOP functions. When a customer provided the desired SOP expression, the connections were added as the final step of fabrication. This configuration technique was similar to an MROM approach. Figure 11.1 shows the basic architecture of a PLA.

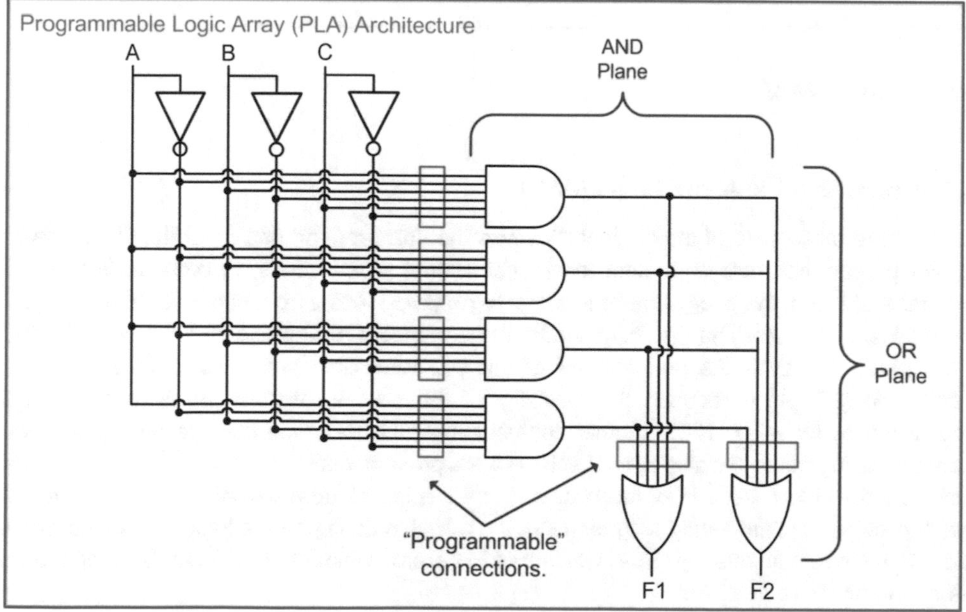

Fig. 11.1
Programmable logic array (PLA) architecture

© Springer Nature Switzerland AG 2019
B. J. LaMeres, *Introduction to Logic Circuits & Logic Design with Verilog*,
https://doi.org/10.1007/978-3-030-13605-5_11

A more compact schematic for the PLA is drawn by representing all of the inputs into the AND and OR gates with a single wire. Connections are indicated by inserting X's at the intersections of wires. Figure 11.2 shows this simplified PLA schematic implementing two different SOP logic expressions.

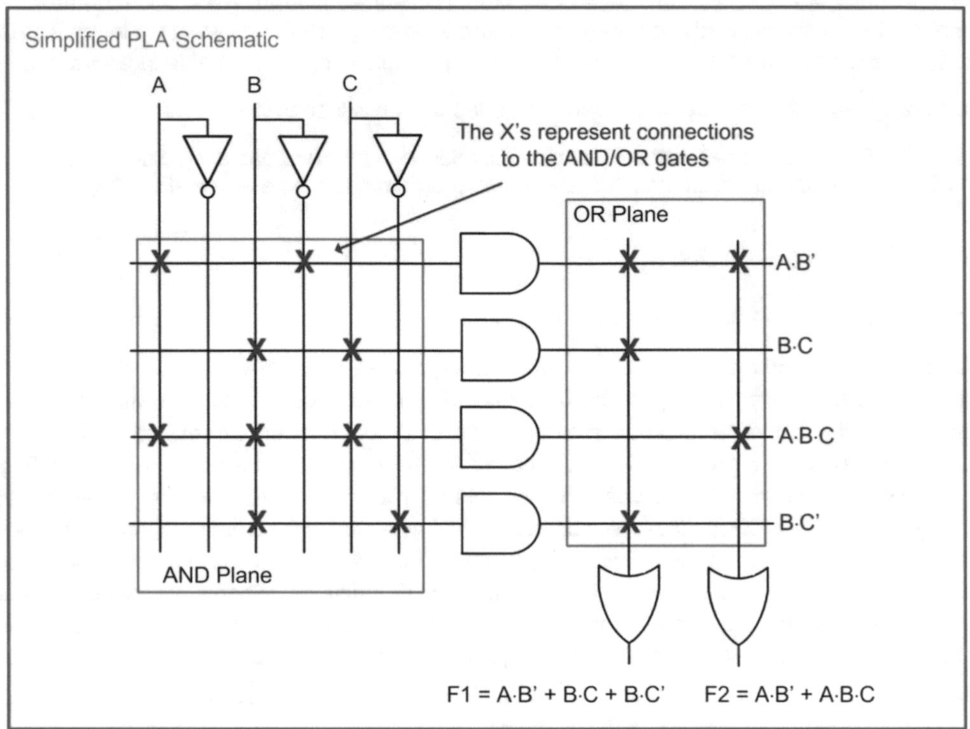

Fig. 11.2
Simplified PLA schematic

11.1.2 Programmable Array Logic (PAL)

One of the drawbacks of the original PLA was that the programmability of the OR plane caused significant propagation delays through the combinational logic circuits. In order to improve on the performance of PLAs, the **programmable array logic (PAL)** was introduced in 1978 by the company *Monolithic Memories, Inc*. The PAL contained a programmable AND plane and a *fixed-OR* plane. The fixed-OR plane improved the performance of this programmable architecture. While not having a programmable OR plane reduced the flexibility of the device, most SOP expressions could be manipulated to work with a PAL. Another contribution of the PAL was that the AND plane could be programmed using fuses. Initially, all connections were present in the AND plane. An external programmer was used to blow fuses in order to disconnect the inputs from the AND gates. While the fuse approach provided one-time-only programming, the ability to configure the logic post-fabrication was a significant advancement over the PLA, which had to be programmed at the manufacturer. Figure 11.3 shows the architecture of a PAL.

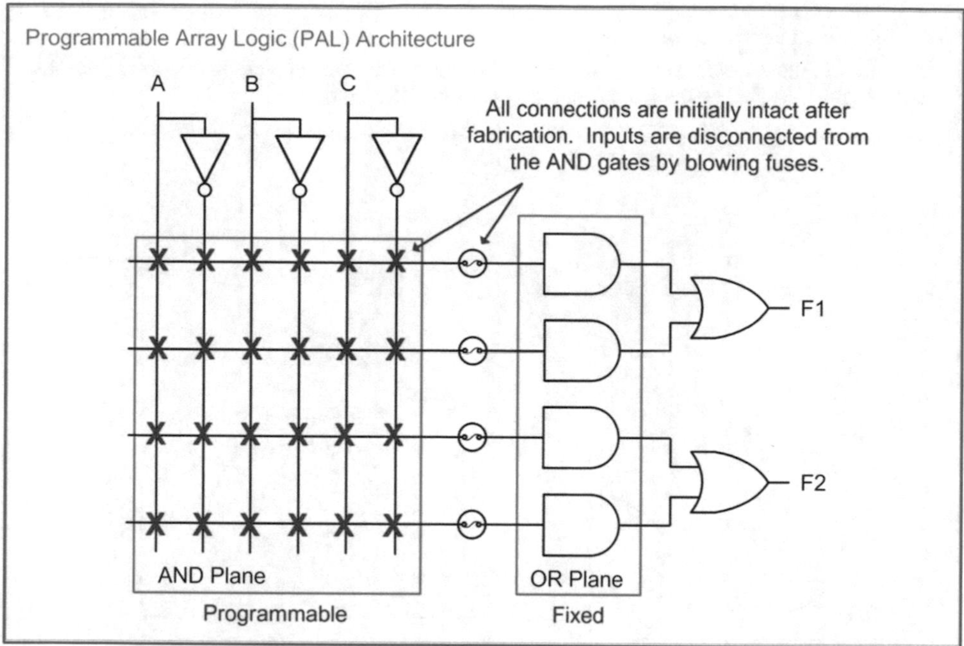

Fig. 11.3
Programmable array logic (PAL) architecture

11.1.3 Generic Array Logic (GAL)

As the popularity of the PAL grew, additional functionality was implemented to support more sophisticated designs. One of the most significant improvements was the addition of an *output logic macrocell (OLMC)*. An OLMC provided a D-flip-flop and a selectable mux so that the output of the SOP circuit from the PAL could be used either as the system output or the input to a D-flip-flop. This enabled the implementation of sequential logic and finite-state machines. The OLMC also could be used to route the I/O pin back into the PAL to increase the number of inputs possible in the SOP expressions. Finally, the OLMC provided a multiplexer to allow feedback from either the PAL output or the output of the D-flip-flop. This architecture was named a **generic array logic (GAL)** to distinguish its features from a standard PAL. Figure 11.4 shows the architecture of a GAL consisting of a PAL and an OLMC.

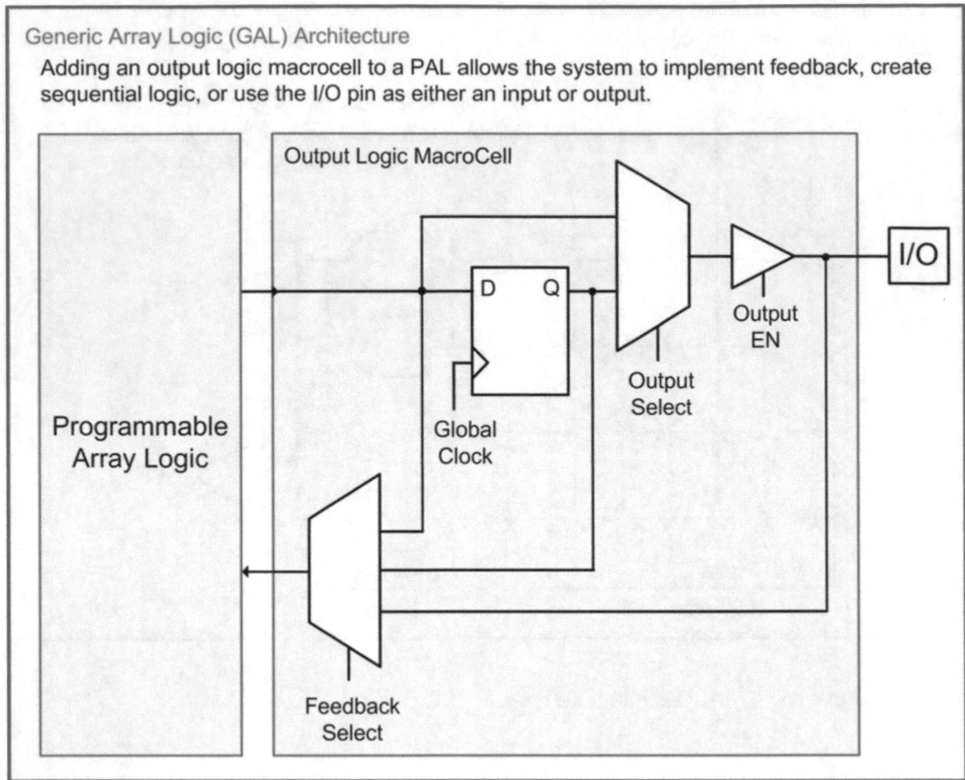

Fig. 11.4
Generic array logic (GAL) architecture

11.1.4 Hard Array Logic (HAL)

For mature designs, PALs and GALs could be implemented as a **hard array logic (HAL)** device. A HAL was a version of a PAL or GAL that had the AND plane connections implemented during fabrication instead of through blowing fuses. This architecture was more efficient for high-volume applications as it eliminated the programming step post-fabrication, and the device did not need to contain the additional programming circuitry.

In 1983, *Altera Inc.* was founded as a programmable logic device company. In 1984, Altera released its first version of a PAL with a unique feature that it could be programmed and erased multiple times using a programmer and an UV light source similar to an EEPROM.

11.1.5 Complex Programmable Logic Devices (CPLD)

As the demand for larger programmable devices grew, the PAL's architecture was not able to scale efficiently due to a number of reasons: first, as the size of combinational logic circuits increased, the PAL encountered fan-in issues in its AND plane, and second, for each input that was added to the PAL, the amount of circuitry needed on the chip grew geometrically due to requiring a connection to each AND gate in addition to the area associated with the additional OLMC. This led to a new PLD architecture in which the on-chip interconnect was partitioned across multiple PALs on a single chip. This partitioning meant that not all inputs to the device could be used by each PAL so the design complexity increased; however, the additional programmable resources outweighed this drawback, and this architecture was broadly adopted. This new architecture was called a **complex programmable logic device (CPLD)**.

The term ***simple* programmable logic device (SPLD)** was created to describe all of the previous PLD architectures (i.e., PLA, PAL, GAL, and HAL). Figure 11.5 shows the architecture of the CPLD.

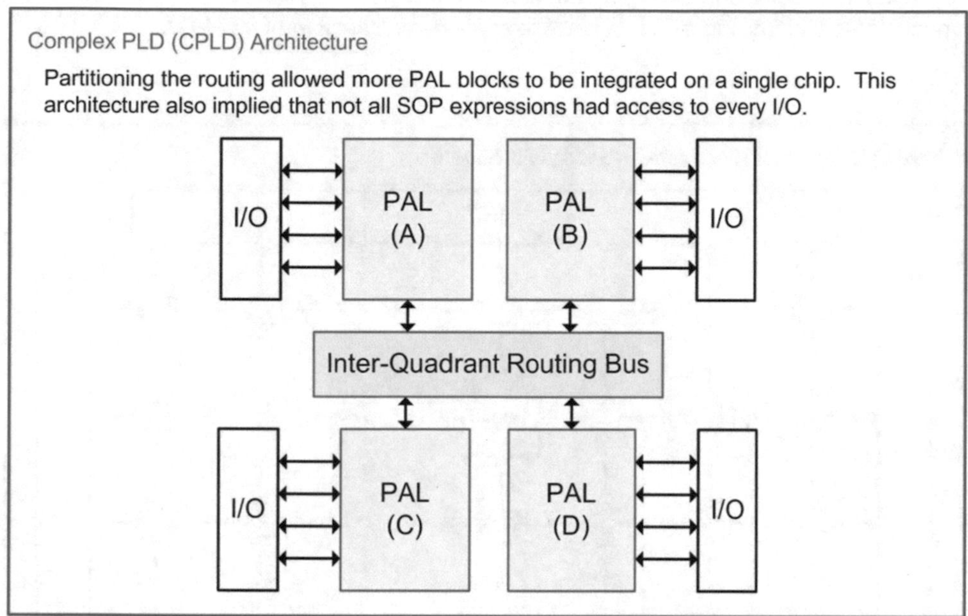

Fig. 11.5
Complex PLD (CPLD) architecture

CONCEPT CHECK

CC11.1 What is the only source of delay mismatch from the inputs to the outputs in a programmable array?

A) The AND gates will have different delays due to having different numbers of inputs.

B) The OR gates will have different delays due to having different numbers of inputs.

C) An input may or may not go through an inverter before reaching the AND gates.

D) None. All paths through the programmable array have identical delay.

11.2 Field-Programmable Gate Arrays (FPGAs)

To address the need for even more programmable resources, a new architecture was developed by *Xilinx Inc.* in 1985. This new architecture was called a **field-programmable gate array (FPGA)**. An FPGA consists of an array of programmable logic blocks (or logic elements) and a network of programmable interconnect that can be used to connect any logic element to any other logic element. Each logic block contained circuitry to implement arbitrary combinational logic circuits in addition to a D-flip-flop and a multiplexer for signal steering. This architecture effectively implemented an OLMC within each block, thus providing ultimate flexibility and providing significantly more resources for sequential logic. Today, FPGAs are the most commonly used programmable logic device with Altera Inc. and Xilinx Inc. being the two largest manufacturers. Figure 11.6 shows the generic architecture of an FPGA.

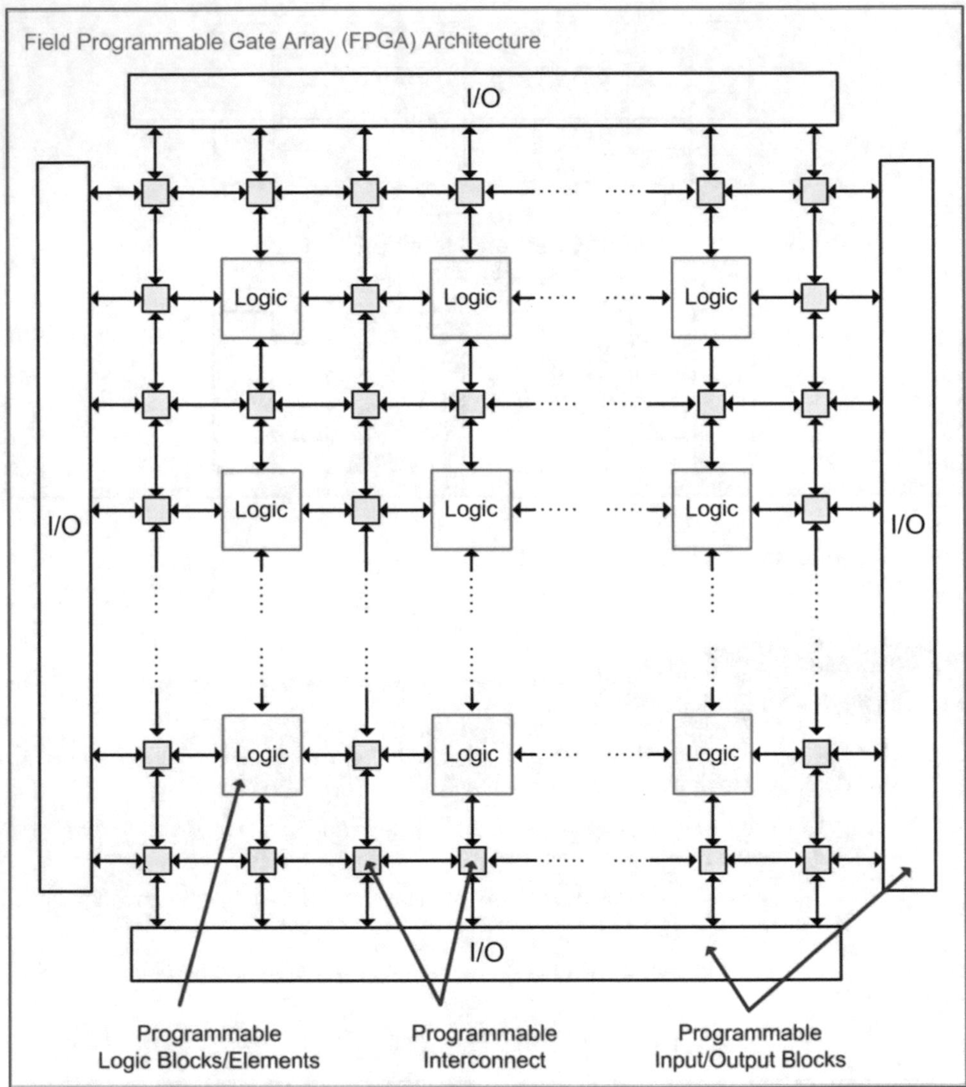

Fig. 11.6
Field-programmable gate array (FPGA) architecture

11.2.1 Configurable Logic Block (or Logic Element)

The primary reconfigurable structure in the FPGA is the **configurable logic block (CLB)** or **logic element (LE)**. Xilinx Inc. uses the term CLB, while Altera uses LE. Combinational logic is implemented using a circuit called a **look-up table (LUT)**, which can implement any arbitrary truth table. The details of a LUT are given in the next section. The CLB/LE also contains a D-flip-flop for sequential logic. A signal steering multiplexer is used to select whether the output of the CLB/LE comes from the LUT or from the D-flip-flop. The LUT can be used to drive a combinational logic expression into the D input of the D-flip-flop, thus creating a highly efficient topology for finite-state machines. A global routing network is used to provide common signals to the CLB/LE such as clock, reset, and enable. This global routing network can provide these common signals to the entire FPGA or local groups of CLB/LEs. Figure 11.7 shows the topology of a simple CLB/LE.

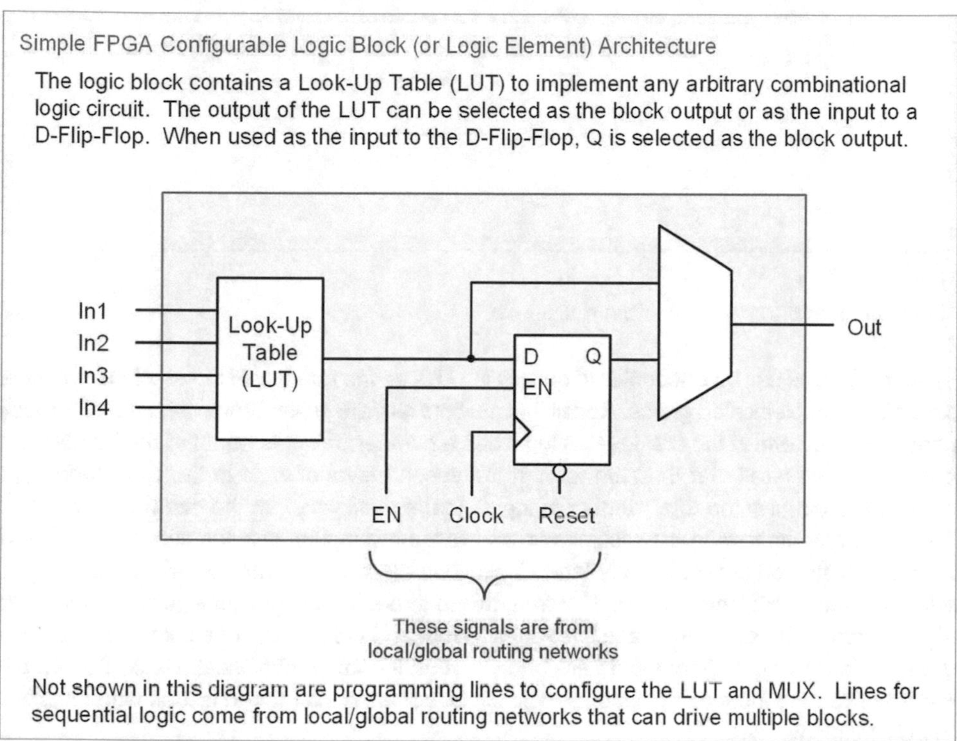

Fig. 11.7
Simple FPGA configurable logic block (or logic element)

CLB/LEs have evolved to include numerous other features such as carry-in/carry-out signals so that arithmetic operations can be cascaded between multiple blocks in addition to signal feedback and D-flip-flop initialization.

11.2.2 Look-Up Tables (LUTs)

A look-up table is the primary circuit used to implement combinational logic in FPGAs. This topology has also been adopted in modern CPLDs. In a LUT, the desired outputs of a truth table are loaded into a local configuration SRAM memory. The SRAM memory provides these values to the inputs of a multiplexer. The inputs to the combinational logic circuit are then used as the select lines to the multiplexer. For an arbitrary input to the combinational logic circuit, the multiplexer selects the appropriate value held in the SRAM and routes it to the output of the circuit. In this way, the multiplexer *looks up* the appropriate output value based on the input code. This architecture has the advantage that any logic function can be created without creating a custom logic circuit. Also, the delay through the LUT is identical regardless of what logic function is being implemented. Figure 11.8 shows a 2-input combinational logic circuit implemented with a 4-input multiplexer.

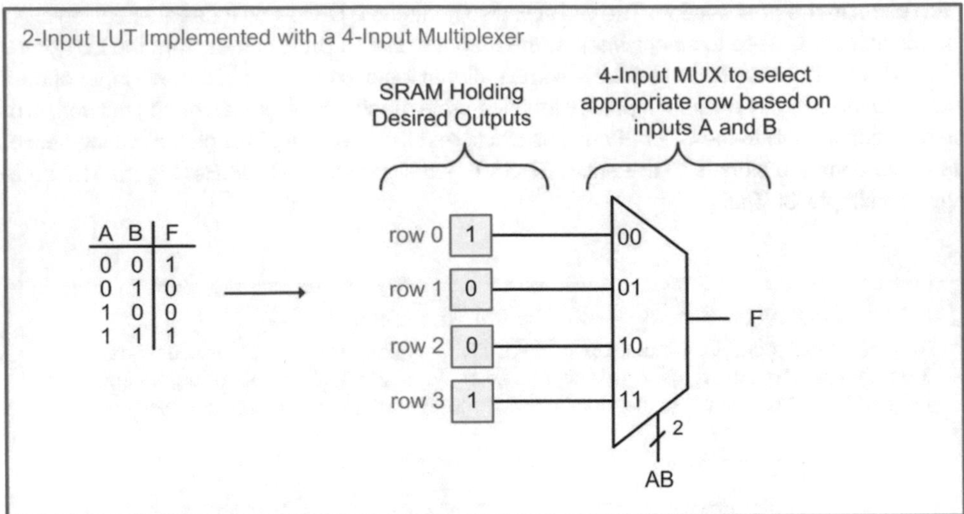

Fig. 11.8
2-input LUT implemented with a 4-input multiplexer

Fan-in limitations can be encountered quickly in LUTs as the number of inputs of the combinational logic circuit being implemented grows. Recall that multiplexers are implemented with an SOP topology in which each product term in the first level of logic has a number of inputs equal to the number of select lines plus one. Also recall that the sum term in the second level of logic in the SOP topology has a number of inputs equal to the total number of inputs to the multiplexer. In the example circuit shown in Fig. 11.8, each product term in the multiplexer will have three inputs, and the sum term will have four inputs. As an illustration of how quickly fan-in limitations are encountered, consider the implication of increasing the number of inputs in Fig. 11.8 from two to three. In this new configuration, the number of inputs in the product terms will increase from three to four and the number of inputs in the sum term will increase from four to eight. Eight inputs are often beyond the fan-in specifications of modern devices, meaning that even a 3-input combinational logic circuit will encounter fan-in issues when implemented using a LUT topology.

To address this issue, multiplexer functionality in LUTs is typically implemented as a series of smaller, cascaded multiplexers. Each of the smaller multiplexers progressively choose which row of the truth table to route to the output of the LUT. This eliminates fan-in issues at the expense of adding additional levels of logic to the circuit. While cascading multiplexers increases the overall circuit delay, this approach achieves a highly consistent delay because regardless of the truth table output value, the number of levels of logic through the multiplexers is always the same. Figure 11.9 shows how the 2-input truth table from Fig. 11.8 can be implemented using a 2-level cascade of 2-input multiplexers.

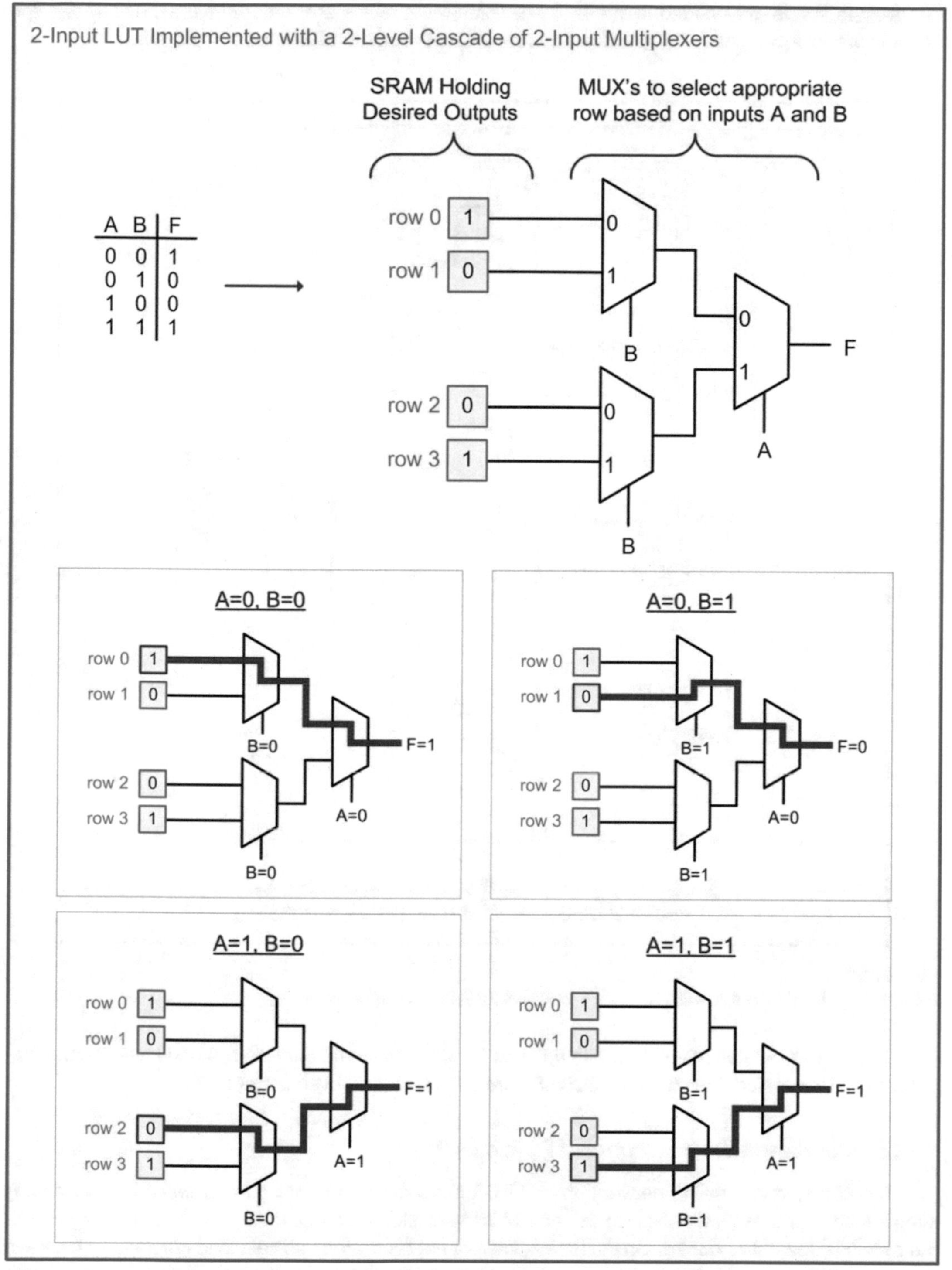

Fig. 11.9
2-input LUT implemented with a 2-level cascade of 2-input multiplexers

If more inputs are needed in the LUT, additional MUX levels are added. Figure 11.10 shows the architecture for a 3-input LUT implemented with a 3-level cascade of 2-input multiplexers.

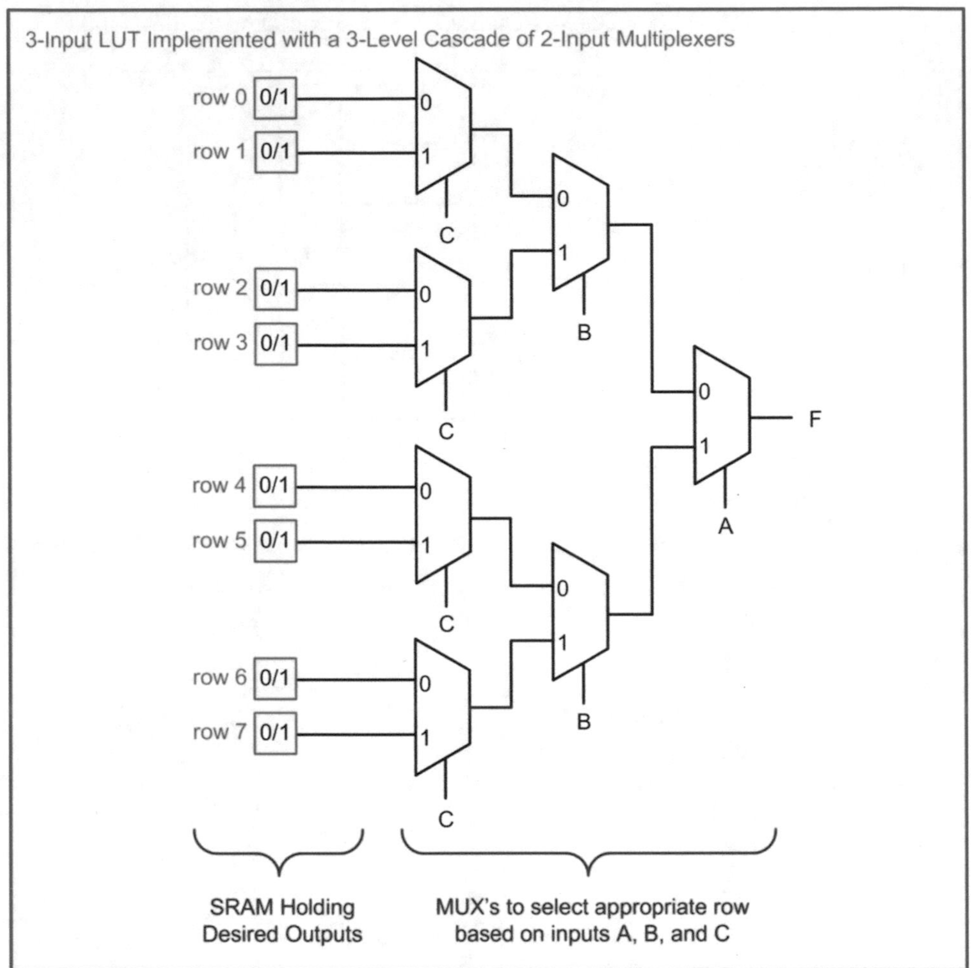

Fig. 11.10
3-input LUT implemented with a 3-level cascade of 2-input multiplexers

Modern FPGAs can have LUTs with up to six inputs. If even more inputs are needed in a combinational logic expression, then multiple CLB/LEs are used that form even larger LUTs.

11.2.3 Programmable Interconnect Points (PIPs)

The configurable routing network on an FPGA is accomplished using programmable switches. A simple model for these switches is to use an NMOS transistor. A configuration SRAM bit stores whether the switch is opened or closed. On the FPGA, interconnect is routed vertically and horizontally between the CLB/LEs with switching points placed throughout the FPGA to facilitate any arbitrary routing configuration. Figure 11.11 shows how the routing can be configured into a full cross-point configuration using programmable switches.

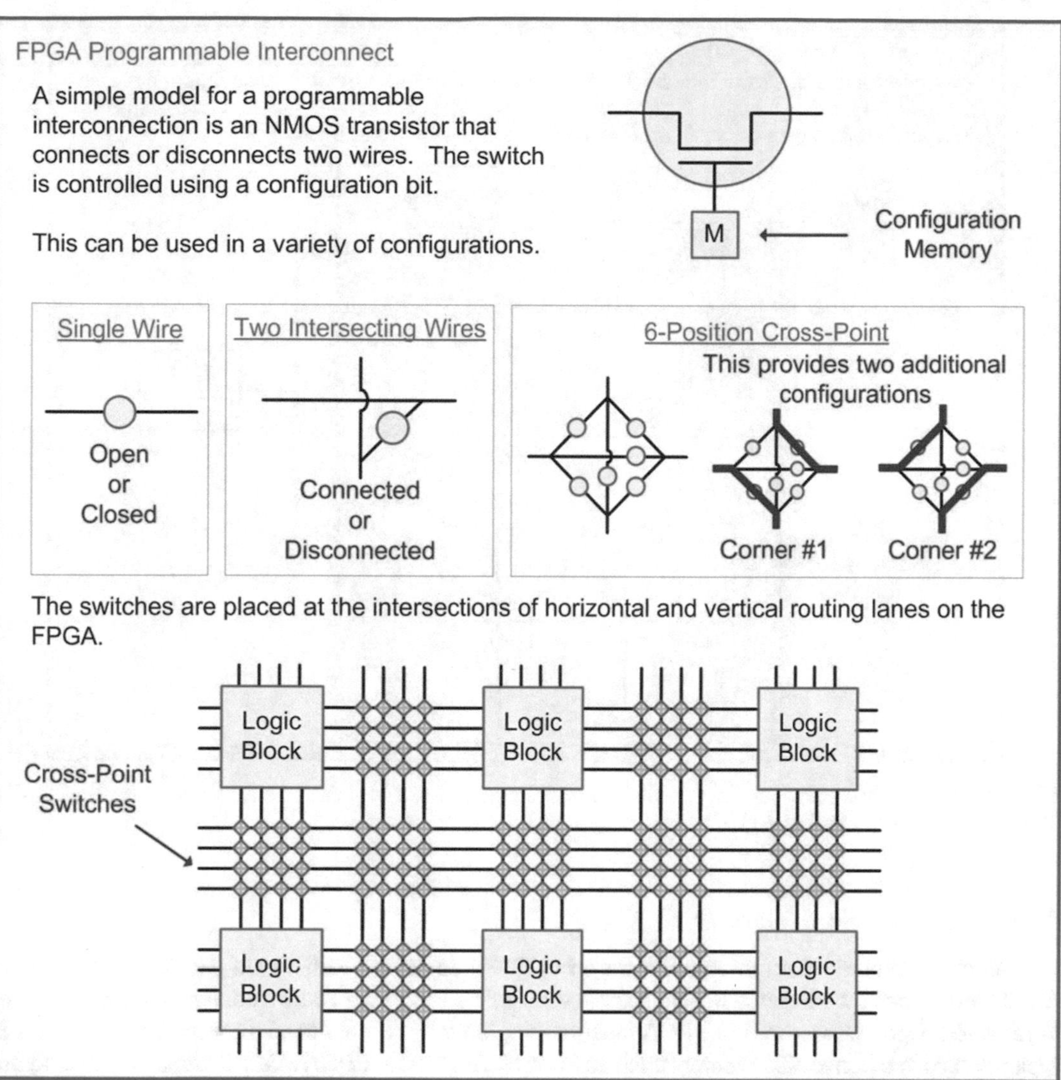

Fig. 11.11
FPGA programmable interconnect

11.2.4 Input/Output Block (IOB)

FPGAs also contain **input/output blocks (IOBs)** that provide programmable functionality for interfacing to external circuitry. The IOBs contain both driver and receiver circuitry so that they can be programmed to be either inputs or outputs. D-flip-flops are included in both the input and output circuitry to support synchronous logic. Figure 11.12 shows the architecture of an FPGA IOB.

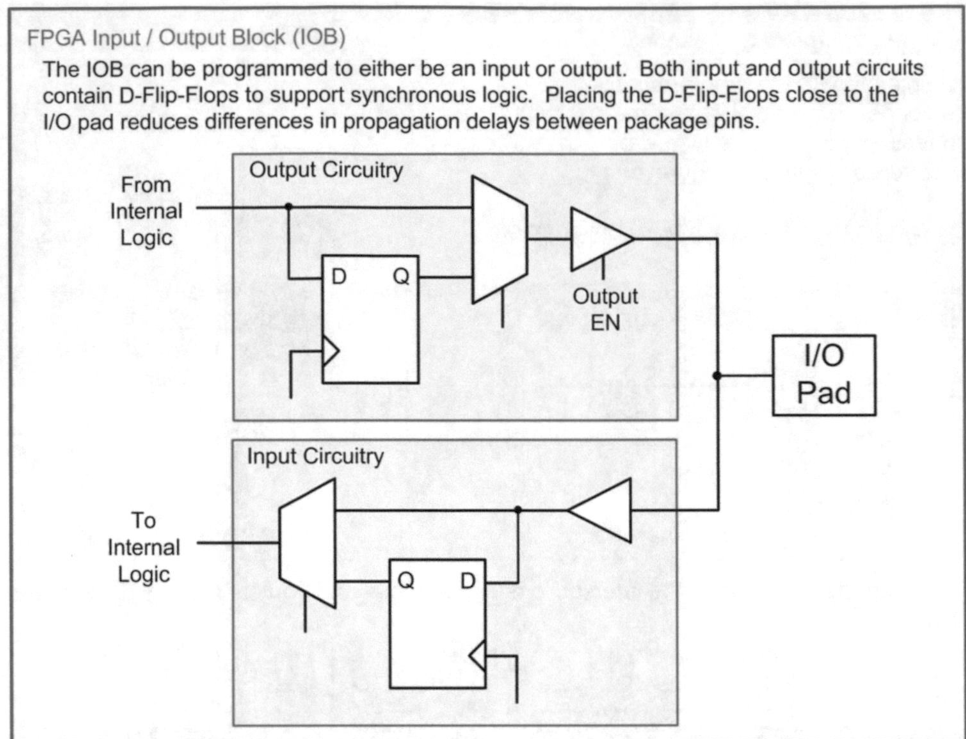

Fig. 11.12
FPGA input/output block (IOB)

11.2.5 Configuration Memory

All of the programming information for an FPGA is contained within configuration SRAM that is distributed across the IC. Since this memory is volatile, the FPGA will lose its configuration when power is removed. Upon power-up, the FPGA must be programmed with its configuration data. This data is typically held in a nonvolatile memory such as FLASH. The “FP” in FPGA refers to the ability to program the device in the *field*, or post-fabrication. The “GA” in FPGA refers to the array topology of the programmable logic blocks or elements.

Concept Check

CC11.2 What is the primary difference between an FPGA and a CPLD?

A) The ability to create arbitrary SOP logic expressions.

B) The abundance of configurable routing.

C) The inclusion of D-flip-flops.

D) The inclusion of programmable I/O pins.

Summary

- A *programmable logic device* (PLD) is a generic term for a circuit that can be configured to implement arbitrary logic functions.
- There are a variety of PLD architectures that have been used to implement combinational logic. These include the PLA and PAL. These devices contain an AND plane and an OR plane. The AND plane is configured to implement the product terms of a SOP expression. The OR plane is configured to implement the sum term of a SOP expression.
- A GAL increases the complexity of logic arrays by adding sequential logic storage and programmable I/O capability.
- A CPLD significantly increases the density of PLDs by connecting an array of PALs together and surrounding the logic with I/O drivers.
- FPGAs contain an array of programmable logic elements that each consist of combinational logic capability and sequential logic storage. FPGAs also contain a programmable interconnect network that provides the highest level of flexibility in programmable logic.
- A look-up table (LUT) is a simple method to create a programmable combinational logic circuit. A LUT is simply a multiplexer with the inputs to the circuit connected to the select lines of the MUX. The desired outputs of the truth table are connected to the MUX inputs. As different input codes arrive on the select lines of the MUX, they *select* the corresponding logic value to be routed to the system output.

Exercise Problems

Section 11.1: Programmable Arrays

11.1.1 Name the type of programmable logic described by the characteristic: *this device adds an output logic macrocell to a traditional PAL.*

11.1.2 Name the type of programmable logic described by the characteristic: *this device combines multiple PALs on a single chip with a partitioned interconnect system.*

11.1.3 Name the type of programmable logic described by the characteristic: *this device has a programmable AND plane and programmable OR plane.*

11.1.4 Name the type of programmable logic described by the characteristic: *this device has a programmable AND plane and fixed-OR plane.*

11.1.5 Name the type of programmable logic described by the characteristic: *this device is a PAL or GAL that is programmed during manufacturing.*

11.1.6 For the following unconfigured PAL schematic in Fig. 11.13, draw in the connection points (i.e., the X's) to implement the two SOP logic expressions shown on the outputs.

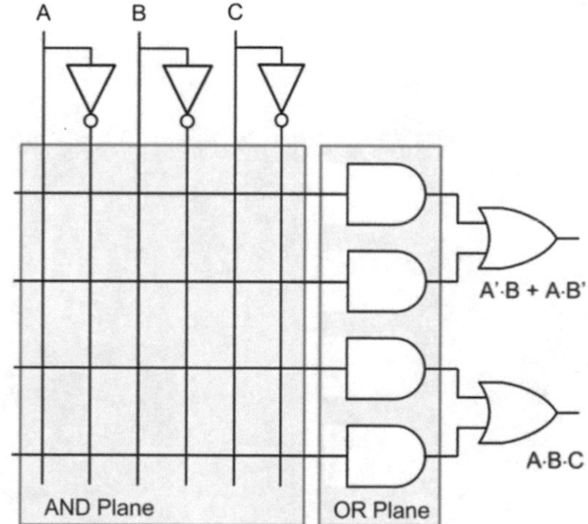

Fig. 11.13
Blank PAL schematic

Section 11.2: Field-Programmable Gate Arrays (FPGAs)

11.2.1 Give a general description of a field-programmable gate array that differentiates it from other programmable logic devices.

11.2.2 Which part of an FPGA is described by the following characteristic: *this is used to interface between the internal logic and external circuitry.*

11.2.3 Which part of an FPGA is described by the following characteristic: *this is used to configure the on-chip routing.*

11.2.4 Which part of an FPGA is described by the following characteristic: *this is the primary programmable element that makes up the array.*

11.2.5 Which part of an FPGA is described by the following characteristic: *this part is used to implement the combinational logic within the array.*

11.2.6 Draw the logic diagram of a 4-input look-up table (LUT) to implement the truth table provided in Fig. 11.14. Implement the LUT with only 2-input multiplexers. Be sure to label the exact location of the inputs (A, B, C, and D), the desired value for each row of the truth table, and the output (F) in the diagram.

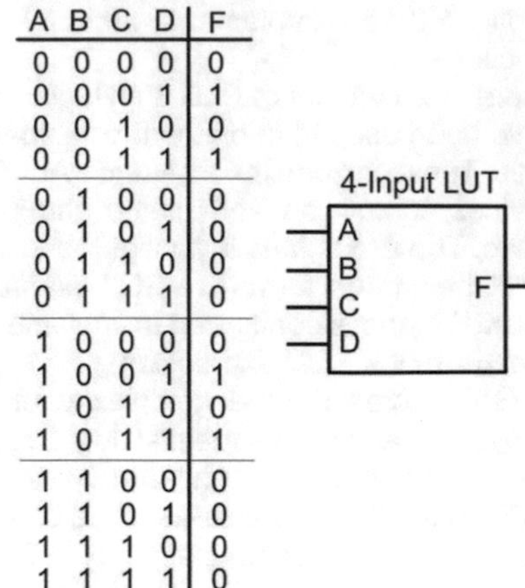

A	B	C	D	F
0	0	0	0	0
0	0	0	1	1
0	0	1	0	0
0	0	1	1	1
0	1	0	0	0
0	1	0	1	0
0	1	1	0	0
0	1	1	1	0
1	0	0	0	0
1	0	0	1	1
1	0	1	0	0
1	0	1	1	1
1	1	0	0	0
1	1	0	1	0
1	1	1	0	0
1	1	1	1	0

Fig. 11.14
4-input LUT exercise

Chapter 12: Arithmetic Circuits

This chapter presents the design and timing considerations of circuits to perform basic arithmetic operations including addition, subtraction, multiplication, and division. A discussion is also presented on how to model arithmetic circuits in Verilog. The goal of this chapter is to provide an understanding of the basic principles of binary arithmetic circuits.

Learning Outcomes—After completing this chapter, you will be able to:

12.1 Design a binary adder using both the classical digital design approach and the modern HDL-based approach.
12.2 Design a binary subtractor using both the classical digital design approach and the modern HDL-based approach.
12.3 Design a binary multiplier using both the classical digital design approach and the modern HDL-based approach.
12.4 Design a binary divider using both the classical digital design approach and the modern HDL-based approach.

12.1 Addition

Binary addition is performed in a similar manner to performing decimal addition by hand. The addition begins in the least significant position of the number ($p = 0$). The addition produces the sum for this position. In the event that this positional sum cannot be represented by a single symbol, then the higher-order symbol is *carried* to the subsequent position ($p = 1$). The addition in the next higher position must include the number that was carried in from the lower positional sum. This process continues until all of the symbols in the number have been operated on. The final positional sum can also produce a carry, which needs to be accounted for in a separate system.

Designing a binary adder involves creating a combinational logic circuit to perform the positional additions. Since a combinational logic circuit can only produce a scalar output, circuitry is needed to produce the sum and the carry at each position. The binary adder size is predetermined and fixed prior to implementing the logic (i.e., an *n*-bit adder). Both inputs to the adder must adhere to the fixed size, regardless of their value. Smaller numbers simply contain leading zeros in their higher-order positions. For an *n*-bit adder, the largest sum that can be produced will require $n + 1$ bits. To illustrate this, consider a 4-bit adder. The largest numbers that the adder will operate on are $1111_2 + 1111_2$ (or $15_{10} + 15_{10}$). The result of this addition is 11110_2 (or 30_{10}). Notice that the largest sum produced fits within 5 bits or $n + 1$. When constructing an adder circuit, the sum is always recorded using *n*-bits with a separate carry out bit. In our 4-bit example, the sum would be expressed as "1110" with a carry out. The carry out bit can be used in multiple word additions, used as part of the number when being decoded for a display, or simply discarded as in the case when using two's complement numbers.

12.1.1 Half Adders

When creating an adder, it is desirable to design incremental sub-systems that can be re-used. This reduces design effort and minimizes troubleshooting complexity. The most basic component in the adder is called a *half adder*. This circuit computes the sum and carry out on two input arguments. The reason it is called a half adder instead of a full adder is because it does not accommodate a *carry in* during the computation; thus it does not provide all of the necessary functionality required for the positional adder. Example 12.1 shows the design of a half adder. Notice that two combinational logic circuits are required

© Springer Nature Switzerland AG 2019

B. J. LaMeres, *Introduction to Logic Circuits & Logic Design with Verilog*,
https://doi.org/10.1007/978-3-030-13605-5_12

in order to produce the sum (the XOR gate) and the carry out (the AND gate). These two gates are in parallel to each other; thus the delay through the half adder is due to only one level of logic.

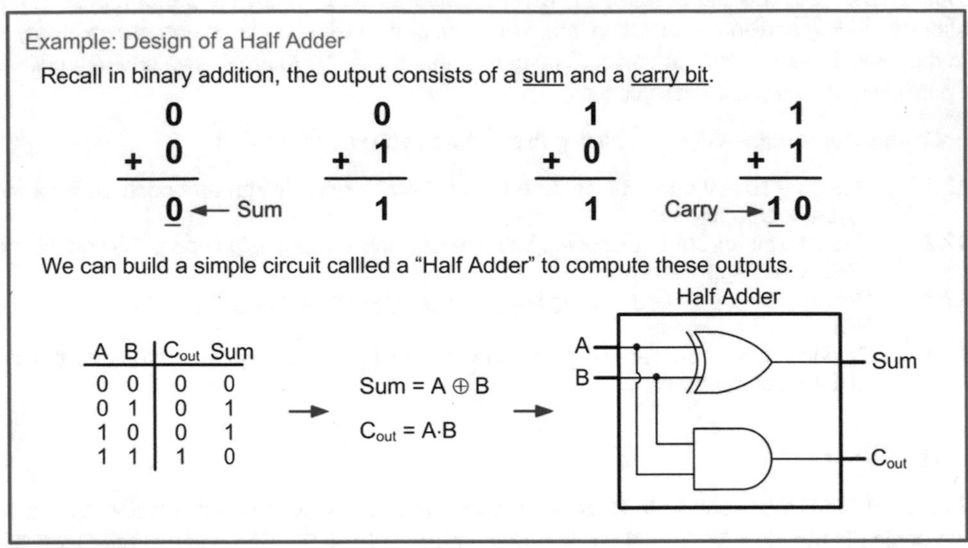

Example 12.1
Design of a half adder

12.1.2 Full Adders

A full adder is a circuit that still produces a sum and carry out but considers three inputs in the computations (A, B, and C_{in}). Example 12.2 shows the design of a full adder.

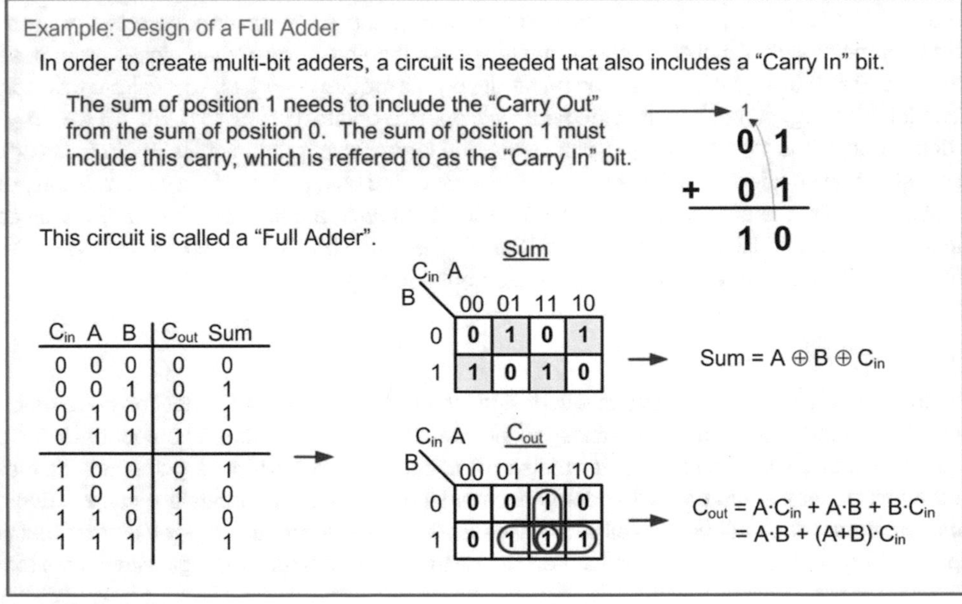

Example 12.2
Design of a full adder

As mentioned before, it is desirable to re-use design components as we construct more complex systems. One such design re-use approach is to create a full adder using two half adders. This is straightforward for the sum output since the logic is simply two cascaded XOR gates (Sum = A ⊕ B ⊕ Cin). The carry out is not as straightforward. Notice that the expression for Cout derived in Example 12.2 contains the term (A + B). If this term could be manipulated to use an XOR gate instead, it would allow the full adder to take advantage of existing circuitry in the system. Figure 12.1 shows a derivation of an equivalency that allows (A + B) to be replaced with (A ⊕ B) in the Cout logic expression.

A Useful Logic Equivalency that can be Exploited in Arithmetic Circuits

The logic expression for the carry out of a full adder was given as: $C_{out} = A{\cdot}B + (A + B){\cdot}C_{in}$. It turns out that the exact same output is produced by the expression $A{\cdot}B + (A \oplus B){\cdot}C_{in}$. Let's examine how this is possible by breaking down the expressions into their individual parts and solving at each step.

FA Inputs			Desired Output	$C_{out} = A{\cdot}B + (A + B){\cdot}C_{in}$			$C_{out} = A{\cdot}B + (A \oplus B){\cdot}C_{in}$		
C_{in}	A	B	C_{out}	A·B	$(A+B){\cdot}C_{in}$	$A{\cdot}B + (A + B){\cdot}C_{in}$	A·B	$(A \oplus B){\cdot}C_{in}$	$A{\cdot}B + (A \oplus B){\cdot}C_{in}$
0	0	0	0	0	0	0	0	0	0
0	0	1	0	0	0	0	0	0	0
0	1	0	0	0	0	0	0	0	0
0	1	1	1	1	0	1	1	0	1
1	0	0	0	0	0	0	0	0	0
1	0	1	1	0	1	1	0	1	1
1	1	0	1	0	1	1	0	1	1
1	1	1	1	1	1	1	1	0	1

$$\mathbf{C_{out} = A{\cdot}B + (A + B){\cdot}C_{in} = A{\cdot}B + (A \oplus B){\cdot}C_{in}}$$

Equivalent !

Fig. 12.1
A useful logic equivalency that can be exploited in arithmetic circuits

The ability to implement the carry out logic using the expression $C_{out} = A{\cdot}B + (A \oplus B){\cdot}C_{in}$ allows us to implement a full adder with two half adders and the addition of a single OR gate. Example 12.3 shows this approach. In this new configuration, the sum is produced in two levels of logic, while the carry out is produced in three levels of logic.

Example – Design of a Full Adder Out of Two Half Adders

It is often desirable to create a full adder out of two half adders in order to re-use existing design components. The "Sum" of the full adder can be created by using two cascaded XOR gates provided by the half adders.

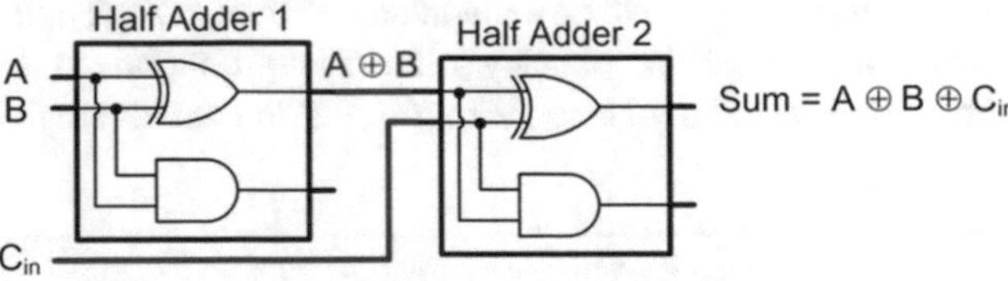

The expression for the "Carry Out" of the full adder is:

$$C_{out} = A \cdot B + (A + B) \cdot C_{in}$$

or

$$C_{out} = A \cdot B + (A \oplus B) \cdot C_{in}$$

Notice that the carry out of Half Adder 1 produces the A·B term in this expression. Also notice that the carry out of Half Adder 2 produces the $(A \oplus B) \cdot C_{in}$ term. The only remaining logic needed to create the carry out of the full adder is an OR gate. The final logic diagram for the full adder is as follows:

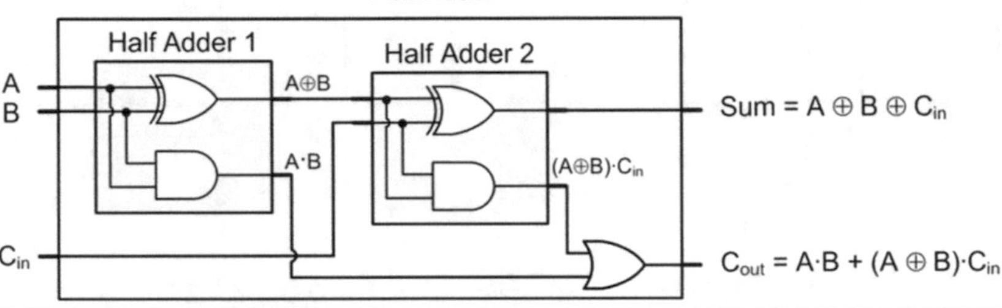

Example 12.3
Design of a full adder out of half adders

12.1.3 Ripple Carry Adder (RCA)

The full adder can now be used in the creation of multi-bit adders. The simplest topology exploiting the full adder is called a *ripple carry adder* (RCA). In this approach, full adders are used to create the sum and carry out of each bit position. The carry out of each full adder is used as the carry in for the next higher position. Since each subsequent full adder needs to wait for the carry to be produced by the preceding stage, the carry is said to *ripple* through the circuit, thus giving this approach its name. Example 12.4 shows how to design a 4-bit ripple carry adder using a chain of full adders. Notice that the carry in for the full adder in position 0 is tied to a logic 0. The 0 input has no impact on the result of the sum but enables a full adder to be used in the 0th position.

Example: Design of a 4-Bit Ripple Carry Adder (RCA)

Full adders can be cascaded together to form a multi-bit adder. The symbols are typically drawn in the following fashion to mirror a positional number system.

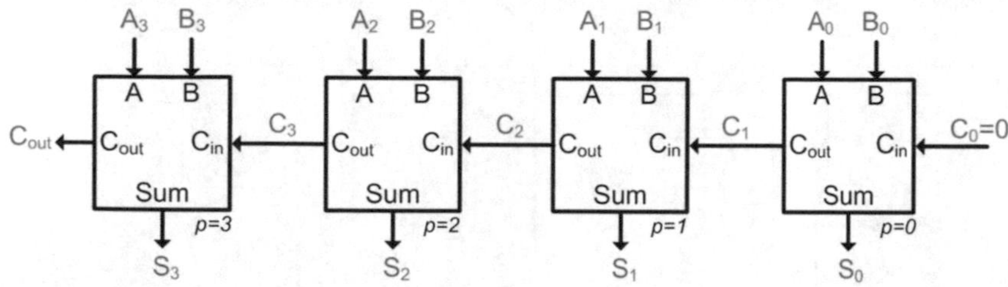

The sum of position 1 cannot complete until it receives the carry in (C_1) from the sum in position 0. The position 2 sum cannot complete until it receives the carry in (C_2) from the sum in position 1, etc. In this way, the carry "ripples" through the circuit from right to left. This configuration is known as a Ripple Carry Adder (RCA).

Example 12.4
Design of a 4-bit ripple carry adder (RCA)

While the ripple carry adder provides a simple architecture based on design re-use, its delay can become considerable when scaling to larger input sizes (e.g., $n = 32$ or $n = 64$). A simple analysis of the timing can be stated such that if the time for a full adder to complete its positional sum is t_{FA}, then the time for an n-bit ripple carry adder to complete its computation is $t_{RCA} = n \cdot t_{FA}$.

If we examine the RCA in more detail, we can break down the delay in terms of the levels of logic necessary for the computation. Example 12.5 shows the timing analysis of the 4-bit RCA. This analysis determines the number of logic levels in the adder. The actual gate delays can then be plugged in to find the final delay. The inputs to the adder are A, B, and C_{in} and are always assumed to update at the same time. The first full adder requires two levels of logic to produce its sum and three levels to produce its carry out. Since the timing of a circuit is always stated as its worst-case delay, we say that the first full adder takes three levels of logic. When the carry (C_1) ripples to the next full adder (FA1), it must propagate through two additional levels of logic in order to produce C_2. Notice that the first half adder in FA1 only depends on A_1 and B_1; thus it is able to perform this computation immediately. This half adder can be considered as first level logic. More importantly, it means that when the carry in arrives (C_1), only two additional levels of logic are needed, not three. The levels of logic for the RCA can be expressed as $3 + 2 \cdot (n-1)$. If each level of logic has a delay of t_{gate}, then a more accurate expression for the RCA delay is $t_{RCA} = (3 + 2 \cdot (n-1)) \cdot t_{gate}$.

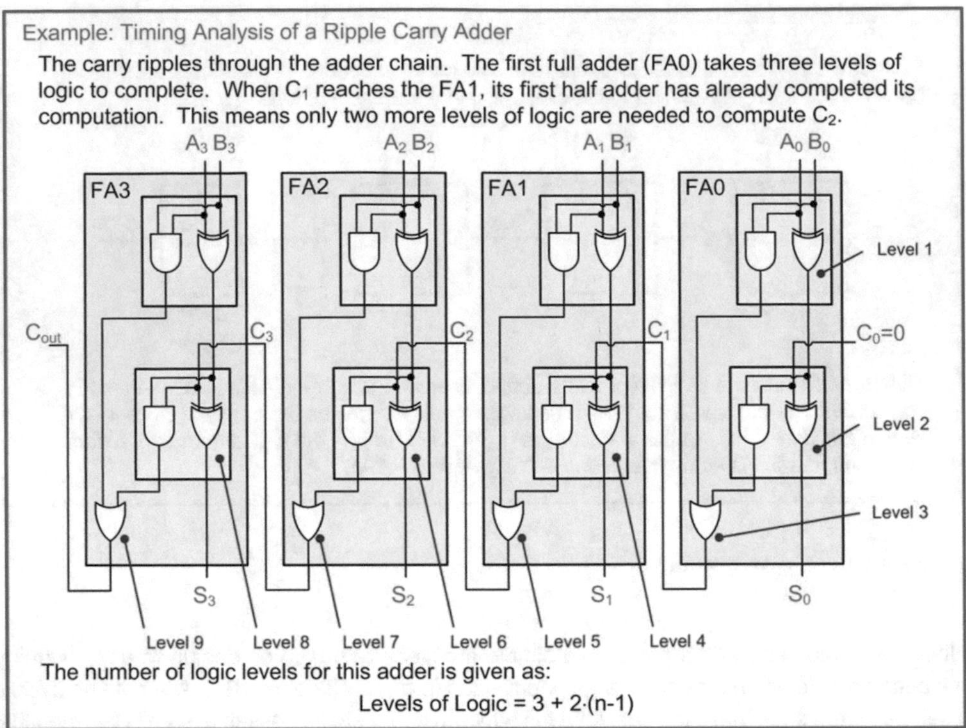

Example 12.5
Timing analysis of a 4-bit ripple carry adder

12.1.4 Carry Look Ahead Adder (CLA)

In order to address the potentially significant delay of a ripple carry adder, a *carry look ahead* (CLA) adder was created. In this approach, additional circuitry is included that produces the intermediate carry in signals immediately instead of waiting for them to be created by the preceding full adder stage. This allows the adder to complete in a fixed amount of time instead of one that scales with the number of bits in the adder. Example 12.6 shows an overview of the design approach for a CLA.

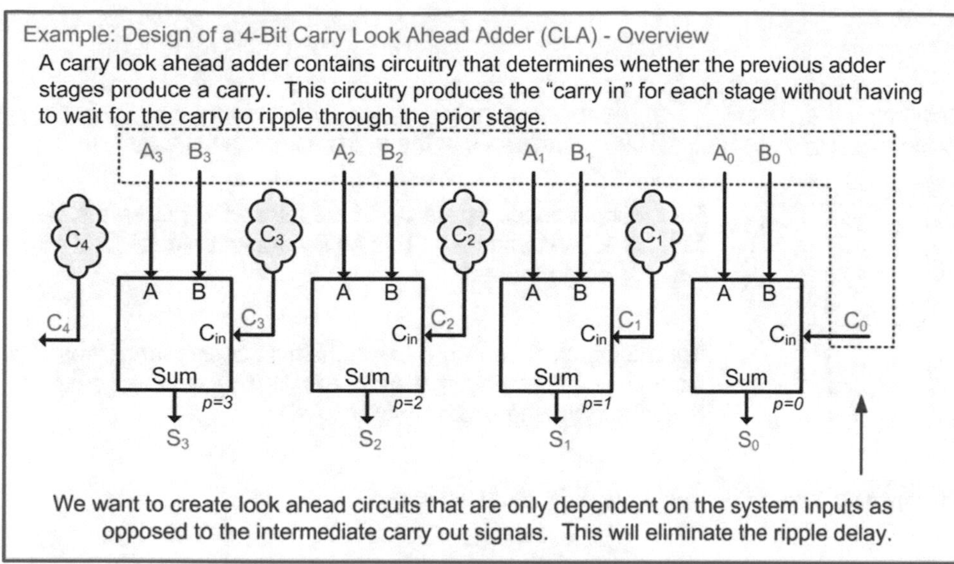

Example 12.6
Design of a 4-bit carry look ahead adder (CLA) – Overview

For the CLA architecture to be effective, the look ahead circuitry needs to be dependent only on the system inputs A, B, and C_{in} (i.e., C_0). A secondary characteristic of the CLA is that it should exploit as much design re-use as possible. In order to examine the design re-use aspects of a multi-bit adder, the concepts of carry **generation** (g) and **propagation** (p) are used. A full adder is said to *generate* a carry if its inputs A and B result in $C_{out} = 1$ when $C_{in} = 0$. A full adder is said to *propagate* a carry if its inputs A and B result in $C_{out} = 1$ when $C_{in} = 1$. These simple statements can be used to derive logic expressions for each stage of the adder that can take advantage of existing logic terms from prior stages. Example 12.7 shows the derivation of these terms and how algebraic substitutions can be exploited to create look ahead circuitry for each full adder that is only dependent on the system inputs. In these derivations, the variable *i* is used to represent position since *p* is used to represent the propagate term.

Example: Design of a 4-Bit Carry Look Ahead Adder (CLA) – Algebraic Formation

The look ahead circuitry considers whether the prior adder stages create a carry by considering two conditions: 1) whether a stage will **generate** (*g*) a carry; and 2) whether the stage will **propagate** (*p*) a carry. Let's look at the truth table for a full adder.

C_{in}	A	B	C_{out}
0	0	0	0
0	0	1	0
0	1	0	0
0	1	1	1
1	0	0	0
1	0	1	1
1	1	0	1
1	1	1	1

For the input codes where C_{in}=0, the full adder "generates" a new carry when A=1 and B=1. This behavior can be described with the expression: $g = A\cdot B$

For the input codes where C_{in}=1, the full adder "propagates" the incoming carry when either A=1 or B=1. This behavior can be described with the expression: $p = A+B$

The entire expression for the carry out can be written as:

$$C_{out} = g + p\cdot C_{in}$$
$$C_{out} = A\cdot B + (A+B)\cdot C_{in}$$

Let's see how this can be used to our advantage in a multiple bit adder. Recall that for any arbitrary adder position, the generate, propagate, and carry out terms are:

$$g_i = A_i\cdot B_i$$
$$p_i = A_i + B_i$$
$$C_{i+1} = g_i + p_i\cdot C_i$$

Note: We'll use the subscript "i" to denote position since we're using "p" for *propagate*.

We can now write expressions for the subsequent carry terms as:

$$C_1 = g_0 + p_0\cdot C_0$$

The C_1 expression only depends on the inputs A, B, and C_0.

$$C_2 = g_1 + p_1\cdot C_1$$
$$C_2 = g_1 + p_1\cdot(g_0 + p_0\cdot C_0)$$
$$C_2 = g_1 + p_1\cdot g_0 + p_1\cdot p_0\cdot C_0$$

For C_2, we can plug in the expression for C_1 to create an expression that only depends on A, B, and C_0...

$$C_3 = g_2 + p_2\cdot C_2$$
$$C_3 = g_2 + p_2\cdot(g_1 + p_1\cdot g_0 + p_0\cdot p_1\cdot C_0)$$
$$C_3 = g_2 + p_2\cdot g_1 + p_2\cdot p_1\cdot g_0 + p_2\cdot p_1\cdot p_0\cdot C_0$$

and again for C_3…

$$C_4 = g_3 + p_3\cdot C_3$$
$$C_4 = g_3 + p_3\cdot(g_2 + p_2\cdot g_1 + p_2\cdot p_1\cdot g_0 + p_2\cdot p_1\cdot p_0\cdot C_0)$$

and again for C_4…

All of these expressions only depend on the inputs A, B, and C_0. Also notice that each expression is in a 2-level sum of products form.

Example 12.7
Design of a 4-bit carry look ahead adder (CLA) – Algebraic formation

Example 12.8 shows a timing analysis of the 4-bit carry look ahead adder. Notice that the full adders are modified to add the logic for the generate and propagate bits in addition to removing the unnecessary gates associated with creating the carry out.

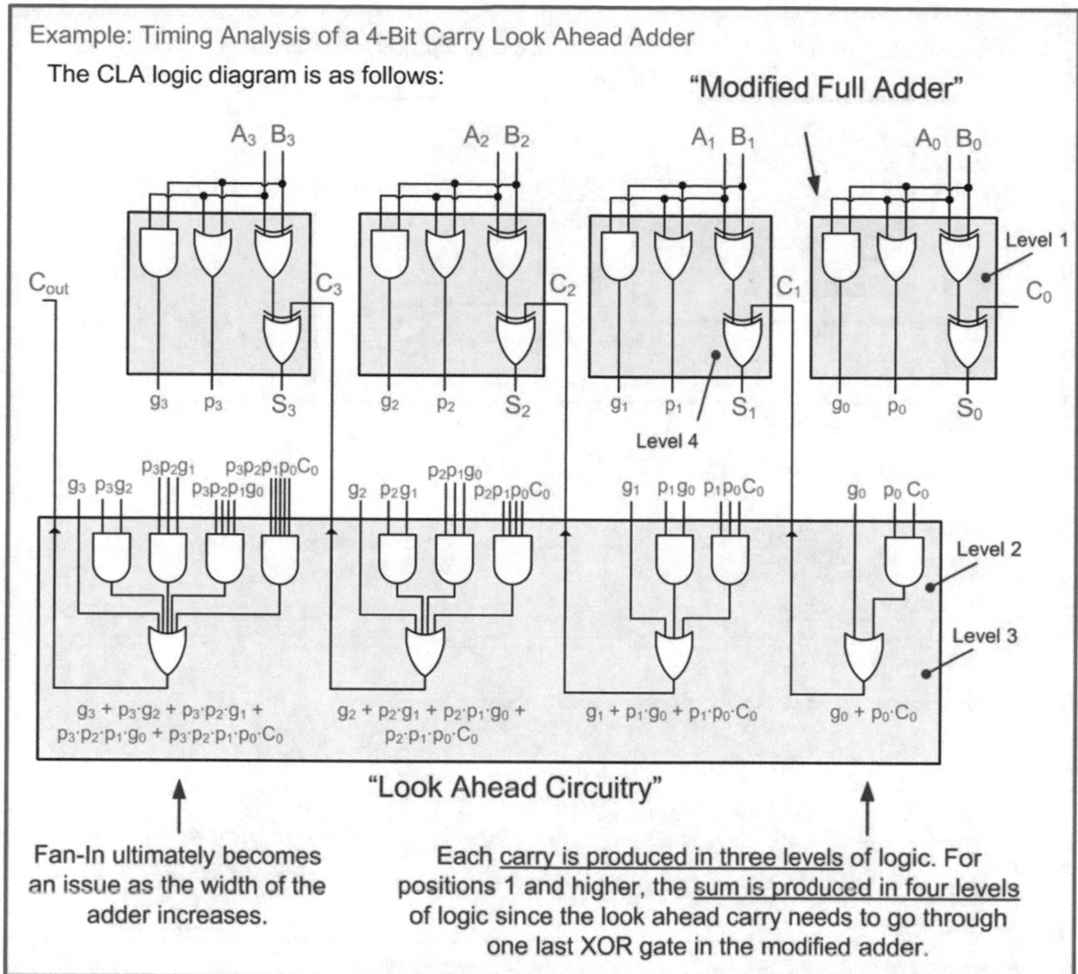

Example 12.8
Timing analysis of a 4-bit carry look ahead adder

The 4-bit CLA can produce the sum in four levels of logic as long as fan-in specifications are met. As the CLA width increases, the look ahead circuitry will become fan-in limited, and additional stages will be required to address the fan-in. Regardless, the CLA has considerably less delay than a RCA as the width of the adder is increased.

12.1.5 Adders in Verilog

12.1.5.1 Structural Model of a Ripple Carry Adder in Verilog

A structural model of a ripple carry adder is useful to visualize the propagation delay of the circuit in addition to the impact of the carry rippling through the chain. Example 12.9 shows the structural model for a full adder in Verilog consisting of two half adders. The half adders are created using two gate-level primitives for the XOR and AND operations, each with a delay of 1 ns. The full adder is created by instantiating two versions of the half adder as sub-systems plus one additional gate-level primitive for the OR gate.

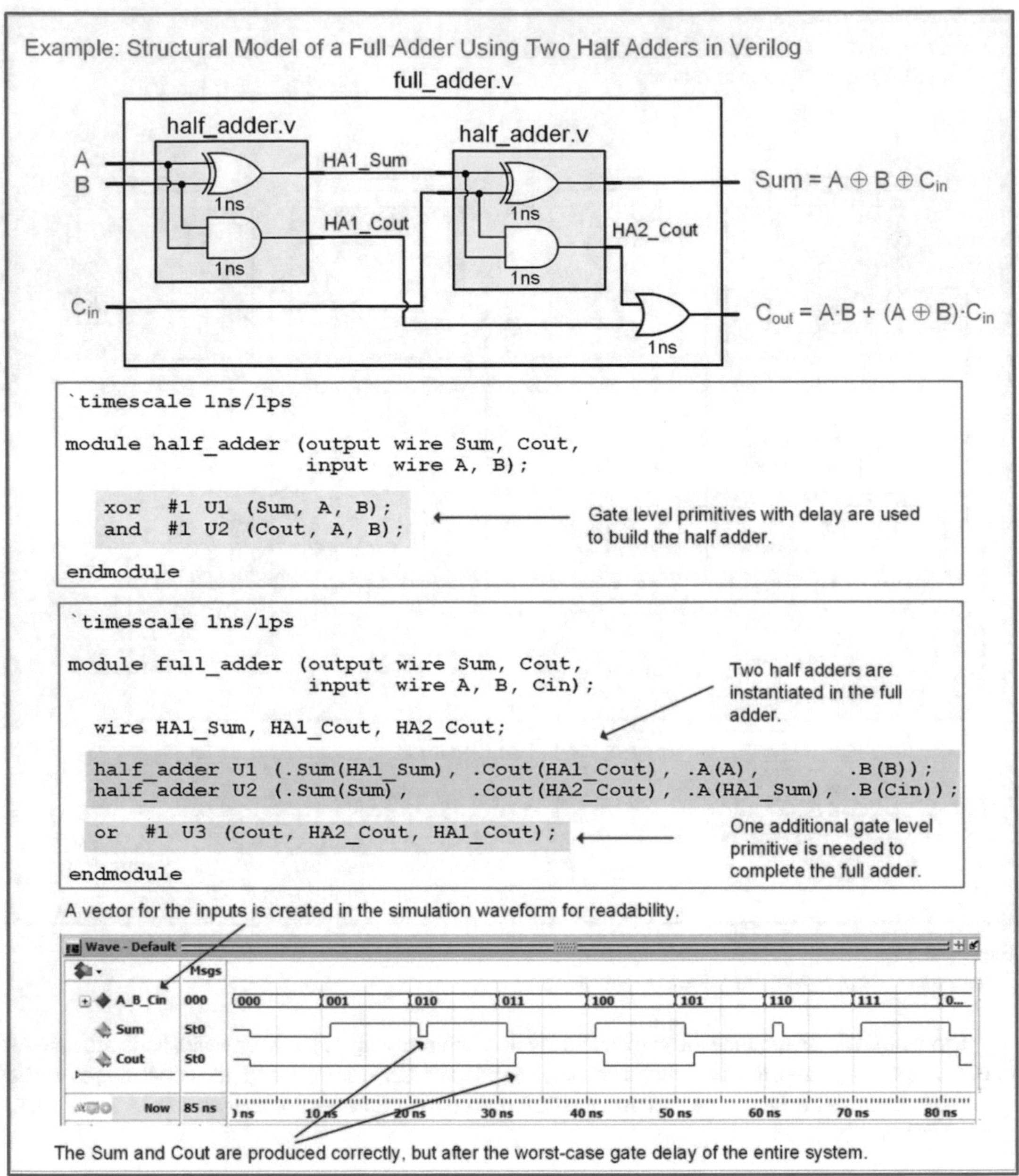

Example 12.9
Structural model of a full adder using two half adders in Verilog

Example 12.10 shows the structural model of a 4-bit ripple carry adder in Verilog. The RCA is created by instantiating four full adders. Notice that a logic 1'b0 can be directly inserted into the port map of the first full adder to model the behavior of $C_0 = 0$.

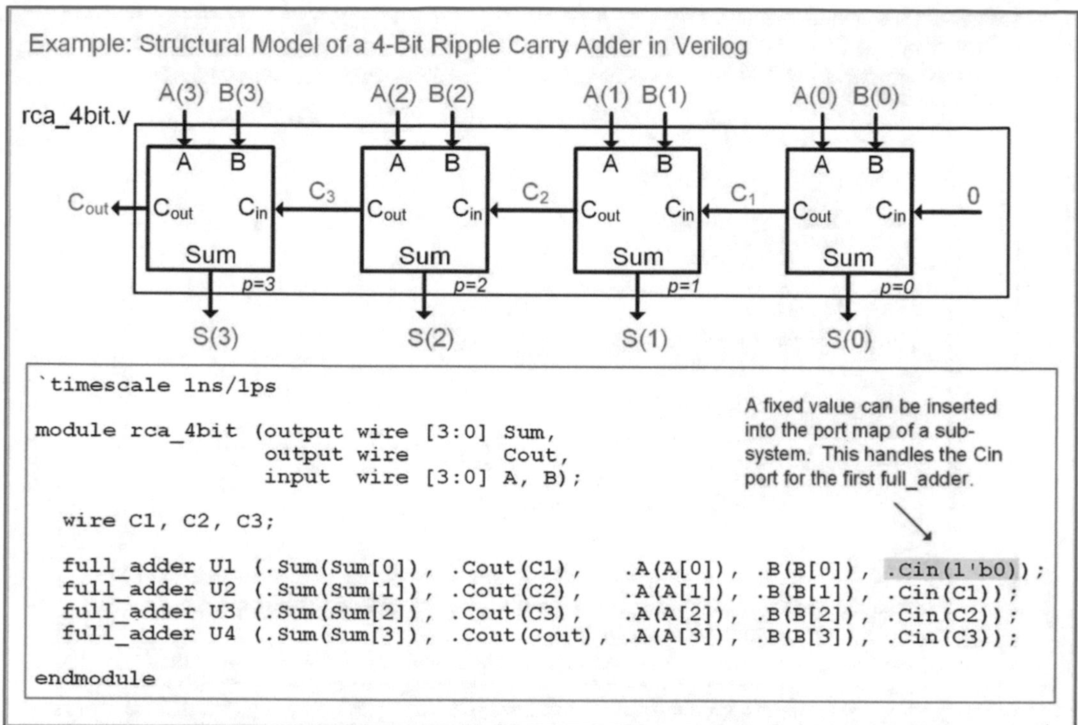


```
`timescale 1ns/1ps

module rca_4bit (output wire [3:0] Sum,
                 output wire       Cout,
                 input  wire [3:0] A, B);

  wire C1, C2, C3;

  full_adder U1 (.Sum(Sum[0]), .Cout(C1),    .A(A[0]), .B(B[0]), .Cin(1'b0));
  full_adder U2 (.Sum(Sum[1]), .Cout(C2),    .A(A[1]), .B(B[1]), .Cin(C1));
  full_adder U3 (.Sum(Sum[2]), .Cout(C3),    .A(A[2]), .B(B[2]), .Cin(C2));
  full_adder U4 (.Sum(Sum[3]), .Cout(Cout),  .A(A[3]), .B(B[3]), .Cin(C3));

endmodule
```

Example 12.10
Structural model of a 4-bit ripple carry adder in Verilog

When creating arithmetic circuitry, testing under all input conditions is necessary to verify functionality. Testing under each and every input condition can require a large number of input conditions. To test an *n*-bit adder under each and every numeric input condition will take $(2^n)^2$ test vectors. For our simple 4-bit adder example, this equates to 256 input patterns. The large number of input patterns precludes the use of manual signal assignments in the test bench to stimulate the circuit. One approach to generating the input test patterns is to use nested for loops. Example 12.11 shows a test bench that uses two nested for loops to generate the 256 unique input conditions for the 4-bit ripple carry adder. Note that the loop variables *i* and *j* are declared as type integer and then automatically incremented within the for loops. Within the loops, the loop variables i and j are assigned to the DUT inputs A_TB and B_TB. The truncation to 4-bits is automatically handled in Verilog. The simulation waveform illustrates how the ripple carry adder has a noticeable delay before the output sum is produced. During the time the carry is rippling through the adder chain, glitches can appear on each of the sum bits in addition to the carry out signal. The values in this waveform are displayed as unsigned decimal symbols to make the results easier to interpret.

Example: Test Bench for a 4-Bit Ripple Carry Adder Using Nested for Loops in Verilog

Nested for loops can be used in order to generate an exhaustive set of test vectors to stimulate the adder.

```
`timescale 1ns/1ps

module rca_4bit_TB ();

  reg  [3:0] A_TB, B_TB;
  wire [3:0] Sum_TB;
  wire       Cout_TB;

  integer    i, j;

  rca_4bit DUT (Sum_TB, Cout_TB, A_TB, B_TB);

  always
    begin
      for (i=0; i<16; i=i+1)
        for (j=0; j<16; j=j+1)
          begin
            A_TB = i; B_TB = j; #30;
          end
    end

endmodule
```

Nested for loops handled created all possible input vectors.

The simulation waveform for the ripple carry adder is as follows. The numbers are shown in unsigned decimal format for readability.

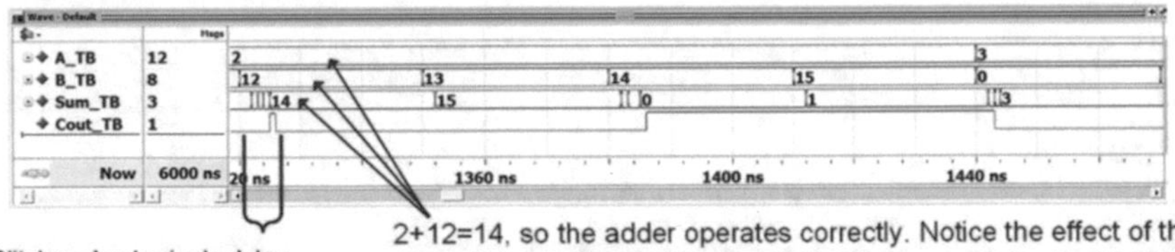

Glitches due to ripple delay.

2+12=14, so the adder operates correctly. Notice the effect of the ripple through the circuit. In addition to the correct output being delayed, there are glitches on both the Sum and C_{out} ports.

Example 12.11
Test bench for a 4-bit ripple carry adder using nested for loops in Verilog

12.1.5.2 Structural Model of a Carry Look Ahead Adder in Verilog

A carry look ahead adder can also be modeled using procedural assignments and modified full adder sub-systems. Example 12.12 shows a structural model for a 4-bit CLA in Verilog. In this example, the gate delay is modeled at 1 ns. The delay due to multiple levels of logic is entered manually to simplify the model. The two cascaded XOR gates in the modified full adder are modeled using a single, 3-input gate primitive with 2 ns of delay.

Example: Structural Model of a 4-Bit Carry Look Ahead Adder in Verilog

```
`timescale 1ns/1ps

module mod_full_adder (output wire Sum, p, g,
                       input  wire A, B, Cin);

   xor  #2 U1 (Sum, A, B, Cin);
   or   #1 U2 (p, A, B);
   and  #1 U3 (g, A, B);

endmodule
```

A modified full adder creates the propagate (p) and generate (g) signals instead of Cout.

```
`timescale 1ns/1ps

module cla_4bit (output wire [3:0] Sum,
                 output wire       Cout,
                 input  wire [3:0] A, B);

   wire       C0, C1, C2, C3;
   wire [3:0] p, g;

   assign  C0   = 1'b0;
   assign  C1   = g[0] | (p[0] & C0);
   assign  C2   = g[1] | (p[1] & C1);
   assign  C3   = g[2] | (p[2] & C2);
   assign  Cout = g[3] | (p[3] & C3);

  mod_full_adder U0 (.Sum(Sum[0]), .p(p[0]), .g(g[0]), .A(A[0]), .B(B[0]), .Cin(C0));
  mod_full_adder U1 (.Sum(Sum[1]), .p(p[1]), .g(g[1]), .A(A[1]), .B(B[1]), .Cin(C1));
  mod_full_adder U2 (.Sum(Sum[2]), .p(p[2]), .g(g[2]), .A(A[2]), .B(B[2]), .Cin(C2));
  mod_full_adder U3 (.Sum(Sum[3]), .p(p[3]), .g(g[3]), .A(A[3]), .B(B[3]), .Cin(C3));

endmodule
```

These continuous assignments model the combinational logic for the propagate and generate signals.

Example 12.12
Structural model of a 4-bit carry look ahead adder in Verilog

Example 12.13 shows the simulation waveform for the 4-bit carry look ahead adder. The outputs still have intermediate transitions, while the combinational logic is computing the results; however, the overall delay of the adder is bound to $\leq 4{*}t_{gate}$.

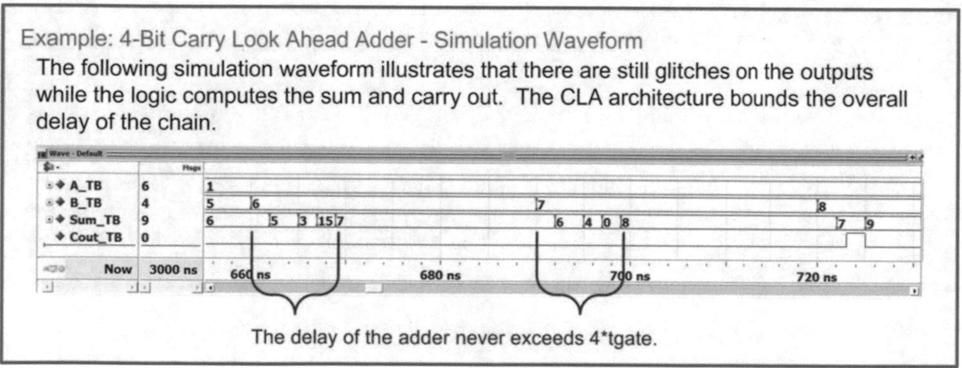

Example 12.13
4-bit carry look ahead adder – Simulation waveform

12.1.5.3 Behavior Model of an Adder Using Arithmetic Operators in Verilog

Verilog also supports adder models at a higher level of abstraction using the "+" operator. Note that when adding two *n*-bit arguments, the sum produced will be *n* + 1 bits. This can be handled in Verilog by concatenating the Cout and Sum outputs on the LHS of the assignment. The entire add operation can be accomplished in a single continuous assignment that contains both the concatenation and addition operators. When using continuous assignment, the LHS must be a net data type. This means the outputs Cout and Sum need to be declared as type wire. If it was desired to have the outputs declared of type reg, a procedural assignment could be used instead. Example 12.14 shows the behavioral model for a 4-bit adder in Verilog.

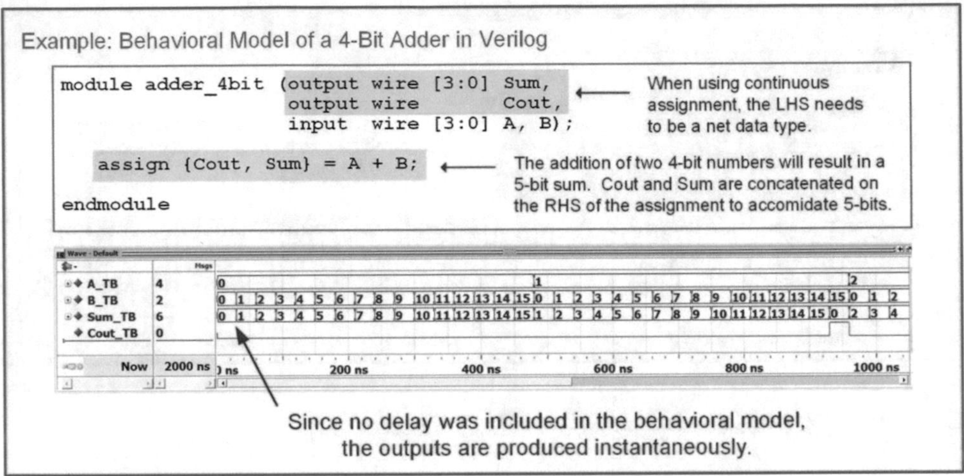

Example 12.14
Behavioral model of a 4-bit adder in Verilog

CONCEPT CHECK

CC12.1 Does a binary adder behave differently when it's operating on unsigned vs. two's complement numbers? Why or why not?

A) Yes. The adder needs to keep track of the sign bit; thus extra circuitry is needed.

B) No. The binary addition is identical. It is up to the designer to handle how the two's complement codes are interpreted and whether two's complement overflow occurred using a separate system.

12.2 Subtraction

Binary subtraction can be accomplished by building a dedicated circuit using a similar design approach as just described for adders. A more effective approach is to take advantage of two's complement representation in order to re-use existing adder circuitry. Recall that taking the two's complement of a number will produce an equivalent magnitude number, but with the opposite sign (i.e., positive to negative or negative to positive). This means that all that is required to create a subtractor from an adder is to first take the two's complement of the subtrahend input. Since the steps to take the two's complement of a number involve complementing each of the bits in the number and then adding 1, the logic required is relatively simple. Example 12.15 shows a 4-bit subtractor using full adders. The subtrahend B is inverted prior to entering the full adders. Also, the carry in bit C_0 is set to 1. This handles the "adding 1" step of the two's complement. All of the carries in the circuit are now treated as *borrows*, and the sum is now treated as the *difference*.

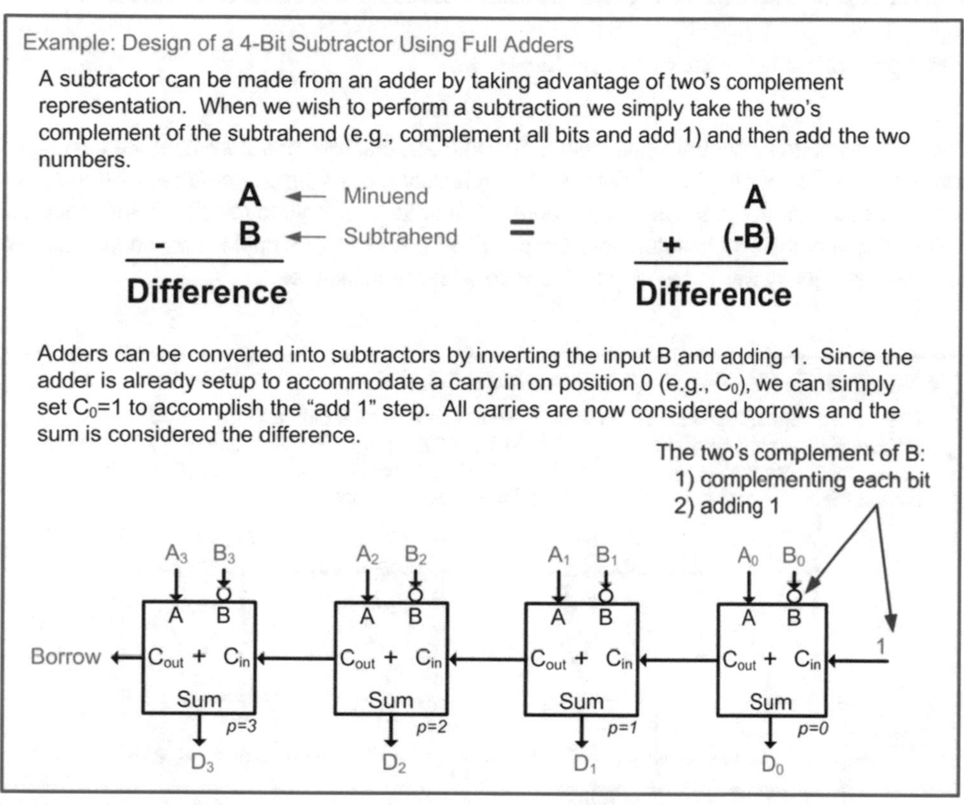

Example 12.15
Design of a 4-bit subtractor using full adders

A programmable adder/subtractor can be created with the use of a programmable inverter and a control signal. The control signal will selectively invert B and also change the C_0 bit between a 0 (for adding) and a 1 (for subtracting). Example 12.16 shows how an XOR gate can be used to create a programmable inverter for use in a programmable adder/subtractor circuit.

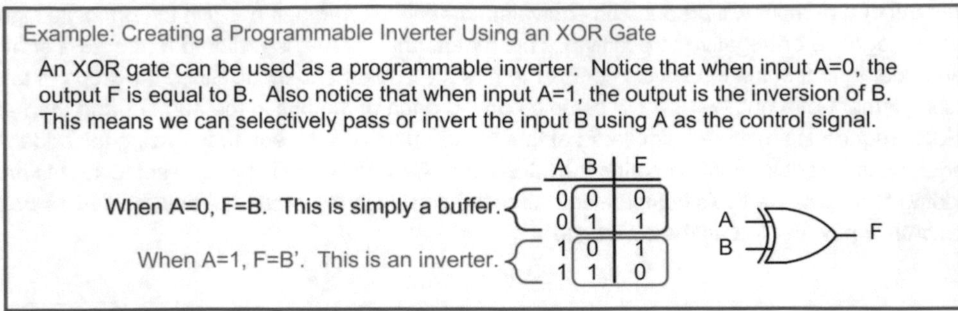

Example 12.16
Creating a programmable inverter using an XOR gate

We can now define a control signal called (ADDn/SUB) that will control whether the circuit performs addition or subtraction. Example 12.17 shows the architecture of a 4-bit programmable adder/subtractor. It should be noted that this programmability adds another level of logic to the circuit, thus increasing its delay. The programmable architecture in Example 12.17 is shown for a ripple carry adder; however, this approach works equally well for a carry look ahead adder architecture.

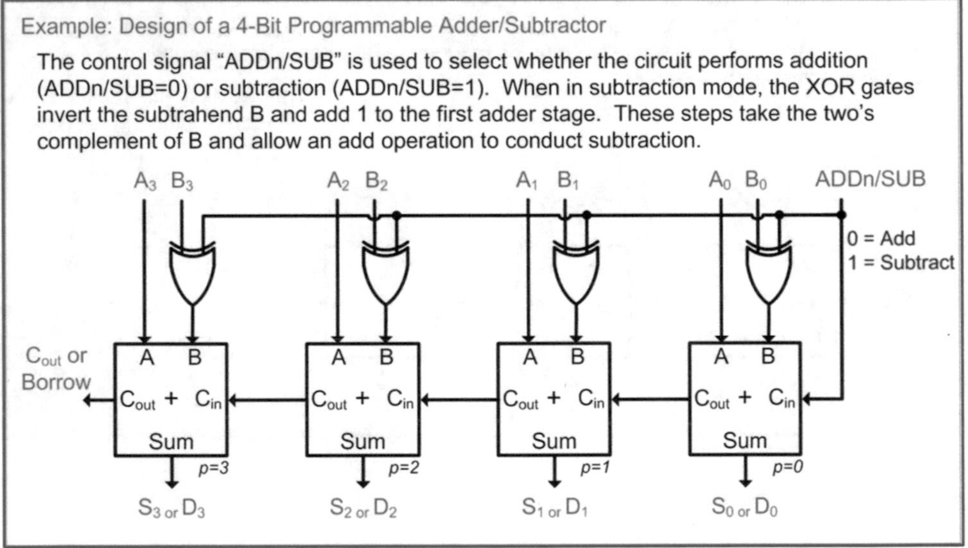

Example 12.17
Design of a 4-bit programmable adder/subtractor

When using two's complement representation in arithmetic, care must be taken to monitor for two's complement overflow. Recall that when using two's complement representation, the number of bits of the numbers is fixed (e.g., 4-bits), and if a carry/borrow out is generated, it is ignored. This means that the Cout bit does not indicate whether two's complement overflow occurred. Instead, we must construct additional circuitry to monitor the arithmetic operations for overflow. Recall from Chap. 2 that two's complement overflow occurs in any of these situations:

- The sum of like signs results in an answer with opposite sign (i.e., positive + positive = negative or negative + negative = positive).
- The subtraction of a positive number from a negative number results in a positive number (i.e., negative – positive = positive).
- The subtraction of a negative number from a positive number results in a negative number (i.e., positive – negative = negative).

The construction of circuitry for these conditions is straightforward since the sign bit of all numbers involved in the operation indicates whether the number is positive or negative. The sign bits of the input arguments and the output are fed into combinational logic circuitry that will assert for any of the above conditions. These signals are then logically combined to create two's complement overflow signal.

CONCEPT CHECK

CC12.2 What modifications can be made to the programmable adder/subtractor architecture so that it can be used to take the two's complement of a number?

A) Remove the input A.

B) Add an additional control signal that will cause the circuit to ignore A and just perform a complement on B, and then add 1.

C) Add an additional 1 to the original number using an OR gate on Cin.

D) Set A to 0, put the number to be manipulated on B, and put the system into subtraction mode. The system will then complement the bits on B, and then add 1, thus performing two's complement negation.

12.3 Multiplication

12.3.1 Unsigned Multiplication

Binary multiplication is performed in a similar manner to performing decimal multiplication by hand. Recall the process for long multiplication. First, the two numbers are placed vertically over one another with their least significant digits aligned. The upper number is called the *multiplicand* and the lower number is called the *multiplier*. Next, we multiply each individual digit within multiplier with the entire multiplicand, starting with the least position. The result of this interim multiplication is called the *partial product*. The partial product is recorded with its least significant digit aligned with the corresponding position of the multiplier digit. Finally, all partial products are summed to create the final product of the multiplication. This process is often called the *shift and add* approach. Example 12.18 shows the process for performing long multiplication on decimal numbers highlighting the individual steps.

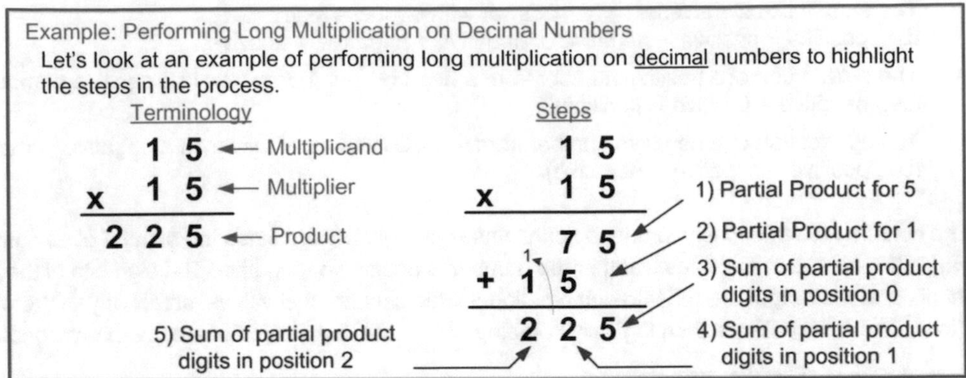

Example 12.18
Performing long multiplication on decimal numbers

Binary multiplication follows this same process. Example 12.19 shows the process for performing long multiplication on binary numbers. Note that the inputs represent the largest unsigned numbers possible using 4-bits, thus producing the largest possible product. The largest product will require 8-bits to be represented. This means that for any multiplication of *n*-bit inputs, the product will require 2·n bits for the result.

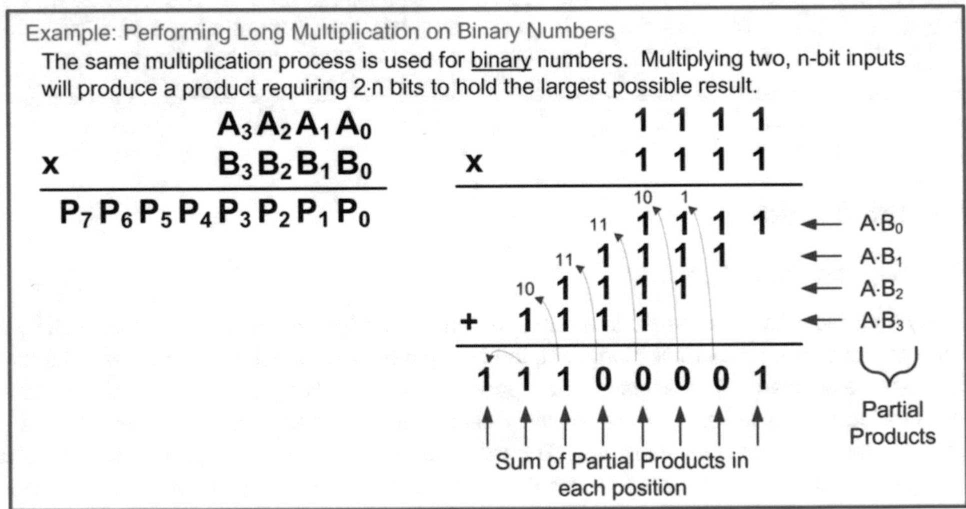

Example 12.19
Performing long multiplication on binary numbers

The first step in designing a binary multiplier is to create circuitry that can compute the product on individual bits. Example 12.20 shows the design of a single-bit multiplier.

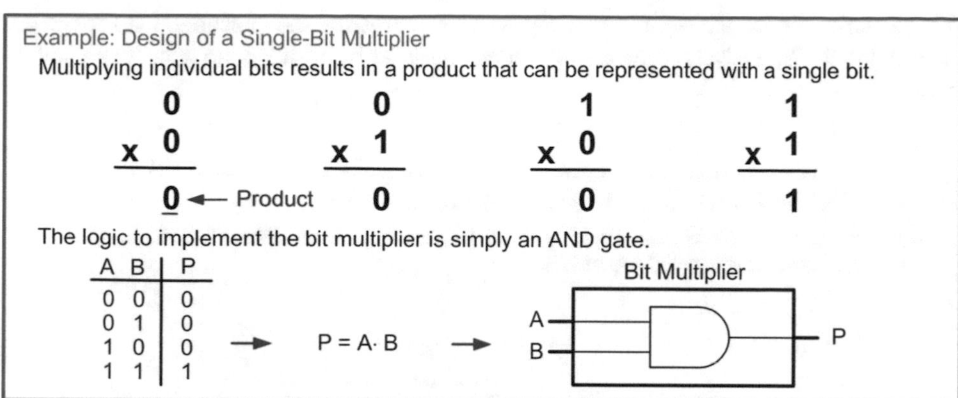

Example 12.20
Design of a single-bit multiplier

We can create all of the partial products in one level of logic by placing an AND gate between each bit pairing in the two input numbers. This will require n^2 AND gates. The next step involves creating adders that can perform the sum of the columns of bits within the partial products. This step is not as straightforward. Notice that in our 4-bit example in Example 12.19 that the number of input bits in the column addition can reach up to 6 (in position 3). It would be desirable to re-use the full adders previously created; however, the existing full adders could only accommodate 3 inputs (A, B, C_{in}). We can take advantage of the associative property of addition to form the final sum incrementally. Example 12.21 shows the architecture of this multiplier. This approach implements a shift and add process to compute the product and is known as a *combinational multiplier* because it is implemented using only combinational logic. Note that this multiplier only handles unsigned numbers.

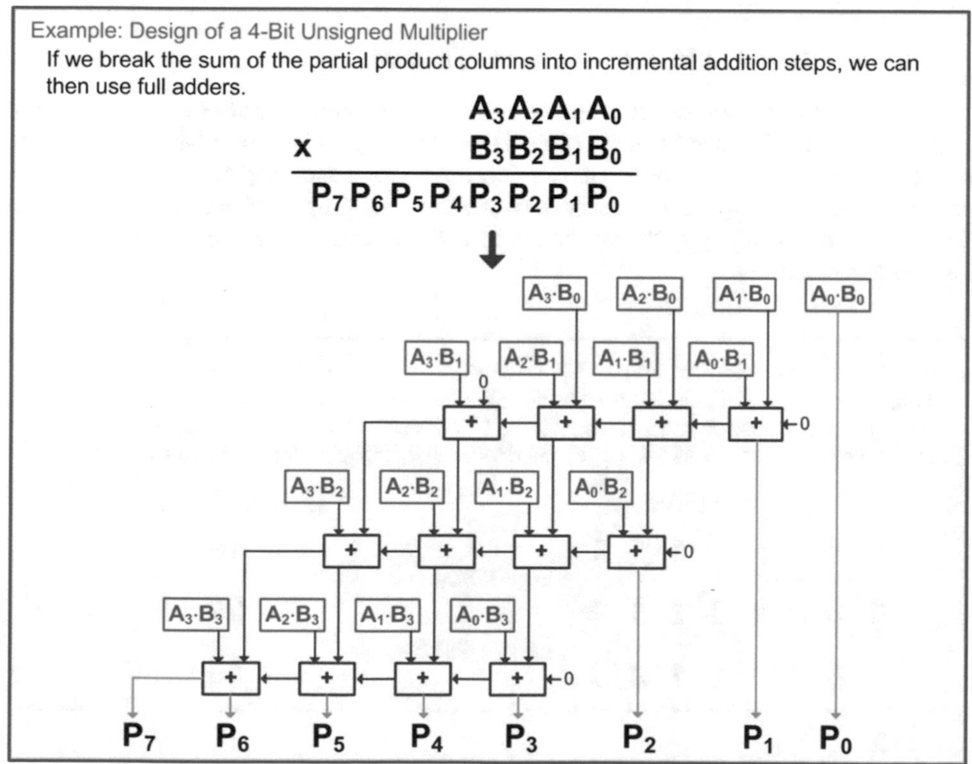

Example 12.21
Design of a 4-bit unsigned multiplier

This multiplier can have a significant delay, which is caused by the cascaded full adders. Example 12.22 shows the timing analysis of the combinational multiplier highlighting the worst-case path through the circuit.

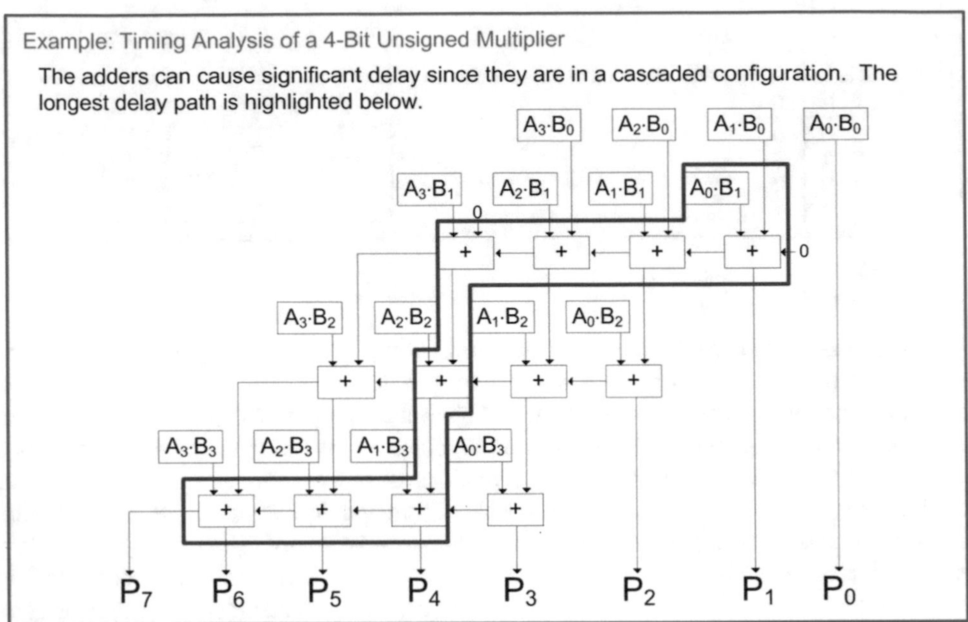

Example 12.22
Timing analysis of a 4-bit unsigned multiplier

12.3.2 A Simple Circuit to Multiply by Powers of Two

In digital systems, a common operation is to multiply numbers by powers of two. For unsigned numbers, multiplying by two can be accomplished by performing a logical shift left. In this operation, all bits are moved to the next higher position (i.e., left) by one position and filling the 0th position with a zero. This has the effect of doubling the value of the number. This can be repeated to achieve higher powers of two. This process works as long as the resulting product fits within the number of bits available. Example 12.23 shows this procedure.

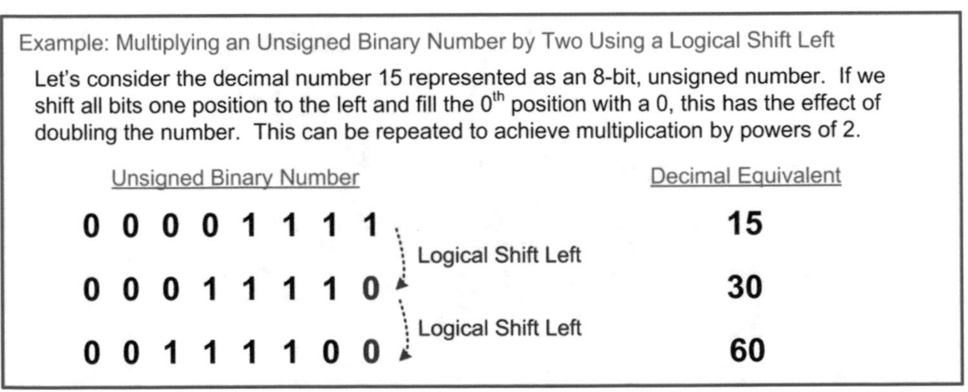

Example 12.23
Multiplying an unsigned binary number by two using a logical shift left

12.3.3 Signed Multiplication

When performing multiplication on signed numbers, it is desirable to re-use the unsigned multiplier in Example 12.21. Let's examine if this is possible. Recall in decimal multiplication that the inputs are multiplied together independent of their sign. The sign of the product is handled separately following these rules:

- A positive number times a positive number produces a positive number.
- A negative number times a negative number produces a positive number.
- A positive number times a negative number produces a negative number.

This process does not work properly in binary due to the way that negative numbers are represented with two's complement. Example 12.24 illustrates how an unsigned multiplier incorrectly handles signed numbers.

Example: Illustrating How an Unsigned Multiplier Incorrectly Handles Signed Numbers

In decimal, the process for multiplying signed numbers is to treat both numbers as unsigned, perform the multiplication, and then apply the correct sign to the product.

```
    -7
x    7
------
  - 4 9
```

The product is formed using the traditional long multiplication process treating the inputs as unsigned (e.g., 7x7=49).

The sign is applied to the product as the final step (neg x pos = neg).

This process does not work directly in binary due to the way that negative numbers are represented using two's complement. Consider the same multiplication using 4-bit, signed numbers.

```
              1 0 0 1     <- -7_10 in 4-bit, two's complement
x             0 1 1 1     <- +7_10
-------------------------
              1 0 0 1
            1 0 0 1
          1 0 0 1
+       0 0 0 0
-------------------------
  0 0 1 1 1 1 1 1         <- +63_10 INCORRECT!
```

Example 12.24
Illustrating how an unsigned multiplier incorrectly handles signed numbers

Instead of building a dedicated multiplier for signed numbers, we can add functionality to the unsigned multiplier previously presented to handle negative numbers. The process involves first identifying any negative numbers. If a negative number is present, the two's complement is taken on it to produce its equivalent magnitude, positive representation. The multiplication is then performed on the positive values. The final step is to apply the correct sign to the product. If the product should be negative due to one of the inputs being negative, the sign is applied by taking the two's complement on the final result. This creates a number that is now in 2·n two's complement format. Example 12.25 shows an illustration of the process to correctly handle signed numbers using an unsigned multiplier.

Example: Process to Correctly Handle Signed Numbers Using an Unsigned Multiplier

The process for handling negative numbers in binary multiplication involves taking the two's complement of any negative numbers to get their positive magnitude equivalents. The unsigned multiplier is then used to create a positive product. If the signs of the inputs should produce a negative product, then the last step is to take the two's complement of the product. Let's do an example of this process on $(-7_{10})\text{x}(+7_{10})=(-49_{10})$.

Step 1 – Take the two's complement of any negative inputs.

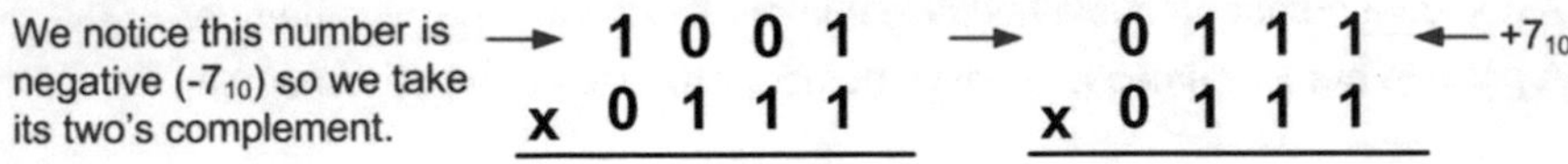

Step 2 – Perform the multiplication.

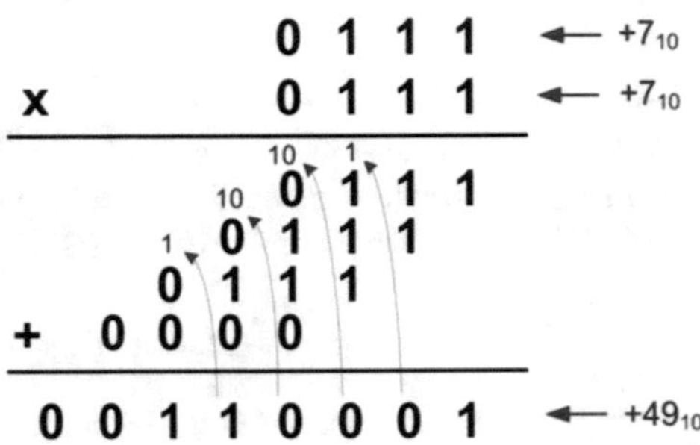

Step 3 – Apply the sign to the product (if applicable).

Since we had a (neg)x(pos), the product should be a negative, so we need to apply the sign by taking the two's complement.

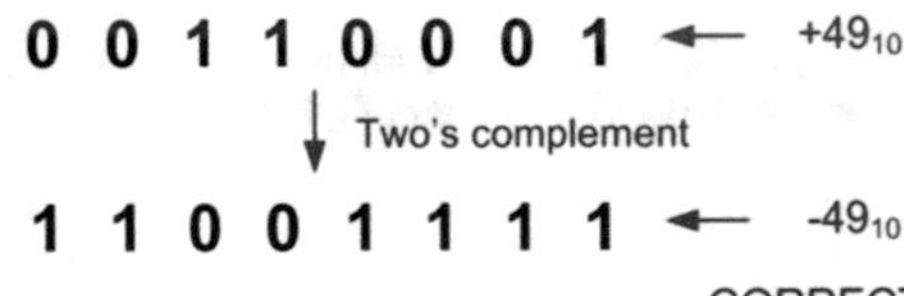

CORRECT!

Notice the result is now in 8-bit two's complement representation.

Example 12.25
Process to correctly handle signed numbers using an unsigned multiplier

CONCEPT CHECK

CC12.3 Will the AND gates used to compute the partial products in a binary multiplier ever experience an issue with fan-in as the size of the multiplier increases?

A) Yes. When the number of bits of the multiplier arguments exceed the fan-in specification of the AND gates used for the partial products, a fan-in issue has occurred.

B) No. The number of inputs of the AND gates performing the partial products will always be two, regardless of the size of the input arguments to the multiplier.

12.4 Division

12.4.1 Unsigned Division

There are a variety of methods to perform division, each with trade-offs between area, delay, and accuracy. To understand the general approach to building a divider circuit, let's focus on how a simple iterative divider can be built. Basic division yields a *quotient* and a *remainder*. The process begins by checking whether the *divisor* goes into the highest position digit in the *dividend*. The number of times this dividend digit can be divided is recorded as the highest position value of the quotient. Note that when performing division by hand, we typically skip over the condition when the result of these initial operations are zero, but when breaking down the process into steps that can be built with logic circuits, each step needs to be highlighted. The first quotient digit is then multiplied with the divisor and recorded below the original dividend. The next lower position digit of the dividend is brought down and joined with the product from the prior multiplication. This forms a new number to be divided by the divisor to create the next quotient value. This process is repeated until each of the quotient digits have been created. Any value that remains after the last subtraction is recorded as the remainder. Example 12.26 shows the long division process on decimal numbers highlight each incremental step.

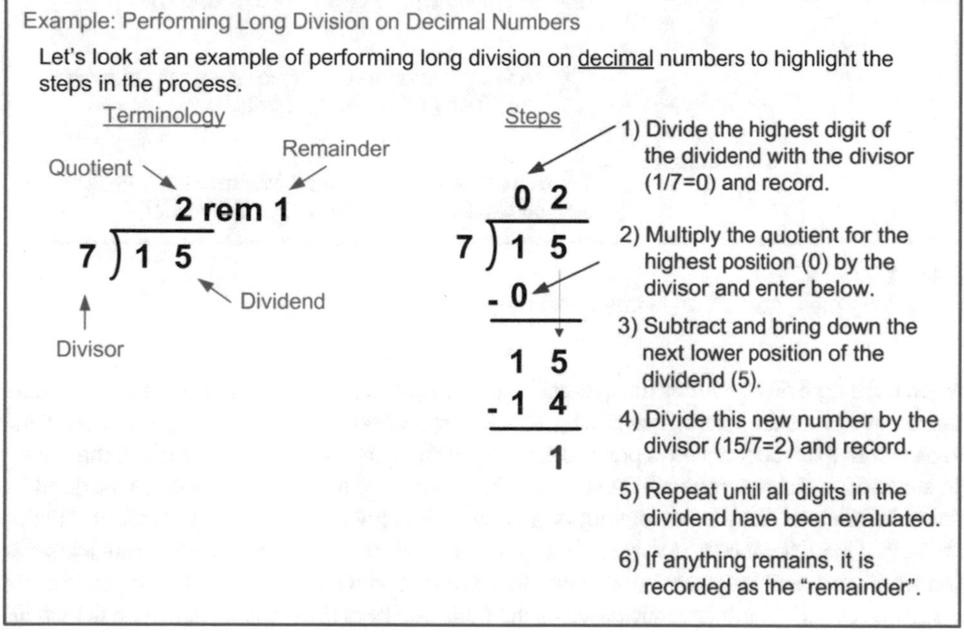

Example 12.26
Performing long division on decimal numbers

Long division in binary follows this same process. Example 12.27 shows the long division process on two 4-bit, unsigned numbers. This division results in a 4-bit quotient and a 4-bit remainder.

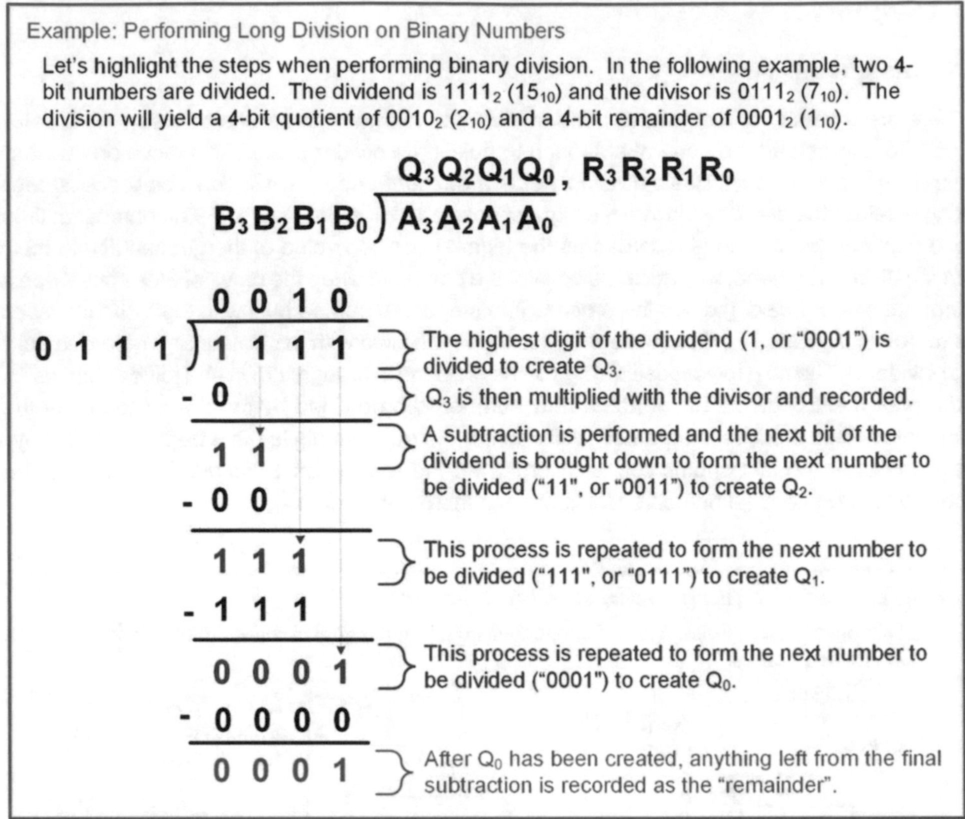

Example 12.27
Performing long multiplication on binary numbers

When building a divider circuit using combinational logic, we can accomplish the computation using a series of iterative subtractors. Performing division is equivalent to subtracting the divisor from the interim dividend. If the subtraction is positive, then the divisor went into the dividend and the quotient is a 1. If the subtraction yields a negative number, then the divisor did not go into the interim dividend, and the quotient is 0. We can use the borrow out of a subtraction chain to provide the quotient. This has the advantage that the difference has already been calculated for the next subtraction. A multiplexer is used to select whether the difference is used in the next subtraction ($Q = 0$) or if the interim divisor is simply brought down ($Q = 1$). This inherently provides the functionality of the multiplication step in long division. Example 12.28 shows the architecture of a 4-bit, unsigned divider based on the iterative subtraction approach. Notice that when the borrow out of the 4-bit subtractor chain is a 0, it indicates that the subtraction yielded a positive number. This means that the divisor went into the interim dividend once. In this case, the quotient for this position is a 1. An inverter is required to produce the correct polarity of the quotient. The borrow out is also fed into the multiplexer stage as the select line to pass the difference to the next stage of subtractors. If the borrow out of the 4-bit subtractor chain is a 1, it indicates that the subtraction yielded a negative number. In this case, the quotient is a 0. This also means that the difference calculated is garbage and should not be used. The multiplexer stage instead selects the interim dividend as the input to the next stage of subtractors.

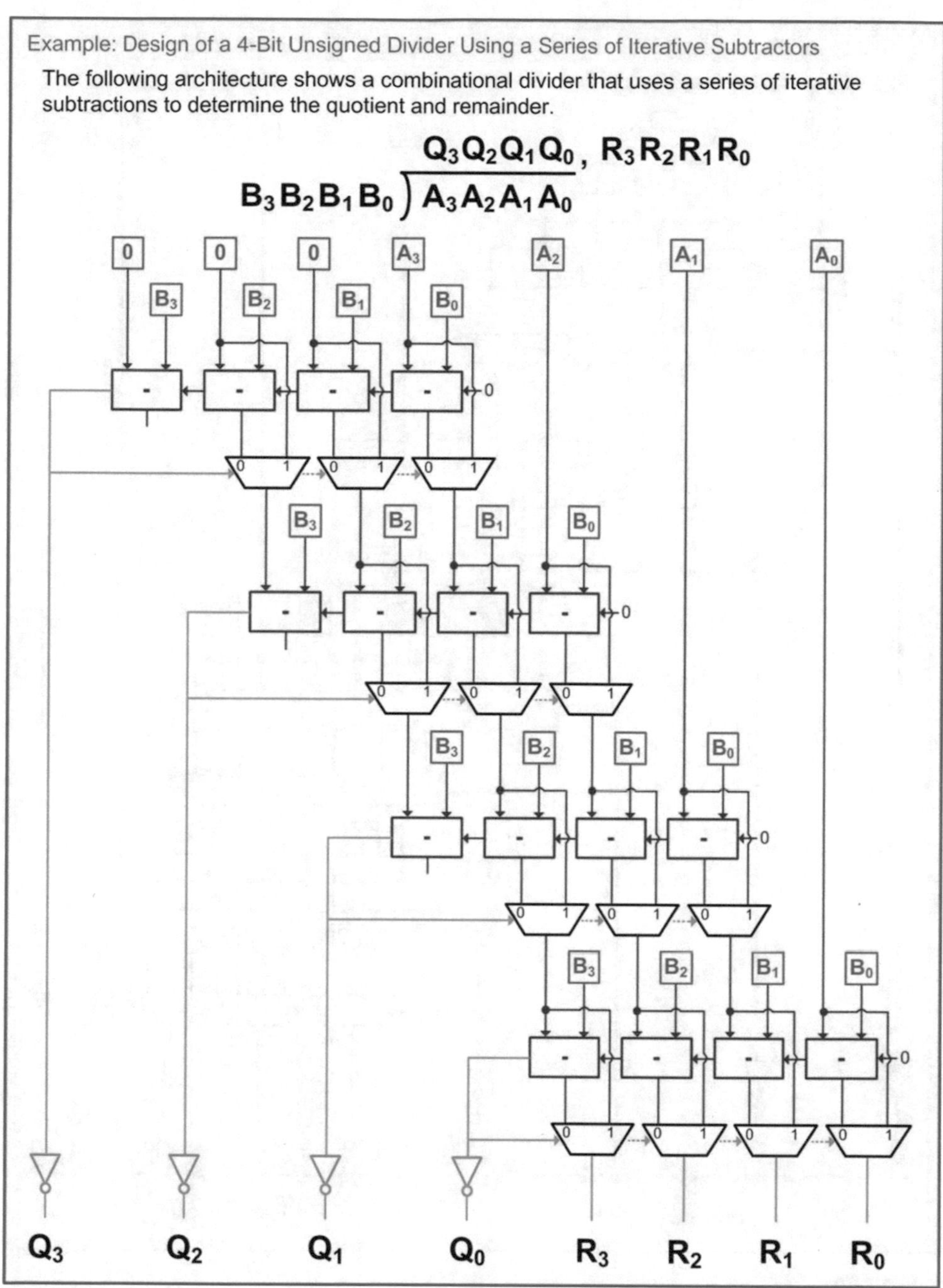

Example 12.28
Design of a 4-bit unsigned divider using a series of iterative subtractors

To illustrate how this architecture works, Example 12.29 walks through each step in the process where 1111_2 (15_{10}) is divided by 0111_2 (7_{10}). In this example, the calculations propagate through the logic stages from top to bottom in the diagram.

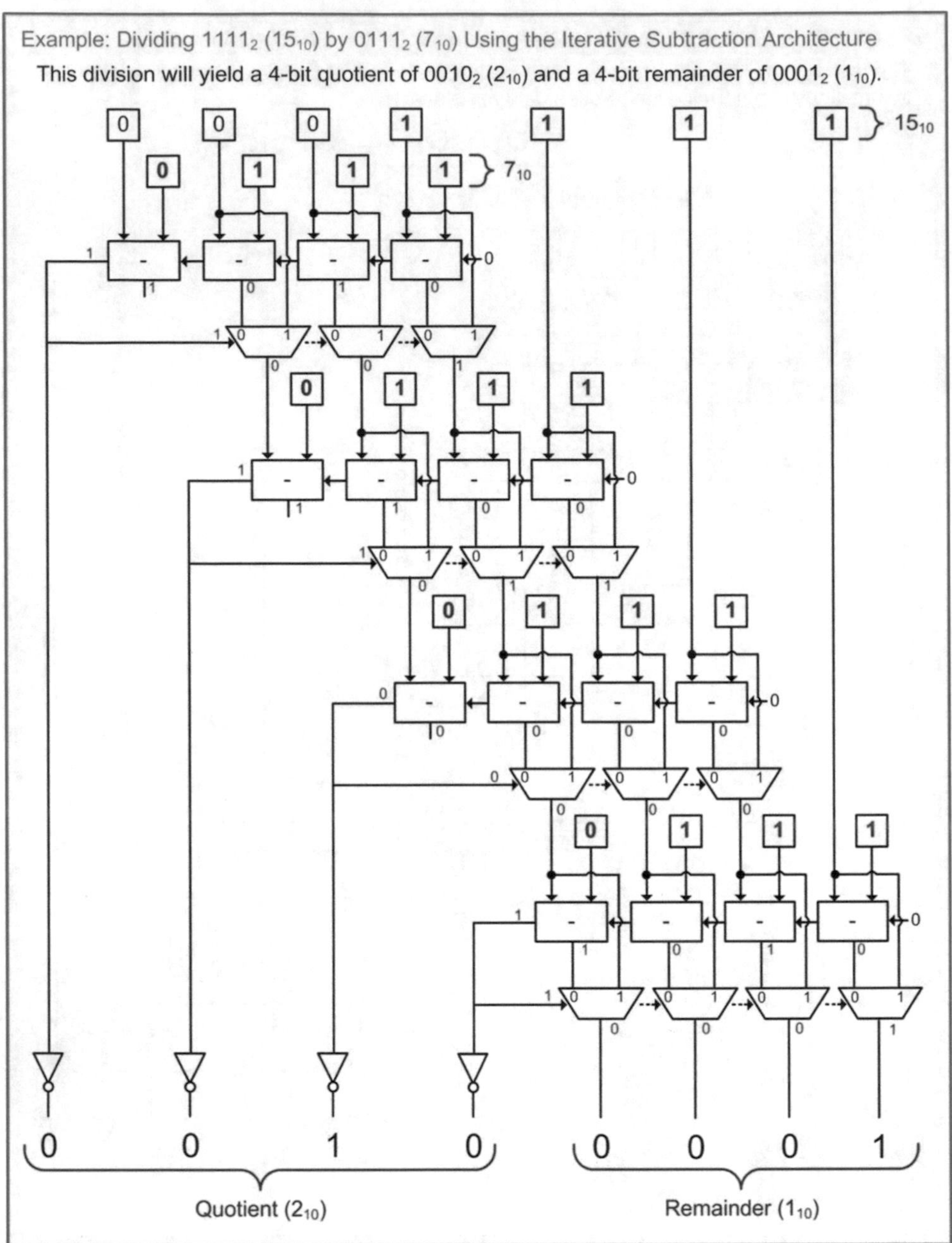

Example 12.29
Dividing 1111_2 (15_{10}) by 0111_2 (7_{10}) using the iterative subtraction architecture

12.4.2 A Simple Circuit to Divide by Powers of Two

For unsigned numbers, dividing by two can be accomplished by performing a logical shift right. In this operation, all bits are moved to the next lower position (i.e., right) by one position and then filling the highest position with a zero. This has the effect of halving the value of the number. This can be repeated to achieve higher powers of two. This process works until no more ones exist in the number, and the result is simply all zeros. Example 12.30 shows this process.

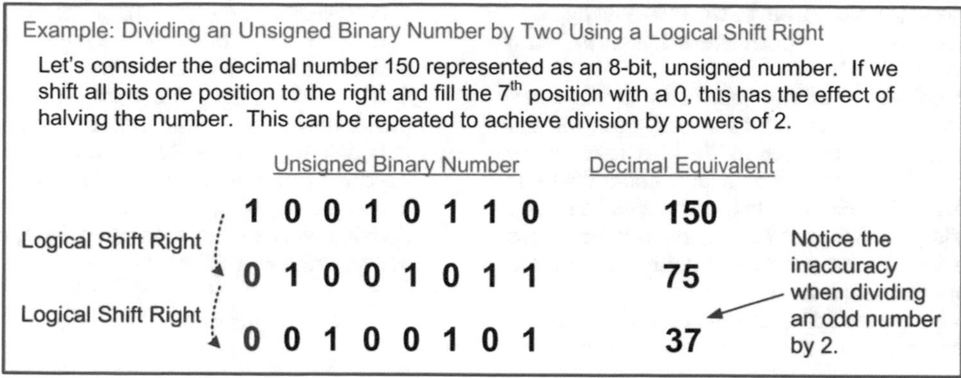

Example 12.30
Dividing an unsigned binary numbers by two using a logical shift right

12.4.3 Signed Division

When performing division on signed numbers, a similar strategy as in signed multiplication is used. The process involves first identifying any negative numbers. If a negative number is present, the two's complement is taken on it to produce its equivalent magnitude, positive representation. The division is then performed on the positive values. The final step is to apply the correct sign to the divisor and quotient. This is accomplished by taking the two's complement if a negative number is required. The rules governing the polarities of the quotient and remainders are:

- The quotient will be negative if the input signs are different (i.e., pos/neg or neg/pos).
- The remainder has the same sign as the dividend.

CONCEPT CHECK

CC12.4 Could a shift register help reduce the complexity of a combinational divider circuit? How?

A) Yes. Instead of having redundant circuits holding the different shifted versions of the divisor, a shift register could be used to hold and shift the divisor after each subtraction.

B) No. A state machine would then be needed to control the divisor shifting, which would make the system even more complex.

Summary

- Binary arithmetic is accomplished using combinational logic circuitry. These circuits tend to be the largest circuits in a system and have the longest delay. Arithmetic circuits are often broken up into interim calculations in order to reduce the overall delay of the computation.
- A *ripple carry adder* performs addition by re-using lower level components that each performs a small part of the computation. A full adder is made from two half adders, and a ripple carry adder is made from a chain of full adders. This approach simplifies the design of the adder, but leads to long delay times

since the carry from each sum must ripple to the next higher position's addition before it can complete.

- A *carry look ahead adder* attempts to eliminate the linear dependence of delay on the number of bits that exists in a ripple carry adder. The carry look ahead adder contains dedicated circuitry that calculates the carry bits for each position of the addition. This leads to a more constant delay as the width of the adder increases.
- A binary multiplier can be created in a similar manner to the way multiplication is accomplished by hand using the *shift and add* approach. The partial products of the multiplication can be performed using 2-input AND gates. The sum of the partial products can have more inputs than the typical ripple carry adder can accommodate. To handle this, the additions are performed two bits at a time using a series of adders.
- Division can be accomplished using an iterative subtractor architecture.

Exercise Problems

Section 12.1: Addition

12.1.1 Give the total delay of the full adder shown in Fig. 12.2 if all gates have a delay of 1 ns.

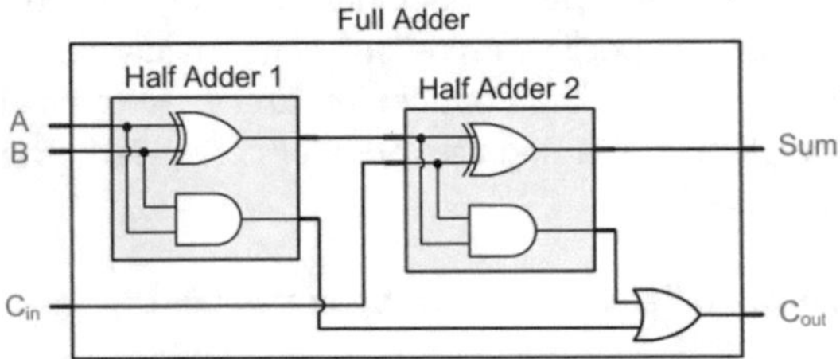

Fig. 12.2
Full adder timing exercise

12.1.2 Give the total delay of the full adder shown in Fig. 12.2 if the XOR gates have delays of 5 ns while the AND and OR gates have delays of 1 ns.

12.1.3 Give the total delay of the 4-bit ripple carry adder shown in Fig. 12.3 if all gates have a delay of 2 ns.

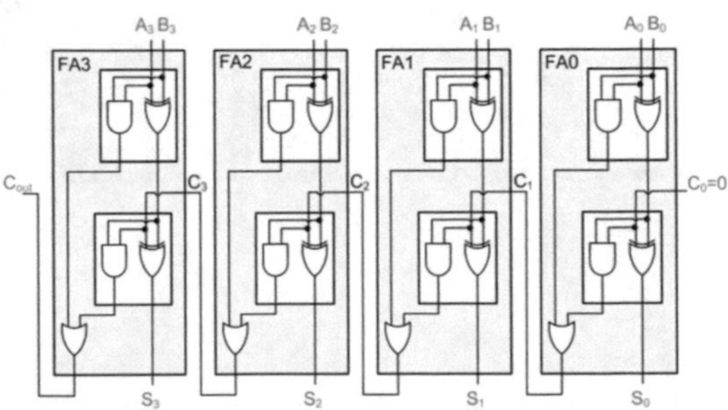

Fig. 12.3
4-bit RCA timing exercise

12.1.4 Give the total delay of the 4-bit ripple carry adder shown in Fig. 12.3 if the XOR gates have delays of 10 ns while the AND and OR gates have delays of 2 ns.

12.1.5 Design a Verilog model for an 8-bit ripple carry adder (RCA) using a structural design approach. This involves creating a half adder (half_adder.v), full adder (full_adder.v), and then finally a top-level adder (rca.v) by instantiating eight full adder sub-systems. Model the logic operations using gate-level primitives. Give each gate primitive a delay of 1 ns. The general topology and module definition for the design are shown in Fig. 12.4. Create a test bench to exhaustively verify your design under all input conditions. The test bench should drive in different values every 30 ns in order to give sufficient time for the results to ripple through the adder.

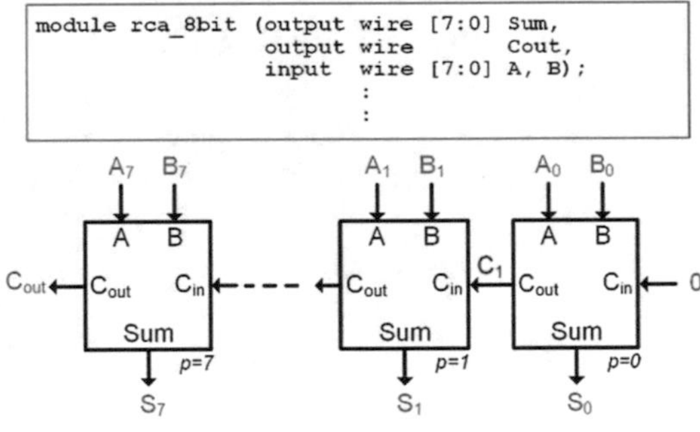

```
module rca_8bit (output wire [7:0] Sum,
                 output wire       Cout,
                 input  wire [7:0] A, B);
                 :
                 :
```


Fig. 12.4
4-bit RCA module definition

12.1.6 Give the total delay of the 4-bit carry look ahead adder shown in Fig. 12.5 if all gates have a delay of 2 ns.

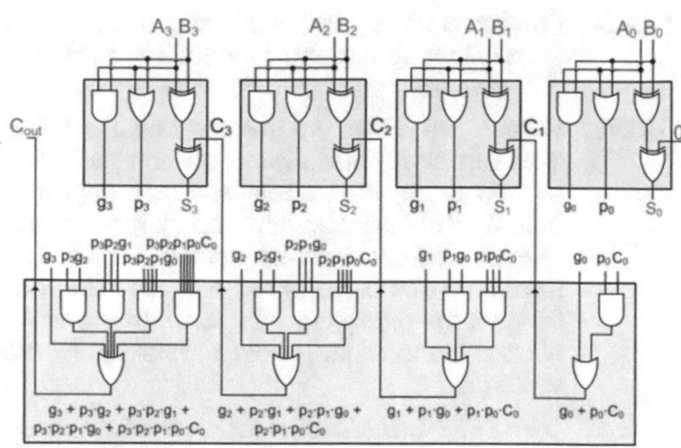

Fig. 12.5
4-bit CLA timing exercise

12.1.7 Give the total delay of the 4-bit carry look ahead adder shown in Fig. 12.5 if the XOR gates have delays of 10 ns while the AND and OR gates have delays of 2 ns.

12.1.8 Design a Verilog model for an 8-bit carry look ahead adder (cla.v). The model should instantiate eight instances of a modified full adder (mod_full_adder.v), which is implemented with gate-level primitives. The carry look ahead logic should be implemented using continuous assignment with logical operators within the cla.v module. All logic operations should have 1 ns of delay. The topology and port definition for the design are shown in Fig. 12.6. Create a test bench to exhaustively verify this design under all input conditions. The test bench should drive in different values every 30 ns in order to give sufficient time for the signals to propagate through the adder.

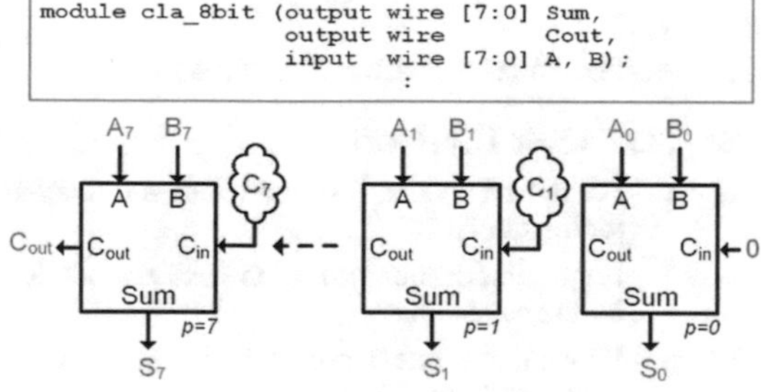

```
module cla_8bit (output wire [7:0] Sum,
                 output wire       Cout,
                 input  wire [7:0] A, B);
                        :
```


Fig. 12.6
4-bit CLA module definition

Section 12.2: Subtraction

12.2.1 How is the programmable add/subtract topology shown in Fig. 12.7 analogous to 2's complement arithmetic?

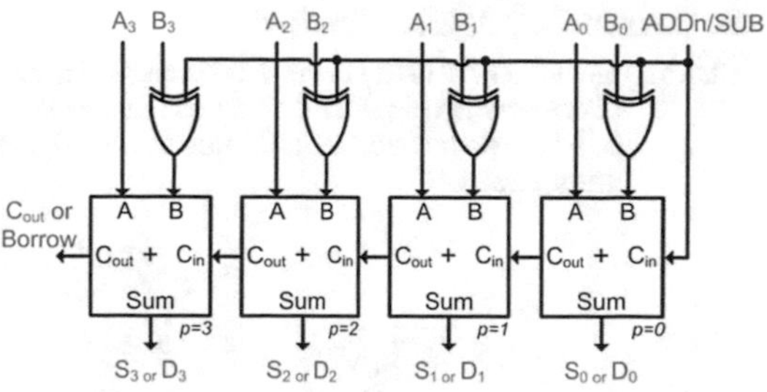

Fig. 12.7
Programmable adder/subtractor block diagram

12.2.2 Will the programmable adder/subtractor architecture shown in Fig. 12.7 work for negative numbers encoded using signed magnitude or 1's complement?

12.2.3 When calculating the delay of the programmable adder/subtractor architecture shown in Fig. 12.7 does the delay of the XOR gate that acts as the programmable inverter need to be considered?

12.2.4 Design a Verilog model for an 8-bit, programmable adder/subtractor. The design will have an input called "ADDn_SUB" that will control whether the system behaves as an adder (0) or as a subtractor (1). The design should operate on two's complement signed numbers. The result of the operation(s) will appear on the port called "Sum_Diff." The model should assert the output "Cout_Bout" when an addition creates a carry or when a subtraction creates a borrow. The circuit will also assert the output Vout when either operation results in two's complement overflow. The port definition and block diagram for the system is shown in Fig. 12.8. Create a test bench to exhaustively verify this design under all input conditions.

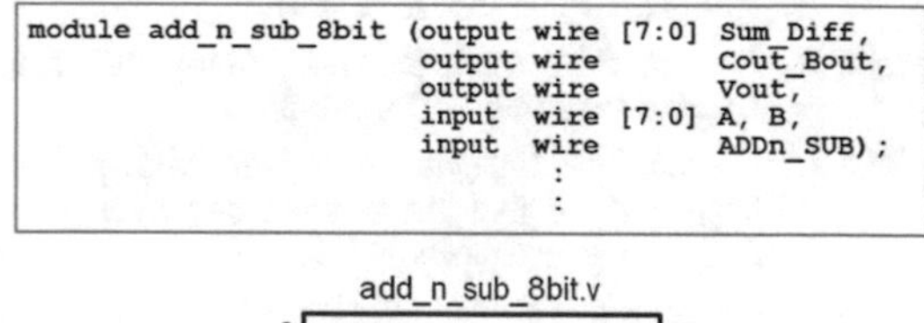

```
module add_n_sub_8bit (output wire [7:0] Sum_Diff,
                       output wire       Cout_Bout,
                       output wire       Vout,
                       input  wire [7:0] A, B,
                       input  wire       ADDn_SUB);
                              :
                              :
```

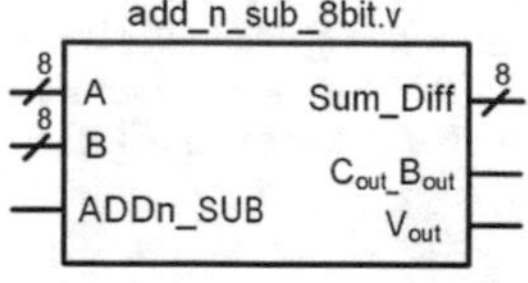

Fig. 12.8
Programmable adder/subtractor module definition

Section 12.3: Multiplication

12.3.1 Give the total delay of the 4-bit unsigned multiplier shown in Fig. 12.9 if all gates have a delay of 1 ns. The addition is performed using a ripple carry adder.

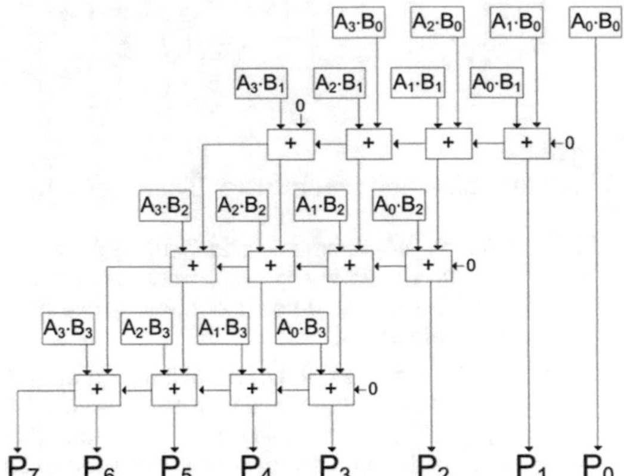

Fig. 12.9
4-bit unsigned multiplier block diagram

12.3.2 For the 4-bit unsigned multiplier shown in Fig. 12.9, how many levels of logic does it take to compute all of the partial products?

12.3.3 For the 4-bit unsigned multiplier shown in Fig. 12.9, how many AND gates are needed to compute the partial products?

12.3.4 For the 4-bit unsigned multiplier shown in Fig. 12.9, how many total AND gates are used if the additions are implemented using full adders made of half adders?

12.3.5 Based on the architecture of an unsigned multiplier in Fig. 12.9, how many AND gates are needed to compute the partial products if the inputs are increased to 8 bits?

12.3.6 For an 8-bit multiplier, how many bits are needed to represent the product?

12.3.7 For an 8-bit *unsigned* multiplier, what is the largest value that the product can ever take on? Give your answer in decimal.

12.3.8 For an 8-bit *signed* multiplier, what is the largest value that the product can ever take on? Give your answer in decimal.

12.3.9 For an 8-bit *signed* multiplier, what is the smallest value that the product can ever take on? Give your answer in decimal.

12.3.10 What is the maximum number of times that a 4-bit unsigned multiplicand can be multiplied by two using the *logical shift left* approach before the product is too large to be represented by an 8-bit product? Hint: The maximum number of times this operation can be performed corresponds to when the multiplicand starts at its lowest possible non-zero value (i.e., 1).

12.3.11 Design a Verilog model for an 8-bit unsigned multiplier using whatever modeling approach you wish. Create a test bench to exhaustively verify this design under all input conditions. The port definition for this multiplier is given in Fig. 12.10.

```
module mul_unsigned_8bit (output wire [15:0] P,
                          input  wire [7:0] A, B);
                                   :
                                   :
```

Fig. 12.10
4-bit unsigned multiplier module definition

12.3.12 Design a Verilog model for an 8-bit signed multiplier using whatever modeling approach you wish. Create a test bench to exhaustively verify this design under all input conditions. The port definition for this multiplier is given in Fig. 12.11.

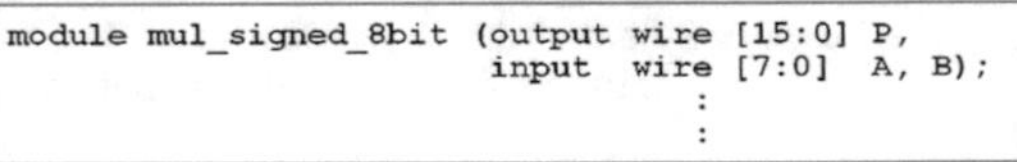

```
module mul_signed_8bit (output wire [15:0] P,
                        input  wire [7:0]  A, B);
                                 :
                                 :
```

Fig. 12.11
4-bit signed multiplier module definition

Section 12.4: Division

12.4.1 For a 4-bit divider, how many bits are needed for the quotient?

12.4.2 For a 4-bit divider, how many bits are needed for the remainder?

12.4.3 Explain the basic concept of the iterative-subtractor approach to division.

12.4.4 For the 4-bit divider shown in Example 12.28, estimate the total delay assuming all gates have a delay of 1 ns.

Chapter 13: Computer System Design

One of the most common digital systems in use today is the computer. A computer accomplishes tasks through an architecture that uses both *hardware* and *software*. The hardware in a computer consists of many of the elements that we have covered so far. These include registers, arithmetic and logic circuits, finite-state machines, and memory. What makes a computer so useful is that the hardware is designed to accomplish a predetermined set of **instructions**. These instructions are relatively simple, such as moving data between memory and a register or performing arithmetic on two numbers. The instructions are comprised of binary codes that are stored in a memory device and represent the sequence of operations that the hardware will perform to accomplish a task. This sequence of instructions is called a computer **program**. What makes this architecture so useful is that the preexisting hardware can be *programmed* to perform an almost unlimited number of tasks by simply defining the sequence of instructions to be executed. The process of designing the sequence of instructions, or program, is called *software development* or *software engineering*.

The idea of a general-purpose computing machine dates back to the nineteenth century. The first computing machines were implemented with mechanical systems and were typically analog in nature. As technology advanced, computer hardware evolved from electromechanical switches to vacuum tubes and ultimately to integrated circuits. These newer technologies enabled switching circuits and provided the capability to build binary computers. Today's computers are built exclusively with semiconductor materials and integrated circuit technology. The term *microcomputer* is used to describe a computer that has its processing hardware implemented with integrated circuitry. Nearly all modern computers are binary. Binary computers are designed to operate on a fixed set of bits. For example, an 8-bit computer would perform operations on 8 bits at a time. This means it moves data between registers and memory and performs arithmetic and logic operations in groups of 8 bits.

This chapter will cover the basics of a simple computer system and present the design of an 8-bit system to illustrate the details of instruction execution. The goal of this chapter is to provide an understanding of the basic principles of computer systems.

Learning Outcomes—After completing this chapter, you will be able to:

13.1 Describe the basic components and operation of computer hardware.
13.2 Describe the basic components and operation of computer software.
13.3 Design a fully operational computer system using Verilog.
13.4 Describe the difference between the Von Neumann and Harvard computer architectures.

13.1 Computer Hardware

Computer hardware refers to all of the physical components within the system. This hardware includes all circuit components in a computer such as the memory devices, registers, and finite-state machines. Figure 13.1 shows a block diagram of the basic hardware components in a computer.

© Springer Nature Switzerland AG 2019
B. J. LaMeres, *Introduction to Logic Circuits & Logic Design with Verilog*,
https://doi.org/10.1007/978-3-030-13605-5_13

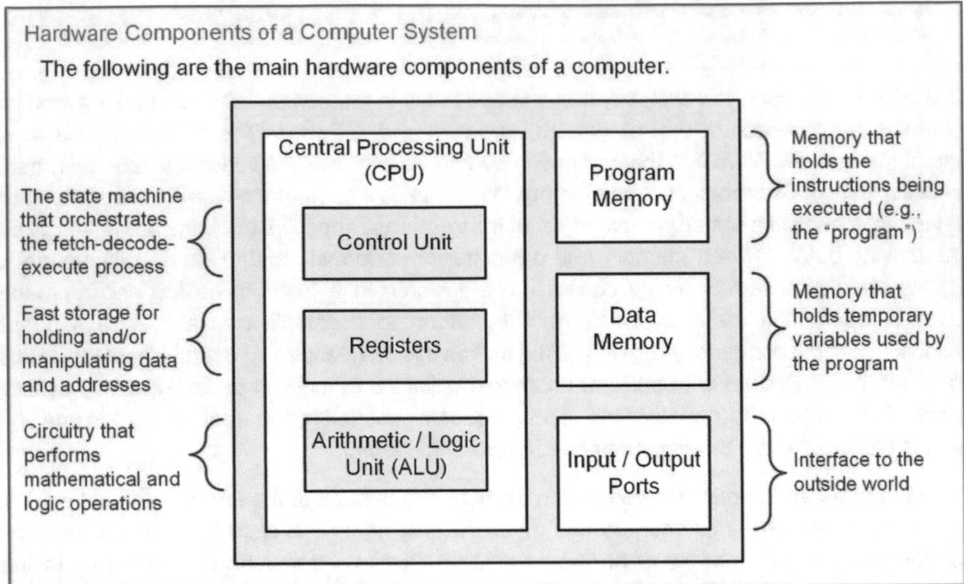

Fig. 13.1
Hardware components of a computer system

13.1.1 Program Memory

The instructions that are executed by a computer are held in *program memory*. Program memory is treated as read-only memory during execution in order to prevent the instructions from being overwritten by the computer. Some computer systems will implement the program memory on a true ROM device (MROM or PROM), while others will use a EEPROM that can be read from during normal operation but can only be written to using a dedicated write procedure. Programs are typically held in non-volatile memory so that the computer system does not lose its program when power is removed. Modern computers will often copy a program from non-volatile memory (e.g., a hard disk drive) to volatile memory after start-up in order to speed up instruction execution. In this case, care must be taken that the program does not overwrite itself.

13.1.2 Data Memory

Computers also require *data memory*, which can be written to and read from during normal operation. This memory is used to hold temporary variables that are created by the software program. This memory expands the capability of the computer system by allowing large amounts of information to be created and stored by the program. Additionally, computations can be performed that are larger than the width of the computer system by holding interim portions of the calculation (e.g., performing a 128-bit addition on a 32-bit computer). Data memory is implemented with R/W memory, most often SRAM or DRAM.

13.1.3 Input/Output Ports

The term *port* is used to describe the mechanism to get information from the output world into or out of the computer. Ports can be input, output, or bidirectional. I/O ports can be designed to pass information in a serial or parallel format.

13.1.4 Central Processing Unit

The *central processing unit* (CPU) is considered the *brains* of the computer. The CPU handles reading instructions from memory, decoding them to understand which instruction is being performed, and executing the necessary steps to complete the instruction. The CPU also contains a set of registers that are used for general-purpose data storage, operational information, and system status. Finally, the CPU contains circuitry to perform arithmetic and logic operations on data.

13.1.4.1 Control Unit

The *control unit* is a finite-state machine that controls the operation of the computer. This FSM has states that perform fetching the instruction (i.e., reading it from program memory), decoding the instruction, and executing the appropriate steps to accomplish the instruction. This process is known as *fetch, decode, and execute* and is repeated each time an instruction is performed by the CPU. As the control unit state machine traverses through its states, it asserts control signals that move and manipulate data in order to achieve the desired functionality of the instruction.

13.1.4.2 Data Path: Registers

The CPU groups its registers and ALU into a sub-system called the *data path*. The data path refers to the fast storage and data manipulations within the CPU. All of these operations are initiated and managed by the control unit state machine. The CPU contains a variety of registers that are necessary to execute instructions and hold status information about the system. Basic computers have the following registers in their CPU:

- **Instruction Register (IR)** – The instruction register holds the current binary code of the instruction being executed. This code is read from program memory as the first part of instruction execution. The IR is used by the control unit to decide which states in its FSM to traverse in order to execute the instruction.
- **Memory Address Register (MAR)** – The memory address register is used to hold the current address being used to access memory. The MAR can be loaded with addresses in order to fetch instructions from program memory or with addresses to access data memory and/or I/O ports.
- **Program Counter (PC)** – The program counter holds the address of the current instruction being executed in program memory. The program counter will increment sequentially through the program memory reading instructions until a dedicated instruction is used to set it to a new location.
- **General-Purpose Registers** – These registers are available for temporary storage by the program. Instructions exist to move information from memory into these registers and to move information from these registers into memory. Instructions also exist to perform arithmetic and logic operations on the information held in these registers.
- **Condition Code Register (CCR)** – The condition code register holds status flags that provide information about the arithmetic and logic operations performed in the CPU. The most common flags are *negative* (N), zero (Z), two's complement overflow (V), and carry (C). This register can also contain flags that indicate the status of the computer, such as if an interrupt has occurred or if the computer has been put into a low-power mode.

13.1.4.3 Data Path: Arithmetic Logic Unit (ALU)

The *arithmetic logic unit* is the system that performs all mathematical (i.e., addition, subtraction, multiplication, and division) and logic operations (i.e., and, or, not, shifts, etc.). This system operates on data being held in CPU registers. The ALU has a unique symbol associated with it to distinguish it from other functional units in the CPU.

Figure 13.2 shows the typical organization of a CPU. The registers and ALU are grouped into the data path. In this example, the computer system has two general-purpose registers called A and B. This CPU organization will be used throughout this chapter to illustrate the detailed execution of instructions.

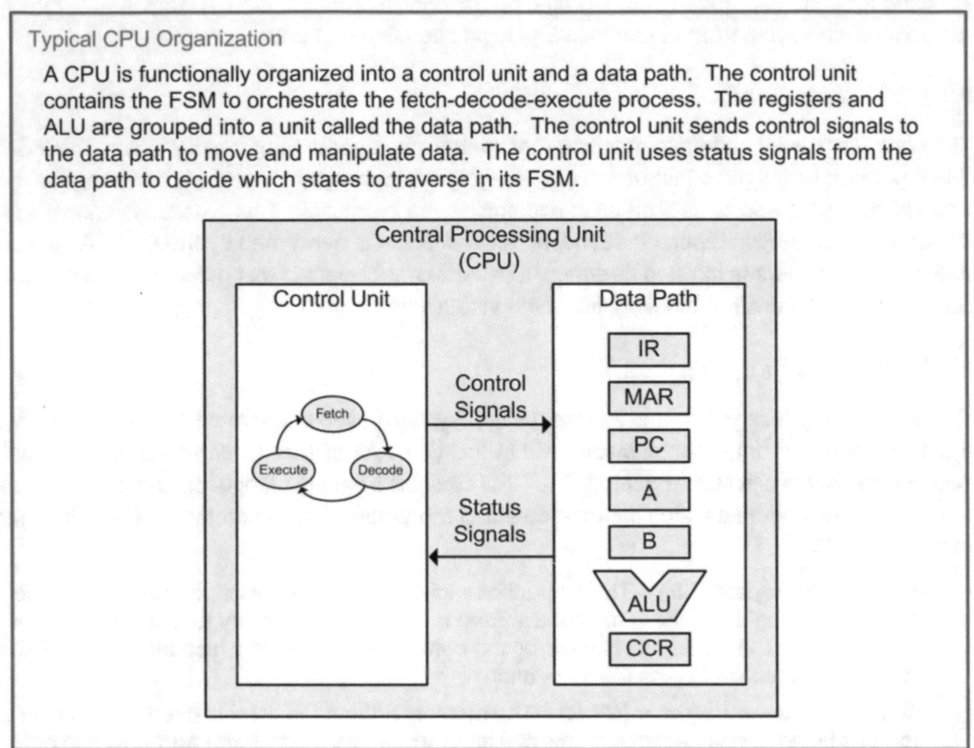

Fig. 13.2
Typical CPU organization

13.1.5 A Memory-Mapped System

A common way to simplify moving data in or out of the CPU is to assign a unique address to all hardware components in the memory system. Each input/output port and each location in both program and data memory are assigned a unique address. This allows the CPU to access everything in the memory system with a dedicated address. This reduces the number of lines that must pass into the CPU. A *bus system* facilitates transferring information within the computer system. An address bus is driven by the CPU to identify which location in the memory system is being accessed. A data bus is used to transfer information to/from the CPU and the memory system. Finally, a control bus is used to provide other required information about the transactions such as *read* or *write* lines. Figure 13.3 shows the computer hardware in a memory-mapped architecture.

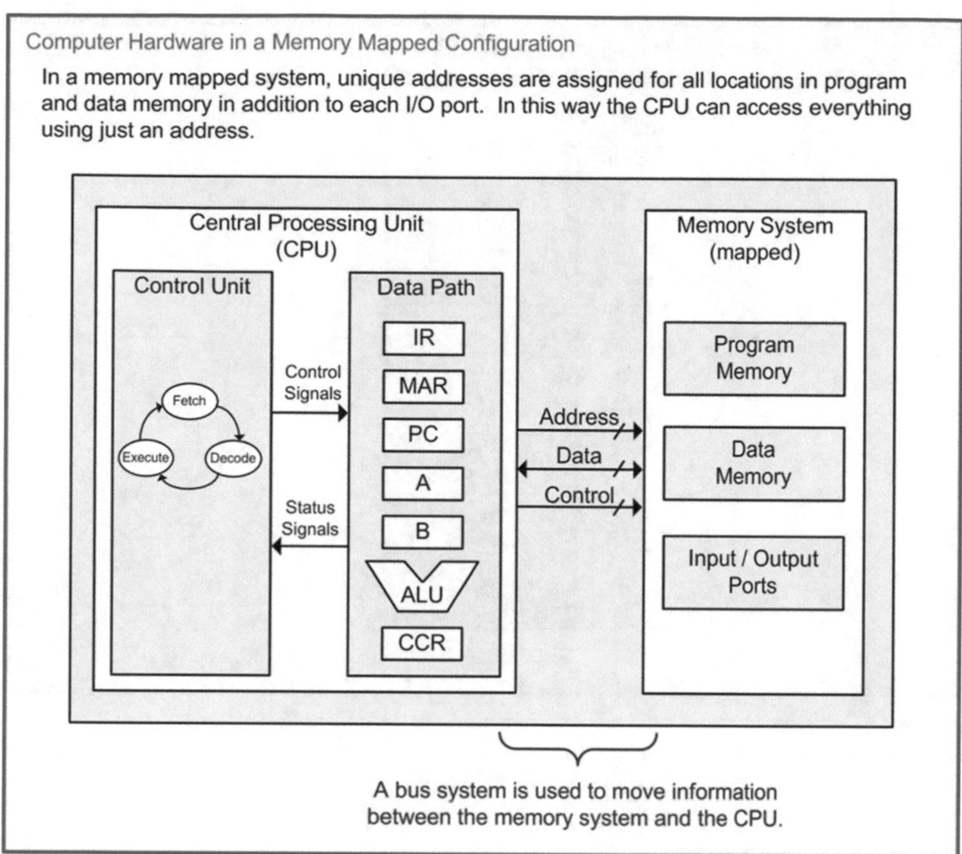

Fig. 13.3
Computer hardware in a memory-mapped configuration

To help visualize how the memory addresses are assigned, a *memory map* is used. This is a graphical depiction of the memory system. In the memory map, the ranges of addresses are provided for each of the main subsections of memory. This gives the programmer a quick overview of the available resources in the computer system. Example 13.1 shows a representative memory map for a computer system with an address bus with a width of 8 bits. This address bus can provide 256 unique locations. For this example, the memory system is also 8 bits wide; thus, the entire memory system is 256 $\times$ 8 in size. In this example 128 bytes are allocated for program memory; 96 bytes are allocated for data memory; 16 bytes are allocated for output ports; and 16 bytes are allocated for input ports.

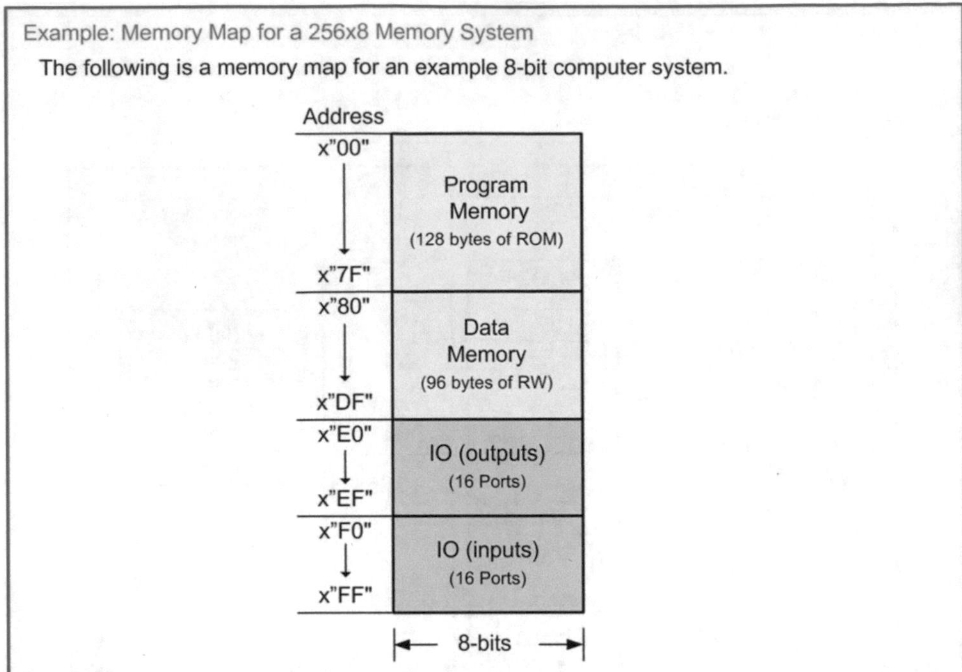

Example 13.1
Memory map for a 256 × 8 memory system

Concept Check

CC13.1 Is the hardware of a computer programmed in a similar way to a programmable logic device?

A) Yes. The control unit is reconfigured to produce the correct logic for each unique instruction just like a logic element in an FPGA is reconfigured to produce the desired logic expression.

B) No. The instruction code from program memory simply tells the state machine in the control unit which path to traverse in order to accomplish the desired task.

13.2 Computer Software

Computer software refers to the instructions that the computer can execute and how they are designed to accomplish various tasks. The specific group of instructions that a computer can execute is known as its **instruction set**. The instruction set of a computer needs to be defined first before the computer hardware can be implemented. Some computer systems have a very small number of instructions in order to reduce the physical size of the circuitry needed in the CPU. This allows the CPU to execute the instructions very quickly but requires a large number of operations to accomplish a given task. This architectural approach is called a **reduced instruction set computer** (RISC).

The alternative to this approach is to make an instruction set with a large number of dedicated instructions that can accomplish a given task in fewer CPU operations. The drawback of this approach is that the physical size of the CPU must be larger in order to accommodate the various instructions. This architectural approach is called a **complex instruction set computer** (CISC).

13.2.1 Opcodes and Operands

A computer instruction consists of two fields, an *opcode* and an *operand*. The opcode is a unique binary code given to each instruction in the set. The CPU decodes the opcode in order to know which instruction is being executed and then takes the appropriate steps to complete the instruction. Each opcode is assigned a **mnemonic**, which is a descriptive name for the opcode that can be used when discussing the instruction functionally. An operand is additional information for the instruction that may be required. An instruction may have any number of operands including zero. Figure 13.4 shows an example of how the instruction opcodes and operands are placed into program memory.

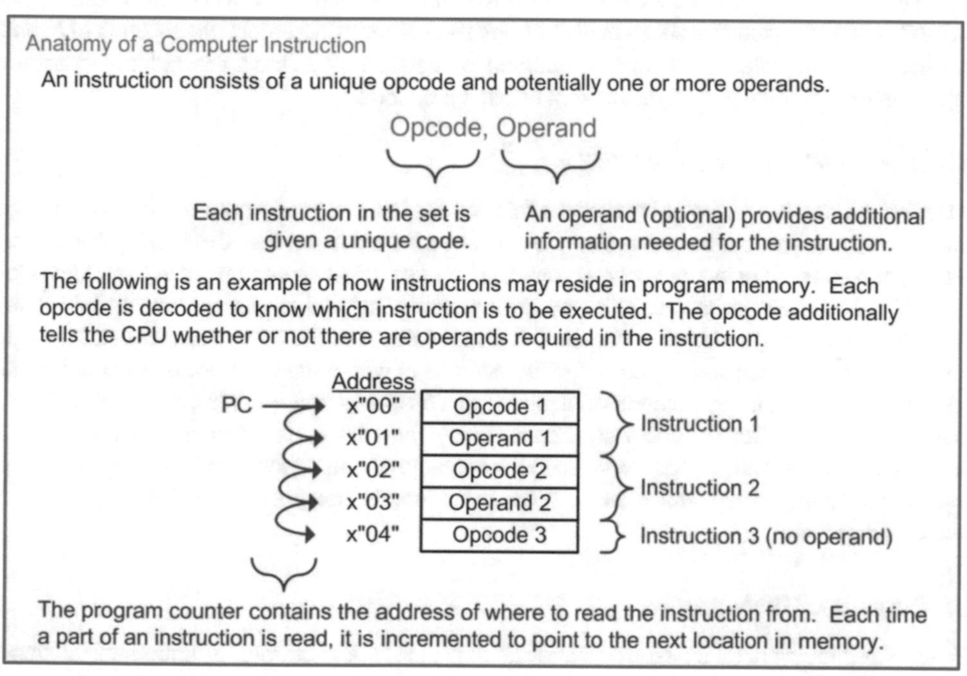

Fig. 13.4
Anatomy of a computer instruction

13.2.2 Addressing Modes

An *addressing mode* describes the way in which the operand of an instruction is used. While modern computer systems may contain numerous addressing modes with varying complexities, we will focus on just a subset of basic addressing modes. These modes are immediate, direct, inherent, and indexed.

13.2.2.1 Immediate Addressing (IMM)

Immediate addressing is when the operand of an instruction *is* the information to be used by the instruction. For example, if an instruction existed to put a constant into a register within the CPU using immediate addressing, the operand would *be* the constant. When the CPU reads the operand, it simply inserts the contents into the CPU register, and the instruction is complete.

13.2.2.2 Direct Addressing (DIR)

Direct addressing is when the operand of an instruction contains the *address* of where the information to be used is located. For example, if an instruction existed to put a constant into a register within the CPU using direct addressing, the operand would contain the address of *where* the constant was located in memory. When the CPU reads the operand, it puts this value out on the address bus and performs an additional read to retrieve the contents located at that address. The value read is then put into the CPU register and the instruction is complete.

13.2.2.3 Inherent Addressing (INH)

Inherent addressing refers to an instruction that does not require an operand because the opcode itself contains all of the necessary information for the instruction to complete. This type of addressing is used on instructions that perform manipulations on data held in CPU registers without the need to access the memory system. For example, if an instruction existed to increment the contents of a register (A), then once the opcode is read by the CPU, it knows everything it needs to know in order to accomplish the task. The CPU simply asserts a series of control signals in order to increment the contents of A, and then the instruction is complete. Notice that no operand is needed for this task. Instead, the location of the register to be manipulated (i.e., A) is inherent within the opcode.

13.2.2.4 Indexed Addressing (IND)

Indexed addressing refers to instructions that will access information at an address in memory to complete the instruction, but the address to be accessed is held in another CPU register. In this type of addressing, the operand of the instruction is used as an *offset* that can be applied to the address located in the CPU register. For example, let's say an instruction existed to put a constant into a register (A) within the CPU using indexed addressing. Let's also say that the instruction was designed to use the contents of another register (B) as part of the address of where the constant was located. When the CPU reads the opcode, it understands what the instruction is and that B holds part of the address to be accessed. It also knows that the operand is applied to B to form the actual address to be accessed. When the CPU reads the operand, it adds the value to the contents of B and then puts this new value out on the address bus and performs an additional read. The value read is then put into the CPU register A and the instruction is complete.

13.2.3 Classes of Instructions

There are three general classes of instructions: (1) loads and stores, (2) data manipulations, and (3) branches. To illustrate how these instructions are executed, examples will be given based on the computer architecture shown in Fig. 13.3.

13.2.3.1 Loads and Stores

This class of instructions accomplishes moving information between the CPU and memory. A **load** is an instruction that moves information from memory *into* a CPU register. When a load instruction uses immediate addressing, the operand of the instruction *is* the data to be loaded into the CPU register. As an example, let's look at an instruction to load the general-purpose register A using immediate addressing. Let's say that the opcode of the instruction is x"86", has a mnemonic LDA_IMM, and is inserted into program memory starting at x"00". Example 13.2 shows the steps involved in executing the LDA_IMM instruction.

Example: Execution of an Instruction to "Load Register A Using Immediate Addressing"

A load instruction using immediate addressing will put the value of the operand into a CPU register. Let's create a program that will load register A in the CPU with the value x"AA". The program is as follows:

Using Mnemonics		Using Hex Values
LDA_IMM x"AA"	or	x"86" x"AA"

When the opcode and operand are put into program memory at x"00", they look like this:

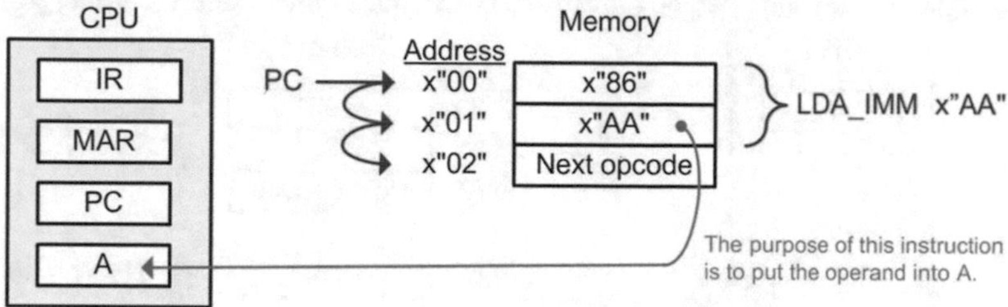

When the CPU begins executing the program, it will perform the following steps:

Step 1 – Fetch the opcode

The program counter begins at x"00", meaning that this address is the location of the first instruction opcode. The PC address is put on the address bus using the MAR and a read is performed. The information read from memory (e.g., the opcode) is placed into the instruction register. The PC is then incremented to point to the next address in program memory. After this step, the IR holds x"86" and the PC holds x"01".

Step 2 – Decode the instruction

The CPU decodes x"86" and understands that it is a "load A with immediate addressing". It also knows from the opcode that the instruction has an operand that exists at the next address location.

Step 3 – Execute the instruction

The CPU now needs to read the operand. It places the PC address (x"01") on the address bus using the MAR and a read is performed. The information read from memory (e.g., the operand) is placed into register A. After this step, A=x"AA". Also in this step, the PC is incremented to point to the next location in memory (x"02"), which holds the opcode of the next instruction to be executed.

Example 13.2
Execution of an instruction to "load register A using immediate addressing"

Now let's look at a load instruction using direct addressing. In direct addressing, the operand of the instruction is the *address* of where the data to be loaded resides. As an example, let's look at an instruction to load the general-purpose register A. Let's say that the opcode of the instruction is x"87", has a mnemonic LDA_DIR, and is inserted into program memory starting at x"08". The value to be loaded into A resides at address x"80", which has already been initialized with x"AA" before this instruction. Example 13.3 shows the steps involved in executing the LDA_DIR instruction.

Example: Execution of an Instruction to "Load Register A Using Direct Addressing"

A load instruction using direct addressing will put the value located at the address provided by the operand into a CPU register. Let's create a program that will load register A in the CPU with the contents located at address x"80", which has already been initialized to x"AA". The program is as follows:

Using Mnemonics		Using Hex Values
LDA_DIR x"80"	or	x"87" x"80"

When the opcode and operand are put into program memory at x"08", they look like this:

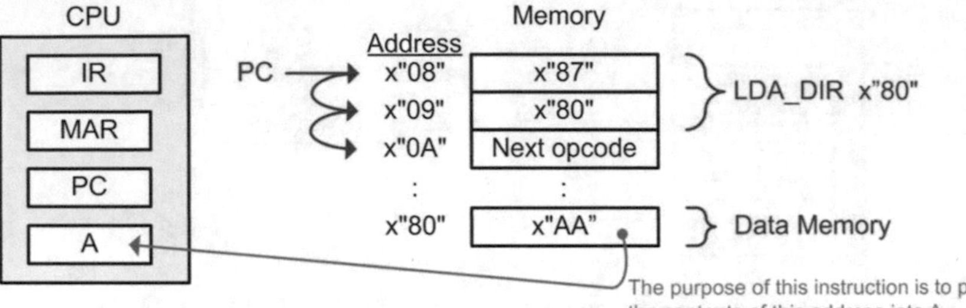

When the CPU begins executing the program, it will perform the following steps:

Step 1 – Fetch the opcode

The program counter begins at x"08", meaning that this address is the location of the instruction opcode. The PC address is put on the address bus using the MAR and a read is performed. The information read from memory (e.g., the opcode) is placed into the instruction register. The PC is then incremented to point to the next address in program memory. After this step, the IR holds x"87" and the PC holds x"09".

Step 2 – Decode the instruction

The CPU decodes x"87" and understands that it is a "load A with direct addressing". It also knows from the opcode that the instruction has an operand that exists at the next address location.

Step 3 – Execute the instruction

The CPU now needs to read the operand. It places the PC address (x"09") on the address bus using the MAR and a read is performed. The information read from memory (e.g., the operand) is the address that contains the value to be put into A. The operand is immediately put on the address bus using the MAR and another read is performed. The value read from address x"80" is placed into register A. After this step, A=x"AA". Also in this step, the PC is incremented to point to the next location in memory (x"0A"), which holds the opcode of the next instruction to be executed.

Example 13.3
Execution of an instruction to "load register A using direct addressing"

A **store** is an instruction that moves information from a CPU register *into* memory. The operand of a store instruction indicates the address of where the contents of the CPU register will be written. As an example, let's look at an instruction to store the general-purpose register A into memory address x"E0". Let's say that the opcode of the instruction is x"96", has a mnemonic STA_DIR, and is inserted into program memory starting at x"04". The initial value of A is x"CC" before the instruction is executed. Example 13.4 shows the steps involved in executing the STA_DIR instruction.

Example: Execution of an Instruction to "Store Register A Using Direct Addressing"

A store instruction using direct addressing will put the value held in a CPU register into memory at the address provided by the operand. Let's create a program that will store register A in the CPU to address location x"E0". We can assume A holds x"CC" prior to this instruction. The program is as follows:

Using Mnemonics		Using Hex Values
STA_DIR x"E0"	or	x"96" x"E0"

When the opcode and operand are put into program memory at x"04", they look like this:

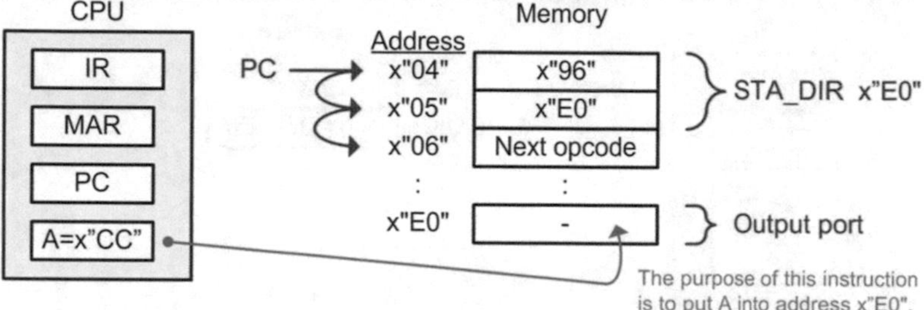

When the CPU begins executing the program, it will perform the following steps:

Step 1 – Fetch the opcode

The program counter begins at x"04", meaning that this address is the location of the instruction opcode. The PC address is put on the address bus using the MAR and a read is performed. The information read from memory (e.g., the opcode) is placed into the instruction register. The PC is then incremented to point to the next address in program memory. After this step, the IR holds x"96" and the PC holds x"05".

Step 2 – Decode the instruction

The CPU decodes x"96" and understands that it is a "store A with direct addressing". It also knows from the opcode that the instruction has an operand that exists at the next address location.

Step 3 – Execute the instruction

The CPU now needs to read the operand. It places the PC address (x"05") on the address bus using the MAR and a read is performed. The information read from memory (e.g., the operand) is the address of where A will be written. The operand is immediately put on the address bus using the MAR, A is put on the data bus, and a write is performed. After this step, location x"E0" in memory contains x"CC". Also in this step, the PC is incremented to point to the next location in memory (x"06"), which holds the opcode of the next instruction to be executed. The write did not effect register A so it still contains x"CC" after the instruction completes.

Example 13.4
Execution of an instruction to "store register A using direct addressing"

13.2.3.2 Data Manipulations

This class of instructions refers to ALU operations. These operations act on data that resides in the CPU registers. These instructions include arithmetic, logic operators, shifts and rotates, and tests and compares. Data manipulation instructions typically use inherent addressing because the operations are conducted on the contents of CPU registers and don't require additional memory access. As an example, let's look at an instruction to perform addition on registers A and B. The sum will be placed back in A. Let's say that the opcode of the instruction is x"42", has a mnemonic ADD_AB, and is inserted into program memory starting at x"04". Example 13.5 shows the steps involved in executing the ADD_AB instruction.

Example: Execution of an Instruction to "Add Registers A and B"

This instruction adds A and B and puts the sum back into A (A = A+B). This instruction does not require an operand because the inputs and output of the operation reside completely within the CPU. This type of instruction uses inherent addressing, meaning that the location of the information impacted is inherent in the opcode. Let's create a program to perform this addition. The program is as follows:

Using Mnemonics		Using Hex Values
ADD_AB	or	x"42"

When the opcode is put into program memory at x"04", it looks like this:

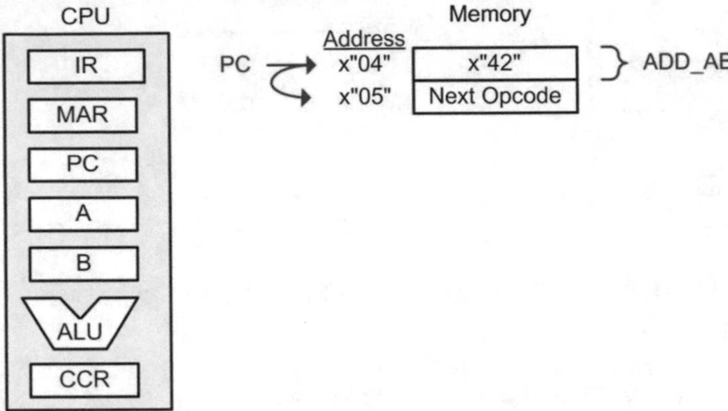

When the CPU begins executing the program, it will perform the following steps:

Step 1 – Fetch the opcode

The program counter begins at x"04", meaning that this address is the location of the instruction opcode. The PC address is put on the address bus using the MAR and a read is performed. The information read from memory (e.g., the opcode) is placed into the instruction register. The PC is then incremented to point to the next address in program memory. After this step, the PC holds x"05" and the IR holds x"42".

Step 2 – Decode the instruction

The CPU decodes x"42" and understands that it is an "Add A and B". It also knows that there is no operand associated with this instruction.

Step 3 – Execute the instruction

The CPU asserts the necessary control signals to route A and B to the ALU, performs the addition, and places the sum back into A. The CCR is also updated to provide additional status information about the operation.

Example 13.5
Execution of an instruction to "add registers A and B"

13.2.3.3 Branches

In the previous examples, the program counter was always incremented to point to the address of the next instruction in program memory. This behavior only supports a linear execution of instructions. To provide the ability to specifically set the value of the program counter, instructions called *branches* are used. There are two types of branches: **unconditional** and **conditional**. In an unconditional branch, the program counter is always loaded with the value provided in the operand. As an example, let's look at an instruction to *branch always* to a specific address. This allows the program to perform loops. Let's say that the opcode of the instruction is x"20", has a mnemonic BRA, and is inserted into program memory starting at x"06". Example 13.6 shows the steps involved in executing the BRA instruction.

Example: Execution of an Instruction to "Branch Always"

A *branch always* instruction will set the program counter to the value provided by the operand. Let's create a program that will set the program counter to x"00". The program is as follows:

Using Mnemonics		Using Hex Values
BRA x"00"	or	x"20" x"00"

When the opcode and operand are put into program memory at x"06", they look like this:

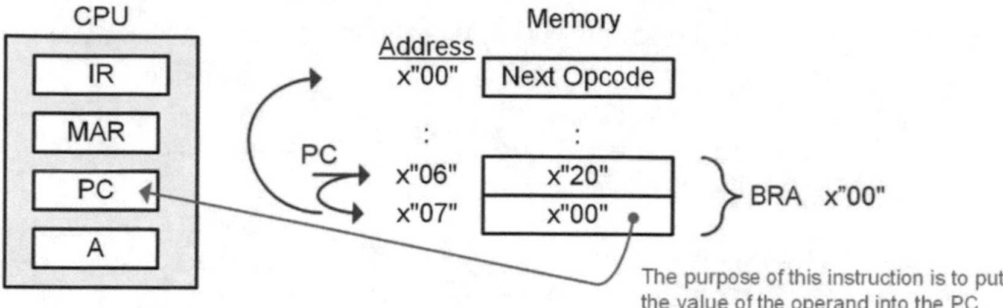

When the CPU begins executing the program, it will perform the following steps:

Step 1 – Fetch the opcode

The program counter begins at x"06", meaning that this address is the location of the instruction opcode. The PC address is put on the address bus using the MAR and a read is performed. The information read from memory (e.g., the opcode) is placed into the instruction register. The PC is then incremented to point to the next address in program memory. After this step, the PC holds x"07" and the IR holds x"20".

Step 2 – Decode the instruction

The CPU decodes x"20" and understands that it is a "branch always". It also knows from the opcode that the instruction has an operand that exists at the next address location.

Step 3 – Execute the instruction

The CPU now needs to read the operand. It places the PC address (x"07") on the address bus using the MAR and a read is performed. The information read from memory (e.g., the operand) is the address to load into the PC. The operand is latched into the PC and the instruction is complete. After this instruction, the PC=x"00" and the program will begin executing instructions at that address.

Example 13.6
Execution of an instruction to "branch always"

In a conditional branch, the program counter is only updated if a particular condition is true. The conditions come from the status flags in the condition code register (NZVC). This allows a program to selectively execute instructions based on the result of a prior operation. Let's look at an example instruction that will branch only if the Z flag is asserted. This instruction is called a *branch if equal to zero*. Let's say that the opcode of the instruction is x"23", has a mnemonic BEQ, and is inserted into program memory starting at x"05". Example 13.7 shows the steps involved in executing the BEQ instruction.

Example: Execution of an Instruction to "Branch if Equal to Zero"

This instruction will update the program counter with the address in the operand if the zero flag (Z) in the condition code register is asserted (Z=1). If Z=0, the program counter will simply increment to the next location in program memory. Let's look at how this program is executed. The instruction resides in program memory at addresses x"05" and x"06".

Using Mnemonics		Using Hex Values
BEQ x"00"	or	x"23" x"00"

When the opcode and operand are put into program memory at x"02", they look like this:

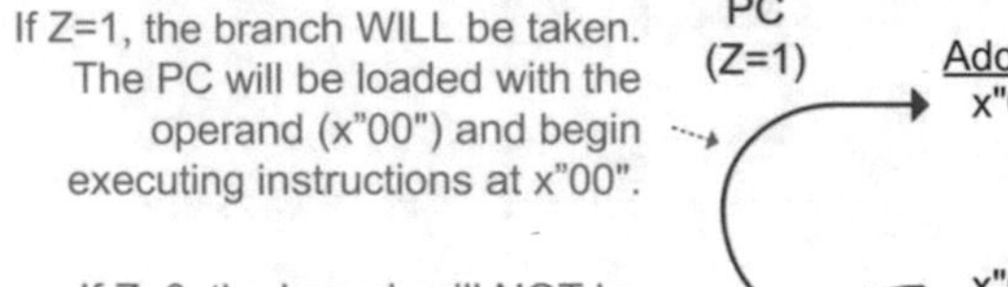

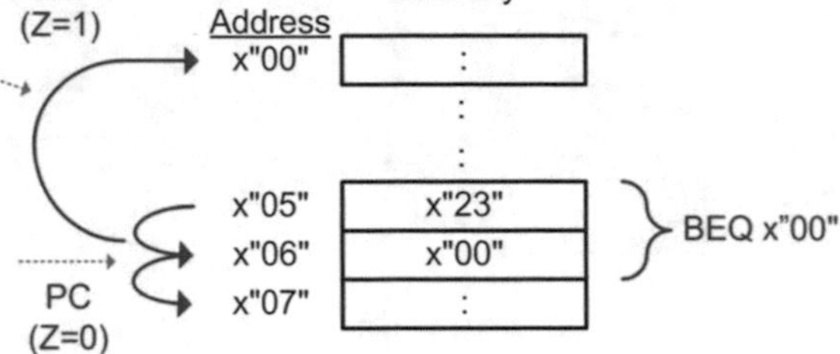

When the CPU begins executing the program, it will perform the following steps:

Step 1 – Fetch the opcode

The program counter begins at x"05", meaning that this address is the location of the instruction opcode. The PC address is put on the address bus using the MAR and a read is performed. The information read from memory (e.g., the opcode) is placed into the instruction register. The PC is then incremented to point to the next address in program memory. After this step, the PC holds x"06" and the IR holds x"23".

Step 2 – Decode the instruction

The CPU decodes x"23" and understands that it is a "branch if equal to zero". It also knows from the opcode that the instruction has an operand that exists at the next address location. The FSM now looks at the Z flag and decides which path in the FSM to take in order to execute the instruction properly.

Step 3 – Execute the instruction

Z=1 – The branch will be taken by loading the PC with the operand. It places the PC address (x"06") on the address bus using the MAR and a read is performed. The information read from memory (e.g., the operand) is then loaded into the PC. If this action is taken, the PC=x"00".

Z=0 – The branch will not be taken. Instead, the PC is simply incremented to point to the next location in memory, bypassing the operand. If this action is taken, the PC=x"07".

Example 13.7
Execution of an instruction to "branch if equal to zero"

Conditional branches allow computer programs to make *decisions* about which instructions to execute based on the results of previous instructions. This gives computers the ability to react to input signals or act based on the results of arithmetic or logic operations. Computer instruction sets typically contain conditional branches based on the NZVC flags in the condition code registers. The following instructions are based on the values of the NZVC flags.

- BMI – Branch if minus (N = 1)
- BPL – Branch if plus (N = 0)
- BEQ – Branch if equal to zero (Z = 1)
- BNE – Branch if not equal to zero (Z = 0)

- BVS – Branch if two's complement overflow occurred, or V is set (V = 1)
- BVC – Branch if two's complement overflow did not occur, or V is clear (V = 0)
- BCS – Branch if a carry occurred, or C is set (C = 1)
- BCC – Branch if a carry did not occur, or C is clear (C = 0)

Combinations of these flags can be used to create more conditional branches.

- BHI – Branch if higher (C = 1 and Z = 0)
- BLS – Branch if lower or the same (C = 0 and Z = 1)
- BGE – Branch if greater than or equal ((N = 0 and V = 0) or (N = 1 and V = 1)), only valid for signed numbers
- BLT – Branch if less than ((N = 1 and V = 0) or (N = 0 and V = 1)), only valid for signed numbers
- BGT – Branch if greater than ((N = 0 and V = 0 and Z = 0) or (N = 1 and V = 1 and Z = 0)), only valid for signed numbers
- BLE – Branch if less than or equal ((N = 1 and V = 0) or (N = 0 and V = 1) or (Z = 1)), only valid for signed numbers

Concept Check

CC13.2 Software development consists of choosing which instructions, and in what order, will be executed to accomplish a certain task. The group of instructions is called the *program* and is inserted into program memory. Which of the following might a software developer care about?

A) Minimizing the number of instructions that need to be executed to accomplish the task in order to increase the computation rate

B) Minimizing the number of registers used in the CPU to save power

C) Minimizing the overall size of the program to reduce the amount of program memory needed

D) Both A and C

13.3 Computer Implementation: An 8-Bit Computer Example

13.3.1 Top-Level Block Diagram

Let's now look at the detailed implementation and instruction execution of a computer system. In order to illustrate the detailed operation, we will use a simple 8-bit computer system design. Example 13.8 shows the block diagram for the 8-bit computer system. This block diagram also contains the Verilog file and module names, which will be used when the behavioral model is implemented.

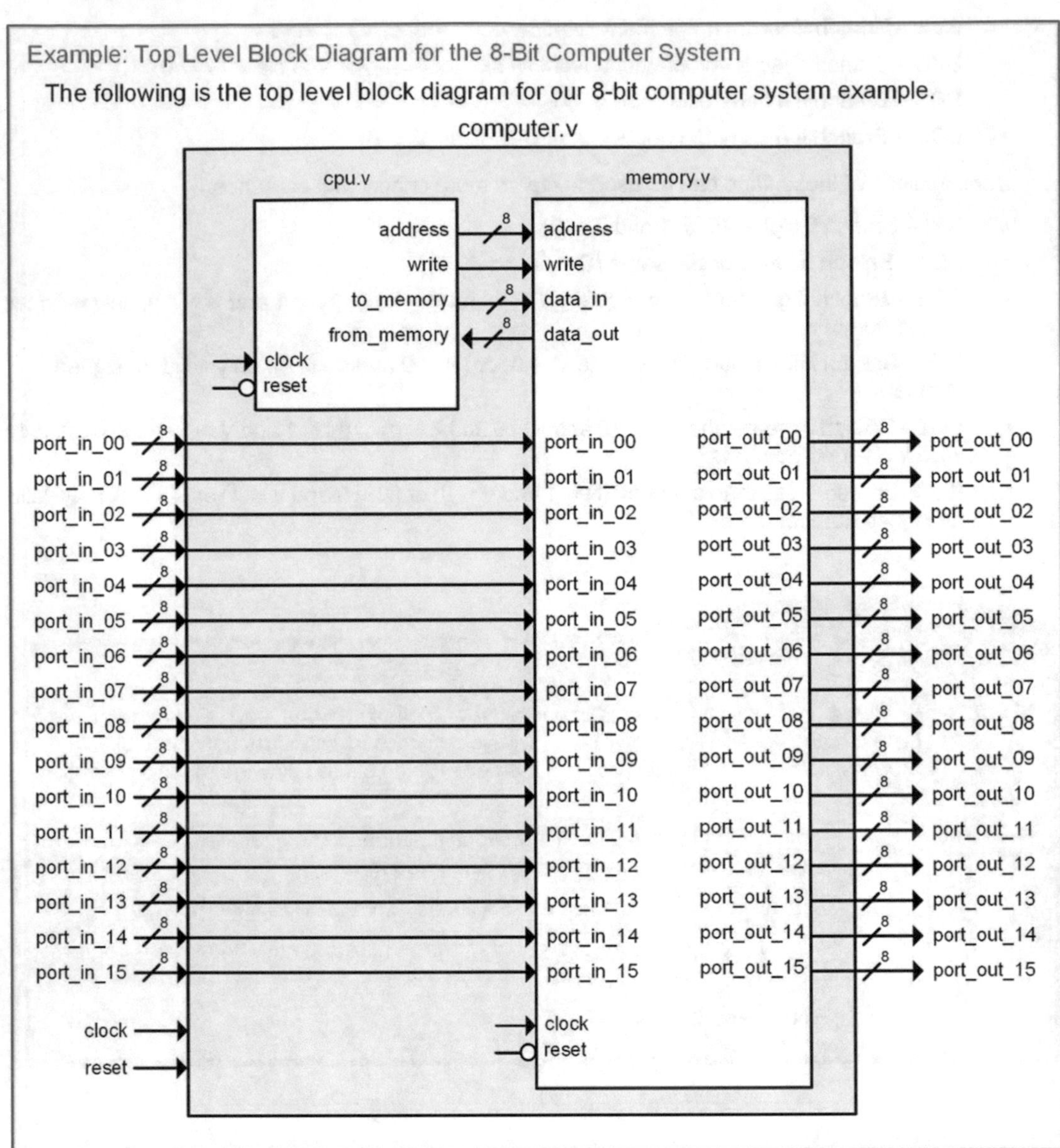

Example 13.8
Top-level block diagram for the 8-bit computer system

We will use the memory map shown in Example 13.1 for our example computer system. This mapping provides 128 bytes of program memory, 96 bytes of data memory, 16× output ports, and 16× input ports. To simplify the operation of this example computer, the address bus is limited to 8 bits. This only provides 256 locations of memory access but allows an entire address to be loaded into the CPU as a single operand of an instruction.

13.3.2 Instruction Set Design

Example 13.9 shows a basic instruction set for our example computer system. This set provides a variety of loads and stores, data manipulations, and branch instructions that will allow the computer to be programmed to perform more complex tasks through software development. These instructions are sufficient to provide a baseline of functionality in order to get the computer system operational. Additional instructions can be added as desired to increase the complexity of the system.

Example: Instruction Set for the 8-Bit Computer System

The following is a base set of instructions that the 8-bit computer system will be able to perform. Each instruction is given a descriptive mnemonic, which allows the system implementation and the programming to be more intuitive. Each instruction is also provided with a unique binary opcode. Some instructions have an operand, which provides additional information necessary for the instruction. If an instruction contains an operand, a description is provided as to how it is used (e.g., as data or as an address).

Mnemonic	Opcode, Operand	Description
"Loads and Stores"		
LDA_IMM	x"86", *<data>*	Load Register A using Immediate Addressing
LDA_DIR	x"87", *<addr>*	Load Register A using Direct Addressing
LDB_IMM	x"88", *<data>*	Load Register B with Immediate Addressing
LDB_DIR	x"89", *<addr>*	Load Register B with Direct Addressing
STA_DIR	x"96", *<addr>*	Store Register A to Memory using Direct Addressing
STB_DIR	x"97", *<addr>*	Store Register B to Memory using Direct Addressing
"Data Manipulations"		
ADD_AB	x"42"	A = A + B (plus)
SUB_AB	x"43"	A = A - B (minus)
AND_AB	x"44"	A = A · B (AND)
OR_AB	x"45"	A = A + B (OR)
INCA	x"46"	A = A + 1 (plus)
INCB	x"47"	B = B + 1 (plus)
DECA	x"48"	A = A - 1 (minus)
DECB	x"49"	B = B - 1 (minus)
"Branches"		
BRA	x"20", *<addr>*	Branch Always to Address Provided
BMI	x"21", *<addr>*	Branch to Address Provided if N=1
BPL	x"22", *<addr>*	Branch to Address Provided if N=0
BEQ	x"23", *<addr>*	Branch to Address Provided if Z=1
BNE	x"24", *<addr>*	Branch to Address Provided if Z=0
BVS	x"25", *<addr>*	Branch to Address Provided if V=1
BVC	x"26", *<addr>*	Branch to Address Provided if V=0
BCS	x"27", *<addr>*	Branch to Address Provided if C=1
BCC	x"28", *<addr>*	Branch to Address Provided if C=0

Example 13.9
Instruction set for the 8-bit computer system

13.3.3 Memory System Implementation

Let's now look at the memory system details. The memory system contains program memory, data memory, and input/output ports. Example 13.10 shows the block diagram of the memory system. The program and data memory will be implemented using lower-level components (rom_128x8_sync.v and rw_96x8_sync.v), while the input and output ports can be modeled using a combination of RTL blocks and combinational logic. The program and data memory sub-systems contain dedicated circuitry to handle their addressing ranges. Each output port also contains dedicated circuitry to handle its unique address. A multiplexer is used to handle the signal routing back to the CPU based on the address provided.

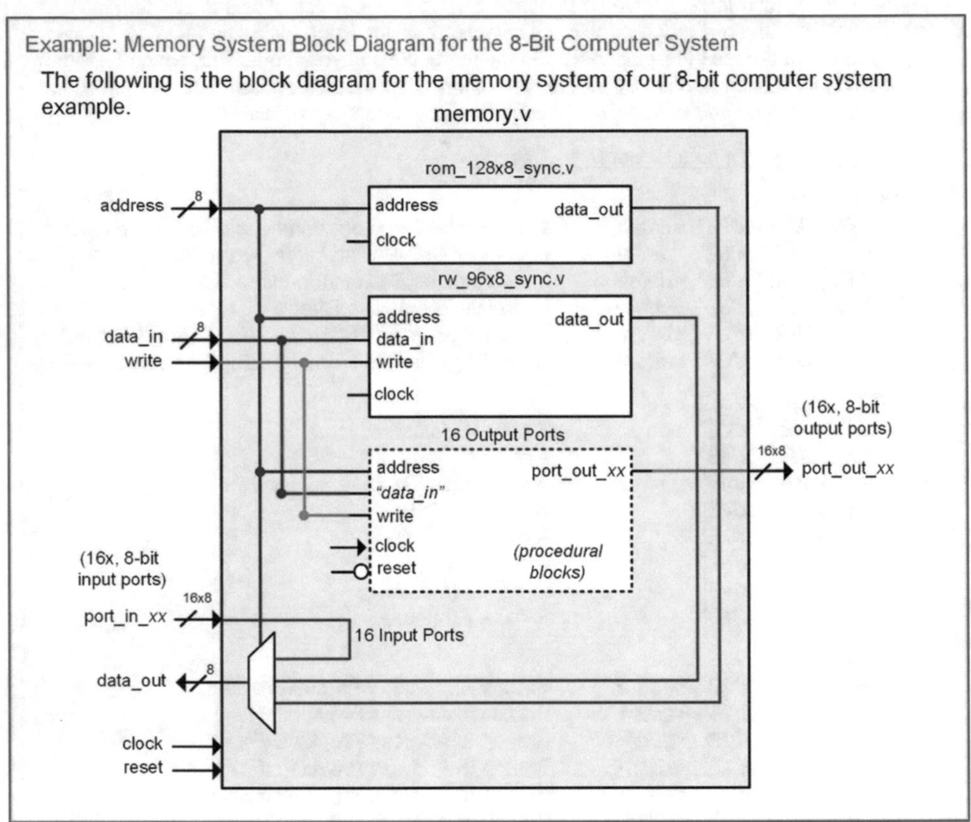

Example 13.10
Memory system block diagram for the 8-bit computer system

13.3.3.1 Program Memory Implementation in Verilog

The program memory can be implemented in Verilog using the modeling techniques presented in Chap. 12. To make the Verilog more readable, the instruction mnemonics can be declared as parameters. This allows the mnemonic to be used when populating the program memory array. The following Verilog shows how the mnemonics for our basic instruction set can be defined as parameters.

```
parameter LDA_IMM = 8'h86;   //-- Load Register A with Immediate Addressing
parameter LDA_DIR = 8'h87;   //-- Load Register A with Direct Addressing
parameter LDB_IMM = 8'h88;   //-- Load Register B with Immediate Addressing
parameter LDB_DIR = 8'h89;   //-- Load Register B with Direct Addressing
parameter STA_DIR = 8'h96;   //-- Store Register A to memory (RAM or IO)
parameter STB_DIR = 8'h97;   //-- Store Register B to memory (RAM or IO)
parameter ADD_AB  = 8'h42;   //-- A <= A + B
parameter SUB_AB  = 8'h43;   //-- A <= A - B
parameter AND_AB  = 8'h44;   //-- A <= A and B
parameter OR_AB   = 8'h45;   //-- A <= A or B
parameter INCA    = 8'h46;   //-- A <= A + 1
parameter INCB    = 8'h47;   //-- B <= B + 1
parameter DECA    = 8'h48;   //-- A <= A - 1
parameter DECB    = 8'h49;   //-- B <= B - 1
parameter BRA     = 8'h20;   //-- Branch Always
parameter BMI     = 8'h21;   //-- Branch if N=1
parameter BPL     = 8'h22;   //-- Branch if N=0
parameter BEQ     = 8'h23;   //-- Branch if Z=1
parameter BNE     = 8'h24;   //-- Branch if Z=0
parameter BVS     = 8'h25;   //-- Branch if V=1
parameter BVC     = 8'h26;   //-- Branch if V=0
parameter BCS     = 8'h27;   //-- Branch if C=1
parameter BCC     = 8'h28;   //-- Branch if C=0
```

Now the program memory can be declared as an array type with initial values to define the program. The following Verilog shows how to declare the program memory and an example program to perform a load, a store, and a branch always. This program will continually write x"AA" to port_out_00.

```
reg[7:0] ROM[0:127];

initial
  begin
    ROM[0] = LDA_IMM;
    ROM[1] = 8'hAA;
    ROM[2] = STA_DIR;
    ROM[3] = 8'hE0;
    ROM[4] = BRA;
    ROM[5] = 8'h00;
  end
```

The address mapping for the program memory is handled in two ways. First, notice that the array type defined above uses indices from 0 to 127. This provides the appropriate addresses for each location in the memory. The second step is to create an internal enable line that will only allow assignments from ROM to data_out when a valid address is entered. Consider the following Verilog to create an internal enable (EN) that will only be asserted when the address falls within the valid program memory range of 0 to 127.

```
always @ (address)
  begin
    if ((address >= 0) && (address <= 127))
      EN = 1'b1;
    else
      EN = 1'b0;
  end
```

If this enable signal is not created, the simulation and synthesis will fail because data_out assignments will be attempted for addresses outside of the defined range of the ROM array. This enable line can now be used in the behavioral model for the ROM as follows:

```
always @ (posedge clock)
  begin
    if (EN)
      data_out = ROM[address];
  end
```

13.3.3.2 Data Memory Implementation in Verilog

The data memory is created using a similar strategy as the program memory. An array signal is declared with an address range corresponding to the memory map for the computer system (i.e., 128 to 223). An internal enable is again created that will prevent data_out assignments for addresses outside of this valid range. The following is the Verilog to declare the R/W memory array:

```
reg[7:0] RW[128:223];
```

The following is the Verilog to model the local enable and signal assignments for the R/W memory:

```
always @ (address)
  begin
    if ((address >= 128) && (address <= 223))
      EN = 1'b1;
    else
      EN = 1'b0;
    end

always @ (posedge clock)
  begin
    if (write && EN)
      RW[address] = data_in;
    else if (!write && EN)
      data_out = RW[address];
  end
```

13.3.3.3 Implementation of Output Ports in Verilog

Each output port in the computer system is assigned a unique address. Each output port also contains storage capability. This allows the CPU to update an output port by writing to its specific address. Once the CPU is done storing to the output port address and moves to the next instruction in the program, the output port holds its information until it is written to again. This behavior can be modeled using an RTL procedural block that uses the address bus and the write signal to create a synchronous enable condition. Each output port is modeled with its own block. The following Verilog shows how the output ports at x"E0" and x"E1" are modeled using address-specific procedural blocks:

```
//-- port_out_00 (address E0)
  always @ (posedge clock or negedge reset)
    begin
      if (!reset)
        port_out_00 <= 8'h00;
      else
        if ((address == 8'hE0) && (write))
           port_out_00 <= data_in;
  end

//-- port_out_01 (address E1)
  always @ (posedge clock or negedge reset)
    begin
      if (!reset)
        port_out_01 <= 8'h00;
       else
      if ((address == 8'hE1) && (write))
           port_out_01 <= data_in;
      end

                    :
      "the rest of the output port models go here..."
                    :
```

13.3.3.4 Implementation of Input Ports in Verilog

The input ports do not contain storage but do require a mechanism to selectively route their information to the data_out port of the memory system. This is accomplished using the multiplexer shown in Example 13.10. The only functionality that is required for the input ports is connecting their ports to the multiplexer.

13.3.3.5 Memory data_out Bus Implementation in Verilog

Now that all of the memory functionality has been designed, the final step is to implement the multiplexer that handles routing the appropriate information to the CPU on the data_out bus based on the incoming address. The following Verilog provides a model for this behavior. Recall that a multiplexer is combinational logic, so if the behavior is to be modeled using a procedural block, all inputs must be listed in the sensitivity list, and blocking assignments are used. These inputs include the outputs from the program and data memory in addition to all of the input ports. The sensitivity list must also include the address bus as it acts as the select input to the multiplexer. Within the block, an if-else statement is used to determine which sub-system drives data_out. Program memory will drive data_out when the incoming address is in the range of 0 to 127 (x"00" to x"7F"). Data memory will drive data_out when the address is in the range of 128 to 223 (x"80" to x"DF"). An input port will drive data_out when the address is in the range of 240 to 255 (x"F0" to x"FF"). Each input port has a unique address, so the specific addresses are listed as nested if-else clauses.

```
always @ (address, rom_data_out, rw_data_out,
          port_in_00, port_in_01, port_in_02, port_in_03,
          port_in_04, port_in_05, port_in_06, port_in_07,
          port_in_08, port_in_09, port_in_10, port_in_11,
          port_in_12, port_in_13, port_in_14, port_in_15)

  begin: MUX1

    if ((address >= 0) && (address <= 127))
      data_out = rom_data_out;
    else if ((address >= 128) && (address <= 223))
      data_out = rw_data_out;
    else if (address == 8'hF0) data_out = port_in_00;
    else if (address == 8'hF1) data_out = port_in_01;
    else if (address == 8'hF2) data_out = port_in_02;
    else if (address == 8'hF3) data_out = port_in_03;
    else if (address == 8'hF4) data_out = port_in_04;
    else if (address == 8'hF5) data_out = port_in_05;
    else if (address == 8'hF6) data_out = port_in_06;
    else if (address == 8'hF7) data_out = port_in_07;
    else if (address == 8'hF8) data_out = port_in_08;
    else if (address == 8'hF9) data_out = port_in_09;
    else if (address == 8'hFA) data_out = port_in_10;
    else if (address == 8'hFB) data_out = port_in_11;
    else if (address == 8'hFC) data_out = port_in_12;
    else if (address == 8'hFD) data_out = port_in_13;
    else if (address == 8'hFE) data_out = port_in_14;
    else if (address == 8'hFF) data_out = port_in_15;

  end
```

13.3.4 CPU Implementation

Let's now look at the central processing unit details. The CPU contains two components, the control unit (control_unit.v) and the data path (data_path.v). The data path contains all of the registers and the ALU. The ALU is implemented as a sub-system within the data path (alu.v). The data path also contains a bus system in order to facilitate data movement between the registers and memory. The bus system is implemented with two multiplexers that are controlled by the control unit. The control unit contains the finite-state machine that generates all control signals for the data path as it performs the fetch-decode-execute steps of each instruction. Example 13.11 shows the block diagram of the CPU in our 8-bit microcomputer example.

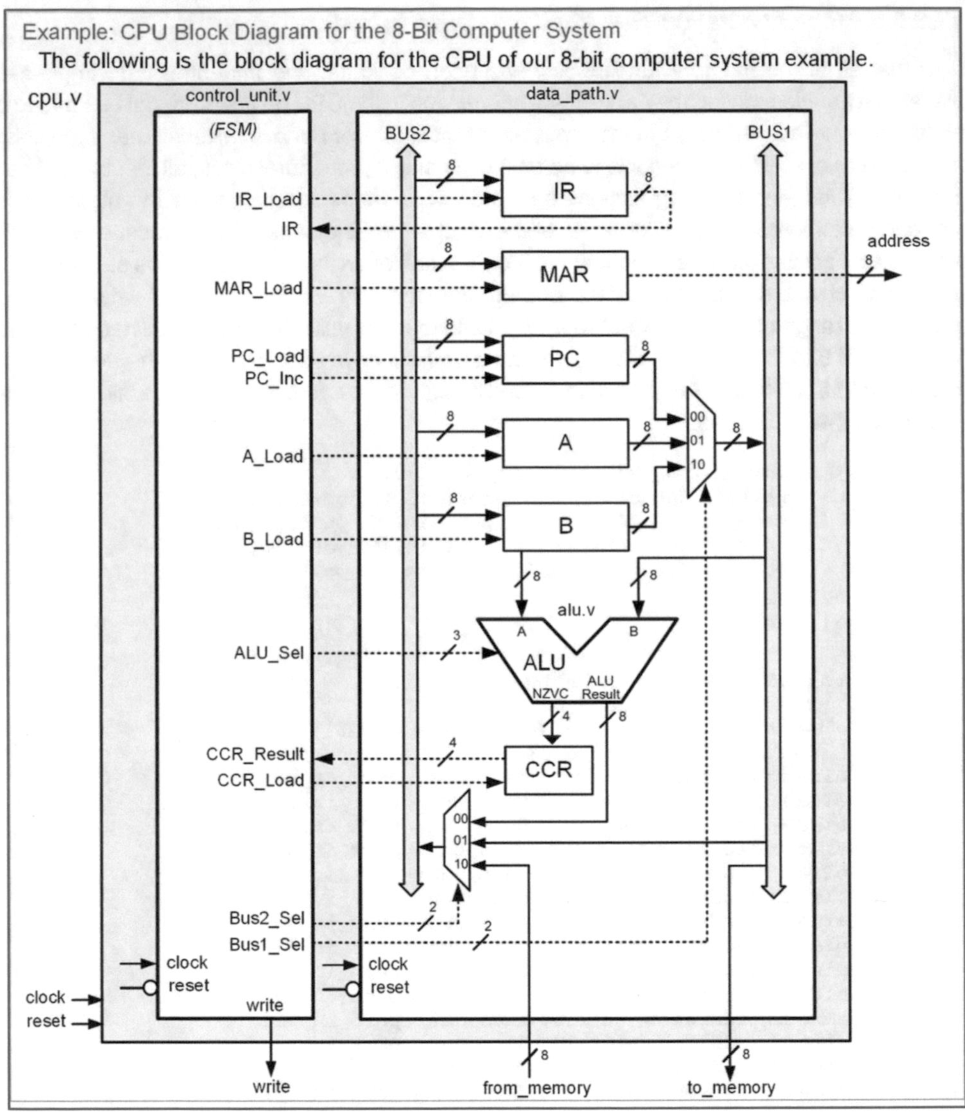

Example 13.11
CPU block diagram for the 8-bit computer system

13.3.4.1 Data Path Implementation in Verilog

Let's first look at the data path bus system that handles internal signal routing. The system consists of two 8-bit busses (Bus1 and Bus2) and two multiplexers. Bus1 is used as the destination of the PC, A, and B register outputs, while Bus2 is used as the input to the IR, MAR, PC, A, and B registers. Bus1 is connected directly to the *to_memory* port of the CPU to allow registers to write data to the memory system. Bus2 can be driven by the *from_memory* port of the CPU to allow the memory system to provide data for the CPU registers. The two multiplexers handle all signal routing and have their select lines (Bus1_Sel and Bus2_Sel) driven by the control unit. The following Verilog shows how the multiplexers are implemented. Again, a multiplexer is combinational logic, so all inputs must be listed in the sensitivity list of its procedural block, and blocking assignments are used. Two additional signal assignments are also required to connect the MAR to the address port and to connect Bus1 to the to_memory port.

```
always @ (Bus1_Sel, PC, A, B)
  begin: MUX_BUS1
    case (Bus1_Sel)
      2'b00    : Bus1 = PC;
      2'b01    : Bus1 = A;
      2'b10    : Bus1 = B;
      default  : Bus1 = 8'hXX;
    endcase
  end

always @ (Bus2_Sel, ALU_Result, Bus1, from_memory)
  begin: MUX_BUS2
    case (Bus2_Sel)
      2'b00    : Bus2 = ALU_Result;
      2'b01    : Bus2 = Bus1;
      2'b10    : Bus2 = from_memory;
    default    : Bus1 = 8'hXX;
  endcase
end

always @ (Bus1, MAR)
  begin
    to_memory = Bus1;
    address = MAR;
  end
```

Next, let's look at implementing the registers in the data path. Each register is implemented using a dedicated procedural block that is sensitive to clock and reset. This models the behavior of synchronous latches, or registers. Each register has a synchronous enable line that dictates when the register is updated. The register output is only updated when the enable line is asserted and a rising edge of the clock is detected. The following Verilog shows how to model the instruction register (IR). Notice that the signal IR is only updated if IR_Load is asserted and there is a rising edge of the clock. In this case, IR is loaded with the value that resides on Bus2.

```
always @ (posedge clock or negedge reset)
  begin: INSTRUCTION_REGISTER
    if (!reset)
      IR <= 8'h00;
    else
      if (IR_Load)
        IR <= Bus2;
    end
```

A nearly identical block is used to model the memory address register. A unique signal is declared called *MAR* in order to make the Verilog more readable. MAR is always assigned to address in this system.

```
always @ (posedge clock or negedge reset)
  begin: MEMORY_ADDRESS_REGISTER
    if (!reset)
      MAR <= 8'h00;
    else
      if (MAR_Load)
          MAR <= Bus2;
    end
```

Now let's look at the program counter block. This register contains additional functionality beyond simply latching in the value of Bus2. The program counter also has an increment feature that will take place synchronously when the signal PC_Inc coming from the control unit is asserted. This is handled using an additional nested if-else clause under the portion of the block handling the rising edge of the clock condition.

```
always @ (posedge clock or negedge reset)
  begin: PROGRAM_COUNTER
    if (!reset)
      PC <= 8'h00;
    else
      if (PC_Load)
        PC <= Bus2;
      else if (PC_Inc)
        PC <= MAR + 1;
    end
```

The two general-purpose registers A and B are modeled using individual procedural blocks as follows:

```
always @ (posedge clock or negedge reset)
  begin: A_REGISTER
    if (!reset)
      A <= 8'h00;
    else
      if (A_Load)
        A <= Bus2;
    end

always @ (posedge clock or negedge reset)
  begin: B_REGISTER
    if (!reset)
      B <= 8'h00;
    else
      if (B_Load)
        B <= Bus2;
  end
```

The condition code register latches in the status flags from the ALU (NZVC) when the CCR_Load line is asserted. This behavior is modeled using a similar approach as follows:

```
always @ (posedge clock or negedge reset)
  begin: CONDITION_CODE_REGISTER
    if (!reset)
      CCR_Result <= 8'h00;
    else
      if (CCR_Load)
        CCR_Result <= NZVC;
  end
```

13.3.4.2 ALU Implementation in Verilog

The ALU is a set of combinational logic circuitry that performs arithmetic and logic operations. The output of the ALU operation is called *Result*. The ALU also outputs four status flags as a 4-bit bus called *NZVC.* The ALU behavior can be modeled using case and if-else statements that decide which operation to perform based on the input control signal *ALU_Sel*. The following Verilog shows an example of how to implement the ALU addition functionality. A case statement is used to decide which operation is being performed based on the ALU_Sel input. Under each operation clause, a series of procedural statements are used to compute the result and update the NZVC flags. Each of these flags is updated individually. The N flag can be simply driven with position 7 of the ALU result since this bit is the sign bit for signed numbers. The Z flag can be driven using an if-else condition that checks whether the result was x"00". The V flag is updated based on the type of the operation. For the addition operation, the V flag will be asserted if a POS + POS=NEG or a NEG + NEG = POS. These conditions can be checked by looking at the sign bits of the inputs and the sign bit of the result. Finally, the C flag can be computed as the 8th bit in the addition of A + B.

```
always @ (A, B, ALU_Sel)
  begin
    case (ALU_Sel)
       3'b000  : begin //-- Addition

                  //-- Sum and Carry Flag
                    {NZVC[0], Result} = A + B;

                  //-- Negative Flag
                    NZVC[3] = Result[7];

                  //-- Zero Flag
                    if (Result == 0)
                      NZVC[2] = 1;
                    else
                      NZVC[2] = 0;

                  //-- Two's Comp Overflow Flag
                    if (((A[7]==0) && (B[7]==0) && (Result[7] == 1)) ||
                        ((A[7]==1) && (B[7]==1) && (Result[7] == 0)))
                      NZVC[1] = 1;
                    else
                      NZVC[1] = 0;

                  end

                          :
          //-- other ALU operations go here...
                          :

          default : begin
                      Result  = 8'hXX;
                      NZVC    = 4'hX;
                    end
    endcase

  end
```

13.3.4.3 Control Unit Implementation in Verilog

Let's now look at how to implement the control unit state machine. We'll first look at the formation of the Verilog to model the FSM and then turn to the detailed state transitions in order to accomplish a variety of the most common instructions. The control unit sends signals to the data path in order to move data in and out of registers and into the ALU to perform data manipulations. The finite-state machine is implemented with the behavioral modeling techniques presented in Chap. 9. The model contains three processes in order to implement the state memory, next state logic, and output logic of the FSM. Parameters are created for each of the states defined in the state diagram of the FSM. The states associated with fetching (S_FETCH_0, S_FETCH_1, S_FETCH_2) and decoding the opcode (S_DECODE_3) are performed each time an instruction is executed. A unique path is then added after the decode state to perform the steps associated with executing each individual instruction. The FSM can be created one instruction at a time by adding additional state paths after the decode state. The following Verilog code shows how the user-defined state names are created for nine basic instructions (LDA_IMM, LDA_DIR, STA_DIR, LDB_IMM, LDB_DIR, STB_DIR, ADD_AB, BRA, and BEQ). Eight-bit state variables are created for current_state and next_state to accommodate future state codes. The state codes are assigned in binary using integer format to allow additional states to be easily added.

```
reg       [7:0] current_state, next_state;
parameter S_FETCH_0   = 0,  //-- Opcode fetch states
          S_FETCH_1   = 1,
          S_FETCH_2   = 2,

          S_DECODE_3  = 3,  //-- Opcode decode state

          S_LDA_IMM_4 = 4,  //-- Load A (Immediate) states
          S_LDA_IMM_5 = 5,
          S_LDA_IMM_6 = 6,

          S_LDA_DIR_4 = 7,  //-- Load A (Direct) states
          S_LDA_DIR_5 = 8,
          S_LDA_DIR_6 = 9,
          S_LDA_DIR_7 = 10,
          S_LDA_DIR_8 = 11,

          S_STA_DIR_4 = 12, //-- Store A (Direct) States
          S_STA_DIR_5 = 13,
          S_STA_DIR_6 = 14,
          S_STA_DIR_7 = 15,

          S_LDB_IMM_4 = 16, //-- Load B (Immediate) states
          S_LDB_IMM_5 = 17,
          S_LDB_IMM_6 = 18,

          S_LDB_DIR_4 = 19, //-- Load B (Direct) states
          S_LDB_DIR_5 = 20,
          S_LDB_DIR_6 = 21,
          S_LDB_DIR_7 = 22,
          S_LDB_DIR_8 = 23,

          S_STB_DIR_4 = 24, //-- Store B (Direct) States
          S_STB_DIR_5 = 25,
          S_STB_DIR_6 = 26,
          S_STB_DIR_7 = 27,

          S_BRA_4     = 28, //-- Branch Always States
          S_BRA_5     = 29,
          S_BRA_6     = 30,

          S_BEQ_4     = 31, //-- Branch if Equal States
          S_BEQ_5     = 32,
          S_BEQ_6     = 33,
          S_BEQ_7     = 34,

          S_ADD_AB_4  = 35; //-- Addition States
```

Within the control unit module, the state memory is implemented as a separate procedural block that will update the current state with the next state on each rising edge of the clock. The reset state will be the first fetch state in the FSM (i.e., S_FETCH_0). The following Verilog shows how the state memory in the control unit can be modeled. Note that this block models sequential logic so non-blocking assignments are used.

```
always @ (posedge clock or negedge reset)
  begin: STATE_MEMORY
    if (!reset)
      current_state <= S_FETCH_0;
    else
      current_state <= next_state;
  end
```

The next state logic is also implemented as a separate procedural block. The next state logic depends on the current state, the instruction register (IR), and the condition code register (CCR_Result). The following Verilog gives a portion of the next state logic process showing how the state transitions can be modeled:

```
always @ (current_state, IR, CCR_Result)
  begin: NEXT_STATE_LOGIC
    case (current_state)
      S_FETCH_0  : next_state = S_FETCH_1;  //-- Path for FETCH instruction
      S_FETCH_1  : next_state = S_FETCH_2;
      S_FETCH_2  : next_state = S_DECODE_3;

      S_DECODE_3 : if      (IR == LDA_IMM) next_state = S_LDA_IMM_4;  //-- Register A
                   else if (IR == LDA_DIR) next_state = S_LDA_DIR_4;
                   else if (IR == STA_DIR  next_state = S_STA_DIR_4;
                   else if (IR == LDB_IMM) next_state = S_LDB_IMM_4;  //-- Register B
                   else if (IR == LDB_DIR) next_state = S_LDB_DIR_4;
                   else if (IR == STB_DIR) next_state = S_STB_DIR_4;
                   else if (IR == BRA)     next_state = S_BRA_4;      //-- Branch Always
                   else if (IR == ADD_AB)  next_state = S_ADD_AB_4;   //-- ADD
                   else                    next_state = S_FETCH_0;    //-- others go here

      S_LDA_IMM_4 : next_state = S_LDA_IMM_5;    //-- Path for LDA_IMM instruction
      S_LDA_IMM_5 : next_state = S_LDA_IMM_6;
      S_LDA_IMM_6 : next_state = S_FETCH_0;

                              :
         Next state logic for other states goes here...
                              :

    endcase
  end
```

Finally, the output logic is modeled as a third, separate procedural block. It is useful to explicitly state the outputs of the control unit for each state in the machine to allow easy debugging and avoid synthesizing latches. Our example computer system has Moore-type outputs, so the process only depends on the current state. The following Verilog shows a portion of the output logic process:

```
always @ (current_state)
  begin: OUTPUT_LOGIC
    case (current_state)

      S_FETCH_0  : begin          //-- Put PC onto MAR to provide address of Opcode
                     IR_Load  = 0;
                     MAR_Load = 1;
                     PC_Load  = 0;
                     PC_Inc   = 0;
                     A_Load   = 0;
                     B_Load   = 0;
                     ALU_Sel  = 3'b000;
                     CCR_Load = 0;
                     Bus1_Sel = 2'b00; //-- "00"=PC,  "01"=A,    "10"=B
                     Bus2_Sel = 2'b01; //-- "00"=ALU, "01"=Bus1, "10"=from_memory
                     write    = 0;
                   end
```

```
          S_FETCH_1  : begin     //-- Increment PC, Opcode will be available next state
                           IR_Load  = 0;
                           MAR_Load= 0;
                           PC_Load  = 0;
                           PC_Inc   = 1;
                           A_Load   = 0;
                           B_Load   = 0;
                           ALU_Sel  = 3'b000;
                           CCR_Load= 0;
                           Bus1_Sel= 2'b00; //-- "00"=PC,  "01"=A,    "10"=B
                           Bus2_Sel= 2'b00; //-- "00"=ALU, "01"=Bus1, "10"=from_memory
                           write    = 0;
                         end;

                                      :
              Output logic for other states goes here...
                                      :

          endcase
       end
```

13.3.4.3.1 Detailed Execution of LDA_IMM

Now let's look at the details of the state transitions and output signals in the control unit FSM when executing a few of the most common instructions. Let's begin with the instruction to load register A using immediate addressing (LDA_IMM). Example 13.12 shows the state diagram for this instruction. The first three states (S_FETCH_0, S_FETCH_1, S_FETCH_2) handle fetching the opcode. The purpose of these states is to read the opcode from the address being held by the program counter and put it into the instruction register. Multiple states are needed to handle putting PC into MAR to provide the address of the opcode, waiting for the memory system to provide the opcode, latching the opcode into IR, and incrementing PC to the next location in program memory. Another state is used to decode the opcode (S_DECODE_3) in order to decide which path to take in the state diagram based on the instruction being executed. After the decode state, a series of three more states are needed (S_LDA_IMM_4, S_LDA_IMM_5, S_LDA_IMM_6) to execute the instruction. The purpose of these states is to read the operand from the address being held by the program counter and put it into A. Multiple states are needed to handle putting PC into MAR to provide the address of the operand, waiting for the memory system to provide the operand, latching the operand into A, and incrementing PC to the next location in program memory. When the instruction completes, the value of the operand resides in A, and PC is pointing to the next location in program memory, which is the opcode of the next instruction to be executed.

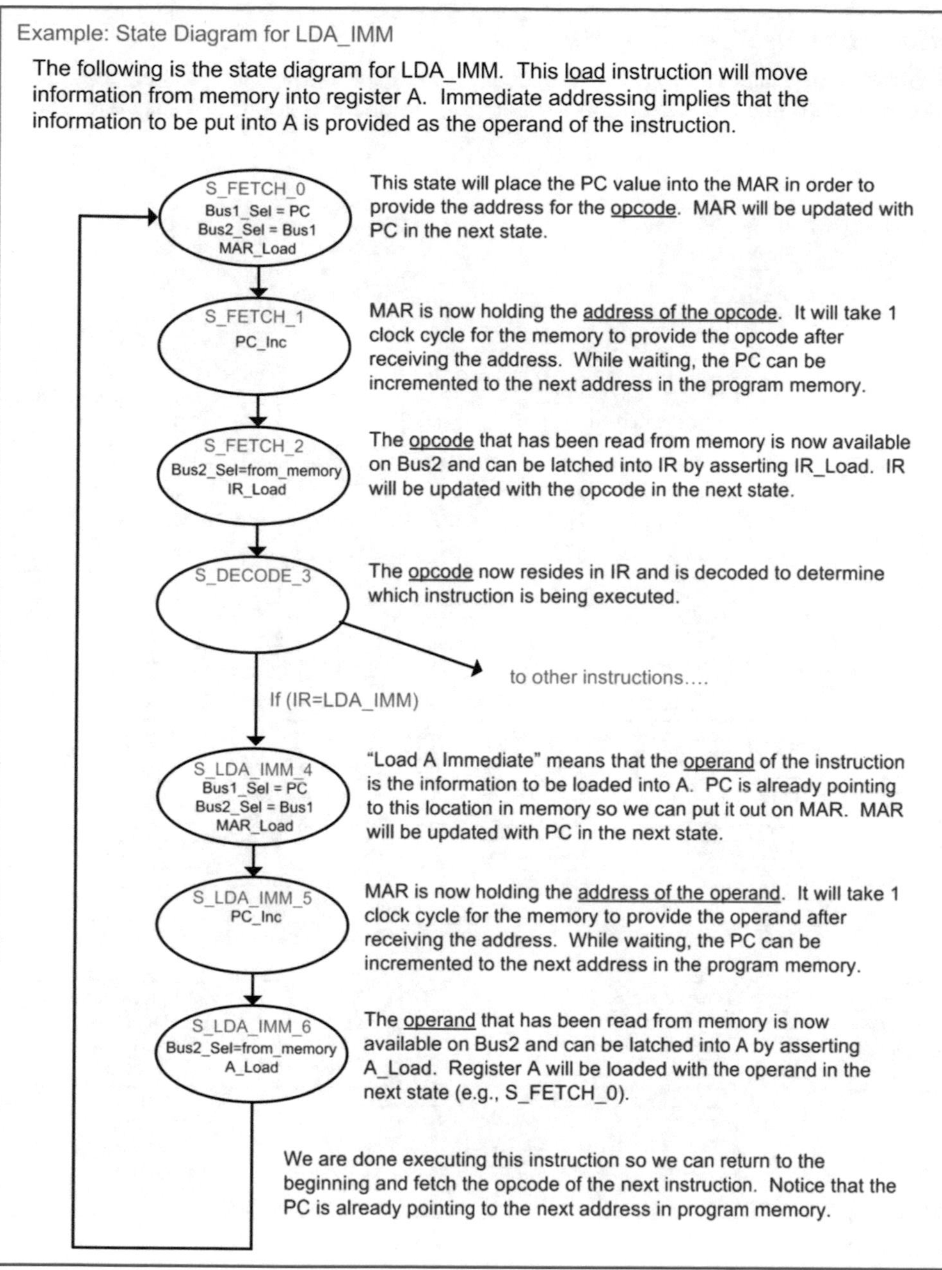

Example 13.12
State diagram for LDA_IMM

Example 13.13 shows the simulation waveform for executing LDA_IMM. In this example, register A is loaded with the operand of the instruction, which holds the value x"AA".

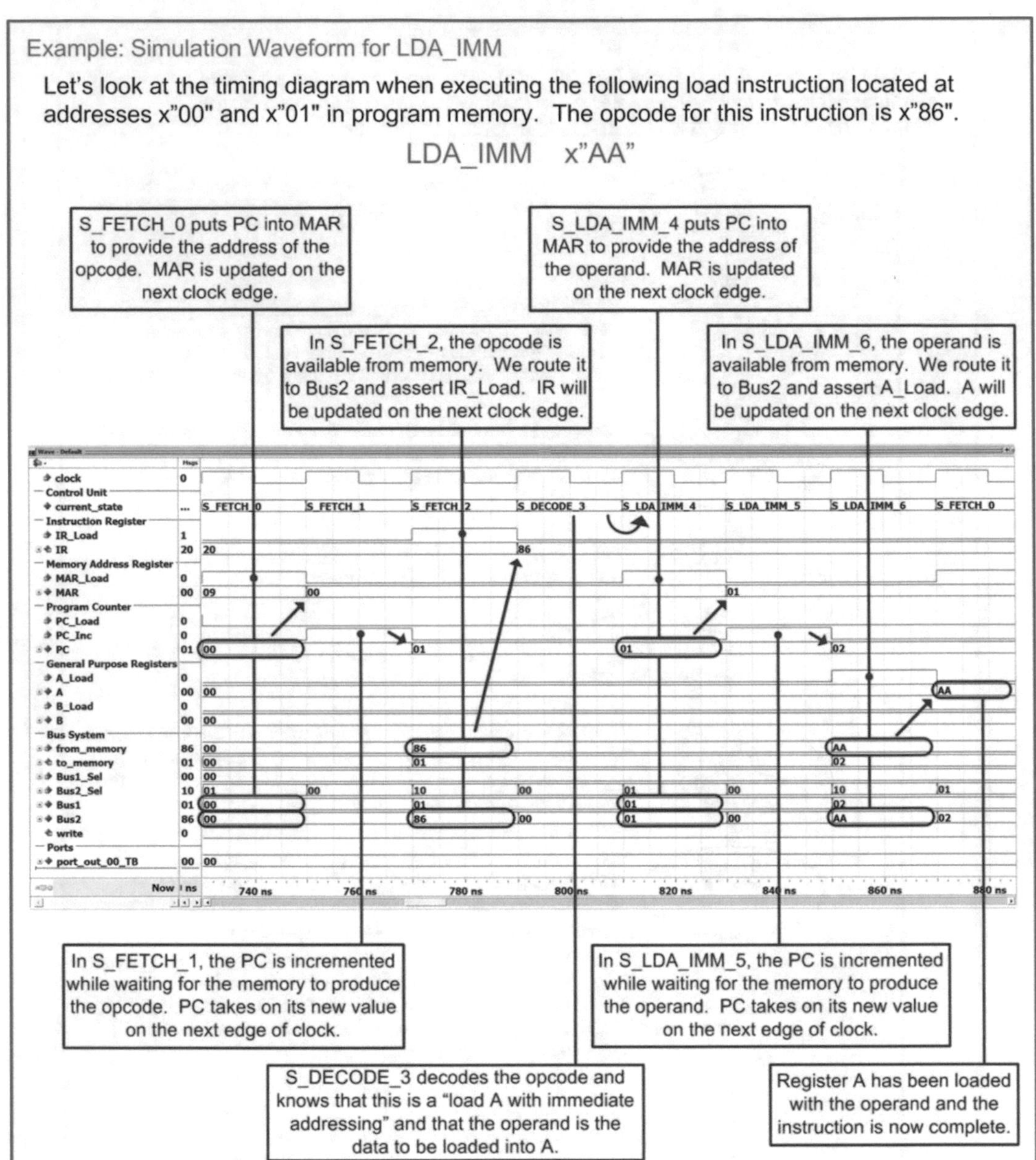

Example 13.13
Simulation waveform for LDA_IMM

13.3.4.3.2 Detailed Execution of LDA_DIR

Now let's look at the details of the instruction to load register A using direct addressing (LDA_DIR). Example 13.14 shows the state diagram for this instruction. The first four states to fetch and decode the opcode are the same states as in the previous instruction and are performed each time a new instruction is executed. Once the opcode is decoded, the state machine traverses five new states to execute the instruction (S_LDA_DIR_4, S_LDA_DIR_5, S_LDA_DIR_6, S_LDA_DIR_7, S_LDA_DIR_8). The purpose of these states is to read the operand and then use it as the address of where to read the contents to put into A.

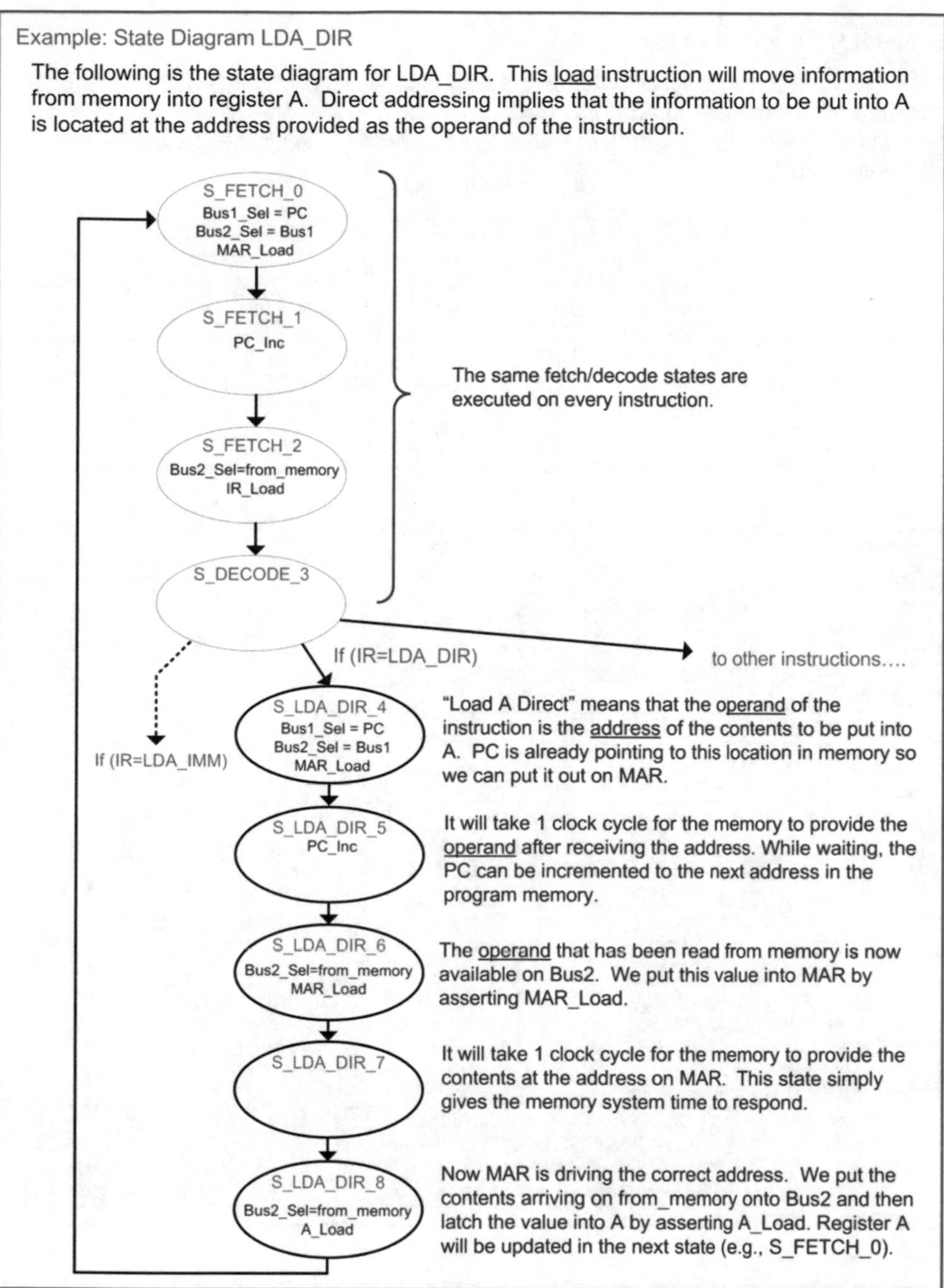

Example 13.14
State diagram for LDA_DIR

Example 13.15 shows the simulation waveform for executing LDA_DIR. In this example, register A is loaded with the contents located at address x"80", which has already been initialized to x"AA".

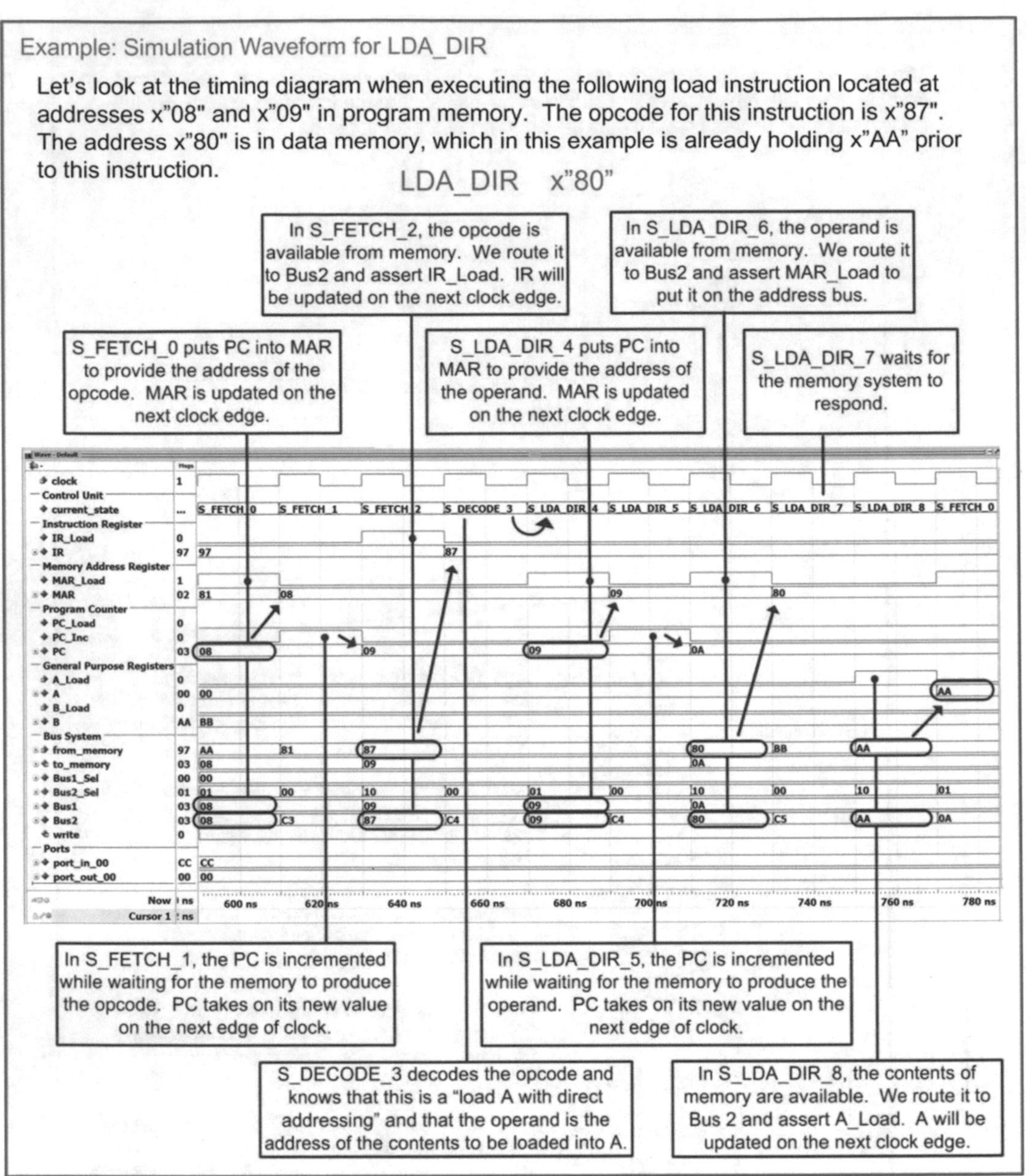

Example 13.15
Simulation waveform for LDA_DIR

13.3.4.3.3 Detailed Execution of STA_DIR

Now let's look at the details of the instruction to store register A to memory using direct addressing (STA_DIR). Example 13.16 shows the state diagram for this instruction. The first four states are again the same as prior instructions in order to fetch and decode the opcode. Once the opcode is decoded, the state machine traverses four new states to execute the instruction (S_STA_DIR_4, S_STA_DIR_5, S_STA_DIR_6, S_STA_DIR_7). The purpose of these states is to read the operand and then use it as the address of where to write the contents of A to.

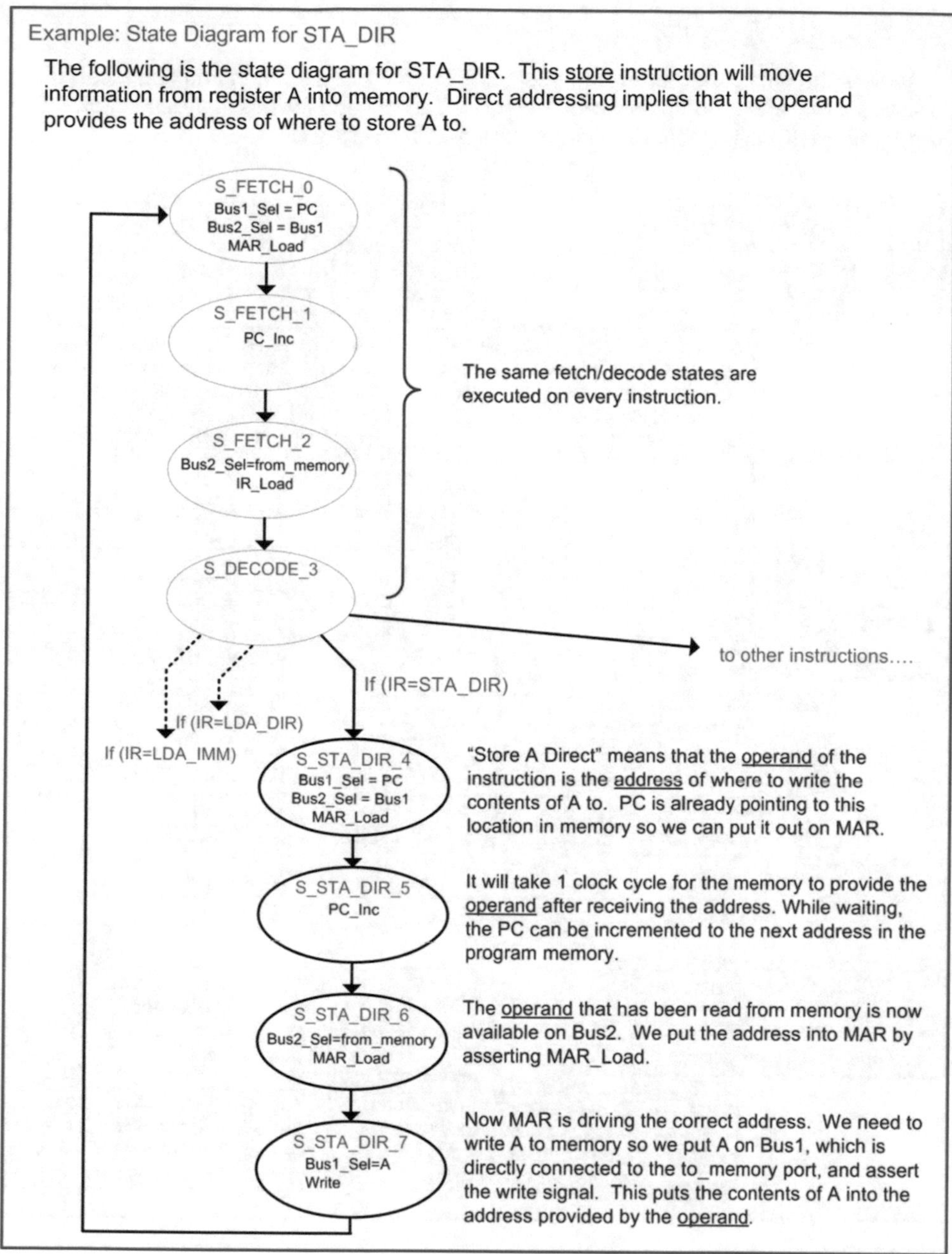

Example 13.16
State diagram for STA_DIR

Example 13.17 shows the simulation waveform for executing STA_DIR. In this example, register A already contains the value x"CC" and will be stored to address x"E0". The address x"E0" is an output port (port_out_00) in our example computer system.

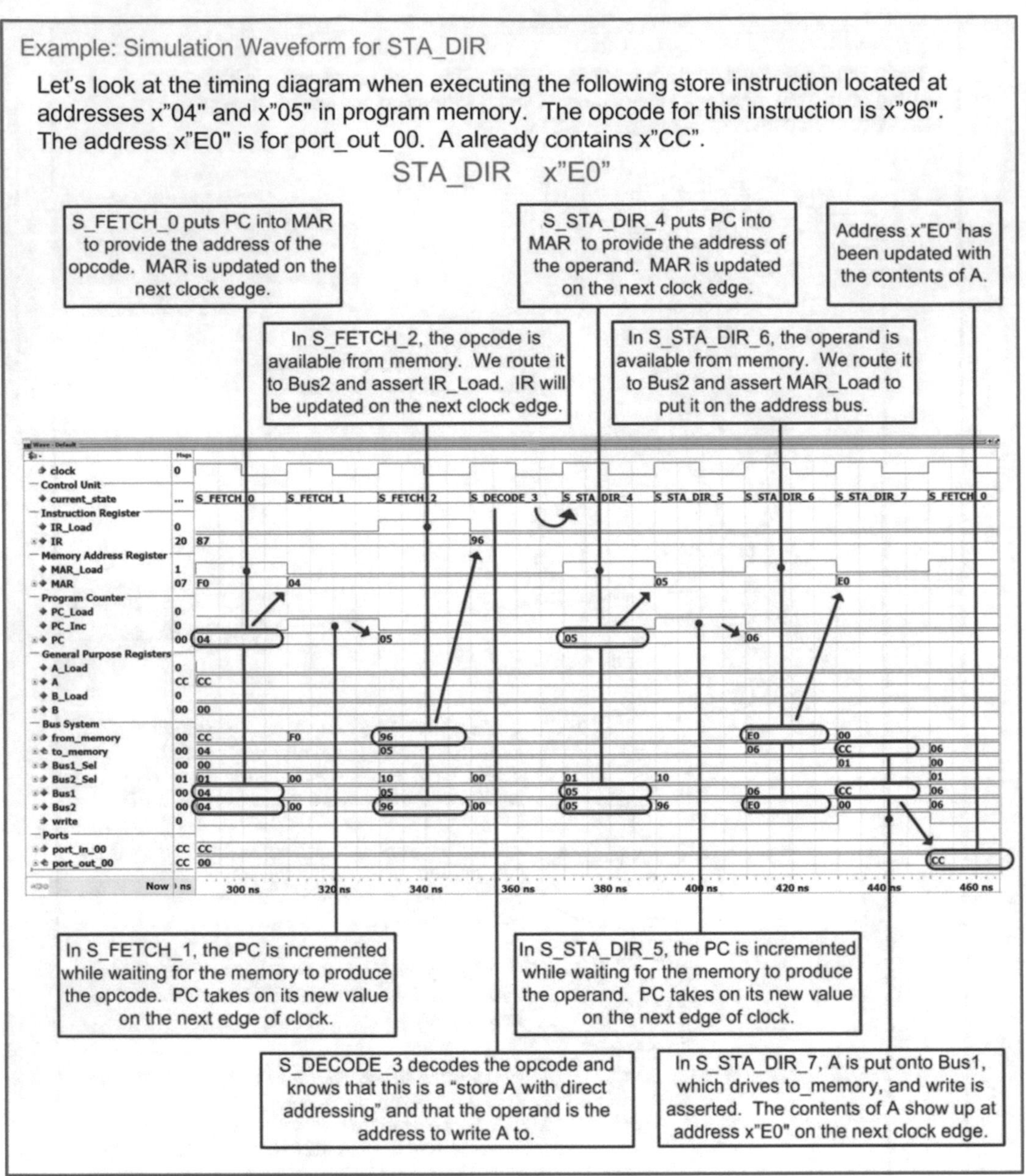

Example 13.17
Simulation waveform for STA_DIR

13.3.4.3.4 Detailed Execution of ADD_AB

Now let's look at the details of the instruction to add A to B and store the sum back in A (ADD_AB). Example 13.18 shows the state diagram for this instruction. The first four states are again the same as prior instructions in order to fetch and decode the opcode. Once the opcode is decoded, the state machine only requires one more state to complete the operation (S_ADD_AB_4). The ALU is combinational logic so it will begin to compute the sum immediately as soon as the inputs are updated. The inputs to the ALU are Bus1 and register B. Since B is directly connected to the ALU, all that is required to start the addition is to put A onto Bus1. The output of the ALU is put on Bus2 so that it can be latched into A on the next clock edge. The ALU also outputs the status flags NZVC, which are directly connected to the condition code register. A_Load and CCR_Load are asserted in this state. A and CCR_Result will be updated in the next state (i.e., S_FETCH_0).

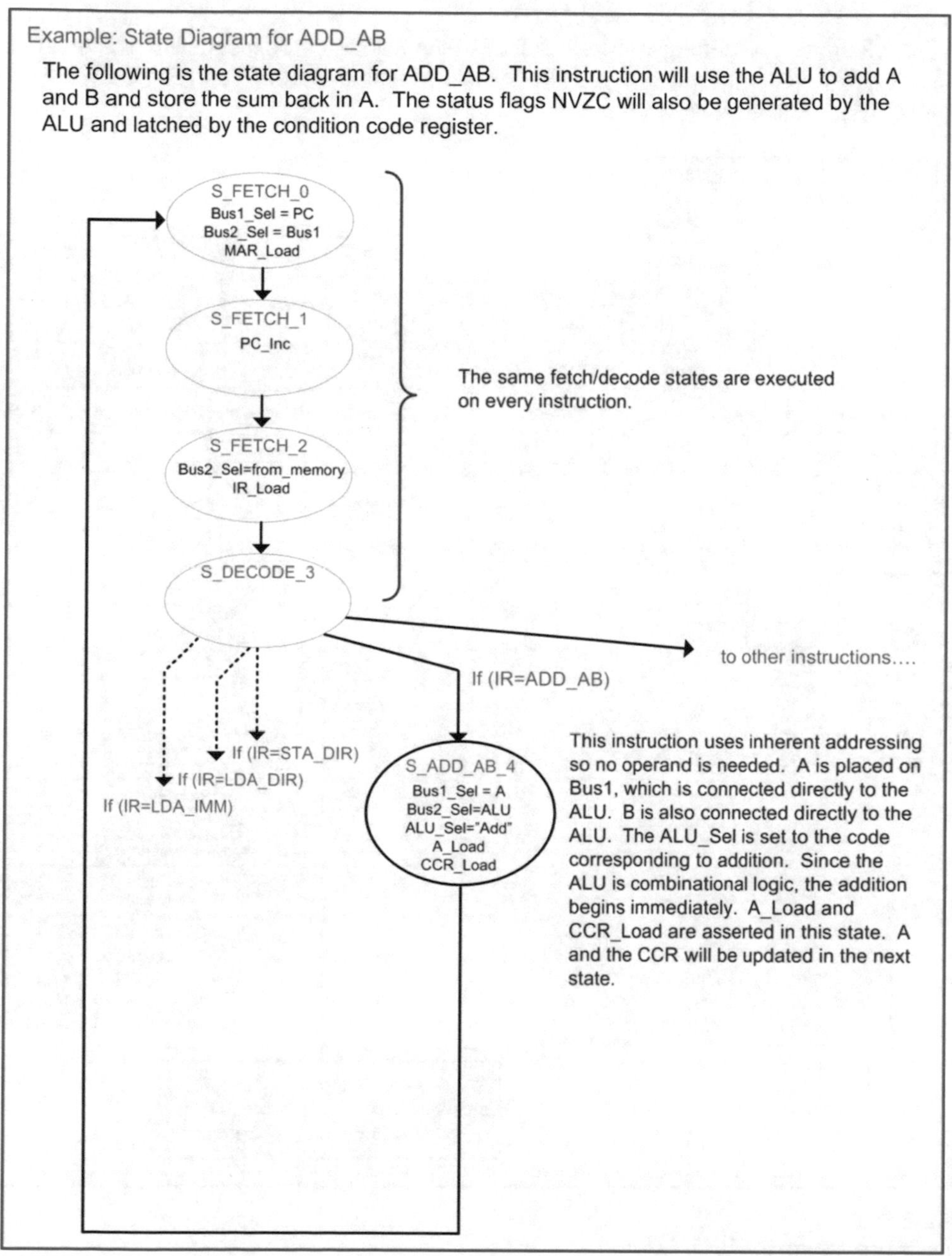

Example 13.18
State diagram for ADD_AB

Example 13.19 shows the simulation waveform for executing ADD_AB. In this example, two load immediate instructions were used to initialize the general-purpose registers to A = x"FF" and B = x"01" prior to the addition. The addition of these values will result in a sum of x"00" and assert the carry (C) and zero (Z) flags in the condition code register.

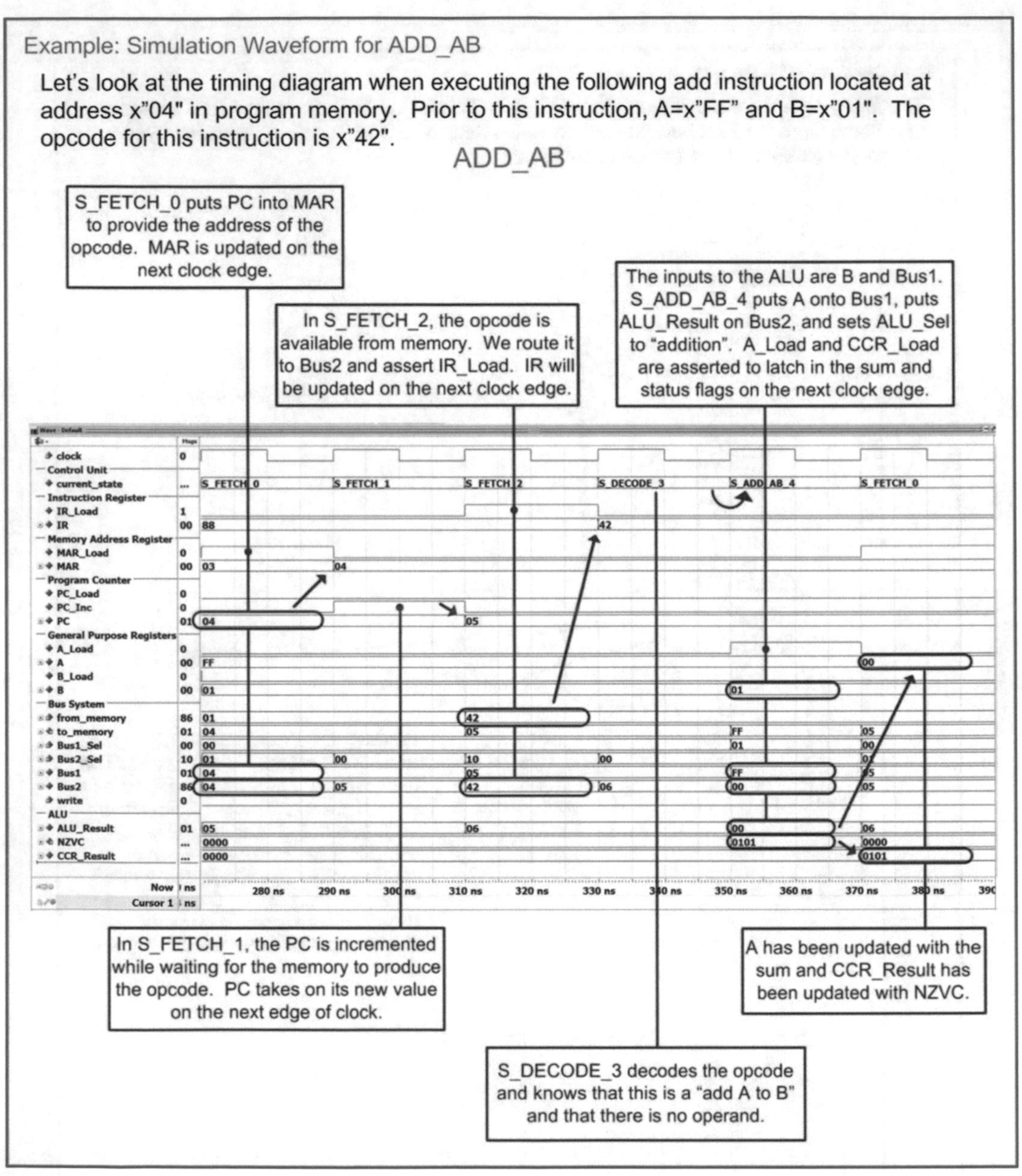

Example 13.19
Simulation waveform for ADD_AB

13.3.4.3.5 Detailed Execution of BRA

Now let's look at the details of the instruction to branch always (BRA). Example 13.20 shows the state diagram for this instruction. The first four states are again the same as prior instructions in order to fetch and decode the opcode. Once the opcode is decoded, the state machine traverses four new states to execute the instruction (S_BRA_4, S_BRA_5, S_BRA_6). The purpose of these states is to read the operand and put its value into PC to set the new location in program memory to execute instructions.

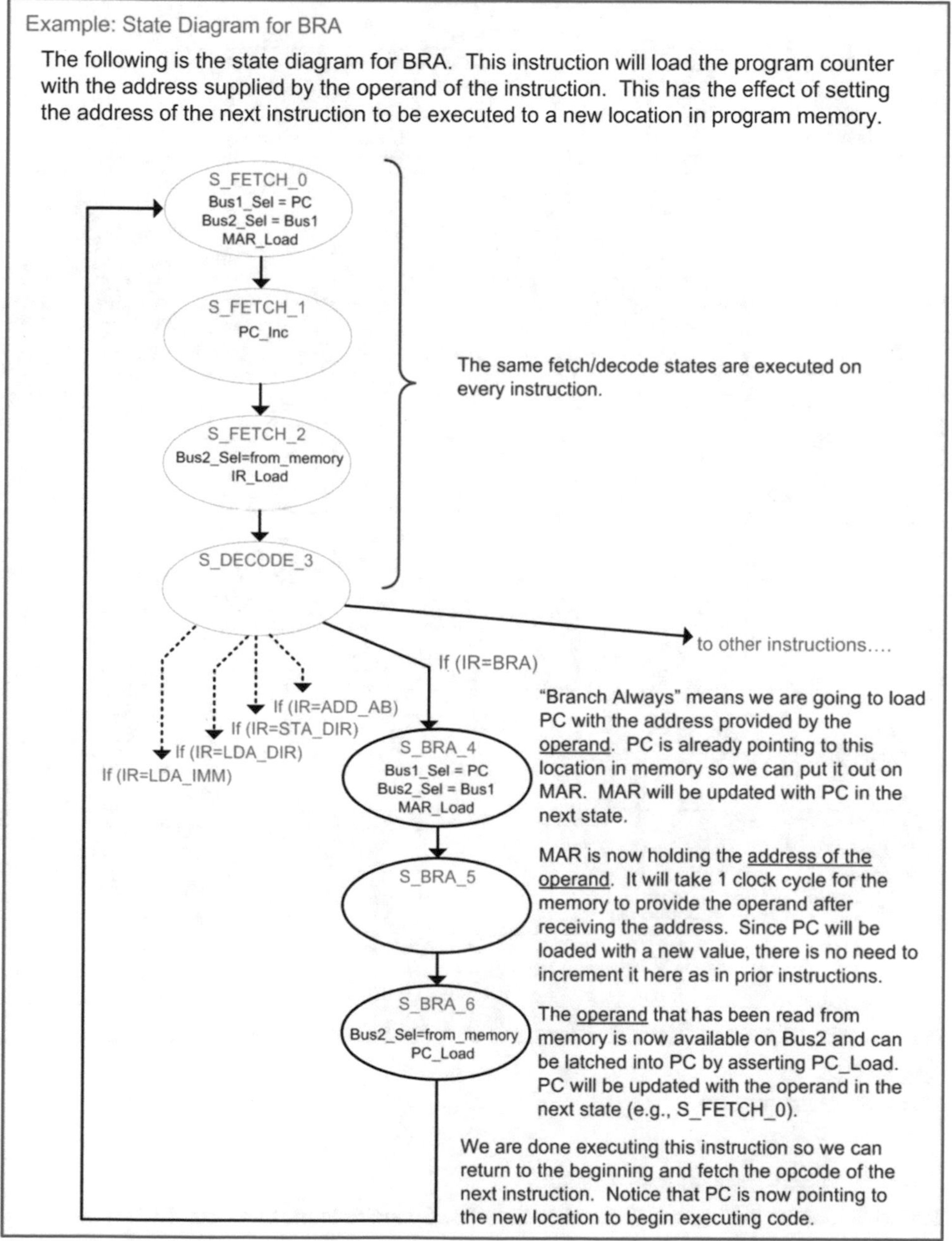

Example 13.20
State diagram for BRA

Example 13.21 shows the simulation waveform for executing BRA. In this example, PC is set back to address x"00".

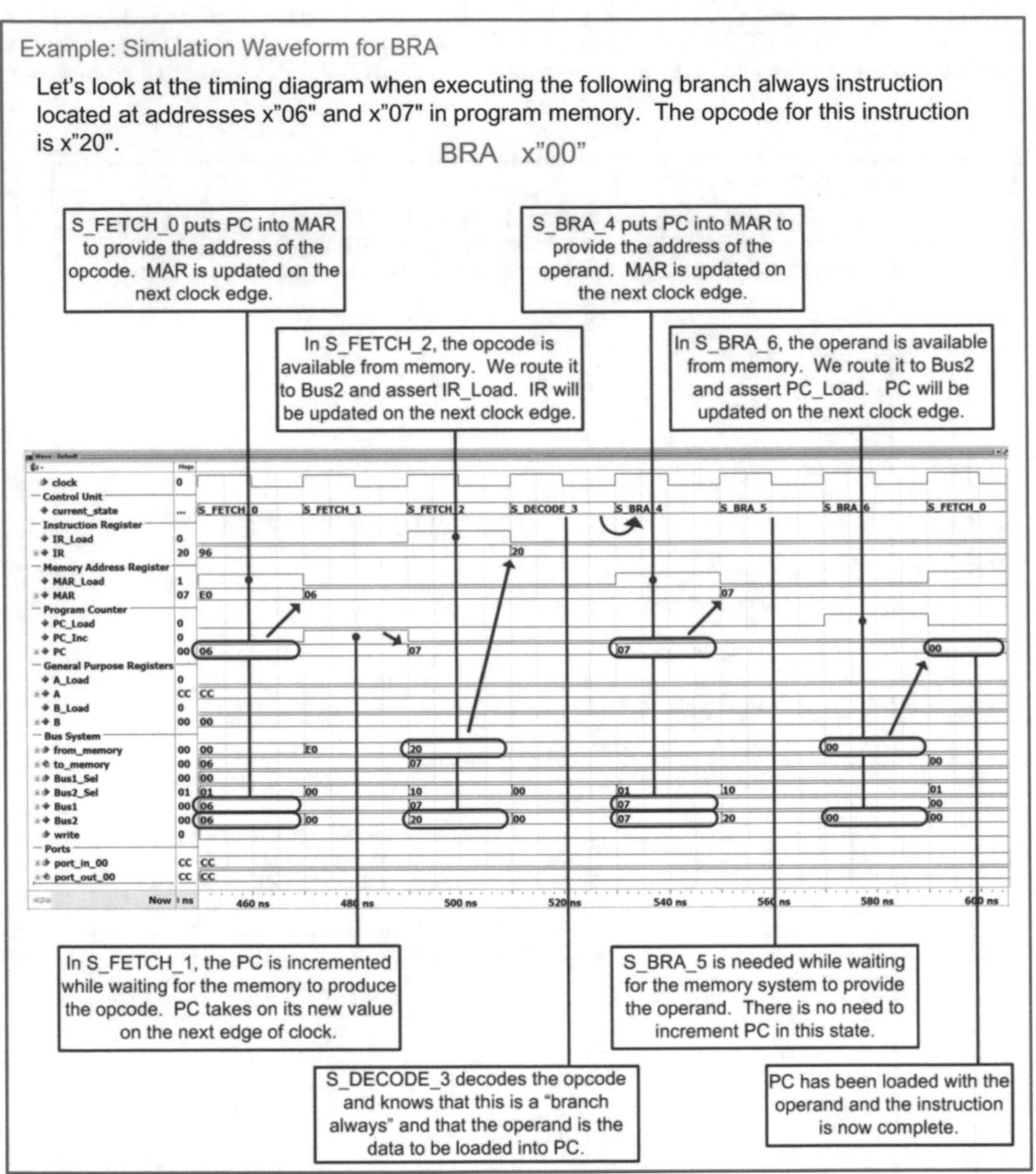

Example 13.21
Simulation waveform for BRA

13.3.4.3.6 Detailed Execution of BEQ

Now let's look at the branch if equal to zero (BEQ) instruction. Example 13.22 shows the state diagram for this instruction. Notice that in this conditional branch, the path that is taken through the FSM depends on both IR and CCR. In the case that Z = 1, the branch is taken, meaning that the operand is loaded into PC. In the case that Z = 0, the branch is not taken, meaning that PC is simply incremented to bypass the operand and point to the beginning of the next instruction in program memory.

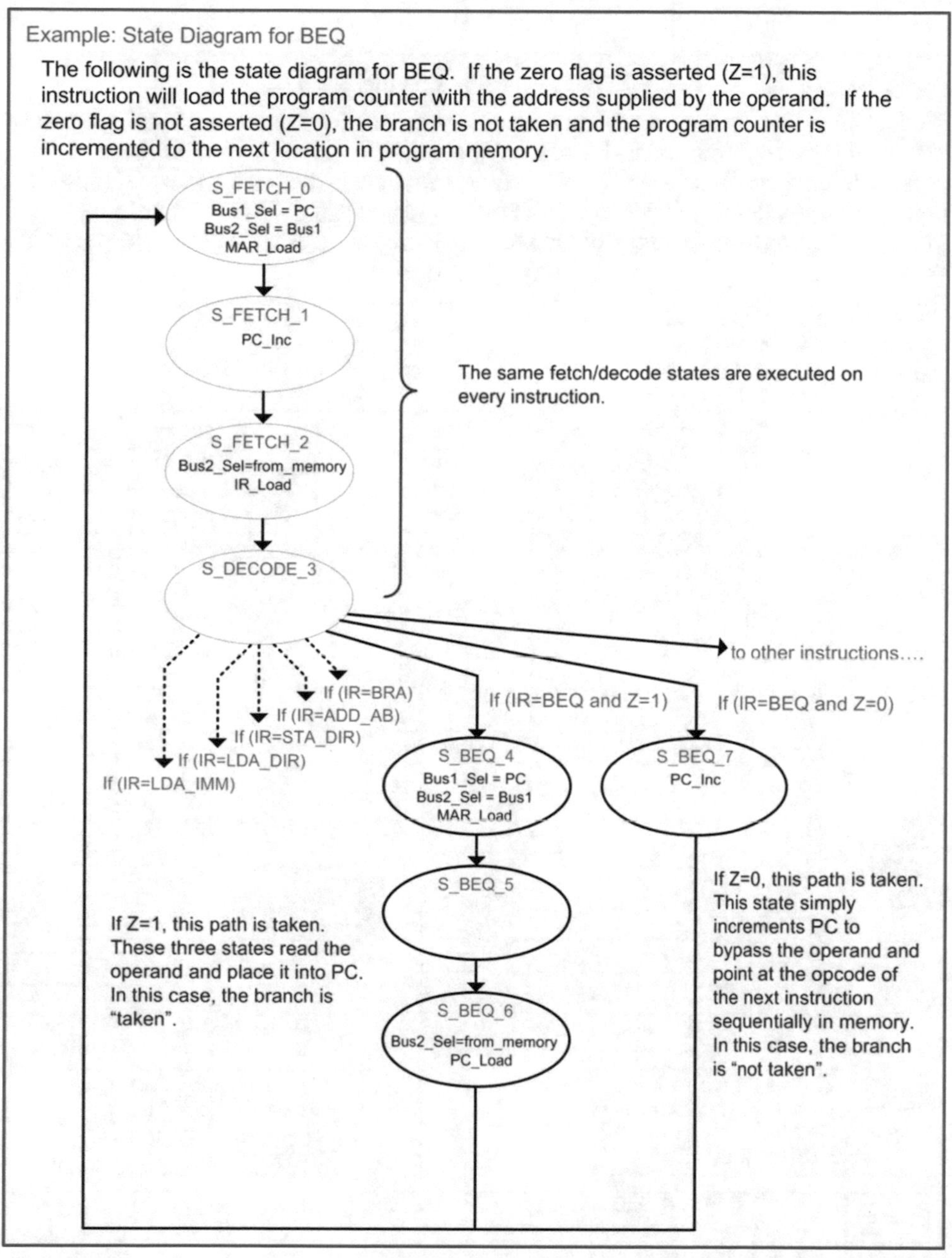

Example 13.22
State diagram for BEQ

Example 13.23 shows the simulation waveform for executing BEQ when the branch *is taken*. Prior to this instruction, an addition was performed on x"FF" and x"01". This resulted in a sum of x"00", which asserted the Z and C flags in the condition code register. Since Z = 1 when BEQ is executed, the branch is taken.

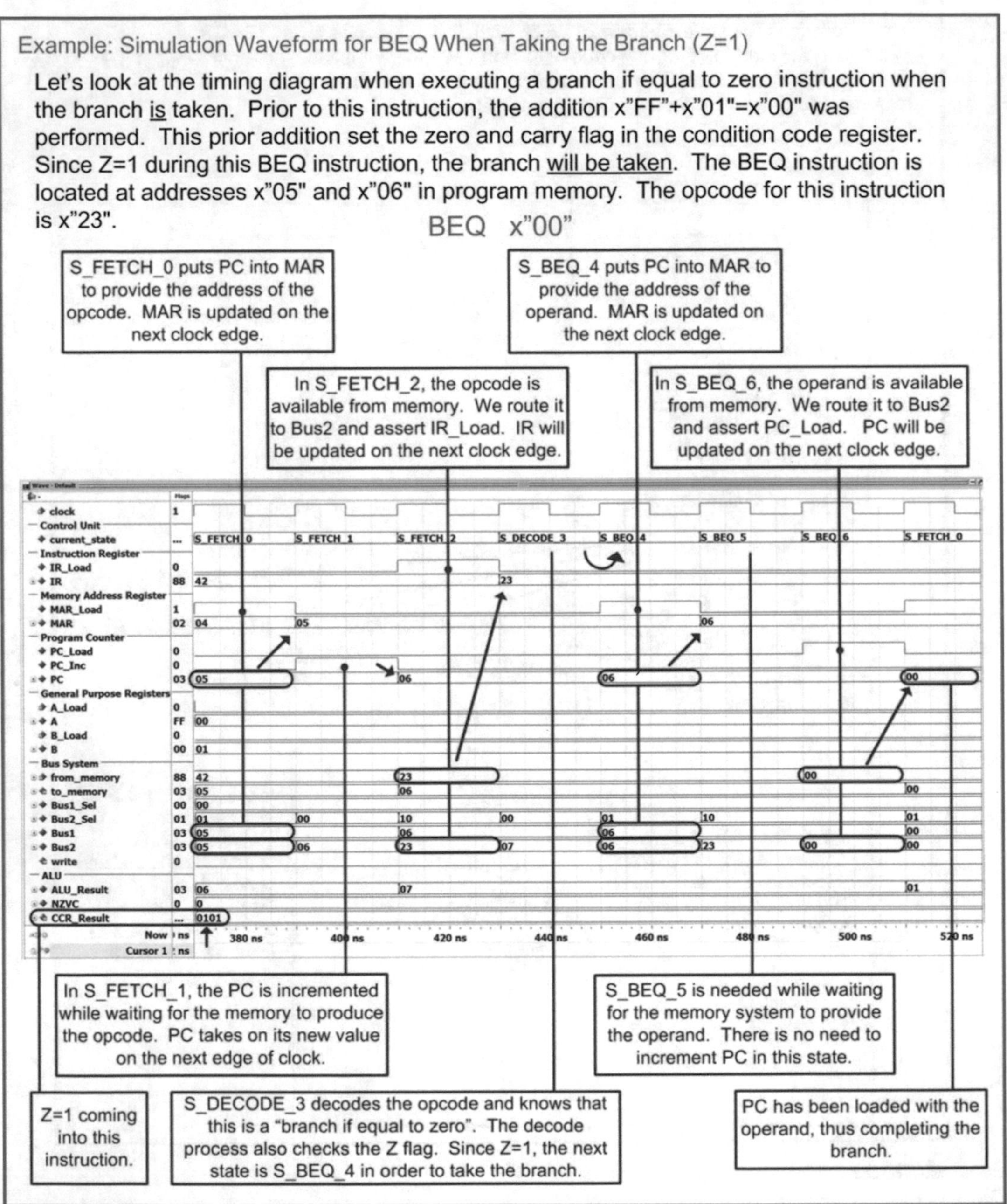

Example 13.23
Simulation waveform for BEQ when taking the branch (Z = 1)

Example 13.24 shows the simulation waveform for executing BEQ when the branch *is not taken*. Prior to this instruction, an addition was performed on x"FE" and x"01". This resulted in a sum of x"FF", which did not assert the Z flag. Since Z = 0 when BEQ is executed, the branch is not taken. When not taking the branch, PC must be incremented again in order to bypass the operand and point to the next location in program memory.

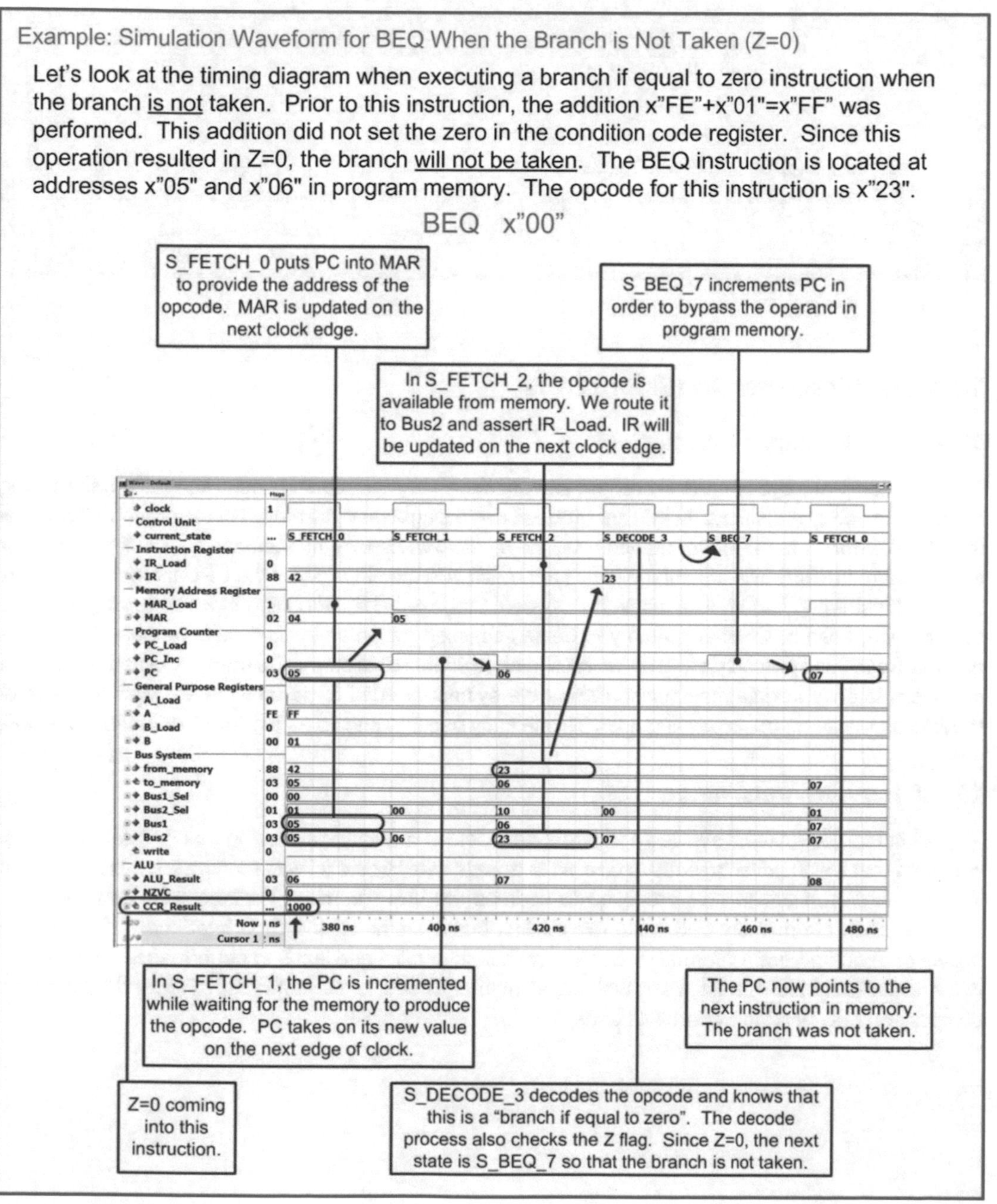

Example 13.24
Simulation waveform for BEQ when the branch is not taken (Z = 0)

Concept Check

CC13.3 The 8-bit microcomputer example presented in this section is a very simple architecture used to illustrate the basic concepts of a computer. If we wanted to keep this computer an 8-bit system but increase the depth of the memory, it would require adding more address lines to the address bus. What changes to the computer system would need to be made to accommodate the wider address bus?

A) The width of the program counter would need to be increased to support the wider address bus.

B) The size of the memory address register would need to be increased to support the wider address bus.

C) Instructions that use direct addressing would need additional bytes of operand to pass the wider address into the CPU 8 bits at a time.

D) All of the above.

13.4 Architecture Considerations

13.4.1 Von Neumann Architecture

The computer system just presented represents a very simple architecture in which all memory devices (i.e., program, data, and I/O) are grouped into a single memory map. This approach is known as the *Von Neumann architecture*, named after the nineteenth-century mathematician that first described this structure in 1945. The advantage of this approach is in the simplicity of the CPU interface. The CPU can be constructed based on a single bus system that executes everything in a linear progression of states, regardless of whether memory is being accessed for an instruction or a variable. One of the drawbacks of this approach is that an instruction and variable data cannot be read at the same time. This creates a latency in data manipulation since the system needed to be constantly switching between reading instructions and accessing data. This latency became known as the *Von Neumann bottleneck*.

13.4.2 Harvard Architecture

As computer systems evolved and larger data sets in memory were being manipulated, it became apparent that it was advantageous to be able to access data in parallel with reading the next instruction. The *Harvard* architecture was proposed to address the Von Neumann bottleneck by separating the program and data memory and using two distinct bus systems for the CPU interface. This approach allows data and program information to be accessed in parallel and leads to performance improvement when large numbers of data manipulations in memory need to be performed. Figure 13.5 shows a comparison between the two architectures.

Von Neumann vs. Harvard Architecture

A Von Neumann architecture maps both program and data memory into a single memory system. A single bus system is used to interface the CPU to all memory. This creates a simple CPU interface but leads to latency due to everything being accessed in a serial manner. This latency is known as the "Von Neumann bottleneck".

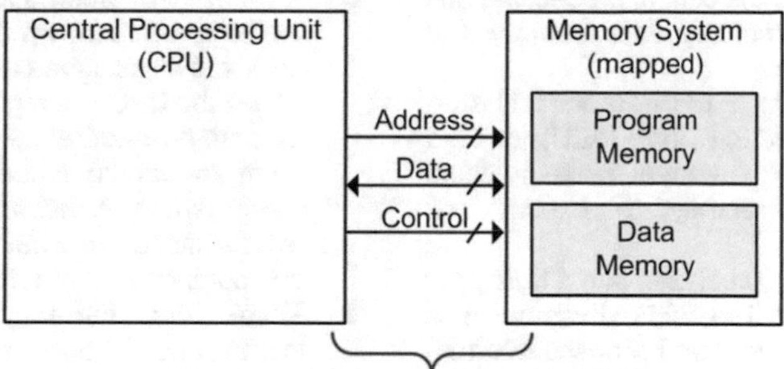

A single bus system is used to access both program and data memory.

A Harvard architecture eliminates this latency by using two separate bus systems to access program and data memory individually. This allows the CPU to read instructions in parallel with accessing variables.

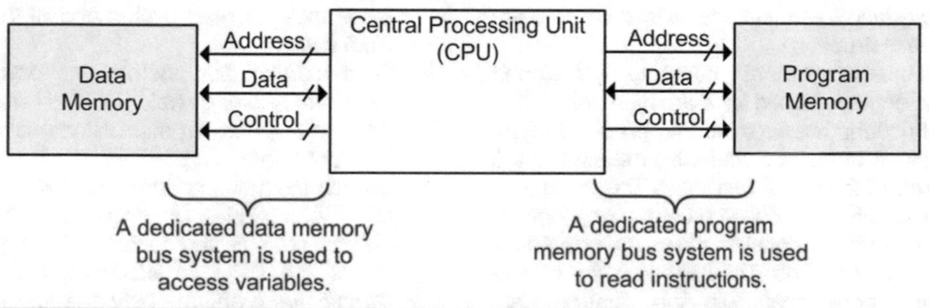

A dedicated data memory bus system is used to access variables.

A dedicated program memory bus system is used to read instructions.

Fig. 13.5
Von Neumann vs. Harvard architecture

Concept Check

CC13.4 Does a computer with a Harvard architecture require two control unit state machines?

A) Yes. It has two bus systems that need to be managed separately, so two finite-state machines are required.

B) No. A single state machine is still used to fetch, decode, and execute the instruction. The only difference is that if data is required for the execute stage, it can be retrieved from data memory at the same time the state machine fetches the opcode of the next instruction from program memory.

Summary

- A computer is a collection of hardware components that are constructed to perform a specific set of instructions to process and store data. The main hardware components of a computer are the central processing unit (CPU), program memory, data memory, and input/output ports.
- The CPU consists of registers for fast storage, an arithmetic logic unit (ALU) for data manipulation, and a control state machine that directs all activity to execute an instruction.
- A CPU is typically organized into a *data path* and a *control unit*. The data path contains all circuitry used to store and process information. The data path includes the registers and the ALU. The control unit is a large state machine that sends control signals to the data path in order to facilitate instruction execution.
- The control unit continuously performs a *fetch-decode-execute* cycle in order to complete instructions.
- The instructions that a computer is designed to execute is called its *instruction set*.
- Instructions are inserted into *program memory* in a sequence that when executed will accomplish a particular task. This sequence of instructions is called a computer *program*.
- An instruction consists of an *opcode* and a potential *operand*. The opcode is the unique binary code that tells the control state machine which instruction is being executed. An operand is additional information that may be needed for the instruction.
- An *addressing mode* refers to the way that the operand is treated. In *immediate* addressing, the operand is the actual data to be used. In *direct* addressing, the operand is the address of where the data is to be retrieved or stored. In *inherent* addressing, all of the information needed to complete the instruction is contained within the opcode, so no operand is needed.
- A computer also contains *data memory* to hold temporary variables during run time.
- A computer also contains input and output ports to interface with the outside world.
- A *memory-mapped* system is one in which the program memory, data memory, and I/O ports are all assigned a unique address. This allows the CPU to simply process information as data and addresses and allows the program to handle where the information is being sent to. A *memory map* is a graphical representation of what address ranges various components are mapped to.
- There are three primary classes of instructions. These are loads and stores, data manipulations, and branches.
- Load instructions move information from memory into a CPU register. A load instruction takes multiple read cycles. Store instructions move information from a CPU register into memory. A store instruction takes multiple read cycles and at least one write cycle.
- Data manipulation instructions operate on information being held in CPU registers. Data manipulation instructions often use inherent addressing.
- Branch instructions alter the flow of instruction execution. *Unconditional branches* always change the location in memory of where the CPU is executing instructions. *Conditional branches* only change the location of instruction execution if a status flag is asserted.
- Status flags are held in the condition code register and are updated by certain instructions. The most commonly used flags are the negative flag (N), zero flag (Z), two's complement overflow flag (V), and carry flag (C).

Exercise Problems

Section 13.1: Computer Hardware

13.1.1 What computer hardware sub-system holds the temporary variables used by the program?

13.1.2 What computer hardware sub-system contains fast storage for holding and/or manipulating data and addresses?

13.1.3 What computer hardware sub-system allows the computer to interface to the outside world?

13.1.4 What computer hardware sub-system contains the state machine that orchestrates the fetch-decode-execute process?

13.1.5 What computer hardware sub-system contains the circuitry that performs mathematical and logic operations?

13.1.6 What computer hardware sub-system holds the instructions being executed?

Section 13.2: Computer Software

13.2.1 In computer software, what are the names of the most basic operations that a computer can perform?

13.2.2 Which element of computer software is the binary code that tells the CPU which instruction is being executed?

13.2.3 Which element of computer software is a collection of instructions that perform a desired task?

13.2.4 Which element of computer software is the supplementary information required by an instruction such as constants or which registers to use?

13.2.5 Which class of instructions handles moving information between memory and CPU registers?

13.2.6 Which class of instructions alters the flow of program execution?

13.2.7 Which class of instructions alters data using either arithmetic or logical operations?

Section 13.3: Computer Implementation: An 8-Bit Computer Example

13.3.1 Design the example 8-bit computer system presented in this chapter in Verilog with the ability to execute the three instructions LDA_IMM, STA_DIR, and BRA. Simulate your computer system using the following program that will continually write the patterns x"AA" and x"BB" to output ports port_out_00 and port_out_01:

```
initial
 begin
   ROM[0]  = LDA_IMM;
   ROM[1]  = 8'hAA;
   ROM[2]  = STA_DIR;
   ROM[3]  = 8'hE0;
   ROM[4]  = STA_DIR;
   ROM[5]  = 8'hE1;
   ROM[6]  = LDB_IMM;
   ROM[7]  = 8'hBB;
   ROM[8]  = STB_DIR;
   ROM[9]  = 8'hE0;
   ROM[10] = STB_DIR;
   ROM[11] = 8'hE1;
   ROM[12] = BRA;
   ROM[13] = 8'h00;
 end
```

13.3.2 Add the functionality to the computer model from 13.3.1 the ability to perform the LDA_DIR instruction. Simulate your computer system using the following program that will continually read from port_in_00 and write its contents to port_out_00:

```
initial
  begin
   ROM[0]  = LDA_DIR;
   ROM[1]  = 8'hF0;
   ROM[2]  = STA_DIR;
   ROM[3]  = 8'hE0;
   ROM[4]  = BRA;
   ROM[5]  = 8'h00;
  End
```

13.3.3 Add the functionality to the computer model from 13.3.2 the ability to perform the instructions LDB_IMM, LDB_DIR, and STB_DIR. Modify the example programs given in exercises 13.3.1 and 13.3.2 to use register B in order to simulate your implementation.

13.3.4 Add the functionality to the computer model from 13.3.3 the ability to perform the addition instruction ADD_AB. Test your addition instruction by simulating the following program. The first addition instruction will perform x"FE" + x"01" = x"FF" and assert the negative (N) flag. The second addition instruction will perform x"01" + x"FF" = x"00" and assert the carry (C) and zero (Z) flags. The third addition instruction will perform x"7F" + x"7F" = x"FE" and assert the two's complement overflow (V) and negative (N) flags.

```
initial
  begin
    ROM[0]  = LDA_IMM; //-- test 1
    ROM[1]  = 8'hFE;
    ROM[2]  = LDB_IMM;
    ROM[3]  = 8'h01;
    ROM[4]  = ADD_AB;
    ROM[5]  = LDA_IMM; //-- test 2
    ROM[6]  = 8'h01;
    ROM[7]  = LDB_IMM;
    ROM[8]  = 8'hFF;
    ROM[9]  = ADD_AB;
    ROM[10] = LDA_IMM; //-- test 3
    ROM[11] = 8'h7F;
    ROM[12] = LDB_IMM;
    ROM[13] = 8'h7F;
    ROM[14] = ADD_AB;
    ROM[15] = BRA;
    ROM[16] = 8'h00;
  end
```

13.3.5 Add the functionality to the computer model from 13.3.4 the ability to perform the *branch if equal to zero* instruction BEQ. Simulate your implementation using the following program. The first addition in this program will perform x"FE" + x"01" = x"FF" (Z = 0). The subsequent BEQ instruction should NOT take the branch. The second addition in this program will perform x"FF" + x"01" = x"00" (Z = 1) and SHOULD take the branch. The final instruction in this program is a BRA that is inserted for safety. In the event that the BEQ is not operating properly, the BRA will set the program counter back to x"00" and prevent the program from running away.

```
initial
  begin
    ROM[0]  = LDA_IMM; //-- test 1
    ROM[1]  = 8'hFE;
    ROM[2]  = LDB_IMM;
    ROM[3]  = 8'h01;
    ROM[4]  = ADD_AB;
    ROM[5]  = BEQ;     //--NO branch
    ROM[6]  = 8'h00;

    ROM[7]  = LDA_IMM; //-- test 2
    ROM[8]  = 8'h01;
    ROM[9]  = LDB_IMM;
    ROM[10] = 8'hFF;
    ROM[11] = ADD_AB;
    ROM[12] = BEQ;     //-- Branch
    ROM[13] = 8'h00;

    ROM[14] = BRA;
    ROM[15] = 8'h00;
  end
```

13.3.6 Add the functionality for the **SUB_AB** instruction described in Example 13.9 to the computer model from 13.3.4. You will need to create a test program to verify the execution of the instruction.

13.3.7 Add the functionality for the **AND_AB** instruction described in Example 13.9 to the computer model from 13.3.4. You will need to create a test program to verify the execution of the instruction.

13.3.8 Add the functionality for the **OR_AB** instruction described in Example 13.9 to the computer model from 13.3.4. You will need to create a test program to verify the execution of the instruction.

13.3.9 Add the functionality for the **INCA** instruction described in Example 13.9 to the computer model from 13.3.4. You will need to create a test program to verify the execution of the instruction.

13.3.10 Add the functionality for the **INCB** instruction described in Example 13.9 to the computer model from 13.3.4. You will need to create a test program to verify the execution of the instruction.

13.3.11 Add the functionality for the **DECA** instruction described in Example 13.9 to the computer model from 13.3.4. You will need to create a test program to verify the execution of the instruction.

13.3.12 Add the functionality for the **DECB** instruction described in Example 13.9 to the computer model from 13.3.4. You will need to create a test program to verify the execution of the instruction.

13.3.13 Add the functionality for the **BMI** instruction described in Example 13.9 to the computer model from 13.3.4. You will need to create a test program to verify the execution of the instruction.

13.3.14 Add the functionality for the **BPL** instruction described in Example 13.9 to the computer model from 13.3.4. You will need to create a test program to verify the execution of the instruction.

13.3.15 Add the functionality for the **BNE** instruction described in Example 13.9 to the computer model from 13.3.4. You will need to create a test program to verify the execution of the instruction.

13.3.16 Add the functionality for the **BVS** instruction described in Example 13.9 to the computer model from 13.3.4. You will need to create a test program to verify the execution of the instruction.

13.3.17 Add the functionality for the **BVC** instruction described in Example 13.9 to the computer model from 13.3.4. You will need to create a test program to verify the execution of the instruction.

13.3.18 Add the functionality for the **BCS** instruction described in Example 13.9 to the computer model from 13.3.4. You will need to create a test program to verify the execution of the instruction.

13.3.19 Add the functionality for the **BCC** instruction described in Example 13.9 to the computer model from 13.3.4. You will need to create a test program to verify the execution of the instruction.

Section 13.4: Architectural Considerations

13.4.1 Would the instruction set need to be different between a Von Neumann versus a Harvard architecture? Why or why not?

13.4.2 Which of the three classes of computer instructions (loads/stores, data manipulations, and branches) are sped up by moving from the Von Neumann architecture to the Harvard architecture.

13.4.3 In a memory-mapped, Harvard architecture, would the I/O system be placed in the program memory or data memory block?

13.4.4 A Harvard architecture requires two memory address registers to handle two separate memory systems. Does it also require two instruction registers? Why or why not?

13.4.5 A Harvard architecture requires two memory address registers to handle two separate memory systems. Does it also require two program counters? Why or why not?

Appendix A: List of Worked Examples

© Springer Nature Switzerland AG 2019
B. J. LaMeres, *Introduction to Logic Circuits & Logic Design with Verilog*,
https://doi.org/10.1007/978-3-030-13605-5

Appendix B: Concept Check Solutions

- ❖ CC1.1 B
- ❖ CC1.2 C

- ❖ CC2.1 C
- ❖ CC2.2 D
- ❖ CC2.3 D
- ❖ CC2.4 A

- ❖ CC3.1 A
- ❖ CC3.4(a) B
- ❖ CC3.4(b) C
- ❖ CC3.4(c) A
- ❖ CC3.3 A
- ❖ CC3.4 D

- ❖ CC4.1 B
- ❖ CC4.2 B
- ❖ CC4.3 D
- ❖ CC4.4(a) B
- ❖ CC4.4(b) A
- ❖ CC4.5 D

- ❖ CC5.1 D
- ❖ CC5.2 C
- ❖ CC5.3 A
- ❖ CC5.4(a) A
- ❖ CC5.4(b) B
- ❖ CC5.5(a) D
- ❖ CC5.5(b) A
- ❖ CC5.6 B
- ❖ CC5.7 C

- ❖ CC6.1 C
- ❖ CC6.2 D
- ❖ CC6.3 C
- ❖ CC6.4 C

- ❖ CC7.1(a) B
- ❖ CC7.1(b) D
- ❖ CC7.2 A
- ❖ CC7.3 C
- ❖ CC7.4(a) A
- ❖ CC7.4(b) C
- ❖ CC7.4(c) B
- ❖ CC7.4(d) D
- ❖ CC7.4(e) A
- ❖ CC7.4(f) C
- ❖ CC7.5 A
- ❖ CC7.6 D
- ❖ CC7.7 B

- ❖ CC8.1 A
- ❖ CC8.2 B
- ❖ CC8.3 B
- ❖ CC8.4 B

- ❖ CC9.1 D
- ❖ CC9.2 D
- ❖ CC9.3 C
- ❖ CC9.4 A
- ❖ CC9.5 C

- ❖ CC10.1 D
- ❖ CC10.2 C
- ❖ CC10.3 B
- ❖ CC10.4 A

- ❖ CC11.1 C
- ❖ CC11.2 B

- ❖ CC12.1 B
- ❖ CC12.2 D
- ❖ CC12.3 B
- ❖ CC12.4 A

- ❖ CC13.1 B
- ❖ CC13.2 D
- ❖ CC13.3 D
- ❖ CC13.4 B

© Springer Nature Switzerland AG 2019
B. J. LaMeres, *Introduction to Logic Circuits & Logic Design with Verilog*,
https://doi.org/10.1007/978-3-030-13605-5

Index

© Springer Nature Switzerland AG 2019
B. J. LaMeres, *Introduction to Logic Circuits & Logic Design with Verilog*,
https://doi.org/10.1007/978-3-030-13605-5

W

X

Y

Printed in the United States
By Bookmasters